Band Structure of Cubic Hydrides

Dimitrios A. Papaconstantopoulos

Band Structure of Cubic Hydrides

Springer

Dimitrios A. Papaconstantopoulos
George Mason University
Fairfax, VA, USA

ISBN 978-3-031-06877-5 ISBN 978-3-031-06878-2 (eBook)
https://doi.org/10.1007/978-3-031-06878-2

This Springer imprint is published by the registered company Springer Nature Switzerland AG
The registered company address is: Gewerbestrasse 11, 6330 Cham, Switzerland

To Gail with gratitude

Preface

This Handbook is a study of the band structure of cubic monohydrides, dihydrides, and several trihydrides related to superconductivity. The data presented correspond to the NaCl, CsCl, CaF2, and the bcc-like Im3m structures and cover the whole periodic table up to the element Lawrencium. Results are given for the equation of state of each compound with the parameters of a Birch fit so that the user can regenerate the results and also derive other information such as pressure-volume relations and the variation of the bulk modulus with pressure. For each compound in addition to the equation of state, the energy bands, the densities of states, and a set of tight-binding parameters are given. The tight-binding parameters are presented in a two-center and three-center approximation, and in both orthogonal and non-orthogonal forms. Also included in this book is a discussion of the McMillan-Gaspari-Gyorffy theories and a tabulation of the electron-ion interaction matrix elements. The evaluation of the Stoner criterion for ferromagnetism is discussed and results are tabulated.

The accumulation of these results together in one source and their generation by the same methodology will be very useful for researchers and students to be able to quickly obtain the basic information of the electronic structure of all cubic hydrides in the periodic table and provide plenty of homework problems for graduate students.

The author is indebted to Michael J. Mehl for his assistance in resolving several problems that occurred in the course of this work and his continuous advice over the years. I would also like to thank my various collaborators and colleagues and, in particular, B. M. Klein, W. E. Pickett, L. L. Boyer, A. C. Switendick, L. F. Mattheiss, J. L. Feldman, M. R. Pederson, E. N. Economou, D. J. Nagel, N. Bernstein, J. Broughton and E. Kaxiras for valuable comments and suggestions. I also wish to acknowledge contributions from my students M. Keegan, B. Akdim, Lei Shi, A. Koufos, J. McGrady, L. Nixon, M. Assadullah, J. Durgavich, M. Chellathurai, S. Sayed, and S. Silayi.

The editorial and technical assistance from Gail, Lia, Rea, Ben, and Eleni is gratefully acknowledged.

Fairfax, VA, USA　　　　Dimitrios A. Papaconstantopoulos
January 2023

Contents

Chapter 1
Introduction

1.1 Introduction

This Handbook follows the same format the author used in his other book on the band structure of the elements [1]. The book has 16 chapters with a short narrative in each describing trends and findings. Hydrides have been used over the years as hydrogen storage materials, and they have also been used to achieve metallization of atomic hydrogen. In addition, hydrides play an important role in the field of superconductivity. Back in the seventies Stritzker [2] discovered that PdH was a superconductor at a transition temperature Tc = 10 K, and also displayed an inverse isotope effect in the NaCl structure. At that time the PdH-PdD system became a classic example of a hydrogenated material that was a superconductor and band-theory based calculations emerged [3, 4].

Recently a new development occurred in this field by the prediction [5], followed by experimental confirmation [6], that SH3 (bcc-like structure) reaches Tc = 200 K at a high pressure of 200GPa. This was a big breakthrough in condensed matter physics which was followed by discovery of other materials such as LaH10 [7] which opened the road to room temperature superconductivity. These advances motivated calculations in the relevant Im3m structure included in the present work.

This Handbook presents results of electronic structure calculations and Tables of three-center and two-center tight-binding (TB) parametrizations for all possible (Z = 1 to Z = 103) cubic-hydrides in the NaCl, CsCl, CaF2 and Im3m structures. These results, which were generated by the Augmented Plane Wave (APW) method and TB fits to the APW data, include diagrams of energy bands and electronic densities of states (DOS) with site and angular momentum decomposition, and tabulations of equilibrium lattice constants, bulk moduli, DOS values at the Fermi level and determinations of the Hopfield-McMillan parameters and the Stoner criterion of magnetism. The TB Hamiltonians were constructed for the NaCl structure by fitting to the APW band structures to a few mRy accuracy.

D. A. Papaconstantopoulos, *Band Structure of Cubic Hydrides*,
https://doi.org/10.1007/978-3-031-06878-2_1

1.2 The Augmented Plane Wave Method

The Augmented Plane Wave (APW) introduced by Slater [8] is one of the most successful band structure methods. The method deals with the central problem in condensed matter physics, namely the solution of the Schrodinger equation in momentum space. Since the details of the method and its development are covered in many publications, we give below the essential features of the code we have used for this work. The APW method is a self-consistent crystal field method which starts with a crystal potential found by solving Poisson's equation assuming a charge density generated from overlapping atomic wave functions. The potential is entered into the Schrodinger equation which computes new wave functions which are entering again in Poisson's equations. This is an iterative procedure that continues until it satisfies an appropriate convergence criterion for the charge density or the total energy. The Hamiltonian, besides the usual kinetic and potential energy terms, contains the relativistic components with mass-velocity and Darwin corrections. This is known as the scalar relativistic approach. In addition, the local density approximation (LDA) [9] is made to treat exchange and correlation. In our code we use the Hedin-Lundqvist [10] form of LDA. The expression for the total energy as a function of the electron charge density follows the Density Functional Theory (DFT) of Hohenberg and Kohn et al. [11]. In this work we applied the muffin-tin approximation (MTA) which assumes a spherically symmetric potential within spheres surrounding each atom and a constant potential in the interstitial region. The MTA works very well for cubic structures as we confirmed using our general potential linearized APW code. It was necessary to perform LAPW [12] calculations only for the Im3m structure where the interstitial volume is too large. Finally we note that our APW code is symmetrized which identifies the eigenvalues separately for each irreducible representation which turns out to be very useful for obtaining accurate tight-binding parameters.

1.3 The Tight-Binding Formalism

In the tight-binding method the one-electron wave function is expressed as a linear combination of Bloch sums. When such an expansion of the wave function is substituted into the Schrödinger equation one obtains a set of simultaneous linear algebraic equations which has a nonzero solution if the determinant of the coefficients vanishes, i.e.,

$$\tilde{H} - E\tilde{S} = 0 \tag{1.1}$$

The matrix elements in Eq. (1.1) have the form:

$$H_{nm} = \sum_{R_j} \exp i\boldsymbol{k} \cdot \left(\boldsymbol{R}_j - \boldsymbol{R}_i\right) \int \varphi_n^*(\boldsymbol{r} - \boldsymbol{R}_i) H \varphi_m\left(\boldsymbol{r} - \boldsymbol{R}_j\right) dv \tag{1.2}$$

$$S_{nm} = \sum_{R_j} \exp i\boldsymbol{k} \cdot \left(\boldsymbol{R}_j - \boldsymbol{R}_i\right) \int \varphi_n^*(\boldsymbol{r} - \boldsymbol{R}_i) \varphi_m\left(\boldsymbol{r} - \boldsymbol{R}_j\right) dv \tag{1.3}$$

where $\mathbf{R}_i$ and R_j denote the positions of atoms located on orbitals φ_n and φ_m respectively. The size of the matrices $\tilde{H}$ and $\tilde{S}$ is determined by the number of atoms per unit cell and the number of atomic orbitals taken on each atomic site. So for NaCl and CsCl structures with two atoms per unit cell, $\tilde{H}$ and $\tilde{S}$ are 13 × 13 matrices representing one s orbital, three p orbitals, and five d orbitals for the first atom, and one s orbital and three p orbitals for the second atom. In this volume since we consider hydrides the matrix size is really 10 × 10. We have a 13 × 13 matrix in our programs to allow for treating carbides and nitrides where the p-orbitals of carbon and nitrogen are absolutely necessary. For some of the hydrides we used the p-orbitals of hydrogen to obtain a better fit. For the fluorite (CaF2) crystal structures since we have three atoms per unit cell the matrix size would be 11 × 11 and 14 × 14 for hydrides and carbides respectively. In Eqs. (1.2) and (1.3) the summations over $\mathbf{R}_j$ indicate summing over all neighbors of the central atom. In the present work for the NaCl we have carried out the summation to two neighboring atoms of the same kind, which means that we have a total of six nearest neighbors included in the fits. The integrals in equations

Equations (1.2) and (1.3) are three-center integrals since they are a product of atomic wavefunctions centered on atoms at Ri and Rj and a potential energy function (included in the Hamilonian H) centered on a third atom. These integrals could be computed from first-principles. However, most computer codes in density functional applications use either augmented plane waves (APW and LAPW methods or plane waves (pseudopotential methods). In this work we replace the above integrals by adjustable parameters that are determined by fitting to APW band structure calculations. In the classic paper of Slater and Koster [13] this method was first proposed with the simplification of using Lowdin [14] functions instead of atomic orbitals. These Lowdin functions are taken to be ortho-normal so that the integral of Eq. (1.3) vanish unless $m = n$. In this case one has to solve the eigenvalue problem:

$$\tilde{H} - E\tilde{I} = 0 \tag{1.4}$$

Here we have performed the fitting to the APW results, both in a nonorthogonal representation (as was done by Mattheiss [15] in a number of materials) and on the more traditional orthogonal basis.

The integrals of Eqs. (1.2) and (1.3) are related by symmetry operations that reduce considerably the number of independent parameters to be determined in these calculations. These symmetry considerations have been worked out by Slater and Koster (SK), who present tables of the matrix elements. Some typographical errors appear in these tables. For this reason and because phase factors may change the definition of these SK parameters by a sign, we have given here the full matrix for each structure as used in our calculations.

There are at least three important virtues in presenting the band structure of solids in the SK form. First, using the SK parameters one can easily diagonalize a small matrix (10×10 for a monohydride) to obtain the energy bands and densities of states for a given material. This avoids the complexity of the APW or KKR techniques, which are used reliably and efficiently only by experts. Hence, the SK method is readily usable for the analysis of experiments by experimentalists themselves. Second, a great variety of theories utilize SK parameters to construct tight-binding (TB) Hamiltonians for the study of, for example, defects, disordered materials, surfaces, interfaces, and phonon spectra. The availability of such parameters is particularly useful to scientists working in these areas. At present, those who study such phenomena spend considerable time creating these TB Hamiltonians. In determining SK parameters they often invoke approximations that limit the accuracy of their results. Third, there is a clear educational value in this approach; it will enable students to easily generate the band structure of a material and gain a better understanding of the concepts involved.

1.4 The Two-Center Approximation

The three-center integrals of Eqs. (1.2) and (1.3) can be reduced to two-center integrals if we assume that the potential energy is a sum of spherical potentials located on the same two atoms where the atomic orbitals are centered. In this case the wavefunctions φ can be expressed as sums of functions that are space-quantized about the vector $\mathbf{R}_j$–R_i, exactly as in a diatomic molecule. For example, if π is a p orbital, it becomes a linear combination of $\mathrm{p}\sigma$, $\mathrm{p}\pi_{+}$, and $\mathrm{p}\pi_{-}$ functions and if φ is a d orbital it becomes a linear combination of $\mathrm{d}\sigma$, $\mathrm{d}\pi_+$, $\mathrm{d}\pi_-$, $\mathrm{d}\delta_+$, and $d\delta_-$ functions. The relationships between three- and two-center integrals have been worked out by Slater and Koster and are given in Table 1 of their paper. For the cubic structures we wrote our computer codes in the three-center approximation, so we have used this table to derive the corresponding expressions for first-, second-, and third-neighbor interactions and incorporated them into our codes. These relationships are given below for the NaCl structure.

First-neighbor A–A interactions:

$$
\begin{aligned}
E_{s,s}(110) &= (\mathrm{ss}\sigma)_1 \\
E_{s,x}(110) &= \frac{1}{\sqrt{2}}(\mathrm{sp}\sigma)_1 \\
E_{s,xy}(110) &= \frac{\sqrt{3}}{2}(\mathrm{sd}\sigma)_1 \\
E_{s,3z^2-r^2}(110) &= -\frac{1}{2}(\mathrm{sd}\sigma)_1 \\
E_{x,x}(110) &= \frac{1}{2}\left[(\mathrm{pp}\sigma)_1 + (\mathrm{pp}\pi)_1\right] \\
E_{x,x}(011) &= (\mathrm{pp}\pi)_1 \\
E_{x,y}(110) &= \frac{1}{2}\left[(\mathrm{pp}\sigma)_1 - (\mathrm{pp}\pi)_1\right] \\
E_{x,xy}(110) &= \frac{\sqrt{3}}{2\sqrt{2}}(\mathrm{pd}\sigma)_1 \\
E_{x,xy}(011) &= \frac{1}{\sqrt{2}}(\mathrm{pd}\pi)_1 \\
E_{z,3z^2-r^2}(011) &= \frac{\sqrt{2}}{4}\left[\frac{(\mathrm{pd}\sigma)_1}{2} + \sqrt{3}(\mathrm{pd}\pi)_1\right] \\
E_{z,x^2-y^2}(011) &= \frac{\sqrt{2}}{4}\left[(\mathrm{pd}\pi)_1 - \frac{\sqrt{3}(\mathrm{pd}\sigma)_1}{2}\right] \\
E_{xy,xy}(110) &= \frac{1}{4}\left[3(\mathrm{dd}\sigma)_1 + (\mathrm{dd}\delta)_1\right] \\
E_{xy,xy}(011) &= \frac{1}{2}\left[(\mathrm{dd}\pi)_1 + (\mathrm{dd}\delta)_1\right] \\
E_{xy,xz}(011) &= \frac{1}{2}\left[(\mathrm{dd}\pi)_1 - (\mathrm{dd}\delta)_1\right] \\
E_{xy,3z^2-r^2}(110) &= \frac{\sqrt{3}}{4}\left[(\mathrm{dd}\delta)_1 + (\mathrm{dd}\sigma)_1\right] \\
E_{3z^2-r^2,3z^2-r^2}(110) &= \frac{1}{4}\left[(\mathrm{dd}\sigma)_1 + 3(\mathrm{dd}\delta)_1\right] \\
E_{x^2-y^2,x^2-y^2}(110) &= (\mathrm{dd}\pi)_1
\end{aligned}
$$

Second-neighbor A–A interactions:

$$\begin{aligned}
\mathrm{Es,s_2(200)} &= (\mathrm{sss})_2 \\
\mathrm{Es,x(200)} &= (\mathrm{sps})_2 \\
\mathrm{Es, 3z^2 - r^2(002)} &= (\mathrm{sds})_2 \\
\mathrm{Ex, x(200)} &= (\mathrm{pps})_2 \\
\mathrm{Ey, y(200)} &= (\mathrm{ppp})_2 \\
\mathrm{Ex, xy(020)} &= (\mathrm{pdp})_2 \\
\mathrm{Ez,3z^2 - r^2(002)} &= (\mathrm{pds})_2 \\
\mathrm{Exy, xy(200)} &= (\mathrm{ddp})_2 \\
\mathrm{Exy, xy(002)} &= (\ \mathrm{ddd}\ )_2 \\
\mathrm{E3z^2 - r^2, 3z2 - r^2(002)} &= (\mathrm{dds})_2 \\
\mathrm{Ex^2 - y^2, x^2 - y^2(002)} &= (\mathrm{ddd})_2
\end{aligned}$$

First-neighbor B–B interactions:

$$\mathrm{Es, s(110)} = (\mathrm{sss})_1$$

Second-neighbor B–B interactions:

$$\mathrm{Es, s(200)} = (\mathrm{sss})_2$$

First-neighbor A–B interactions:

$$\begin{aligned}
\mathrm{Es, s(100)} &= (\mathrm{sss})_1 \\
\mathrm{Ex, s(100)} &= (\mathrm{pss})_1 \\
\mathrm{E3z^2 - r^2, s(001)} &= (\mathrm{dss})_1
\end{aligned}$$

Second-neighbor A–B interactions:

$$\begin{aligned}
\mathrm{Es, s(111)} &= (\mathrm{sss})_2 \\
\mathrm{Ex, s(111)} &= (\mathrm{pss})_2/\mathrm{sqrt(3)} \\
\mathrm{Exy, s(111)} &= (\mathrm{dss})_2/\mathrm{sqrt(3)}
\end{aligned}$$

1.5 Computational Details

The SK matrices for the NaCl lattice were formed by using the tables in the SK article [13]. To ensure correct assignment of states it was necessary to use group theory to reduce the 10×10 secular equation to matrices at high symmetry points

Table 1.1 Block-diagonalization of the 10 × 10 TB Hamiltonian at high-symmetry points of the NaCl lattice. The ordering of the states is s, p(x), p(y), p(z), d(yz), d(zx), d(xy), d($x^2 - y^2$), d ($3z^2 - r^2$), s(H)

State	k-point	Matrix	Deg
Γ_1	(0,0,0)	H(1,1) H(1,10) H(1,10) H(10,10)	1
Γ_{12}	(0, 0, 0)	H(8,8)	2
Γ_{15}	(0,0,0)	H(2,2)	3
Γ'_{25}	(0,0,0)	H(5,5)	3
X_1	(0,0,2π/a)	H(1,1) H (1,9) H(1,10) H(1,9) H(9,9) H(9,10) H(1,10) H(9,10) H(10,10)	1
X_2	(0,0,2π/a)	H(8,8)	1
X_3	(0,0,2π/a)	H(5,5)	1
X'_4	(0,0,2π/a)	H(4,4)	1
X_5	(0,0,2π/a)	H(6,6)	2
X'_5	(0,0,2π/a)	H(2,2)	2
L_1	(π/a,π/a,π/a)	H(1,1) [H(1,5) + H(1,6) + H(1,7)]/sqrt(3) [H(1,5) + H(1,6] + H(1,7)]/sqrt(3) [H(5,5) + H(6,6) + H (7,7) + 2*(H(5,6) + H(5,7) + H(6,7)]/3	1
L'_2	$\left(\frac{\pi}{a}, \frac{\pi}{a}, \frac{\pi}{a}\right)$	[(H(2,2) + H(3,3) + H(4,4)]/3 + 2/3*[H(2,3) + H(2,4) + H(3,4)] [H(2,10) + H(3,10) + H(4,10)]/sqr3 [H(2,10) + H(3,10) + H (4,10)]/sqrt(3) H(10,10)	1
L_3	(π/a,π/a,π/a)	$\begin{bmatrix} (1/2)[H_{55} - 2H_{56} + H_{66}] & (1/\sqrt{2})[H_{58} - H_{68}] \\ (1/\sqrt{2})[H^*_{58} - H^*_{68}] & H_{88} \end{bmatrix}$	2
L'_3	(π/a,π/a,π/a)	[H(2,2)−2.0*H(2,3) + H(3,3)]/2	2
W_1	$\left(0, \frac{2\pi}{a}, \frac{\pi}{a}\right)$	$\begin{bmatrix} H_{11} & H_{19} \\ H^*_{19} & H_{99} \end{bmatrix}$	1
W'_2	(0,2π/a,π/a)	H(4,4) H(4,8) H(4,10) H(4,8) H(8,8) H(8,10) H(4,10) H(8,10) H(10,10)	1
W'_1	(0,π/a,2π/a)	H(7,7)	1
W_3	$\left(0, \frac{\pi}{a}, \frac{2\pi}{a}\right)$	$\begin{bmatrix} H_{22} & H_{27} \\ H^*_{27} & H_{77} \end{bmatrix}$	2

or lines in the irreducible Brillouin zone. In Table 1.1 we show this block-diagonalization of the Hamiltonian. We fit seven bands for each k-point.

The determination of the SK parameters uses a least-squares procedure, and since it is a nonlinear problem it requires a reasonably good set of starting values. As mentioned above, by using symmetry considerations one can reduce the original secular equation to smaller matrices or to a number of linear equations. These linear equations can be solved simultaneously to determine an initial set of starting parameters. We now give the linear equations obtained for the particular crystal structure.

1.5.1 The NaCl Structure

To obtain starting values for the SK parameters in the NaCl structure we use the following linear equations, which involve d-like functions:

$$E(\Gamma_{25'}) = E_{xy,xy}(000) + 4E_{xy,xy}(110) + 8E_{xy,xy}(011) + 4E_{xy,xy}(200) + 2E_{xy,xy}(002) \tag{1.5}$$

$$E(\Gamma_{12}) = E_{d2,d2}(000) + 6E_{d2,d2}(110) + 6E_{d1,d1}(110) + 3E_{d2,d2}(002) + 3E_{d1,d1}(002) \tag{1.6}$$

$$E(X_3) = E_{xy,xy}(000) + 4E_{xy,xy}(110) - 8E_{xy,xy}(011) + 4E_{xy,xy}(200) + 2E_{xy,xy}(002) \tag{1.7}$$

$$E(X_2) = E_{d2,d2}(000) - 6E_{d2,d2}(110) + 2E_{d1,d1}(110) + 3E_{d2,d2}(002) + 3E_{d1,d1}(002) \tag{1.8}$$

$$E(X_5) = E_{xy,xy}(000) - 4E_{xy,xy}(110) + 4E_{xy,xy}(200) + 2E_{xy,xy}(002) \tag{1.9}$$

$$E(W_{1'}) = E_{xy,xy}(000) - 4E_{xy,xy}(110) + 4E_{xy,xy}(200) - 2E_{xy,xy}(002) \tag{1.10}$$

$$E(\Delta_{2'}, k = 0, 0, 4) = E_{xy,xy}(000) + 4E_{xy,xy}(110) + 4E_{xy,xy}(200) - 2E_{xy,xy}(002) \tag{1.11}$$

$$E(\Delta_2, k = 0, 0, 4) = E_{d2,d2}(000) + 4E_{d1,d1}(110) + 3E_{d2,d2}(002) - E_{d1,d1}(002) \tag{1.12}$$

$$\mathrm{E}[\mathrm{Z}_2, (\mathrm{k} = 0, 1, 8), (\mathrm{k} = 0, 2, 8), \mathrm{k} = (0, 3, 8)] = \mathrm{H}(7, 7) \tag{1.13}$$

$$\begin{aligned} &\mathrm{E}[\Sigma_2, (\mathrm{k} = 0, 2, 2), \mathrm{k} = (0, 4, 4), \mathrm{k} = (0, 6, 6)] \\ &= [\mathrm{H}(5, 5) + 2. * \mathrm{H}(5, 7) + \mathrm{H}(7, 7)]/2 \end{aligned} \tag{1.14}$$

Solving the above Eqs. (1.5) through (1.14) we determine most of the d-d parameters of the A–A interactions. Starting values for the rest of the A–A interaction parameters can be found in reference [1]. Then one runs a least-squares minimization code that includes all the spd parameters found in the 10 × 10 matrix. More details are given in the Appendix 1.

1.6 Nonorthogonal Hamiltonian

The use of nonorthogonal orbitals provides a more physically meaningful representation and, since the number of parameters involved is almost double that of the orthogonal SK method, the fit is superior. This approach has been utilized extensively by Mattheiss [15] in the parametrizations of the band structure of many compounds. Following Mattheiss we first set up the Hamiltonian and overlap matrices for a particular wave vector k, and then apply the Löwdin symmetric orthogonalization scheme to convert back to an orthogonal basis. In this scheme the matrices $\tilde{H}$ and $\tilde{S}$ of Eq. (2.1) are replaced by an effective Hamiltonian $\tilde{H}'$ with a unit overlap matrix, where $\tilde{H}' = \tilde{S}^{-1/2}\tilde{H}\tilde{S}^{-1/2}$ and $\tilde{S}' = \tilde{S}^{-1/2}\tilde{S}^{-1/2} = 1$ To determine $\tilde{S}^{-1/2}$ the overlap matrix Sis diagonalized, i.e., $\tilde{U}^{+}\tilde{S}\tilde{U} = \tilde{D}$ Then a new matrix $\tilde{D}^{-1/2}$ is formed by replacing each diagonal element by its inverse square root, which is used to find $\tilde{S}^{-1/2} = \tilde{U}\tilde{D}^{-1/2}\tilde{U}^{+}$.

1.7 Total Energy—Birch Fit

Total Energy calculations provide an evaluation of the equilibrium lattice parameter and the bulk modulus. The total energy computed for a given crystal structure and several volumes is expanded by the Birch-fit formula [16] according to the expression:

$$E(V) = \sum_{i}^{N} a_i V_i^{\frac{2i}{3}}$$

where a_i are the expansion coefficients and N is the order of the fit. A third order fit usually provides an accurate description of the total energy. The second derivative of E with respect to volume gives the bulk modulus:

$$B = -V\frac{d^2E}{dV^2}$$

The Birch fit can also be used to calculate the pressure as a function of volume. In the tables for each hydride we give the values of the coefficients (a_i).

1.8 NRL-Tight-Binding Method

The NRL-Tight-Binding method (TB) is a Slater-Koster-like TB method [17–20], which has been successfully applied to many single element, binary and ternary systems. The method performs very well in metals, insulators, and semiconductors [21, 22], and has been extended to clusters and molecules. It is built on a fitting process to Density Functional Theory results. The NRL-TB uses two-center parameters only and works in both an orthogonal and a nonorthogonal basis and computes both the energy bands and total energies. It has transferability far beyond the fitted DFT database. It should be emphasized here that in the present work we are not using the NRL-TB. The TB parametrizations presented in this book, which include both three and two center parameters, do not evaluate total energies but give an almost exact reproduction of the APW energy bands and densities of states. The NRL-TB formalism with the basic equations is outlined below. First, the total energy is determined using the fact that in DFT the total energy E[n(r)] is given by the expression:

$$E[n(r)] = \sum_i \epsilon_{\mathrm{i}} + F[n(r)] \tag{1.15}$$

where n(r) is the electron density and F[n(r)] contains the remaining parts of the DFT total energy minus the sum of the one-electron eigenvalues.

In other TB approaches F[n(r)] is a sum of pair potentials. However, based on the fact that in DFT one can uniformly shift the energy bands by a constant, in this method a shift $V_0 = F[n(r)]/N$ is applied to the first-principles eigenvalues ϵ_{i}, where N is the number of valence electrons in the system. This shifts the one-electron eigenvalues to new values,

$$\epsilon'_{\mathrm{l}} = \epsilon_{\mathrm{l}} + \mathrm{V}_0 \tag{1.16}$$

and hence the total energy becomes the sum of the shifted DFT eigenvalues over the Brillouin zone, i.e.

$$E[n(r)] = \sum_i \epsilon'_{\mathrm{l}} \tag{1.17}$$

This is a two-center TB scheme, where the on-site terms have a polynomial form as a function of the atomic density. For a single element, the density of atom i is defined as

$$\rho_i = \sum_i \exp\left(-\lambda^2 R_{ij}\right) F_C\left(R_{ij}\right) \tag{1.18}$$

where the sum is over all the neighboring atoms j within a range of cutoff distance R_c of atom i, is a fitting parameter, and F_C (Rij) is a smooth cutoff function. The angular-momentum-dependent on-site terms are defined by

$$h_\ell(R_i) = \alpha_{ii\ell} + \beta_{ii\ell\rho}^{2/3} + \gamma_{ii\ell\rho}^{4/3} \tag{1.19}$$

where ℓ represents the s, p, and d orbitals, and $\alpha\ell$, $\beta\ell$, $\gamma\ell$ and $\delta\ell$ are fitting coefficients.

The two-center s-p-d SK Hamiltonian and overlap integrals contain ten independent SK parameters, which are assumed to all have polynomial times exponential forms in terms of the neighbor distance R, given by the following equation,

$$P_\gamma(R) = \left(e_\gamma + f_\gamma R + g_\gamma R^2\right) \exp\left(\left[-q_\gamma^2\right] F_C(R)\right) \tag{1.20}$$

where γ indicates the type of interactions, including ssσ, ppσ, spσ, ddσ, sdσ, pdσ, ppπ, ddπ, pdπ, and ddδ. e_γ, f_γ, g_γ, and qγ are our fitting coefficients. In the case of a binary compound there are four additional SK parameters representing the A–B interactions i.e. pss, dss, dps and dpp. In a non-orthogonal calculation the overlap functions have the same form as in Eq. (1.20) with different fitting coefficients.

In order to determine the above coefficients a least-squares procedure is used to fit to DFT total energies and energy bands as a function of volume for different crystal structures. The total energy is usually weighed at around 200–300 times over a single band energy. In general, the fitting RMS error is less than 10 mRy and 0.2 mRy for the energy bands and total energy, respectively.

Having determined the above coefficients, the method is used to predict total energies accessible to standard DFT which were not fitted, such as elastic constants, phonon spectra and surface energies. Furthermore, large scale simulations can be performed which are not practical via DFT, such as static calculations for the energetics of systems containing up to 20,000 atoms, or calculations for very large number of k-points needed for mapping Fermi surfaces and evaluating susceptibilities.

The method also has capabilities to perform molecular dynamics (MD) simulations accommodating as many as 10,000 atoms, and 10,000 MD steps, which is an impossible task for standard DFT codes. Such calculations should yield mean square displacements, thermal expansion, and vacancy formation energies [15–17]. An extension of the method to include f-orbitals is discussed in Ref. [23].

1.9 The Gaspari-Gyorffy-McMillan Theory of Superconductivity

McMillan [24] first proposed to write the electron–phonon coupling constant λ as the following ratio

$$\lambda = \frac{\eta}{M\langle\omega^2\rangle} = \frac{N(\epsilon_f)\langle I^2\rangle}{M\langle\omega^2\rangle} \quad (1.21)$$

where $N(\varepsilon_f)$ is the density of electronic states per spin and $<I^2>$ is and electron–ion matrix element, $<\omega^2>$ is an average phonon frequency, and M is the atomic mass.

One calculates the numerator η, known as the Hopfield-McMillan parameter, by the theory known as rigid-muffin-tin approximation. The matrix element $<I^2>$, derived from multiple-scattering theory, is given by the formula due to Gaspari and Gyorffy [25]

$$\langle I^2\rangle = \frac{\epsilon_f}{\pi^2 N^2(\epsilon_f)} \sum_l \frac{2(l+1)\sin^2(\delta_{l+1}-\delta_l)N_l N_{l+1}}{N_l^{(1)} N_{l+1}^{(1)}} \quad (1.22)$$

This formula for $<I^2>$ depends on the results of band structure calculations, specifically angular momentum components (N_l) of the density of states at the Fermi level, the scattering phase shifts (δ_l), calculated from the radial wave functions u_l, spherical Bessel functions j_l and Neumann functions n_l at the muffin-tin radius R_s from the well-known formula

$$\tan\delta_l(R_s, E) = \frac{j'_l(kR_s) - j_l(kR_s)L_l(R_s, E)}{n'_l(kR_s) - n_l(kR_s)L_l(R_s, E)} \quad (1.23)$$

where $L_l = \frac{u'_l}{u_l}$ is the logarithmic derivative of the radial wave function.

The equation for the free-scatterer density of states is

$$N_l^{(1)} = \frac{\sqrt{\epsilon_f}}{\pi}(2l+1)\int_0^{R_s} u_l^2(r, \epsilon_f)r^2 dr \quad (1.24)$$

Since Eq. (1.2) is based on scattering theory, it should be applied using touching muffin-tin spheres, as the excess interstitial volume in the case of non-touching spheres introduces errors. The denominator of Eq. (1.1) is approximated by $<\omega^2> = (1/2)\Theta^2_D$ where Θ_D is the measured value of the Debye temperature. Knowing λ and $<\omega>$ McMillan proposed the following equation to evaluate the superconducting transition temperature T_c [24]:

$$T_c = \frac{\Theta_D}{1.45} \exp\left[\frac{-1.04(1+\lambda)}{\lambda - \mu^*(1+0.62\lambda)}\right] \tag{1.25}$$

The McMillan equation also involves the parameter μ^* called the Coulomb pseudopotential. The value of μ^* is usually taken in the range 0.1–0.15 and introduces non-negligible changes to T_c. The above formalism can also be used to study the pressure variation of T_c. The accuracy of the McMillan equation was improved by Allen and Dynes [26] who changed the prefactor (Θ_D /1.45) to (f1*f2* $\omega_{\log}$/1.2) using an elaborate fitting procedure to the measured Tc of several known superconductors.

The value of the McMillan-Hopfield parameter η is given separately for the A-site and the H-site for each compound in the corresponding tables of this book.

1.10 The Stoner Criterion

Many years ago Stoner [27] proposed that the following inequality:

$$N(E_F)I_s > 1 \tag{1.26}$$

is used as a criterion for the occurrence of ferromagnetism. As was shown by Vosko and Perdew [28], the parameter I_s can be accurately calculated using the results of band theory as follows.

$$I_s = \int dr\gamma^2(r)|K(r)|, \quad \text{and} \quad \gamma = \frac{1}{N(E_F)} \sum_l N_l(E_F)u_l^2(E_F) \tag{1.27}$$

where $u_l(E_F)$ is the Fermi level value of the radial wave function and K(r) is a kernel giving the exchange and correlation enhancement of the external field due to magnetization. We note again the key role of the angular momentum decomposed DOS, $N_l(E_F)$. The first implementation of this theory was made by Janak [29] who demonstrated that the Stoner criterion is satisfied for Fe, Co and Ni with values clearly exceeding one, while for all other elements the values obtained by Janak were well below 1.0.

This method has been extended to binary hydrides by Papaconstantopoulos [30] who performed calculations for all the 3*d,* 4*d,* and 5*d* monohydrides in the NaCl structure. These calculations predicted that CoH exceeds the Stoner criterion, which makes it a ferromagnet in agreement with experiment. In this volume we present Stoner parameters for all hydrides across the periodic table.

1.11 Description of the First-Principles Calculations

The data presented in this book are based on the results of first-principles band structure calculations using the APW/LAPW method. We have performed self-consistent, scalar-relativistic (without spin–orbit) calculations using the local density approximation (LDA) for exchange and correlation as applied by Hedin and Lundqvist [10]. In the APW/LAPW calculations the energy levels are divided into inner and outer levels. The inner levels are treated as atomic-like in each iteration with a modified Liberman et al. relativistic code [31]. The outer levels are computed by the APW/ LAPW method on a uniform mesh of k-points depending on the crystal structure. We carried out our iterations to self-consistency using 89 k-points in the irreducible Brillouin zone for the NaCl and CaF2 structures, 35 k-points for the CsCl and 55 k-points for the Im3m structure. The calculations yield energy bands, densities of states and total energies.

1.12 Accuracy of the Tight-Binding Calculations

The tight-binding Hamiltonians were constructed using a least-squares fit to the APW results for the NaCl structure. In the fit we included 89 k-points for seven bands each with the exception of using ten bands at the high symmetry k-points in order to place correctly the p-orbitals of the metal site. As mentioned earlier block-diagonalization of the 10×10 matrix is applied, which is important in obtaining correctly the angular-momentum character of the states.

The three-center nonorthogonal calculations fit the first seven bands with an accuracy that is better than 1-2mRy. This means that these bands are virtually indistinguishable from the APW results for most compounds. We believe that this is due to the much larger number of disposable parameters that we have available in the nonorthogonal representation and also because such a representation is more meaningful physically. In the other three representations our rms fitting errors are increasing with decreasing number of parameters. Typical rms errors for the first seven bands in transition metals are 3mRy for three-center orthogonal, 2mRy for two-center nonorthogonal, and 5mRy for two-center orthogonal. We should draw the attention of the reader to the fact that the higher bands, eighth through ninth, do not fit as well because these bands contain *f*-character, which is not included in our basis set. However, by fitting the high symmetry points for these bands, we place them correctly and do not allow them to come down and interfere incorrectly with the lower bands, as can happen if we restrict our fit to seven bands for all k-points. This raises the question of the universality of the SK parameters. Our view is that our parameters are really unique for the metals for which we have achieved a high level of accuracy for the first seven bands and the correct positioning of the high-energy p-like states.

1.13 Wavefunctions

The Bloch eigenfunction $\psi_{kn}(r)$, which corresponds to an eigenvalue $E_n(k)$ of the SK Hamiltonian, is given by the following linear combination of LCAOs:

$$\psi_{kn}(r) = \sum_m v_{nm}(k) b_{km}(r)$$

where $v_{nm}(k)$ are the eigenvectors for band n and wave vector k, and the LCAOs are

$$b_{km}(r) = N^{-1/2} \sum_i \exp(ikR_j) \varphi_m(r - R_j)$$

where the sum is over the N atomic sites R_j and $\varphi_m(r - R_j)$ atomic wave-functions for $m = 1, \ldots, 10$. It is clear from the above equations that the wavefunctions of the solid can be calculated using the eigenvectors $v_{nm}(k)$ of the SK Hamiltonian together with atomic wavefunctions $\varphi_m(r - R_j)$. In this text we have used the SK eigenvectors as weights for obtaining the angular momentum components of the DOS. We have not calculated the wave-functions for the solid, but this is a straight-forward task given our SK Hamiltonian and parameters. Work along these lines has been done by Johnston et al. [32] who used the combined interpolation scheme and obtained spherically averaged charge densities that were in good agreement with those of the first-principles calculations.

1.14 Scaling Laws

It is possible to include a lattice constant dependence in the SK parameters and thus obtain the band structure at different lattice spacing without performing new first-principles calculations. This can be done following Andersen et al. [33] who, using muffin-tin orbital theory, found that the distance-dependence of the tight-binding matrix elements is given by the expression

$$V_{ll'm} = C_{ll'm} d^{-(l+l'+1)}$$

where $C_{ll'm}$ is a constant that depends on the particular element and is volume independent, and d is the neighbor distance. This expression suggests that matrix elements for s–s interactions vary as d^{-1}, s-p interactions as d^{-2}, s-d and p–p interactions as d^{-3}, p–d interactions as d^{-4}, and d–d interactions as d^{-5}.

We have checked this scaling law [34], comparing its results to first-principles APW calculations at the same lattice spacing, and found very good agreement for variations of the lattice constant as large as 2%.

Harrison [35] in his elegant theory of the physics of chemical bonds, has reformulated the above law to include only nearest neighbors. His bond-length dependences are d^{-2} for s–s, s-p, and p–p interactions, $d^{-7/2}$ for s-d and p–d interactions, and d^{-5} for d–d interactions. Harrison's theory provides a semi-quantitative picture of the band structure of the elements that is remarkably successful considering the very small number of TB parameters used. Harrison's theory was extended by Shi et al. [1, 36] to give for the elements a very accurate band structure at the expense of introducing a large number of TB parameters. This modification to Harrison's theory could also be extended to the hydrides.

Beyond the scope of this Handbook is the NRL-TB methodology which we describe in Sect. 1.8. This reformulation of the SK approach reproduces the first-principles total energies as well as the band structure and can give the lattice constant dependence near the equilibrium volume as well as having transferability to other structures.

1.15 Calculation of the Densities of States

To calculate the densities of states (DOS) we proceed as follows. Having determined the SK parameters for a particular compound we then diagonalize the corresponding Hamiltonian for a large number of k-points. For the NaCl structure a uniform k-point mesh contains 505 k-points and for the CsCl structure 969 k-points in the irreducible Brillouin zone. The diagonalization of the 10 × **10** matrix gives, in addition to the eigenvalues, a quantity $Q_l = \sum_{j=1}^{9} v_{lj} * v_{lj}$ (with l = s, p, t_{2g}, $e_{g,}$ $_{sH}$), which determines the angular momentum character of each state. The eigenvalues and Q_ls obtained by the above procedure are used to find the total DOS and its l-components by the tetrahedron method [37]. We also calculate the Fermi velocity <v> and the plasmon energy from the formula [38]:

$$\Omega^2 = 4\pi * (e)^2 * N(Ef) * <v>^2$$

For the CsCl, CaF2, Im3m structure results presented in this volume we calculated the DOS from the APW eigenvalues and eigenvectors.

1.16 Coherent Potential Approximation (CPA)

The CPA is a mean field theory [39] designed to calculate the electronic states of disordered materials. The basic idea of the CPA is that the electrons in the solid can be regarded as moving in an effective medium the Hamiltonian of which is determined self-consistently by the condition that the average scattering of electrons by the atomic potentials relative to the medium is zero. This is expressed in terms of a sum of "t" matrices by the following equation:

$$x\frac{\epsilon_A - \Sigma}{1 - (\epsilon_A - \Sigma)\overline{G}} + (1 - x)\frac{\epsilon_B - \Sigma}{1 - (\epsilon_B - \Sigma)\overline{G}} = 0$$

where $\sum$ is the self-energy, ε_A and ε_B are the onsite tight-binding parameters, x is the concentration, and G is the averaged Green's function:

$$\overline{G}(z) = \Omega^{-1} \int d^3k(z - H(k))^{-1}$$

where $H(k)$ is the tight-binding Hamiltonian:

$$H(k) = H_{TB}(k) - \Sigma$$

The density of states is given by:

$$N(E) = -\frac{1}{\pi}\mathrm{ImTr}\overline{G}(z)$$

The above formalism represents a tight-binding form of the CPA treating only diagonal disorder. The underlying tight-binding Hamiltonian is based on 3-center integrals.

1.17 Systematics

The combined APW and TB calculations presented in this volume are used to establish trends of various physical quantities as a function of atomic number **Z** as discussed below:

(a) In Fig. 1.1 the variation of the lattice constant versus Z is shown for the 3d, 4d, and 5d hydrides in the NaCl and CaF2 crystal structures and compared with the corresponding elements in the fcc structure. The expected expansion of the lattice is confirmed. It is also shown that within each row the lattice constant increases with increasing Z. In addition, it is observed that there is a shift of the lattice constants to larger values as we go from the 3d to the 5d hydrides similar to that in the elements.

(b) We have studied as a function of **Z** the quantity $\Delta_l = E_l(000) - E_F$, where $E_l(00)$ denotes the tight-binding on-site parameters for $\boldsymbol{l}$ = *s*, ***p***, d, and s(H) in the orthogonal three-center approximation and E_F is the Fermi energy. The quantity Δ_d represents the approximate location of the d bands with respect to E_F. Within the transition metal series all deltas are positive and slowly varying. Δ_d becomes negative to the right of the 3d and 4d series. Δ_s (H) is negative and nearly constant throughout. The difference $\Delta_{t2g} - \Delta_{eg}$ can be taken as the measure of the crystal-field splitting, has values of less than 0.1. The above observations are depicted in Fig. 1.2a–d.

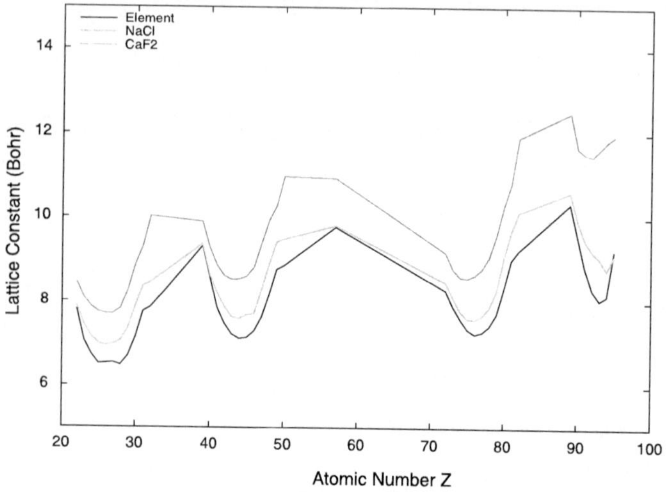

Fig. 1.1 Lattice constant versus atomic number Z

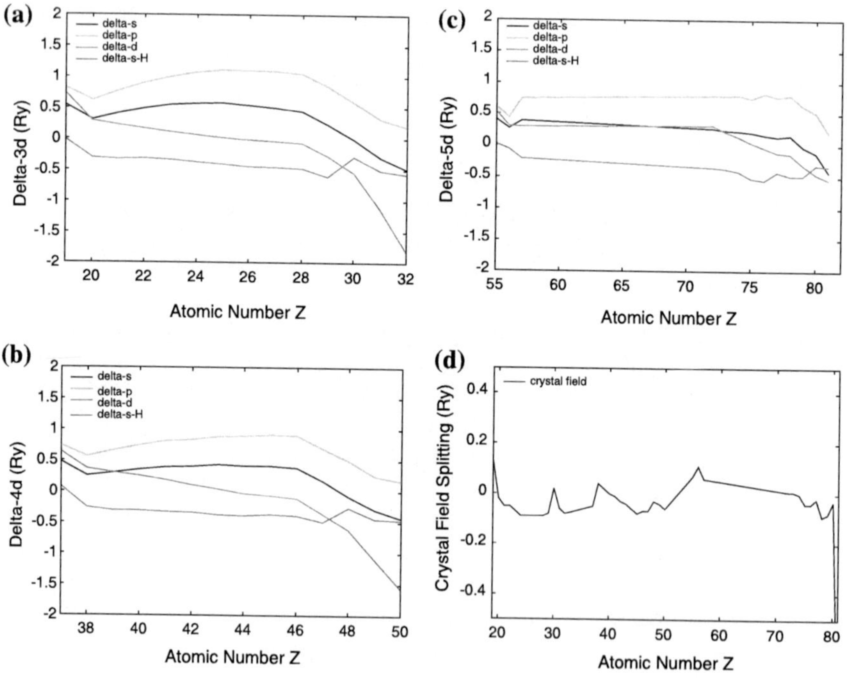

Fig. 1.2 Δ_l versus Z for the NaCl structure

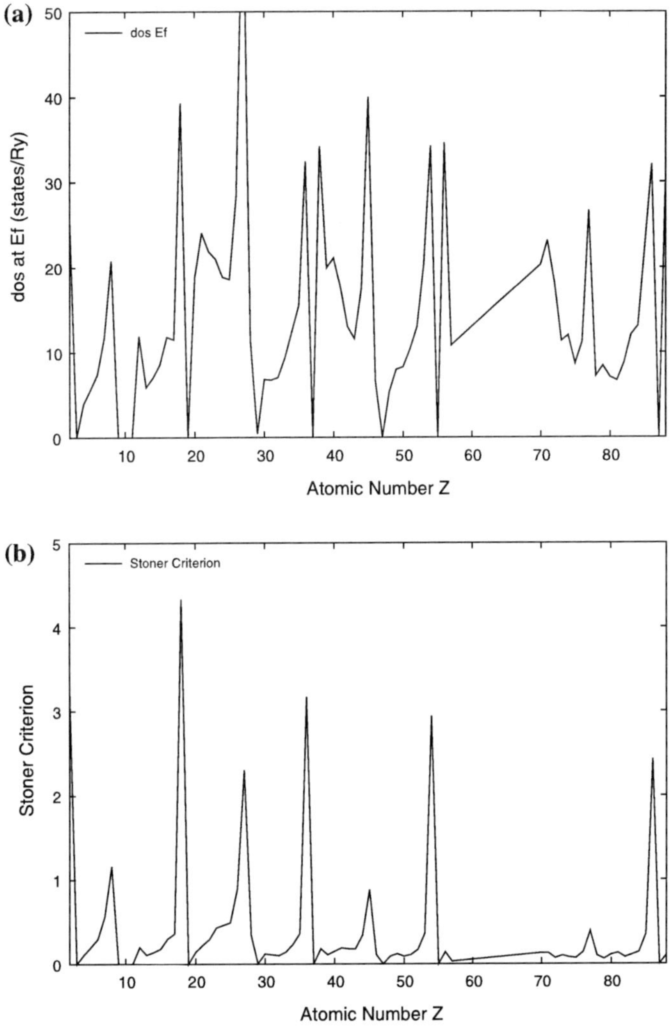

Fig. 1.3 **a** Density of states at E_f v.Z, and **b** Stoner criterion versus Z

(c) In Fig. 1.31.3a, b the Fermi level value of the density of states $N(E_f)$ and the Stoner criterion are shown as a function of Z. The largest $N(E_f)$ is found for CoH which satisfies the Stoner criterion and is known to be ferromagnetic. In the noble gases column large values of the Stoner criterion are found but it is doubtful that these materials can be made in the NaCl structure. In the lanthanides and actinides the f-states create very large $N(E_f)$ which is consistent with experimental evidence that these materials are antiferromagnetic.

(d) In Fig. 1.4a–c the Hopfield parameter η ,the electron-phonon coupling constant λ and the critical temperature Tc for superconductivity. The parameter η is calculated exactly by the Gaspari-Gyorffy theory. However, since we have not performed detailed calculations of the phonon frequencies we have estimated λ from the Debye temperature of the component elements.

1.18 Uses of This Handbook

The total energy results are presented in a graphical form and in terms of the coefficients of a Birch fit. These coefficients can be used to generate the bulk modulus as a function of volume or to obtain pressure versus volume plots. Labeling of the y-axis is not consistent throughout the book in that it ignores the absolute value of the total energy. This, of course, is not an issue since only the energy differences matter. In the energy band and density of states diagrams the Fermi level is set to a value with respect to the muffin-tin zero. The SK parameters given in this book can be used in a great variety of calculations in condensed matter physics. We will discuss a few of their applications here without describing the formalism of these theories. First, from the point of view of band theory of periodic materials, Fermi surface calculations can be performed by using the eigenvalues on a very dense grid of k-points as generated by diagonalizing our SK Hamiltonians, given in Appendix 1. This method also has the flexibility of adjusting a few of the SK parameters (usually the on-site parameters) to obtain a good fit to the Fermi surface for those cases in which the local density theory employed in the first-principles calculation is not in agreement with experiment. The SK Hamiltonian can also be used to obtain the DOS along symmetry directions in k-space, thus facilitating the comparison with photoemission data. Adjustment of SK parameters enables one to fit to various experiments and resolve discrepancies. The SK parameters are particularly useful in the study of disordered materials. The TB form of the coherent-potential approximation (CPA) [39] leads to results that in many cases are nearly equivalent [40] to those obtained by more elaborate techniques such as the KKR-CPA [41]. The TB-CPA has also been used to study the electronic structure of amorphous semiconductors [42]. A computer code using TB-CPA specific to calculate the DOS for substoichiometric hydrides with hydrogen vacancies is listed in Appendix 3. The formalism for this application is given in Ref. [43]. SK parameters have been found to be the essential starting points for theories studying other defects in solids and also for obtaining surface and interface electronic states [44]. Finally, we wish to refer to the work of Varma and Weber [45], who have used SK parametrizations to calculate both the electron–phonon (EP) interaction and phonon dispersion curves. Fry et al. [46] presented a systematic study of the EP interaction in most transition metals using two-center-orthogonal parameters.

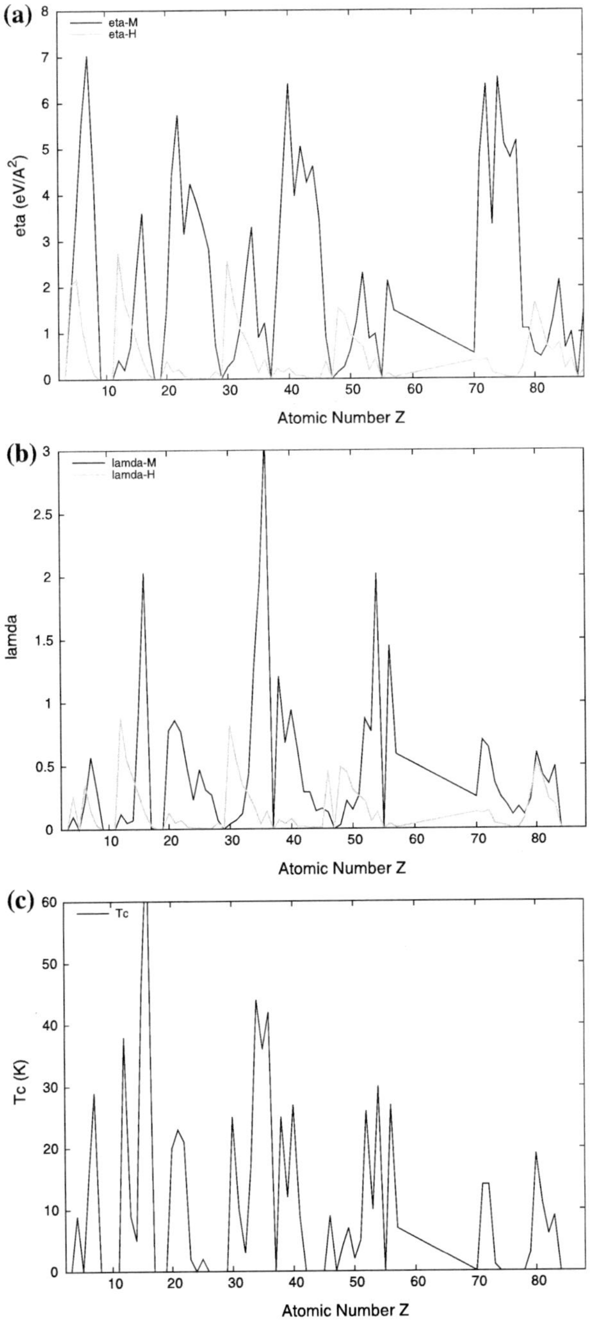

Fig. 1.4 a Hopfield parameter η, **b** Electron–phonon coupling constant λ, and **c** Superconductivity critical temperature T_c

1.19 Description of the Slater-Koster Tables

We tabulate four sets of SK parameters: (1) three-center integrals using an orthogonal basis set, (2) three-center integrals using a nonorthogonal basis set, (3) two-center integrals using an orthogonal basis set, and (4) two-center integrals using a nonorthogonal basis set. For each compound the first page displays the three center orthogonal and nonorthogonal parameters in the three-center approximation. The notation is self-explanatory with the exception of the abbreviations d_1 for $x^2 - y^2$ and d_2 for $3z_2 - r^2$.

On the second page the rms errors and maximum deviations are given per band and a comparison is made of the eigenvalues resulting from the orthogonal and nonorthogonal fits to the APW eigenvalues. At the bottom of the second page a tabulation is made of the Fermi-level quantities, i.e., Fermi energy, total and angular momentum components of the DOS, Fermi velocity, plasmon energy, and integrated DOS up to $\boldsymbol{E_F}$. These quantities correspond to the most accurate of the four calculations presented here, namely the one using a three-center nonorthogonal Hamiltonian. However, the other three sets of SK parameters are also very accurate.

References

1. D.A. Papaconstantopoulos, *Handbook of the Band Structure of Elemental Solids from Z=1 to Z=112*, 2nd edn. (Springer, 2015)
2. B. Stritzker, W. Buckel, Z. Phys **257**, 1 (1972)
3. A.C. Switendick, Ber. Bunsenges, Physik. Chemie **76**, 535 (1972)
4. D.A. Papaconstantopoulos, B.M. Klein, Phys. Rev. Lett. **35**, 110 (1975); B.M. Klein, E.N. Economou, D.A. Papaconstantopoulos, Phys. Rev. Lett. **39**, 574 (1977); D.A. Papaconstantopoulos et al., Phys. Rev. B **17**, 141 (1978)
5. D. Duan, Y. Liu, F. Tian, D. Li, X. Huang, Z. Zhao, H. Yu, B. Liu, W. Tian, T. Cui, Sci. Rep. **4**, 30 (2014)
6. A.P. Drozdov, M.I. Eremets, I.A. Troyan, V. Ksenofontov, S.I. Shylin, Nature **525**, 73 (2015)
7. H. Liu, I.I. Naumov, R. Hoffmann, N.W. Ashcroft, R.J. Hemley, Proc. Natl. Acad. Sci. USA **114**, 6990 (2017); D.A. Papaconstantopoulos, M.J. Mehl, P-H. Chang, Phys. Rev. B **101**, 060506 (R) (2020)
8. J.C. Slater, Phys. Rev. **51**, 846 (1937); L.F. Mattheiss, J. H. Wood, A.C. Switendick, Methods in Computational Physics **8**, 63 (1968)
9. W. Kohn, L. Sham, Phys. Rev. A **140**, 1133 (1965)
10. L. Hedin, B.I. Lundqvist, J. Phys. **C4**, 2064 (1971)
11. P. Hohenberg, W. Kohn, Phys. Rev. B **136**, 864 (1964)
12. O.K. Andersen, Phys. Rev **B12**, 3060 (1975)
13. J.C. Slater, G.F. Koster, Phys. Rev. **94**, 1498 (1954)
14. P.O. Lowdin, J. Chem. Phys. **18**, 1365 (1950); J. Chem. Phys. **19**, 1396 (1951)
15. L.F. Mattheiss, Phys. Rev. B **2**, 3918 (1970)
16. F. Birch, J. Geophys. Res. **83**, 1257 (1978)
17. R.E. Cohen, M.J. Mehl, D.A. Papaconstantopoulos, Phys. Rev. B **50**, 14694 (1994)
18. M.J. Mehl, D.A. Papaconstantopoulos, Phys. Rev. B **54**, 4519 (1996)
19. D.A. Papaconstantopoulos, M.J. Mehl, J. Phys. Condens. Matter **15**, R413 (2003)
20. D.A. Papaconstantopoulos, M.J. Mehl, M.D. Johannes, Phys. Rev. B **82**, 054503 (2010)

21. N. Bernstein, M.J. Mehl, D.A. Papaconstantopoulos, N.I. Papanicolaou, M.Z. Bazant, E. Kaxiras, Phys. Rev. B **62**, 4477 (2000)
22. N. Bernstein, M.J. Mehl, D.A. Papaconstantopoulos, Phys. Rev. B **66**, 075212 (2002)
23. J. Durgavich, S. Sayed, D.A. Papaconstantopoulos, J. Comp. Mat. Sci. **112**, 395 (2016)
24. W. McMillan, Phys. Rev. B **167**, 331 (1968)
25. G.D. Gaspari, B.L. Gyorffy, Phys. Lett. **28**, 801 (1972); for an application see D.A. Papaconstantopoulos, L.L. Boyer, B.M. Klein, A.R. Williams, V.L. Moruzzi, J.F. Janak, Phys. Rev. B **15**, 4221 (1977)
26. P.B. Allen, R.C. Phys, Rev. B **12**, 905 (1975)
27. E.C. Stoner, Proc. R. Soc. A **169**(938), 339 (1939)
28. S.H. Vosko, J.P. Perdew, Can. J. Phys. **53**, 1385 (1975)
29. J.F. Janak, Phys. Rev. B **16**, 255 (1977)
30. D.A. Papaconstantopoulos, Europhys. Lett. **15**, 621 (1991)
31. D.A. Liberman, D.T. Cromer, J.T. Waber, Comp. Phys. Commun. **2**, 107 (1971)
32. N.A. Johnston, C.A. Sholl, P.V. Smith, J. Phys. F **13**, 945 (1983)
33. O.K Andersen, O. Jepsen, Physica **91B**, 317 (1977); O.K Andersen, W. Close, H. Nohl, Phys. Rev. **B17**, 1209 (1978)
34. J.D. Shore, D.A. Papaconstantopoulos, Phys. Rev. B **35**, 1122 (1987)
35. W.A. Harrison, *Electronic Structure and the Properties of Solids* (W. H. Freeman and Co., San Francisco, 1980)
36. L. Shi, D.A. Papaconstantopoulos, Phys. Rev. B **70**, 205101 (2004)
37. G. Lehmann, M. Taut, Phys. Status Solidi B **54**, 469 (1972)
38. B.A. Sanborn, P.B. Allen, D.A. Papaconstantopoulos, Phys. Rev **40**, 6037 (1989)
39. H. Ehrenreich, L.M. Schwartz, Solid State Phys. **31**, 150 (1976); J.S. Faulkner, Prog. Mat. Sci. **27**, 1 (1982)
40. D.A Papaconstantopoulos, P.M. Laufer, A.C. Switendick, in *Hydrogen in Disordered andAmorphous Solids,* ed. by **G.** Bambakidis, R.C. Bowman, Jr. (Plenum Press, New York, 1986), pp. 139–152; P.M. Laufer, D.A. Papaconstantopoulos, *P*hys. Rev. B **33**, 5134 (1986)
41. M. Stocks, W.M. Temmerman, B.L. Gyorffy, Phys. Rev. Lett. **41**, 339 (1978)
42. D.A. Papaconstantopoulos, E.N. Economou, Phys. Rev. B **24**, 7233 (1981)
43. D.A. Papaconstantopoulos, B.M. Klein, J.S. Faulkner, L.L. Boyer, Phys. Rev. B **18**, 2784 (1978)
44. W.E. Pickett, D.A. Papaconstantopoulos, Phys. Rev. B **34**, 8372 (1986)
45. C.M. Varma, W. Weber, Phys. Rev. Lett. **39**, 1094 (1977); C.M. Varma, E.I. Blount, P. Vashista, W. Weber, Phys. *Rev.* **B19**, 6130 (1979)
46. J.L. Fry, G. Fletcher, P.C. Pattnaik, D.A. Papaconstantopoulos, Physica **135B**, 473 (1985)

Chapter 2
Prototype Crystal Structures

In this chapter results for the compounds NaCl, CsCl and CaF2 are given since the crystal structures used for the hydrides in this volume are named after these structures also known as B1, B2 and C1 respectively. These materials are shown to be insulators as known experimentally. However, their energy gap is significantly underestimated as is expected in LDA calculations. The calculated equilibrium lattice parameters differ from experiment by about 3%. It is noted that in both NaCl and CsCl the chlorine p-states are dominant just below the gap. Similarly, in CaF2 the fluorine p-states dominate at the Fermi level. Tight-binding parameters are given only for NaCl in both the two and three-center approximations. The fit to the APW results uses s and p orbitals for both Na and Cl but essentially only the four Cl s–p bands in the valence are included. The RMS errors are very good.

D. A. Papaconstantopoulos, *Band Structure of Cubic Hydrides*,
https://doi.org/10.1007/978-3-031-06878-2_2

2.1 Sodium Chloride (NaCl)

See Figs. 2.1, 2.2 and Tables 2.1, 2.2.

See Figs. 2.3, 2.4, 2.5, 2.6 and Tables 2.3, 2.4.

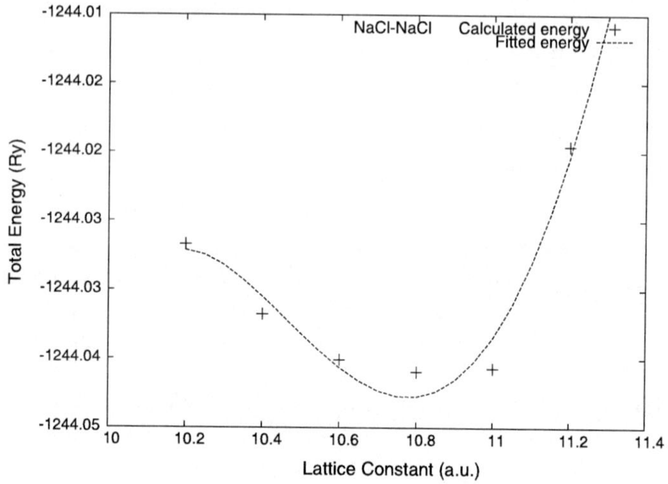

Fig. 2.1 Total energy versus lattice constant for the NaCl structure

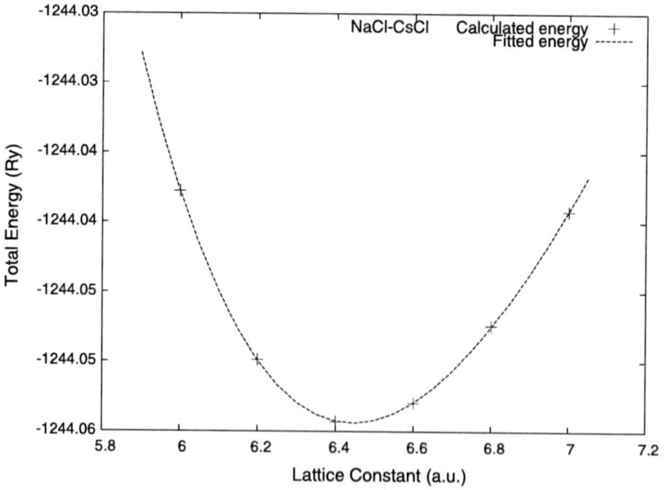

Fig. 2.2 Total energy versus lattice constant for the CsCl structure

Table 2.1 Lattice constant, bulk modulus, gap, total energy

Stru		a (Bohr)	B (MBar)	Gap (Ry)	Total Energy (Ry)
NaCl	insulator	10.74	0.95	0.33	-1244.04114
CsCl	insulator	6.44	0.33	0.29	-1244.05941
exp(NaCl)		10.66	0.24	0.62	

Table 2.2 Birch fit coefficients

	A1	A2	A3	A4
NaCl	-1.229029E+03	-1.972171E+03	8.610900E+04	-1.249290E+06
CsCl	-1.243558E+03	-3.418860E+01	2.469088E+02	1.281111E+04

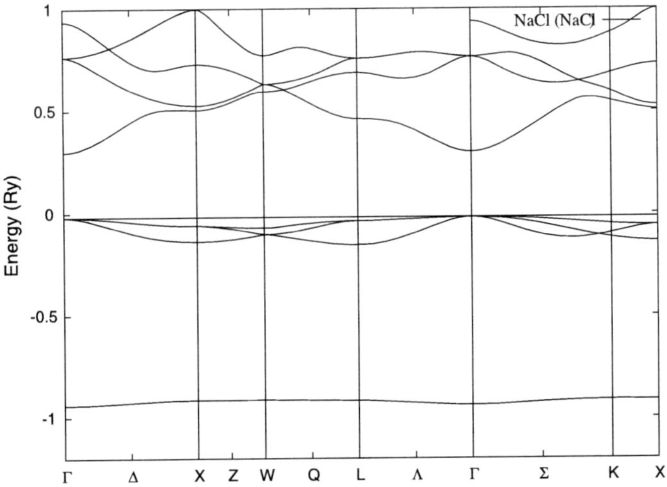

Fig. 2.3 Energy bands of NaCl in the NaCl structure

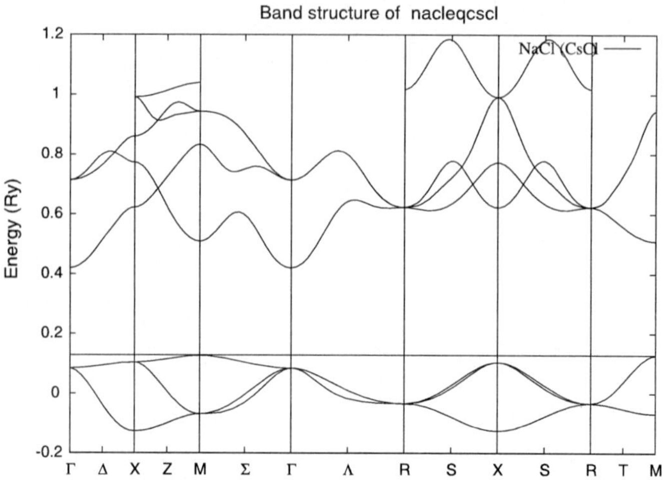

Fig. 2.4 Energy bands of NaCl in the CsCl structure

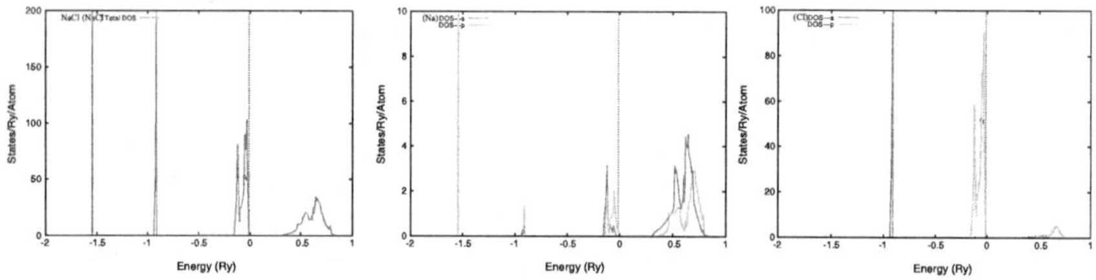

Fig. 2.5 Total, angular momentum and site decomposed densities of states for the NaCl structure

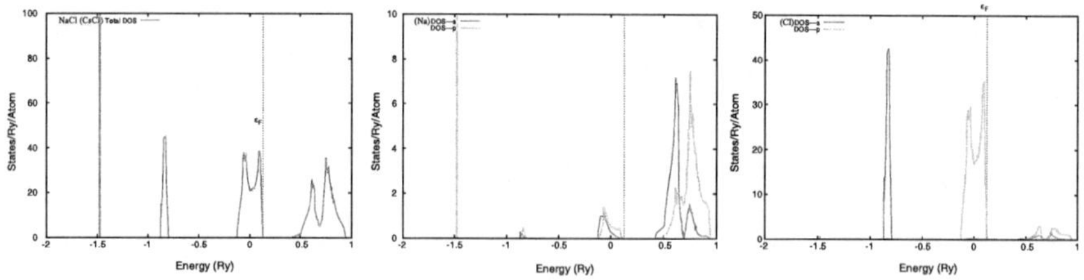

Fig. 2.6 Total, angular momentum and site decomposed densities of states for the CsCl structure

Table 2.3 NaCl (NaCl) a = 10.84 Bohr Slater–Koster 3-center parameters

	ORTHOGONAL	NON-ORTHOGONAL	
	ENERGY INTEGRALS Ry	ENERGY INTEGRALS Ry	OVERLAP INTEGRALS
ON SITE Na-Na			
s,s(000)	0.41733	0.12390	
FIRST NEIGHBOR Na-Na			
s,s(110)	-0.02640	-0.06177	0.04976
SECOND NEIGHBOR Na-Na			
s,s(200)	-0.00222	-0.00868	0.01579
ON SITE Cl-Cl			
s,s(000)	-0.88537	-0.86944	
x,x(000)	-0.05911	-0.06781	
FIRST NEIGHBOR Cl-Cl			
s,s(110)	0.00988	-0.04341	0.06398
s,x(110)	-0.01303	0.00003	0.00410
x,x(110)	0.00789	0.00701	0.00391
x,x(011)	-0.00304	-0.00270	0.01071
x,y(110)	0.00203	0.00934	-0.04897
SECOND NEIGHBOR Cl-Cl			
s,s(200)	0.00571	0.01781	-0.01076
s,x(200)	0.00790	0.00501	-0.01027
x,x(200)	-0.00672	0.00065	-0.04072
y,y(200)	0.00040	0.00049	0.00277
FIRST NEIGHBOR Na-Cl			
s,s(100)	-0.08051	-0.19068	0.10606
s,x(100)	-0.06922	-0.06883	0.13066
SECOND NEIGHBOR Na-Cl			
s,s(111)	0.00139	-0.01104	0.01252
s,x(111)	0.00349	-0.00244	-0.00813

BAND	ORTHOGONAL RMS ERROR mRy	MAX DEVIATION k	MAX DEVIATION mRy	NON-ORTHOGONAL RMS ERROR mRy	MAX DEVIATION k	MAX DEVIATION mRy
1	3.2	(000)	7.4	0.6	(226)	1.5
2	4.3	(005)	13.2	0.8	(444)	2.3
3	2.0	(264)	6.1	0.9	(226)	3.4
4	1.9	(048)	6.8	0.9	(048)	2.5
1-4	2.8			0.4		

ENERGY VALUES IN Ry AT SELECTED k-POINTS

	ORTHOGONAL	APW	NON-ORTHOGONAL
GAMMA 1	-0.94773	-0.94030	-0.93976
GAMMA 1	0.30235	0.29886	0.30020
GAMMA 15	-0.01995	-0.01908	-0.01922
X1 (008)	-0.91146	-0.91496	-0.91371
X1 (008)	0.53046	0.50479	0.50240
X4' (008)	-0.14623	-0.13930	-0.13968
X5' (008)	-0.05877	-0.05916	-0.06062
L1 (444)	-0.16075	-0.15665	-0.15896
L3 (444)	-0.03916	-0.03717	-0.03851
L2' (444)	-0.91966	-0.91847	-0.91714
L2' (444)	0.52790	0.46065	0.46390
W1 (048)	-0.07826	-0.07145	-0.07395
W1 (048)	0.56486	0.59605	0.59700
W2' (048)	-0.91346	-0.91398	-0.91365
W3 (048)	-0.10411	-0.10454	-0.10473

Table 2.4 (NaCl structure) a = 10.84 Bohr Slater–Koster 2-center parameters

	ORTHOGONAL	NON-ORTHOGONAL	
	ENERGY INTEGRALS Ry	ENERGY INTEGRALS Ry	OVERLAP INTEGRALS
ON SITE Na-Na			
s	0.49035	0.48302	
FIRST NEIGHBOR Na-Na			
sss	-0.01535	-0.01594	-0.00655
SECOND NEIGHBOR Na-Na			
sss	-0.00515	-0.00256	0.00410
ON SITE Cl-Cl			
s	-0.91747	-0.91852	
p	-0.07286	-0.07769	
FIRST NEIGHBOR Cl-Cl			
sss	-0.00063	-0.00096	-0.00034
sps	0.00243	-0.00290	0.00214
pps	0.01784	0.02019	-0.02656
ppp	-0.00283	-0.00313	0.00264
SECOND NEIGHBOR Cl-Cl			
sss	0.00019	-0.00031	0.00031
sps	-0.01105	-0.01024	0.00287
pps	-0.00044	0.00209	-0.00235
ppp	0.00159	-0.00008	0.00201
FIRST NEIGHBOR Na-Cl			
sss	0.01911	0.01700	-0.00042
sps	0.00886	0.01424	-0.00821
SECOND NEIGHBOR Na-Cl			
sss	0.00458	-0.00131	-0.00012
sps	-0.02749	-0.00280	0.06455

BAND	ORTHOGONAL RMS ERROR mRy	MAXIMUM DEVIATION k	mRy	NON-ORTHOGONAL RMS ERROR mRy	MAXIMUM DEV k	mRy
1	1.3	(004)	3.0	0.3	(226)	0.7
2	3.6	(033)	9.4	1.6	(002)	3.5
3	2.8	(066)	6.8	1.2	(066)	2.7
4	4.9	(084)	16.0	0.7	(008)	2.6
1-4	3.2			1.0		

ENERGY VALUES IN Ry AT SELECTED k-POINTS

	ORTHOGONAL	APW	NON-ORTHOGONAL
GAMMA 1	-0.94270	-0.94030	-0.94061
GAMMA 1	0.29400	0.29886	0.29951
GAMMA 15	-0.01870	-0.01908	-0.01970
X1 (008)	-0.91385	-0.91496	-0.91496
X1 (008)	0.52080	0.50479	0.50704
X4' (008)	-0.13870	-0.13930	-0.13934
X5' (008)	-0.05600	-0.05916	-0.06179
L1 (444)	-0.16240	-0.15665	-0.15974
L3 (444)	-0.03704	-0.03717	-0.03719
L2' (444)	-0.91859	-0.91847	-0.91836
L2' (444)	0.52260	0.46065	0.51660
W1 (048)	-0.08750	-0.07145	-0.07016
W1 (048)	0.57470	0.59605	0.59996
W2' (048)	-0.91459	-0.91398	-0.91352
W3 (048)	-0.10380	-0.10454	-0.10318

2.2 Cesium Chloride (CsCl)

See Figs. 2.7, 2.8 and Tables 2.5, 2.6.
See Figs. 2.9, 2.10, 2.11 and 2.12.

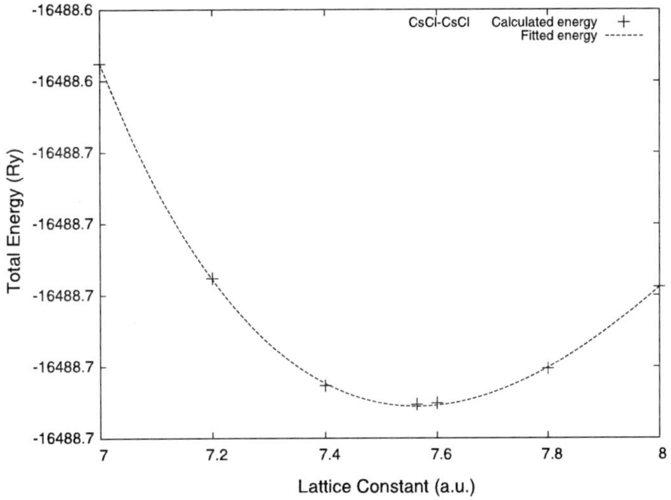

Fig. 2.7 Total energy versus lattice constant for CsCl

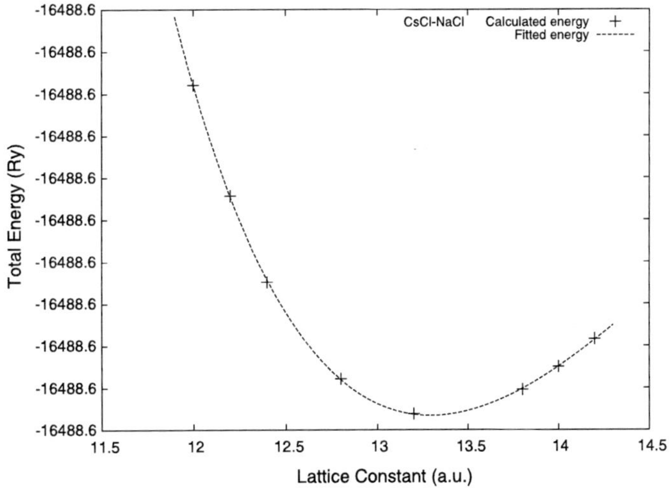

Fig. 2.8 Total energy versus lattice constant for NaCl

Table 2.5 Lattice constant, bulk modulus, gap, total energy

Stru	a (Bohr)	B (MBar)	Energy Gap (eV)	Total Energy (Ry)
CsCl insulator	7.56	0.24	0.38	-16488.67274
NaCl insulator	13.29	0.14	0.32	-16488.63826
exp (CsCl)	7.79		0.61	

Table 2.6 Birch fit coefficients

	A1	A2	A3	A4
CsCl	-1.648834E+04	-1.227769E+01	-1.855384E+03	8.413948E+04
NaCl	-1.648849E+04	1.431256E+01	-4.192655E+03	1.725427E+05

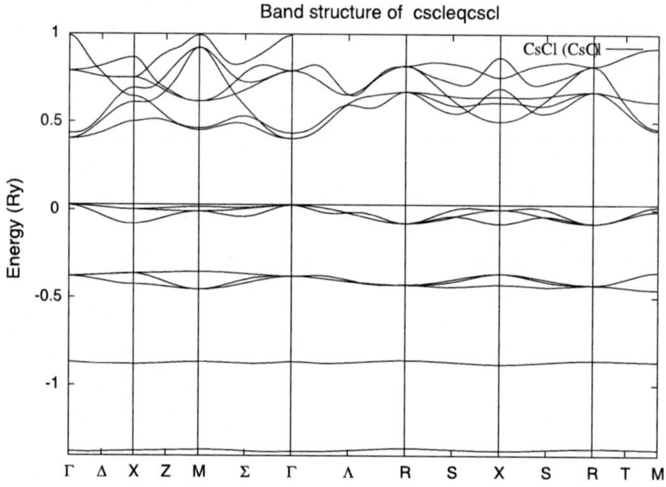

Fig. 2.9 Energy bands of CsCl in the CsCl structure

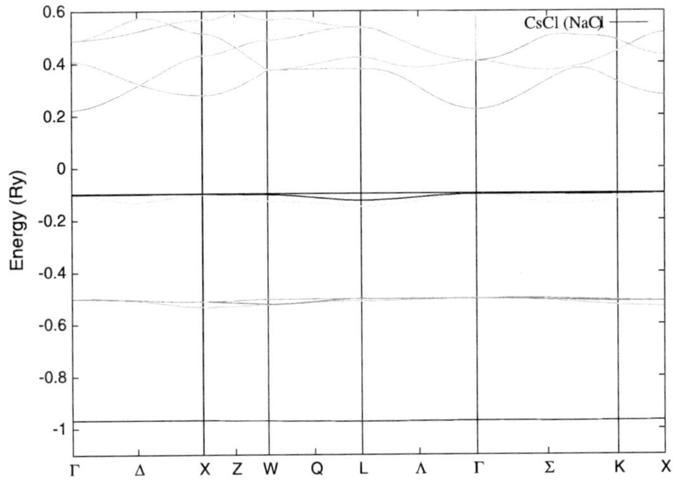

Fig. 2.10 Energy bands of CsCl in the NaCl structure

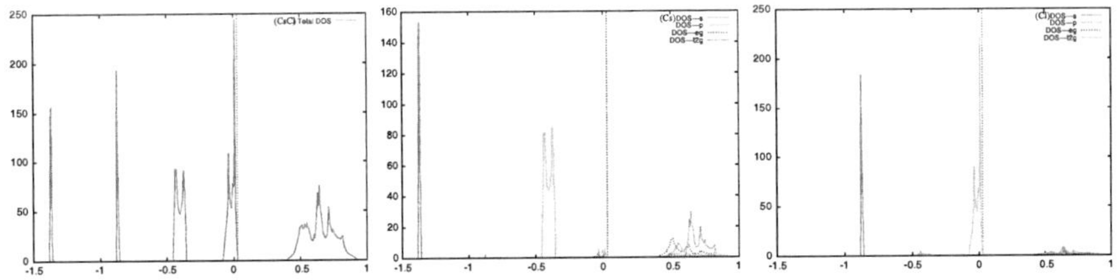

Fig. 2.11 Total, angular momentum and site decomposed densities of states of CsCl in the CsCl structure

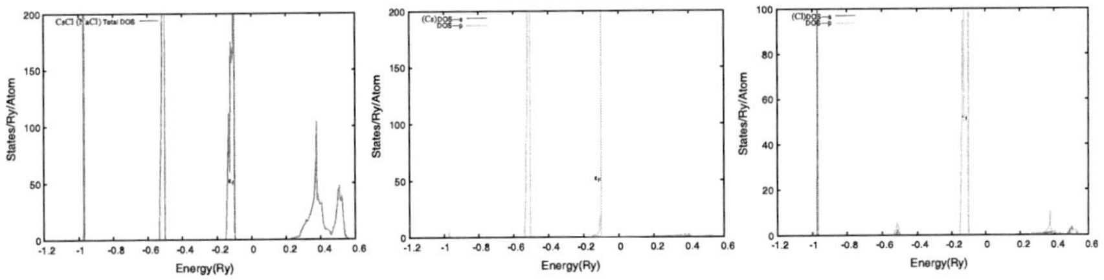

Fig. 2.12 Total, angular momentum and site decomposed density of states of CsCl in the NaCl structure

2.3 Calcium Fluoride (CaF2)

See Fig. 2.13 and Tables 2.7, 2.8.
See Figs. 2.14 and 2.15

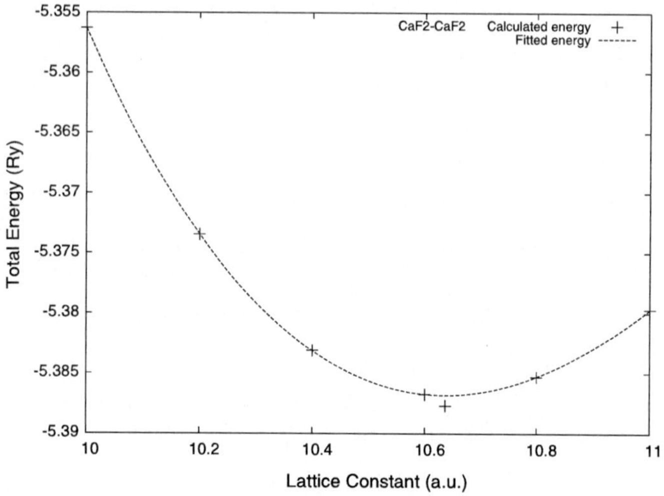

Fig. 2.13 Total energy versus lattice constant for CaF2

Table 2.7 Lattice constant, bulk modulus, gap, total energy

Stru	a (Bohr)	B (MBar)	Gap (Ry)	Total Energy (Ry)
Fluorite	10.64	0.75	0.45	-1755.38769
exp	10.33	0.83	0.87	

Table 2.8 Birch fit coefficients

A1	A2	A3	A4
-4.073252E+00	-9.985728E+01	1.022849E+03	3.648442E+04

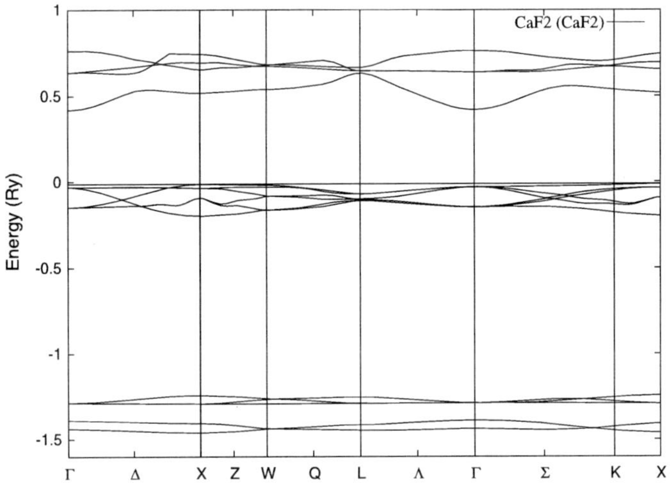

Fig. 2.14 Energy bands of CaF2

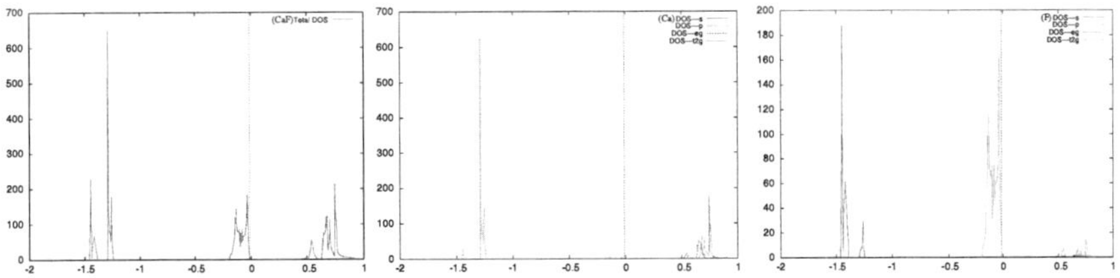

Fig. 2.15 Total, angular momentum and site-decomposed density of states of CaF2

Chapter 3
Alkali Hydrides

This chapter covers the alkali metal hydrides LiH, NaH, KH, RbH, CsH and FrH. Results are presented for the crystal structures NaCl (B1),and CsCl (B2) [1-3]. The ground state of these materials is in the NaCl also known as the rock-salt structure. With the exception of LiH and FrH these hydrides undergo a phase transition to the CsCl structure at the following pressures 29GPa, 4GPa, 2GPa and 0.8GPa for NaH, KH, RbH and CsH respectively. These materials are experimentally found to be insulators which is also confirmed by the present calculations. However, since the calculations were performed in the local density approximation (LDA), the energy gap is underestimated. The lattice parameters are in good agreement with experiment and the bulk moduli in worse agreement which is expected for soft materials. We note the increase of the lattice constant from the lighter to the heavier hydrides. Examining the energy bands and densities of states diagrams we observe that,as expected,the occupied states have predominantly s–H character and the conduction bands have strong p character of the alkali element but a non-negligible d-like contribution occurs especially for the heavier ones. For only the LiH, figures and tables are given for the Im3m structure which under pressure is metallic and presents possible superconducting behavior. Also Tight-binding parameters in the NaCl structure are given based on both orthogonal and non-orthogonal Hamiltonians using three- and two-center integrals.

D. A. Papaconstantopoulos, *Band Structure of Cubic Hydrides*,
https://doi.org/10.1007/978-3-031-06878-2_3

3.1 Lithium Hydride (LiH)

See Fig. 3.1, Tables 3.1, 3.2, and 3.3.

See Figs. 3.2, 3.3, 3.4, 3.5, Tables 3.4, and 3.5.

See Fig. 3.6.

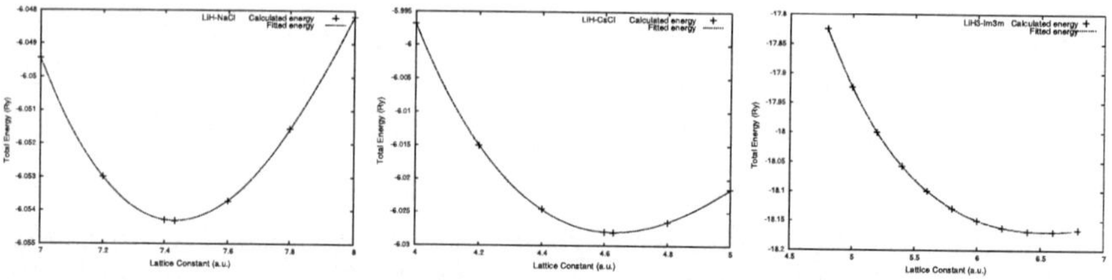

Fig. 3.1 Total energy versus lattice constant of LiH in the NaCl, CsCl and Im3m structures

Table 3.1 Lattice constants, bulk modulus, gap, total energy

Stru	a (Bohr)		B (MBar)	Gap (Ry)	Total Energy (Ry)
NaCl	7.43	Insulator	0.40	0.18	-16.05432
NaCl	5.20	Semimetal	-	-	-
CsCl	4.63	Insulator	0.39	0.026	-
Im3m(LiH3)	6.54	(P=0)	0.53	-	-
Im3m(LiH3)	4.90	(P=2.01MBar)	5.46	-	-
exp(NaCl)	7.72		0.36	0.36	

Table 3.2 Birch fit coefficients

	A1	A2	A3	A4
NaCl	-5.642097E+00	-2.023694E+01	2.930125E+02	-1.040779E+03
CsCl	-5.633271E+00	-1.889874E+01	2.662976E+02	-9.119566E+01
Im3m	-1.749122E+01	-3.962183E+01	6.577760E+02	-2.220277E+03

Table 3.3 LiH (NaCl)

```
   Fermi Energy   Total DOS
      (Ry)        (States/Ry)
    0.358            0.0         Insulator
              LiH  (CsCl)
    Fermi Energy   Total DOS
      (Ry)        (States/Ry)
    0.642            0.0         Insulator
              LiH3             Im3m a=6.543 Bohr
     Density of States at the Fermi Energy
  Energy   Total DOS          Decomposed DOS inside the muffin tins
     (Ry)        (States/Ry)       s          p         d         f
 Li  0.64169     5.03919     0.03492    0.09019    0.01094    0.00112
  H  0.64169     5.03919     0.54144    0.01500    0.00063    0.00004
 Electron-ion interaction (Hopfield parameter) (eV/A^2) Li:0.244 H:0.838
  Li      MT Radius and Charge =     1.4 (Bohr)    2.0274
   H      MT Radius and Charge =     1.0 (Bohr)    0.3733
              LiH3              Im3m a=4.9 Bohr (P=2.01 MBar)
     Density of States at the Fermi Energy
  Energy   Total DOS          Decomposed DOS inside the muffin tins
     (Ry)        (States/Ry)       s          p         d         f
 Li  1.32459     2.34070     0.03436    0.13956    0.03963    0.00567
  H  1.32459     2.34070     0.31186    0.03251    0.00233    0.00020
  Electron-ion interaction (Hopfield parameter) (eV/A^2) Li:1.128 H3:4.962
   Li      MT Radius and Charge     1.4 (Bohr)   2.2805
    H      MT Radius and Charge     1.0 (Bohr)   0.5403
```

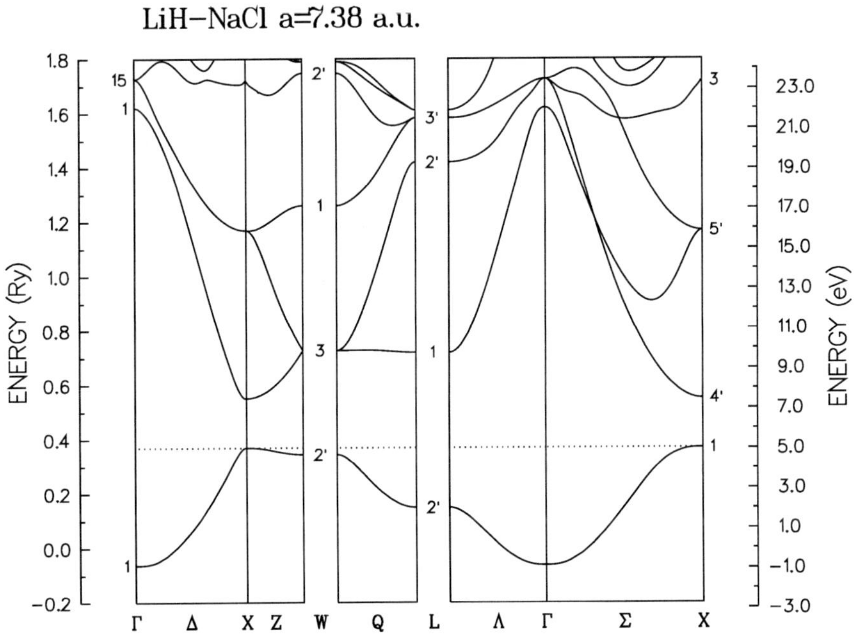

Fig. 3.2 Energy bands of LiH in the NaCl structure (tight binding)

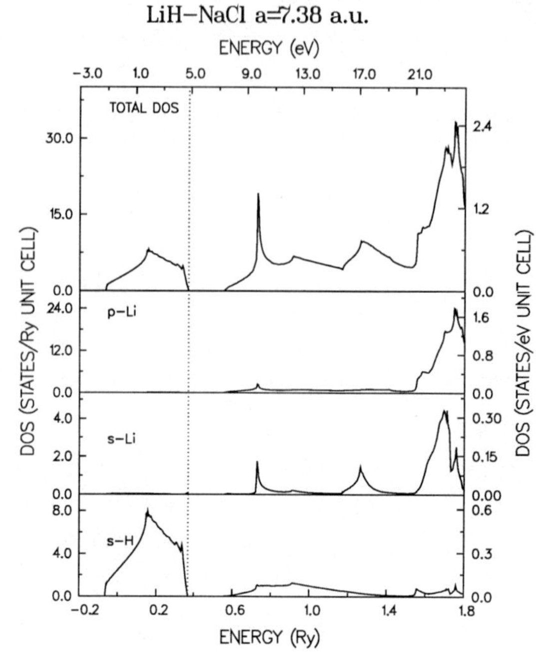

Fig. 3.3 Total, angular momentum and site decomposed densities of states of LiH in the NaCl structure (tight-binding)

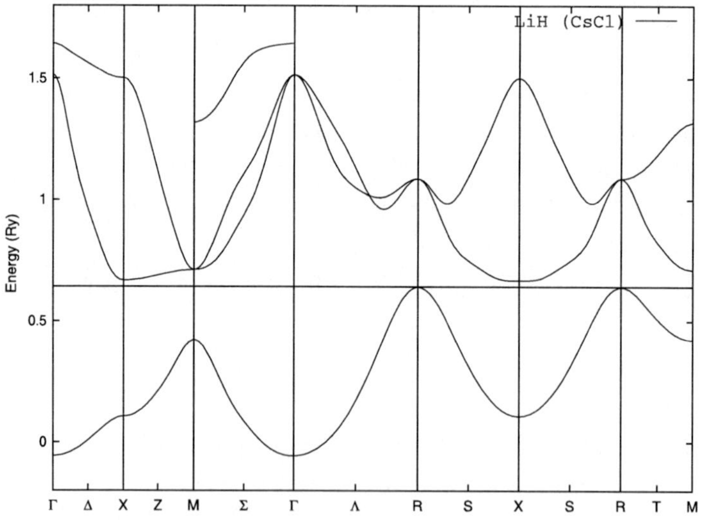

Fig. 3.4 Energy bands of LiH in the CsCl structure

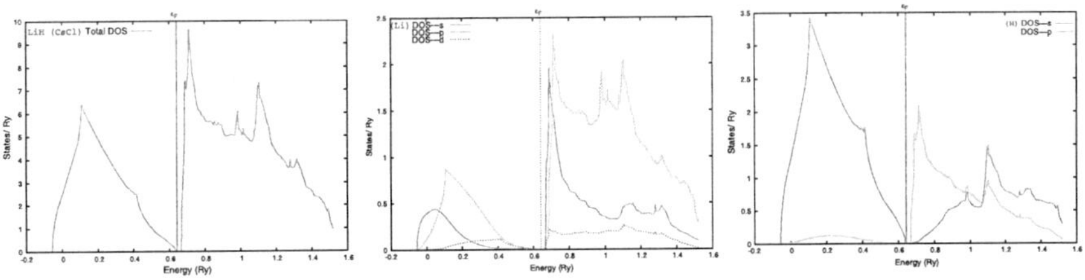

Fig. 3.5 Total, angular momentum and site decomposed densities of states of LiH in the CsCl structure

Table 3.4 LiH (NaCl) a = 7.378 Bohr Slater–Koster 3-center parameters

	ORTHOGONAL	NON-ORTHOGONAL	
	ENERGY INTEGRALS Ry	ENERGY INTEGRALS Ry	OVERLAP INTEGRALS
ON SITE Li-Li			
s,s(000)	2.23673	4.07128	
x,x(000)	3.23361	4.81079	
FIRST NEIGHBOR Li-Li			
s,s(110)	-0.07999	-0.04318	0.05380
s,x(110)	0.13501	0.38617	0.05762
x,x(110)	-0.63067	-1.29812	-0.10849
x,x(011)	0.41712	0.65889	0.06206
x,y(110)	0.07534	0.02343	-0.04511
SECOND NEIGHBOR Li-Li			
s,s(200)	-0.14661	-0.50442	0.00000
s,x(200)	-0.27330	-0.76768	0.00000
x,x(200)	0.47191	1.15088	0.00000
y,y(200)	0.24945	0.32307	0.00000
ON SITE H-H			
s,s(000)	0.70746	0.53983	
FIRST NEIGHBOR H-H			
s,s(110)	0.04930	0.04338	0.03430
s,x(110)	-0.01064	-0.03369	0.00427
x,x(110)	-0.00367	-0.06031	-0.14147
x,x(011)	0.10106	0.02071	0.05736
x,y(110)	-0.05904	-0.00940	-0.01704
SECOND NEIGHBOR H-H			
s,s(200)	-0.02325	-0.00487	0.00000
s,x(200)	0.01144	0.00684	0.00000
x,x(200)	0.02143	0.02133	0.00000
y,y(200)	-0.02817	-0.02071	0.00010
FIRST NEIGHBOR Li-H			
s,s(100)	-0.11228	-0.11864	0.07886
s,x(100)	0.98204	1.38660	-0.17296
x,s(100)	0.17654	0.15908	-0.06126
x,x(100)	1.55587	2.23420	-0.04945
y,y(100)	-0.99096	-1.09528	0.07298
SECOND NEIGHBOR Li-H			
s,s(111)	-0.00923	-0.01871	0.00000
x,s(111)	0.00116	0.01978	0.00000
s,x(111)	0.02256	0.03715	0.00000
x,x(111)	0.03722	0.02090	0.00000
x,y(111)	0.01796	0.03063	0.00000

BAND	ORTHOGONAL RMS ERROR mRy	ORTHOGONAL MAXIMUM DEVIATION k	ORTHOGONAL MAXIMUM DEVIATION mRy	NON-ORTHOGONAL RMS ERROR mRy	NON-ORTHOGONAL MAXIMUM DEVIATION k	NON-ORTHOGONAL MAXIMUM DEVIATION mRy
1	7.5	(048)	20.5	5.5	(000)	16.4
2	7.6	(033)	17.1	4.9	(111)	14.0
3	22.1	(022)	49.8	11.7	(011)	27.0
4	14.1	(004)	33.2	8.5	(003)	20.9
5	26.5	(118)	122.3	24.5	(118)	120.3
1-5	17.3			13.2		

ENERGY VALUES IN Ry AT SELECTED k-POINTS

	ORTHOGONAL	APW	NON-ORTHOGONAL
GAMMA 1	-0.06066	-0.06216	-0.05895
GAMMA 1	1.61747	1.62056	1.60986
GAMMA 15	1.71281	1.72532	1.72008
X1 (008)	0.35360	0.37042	0.36449
X1 (008)	1.69415	1.71915	1.69397
X4'(008)	0.56348	0.55291	0.55633
X5'(008)	1.15659	1.17263	1.16053
L1 (444)	0.71762	0.72117	0.72037
L2'(444)	0.15151	0.14934	0.15367
L2'(444)	1.38470	1.40240	1.39484
L3'(444)	1.59337	1.60983	1.59894
W1 (048)	1.26235	1.27375	1.27522
W2'(048)	0.36920	0.34868	0.35942
W2'(048)	1.71364	1.70980	1.71290
W3 (048)	0.71938	0.73075	0.72493
ODD(224)	1.44118	1.42938	1.43283
EVEN(224)	0.08940	0.09462	0.08881
EVEN(224)	0.91858	0.91132	0.91510
EVEN(224)	1.33195	1.30466	1.32124
EVEN(224)	1.65437	1.65137	1.65547

Table 3.5 LiH (NaCl) a = 7.378 Bohr Slater–Koster 2-center parameters

	ORTHOGONAL ENERGY INTEGRALS	NON-ORTHOGONAL ENERGY INTEGRALS Ry	NON-ORTHOGONAL OVERLAP INTEGRALS Ry
ON SITE Li-Li			
s	1.72342	0.55837	
p	1.58831	0.74158	
FIRST NEIGHBOR Li-Li			
(sss)	-0.07262	0.02218	0.17143
(pps)	0.07919	-0.12548	-0.32669
(ppp)	-0.02479	0.00084	0.06825
(sps)	0.06614	-0.05519	-0.23189
SECOND NEIGHBOR Li-Li			
(sss)	-0.08053	-0.01964	0.01180
(pps)	0.07867	0.02619	-0.03865
(ppp)	0.01166	-0.01510	-0.00596
(sps)	0.07780	0.02434	-0.02233
ON SITE H-H			
s	0.61815	0.23588	
FIRST NEIGHBOR H-H			
(sss)	0.04190	-0.02893	-0.00641
(sps)	-0.02578	0.00000	0.00000
(pps)	-0.28108	0.00000	0.00000
(ppp)	-0.17938	0.00000	0.00000
SECOND NEIGHBOR H-H			
(sss)	0.01153	0.01336	0.01374
(sps)	0.00916	0.00000	0.00000
(pps)	0.03733	0.00000	0.00000
(ppp)	0.14699	0.00000	0.00000

FIRST NEIGHBOR Li-H			
(sss)	0.14461	-0.02504	0.24033
(sps)	0.71382	0.00000	0.00000
(pss)	0.16588	0.00344	0.34752
(pps)	-0.85499	0.00000	0.00000
(ppp)	0.23024	0.00000	0.00000
SECOND NEIGHBOR Li-H			
(sss)	-0.01701	-0.01520	0.00617
(pss)	-0.01182	-0.03905	-0.00047
(sps)	-0.06478	0.00000	0.00000
(pps)	0.05723	0.00000	0.00000
(ppp)	0.01701	0.00000	0.00000

BAND	RMS ERROR	MAXIMUM DEVIATION		RMS ERROR	MAXIMUM DEVIATION	
	mRy	k	mRy	mRy	k	mRy
1	3.5	(226)	7.2	5.7	(008)	10.5
2	14.8	(055)	41.8	6.8	(003)	19.0
3	31.7	(022)	71.7	11.4	(224)	33.4
4	18.4	(005)	45.4	8.7	(224)	23.3
1-4	17.1			8.2		

ENERGY VALUES IN Ry AT SELECTED k-POINTS

	ORTHOGONAL	APW	NON-ORTHOGONAL
GAMMA 1	-0.05955	-0.06216	-0.06464
GAMMA 1	1.61845	1.62056	1.61074
GAMMA 15	1.69869	1.72532	1.72488
X1 (008)	0.36467	0.37042	0.35997
X1 (008)	1.68580	1.71915	1.69399
X4' (008)	0.56005	0.55291	0.56014
X5' (008)	1.16050	1.17263	1.16671
L1 (444)	0.69614	0.72117	0.72777
L2' (444)	0.14697	0.14934	0.13957
L2' (444)	1.37039	1.40240	1.38446
L3' (444)	1.59230	1.60983	1.59683
W1 (048)	1.25369	1.27375	1.27374
W2' (048)	0.35179	0.34868	0.35044
W2' (048)	1.69862	1.70980	1.71297
W3 (048)	0.74044	0.73075	0.72469

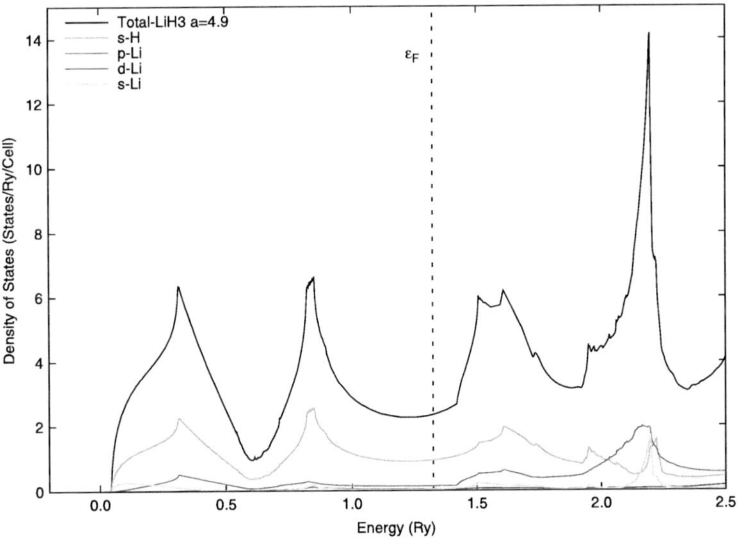

Fig. 3.6 Total, angular momentum and site decomposed densities of states of LiH3 in the Im3m structure

3.2 Sodium Hydride (NaH)

See Fig. 3.7, Tables 3.6, and 3.7.

See Figs. 3.8, 3.9, 3.10, 3.11, Tables 3.8 and 3.9.

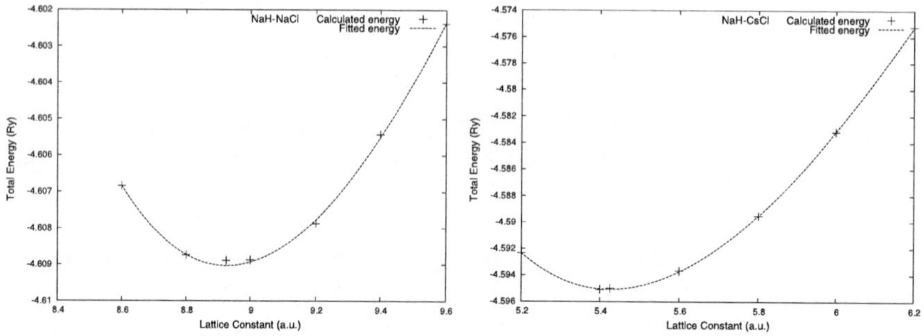

Fig. 3.7 Total energy versus lattice constant for the NaCl and CsCl structures

Table 3.6 Lattice constants, bulk modulus, gap, total energy

	a (Bohr)		B (MBar)	Gap (Ry)	Total Energy (Ry)
NaCl	8.93	Insulator	0.27	0.25	-324.60903
CsCl	5.43	Insulator	0.29	0.46	-
CsCl	4.70	(P=0.27MBar) Ins.	1.14	0.05	-
exp(NaCl)	9.22		0.19	0.41	-

Table 3.7 Birch fit coefficients

	A1	A2	A3	A4
NaCl	-4.270068E+00	-2.055931E+01	2.835999E+02	8.721653E+02
CsCl	-4.198440E+00	-2.470660E+01	4.234917E+02	-1.174396E+03

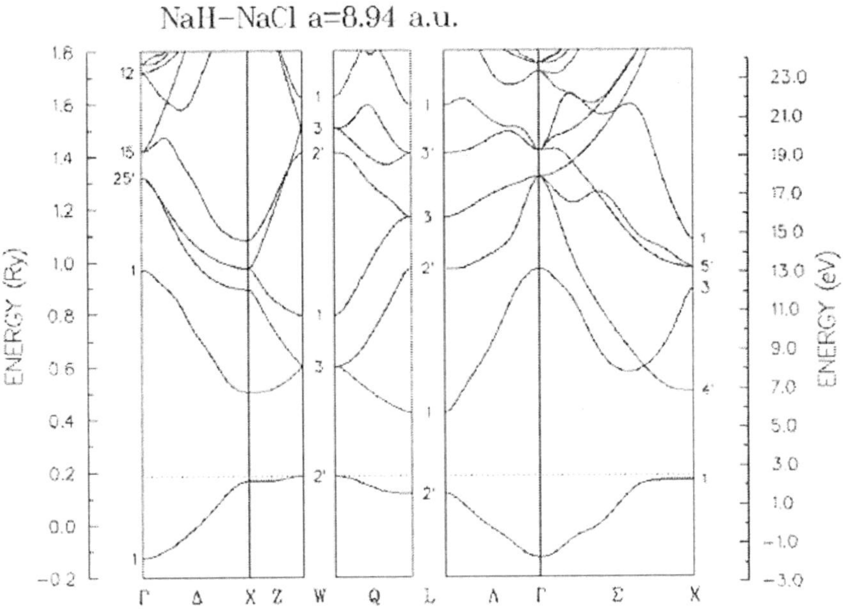

Fig. 3.8 Energy bands of NaH in the NaCl structure

Fig. 3.9 Total, angular momentum and site decomposed densities of states of NaH in the NaCl structure

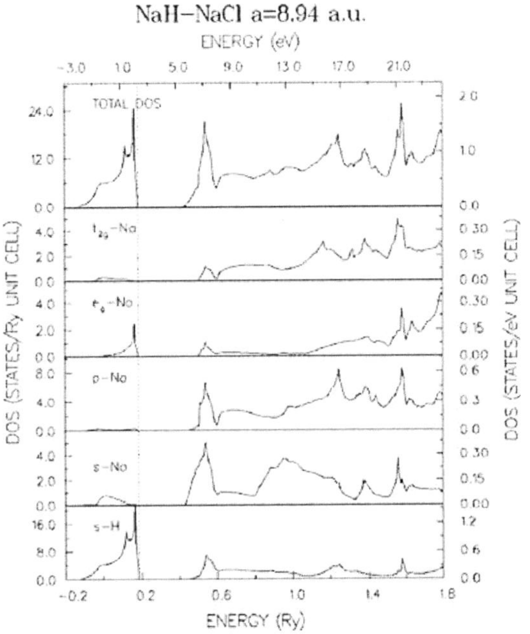

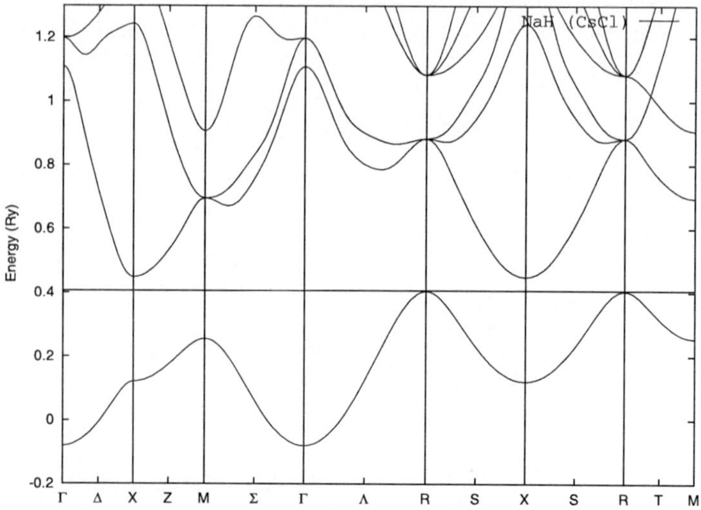

Fig. 3.10 Energy bands of NaH in the CsCl structure

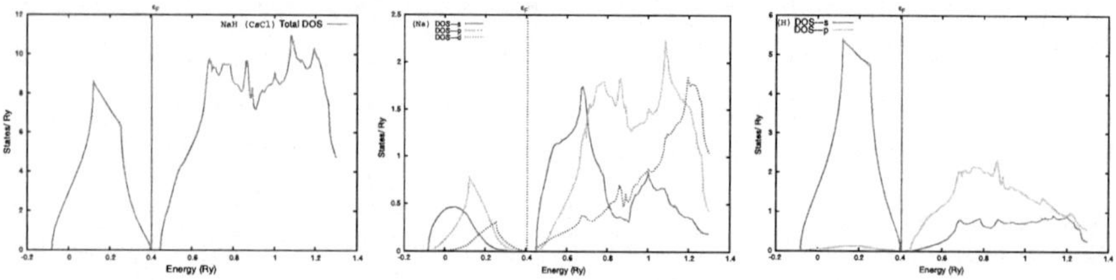

Fig. 3.11 Total, angular momentum and site decomposed densities of states of NaH in the CsCl structure

Table 3.8 NaH (NaCl) a = 8.944 Bohr Slater–Koster 3-center parameters

	ORTHOGONAL	NON-ORTHOGONAL	
	ENERGY INTEGRALS Ry	ENERGY INTEGRALS Ry	OVERLAP INTEGRALS
ON SITE Na-Na			
s,s(000)	0.85745	0.38031	
x,x(000)	1.22519	0.84163	
xy,xy(000)	1.81859	1.19244	
d2,d2(000)	1.67088	1.63266	
FIRST NEIGHBOR Na-Na			
s,s(110)	-0.00240	-0.03045	0.15149
s,x(110)	0.02252	-0.00883	-0.12858
s,xy(110)	-0.12665	-0.04054	-0.10351
s,d2(110)	-0.01271	0.04809	0.06545
x,x(110)	0.03583	-0.03635	-0.09707
x,x(011)	-0.01668	-0.04688	0.01543
x,y(110)	0.03661	-0.05225	-0.15788
x,xy(110)	-0.11317	-0.04256	-0.10844
x,xy(011)	0.02249	0.00935	0.06511
z,d2(011)	-0.00812	0.03046	0.01100
z,d1(011)	0.00368	0.08610	0.08709
xy,xy(110)	-0.13740	0.07827	0.14152
xy,xy(011)	0.02819	-0.10921	-0.13463
xy,xz(011)	-0.05742	-0.07023	-0.09965
xy,d2(110)	-0.04970	-0.12367	-0.10708
d2,d2(110)	-0.04865	0.14815	0.10181
d1,d1(110)	0.05416	0.11518	0.04256
SECOND NEIGHBOR Na-Na			
s,s(200)	0.00386	0.00573	0.06617
s,x(200)	-0.01889	-0.00174	-0.05569
s,d2(002)	0.03940	-0.00290	0.00273
x,x(200)	-0.03534	-0.03813	-0.09051
y,y(200)	0.01234	-0.00819	0.00568
x,xy(020)	-0.00045	-0.00527	0.00693
z,d2(002)	0.01308	-0.01882	-0.01334
xy,xy(200)	-0.04932	-0.00084	-0.00983
xy,xy(002)	0.00674	-0.03444	-0.01399
d2,d2(002)	-0.00212	0.01836	-0.00088
d1,d1(002)	0.00737	-0.02518	-0.00604
ON SITE H-H			
s,s(000)	0.44155	0.10580	
FIRST NEIGHBOR H-H			
s,s(110)	-0.06294	-0.01834	0.03365
s,x(110)	0.03436	0.01045	0.00789
x,x(110)	-0.13618	0.08252	0.04591
x,x(011)	-0.12550	-0.04580	0.00738
x,y(110)	-0.07697	-0.01179	0.00362
SECOND NEIGHBOR H-H			
s,s(200)	0.05207	-0.00471	-0.02382
s,x(200)	-0.03618	0.04061	0.02294
x,x(200)	0.18159	0.30988	0.18059
y,y(200)	0.06864	0.04675	-0.00044
FIRST NEIGHBOR Na-H			
s,s(100)	-0.00540	-0.00090	-0.31516
s,x(100)	0.06589	-0.08203	-0.01791
x,s(100)	-0.03855	-0.06743	-0.36161
x,x(100)	-0.35992	0.17745	-0.04585
y,y(100)	0.12648	0.11925	0.12747
xy,x(010)	0.23828	0.04008	0.09697
d2,s(001)	0.20093	0.06344	0.14397
d2,z(001)	0.12222	-0.14079	-0.21534

SECOND NEIGHBOR Na-H			
s,s(111)	0.04472	0.02348	-0.00390
x,s(111)	0.05535	0.01621	-0.00148
s,x(111)	-0.01128	0.02645	0.01221
x,x(111)	0.02044	-0.01520	-0.00882
x,y(111)	0.01866	0.04150	0.03168
xy,s(111)	0.05956	0.00393	0.01665
xy,x(111)	-0.03410	0.02473	0.00497
yz,x(111)	-0.02788	0.02812	0.02207
d1,x(111)	-0.04434	-0.03685	-0.01833

	ORTHOGONAL			NON-ORTHOGONAL		
BAND	RMS ERROR	MAXIMUM DEVIATION		RMS ERROR	MAXIMUM DEVIATION	
	mRy	k	mRy	mRy	k	mRy
1	8.4	(048)	20.8	2.9	(066)	8.9
2	11.2	(444)	25.1	4.0	(224)	10.7
3	21.1	(022)	53.2	4.0	(003)	9.2
4	21.3	(224)	54.3	6.1	(444)	14.6
5	31.3	(224)	77.2	6.2	(444)	14.6
6	23.9	(224)	62.0	9.1	(042)	29.8
7	30.9	(224)	91.3	11.0	(002)	24.3
8	31.8	(280)	70.4	10.2	(022)	38.6
9	27.1	(444)	73.5	8.0	(042)	20.4
10	39.1	(048)	83.0	11.7	(011)	59.1
1-10	26.2			7.9		

ENERGY VALUES IN Ry AT SELECTED k-POINTS

	ORTHOGONAL	APW	NON-ORTHOGONAL
GAMMA 1	-0.11122	-0.11248	-0.11622
GAMMA 1	0.96172	0.96437	0.96723
GAMMA 12	1.71971	1.73436	1.72972
GAMMA 15	1.41548	1.42974	1.42956
GAMMA 25'	1.31075	1.32906	1.32699
X1 (008)	0.15448	0.16300	0.16795
X1 (008)	1.03092	1.06966	1.06993
X1 (008)	1.97492	1.97872	1.98181
X2 (008)	2.08685	2.10211	2.10292
X3 (008)	0.85967	0.89144	0.89562
X5 (008)	2.18439	2.20542	2.20160
X4' (008)	0.51705	0.49930	0.49401
X4' (008)	4.23423	2.36777	2.33340
X5' (008)	0.99400	0.97433	0.97467
L1 (444)	0.41071	0.43584	0.43743
L1 (444)	1.60758	1.58855	1.57752
L3 (444)	1.18251	1.17794	1.16338
L3 (444)	1.96179	2.03533	2.02518
L2' (444)	0.10806	0.10794	0.11000
L2' (444)	0.97472	0.97047	0.97803
L3' (444)	1.39297	1.40338	1.40790
W1 (048)	0.81999	0.80185	0.80427
W1 (048)	1.64651	1.64527	1.64405
W2' (048)	0.19843	0.17767	0.17799
W2' (048)	1.39393	1.40968	1.41390
W2' (048)	2.05751	2.05500	2.06196
W1' (048)	2.15742	2.24043	2.23605
W3 (048)	0.59865	0.60395	0.60319
W3 (048)	1.49136	1.49976	1.49735
ODD (224)	1.17753	1.11015	1.11380
ODD (224)	1.48233	1.57362	1.59352
ODD (224)	2.00163	1.97506	1.96980
EVEN(224)	0.04686	0.03901	0.03959
EVEN(224)	0.58685	0.59251	0.58177
EVEN(224)	0.91209	0.90117	0.90636
EVEN(224)	1.16442	1.25468	1.26220
EVEN(224)	1.43866	1.37662	1.38090
EVEN(224)	1.68840	1.62223	1.62199
EVEN(224)	1.77812	1.75728	1.75425

FERMI LEVEL QUANTITIES (Non-orthogonal fit)

ENERGY							
	DENSITIES OF STATES						
	Total	s-Na	p-Na	t2g-Na	eg-Na	s-H	p-H
Ry		states/Ry/cell					
0.1850	0.00	0.00	0.00	0.00	0.00	0.00	0.00
	INTEGRATED DENSITIES OF STATES						
	Total	s-Na	p-Na	t2g-Na	eg-Na	s-H	p-H
		electrons					
	2.00	0.09	0.04	0.04	0.09	1.72	0.02

Table 3.9 NaH (NaCl) a = 8.944 Bohr Slater–Koster 2-center parameters

	ORTHOGONAL	NON-ORTHOGONAL	
	ENERGY INTEGRALS Ry	ENERGY INTEGRALS Ry	OVERLAP INTEGRALS
ON SITE Na-Na			
s	1.18921	1.66154	
p	1.55073	1.57669	
t2g	1.63312	1.49354	
eg	2.07400	2.09256	
FIRST NEIGHBOR Na-Na			
(sss)	-0.01537	-0.03509	0.04145
(pps)	0.09299	0.07472	-0.03422
(ppp)	-0.03710	-0.01615	0.00365
(dds)	-0.19465	0.07529	0.14520
(ddp)	0.04820	0.05877	-0.00350
(ddd)	0.01045	-0.02029	-0.02091
(sps)	-0.05148	-0.08399	-0.00853
(sds)	-0.04461	-0.08378	-0.01303
(pds)	0.13059	-0.00871	-0.09511
(pdp)	-0.03585	-0.04972	-0.01290
SECOND NEIGHBOR Na-Na			
(sss)	-0.04989	-0.08212	0.07290
(pps)	0.19909	0.15838	-0.05476
(ppp)	0.00979	0.05720	0.04269
(dds)	-0.13493	-0.50316	-0.18160
(ddp)	-0.00476	-0.10476	-0.04815
(ddd)	0.01604	0.05051	0.02515
(sps)	-0.09928	-0.11135	0.07692
(sds)	-0.05190	-0.18773	-0.01169
(pds)	0.13932	0.36492	0.10733
(pdp)	0.00617	0.10073	0.06358
ON SITE H-H			
s	0.61322	0.43770	
FIRST NEIGHBOR H-H			
(sss)	-0.04741	-0.03353	0.00689
(sps)	0.05578	0.05051	-0.02410
(pps)	0.10412	0.08063	-0.09300
(ppp)	-0.02465	0.03733	0.08366
SECOND NEIGHBOR H-H			
(sss)	0.01940	0.02699	0.01107
(sps)	-0.02811	-0.02902	-0.01908
(pps)	-0.10541	0.04005	0.07398
(ppp)	-0.00163	-0.00649	-0.03553
FIRST NEIGHBOR Na-H			
(sss)	0.08715	0.09072	-0.05482
(sps)	-0.23602	-0.27708	0.11911
(pss)	-0.12723	-0.08242	0.06400
(pps)	-0.34666	-0.41492	0.11142
(ppp)	0.04915	-0.03913	-0.08819
(dpp)	0.06749	-0.06051	-0.11402
(dss)	0.21096	0.05761	-0.09869

SECOND NEIGHBOR Na-H			
(sss)	-0.00733	0.00761	-0.01128
(pss)	-0.00910	0.01013	-0.00602
(sps)	0.02357	0.04904	0.00143
(pps)	0.04446	0.04741	0.01934
(ppp)	0.00017	-0.03811	-0.03022
(dss)	-0.01867	-0.04110	0.00010
(dps)	0.01009	-0.01032	-0.00018
(dpp)	0.00844	0.00739	0.00010

♀NaH

	ORTHOGONAL			NON-ORTHOGONAL		
BAND	RMS ERROR	MAXIMUM DEVIATION		RMS ERROR	MAXIMUM DEVIATION	
	mRy	k	mRy	mRy	k	mRy
1	9.1	(048)	27.6	2.6	(062)	9.2
2	12.7	(055)	27.6	3.5	(006)	8.3
3	24.5	(008)	53.4	11.6	(002)	25.3
4	15.7	(226)	31.0	8.3	(000)	22.3
5	14.8	(022)	36.8	8.7	(000)	22.3
6	23.4	(048)	65.9	4.9	(055)	15.3
7	35.8	(044)	100.5	16.5	(022)	69.4
8	29.6	(033)	59.0	17.8	(111)	53.3
9	37.5	(444)	102.3	18.2	(042)	45.6
10	48.7	(066)	120.8	20.8	(442)	54.9
1-10	27.9			12.9		

ENERGY VALUES IN Ry AT SELECTED k-POINTS

	ORTHOGONAL	APW	NON-ORTHOGONAL
GAMMA 1	-0.10521	-0.11248	-0.11074
GAMMA 1	0.97139	0.96437	0.96465
GAMMA 12	1.76159	1.73436	1.75082
GAMMA 15	1.41669	1.42974	1.43568
GAMMA 25'	1.30727	1.32906	1.30674
X1 (008)	0.15435	0.16300	0.16059
X1 (008)	1.05937	1.06966	1.07000
X1 (008)	2.00332	1.97872	1.97945
X2 (008)	2.05868	2.10211	2.07648
X3 (008)	0.83804	0.89144	0.89577
X5 (008)	2.21967	2.20542	2.18945
X4' (008)	0.47902	0.49930	0.50320
X4' (008)	2.13815	2.36777	2.97999
X5' (008)	0.95639	0.97433	0.98031
L1 (444)	0.45056	0.43584	0.44181
L1 (444)	1.60983	1.58855	1.58487
L3 (444)	1.14897	1.17794	1.17767
L3 (444)	1.93306	2.03533	2.00754
L2' (444)	0.10158	0.10794	0.10853
L2' (444)	0.98826	0.97047	0.97605
L3' (444)	1.37355	1.40338	1.40538
W1 (048)	0.80126	0.80185	0.79111
W1 (048)	1.66647	1.64527	1.63942
W2' (048)	0.20524	0.17767	0.17542
W2' (048)	1.40057	1.40968	1.41484
W2' (048)	2.03631	2.05500	2.05272
W1' (048)	2.15552	2.24043	2.24255
W3 (048)	0.59633	0.60395	0.60748
W3 (048)	1.56561	1.49976	1.50624
ODD (224)	1.11810	1.11015	1.10229
ODD (224)	1.58638	1.57362	1.57512
ODD (224)	1.90297	1.97506	1.95114
EVEN(224)	0.03834	0.03901	0.03968
EVEN(224)	0.58174	0.59251	0.59271
EVEN(224)	0.90075	0.90117	0.90371
EVEN(224)	1.25194	1.25468	1.25452
EVEN(224)	1.35565	1.37662	1.37050
EVEN(224)	1.67537	1.62223	1.64621
EVEN(224)	1.79332	1.75728	1.77869

3.3 Potassium Hydride (KH)

See Fig. 3.12, Tables 3.10, and 3.11.

See Figs. 3.13, 3.14, 3.15, 3.16, Tables 3.12, and 3.13.

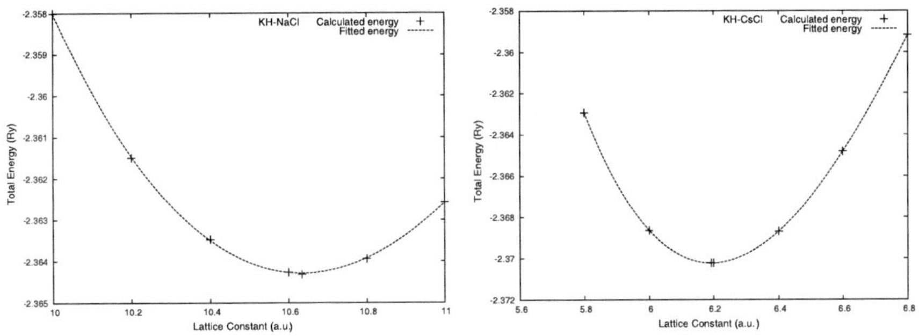

Fig. 3.12 Total energy versus lattice constant for the NaCl and CsCl structures

Table 3.10 Lattice constants, bulk modulus, gap, total energy

	a (Bohr)		B (MBar)	Gap (Ry)	Total Energy (Ry)
NaCl	10.63	Insulator	0.17	0.25	-1202.36429
CsCl	6.19	Insulator	0.21	0.46	-
CsCl	5.90	(P=0.04MBar) Ins.	0.35	0.05	-
exp(NaCl)	10.78		0.16		

Table 3.11 Birch fit coefficients

	A1	A2	A3	A4
NaCl	-1.823783E+00	-5.540066E+01	1.706888E+03	-1.387665E+04
CsCl	-1.972028E+00	-3.141393E+01	6.526044E+02	-1.285615E+03

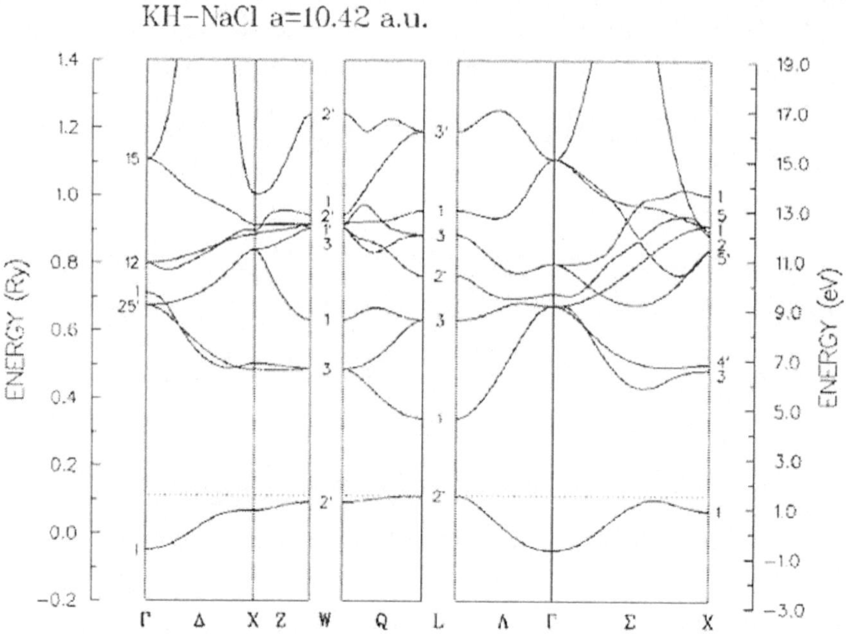

Fig. 3.13 Energy bands of KH in the NaCl structure (tight-binding)

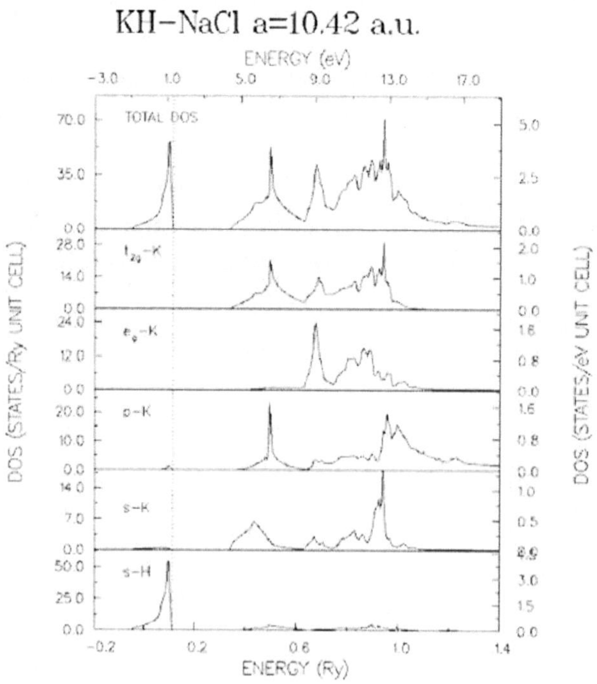

Fig. 3.14 Total, angular momentum and site decomposed densities of states of KH in the NaCl structure (tight-binding)

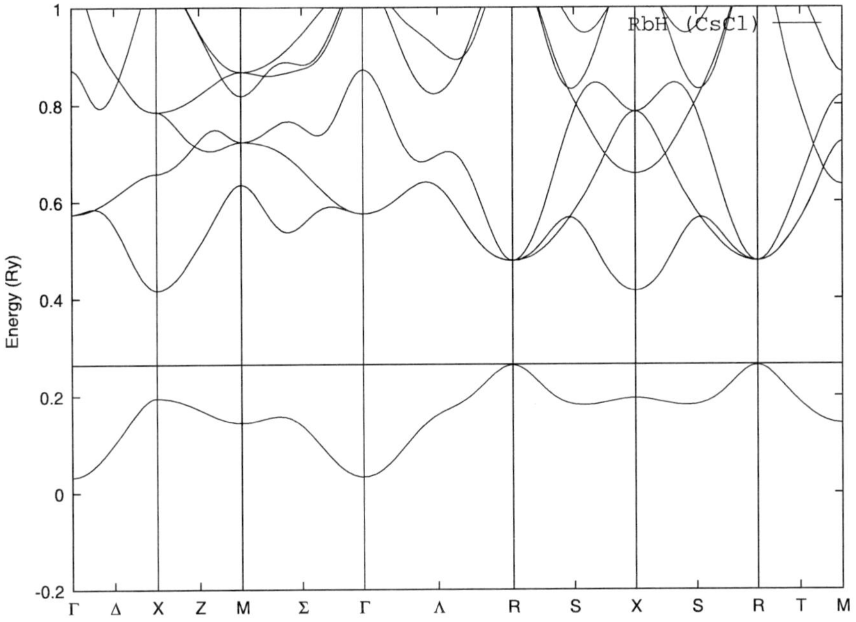

Fig. 3.15 Energy bands of KH in the CsCl structure

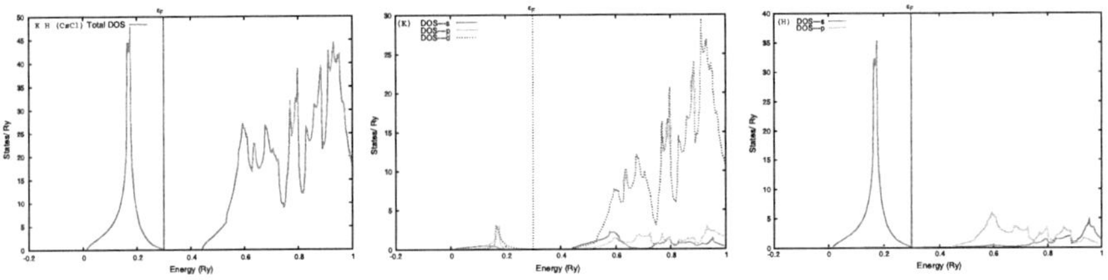

Fig. 3.16 Total, angular momentum and site decomposed densities of states of KH in the CsCl structure

Table 3.12 KH (NaCl) a = 10.417 Bohr Slater–Koster 3-center parameters

	ORTHOGONAL	NON-ORTHOGONAL	
	ENERGY INTEGRALS Ry	ENERGY INTEGRALS Ry	OVERLAP INTEGRALS
ON SITE K- K			
s,s(000)	0.65779	0.58436	
x,x(000)	0.93612	0.66897	
xy,xy(000)	0.79773	0.68896	
d2,d2(000)	0.77310	0.73491	
FIRST NEIGHBOR K- K			
s,s(110)	-0.01865	0.03373	0.08022
s,x(110)	-0.02474	0.00691	0.07748
s,xy(110)	-0.03138	-0.00936	0.06929
s,d2(110)	0.00803	0.03743	0.03212
x,x(110)	0.03770	-0.01651	-0.07399
x,x(011)	-0.00470	0.02299	0.06005
x,y(110)	0.03665	-0.00335	-0.13290
x,xy(110)	0.04635	0.01486	-0.04327
x,xy(011)	-0.00280	0.01753	0.05558
z,d2(011)	0.01180	0.03360	0.09281
z,d1(011)	-0.00479	-0.00517	0.02635
xy,xy(110)	-0.04275	-0.03089	0.02078
xy,xy(011)	0.01195	-0.01332	-0.04536
xy,xz(011)	-0.00209	-0.01115	-0.04383
xy,d2(110)	0.01827	0.00816	0.04513
d2,d2(110)	-0.00468	0.02330	0.03809
d1,d1(110)	-0.00765	0.02501	0.01874
SECOND NEIGHBOR K- K			
s,s(200)	0.03760	0.00323	-0.02078
s,x(200)	0.05381	-0.02878	-0.04157
s,d2(002)	0.03224	-0.00402	0.04683
x,x(200)	-0.07761	0.06737	0.03587
y,y(200)	0.01084	-0.03970	-0.04351
x,xy(020)	0.01081	-0.02091	-0.01704
z,d2(002)	-0.04444	-0.02637	-0.10072
xy,xy(200)	-0.01430	0.00997	0.00407
xy,xy(002)	0.00288	-0.00426	-0.00421
d2,d2(002)	0.03128	-0.03967	-0.07238
d1,d1(002)	0.00184	-0.01717	-0.01861
ON SITE H-H			
s,s(000)	0.10493	0.08931	
FIRST NEIGHBOR H-H			
s,s(110)	-0.00738	-0.00388	0.02832
s,x(110)	0.00000	0.00000	0.00000
x,x(110)	0.00000	0.00000	0.00000
x,x(011)	0.00000	0.00000	0.00000
x,y(110)	0.00000	0.00000	0.00000
SECOND NEIGHBOR H-H			
s,s(200)	-0.00289	-0.00673	0.00891
s,x(200)	0.00000	0.00000	0.00000
x,x(200)	0.00000	0.00000	0.00000
y,y(200)	0.00000	0.00000	0.00000
FIRST NEIGHBOR K-H			
s,s(100)	-0.04142	0.03691	-0.06422
s,x(100)	0.00000	0.00000	0.00000
x,s(100)	0.02136	0.00027	0.28173
x,x(100)	0.00000	0.00000	0.00000
y,y(100)	0.00000	0.00000	0.00000
xy,x(010)	0.00000	0.00000	0.00000
d2,s(001)	-0.03135	-0.00696	-0.11253
d2,z(001)	0.00000	0.00000	0.00000
SECOND NEIGHBOR K-H			
s,s(111)	0.00760	0.00954	0.00520
x,s(111)	-0.01312	-0.00083	0.00950
s,x(111)	0.00000	0.00000	0.00000
x,x(111)	0.00000	0.00000	0.00000
x,y(111)	0.00000	0.00000	0.00000
xy,s(111)	0.00882	-0.00166	-0.00153
xy,x(111)	0.00000	0.00000	0.00000
yz,x(111)	0.00000	0.00000	0.00000
d1,x(111)	0.00000	0.00000	0.00000

KH

BAND	ORTHOGONAL RMS ERROR mRy	MAXIMUM DEVIATION k	mRy	NON-ORTHOGONAL RMS ERROR mRy	MAXIMUM DEVIATION k	mRy
1	1.5	(174)	2.7	1.0	(226)	2.6
2	2.5	(002)	8.5	1.0	(005)	2.8
3	3.9	(003)	11.4	1.0	(354)	3.3
4	3.7	(048)	7.9	1.0	(062)	3.1
5	3.3	(380)	6.5	2.9	(442)	7.5
6	3.3	(055)	7.5	2.6	(042)	8.4
7	4.6	(044)	12.1	2.7	(062)	7.0
1-7	3.4			1.9		

ENERGY VALUES IN Ry AT SELECTED k-POINTS

	ORTHOGONAL	APW	NON-ORTHOGONAL
GAMMA 1	-0.05058	-0.05128	-0.04973
GAMMA 1	0.70915	0.71244	0.71050
GAMMA 12	0.79846	0.79845	0.79981
GAMMA 15	1.10708	1.10769	1.10764
GAMMA 15	10.00000	1.72553	10.00000
GAMMA 25'	0.67094	0.67371	0.67322
X1 (008)	0.07172	0.06909	0.06801
X1 (008)	0.90673	0.89804	0.89679
X1 (008)	1.00571	1.00511	1.00481
X2 (008)	0.88526	0.88337	0.88536
X3 (008)	0.47968	0.48337	0.48378
X5 (008)	0.91729	0.91323	0.91259
X4' (008)	0.50381	0.50143	0.50235
X5' (008)	0.84304	0.84100	0.84098
L1 (444)	0.34201	0.34009	0.34125
L1 (444)	0.95610	0.95746	0.95649
L3 (444)	0.63481	0.63169	0.63127
L3 (444)	0.87971	0.88363	0.88478
L2' (444)	0.11371	0.11263	0.11137
L2' (444)	0.76333	0.76136	0.76384
L3' (444)	1.19460	1.19082	1.19088
W1 (048)	0.62441	0.63227	0.63260
W1 (048)	0.94926	0.94353	0.94331
W2' (048)	0.09195	0.09438	0.09300
W2' (048)	0.90967	0.91547	0.91686
W2' (048)	1.17626	1.24287	1.24218
W1' (048)	0.90576	0.91377	0.91435
W3 (048)	0.48650	0.48715	0.48701
W3 (048)	0.89926	0.90544	0.90482
ODD (224)	0.67333	0.66722	0.66920
ODD (224)	0.86310	0.86047	0.85636
EVEN(224)	0.06357	0.06154	0.06228
EVEN(224)	0.42991	0.43191	0.43196
EVEN(224)	0.58601	0.58605	0.58390
EVEN(224)	0.76559	0.76414	0.76914
EVEN(224)	0.80845	0.81233	0.81000

FERMI LEVEL QUANTITIES (Non-orthogonal fit)

DENSITIES OF STATES (states/Ry/cell)

ENERGY Ry	Total	s- K	p- K	t2g- K	eg- K	s-H	p-H
0.1120	0.00	0.00	0.00	0.00	0.00	0.00	0.00

INTEGRATED DENSITIES OF STATES (electrons)

Total	s- K	p- K	t2g- K	eg- K	s-H	p-H
2.00	0.04	0.03	0.00	0.00	1.92	0.00

VELOCITY cm/s	PLASMON ENERGY eV	EIGENVALUE SUM Ry
0.00x10E8	0.00	-0.0812

Table 3.13 KH (NaCl) a = 10.417 Bohr Slater–Koster 2-center parameters

	ORTHOGONAL	NON-ORTHOGONAL	
	ENERGY INTEGRALS Ry	ENERGY INTEGRALS Ry	OVERLAP INTEGRALS
ON SITE K- K			
s	0.62817	0.56055	
p	0.91488	0.71080	
t2g	0.75185	0.65248	
eg	0.72641	0.63812	
FIRST NEIGHBOR K- K			
(sss)	-0.03351	-0.02906	0.02185
(pps)	0.08366	-0.03258	-0.19466
(ppp)	-0.00654	0.04426	0.08071
(dds)	-0.05729	0.03224	0.12112
(ddp)	0.02176	-0.01245	-0.08833
(ddd)	0.00106	0.00292	0.01531
(sps)	0.05456	-0.00232	-0.11917
(sds)	-0.05013	-0.00177	0.09031
(pds)	-0.07373	0.02296	0.12937
(pdp)	0.00714	-0.02051	-0.07440
SECOND NEIGHBOR K- K			
(sss)	0.00197	0.00677	0.00204
(pps)	-0.04249	0.06085	0.03251
(ppp)	-0.00083	-0.01366	-0.00687
(dds)	0.00723	-0.02779	-0.01422
(ddp)	-0.00224	0.00442	0.00085
(ddd)	0.00076	0.00253	0.00295
(sps)	-0.01888	-0.01228	-0.02553
(sds)	0.01430	-0.01279	-0.01307
(pds)	0.01494	-0.04499	-0.03236
(pdp)	-0.00098	0.00435	-0.00147
ON SITE H-H			
s	0.39599	0.17512	
FIRST NEIGHBOR H-H			
(sss)	-0.01190	0.00126	-0.01846
(sps)	0.00000	0.00000	0.00000
(pps)	0.00000	0.00000	0.00000
(ppp)	0.00000	0.00000	0.00000
SECOND NEIGHBOR H-H			
(sss)	0.02577	0.00960	-0.00429
(sps)	0.00000	0.00000	0.00000
(pps)	0.00000	0.00000	0.00000
(ppp)	0.00000	0.00000	0.00000
FIRST NEIGHBOR K-H			
(sss)	0.08027	0.06317	-0.03421
(sps)	0.00000	0.00000	0.00000
(pss)	0.07988	-0.01688	-0.26524
(pps)	0.00000	0.00000	0.00000
(ppp)	0.00000	0.00000	0.00000
(dpp)	0.00000	0.00000	0.00000
(dss)	0.09473	0.04175	-0.17162
(dps)	0.00000	0.00000	0.00000
SECOND NEIGHBOR K-H			
(sss)	-0.01368	-0.00872	0.00369
(pss)	-0.03221	0.00698	0.00865
(sps)	0.00000	0.00000	0.00000
(pps)	0.00000	0.00000	0.00000
(ppp)	0.00000	0.00000	0.00000
(dss)	-0.02145	0.01197	0.01522
(dps)	0.00000	0.00000	0.00000
(dpp)	0.00000	0.00000	0.00000

♀ KH

BAND	ORTHOGONAL RMS ERROR	ORTHOGONAL MAXIMUM DEVIATION		NON-ORTHOGONAL RMS ERROR	NON-ORTHOGONAL MAXIMUM DEVIATION	
	mRy	k	mRy	mRy	k	mRy
1	6.7	(005)	14.2	1.0	(333)	2.0
2	6.4	(022)	17.2	1.3	(111)	3.3
3	8.5	(264)	16.3	1.2	(003)	3.0
4	8.5	(048)	18.9	1.1	(022)	3.3
5	9.8	(444)	22.2	1.8	(118)	5.2
6	10.0	(044)	31.7	2.1	(174)	7.0
7	9.8	(006)	34.3	2.0	(066)	5.6
1.7	8.6			1.6		

ENERGY VALUES IN Ry AT SELECTED k-POINTS

	ORTHOGONAL	APW	NON-ORTHOGONAL
GAMMA 1	-0.05898	-0.05128	-0.04998
GAMMA 1	0.70456	0.71244	0.71211
GAMMA 12	0.79978	0.79845	0.79949
GAMMA 15	1.10889	1.10769	1.10723
GAMMA 15	10.00000	1.72553	10.00000
GAMMA 25'	0.66491	0.67371	0.67223
X1 (008)	0.06095	0.06909	0.06919
X1 (008)	0.90109	0.89804	0.89476
X1 (008)	1.00302	1.00511	1.00344
X2 (008)	0.87504	0.88337	0.88299
X3 (008)	0.48233	0.48337	0.48390
X5 (008)	0.91523	0.91323	0.91264
X4' (008)	0.49196	0.50143	0.50006
X5' (008)	0.85277	0.84100	0.84068
L1 (444)	0.34411	0.34009	0.34197
L1 (444)	0.94871	0.95746	0.95998
L3 (444)	0.61839	0.63169	0.63171
L3 (444)	0.88473	0.88363	0.88323
L2' (444)	0.10016	0.11263	0.11193
L2' (444)	0.78357	0.76136	0.76073
L3' (444)	1.18358	1.19082	1.19030
W1 (048)	0.61334	0.63227	0.63306
W1 (048)	0.92839	0.94353	0.93930
W2' (048)	0.09428	0.09438	0.09394
W2' (048)	0.91293	0.91547	0.91268
W2' (048)	1.17098	1.24287	1.24296
W1' (048)	0.91220	0.91377	0.91370
W3 (048)	0.49025	0.48715	0.48786
W3 (048)	0.89314	0.90544	0.90437
ODD (224)	0.66437	0.66722	0.66659
ODD (224)	0.86305	0.86047	0.85942
EVEN(224)	0.06686	0.06154	0.06279
EVEN(224)	0.43239	0.43191	0.42982
EVEN(224)	0.59530	0.58605	0.58601
EVEN(224)	0.77320	0.76414	0.76247
EVEN(224)	0.81211	0.81233	0.81594

3.4 Rubidium Hydride (RbH)

See Fig. 3.17, Table 3.14, and 3.15.

See Figs. 3.18, 3.19, 3.20, 3.21, Tables 3.16 and 3.17.

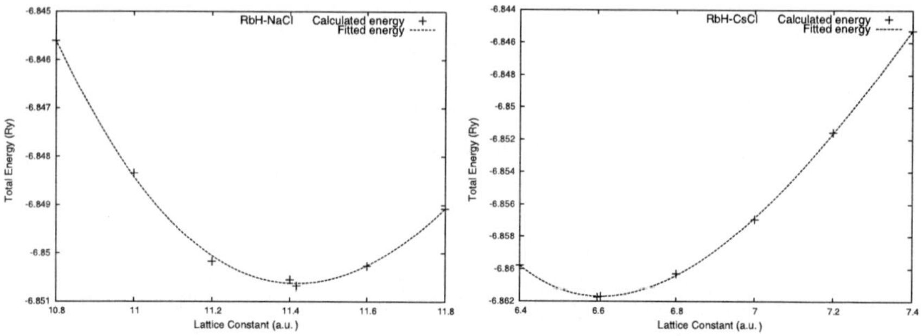

Fig. 3.17 Total energy versus lattice constant of RbH for the NaCl and CsCl structures

Table 3.14 Lattice constants, bulk modulus, gap, total energy

	a (Bohr)		B (MBar)	Gap (Ry)	Total Energy (Ry)
NaCl	11.42	Insulator	0.13	0.19	-5956.84835
CsCl	6.61	Insulator	0.19	0.15	-
exp(NaCl)	11.42		0.10		

Table 3.15 Birch fit coefficients

	A1	A2	A3	A4
NaCl	-6.377465E+00	-5.398885E+01	1.787108E+03	-1.347255E+04
CsCl	-6.553983E+00	-2.178346E+01	1.420787E+02	9.711667E+03

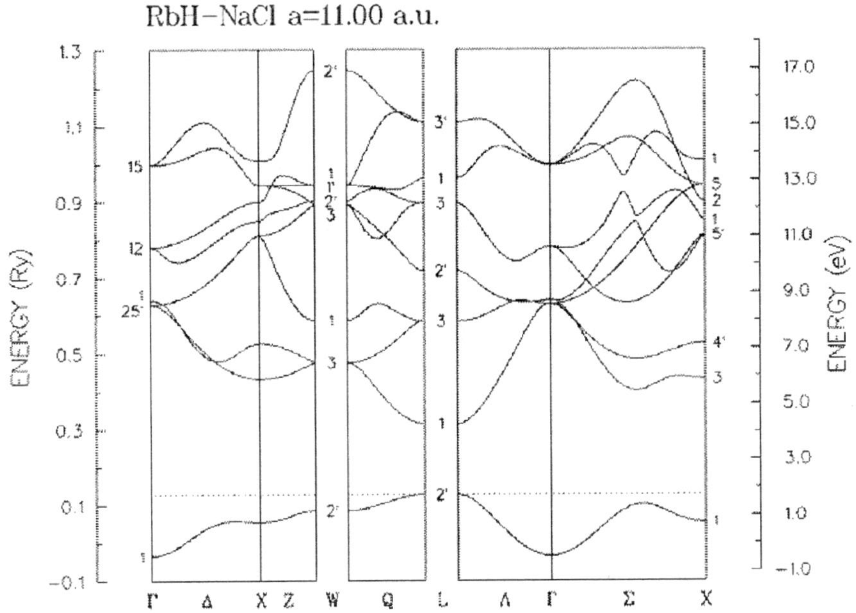

Fig. 3.18 Energy bands of RbH in the NaCl structure (tight-binding)

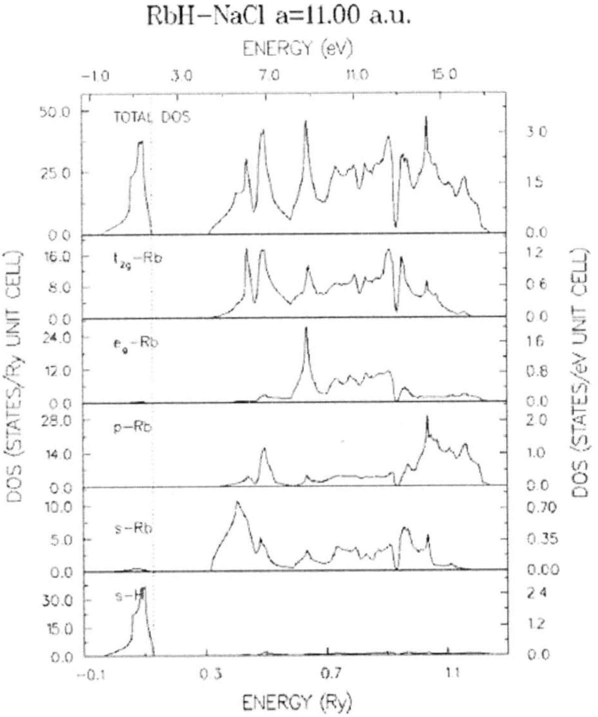

Fig. 3.19 Total, angular momentum and site decomposed densities of states of RbH in the NaCl structure (tight-binding)

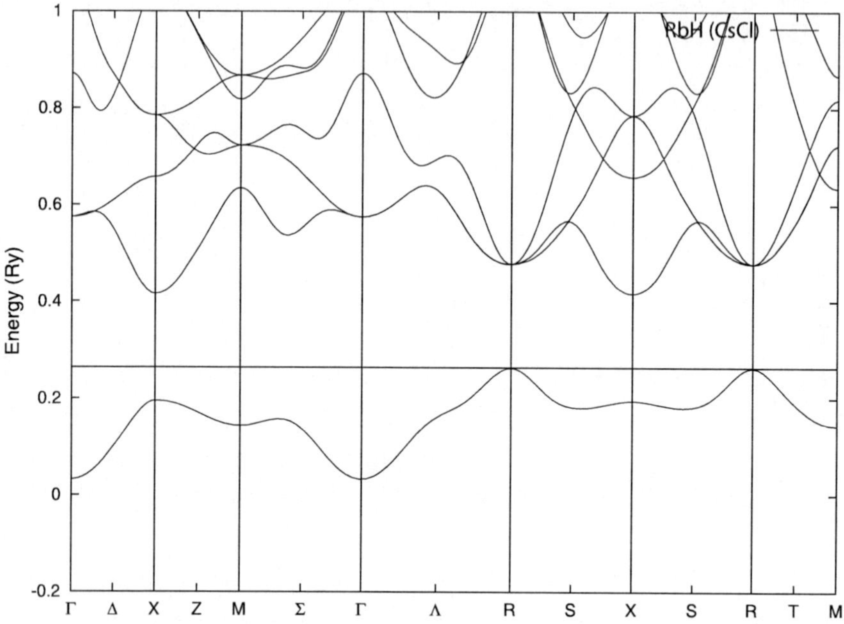

Fig. 3.20 Energy bands of RbH in the CsCl structure

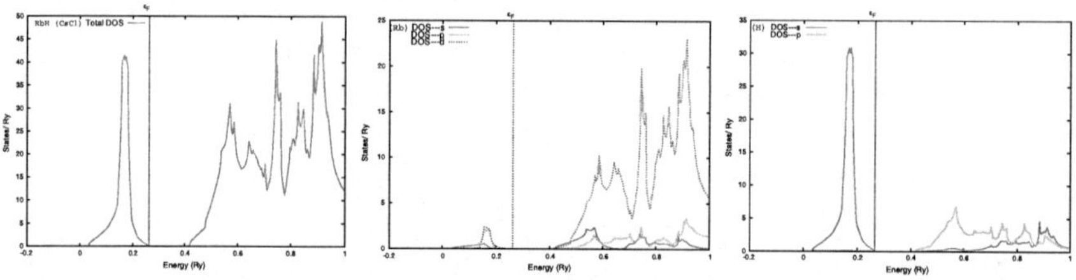

Fig. 3.21 Total, angular momentum and site decomposed densities of states of RbH in the CsCl structure

Table 3.16 RbH (NaCl) a = 11.00 Bohr Slater–Koster 3-center parameters

	ORTHOGONAL	NON-ORTHOGONAL	
	ENERGY INTEGRALS Ry	ENERGY INTEGRALS Ry	OVERLAP INTEGRALS
ON SITE Rb-Rb			
s,s(000)	0.61052	0.55056	
x,x(000)	0.87273	0.76708	
xy,xy(000)	0.74946	0.64936	
d2,d2(000)	0.79880	0.65520	
FIRST NEIGHBOR Rb-Rb			
s,s(110)	-0.00865	0.04090	0.08287
s,x(110)	-0.02331	0.02011	0.07175
s,xy(110)	-0.04379	0.02274	0.12095
s,d2(110)	0.01985	0.01683	0.02557
x,x(110)	0.02889	-0.00295	-0.06667
x,x(011)	-0.00593	0.08158	0.13225
x,y(110)	0.03905	-0.03312	-0.08704
x,xy(110)	0.04566	-0.01576	-0.08728
x,xy(011)	-0.00563	0.03334	0.07694
z,d2(011)	0.03070	0.01784	0.06963
z,d1(011)	-0.02815	-0.00659	0.01005
xy,xy(110)	-0.05257	-0.01324	0.05163
xy,xy(011)	0.01219	-0.01385	-0.05172
xy,xz(011)	0.00191	-0.02425	-0.06631
xy,d2(110)	0.03178	0.02122	0.07109
d2,d2(110)	-0.00765	-0.00268	0.01896
d1,d1(110)	-0.00988	0.02449	-0.01180
SECOND NEIGHBOR Rb-Rb			
s,s(200)	0.01425	0.03404	-0.00125
s,x(200)	0.02087	0.02686	-0.00929
s,d2(002)	0.02002	0.00441	0.04936
x,x(200)	-0.05383	-0.04329	-0.01733
y,y(200)	0.00569	0.02728	0.03370
x,xy(020)	0.00469	0.01982	0.02973
z,d2(002)	-0.02679	-0.02267	-0.09062
xy,xy(200)	-0.00494	-0.01090	-0.02368
xy,xy(002)	0.00459	-0.00419	-0.00443
d2,d2(002)	0.02338	-0.04996	-0.07550
d1,d1(002)	0.00496	0.00053	0.00011
ON SITE H-H			
s,s(000)	0.21117	0.09078	
FIRST NEIGHBOR H-H			
s,s(110)	-0.01737	-0.00434	0.01188
s,x(110)	0.00000	0.00000	0.00000
x,x(110)	0.00000	0.00000	0.00000
x,x(011)	0.00000	0.00000	0.00000
x,y(110)	0.00000	0.00000	0.00000
SECOND NEIGHBOR H-H			
s,s(200)	-0.00023	-0.00672	-0.00356
s,x(200)	0.00000	0.00000	0.00000
x,x(200)	0.00000	0.00000	0.00000
y,y(200)	0.00000	0.00000	0.00000
FIRST NEIGHBOR Rb-H			
s,s(100)	-0.05485	0.03123	0.07526
s,x(100)	0.00000	0.00000	0.00000
x,s(100)	0.05594	0.02473	0.14975
x,x(100)	0.00000	0.00000	0.00000
y,y(100)	0.00000	0.00000	0.00000
xy,x(010)	0.00000	0.00000	0.00000
d2,s(001)	-0.05132	0.00348	-0.15929
d2,z(001)	0.00000	0.00000	0.00000
SECOND NEIGHBOR Rb-H			
s,s(111)	0.02216	0.00172	0.01011
x,s(111)	-0.02630	-0.00351	-0.00668
s,x(111)	0.00000	0.00000	0.00000
x,x(111)	0.00000	0.00000	0.00000
x,y(111)	0.00000	0.00000	0.00000
xy,s(111)	0.02351	0.00037	0.00568
xy,x(111)	0.00000	0.00000	0.00000
yz,x(111)	0.00000	0.00000	0.00000
d1,x(111)	0.00000	0.00000	0.00000

RbH

BAND	ORTHOGONAL RMS ERROR mRy	MAXIMUM DEVIATION k	mRy	NON-ORTHOGONAL RMS ERROR mRy	MAXIMUM DEVIATION k	mRy
1	5.0	(055)	11.1	0.5	(333)	1.3
2	4.7	(022)	14.1	0.7	(033)	2.1
3	5.6	(003)	12.5	0.6	(226)	1.6
4	6.7	(048)	16.1	0.8	(222)	1.4
5	6.8	(224)	13.5	0.8	(000)	1.4
6	6.0	(226)	18.5	0.8	(022)	1.8
7	6.8	(442)	13.3	0.8	(226)	2.4
1-7	6.0			0.7		

ENERGY VALUES IN Ry AT SELECTED k-POINTS

	ORTHOGONAL	APW	NON-ORTHOGONAL
GAMMA 1	-0.03536	-0.03442	-0.03413
GAMMA 1	0.62890	0.64198	0.64060
GAMMA 12	0.77862	0.78089	0.78085
GAMMA 15	0.99521	0.99685	0.99687
GAMMA 25'	0.62611	0.62949	0.62913
X1 (008)	0.06481	0.05490	0.05543
X1 (008)	0.85291	0.84696	0.84851
X1 (008)	1.02000	1.00910	1.00916
X2 (008)	0.90992	0.89876	0.89868
X3 (008)	0.43111	0.43164	0.43201
X5 (008)	0.94917	0.94210	0.94269
X4' (008)	0.53303	0.52741	0.52696
X5' (008)	0.81154	0.81042	0.81122
L1 (444)	0.31223	0.31438	0.31451
L1 (444)	0.95756	0.96128	0.96065
L3 (444)	0.58271	0.58352	0.58451
L3 (444)	0.89875	0.89642	0.89587
L2' (444)	0.13842	0.12894	0.12790
L2' (444)	0.71931	0.71768	0.71803
L3' (444)	1.11384	1.10938	1.10937
W1 (048)	0.57158	0.58772	0.58738
W1 (048)	0.92292	0.94399	0.94336
W2' (048)	0.08499	0.08658	0.08648
W2' (048)	0.89634	0.90068	0.90122
W2' (048)	1.22925	1.24550	1.24551
W1' (048)	0.93082	0.94334	0.94277
W3 (048)	0.47235	0.47529	0.47561
W3 (048)	0.88706	0.89218	0.89290
ODD (224)	0.63925	0.62474	0.62445
ODD (224)	0.86681	0.85921	0.85877
EVEN(224)	0.07949	0.07750	0.07755
EVEN(224)	0.39844	0.39812	0.39891
EVEN(224)	0.54218	0.54419	0.54291
EVEN(224)	0.72987	0.71635	0.71587
EVEN(224)	0.77392	0.78391	0.78519

FERMI LEVEL QUANTITIES (Non-orthogonal fit)

DENSITIES OF STATES (states/Ry/cell)

ENERGY Ry	Total	s-Rb	p-Rb	t2g-Rb	eg-Rb	s-H	p-H
0.1280	0.00	0.00	0.00	0.00	0.00	0.00	0.00

INTEGRATED DENSITIES OF STATES (electrons)

Total	s-Rb	p-Rb	t2g-Rb	eg-Rb	s-H	p-H
2.00	0.03	0.01	0.00	0.02	1.94	0.00

VELOCITY cm/s	PLASMON ENERGY eV	EIGENVALUE SUM Ry
0.00x10E8	0.00	-0.1014

Table 3.17 RbH (NaCl) a = 11.00000 Bohr Slater–Koster 2-center parameters

	ORTHOGONAL	NON-ORTHOGONAL	
	ENERGY INTEGRALS Ry	ENERGY INTEGRALS Ry	OVERLAP INTEGRALS
ON SITE Rb-Rb			
s	0.60725	0.40830	
p	0.87235	0.63618	
t2g	0.73990	0.61004	
eg	0.70679	0.58067	
FIRST NEIGHBOR Rb-Rb			
(sss)	-0.02505	-0.02153	0.06844
(pps)	0.06849	-0.09436	-0.26096
(ppp)	-0.00809	-0.01891	0.02049
(dds)	-0.07012	0.04596	0.15940
(ddp)	0.02155	-0.00607	-0.09313
(ddd)	0.00181	-0.00251	0.00837
(sps)	0.04008	-0.00806	-0.13984
(sds)	-0.05126	0.01530	0.12656
(pds)	-0.07403	0.08045	0.21766
(pdp)	0.00475	-0.00060	-0.05818
SECOND NEIGHBOR Rb-Rb			
(sss)	0.00111	-0.00882	-0.00940
(pps)	-0.04597	0.06774	0.06112
(ppp)	0.00289	-0.00454	-0.00421
(dds)	0.01181	-0.04370	-0.03829
(ddp)	-0.00223	0.00010	-0.00234
(ddd)	0.00153	0.00494	0.00581
(sps)	-0.02249	0.02655	0.02827
(sds)	0.01871	-0.02881	-0.02987
(pds)	0.02147	-0.05926	-0.05288
(pdp)	-0.00074	-0.00211	-0.00477
ON SITE H-H			
s	0.37869	0.13733	
FIRST NEIGHBOR H-H			
(sss)	-0.02232	0.00981	-0.01077
(sps)	0.00000	0.00000	0.00000
(pps)	0.00000	0.00000	0.00000
(ppp)	0.00000	0.00000	0.00000
SECOND NEIGHBOR H-H			
(sss)	0.02885	-0.00140	-0.00271
(sps)	0.00000	0.00000	0.00000
(pps)	0.00000	0.00000	0.00000
(ppp)	0.00000	0.00000	0.00000
FIRST NEIGHBOR Rb-H			
(sss)	0.08050	0.04104	-0.09879
(sps)	0.00000	0.00000	0.00000
(pss)	0.06172	-0.00588	-0.25182
(pps)	0.00000	0.00000	0.00000
(ppp)	0.00000	0.00000	0.00000
(dpp)	0.00000	0.00000	0.00000
(dss)	0.09080	-0.00876	-0.24939
(dps)	0.00000	0.00000	0.00000
SECOND NEIGHBOR Rb-H			
(sss)	-0.01811	-0.00261	0.00884
(pss)	-0.05474	0.00818	0.01173
(sps)	0.00000	0.00000	0.00000
(pps)	0.00000	0.00000	0.00000
(ppp)	0.00000	0.00000	0.00000
(dss)	-0.03589	0.00866	0.01009
(dps)	0.00000	0.00000	0.00000
(dpp)	0.00000	0.00000	0.00000

♀ RbH

BAND	ORTHOGONAL RMS ERROR mRy	MAXIMUM DEVIATION k	mRy	NON-ORTHOGONAL RMS ERROR mRy	MAXIMUM DEVIATION k	mRy
1	7.6	(005)	15.6	0.7	(044)	1.5
2	6.8	(003)	15.1	1.3	(002)	3.2
3	10.1	(444)	21.2	1.5	(022)	5.1
4	12.6	(048)	30.2	1.2	(022)	2.5
5	12.9	(226)	26.9	1.5	(011)	2.9
6	18.2	(118)	32.8	1.8	(055)	3.8
7	12.9	(180)	29.0	1.2	(222)	3.2
1.7	12.1			1.3		

ENERGY VALUES IN Ry AT SELECTED k-POINTS

	ORTHOGONAL	APW	NON-ORTHOGONAL
GAMMA 1	-0.03987	-0.03442	-0.03377
GAMMA 1	0.63709	0.64198	0.64358
GAMMA 12	0.77906	0.78089	0.78249
GAMMA 15	1.00124	0.99685	0.99685
GAMMA 25'	0.61893	0.62949	0.62737
X1 (008)	0.05251	0.05490	0.05403
X1 (008)	0.86128	0.84696	0.84718
X1 (008)	1.02654	1.00910	1.00890
X2 (008)	0.88694	0.89876	0.89771
X3 (008)	0.43205	0.43164	0.43220
X5 (008)	0.94258	0.94210	0.94036
X4' (008)	0.51798	0.52741	0.52619
X5' (008)	0.82431	0.81042	0.81013
L1 (444)	0.32435	0.31438	0.31560
L1 (444)	0.94300	0.96128	0.96071
L3 (444)	0.56235	0.58352	0.58443
L3 (444)	0.88969	0.89642	0.89495
L2' (444)	0.11892	0.12894	0.12861
L2' (444)	0.73308	0.71768	0.71714
L3' (444)	1.10591	1.10938	1.10941
W1 (048)	0.55751	0.58772	0.58808
W1 (048)	0.91644	0.94399	0.94419
W2' (048)	0.08158	0.08658	0.08594
W2' (048)	0.89184	0.90068	0.90193
W2' (048)	1.21491	1.24550	1.24554
W1' (048)	0.93648	0.94334	0.94463
W3 (048)	0.48229	0.47529	0.47568
W3 (048)	0.87354	0.89218	0.89128
ODD (224)	0.61680	0.62474	0.62358
ODD (224)	0.86435	0.85921	0.85884
EVEN(224)	0.08756	0.07750	0.07829
EVEN(224)	0.39144	0.39812	0.39797
EVEN(224)	0.55559	0.54419	0.54415
EVEN(224)	0.73588	0.71635	0.71669
EVEN(224)	0.78418	0.78391	0.78444

3.5 Cesium Hydride (CsH)

See Fig. 3.22, Tables 3.18, and 3.19.

See Figs. 3.23, 3.24, 3.25, 3.26, Tables 3.20 and 3.21.

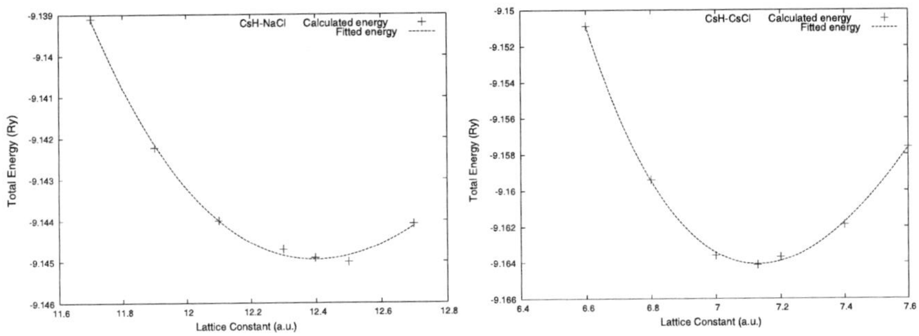

Fig. 3.22 Total energy versus lattice constant of CsH for the NaCl and CsCl structures

Table 3.18 Lattice constants, bulk modulus, gap, total energy

	a (Bohr)		B (MBar)	Gap (Ry)	Total Energy (Ry)
NaCl	12.40	Insulator	0.10	0.14	-15569.14494
CsCl	7.13	Insulator	0.17	0.17	-
exp(NaCl)	12.06		0.08		

Table 3.19 Birch fit coefficients

	A1	A2	A3	A4
NaCl	-8.879348E+00	-2.590967E+01	1.962321E+02	2.415515E+04
CsCl	-8.793205E+00	-3.321724E+01	5.021196E+02	1.159385E+04

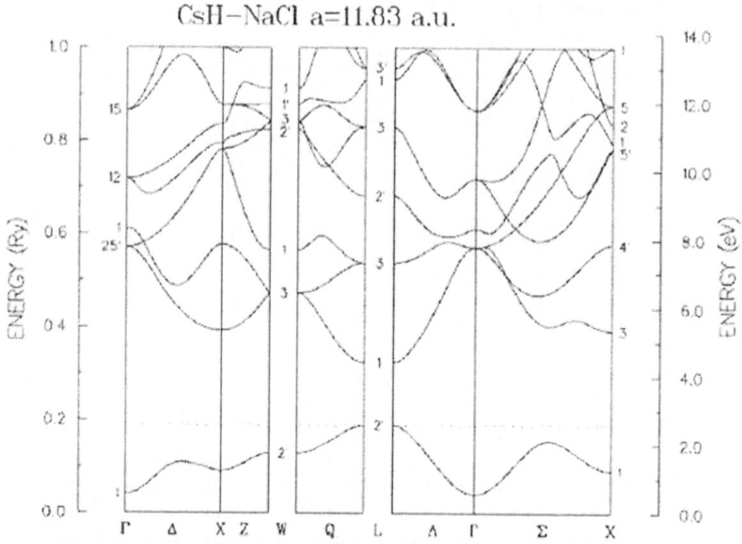

Fig. 3.23 Energy bands of CsH in the NaCl structure (tight-binding)

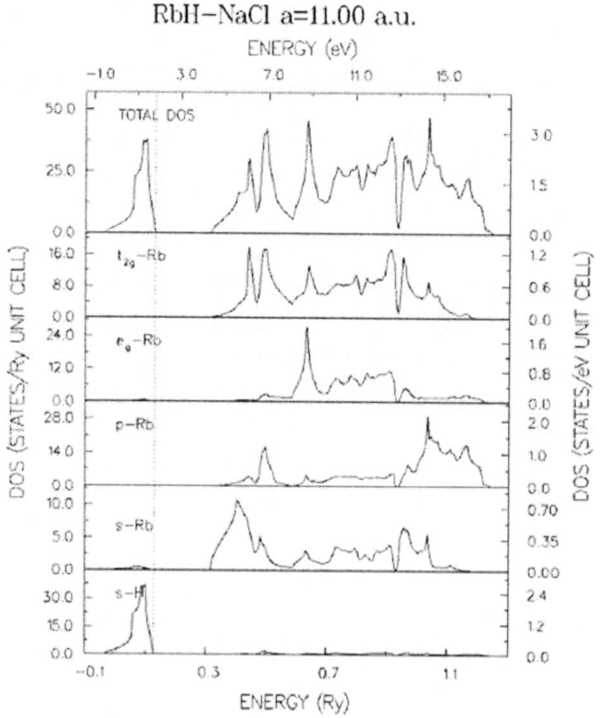

Fig. 3.24 Total, angular momentum and site decomposed densities of states of CsH in the NaCl structure (tight-binding)

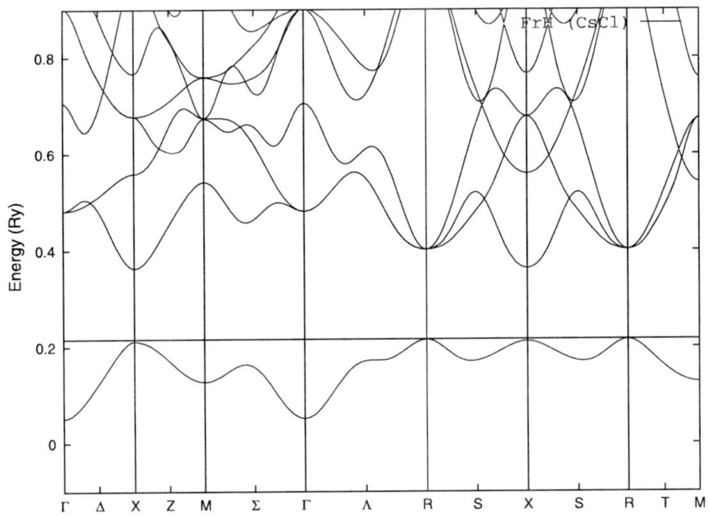

Fig. 3.25 Energy bands of CsH in the CsCl structure

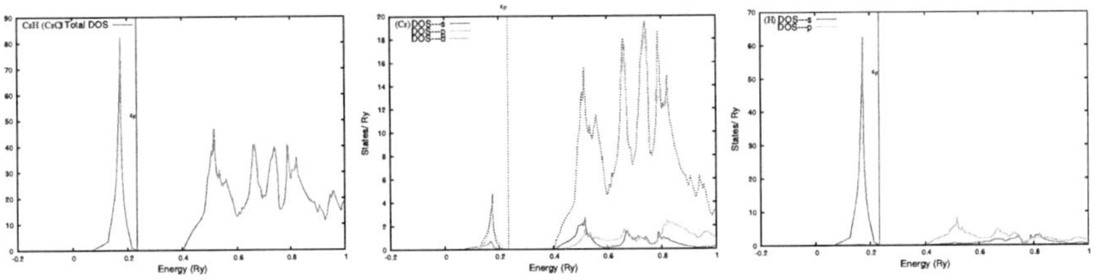

Fig. 3.26 Total, angular momentum and site decomposed densities of states of CsH in the CsCl structure

Table 3.20 CsH (NaCl) a = 11.828 Bohr Slater–Koster 3-center parameters

	ORTHOGONAL	NON-ORTHOGONAL	
	ENERGY INTEGRALS Ry	ENERGY INTEGRALS Ry	OVERLAP INTEGRALS
ON SITE Cs-Cs			
s,s(000)	0.59650	0.55909	
x,x(000)	0.79913	0.75538	
xy,xy(000)	0.74888	0.59564	
d2,d2(000)	0.67540	0.58597	
FIRST NEIGHBOR Cs-Cs			
s,s(110)	-0.02315	-0.01807	0.00034
s,x(110)	-0.02102	0.00386	0.04181
s,xy(110)	-0.01636	0.01117	0.10291
s,d2(110)	0.01516	0.02508	0.03096
x,x(110)	0.01808	-0.03844	-0.09246
x,x(011)	-0.00648	0.02906	0.06965
x,y(110)	0.02631	0.00953	-0.01032
x,xy(110)	0.04073	-0.02483	-0.10179
x,xy(011)	-0.00400	0.01276	0.04772
z,d2(011)	0.00777	0.03768	0.07755
z,d1(011)	-0.01032	-0.00274	-0.01634
xy,xy(110)	-0.05008	-0.00736	0.06733
xy,xy(011)	0.01170	-0.01308	-0.05598
xy,xz(011)	-0.01239	0.00095	-0.03039
xy,d2(110)	0.02046	0.01669	0.06878
d2,d2(110)	-0.01042	0.00492	0.03226
d1,d1(110)	0.00077	0.01259	-0.03725
SECOND NEIGHBOR Cs-Cs			
s,s(200)	0.04210	0.00841	-0.02930
s,x(200)	0.05178	0.00179	-0.04260
s,d2(002)	0.02957	0.00423	0.06206
x,x(200)	-0.05655	0.04800	0.10551
y,y(200)	0.01554	0.01320	0.01839
x,xy(020)	0.01529	0.00685	0.01251
z,d2(002)	-0.03350	-0.07546	-0.12884
xy,xy(200)	-0.01997	-0.00137	-0.00605
xy,xy(002)	0.00395	-0.00070	-0.00058
d2,d2(002)	0.02842	-0.03948	-0.05808
d1,d1(002)	0.00371	-0.00926	-0.01227
ON SITE H-H			
s,s(000)	0.20708	0.19134	
FIRST NEIGHBOR H-H			
s,s(110)	-0.00829	0.01256	0.01821
s,x(110)	0.00000	0.00000	0.00000
x,x(110)	0.00000	0.00000	0.00000
x,x(011)	0.00000	0.00000	0.00000
x,y(110)	0.00000	0.00000	0.00000
SECOND NEIGHBOR H-H			
s,s(200)	-0.00499	-0.00570	-0.00825
s,x(200)	0.00000	0.00000	0.00000
x,x(200)	0.00000	0.00000	0.00000
y,y(200)	0.00000	0.00000	0.00000
FIRST NEIGHBOR Cs-H			
s,s(100)	-0.03123	0.05541	0.03409
s,x(100)	0.00000	0.00000	0.00000
x,s(100)	0.03770	0.03182	0.01297
x,x(100)	0.00000	0.00000	0.00000
y,y(100)	0.00000	0.00000	0.00000
xy,x(010)	0.00000	0.00000	0.00000
d2,s(001)	-0.05994	-0.02059	-0.20072
d2,z(001)	0.00000	0.00000	0.00000
SECOND NEIGHBOR Cs-H			
s,s(111)	0.00561	-0.00249	-0.00556
x,s(111)	-0.00833	-0.00258	-0.00208
s,x(111)	0.00000	0.00000	0.00000
x,x(111)	0.00000	0.00000	0.00000
x,y(111)	0.00000	0.00000	0.00000
xy,s(111)	0.01351	0.00564	0.00928
xy,x(111)	0.00000	0.00000	0.00000
yz,x(111)	0.00000	0.00000	0.00000
d1,x(111)	0.00000	0.00000	0.00000

CsH

BAND	ORTHOGONAL RMS ERROR mRy	MAXIMUM DEVIATION k	mRy	NON-ORTHOGONAL RMS ERROR mRy	MAXIMUM DEVIATION k	mRy
1	3.5	(444)	9.6	0.3	(174)	0.8
2	5.5	(003)	14.1	1.1	(002)	2.9
3	4.1	(044)	9.2	1.0	(004)	3.0
4	5.5	(226)	13.0	0.8	(118)	2.6
5	7.5	(174)	15.5	0.9	(442)	2.9
6	7.6	(444)	17.6	1.0	(444)	1.9
7	6.0	(444)	17.6	1.1	(226)	3.0
1-7	5.8			0.9		

ENERGY VALUES IN Ry AT SELECTED k-POINTS

	ORTHOGONAL	APW	NON-ORTHOGONAL
GAMMA 1	0.03946	0.03992	0.04026
GAMMA 1	0.60958	0.61075	0.61104
GAMMA 12	0.71383	0.71818	0.71779
GAMMA 15	0.86690	0.86567	0.86571
GAMMA 25'	0.57019	0.57247	0.57116
X1 (008)	0.09422	0.09138	0.09106
X1 (008)	0.80432	0.79636	0.79535
X1 (008)	0.99984	0.99950	0.99941
X2 (008)	0.83585	0.83583	0.83587
X3 (008)	0.38300	0.39103	0.39246
X5 (008)	0.87722	0.87614	0.87650
X4' (008)	0.57758	0.57701	0.57664
X5' (008)	0.77410	0.78326	0.78326
L1 (444)	0.32236	0.32361	0.32380
L1 (444)	0.94150	0.93298	0.93167
L3 (444)	0.53625	0.53654	0.53722
L3 (444)	0.81406	0.83161	0.82967
L2' (444)	0.19834	0.18875	0.18894
L2' (444)	0.67825	0.68139	0.68086
L3' (444)	0.95535	0.95637	0.95634
W1 (048)	0.55390	0.56503	0.56502
W1 (048)	0.91919	0.91285	0.91209
W2' (048)	0.12806	0.12817	0.12828
W2' (048)	0.82174	0.82515	0.82485
W1' (048)	0.86143	0.87780	0.87759
W3 (048)	0.47655	0.47280	0.47300
W3 (048)	0.84713	0.84111	0.84177
ODD (224)	0.58514	0.57783	0.57859
ODD (224)	0.79486	0.79100	0.78918
EVEN(224)	0.14196	0.14223	0.14188
EVEN(224)	0.39163	0.39324	0.39388
EVEN(224)	0.50625	0.51140	0.51028
EVEN(224)	0.66553	0.65670	0.65755
EVEN(224)	0.73165	0.72764	0.72941

FERMI LEVEL QUANTITIES (Non-orthogonal fit)

ENERGY Ry	Total	s-Cs	p-Cs	t2g-Cs	eg-Cs	s-H	p-H
	DENSITIES OF STATES (states/Ry/cell)						
0.1890	0.00	0.00	0.00	0.00	0.00	0.00	0.00
	INTEGRATED DENSITIES OF STATES (electrons)						
	2.00	0.18	0.05	0.01	0.03	1.73	0.00

VELOCITY cm/s	PLASMON ENERGY eV	EIGENVALUE SUM Ry
0.00x10E8	0.00	-0.1142

Table 3.21 CsH (NaCl) a = 11.828 Bohr Slater–Koster 2-center parameters

	ORTHOGONAL	NON-ORTHOGONAL	
	ENERGY INTEGRALS Ry	ENERGY INTEGRALS Ry	OVERLAP INTEGRALS
ON SITE Cs-Cs			
s	0.61653	0.41929	
p	0.79466	0.67484	
t2g	0.67632	0.56752	
eg	0.65370	0.54604	
FIRST NEIGHBOR Cs-Cs			
(sss)	-0.02346	0.02530	0.12817
(pps)	0.04636	-0.04392	-0.15734
(ppp)	-0.00832	0.03784	0.07166
(dds)	-0.06672	0.04142	0.15992
(ddp)	0.01995	-0.00656	-0.08800
(ddd)	0.00164	0.00157	0.01018
(sps)	0.03312	-0.04755	-0.16104
(sds)	-0.04818	0.03274	0.14072
(pds)	-0.06500	0.06724	0.19080
(pdp)	0.00554	-0.02004	-0.06840
SECOND NEIGHBOR Cs-Cs			
(sss)	-0.00333	0.01572	0.01988
(pps)	-0.02043	0.10345	0.11448
(ppp)	-0.00035	0.00557	0.00452
(dds)	0.00959	-0.04947	-0.05387
(ddp)	-0.00142	-0.00698	-0.00948
(ddd)	0.00210	0.00594	0.00756
(sps)	-0.01339	0.00987	0.00949
(sds)	0.01713	-0.01714	-0.01507
(pds)	0.01395	-0.07384	-0.08132
(pdp)	0.00028	-0.01251	-0.01519
ON SITE H-H			
s	0.41818	0.22780	
FIRST NEIGHBOR H-H			
(sss)	-0.02047	0.00963	-0.02848
(sps)	0.00000	0.00000	0.00000
(pps)	0.00000	0.00000	0.00000
(ppp)	0.00000	0.00000	0.00000
SECOND NEIGHBOR H-H			
(sss)	0.02642	-0.01339	-0.01505
(sps)	0.00000	0.00000	0.00000
(pps)	0.00000	0.00000	0.00000
(ppp)	0.00000	0.00000	0.00000
FIRST NEIGHBOR Cs-H			
(sss)	0.07248	0.05556	-0.00862
(sps)	0.00000	0.00000	0.00000
(pss)	0.05414	0.04235	-0.10671
(pps)	0.00000	0.00000	0.00000
(ppp)	0.00000	0.00000	0.00000
(dpp)	0.00000	0.00000	0.00000
(dss)	0.07813	-0.00451	-0.22965
(dps)	0.00000	0.00000	0.00000
SECOND NEIGHBOR Cs-H			
(sss)	-0.01886	0.01159	0.04002
(pss)	-0.04594	0.02085	0.05043
(sps)	0.00000	0.00000	0.00000
(pps)	0.00000	0.00000	0.00000
(ppp)	0.00000	0.00000	0.00000
(dss)	-0.03448	0.01277	0.04255
(dps)	0.00000	0.00000	0.00000
(dpp)	0.00000	0.00000	0.00000

♀ CsH

BAND	ORTHOGONAL RMS ERROR mRy	MAXIMUM DEVIATION k	MAXIMUM DEVIATION mRy	NON-ORTHOGONAL RMS ERROR mRy	MAXIMUM DEVIATION k	MAXIMUM DEVIATION mRy
1	5.4	(005)	12.6	1.3	(044)	3.4
2	7.0	(003)	17.4	2.5	(002)	7.3
3	9.4	(444)	22.7	2.2	(022)	7.3
4	11.8	(444)	22.7	2.2	(226)	4.2
5	14.8	(226)	34.5	1.9	(222)	4.9
6	18.5	(044)	39.7	3.0	(066)	6.7
7	15.3	(055)	38.6	2.0	(222)	4.5
1.7	12.5			2.2		

ENERGY VALUES IN Ry AT SELECTED k-POINTS

	ORTHOGONAL	APW	NON-ORTHOGONAL
GAMMA 1	0.03893	0.03992	0.03792
GAMMA 1	0.60721	0.61075	0.61219
GAMMA 12	0.71580	0.71818	0.71963
GAMMA 15	0.87127	0.86567	0.86575
GAMMA 25'	0.56268	0.57247	0.56869
X1 (008)	0.08746	0.09138	0.09091
X1 (008)	0.80594	0.79636	0.79676
X1 (008)	0.99366	0.99950	0.99952
X2 (008)	0.82138	0.83583	0.83427
X3 (008)	0.38992	0.39103	0.39207
X5 (008)	0.87334	0.87614	0.87363
X4' (008)	0.56698	0.57701	0.57542
X5' (008)	0.78569	0.78326	0.78371
L1 (444)	0.33100	0.32361	0.32491
L1 (444)	0.91008	0.93298	0.93178
L3 (444)	0.51382	0.53654	0.53953
L3 (444)	0.81927	0.83161	0.83084
L2' (444)	0.17951	0.18875	0.18873
L2' (444)	0.69838	0.68139	0.68184
L3' (444)	0.94627	0.95637	0.95645
W1 (048)	0.54300	0.56503	0.56470
W1 (048)	0.87290	0.91285	0.91147
W2' (048)	0.13174	0.12817	0.12559
W2' (048)	0.82192	0.82515	0.82745
W1' (048)	0.86495	0.87780	0.87940
W3 (048)	0.48147	0.47280	0.47285
W3 (048)	0.83359	0.84111	0.84025
ODD (224)	0.56804	0.57783	0.57408
ODD (224)	0.79285	0.79100	0.79134
EVEN(224)	0.14748	0.14223	0.14235
EVEN(224)	0.38467	0.39324	0.39237
EVEN(224)	0.51883	0.51140	0.51128
EVEN(224)	0.68585	0.65670	0.65689
EVEN(224)	0.72600	0.72764	0.72741

3.6 Francium Hydride (FrH)

See Fig. 3.27, Tables 3.22, and 3.23.

See Figs. 3.28, 3.29, 3.30, and 3.31.

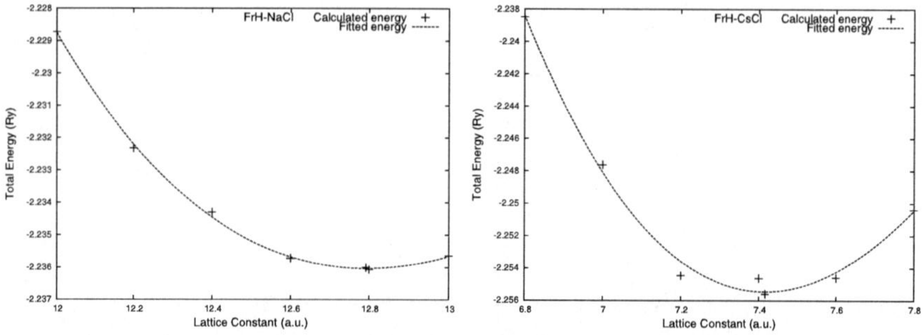

Fig. 3.27 Total energy versus lattice constant of FrH for the NaCl and CsCl structures

Table 3.22 Lattice constants, bulk modulus, gap, total energy

	a (Bohr)		B (MBar)	Gap (Ry)	Total Energy
NaCl	12.79	Insulator	0.09	0.13	-48632.23601
CsCl	7.42	Insulator	0.17	0.15	

Table 3.23 Birch fit coefficients

	A1	A2	A3	A4
NaCl	-1.669450E+00	-9.405588E+01	5.038194E+03	-8.573017E+04
CsCl	-1.496210E+00	-9.661942E+01	3.738054E+03	-3.963887E+04

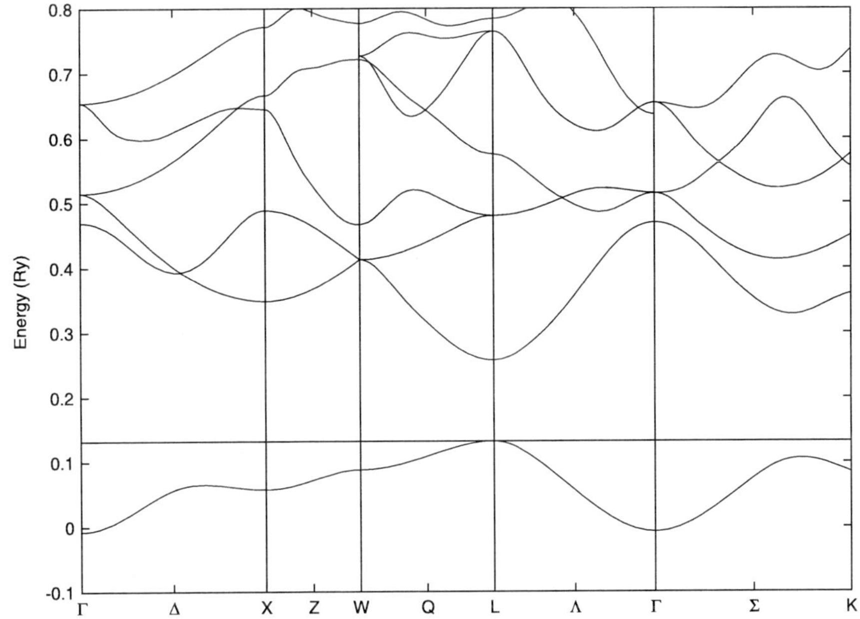

Fig. 3.28 Energy bands of FrH in the NaCl structure

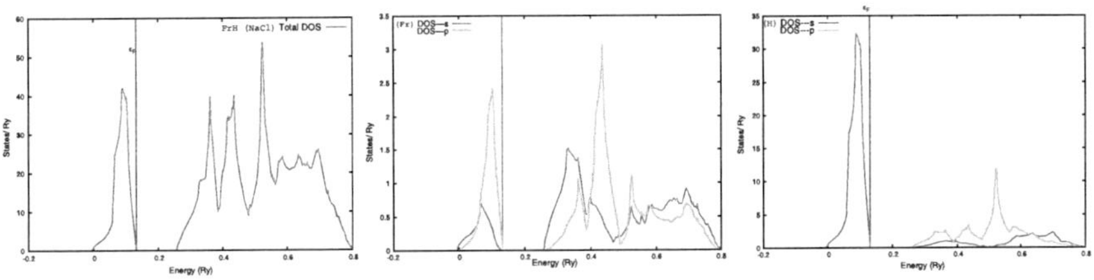

Fig. 3.29 Total, angular momentum and site decomposed densities of states of FrH in the NaCl structure

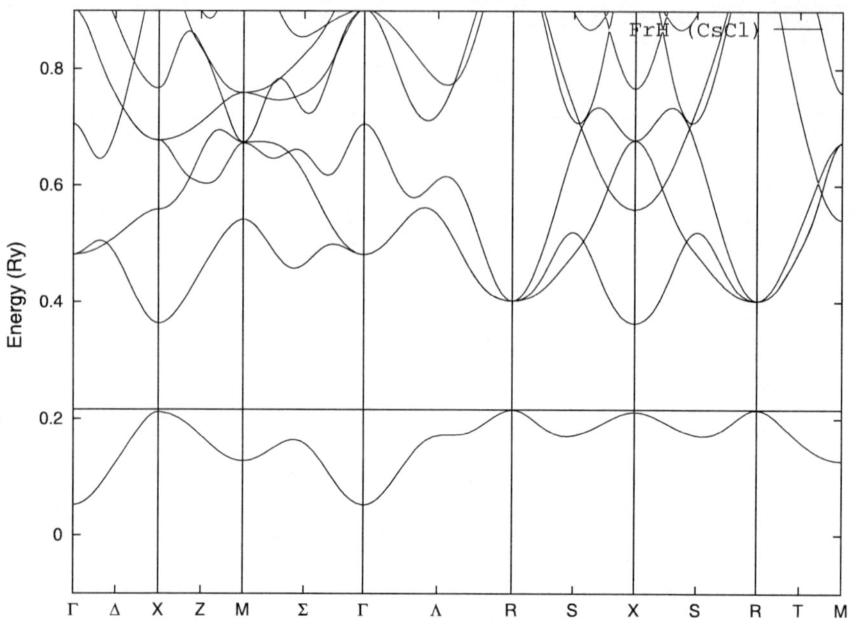

Fig. 3.30 Energy bands of FrH in the CsCl structure

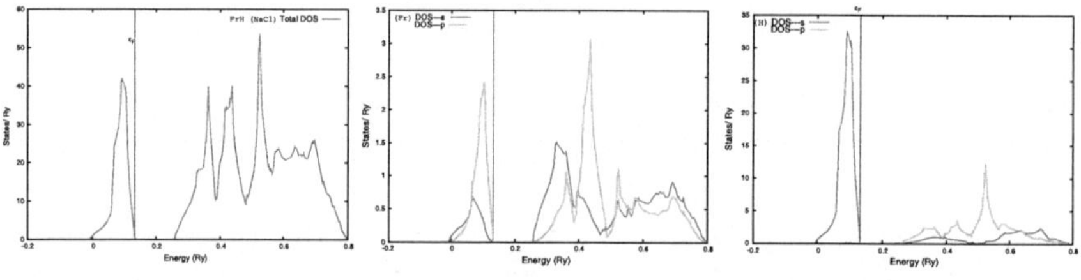

Fig. 3.31 Total, angular momentum and site decomposed densities of states of FrH in the CsCl structure

References

1. H. Smithson, C.A. Marianetti, D. Morgan, A. Van der Ven, A. Predith, G. Ceder, First-Principles study of the stability and electronic structure of metal hydrides. Phys. Rev. B **66**, 144107 (2002)
2. G.D. Barrera, D. Colognesi, P.C.H. Mitchell, A.D. Ramirez-Cuesta, LDA or GGA? A combined experimental inelastic neutron scattering and ab-initio lattice dynamics study of alkali metal hydrides. Chem. Phys. **317**, 119 (2005)
3. L. Fohlmeister, A. Stach, Alkali metal hydride complexes: well-defined molecular species of saline hydrides. Aust. J. Chem. **68**, 1190 (2015)

Chapter 4
Alkaline Earth Hydrides

This chapter covers the alkaline earth hydrides BeH, MgH, CaH, SrH, BaH and RaH) [1-2]. Results are presented for the crystal structures NaCl (B1), CsCl (B2), CaF2 (C1) in line with the scope of the book to study only cubic hydrides. These systems form as dihydrides in non-cubic structures with the exception of MgH2 and BeH2 which under pressure take the CaF2 structure [3]. CaH, SrH and BaH form orthorhombic crystals. Due to the interest on superconductivity under high pressures in light hydrides, results are presented for BeH3 in the Im3m structure. Our results for BeH3 in both the Im3m and the fluorite structures show strong electron–ion interaction coming mainly from the hydrogen sites consistent with the work of Zang et al.

Examining the energy bands and densities of states diagrams we observe that, as in the alkalis, the occupied states have predominantly s–H character and the conduction bands have p character of the alkali-earth element for BeH and MgH and an admixture for p and d-like contributions for the heavier ones. Also Tight-binding parameters are given in the NaCl structure based on both orthogonal and non-orthogonal Hamiltonians using three- and two-center integrals.

D. A. Papaconstantopoulos, *Band Structure of Cubic Hydrides*,
https://doi.org/10.1007/978-3-031-06878-2_4

4.1 Beryllium Hydride (BeH)

See Fig. 4.1.

See Tables 4.1, 4.2 and 4.3.

See Figs. 4.2 and 4.3.

See Tables 4.4 and 4.5.

See Figs. 4.4 and 4.5.

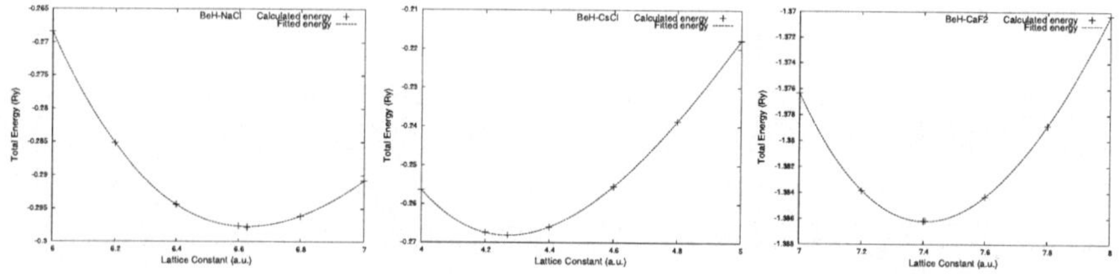

Fig. 4.1 Total energy versus lattice constant in the NaCl, CsCl and CaF2 structures

Table 4.1 Lattice Constant, Bulk Modulus, Gap, Total Energy

Stru	a (Bohr)	B (MBar)	Gap (Ry)	Total Energy (Ry)
NaCl	6.60	1.13	-	-30.29794
NaCl (P=0.57MBar)	6.00	2.92		-30.26839
CsCl	4.27	1.06		-30.26805
CaF2	7.41	0.94		-31.38616
Im3m (P=0.0)	6.38	0.83		-32.46128
Im3m (P=2.02MBar)	5.00	2.06		-32.18568
exp(BC Ortho)				

Table 4.2 Birch Fit Coefficients

	A1	A2	A3	A4
NaCl	4.633983E-01	-2.880345E+01	3.103357E+02	-6.885417E+02
CsCl	5.715413E-01	-3.444768E+01	4.188260E+02	-1.273777E+03
CaF2	-3.365758E-01	-5.264171E+01	7.997820E+02	-3.292197E+03
Im3m	-3.143913E+01	-5.736532E+01	9.265431E+02	-3.273279E+03

Table 4.3 DOS at Ef, Hopfield parameter, Stoner Criterion

```
                 BeH     a=6.60 Bohr          NaCl
--------------------------------------------------------------------------------
 Fermi Energy   Total DOS        Decomposed DOS inside the muffin tins
   (Ry)        (States/Ry)      s          p          eg          t2g      f
--------------------------------------------------------------------------------
Be  0.901         3.953         0.265     0.733     0.022      0.057    0.005
H   0.901         3.953         0.371     0.590     0.004      0.033    0.003
--------------------------------------------------------------------------------
Fermi-Velocity (cm/s): 1.35025 x10E8
Plasmon Energy (eV)   : 11.41339
Electron-ion interaction (Hopfield parameter) (eV/A^2) Be: 1.702  H: 2.017
------------------------------------------------
          Be      MUFFIN-TIN RADIUS and CHARGE =     1.6487     2.6142
           H      MUFFIN-TIN RADIUS and CHARGE =     1.6487     1.1717
                    Be        STONER I = 0.015
                     H        STONER I = 0.011
          STONER PARAMETER (Ry)      I = 0.028
          STONER CRITERION         N*I = 0.110
```

```
                 BeH (P=0.57 Mbar)  a=6.00 Bohr      NaCl
Fermi Energy    Total DOS                 Decomposed DOS inside the muffin tins
    (Ry)        (states/Ry)                    s         p       eg      t2g     f
   1.132           3.283                    0.218     0.675   0.025   0.071   0.005
   1.132           3.283                    0.272     0.408   0.002   0.022   0.002
Fermi-Velocity (cm/s): 1.47044x10E8
Plasmon Energy (eV)   : 13.05461
Electron-ion interaction (Hopfield parameter (eV/A^2)    Be: 3.014  H: 2.330
     Be    MUFFIN-TIN RADIUS and CHARGE    1.5750      2.7001
      H    MUFFIN-TIN RADIUS and CHARGE    1.4250      1.0375
             Be     STONER      I=0.0160
              H     STONER      I=0.0102
     STONER PARAMETER (Ry) I=0.0278
     STONER CRITERION      NI=0.0914
```

```
                 BeH     a=4.27 Bohr        CsCl
-------------------------------------------------------------------------------
  Fermi Energy    Total DOS        Decomposed DOS inside the muffin tins
    (Ry)        (States/Ry)       s          p          eg          t2g          f
--------------------------------------------------------------------------------
 Be  0.797          5.657         0.544    1.974    0.089    0.045    0.023
 H   0.797          5.657         0.383    0.847    0.027    0.042    0.008
--------------------------------------------------------------------------------
 Fermi-Velocity (cm/s): 1.32182 x10E8
 Plasmon Energy (eV)   : 12.82849
 Electron-ion interaction (Hopfield parameter) (eV/A^2) Be: 1.791  H: 1.426
 ------------------------------------------------
          Be      MUFFIN-TIN RADIUS and CHARGE =     1.9414     3.0123
           H      MUFFIN-TIN RADIUS and CHARGE =     1.7565     1.1646
                    Be        STONER I = 0.027
                     H        STONER I = 0.009
          STONER PARAMETER (Ry)      I = 0.036
          STONER CRITERION         N*I = 0.205
 ------------------------------------------------
```

```
                 BeH2     a=7.41 Bohr       CaF2
 ------------------------------------------------------------
  Fermi Energy    Total DOS        Decomposed DOS inside the muffin tins
    (Ry)        (States/Ry)       s          p          eg      t2g        f
 Be   0.734         4.717        0.314    0.384    0.050    0.037    0.007
  H   0.734         4.717        0.990    0.602    0.008    0.011    0.003
 ------------------------------------------------
 Fermi-Velocity (cm/s): 1.34957 x10E8
 Plasmon Energy (eV)   : 10.47169
 Electron-ion interaction (Hopfield parameter) (eV/A^2) Be: 0.716  2H: 7.820
 ------------------------------------------------
          Be      MUFFIN-TIN RADIUS and CHARGE =     1.6836     2.5703
           H      MUFFIN-TIN RADIUS and CHARGE =     1.5232     0.9985
            Be        STONER              I =    0.005
             H        STONER              I =    0.006
               STONER PARAMETER (Ry) I =    0.018
               STONER CRITERION    N*I =    0.086
```

```
                BeH3        a=5.0 Bohr          Im3m
Fermi Energy        Total DOS           Decomposed DOS inside the muffin tins
    (Ry)            (states/Ry)           s      p      d      f
Be  1.403              4.997          0.215  0.625  0.051  0.011
 H  1.403              4.997          0.445  0.081  0.006  0.000
Fermi-Velocity (cm/s): 1.62053 x10E8
Plasmon Energy (eV)   : 16.49726
Electron-ion interaction (Hopfield parameter) (eV/A^2) Be: 3.931   3H: 7.697
```

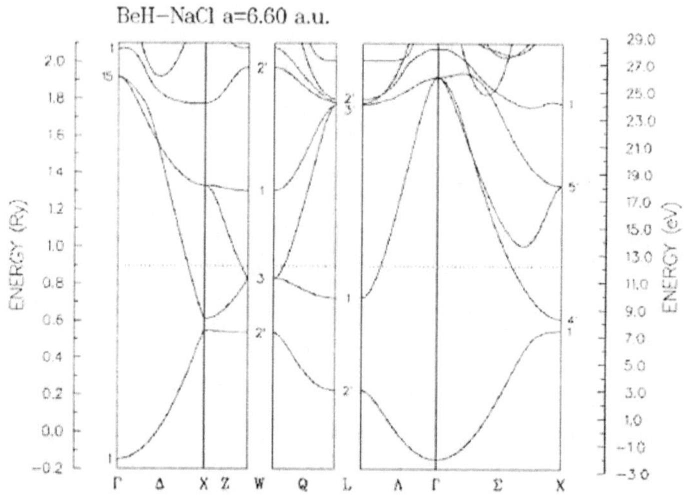

Fig. 4.2 Energy bands of BeH in the NaCl structure (tight-binding)

Fig. 4.3 Total, angular momentum and site decomposed densities of states of BeH in the NaCl structure

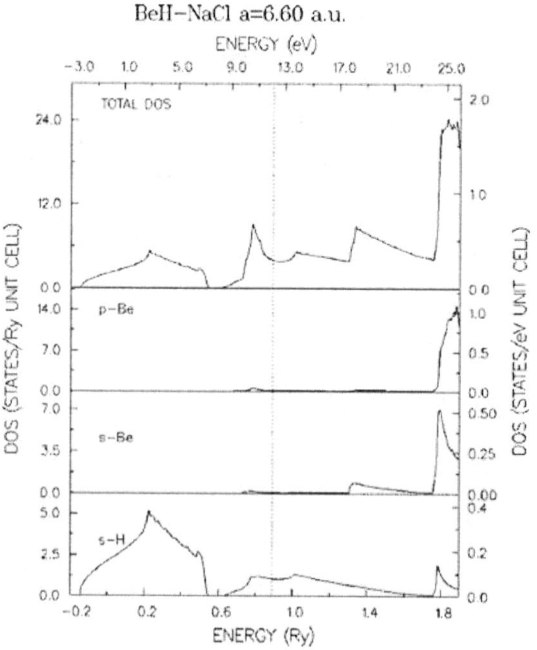

Table 4.4 BeH (NaCl) a = 6.595 Bohr Slater–Koster 3-center parameters

	ORTHOGONAL	NON-ORTHOGONAL	
	ENERGY INTEGRALS Ry	ENERGY INTEGRALS Ry	OVERLAP INTEGRALS
ON SITE Be-Be			
s,s(000)	1.41150	3.31209	
x,x(000)	2.11739	4.06080	
FIRST NEIGHBOR Be-Be			
s,s(110)	-0.08009	0.00362	0.11491
s,x(110)	-0.00090	0.32959	0.11666
x,x(110)	-0.16730	-0.87772	-0.01331
x,x(011)	0.14210	0.45694	0.06272
x,y(110)	0.06685	-0.09451	-0.18568
SECOND NEIGHBOR Be-Be			
s,s(200)	-0.00390	-0.46048	-0.03290
s,x(200)	-0.03646	-0.57614	0.04046
x,x(200)	0.07817	0.72700	-0.16976
y,y(200)	0.11579	0.12344	-0.09520
ON SITE H-H			
s,s(000)	0.93219	0.85648	
FIRST NEIGHBOR H-H			
s,s(110)	0.05624	0.01267	0.01112
s,x(110)	0.01500	-0.00815	-0.00474
x,x(110)	-0.05549	0.00005	-0.03198
x,x(011)	0.14674	-0.06625	-0.01464
x,y(110)	-0.13459	0.04273	0.05102
SECOND NEIGHBOR H-H			
s,s(200)	-0.01829	-0.02520	0.00215
s,x(200)	0.00136	0.00238	0.02249
x,x(200)	0.07529	0.01885	0.02520
y,y(200)	-0.06595	-0.00103	-0.00320
FIRST NEIGHBOR Be-H			
s,s(100)	-0.15536	-0.15551	0.09006
s,x(100)	0.57353	1.36434	0.01340
x,s(100)	0.22429	0.37070	-0.02996
x,x(100)	0.92164	1.80168	-0.15779
y,y(100)	-0.62986	-0.91477	0.09324
SECOND NEIGHBOR Be-H			
s,s(111)	-0.00503	-0.01676	0.01529
x,s(111)	0.00678	0.02994	-0.01558
s,x(111)	0.00843	0.00128	-0.00538
x,x(111)	0.03115	-0.01861	-0.01736
x,y(111)	0.01375	-0.00203	-0.01565

	ORTHOGONAL			NON-ORTHOGONAL		
BAND	RMS ERROR	MAXIMUM DEVIATION		RMS ERROR	MAXIMUM DEVIATION	
	mRy	k	mRy	mRy	k	mRy
1	6.8	(048)	18.3	1.8	(007)	4.8
2	10.7	(111)	23.2	2.3	(005)	5.7
3	17.8	(022)	49.1	2.7	(180)	7.8
4	11.7	(004)	27.5	2.5	(224)	11.6
5	16.8	(066)	37.5	5.2	(118)	20.9
1-5	13.4			3.1		

ENERGY VALUES IN Ry AT SELECTED k-POINTS

	ORTHOGONAL	APW	NON-ORTHOGONAL
GAMMA 1	-0.14778	-0.14894	-0.14778
GAMMA 1	2.07214	2.06896	2.07044
GAMMA 15	1.90849	1.91719	1.91908
X1 (008)	0.53514	0.54600	0.54666
X1 (008)	1.77078	1.77561	1.77335
X4' (008)	0.62493	0.61407	0.61169
X5' (008)	1.31668	1.33282	1.33203
L1 (444)	0.71407	0.72157	0.72058
L2' (444)	0.22456	0.22315	0.22049

L2' (444)	1.78049	1.79512	1.78645
L3' (444)	1.76529	1.77602	1.77476
W1 (048)	1.29323	1.30189	1.30293
W2' (048)	0.55141	0.53316	0.53290
W2' (048)	1.97504	1.96940	1.97200
W3 (048)	0.82540	0.83153	0.83090

FERMI LEVEL QUANTITIES (Non-orthogonal fit)

ENERGY		DENSITIES OF STATES					
	Total	s-Be	p-Be	t2g-Be	eg-Be	s-H	p-H
Ry		states/Ry/cell					
0.8949	3.99	0.09	0.18	0.00	0.00	1.05	2.67
		INTEGRATED DENSITIES OF STATES					
	Total	s-Be	p-Be	t2g-Be	eg-Be	s-H	p-H
		electrons					
	3.00	0.03	0.06	0.00	0.00	2.15	0.76

VELOCITY	PLASMON ENERGY	EIGENVALUE SUM
cm/s	eV	Ry
1.44x10E8	12.20	-1.4168

Table 4.5 BeH (NaCl) a = 6.595 Bohr Slater–Koster 2-center parameters

	ORTHOGONAL	NON-ORTHOGONAL	
	ENERGY INTEGRALS Ry	ENERGY INTEGRALS Ry	OVERLAP INTEGRALS
ON SITE Be-Be			
s	3.35111	0.55571	
p	2.65235	0.87371	
FIRST NEIGHBOR Be-Be			
(sss)	-0.00857	-0.03856	0.12629
(pps)	-0.03144	-0.10030	-0.29062
(ppp)	-0.03255	-0.00853	0.05732
(sps)	-0.02896	-0.00128	-0.18953
SECOND NEIGHBOR Be-Be			
(sss)	-0.28536	-0.02270	0.00757
(pps)	0.43100	0.03560	-0.02889
(ppp)	0.00842	-0.01541	-0.00564
(sps)	0.35128	0.02898	-0.01593
ON SITE H-H			
s	0.40539	0.33076	
FIRST NEIGHBOR H-H			
(sss)	-0.01587	-0.04835	0.01138
(sps)	0.02797	0.00000	0.00000
(pps)	0.00508	0.00000	0.00000
(ppp)	-0.02524	0.00000	0.00000
SECOND NEIGHBOR H-H			
(sss)	0.02743	0.01621	0.00985
(sps)	-0.04696	0.00000	0.00000
(pps)	-0.10149	0.00000	0.00000
(ppp)	0.07105	0.00000	0.00000
FIRST NEIGHBOR Be-H			
(sss)	0.15881	-0.03276	0.25860
(sps)	1.21969	0.00000	0.00000
(pss)	0.06315	0.03197	0.36462
(pps)	-1.45691	0.00000	0.00000
(ppp)	0.21727	0.00000	0.00000
SECOND NEIGHBOR Be-H			
(sss)	-0.00050	-0.01813	0.01170
(pss)	0.02195	-0.03901	0.01541
(sps)	-0.02501	0.00000	0.00000
(pps)	0.01783	0.00000	0.00000
(ppp)	0.01042	0.00000	0.00000

BeH

BAND	ORTHOGONAL RMS ERROR mRy	MAXIMUM DEVIATION k	MAXIMUM DEVIATION mRy	NON-ORTHOGONAL RMS ERROR mRy	MAXIMUM DEVIATION k	MAXIMUM DEVIATION mRy
1	5.3	(066)	13.5	6.1	(444)	11.7
2	15.6	(055)	41.7	5.4	(062)	12.9
3	26.6	(022)	68.9	8.6	(118)	19.0
4	17.0	(005)	45.0	6.6	(224)	27.8
5	16.6	(066)	48.4	7.5	(118)	27.4
1-5	17.5			6.9		

ENERGY VALUES IN Ry AT SELECTED k-POINTS

	ORTHOGONAL	APW	NON-ORTHOGONAL
GAMMA 1	-0.15337	-0.14894	-0.14967
GAMMA 1	2.06901	2.06896	2.06968
GAMMA 15	1.90358	1.91719	1.91782
X1 (008)	0.54199	0.54600	0.53697
X1 (008)	1.76471	1.77561	1.77183
X4' (008)	0.62273	0.61407	0.61689
X5' (008)	1.32097	1.33282	1.32884
L1 (444)	0.70316	0.72157	0.72482
L2' (444)	0.20980	0.22315	0.21149
L2' (444)	1.78324	1.79512	1.78135
L3' (444)	1.75888	1.77602	1.77044
W1 (048)	1.29443	1.30189	1.30313
W2' (048)	0.52328	0.53316	0.53603
W2' (048)	1.95469	1.96940	1.97003
W3 (048)	0.83855	0.83153	0.82521

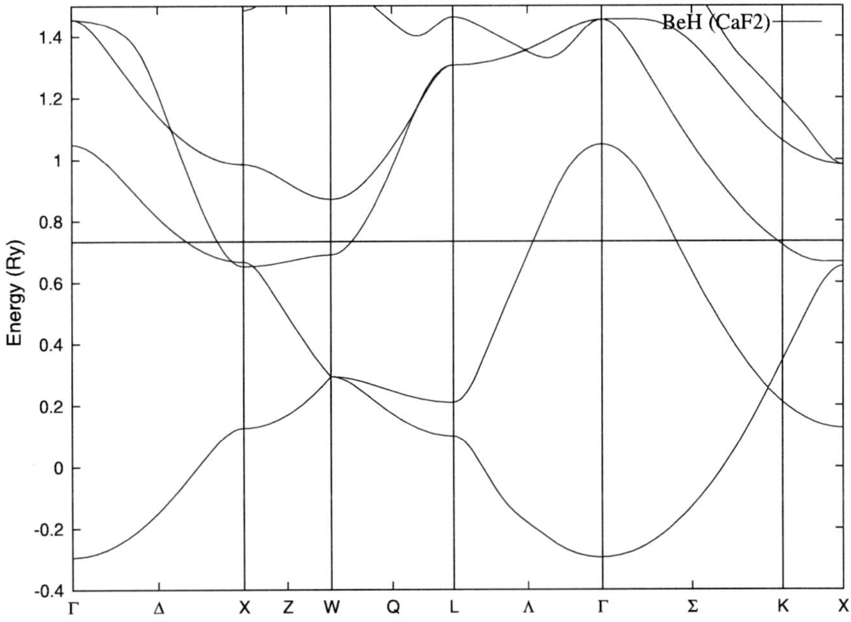

Fig. 4.4 Energy bands of BeH2 in the CaF2 structure

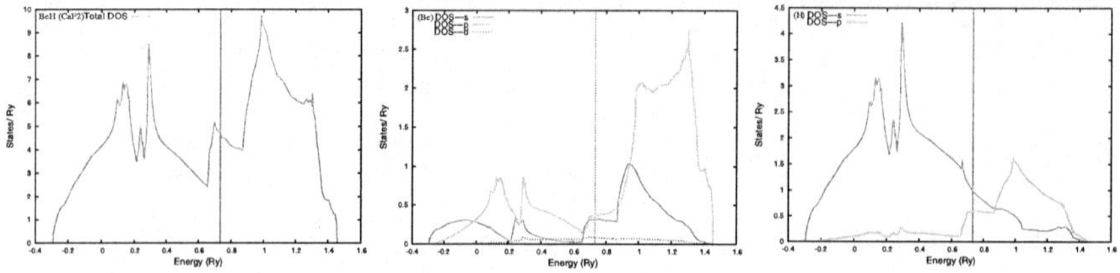

Fig. 4.5 Total, angular momentum and site decomposed densities of states of BeH2 in the CaF2 structure

4.2 Magnesium Hydride (MgH)

See Fig. 4.6.

See Tables 4.6, 4.7 and 4.8.

See Figs. 4.7, 4.8, 4.9 and 4.10.

See Tables 4.9 and 4.10.

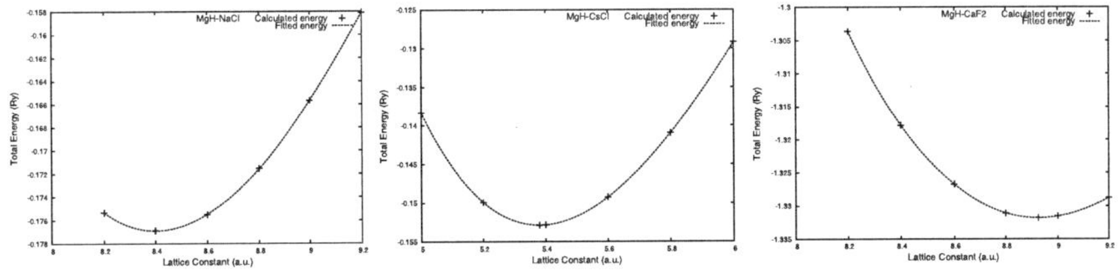

Fig. 4.6 Total energy versus lattice constant in the NaCl, CsCl and CaF2 structures

Table 4.6 Lattice Constant, Bulk Modulus, Gap, Total Energy

Stru	a (Bohr)	B (MBar)	Gap (Ry)	Total Energy (Ry)
NaCl	8.40	0.58		-400.17688
CsCl	5.38	0.50		-400.15285
CaF2	8.93	0.64		-401.33180
CaF2 (P=0.25MBar)	8.20	1.40		-401.30362
exp(bc tetragonal)	a=8.54	c=5.71		

Table 4.7 Birch Fit Coefficients

	A1	A2	A3	A4
NaCl	5.800398E-01	-4.505727E+01	7.431825E+02	-2.100478E+03
CsCl	5.333740E-01	-4.227267E+01	7.224455E+02	-2.135852E+03
CaF2	-2.383832E-01	-7.632023E+01	1.546942E+03	-7.175319E+03

Table 4.8 DOS at Ef, Hopfield parameter, Stoner Criterion

MgH a=8.40 Bohr NaCl

Fermi Energy (Ry)	Total DOS (States/Ry)	Decomposed DOS inside the muffin tins s	p	eg	t2g	f
Mg 0.625	11.931	1.209	0.907	0.102	0.224	0.024
H 0.625	11.931	1.225	2.009	0.026	0.073	0.007

Fermi-Velocity (cm/s): 0.73639 x10E8
Plasmon Energy (eV) : 7.52517
Electron-ion interaction (Hopfield parameter) (eV/A^2) Mg: 0.415 H: 2.724

Mg MUFFIN-TIN RADIUS and CHARGE = 2.0996 10.4154
H MUFFIN-TIN RADIUS and CHARGE = 2.0996 1.3965
Mg STONER I = 0.007
H STONER I = 0.010
STONER PARAMETER (Ry) I = 0.019
STONER CRITERION N*I = 0.2011
****

MgH a=5.38 Bohr CsCl

Fermi Energy (Ry)	Total DOS (States/Ry)	Decomposed DOS inside the muffin tins s	p	eg	t2g	f
Mg 0.593	8.164	1.813	1.127	0.189	0.140	0.048
H 0.593	8.164	0.944	1.241	0.037	0.111	0.009

Fermi-Velocity (cm/s): 1.20137 x10E8
Plasmon Energy (eV) : 9.90381
Electron-ion interaction (Hopfield parameter) (eV/A^2) Mg: 0.208 H: 1.389

Mg MUFFIN-TIN RADIUS and CHARGE = 2.4460 10.7991
H MUFFIN-TIN RADIUS and CHARGE = 2.2131 1.4139
Mg STONER I = 0.014
H STONER I = 0.010
STONER PARAMETER (Ry) I = 0.024
STONER CRITERION N*I = 0.1965

MgH2 a=8.93 Bohr CaF2

Fermi Energy (Ry)	Total DOS (States/Ry)	Decomposed DOS inside the muffin tins s	p	eg	t2g	f
Mg 0.512	4.104	0.240	0.045	0.062	0.028	0.011
H 0.512	4.104	1.065	0.438	0.003	0.008	0.001

Fermi-Velocity (cm/s): 0.78748 x10E8
Plasmon Energy (eV) : 4.30741
Electron-ion interaction (Hopfield parameter) (eV/A^2) Mg: 0.017 2H: 2.990

Mg MUFFIN-TIN RADIUS and CHARGE = 2.0291 10.3025
H MUFFIN-TIN RADIUS and CHARGE = 1.8359 1.2076
Mg STONER I = 0.001
H STONER I = 0.005
STONER PARAMETER (Ry) I = 0.012
STONER CRITERION N*I = 0.0511

MgH2 a=8.20 Bohr CaF2

Fermi Energy (Ry)	Total DOS (States/Ry)	Decomposed DOS inside the muffin tins s	p	eg	t2g	f

```
-------------------------------------------------------------------------------------------
Mg  0.659       4.350          0.140     0.066      0.078      0.037     0.010
 H  0.659       4.350          1.027     0.467      0.005      0.016     0.002
-------------------------------------------------------------------------------------------
Fermi-Velocity (cm/s): 0.99155 x10E8
Plasmon Energy (eV)   : 6.34132
Electron-ion interaction (Hopfield parameter) (eV/A^2) Mg: 0.062 2H: 4.081
--------------------------------------------------
         Mg      MUFFIN-TIN RADIUS and CHARGE =     1.8641    10.2545
          H      MUFFIN-TIN RADIUS and CHARGE =     1.6866     1.1789
                    Mg         STONER I = 0.0007
                     H         STONER I = 0.0053
         STONER PARAMETER (Ry)        I =  0.0120
         STONER CRITERION           N*I =  0.0520
```

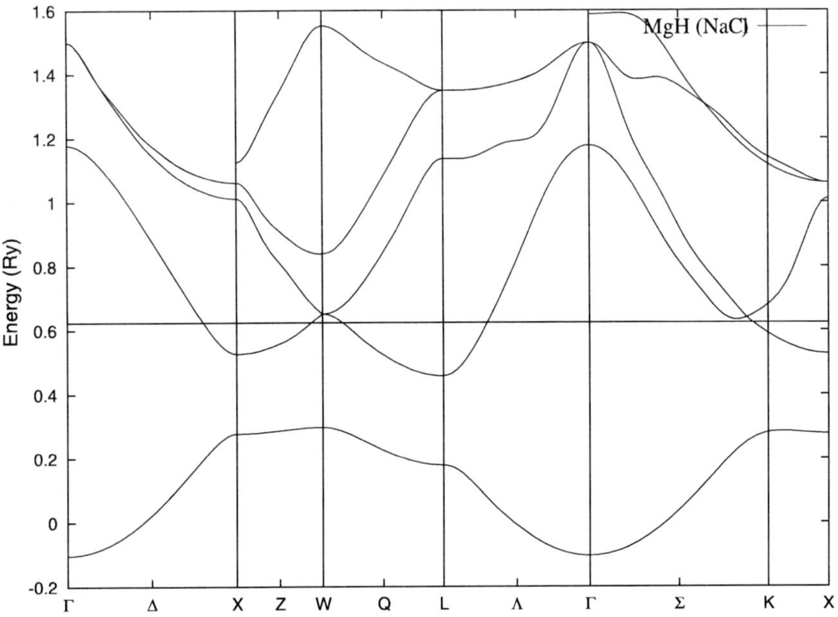

Fig. 4.7 Energy bands of MgH in the NaCl structure

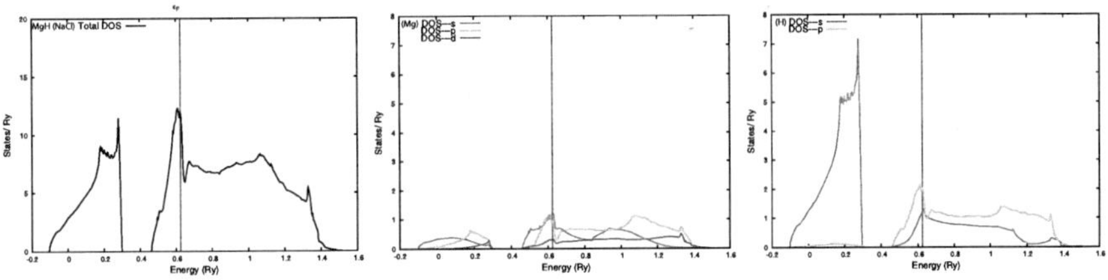

Fig. 4.8 Total, angular momentum and site decomposed densities of states of MgH in the NaCl structure

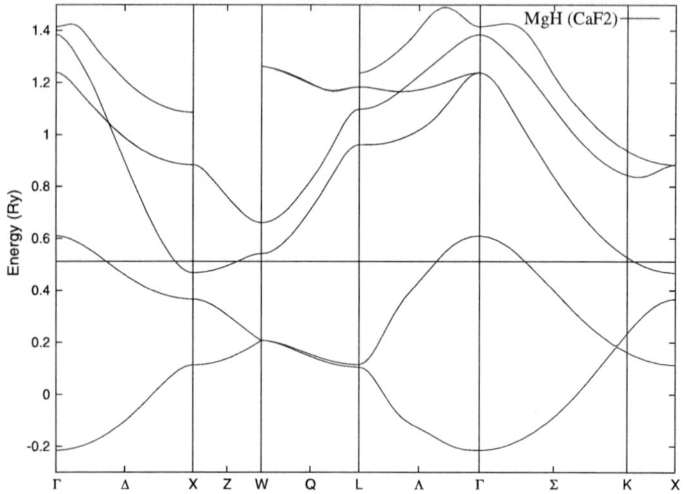

Fig. 4.9 Energy bands of MgH2 in the CaF2 structure

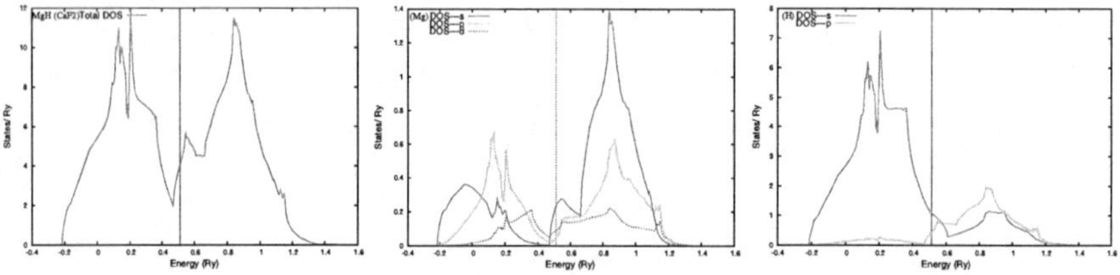

Fig. 4.10 Total, angular momentum and site decomposed densities of states of MgH2 in the CaF2 structure

Table 4.9 MgH (NaCl) a=8.40 Bohr Slater–Koster 3-center parameters

	ORTHOGONAL	NON-ORTHOGONAL	
	ENERGY INTEGRALS Ry	ENERGY INTEGRALS Ry	OVERLAP INTEGRALS
ON SITE Mg-Mg			
s,s(000)	0.84742	0.81150	
x,x(000)	1.21622	1.12014	
xy,xy(000)	1.66832	1.71437	
d2,d2(000)	1.82014	1.37407	
FIRST NEIGHBOR Mg-Mg			
s,s(110)	0.00973	-0.04938	-0.04924
s,x(110)	0.02162	0.05789	0.02543
s,xy(110)	0.06502	0.02568	0.00000
s,d2(110)	0.02031	0.04195	0.00000
x,x(110)	0.06631	0.06112	0.00000
x,x(011)	-0.00063	-0.05235	-0.04882
x,y(110)	0.02974	0.06723	0.00000
x,xy(110)	0.07207	0.07147	0.00000
x,xy(011)	-0.00676	-0.02295	0.00000
z,d2(011)	-0.03927	0.00293	0.00806
z,d1(011)	0.01226	0.02093	-0.03925
xy,xy(110)	-0.14857	-0.08314	0.02624
xy,xy(011)	0.02908	-0.04546	-0.05801
xy,xz(011)	0.02997	0.01254	0.00247
xy,d2(110)	-0.03584	-0.00575	0.03190
d2,d2(110)	-0.01476	-0.08836	-0.01962
d1,d1(110)	-0.04671	0.03627	-0.05515
SECOND NEIGHBOR Mg-Mg			
s,s(200)	0.02287	0.07212	0.03628
s,x(200)	-0.04517	-0.06526	-0.00669
s,d2(002)	0.01449	0.00631	-0.02544
x,x(200)	-0.09485	-0.01903	0.03697
y,y(200)	0.00751	0.02565	0.01076
x,xy(020)	0.00566	0.03758	0.01566
z,d2(002)	0.09048	-0.05881	-0.06204
xy,xy(200)	0.03205	-0.06910	-0.03348
xy,xy(002)	0.02391	-0.04973	-0.03892
d2,d2(002)	0.11137	-0.12601	-0.05974
d1,d1(002)	0.00761	-0.02675	-0.03093
ON SITE H -H			
s,s(000)	0.26930	0.25397	
FIRST NEIGHBOR H -H			
s,s(110)	-0.03084	-0.01726	-0.02062
s,x(110)	0.00000	0.00000	0.00000
x,x(110)	0.00000	0.00000	0.00000
x,x(011)	0.00000	0.00000	0.00000
x,y(110)	0.00000	0.00000	0.00000
SECOND NEIGHBOR H -H			
s,s(200)	0.01407	0.00730	0.01594
s,x(200)	0.00000	0.00000	0.00000
x,x(200)	0.00000	0.00000	0.00000
y,y(200)	0.00000	0.00000	0.00000
FIRST NEIGHBOR Mg-H			
s,s(100)	0.00430	-0.05812	0.02216
s,x(100)	0.00000	0.00000	0.00000
x,s(100)	-0.01804	-0.01107	0.15046
x,x(100)	0.00000	0.00000	0.00000
y,y(100)	0.00000	0.00000	0.00000
xy,x(010)	0.00000	0.00000	0.00000
d2,s(001)	0.02641	0.08031	0.26790
d2,z(001)	0.00000	0.00000	0.00000

SECOND NEIGHBOR Mg-H

s,s(111)	-0.04238	-0.00049	-0.02414
x,s(111)	-0.05002	-0.01264	-0.02568
s,x(111)	0.00000	0.00000	0.00000
x,x(111)	0.00000	0.00000	0.00000
x,y(111)	0.00000	0.00000	0.00000
xy,s(111)	0.04420	0.02226	0.01650
xy,x(111)	0.00000	0.00000	0.00000
yz,x(111)	0.00000	0.00000	0.00000
d1,x(111)	0.00000	0.00000	0.00000

MgH

BAND	ORTHOGONAL RMS ERROR mRy	MAXIMUM DEVIATION k	mRy	NON-ORTHOGONAL RMS ERROR mRy	MAXIMUM DEVIATION k	mRy
1	3.1	(062)	7.0	0.8	(226)	2.7
2	8.4	(055)	20.7	0.8	(111)	2.2
3	10.2	(011)	28.0	1.8	(003)	4.4
4	8.8	(022)	22.9	1.5	(001)	3.5
5	16.6	(042)	47.0	3.8	(022)	15.1
1-5	10.4			2.0		

ENERGY VALUES IN Ry AT SELECTED k-POINTS

	ORTHOGONAL	APW	NON-ORTHOGONAL
GAMMA 1	-0.09824	-0.10330	-0.10482
GAMMA 1	1.18316	1.17834	1.17905
GAMMA 15	1.58452	1.58699	1.58868
GAMMA 25'	1.48268	1.49917	1.49670
X1 (008)	0.27869	0.27779	0.27811
X1 (008)	1.12408	1.12643	1.12454
X3 (008)	1.01744	1.01123	1.00911
X5 (008)	2.43863	2.44568	2.44567
X4' (008)	0.52355	0.52679	0.52767
X5' (008)	1.05909	1.06100	1.06235
L1 (444)	0.45561	0.45780	0.45619
L1 (444)	1.50718	0.00000	1.69740
L3 (444)	1.34698	1.34919	1.34924
L2' (444)	0.18081	0.17917	0.17907
L2' (444)	1.14204	1.13687	1.13764
L3' (444)	1.49483	1.50015	1.49995
W1 (048)	0.83400	0.83947	0.84013
W1 (048)	1.81091	1.81484	1.81481
W2' (048)	0.29724	0.29826	0.29837
W2' (048)	1.54993	1.55220	1.55132
W3 (048)	0.65594	0.65242	0.65265
W3 (048)	1.70518	1.68379	1.68400
ODD (224)	1.24009	1.23370	1.23267
EVEN(224)	0.08511	0.08074	0.08019
EVEN(224)	0.65228	0.66990	0.67171
EVEN(224)	1.03799	1.04026	1.04061
EVEN(224)	1.36834	1.36936	1.36999
EVEN(224)	1.84026	0.00000	1.90979

Table 4.10 MgH (NaCl) a=8.40 Bohr Slater–Koster 2-center parameters

	ORTHOGONAL	NON-ORTHOGONAL	
	ENERGY INTEGRALS Ry	ENERGY INTEGRALS Ry	OVERLAP INTEGRALS
ON SITE Mg-Mg			
s	0.69591	0.49035	
p	1.12518	1.16980	
t2g	1.71738	1.69383	
eg	1.61382	1.48010	
FIRST NEIGHBOR Mg-Mg			
(sss)	-0.04242	0.00087	0.11364
(pps)	0.13229	0.01599	-0.08733
(ppp)	0.00084	0.02459	0.03596
(dds)	-0.18848	0.04709	0.09404
(ddp)	0.07866	0.07899	0.03468
(ddd)	-0.01939	-0.14687	-0.11897
(sps)	-0.09466	0.00664	-0.01112
(sds)	-0.08373	0.07178	0.20553
(pds)	0.12373	0.00931	0.07165
(pdp)	-0.00608	-0.00119	-0.00136
SECOND NEIGHBOR Mg-Mg			
(sss)	-0.01378	-0.01592	-0.00870
(pps)	-0.04348	-0.32211	-0.21397
(ppp)	0.00345	0.00960	0.00208
(dds)	0.00553	0.18967	0.11923
(ddp)	0.01935	-0.23056	-0.12214
(ddd)	0.01068	-0.04631	-0.03419
(sps)	-0.01156	0.09880	0.03080
(sds)	-0.02403	-0.03744	0.00509
(pds)	-0.03273	0.16896	0.07583
(pdp)	-0.00226	-0.08610	-0.06848
ON SITE H-H			
s	0.83442	0.23300	
FIRST NEIGHBOR H-H			
(sss)	-0.00420	-0.00398	0.04061
(sps)	0.00000	0.00000	0.00000
(pps)	0.00000	0.00000	0.00000
(ppp)	0.00000	0.00000	0.00000
SECOND NEIGHBOR H-H			
(sss)	0.02940	-0.00192	-0.04147
(sps)	0.00000	0.00000	0.00000
(pps)	0.00000	0.00000	0.00000
(ppp)	0.00000	0.00000	0.00000
FIRST NEIGHBOR Mg-H			
(sss)	-0.08915	0.03852	-0.18233
(sps)	0.00000	0.00000	0.00000
(pss)	0.13893	-0.10199	-0.25823
(pps)	0.00000	0.00000	0.00000
(ppp)	0.00000	0.00000	0.00000
(dpp)	0.00000	0.00000	0.00000
(dss)	-0.11173	-0.07030	-0.23712
(dps)	0.00000	0.00000	0.00000
SECOND NEIGHBOR Mg-H			
(sss)	0.00971	0.03815	0.01053
(pss)	-0.04937	-0.06330	-0.03984
(sps)	0.00000	0.00000	0.00000
(pps)	0.00000	0.00000	0.00000
(ppp)	0.00000	0.00000	0.00000
(dss)	0.07268	-0.05467	-0.10895
(dps)	0.00000	0.00000	0.00000
(dpp)	0.00000	0.00000	0.00000

MgH

BAND	ORTHOGONAL RMS ERROR mRy	ORTHOGONAL MAXIMUM DEVIATION k	ORTHOGONAL MAXIMUM DEVIATION mRy	NON-ORTHOGONAL RMS ERROR mRy	NON-ORTHOGONAL MAXIMUM DEVIATION k	NON-ORTHOGONAL MAXIMUM DEVIATION mRy
1	8.0	(062)	17.5	4.2	(004)	10.4
2	15.1	(004)	38.1	8.8	(004)	21.2
3	15.3	(004)	34.1	9.5	(118)	32.7
4	18.7	(022)	45.1	14.0	(174)	41.7
5	25.1	(042)	78.4	20.0	(042)	85.2
1-5	17.4			12.5		

ENERGY VALUES IN Ry AT SELECTED k-POINTS

	ORTHOGONAL	APW	NON-ORTHOGONAL
GAMMA 1	-0.09404	-0.10330	-0.10006
GAMMA 1	1.15862	1.17834	1.17787
GAMMA 15	1.58785	1.58699	1.58976
GAMMA 25'	1.46839	1.49917	1.49365
X1 (008)	0.26370	0.27779	0.28349
X1 (008)	1.11123	1.12643	1.12635
X3 (008)	0.99422	1.01123	1.00163
X5 (008)	2.40097	2.44568	2.44587
X4' (008)	0.52285	0.52679	0.53783
X5' (008)	1.04864	1.06100	1.06664
L1 (444)	0.45252	0.45780	0.45777
L3 (444)	1.31019	1.34919	1.33570
L2' (444)	0.18396	0.17917	0.17474
L2' (444)	1.14660	1.13687	1.14473
L3' (444)	1.46126	1.50015	1.49903
W1 (048)	0.83320	0.83947	0.85675
W1 (048)	1.89182	1.81484	1.81366
W2' (048)	0.28905	0.29826	0.29169
W2' (048)	1.54124	1.55220	1.52513
W3 (048)	0.66461	0.65242	0.64693
W3 (048)	1.72755	1.68379	1.67856
ODD (224)	1.23669	1.23370	1.23846
EVEN(224)	0.07895	0.08074	0.08261
EVEN(224)	0.66147	0.66990	0.66781
EVEN(224)	1.05020	1.04026	1.05072
EVEN(224)	1.35968	1.36936	1.37322

4.3 Calcium Hydride (CaH)

See Fig. 4.11.

See Tables 4.11, 4.12 and 4.13.

See Figs. 4.12, 4.13, 4.14 and 4.15.

See Tables 4.14 and 4.15.

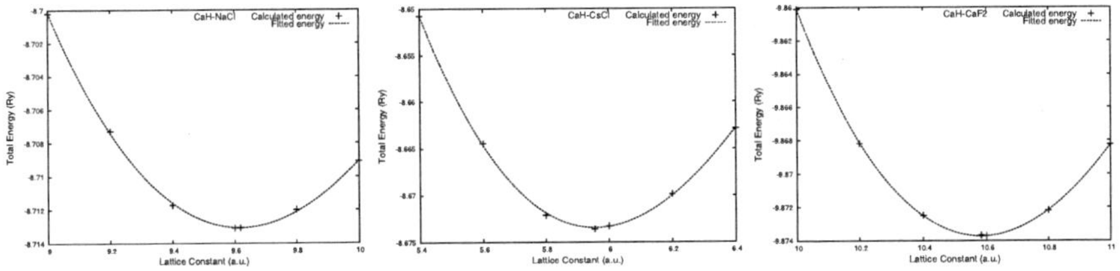

Fig. 4.11 Total energy versus lattice constant in the NaCl, CsCl and CaF2 structures

Table 4.11 Lattice Constant, Bulk Modulus, Gap, Total Energy

Stru	a (Bohr)	B (MBar)	Gap	Total Energy
NaCl	9.62	0.40		-1358.71308
CsCl	5.96	0.34		-1358.67344
CaF2	10.58	0.43		-1359.87373
exp(Orthorhombic)	a=12.95	b=11.24	c=6.82	

Table 4.12 Birch Fit Coefficients

	A1	A2	A3	A4
NaCl	-7.702272E+00	-8.611275E+01	2.236147E+03	-1.604652E+04
CsCl	-7.816296E+00	-7.161355E+01	1.845223E+03	-1.360224E+04
CaF2	-8.702623E+00	-1.129493E+02	3.100307E+03	-1.749504E+04

Table 4.13 DOS at Ef, Hopfield parameter, Stoner Criterion

```
                    CaH      a=9.62 Bohr          NaCl
 ------------------------------------------------------------------------------
  Fermi Energy   Total DOS         Decomposed DOS inside the muffin tins
    (Ry)        (States/Ry)      s          p         eg          t2g        f
 ------------------------------------------------------------------------------
 Ca 0.529          18.798      0.639     0.627      0.521       5.243      0.040
 H  0.529          18.798      0.481     1.614      0.006       0.295      0.028
 ------------------------------------------------------------------------------
 Fermi-Velocity (cm/s): 0.44728 x10E8
 Plasmon Energy (eV)   : 4.68083
 Electron-ion interaction (Hopfield parameter) (eV/A^2) Ca: 1.466 H: 0.392
-----------------------------------------------
  Ca      MUFFIN-TIN RADIUS and CHARGE =    2.5245   18.3760
   H      MUFFIN-TIN RADIUS and CHARGE =    2.2841    1.4235
           Ca        STONER I = 0.005
            H        STONER I = 0.002
          STONER PARAMETER (Ry) I =  0.007
          STONER CRITERION    N*I =  0.1360
-----------------------------------------------
                    CaH      a=5.96 Bohr          CsCl
 ------------------------------------------------------------------------------
  Fermi Energy   Total DOS         Decomposed DOS inside the muffin tins
    (Ry)        (States/Ry)      s          p         eg          t2g        f
 ------------------------------------------------------------------------------
 Ca 0.571        23.468          2.096     0.891      4.184       4.668      0.105
 H  0.571        23.468          0.375     3.799      0.178       0.156      0.019|
 ------------------------------------------------------------------------------
Fermi-Velocity (cm/s): 0.49576 x10E8
Plasmon Energy (eV)   : 5.94963
Electron-ion interaction (Hopfield parameter) (eV/A^2) Ca: 2.093  H: 0.398
-----------------------------------------------
        Ca      MUFFIN-TIN RADIUS and CHARGE =    2.7077   18.6386
        H       MUFFIN-TIN RADIUS and CHARGE =    2.4498    1.5596
                  Ca         STONER I = 0.008
                   H         STONER I = 0.005
          STONER PARAMETER (Ry)   I =   0.013
          STONER CRITERION       N*I =  0.309
-----------------------------------------------
                    CaH2      a=10.58 Bohr          CaF2
 ------------------------------------------------------------------------------
  Fermi Energy   Total DOS         Decomposed DOS inside the muffin tins

    (Ry)          (States/Ry)      s          p         eg          t2g        f
 ------------------------------------------------------------------------------
 Ca 0.323          2.216       0.038      0.003      0.258       0.016      0.006
  H 0.323          2.216       0.394      0.188      0.001       0.014      0.000|
 ------------------------------------------------------------------------------
Fermi-Velocity (cm/s): 0.22541 x10E8
Plasmon Energy (eV)   : 0.70138
Electron-ion interaction (Hopfield parameter) (eV/A^2) Ca: 0.036  H: 0.405
-----------------------------------------------
          Ca      MUFFIN-TIN RADIUS and CHARGE =    2.4057   18.1109
          H       MUFFIN-TIN RADIUS and CHARGE =    2.1766    1.3503
                  Ca         STONER I = 0.001
                   H         STONER I = 0.002
          STONER PARAMETER (Ry)      I = 0.005
          STONER CRITERION         N*I = 0.0124
```

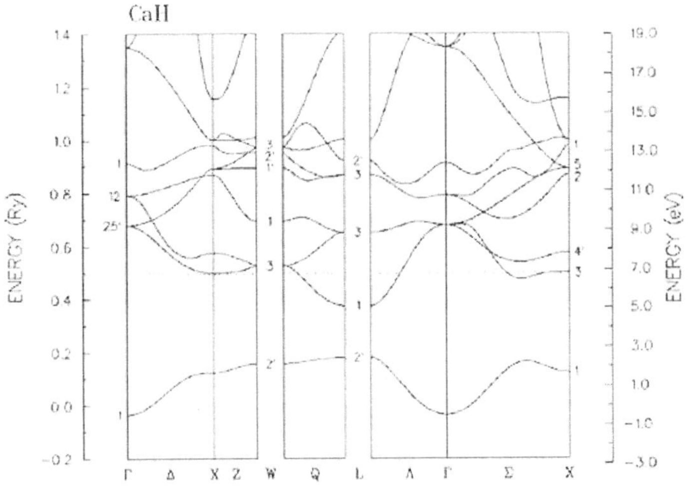

Fig. 4.12 Energy bands of CaH in the NaCl structure (tight-binding)

Fig. 4.13 Total, angular momentum and site-decomposed densities of states of CaH in the NaCl structure (tight-binding)

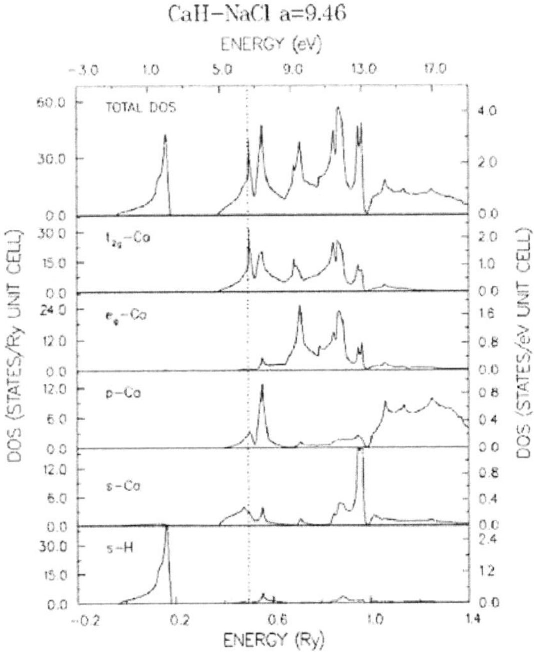

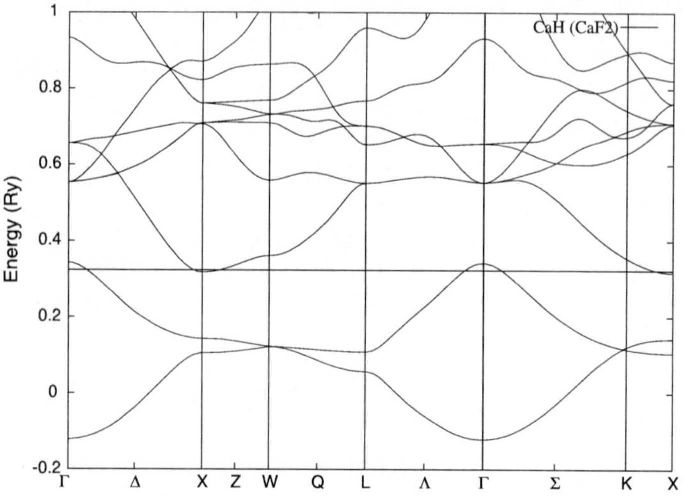

Fig. 4.14 Energy bands of CaH2 in the CaF2 structure

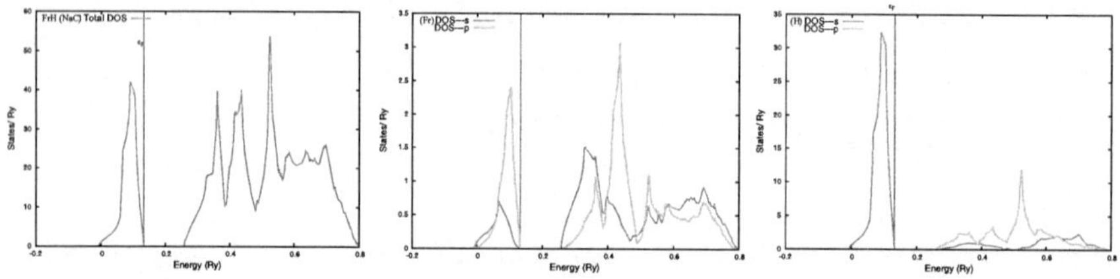

Fig. 4.15 Total, angular momentum and site decomposed densities of states of CaH2 in the CaF2 structure

Table 4.14 CaH (NaCl) a = 9.46 Bohr Slater–Koster 3-center parameters

	ORTHOGONAL	NON-ORTHOGONAL	
	ENERGY INTEGRALS Ry	ENERGY INTEGRALS Ry	OVERLAP INTEGRALS
ON SITE Ca-Ca			
s,s(000)	0.80800	0.74984	
x,x(000)	1.11646	0.92222	
xy,xy(000)	0.77509	0.70559	
d2,d2(000)	0.79780	0.74867	
FIRST NEIGHBOR Ca-Ca			
s,s(110)	-0.01080	0.04390	0.06286
s,x(110)	-0.03048	0.01077	0.07061
s,xy(110)	-0.04320	-0.01635	0.05026
s,d2(110)	0.00228	0.04790	0.04263
x,x(110)	0.04804	0.01426	-0.05639
x,x(011)	-0.00593	0.05415	0.08247
x,y(110)	0.04304	-0.03772	-0.10050
x,xy(110)	0.05040	0.01572	-0.03821
x,xy(011)	-0.00391	0.01195	0.04490
z,d2(011)	0.01437	0.03507	0.08487
z,d1(011)	-0.00094	-0.02565	-0.00536
xy,xy(110)	-0.03836	-0.02768	0.01156
xy,xy(011)	0.01125	0.00256	-0.01414
xy,xz(011)	0.00259	0.00390	-0.01997
xy,d2(110)	0.01880	0.00972	0.04386
d2,d2(110)	-0.00132	0.01545	0.02821
d1,d1(110)	-0.01524	0.02299	0.01939
SECOND NEIGHBOR Ca-Ca			
s,s(200)	0.03565	0.02048	-0.00412
s,x(200)	0.05121	-0.00029	-0.00427
s,d2(002)	0.03324	-0.00787	0.03991
x,x(200)	-0.08496	0.05621	0.06648
y,y(200)	0.01053	-0.01457	-0.01097
x,xy(020)	0.00845	0.00062	0.00630
z,d2(002)	-0.04240	-0.05583	-0.12477
xy,xy(200)	-0.00921	-0.01085	-0.02223
xy,xy(002)	0.00197	0.00581	0.00707
d2,d2(002)	0.02999	-0.01902	-0.04260
d1,d1(002)	0.00124	-0.01837	-0.02109
ON SITE H-H			
s,s(000)	0.18563	0.15410	
FIRST NEIGHBOR H-H			
s,s(110)	-0.01541	-0.00340	0.00989
s,x(110)	0.00000	0.00000	0.00000
x,x(110)	0.00000	0.00000	0.00000
x,x(011)	0.00000	0.00000	0.00000
x,y(110)	0.00000	0.00000	0.00000
SECOND NEIGHBOR H-H			
s,s(200)	-0.00243	-0.00592	-0.00385
s,x(200)	0.00000	0.00000	0.00000
x,x(200)	0.00000	0.00000	0.00000
y,y(200)	0.00000	0.00000	0.00000
FIRST NEIGHBOR Ca-H			
s,s(100)	-0.05401	0.06075	-0.00416
s,x(100)	0.00000	0.00000	0.00000
x,s(100)	0.03891	0.01710	0.19270
x,x(100)	0.00000	0.00000	0.00000
y,y(100)	0.00000	0.00000	0.00000
xy,x(010)	0.00000	0.00000	0.00000
d2,s(001)	-0.00501	-0.03782	-0.14670
d2,z(001)	0.00000	0.00000	0.00000
SECOND NEIGHBOR Ca-H			
s,s(111)	0.02449	0.00562	0.00312
x,s(111)	-0.03095	-0.00155	0.00573
s,x(111)	0.00000	0.00000	0.00000
x,x(111)	0.00000	0.00000	0.00000
x,y(111)	0.00000	0.00000	0.00000
xy,s(111)	0.00938	0.00082	0.00613
xy,x(111)	0.00000	0.00000	0.00000
yz,x(111)	0.00000	0.00000	0.00000
d1,x(111)	0.00000	0.00000	0.00000

CaH

BAND	ORTHOGONAL RMS ERROR	ORTHOGONAL MAXIMUM DEVIATION		NON-ORTHOGONAL RMS ERROR	NON-ORTHOGONAL MAXIMUM DEVIATION	
	mRy	k	mRy	mRy	k	mRy
1	5.1	(004)	10.6	0.8	(333)	1.8
2	3.3	(003)	8.4	0.9	(002)	2.0
3	3.6	(055)	9.9	0.6	(062)	1.5
4	2.4	(048)	6.2	0.8	(008)	2.1
5	2.9	(066)	7.6	1.0	(055)	2.8
6	2.9	(174)	6.0	1.3	(174)	4.4
7	4.4	(044)	10.5	2.6	(226)	8.7
1-7	3.6			1.3		

ENERGY VALUES IN Ry AT SELECTED k-POINTS

	ORTHOGONAL	APW	NON-ORTHOGONAL
GAMMA 1	-0.03164	-0.03424	-0.03396
GAMMA 1	0.91001	0.91475	0.91404
GAMMA 12	0.79212	0.79272	0.79223
GAMMA 15	1.34929	1.35041	1.35042
GAMMA 25'	0.67874	0.68089	0.67988
X1 (008)	0.13278	0.12529	0.12492
X1 (008)	0.98029	0.97726	0.97889
X1 (008)	1.16494	1.15591	1.15612
X2 (008)	0.86894	0.86719	0.86928
X3 (008)	0.49870	0.49906	0.50030
X5 (008)	0.89564	0.89224	0.89251
X4' (008)	0.58059	0.57708	0.57692
X5' (008)	1.01236	1.00077	1.00068
L1 (444)	0.37625	0.37353	0.37443
L1 (444)	1.00519	1.00173	1.00391
L3 (444)	0.65285	0.65225	0.65295
L3 (444)	0.86962	0.86843	0.86866
L2' (444)	0.17514	0.18210	0.18103
L2' (444)	0.92499	0.92051	0.92143
L3' (444)	1.41644	1.40937	1.40935
W1 (048)	0.68999	0.69617	0.69547
W1 (048)	1.00932	1.01403	1.01185
W2' (048)	0.15356	0.15896	0.15751
W2' (048)	0.94789	0.95253	0.95211
W2' (048)	1.44063	1.46398	1.46523
W1' (048)	0.88778	0.89255	0.89486
W3 (048)	0.52669	0.52851	0.52890
W3 (048)	0.96168	0.97121	0.97192
ODD (224)	0.69137	0.68919	0.69027
ODD (224)	0.85473	0.85122	0.85157
EVEN(224)	0.11581	0.10908	0.11010
EVEN(224)	0.48324	0.48320	0.48289
EVEN(224)	0.62323	0.62384	0.62307
EVEN(224)	0.79510	0.79578	0.79710
EVEN(224)	0.89115	0.89167	0.89143

FERMI LEVEL QUANTITIES (Non-orthogonal fit)

DENSITIES OF STATES (states/Ry/cell)

ENERGY Ry	Total	s-Ca	p-Ca	t2g-Ca	eg-Ca	s-H	p-H
0.4988	20.45	2.96	2.70	13.36	0.61	0.83	0.00

INTEGRATED DENSITIES OF STATES (electrons)

Total	s-Ca	p-Ca	t2g-Ca	eg-Ca	s-H	p-H
3.00	0.35	0.10	0.54	0.03	1.99	0.00

VELOCITY cm/s	PLASMON ENERGY eV
0.47x10E8	5.29

Table 4.15 CaH (NaCl) a = 9.46 Bohr Slater–Koster 2-center parameters

	ORTHOGONAL	NON-ORTHOGONAL	
	ENERGY INTEGRALS Ry	ENERGY INTEGRALS Ry	OVERLAP INTEGRALS
ON SITE Ca-Ca			
s	0.73111	0.68099	
p	1.08823	0.83548	
t2g	0.75191	0.67822	
eg	0.73354	0.66794	
FIRST NEIGHBOR Ca-Ca			
(sss)	-0.04524	-0.03390	0.01622
(pps)	0.10492	-0.03397	-0.19257
(ppp)	-0.00782	0.00960	0.04397
(dds)	-0.05159	0.01482	0.08917
(ddp)	0.02003	-0.00532	-0.06625
(ddd)	0.00113	0.00131	0.01090
(sps)	0.07390	0.01747	-0.07420
(sds)	-0.05272	-0.01553	0.05796
(pds)	-0.07716	-0.00065	0.08830
(pdp)	0.00810	-0.00210	-0.04635
SECOND NEIGHBOR Ca-Ca			
(sss)	0.00770	0.01851	0.01983
(pps)	-0.05962	0.07610	0.03018
(ppp)	0.00583	-0.01805	-0.00193
(dds)	0.00383	-0.02728	-0.01723
(ddp)	-0.00275	0.00170	-0.00128
(ddd)	0.00085	-0.00070	-0.00124
(sps)	-0.02505	-0.00906	-0.02187
(sds)	0.01171	-0.02160	-0.02556
(pds)	0.01092	-0.05276	-0.03835
(pdp)	-0.00438	0.00585	-0.00008
ON SITE H-H			
s	0.53939	0.32357	
FIRST NEIGHBOR H-H			
(sss)	-0.00380	0.00109	-0.00763
(sps)	0.00000	0.00000	0.00000
(pps)	0.00000	0.00000	0.00000
(ppp)	0.00000	0.00000	0.00000
SECOND NEIGHBOR H-H			
(sss)	0.02428	0.02240	-0.00327
(sps)	0.00000	0.00000	0.00000
(pps)	0.00000	0.00000	0.00000
(ppp)	0.00000	0.00000	0.00000
FIRST NEIGHBOR Ca-H			
(sss)	0.09142	0.08389	-0.01398
(sps)	0.00000	0.00000	0.00000
(pss)	0.09937	-0.03440	-0.28221
(pps)	0.00000	0.00000	0.00000
(ppp)	0.00000	0.00000	0.00000
(dpp)	0.00000	0.00000	0.00000
(dss)	0.09255	0.06367	-0.12694
(dps)	0.00000	0.00000	0.00000
SECOND NEIGHBOR Ca-H			
(sss)	-0.01471	-0.00479	0.00170
(pss)	-0.03652	0.00637	0.00607
(sps)	0.00000	0.00000	0.00000
(pps)	0.00000	0.00000	0.00000
(ppp)	0.00000	0.00000	0.00000
(dss)	-0.01097	-0.00231	-0.00760
(dps)	0.00000	0.00000	0.00000
(dpp)	0.00000	0.00000	0.00000

♠CaH

	ORTHOGONAL			NON-ORTHOGONAL		
BAND	RMS ERROR	MAXIMUM DEVIATION		RMS ERROR	MAXIMUM DEVIATION	
	mRy	k	mRy	mRy	k	mRy
1	4.7	(005)	12.5	1.1	(224)	2.2
2	7.2	(222)	18.9	1.0	(222)	2.6
3	7.6	(264)	15.6	1.0	(003)	3.0
4	6.1	(444)	11.7	1.2	(055)	3.1
5	6.2	(002)	15.5	1.2	(022)	2.6
6	7.9	(044)	18.7	1.4	(048)	3.4
7	11.4	(444)	26.7	4.1	(354)	17.4
1-7	7.5			1.9		

ENERGY VALUES IN Ry AT SELECTED k-POINTS

	ORTHOGONAL	APW	NON-ORTHOGONAL
GAMMA 1	-0.03915	-0.03424	-0.03268
GAMMA 1	0.91304	0.91475	0.91572
GAMMA 12	0.79546	0.79272	0.79493
GAMMA 15	1.34946	1.35041	1.35040
GAMMA 25'	0.67359	0.68089	0.67978
X1 (008)	0.12431	0.12529	0.12653
X1 (008)	0.97851	0.97726	0.97778
X1 (008)	1.15903	1.15591	1.15530
X2 (008)	0.85997	0.86719	0.86643
X3 (008)	0.50433	0.49906	0.49955
X5 (008)	0.89625	0.89224	0.89339
X4' (008)	0.57267	0.57708	0.57673
X5' (008)	1.02360	1.00077	1.00034
L1 (444)	0.36894	0.37353	0.37393
L1 (444)	1.00156	1.00173	1.00385
L3 (444)	0.64057	0.65225	0.65327
L3 (444)	0.87795	0.86843	0.86804
L2' (444)	0.17962	0.18210	0.18012
L2' (444)	0.94726	0.92051	0.92063
L3' (444)	1.40959	1.40937	1.40928
W1 (048)	0.68489	0.69617	0.69644
W1 (048)	1.02305	1.01403	1.00077
W2' (048)	0.15502	0.15896	0.15714
W2' (048)	0.94184	0.95253	0.94909
W2' (048)	1.43243	1.46398	1.46392
W1' (048)	0.89286	0.89255	0.89109
W3 (048)	0.52878	0.52851	0.52767
W3 (048)	0.95730	0.97121	0.97204
ODD (224)	0.68421	0.68919	0.68803
ODD (224)	0.85731	0.85122	0.84960
EVEN(224)	0.11110	0.10908	0.11129
EVEN(224)	0.49026	0.48320	0.48214
EVEN(224)	0.62856	0.62384	0.62415
EVEN(224)	0.80447	0.79578	0.79730
EVEN(224)	0.88831	0.89167	0.89303

4.4 Strontium Hydride (SrH)

See Fig. 4.16.

See Tables 4.16, 4.17 and 4.18.

See Figs. 4.17, 4.18, 4.19 and 4.20.

See Tables 4.19 and 4.20.

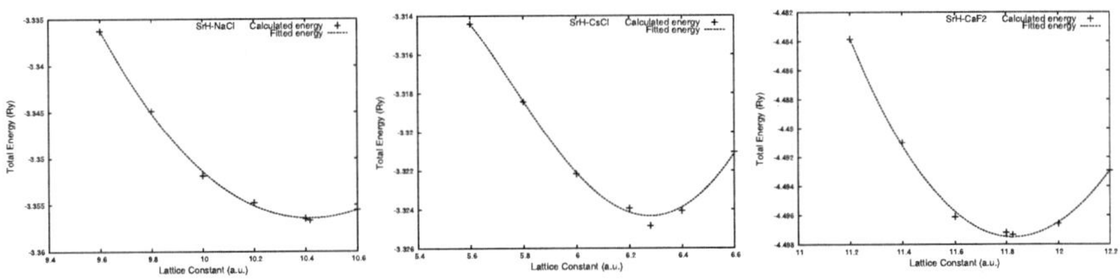

Fig. 4.16 Total energy versus lattice constant of SrH in the NaCl, CsCl and CaF2 structures

Table 4.16 Lattice Constant, Bulk Modulus, Gap, Total Energy

Stru	a (Bohr)	B (MBar)	Gap (Ry)	Total Energy (Ry)
NaCl	10.42	0.34		-6353.35642
CsCl	6.28	0.16		-6353.32434
CaF2 (Insulator)	11.83	0.37		-6354.49749
exp (Orthorhombic)	a=13.91 b=12.05 c=7.34			

Table 4.17 Birch Fit Coefficients

	A1	A2	A3	A4
NaCl	-2.071790E+00	-1.341498E+02	4.406967E+03	-4.359372E+04
CsCl	-2.427216E+00	-9.418692E+01	3.243233E+03	-3.644198E+04
CaF2	-2.081386E+00	-3.368899E+02	1.506804E+04	-2.116131E+05

Table 4.18 DOS at Ef, Hopfield parameter, Stoner Criterion

```
                 SrH      a=10.42 Bohr         NaCl
--------------------------------------------------------------------------------
 Fermi Energy   Total DOS        Decomposed DOS inside the muffin tins
   (Ry)        (States/Ry)      s         p         eg         t2g        f
--------------------------------------------------------------------------------
Sr 0.486         34.197       0.812     0.800     0.384      9.443      0.063
 H 0.486         34.197       0.555     2.066     0.013      0.985      0.053
--------------------------------------------------------------------------------
Fermi-Velocity (cm/s): 0.34607 x10E8
Plasmon Energy (eV)   : 4.33335
Electron-ion interaction (Hopfield parameter) (eV/A^2) Sr: 2.029  H: 0.225
-------------------------------------------------
  Sr      MUFFIN-TIN RADIUS and CHARGE =    2.7345    36.1895
   H      MUFFIN-TIN RADIUS and CHARGE =    2.4740     1.4986
           Sr        STONER I =    0.003
            H        STONER I =    0.002
          STONER PARAMETER (Ry) I = 0.005
          STONER CRITERION    N*I = 0.182
-------------------------------------------------
                 SrH      a=6.28 Bohr          CsCl
--------------------------------------------------------------------------------
 Fermi Energy   Total DOS        Decomposed DOS inside the muffin tins
   (Ry)        (States/Ry)      s         p         eg         t2g        f
--------------------------------------------------------------------------------
Sr 0.582         21.292       1.912     0.688     3.198      3.503      0.147
 H 0.582         21.292       0.425     3.894     0.22  6    0.144      0.028
--------------------------------------------------------------------------------
Fermi-Velocity (cm/s): 0.58761 x10E8
Plasmon Energy (eV)   : 6.20386
Electron-ion interaction (Hopfield parameter) (eV/A^2) Sr: 2.271  H: 0.417
-------------------------------------------------
  Sr      MUFFIN-TIN RADIUS and CHARGE =    2.8550    36.4351
   H      MUFFIN-TIN RADIUS and CHARGE =    2.5831     1.6067
                    Sr        STONER I =  0.006
                     H        STONER I =  0.006
          STONER PARAMETER (Ry) I = 0.012
          STONER CRITERION    N*I = 0.251
-------------------------------------------------
                 SrH2     a=11.83 Bohr         CaF2
--------------------------------------------------------------------------------
 Fermi Energy   Total DOS        Decomposed DOS inside the muffin tins
   (Ry)        (States/Ry)      s         p         eg         t2g        f
--------------------------------------------------------------------------------
Sr 0.226        0.0     Insulator
 H 0.226        0.0
--------------------------------------------------------------------------------
         Sr      MUFFIN-TIN RADIUS and CHARGE =    2.6884    35.9999
          H      MUFFIN-TIN RADIUS and CHARGE =    2.4324     1.4123
```

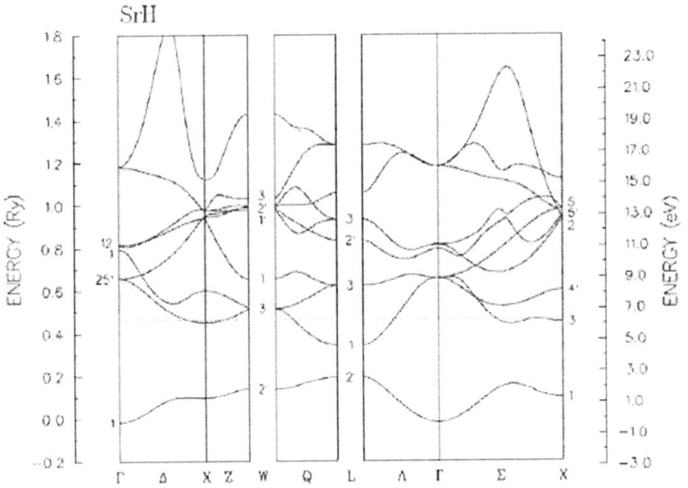

Fig. 4.17 Energy bands of SrH in the NaCl structure (tight-binding)

Fig. 4.18 Total, angular momentum and site-decomposed densities of states of SrH in the NaCl structure (tight-binding)

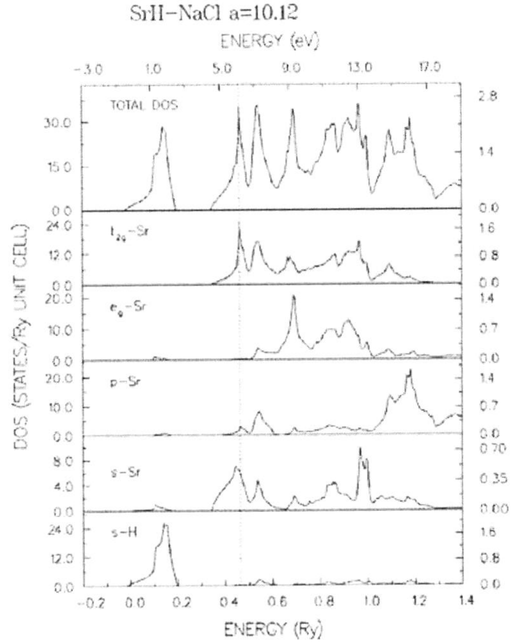

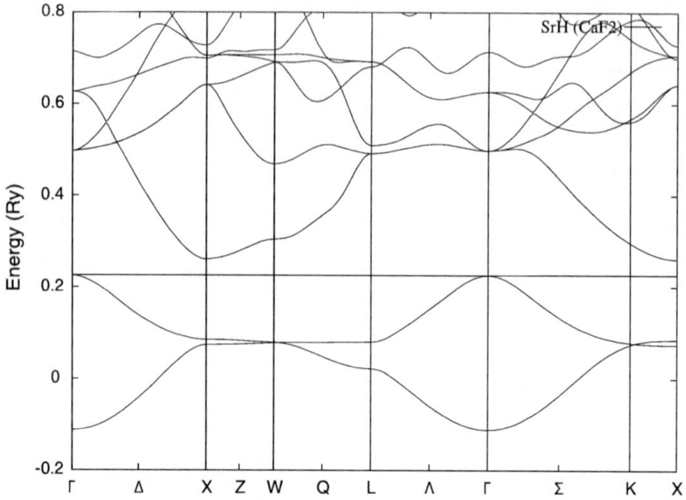

Fig. 4.19 Energy bands of SrH2 in the CaF2 structure

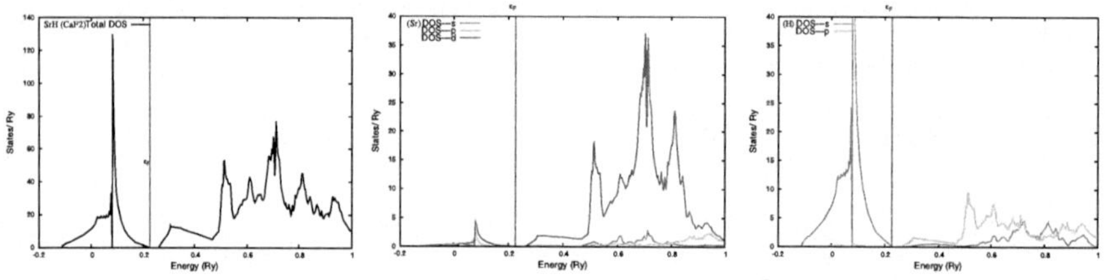

Fig. 4.20 Total, angular momentum and site decomposed densities of states of SrH2 in the CaF2 structure

Table 4.19 SrH (NaCl) a = 10.12 Bohr Slater–Koster 3-center parameters

	ORTHOGONAL	NON-ORTHOGONAL	
	ENERGY INTEGRALS Ry	ENERGY INTEGRALS Ry	OVERLAP INTEGRALS
ON SITE Sr-Sr			
s,s(000)	0.70880	0.66184	
x,x(000)	1.02349	0.90435	
xy,xy(000)	0.83577	0.69306	
d2,d2(000)	0.79689	0.71349	
FIRST NEIGHBOR Sr-Sr			
s,s(110)	-0.02236	0.06110	0.08710
s,x(110)	-0.02663	0.01632	0.05990
s,xy(110)	-0.03286	0.00461	0.09498
s,d2(110)	0.01128	0.03618	0.04007
x,x(110)	0.03559	0.01553	-0.04337
x,x(011)	-0.00734	0.07300	0.10141
x,y(110)	0.04016	-0.03129	-0.08249
x,xy(110)	0.05626	-0.00050	-0.06711
x,xy(011)	-0.00268	0.02378	0.05942
z,d2(011)	0.01248	0.03582	0.08935
z,d1(011)	-0.00708	0.00020	0.01977
xy,xy(110)	-0.05349	-0.01655	0.04619
xy,xy(011)	0.01324	-0.00409	-0.03313
xy,xz(011)	-0.00867	-0.00949	-0.04378
xy,d2(110)	0.02355	0.01254	0.05853
d2,d2(110)	-0.00687	-0.00149	0.01746
d1,d1(110)	-0.00979	0.03560	0.01043
SECOND NEIGHBOR Sr-Sr			
s,s(200)	0.04743	0.03044	-0.00784
s,x(200)	0.06338	0.00098	-0.03064
s,d2(002)	0.03723	0.00493	0.05219
x,x(200)	-0.08137	0.01056	0.04520
y,y(200)	0.01449	0.00353	-0.00004
x,xy(020)	0.01321	0.01147	0.01586
z,d2(002)	-0.04631	-0.05814	-0.12994
xy,xy(200)	-0.01909	-0.01119	-0.02102
xy,xy(002)	0.00299	0.00216	0.00231
d2,d2(002)	0.03680	-0.04717	-0.07257
d1,d1(002)	0.00231	-0.01032	-0.01190
ON SITE H-H			
s,s(000)	0.19860	0.15771	
FIRST NEIGHBOR H-H			
s,s(110)	-0.00953	-0.00569	0.01985
s,x(110)	0.00000	0.00000	0.00000
x,x(110)	0.00000	0.00000	0.00000
x,x(011)	0.00000	0.00000	0.00000
x,y(110)	0.00000	0.00000	0.00000
SECOND NEIGHBOR H-H			
s,s(200)	-0.00544	-0.00847	-0.00491
s,x(200)	0.00000	0.00000	0.00000
x,x(200)	0.00000	0.00000	0.00000
y,y(200)	0.00000	0.00000	0.00000
FIRST NEIGHBOR Sr-H			
s,s(100)	-0.05364	0.04477	0.07148
s,x(100)	0.00000	0.00000	0.00000
x,s(100)	0.04097	0.06444	0.19984
x,x(100)	0.00000	0.00000	0.00000
y,y(100)	0.00000	0.00000	0.00000
xy,x(010)	0.00000	0.00000	0.00000
d2,s(001)	-0.04987	-0.00266	-0.14224
d2,z(001)	0.00000	0.00000	0.00000
SECOND NEIGHBOR Sr-H			
s,s(111)	0.01246	0.00396	0.00797
x,s(111)	-0.01980	-0.00583	0.00298
s,x(111)	0.00000	0.00000	0.00000
x,x(111)	0.00000	0.00000	0.00000
x,y(111)	0.00000	0.00000	0.00000
xy,s(111)	0.01203	0.00286	0.00904
xy,x(111)	0.00000	0.00000	0.00000
yz,x(111)	0.00000	0.00000	0.00000
d1,x(111)	0.00000	0.00000	0.00000

SrH

BAND	ORTHOGONAL RMS ERROR mRy	MAXIMUM DEVIATION k	MAXIMUM DEVIATION mRy	NON-ORTHOGONAL RMS ERROR mRy	MAXIMUM DEVIATION k	MAXIMUM DEVIATION mRy
1	3.0	(048)	5.8	0.9	(044)	2.2
2	4.6	(003)	14.3	1.3	(111)	2.9
3	3.9	(055)	10.9	0.8	(007)	1.6
4	5.8	(048)	14.4	0.7	(008)	2.1
5	5.2	(048)	10.1	1.2	(224)	2.7
6	4.5	(048)	10.4	1.5	(224)	4.7
7	6.1	(044)	15.5	1.3	(066)	4.1
1-7	4.8			1.1		

ENERGY VALUES IN Ry AT SELECTED k-POINTS

	ORTHOGONAL	APW	NON-ORTHOGONAL
GAMMA 1	-0.01508	-0.01727	-0.01700
GAMMA 1	0.79178	0.79560	0.79403
GAMMA 12	0.81428	0.81617	0.81589
GAMMA 15	1.17406	1.18039	1.18015
GAMMA 25'	0.65740	0.65996	0.65895
X1 (008)	0.10395	0.09820	0.09880
X1 (008)	0.99866	0.98375	0.98364
X1 (008)	1.14352	1.12442	1.12323
X2 (008)	0.93582	0.93524	0.93734
X3 (008)	0.44549	0.45099	0.45182
X5 (008)	0.97936	0.97664	0.97702
X4' (008)	0.60465	0.60124	0.60088
X5' (008)	0.94804	0.94623	0.94598
L1 (444)	0.34236	0.34115	0.34398
L1 (444)	1.05736	1.05585	1.05701
L3 (444)	0.62545	0.62271	0.62345
L3 (444)	0.92557	0.93432	0.93444
L2' (444)	0.19814	0.19481	0.19305
L2' (444)	0.84010	0.83777	0.83686
L3' (444)	1.28892	1.28300	1.28305
W1 (048)	0.64356	0.65798	0.65691
W1 (048)	1.04398	1.03484	1.03280
W2' (048)	0.13302	0.13880	0.13860
W2' (048)	0.98352	0.99390	0.99112
W2' (048)	1.32700	1.43015	1.42992
W1' (048)	0.96741	0.97753	0.97753
W3 (048)	0.51524	0.51558	0.51584
W3 (048)	0.99192	0.99669	0.99596
ODD (224)	0.67584	0.66851	0.66846
ODD (224)	0.90465	0.90158	0.90174
EVEN(224)	0.12594	0.12247	0.12249
EVEN(224)	0.44098	0.44334	0.44318
EVEN(224)	0.58845	0.58924	0.58799
EVEN(224)	0.79529	0.79187	0.78914
EVEN(224)	0.85979	0.85966	0.86432

FERMI LEVEL QUANTITIES (Non-orthogonal fit)

ENERGY		DENSITIES OF STATES					
	Total	s-Sr	p-Sr	t2g-Sr	eg-Sr	s-H	p-H
Ry		states/Ry/cell					
0.4637	35.29	6.55	3.22	24.79	0.47	0.27	0.00
		INTEGRATED DENSITIES OF STATES					
	Total	s-Sr	p-Sr	t2g-Sr	eg-Sr	s-H	p-H
		electrons					
	3.00	0.53	0.10	0.45	0.06	1.86	0.00

VELOCITY	PLASMON ENERGY	EIGENVALUE SUM
cm/s	eV	Ry
0.39x10E8	5.22	-0.7132

Table 4.20 SrH (NaCl) a = 10.12 Bohr Slater–Koster 2-center parameters

	ORTHOGONAL	NON-ORTHOGONAL	
	ENERGY INTEGRALS Ry	ENERGY INTEGRALS Ry	OVERLAP INTEGRALS
ON SITE Sr-Sr			
s	0.70609	0.62611	
p	1.01112	0.80464	
t2g	0.77353	0.65134	
eg	0.74857	0.63988	
FIRST NEIGHBOR Sr-Sr			
(sss)	-0.03147	-0.03927	0.00431
(pps)	0.08406	-0.01323	-0.15805
(ppp)	-0.00778	0.07532	0.10011
(dds)	-0.07177	0.03201	0.13414
(ddp)	0.02184	-0.01198	-0.09391
(ddd)	0.00288	0.00626	0.01883
(sps)	0.05366	0.01807	-0.08428
(sds)	-0.05781	-0.00062	0.09800
(pds)	-0.08676	0.03363	0.14340
(pdp)	0.00567	-0.02714	-0.07901
SECOND NEIGHBOR Sr-Sr			
(sss)	0.00375	0.02169	0.01451
(pps)	-0.04103	0.11476	0.07624
(ppp)	-0.00031	-0.00169	0.00393
(dds)	0.00974	-0.03341	-0.01876
(ddp)	-0.00298	-0.00426	-0.00854
(ddd)	0.00157	0.00332	0.00362
(sps)	-0.02350	-0.00614	-0.01341
(sds)	0.01862	-0.02067	-0.02473
(pds)	0.01353	-0.06751	-0.05271
(pdp)	-0.00127	-0.00680	-0.01157
ON SITE H-H			
s	0.48670	0.22448	
FIRST NEIGHBOR H-H			
(sss)	-0.02107	0.00937	-0.01939
(sps)	0.00000	0.00000	0.00000
(pps)	0.00000	0.00000	0.00000
(ppp)	0.00000	0.00000	0.00000
SECOND NEIGHBOR H-H			
(sss)	0.02934	0.00113	-0.01203
(sps)	0.00000	0.00000	0.00000
(pps)	0.00000	0.00000	0.00000
(ppp)	0.00000	0.00000	0.00000
FIRST NEIGHBOR Sr-H			
(sss)	0.08918	0.06763	-0.01950
(sps)	0.00000	0.00000	0.00000
(pss)	0.07731	-0.03336	-0.26439
(pps)	0.00000	0.00000	0.00000
(ppp)	0.00000	0.00000	0.00000
(dpp)	0.00000	0.00000	0.00000
(dss)	0.09937	0.00604	-0.22387
(dps)	0.00000	0.00000	0.00000
SECOND NEIGHBOR Sr-H			
(sss)	-0.01642	-0.00954	0.00980
(pss)	-0.04395	0.00066	0.00418
(sps)	0.00000	0.00000	0.00000
(pps)	0.00000	0.00000	0.00000
(ppp)	0.00000	0.00000	0.00000
(dss)	-0.02238	0.00768	0.01056
(dps)	0.00000	0.00000	0.00000
(dpp)	0.00000	0.00000	0.00000

♠SrH

BAND	ORTHOGONAL RMS ERROR mRy	ORTHOGONAL MAXIMUM DEVIATION k	ORTHOGONAL MAXIMUM DEVIATION mRy	NON-ORTHOGONAL RMS ERROR mRy	NON-ORTHOGONAL MAXIMUM DEVIATION k	NON-ORTHOGONAL MAXIMUM DEVIATION mRy
1	8.4	(005)	18.8	0.9	(062)	2.1
2	7.7	(004)	17.9	1.7	(111)	4.4
3	9.8	(444)	22.2	1.1	(354)	2.8
4	11.6	(048)	25.7	1.3	(022)	3.3
5	14.7	(226)	36.2	1.8	(226)	4.7
6	11.9	(044)	35.6	2.6	(174)	6.5
7	11.3	(006)	26.7	2.5	(066)	8.2
1-7	11.0			1.8		

ENERGY VALUES IN Ry AT SELECTED k-POINTS

	ORTHOGONAL	APW	NON-ORTHOGONAL
GAMMA 1	-0.02439	-0.01727	-0.01684
GAMMA 1	0.78518	0.79560	0.79511
GAMMA 12	0.81885	0.81617	0.81805
GAMMA 15	1.20181	1.18039	1.17968
GAMMA 25'	0.65122	0.65996	0.65842
X1 (008)	0.09258	0.09820	0.09939
X1 (008)	0.98434	0.98375	0.97799
X1 (008)	1.14442	1.12442	1.11609
X2 (008)	0.92087	0.93524	0.93479
X3 (008)	0.45345	0.45099	0.45208
X5 (008)	0.97719	0.97664	0.97608
X4' (008)	0.59156	0.60124	0.60004
X5' (008)	0.95894	0.94623	0.94577
L1 (444)	0.34796	0.34115	0.34226
L1 (444)	1.04211	1.05585	1.06093
L3 (444)	0.60050	0.62271	0.62456
L3 (444)	0.93435	0.93432	0.93339
L2' (444)	0.17963	0.19481	0.19632
L2' (444)	0.85808	0.83777	0.83622
L3' (444)	1.27811	1.28300	1.28301
W1 (048)	0.63226	0.65798	0.65889
W1 (048)	1.01388	1.03484	1.02812
W2' (048)	0.13566	0.13880	0.13876
W2' (048)	0.99094	0.99390	0.99003
W2' (048)	1.31496	1.43015	1.42996
W1' (048)	0.97091	0.97753	0.97769
W3 (048)	0.52199	0.51558	0.51520
W3 (048)	0.98172	0.99669	0.99608
ODD (224)	0.66170	0.66851	0.66663
ODD (224)	0.90276	0.90158	0.90055
EVEN(224)	0.12937	0.12247	0.12267
EVEN(224)	0.44051	0.44334	0.44246
EVEN(224)	0.59776	0.58924	0.58918
EVEN(224)	0.81620	0.79187	0.79057
EVEN(224)	0.85639	0.85966	0.86376

4.5 Barium Hydride (BaH)

See Fig. 4.21.

See Tables 4.21, 4.22 and 4.23.

See Figs. 4.22, 4.23, 4.24 and 4.25.

See Tables 4.24 and 4.25.

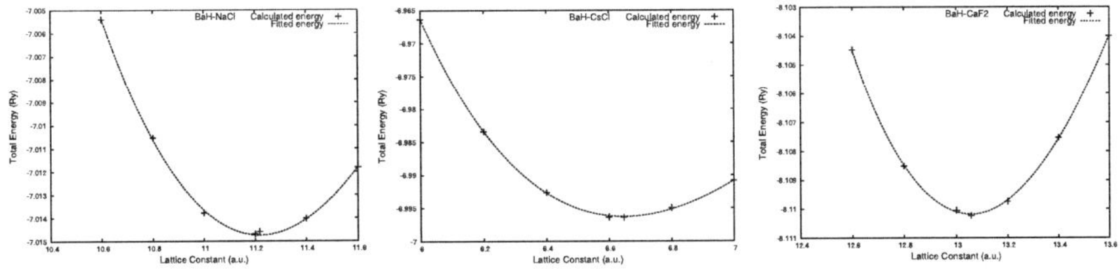

Fig. 4.21 Total energy versus lattice constant of BaH in the NaCl, CsCl and CaF2 structures

Table 4.21 Lattice Constant, Bulk Modulus, Gap, Total Energy

	a (Bohr)	B (MBar)	Gap (Ry)	Total Energy (Ry)
NaCl	11.22	0.25		-16267.01474
CsCl	6.65	0.25		-16266.99626
CaF2 (Insulator)	13.06	0.24		-16268.11017
exp (Orthorhombic)	a=14.83	b=12.86	c=7.89	

Table 4.22 Birch Fit Coefficients

	A1	A2	A3	A4
NaCl	-6.057597E+00	-1.093096E+02	3.756255E+03	-3.418192E+04
CsCl	-6.469137E+00	-4.471803E+01	8.644188E+02	3.638223E+03
CaF2	-7.411953E+00	-7.218233E+01	1.816226E+02	1.018869E+05

Table 4.23 DOS at Ef, Hopfield parameter, Stoner Criterion

```
                    BaH     a=11.22 Bohr        NaCl
 ----------------------------------------------------------------------------
  Fermi Energy   Total DOS        Decomposed DOS inside the muffin tins
    (Ry)        (States/Ry)       s         p         eg         t2g        f
 ----------------------------------------------------------------------------
  Ba 0.437        34.625       0.823     0.882     0.368      9.121      0.199
   H 0.437        34.625       0.412     2.225     0.016      0.834      0.111
 ----------------------------------------------------------------------------
 Fermi-Velocity (cm/s): 0.35697 x10E8
 Plasmon Energy (eV)   : 4.01702
 Electron-ion interaction (Hopfield parameter) (eV/A^2) Ba: 2.137 H: 0.117
----------------------------------------------
   Ba      MUFFIN-TIN RADIUS and CHARGE =     2.9448    53.9375
    H      MUFFIN-TIN RADIUS and CHARGE =     2.6643     1.6273
            Ba        STONER I =    0.002
             H        STONER I =    0.001
           STONER PARAMETER (Ry) I =  0.004
           STONER CRITERION    N*I =  0.137
----------------------------------------------
                    BaH     a=6.65 Bohr      CsCl
 ----------------------------------------------------------------------------
  Fermi Energy   Total DOS        Decomposed DOS inside the muffin tins
    (Ry)        (States/Ry)       s         p         eg         t2g        f
----------------------------------------------------------------------------
 Ba  0.557        34.386       1.741     0.579    10.534      3.641      0.535
  H  0.557        34.386       0.516     4.096     0.847      0.150      0.123
 ----------------------------------------------------------------------------
Fermi-Velocity (cm/s): 0.54206 x10E8
Plasmon Energy (eV)   : 6.67773
Electron-ion interaction (Hopfield parameter) (eV/A^2) Ba: 2.577 H: 0.259
----------------------------------------------
   Ba      MUFFIN-TIN RADIUS and CHARGE =     3.0222    54.2247
    H      MUFFIN-TIN RADIUS and CHARGE =     2.7344     1.7354
                     Ba        STONER I =  0.005
                      H        STONER I =  0.003
           STONER PARAMETER (Ry) I = 0.009
           STONER CRITERION    N*I = 0.293
----------------------------------------------
                    BaH2    a=13.06 Bohr     CaF2
 ----------------------------------------------------------------------------
  Fermi Energy   Total DOS        Decomposed DOS inside the muffin tins

    (Ry)        (States/Ry)       s         p         eg         t2g        f
 ----------------------------------------------------------------------------
 Ba 0.146       0.0    Insulator
  H 0.146       0.0
 ----------------------------------------------------------------------------
          Ba      MUFFIN-TIN RADIUS and CHARGE =     2.9679    53.7986
           H      MUFFIN-TIN RADIUS and CHARGE =     2.6852     1.5194
```

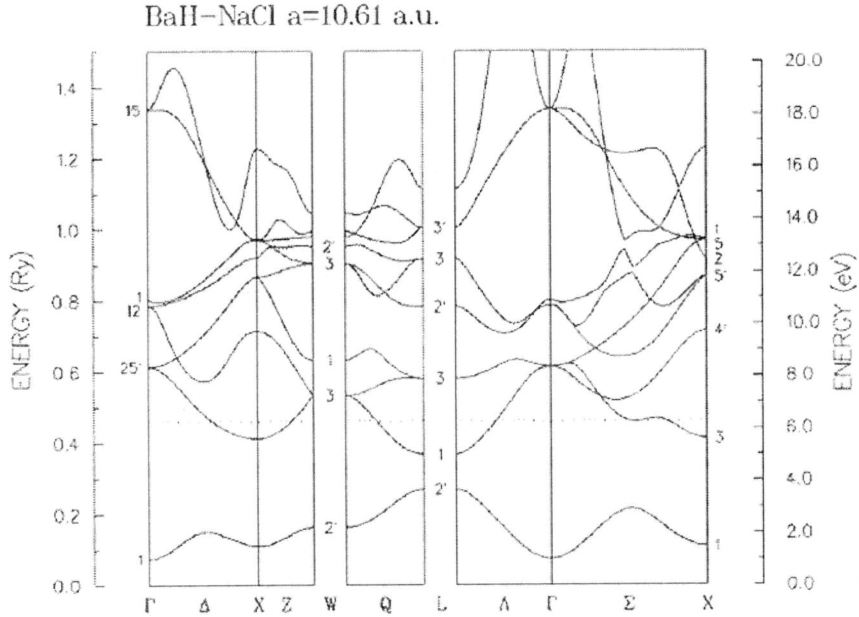

Fig. 4.22 Energy bands of BaH in the NaCl structure (tight-binding)

Fig. 4.23 Total, angular momentum and site-decomposed densities of states of BaH in the NaCl structure (tight-binding)

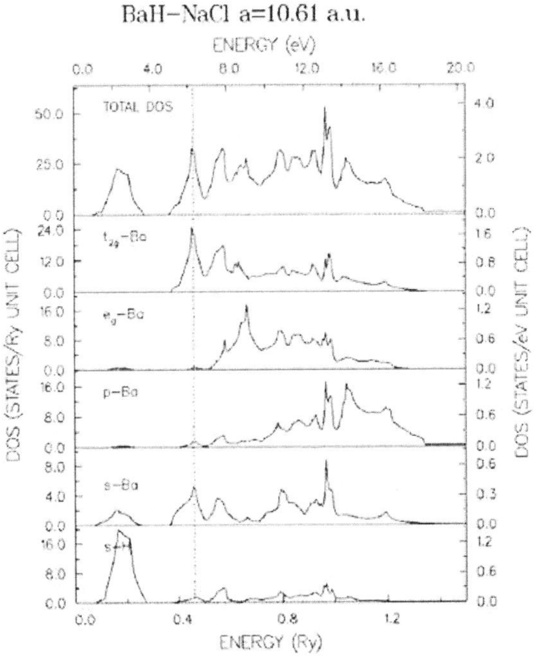

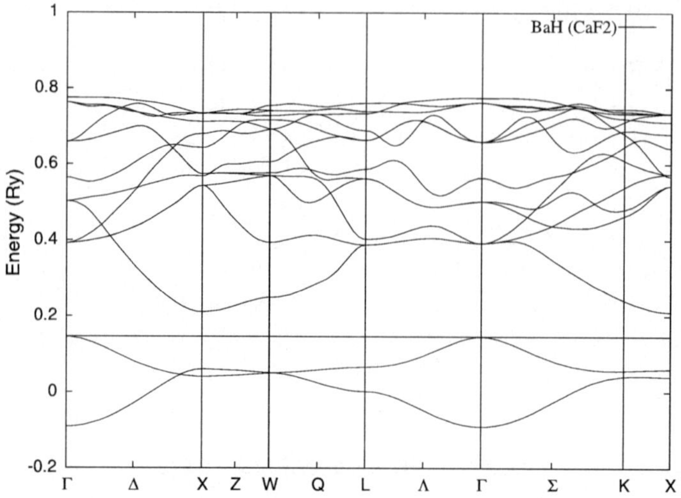

Fig. 4.24 Energy bands of BaH2 in the CaF2 structure

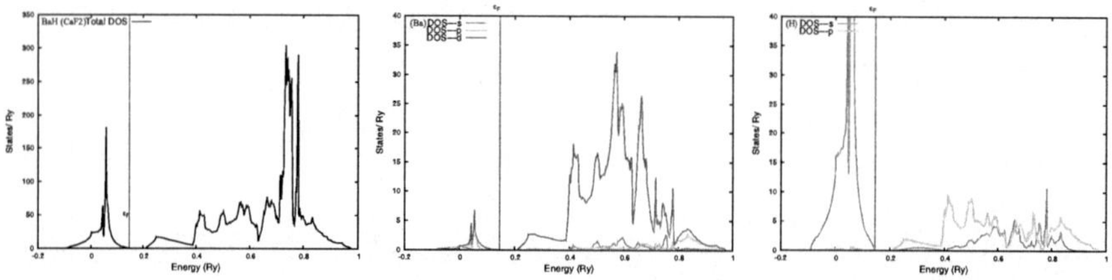

Fig. 4.25 Total, angular momentum and site decomposed densities of states of BaH2 in the CaF2 structure

Table 4.24 BaH (NaCl) a = 10.61 Bohr Slater–Koster 3-center parameters

	ORTHOGONAL	NON-ORTHOGONAL	
	ENERGY INTEGRALS Ry	ENERGY INTEGRALS Ry	OVERLAP INTEGRALS
ON SITE Ba-Ba			
s,s(000)	0.73292	0.67096	
x,x(000)	0.90299	0.83453	
xy,xy(000)	0.81238	0.63717	
d2,d2(000)	0.70408	0.64137	
FIRST NEIGHBOR Ba-Ba			
s,s(110)	-0.03001	-0.01259	-0.00150
s,x(110)	-0.02190	-0.02553	-0.00373
s,xy(110)	-0.01171	0.02835	0.12199
s,d2(110)	0.02168	0.03148	0.03978
x,x(110)	0.03778	-0.09110	-0.14345
x,x(011)	0.01935	0.07112	0.10060
x,y(110)	0.02198	0.00573	-0.02749
x,xy(110)	0.04922	-0.03488	-0.10034
x,xy(011)	0.00401	0.01574	0.04606
z,d2(011)	0.00734	0.03975	0.09387
z,d1(011)	-0.01796	-0.00445	0.01137
xy,xy(110)	-0.05712	0.00151	0.08114
xy,xy(011)	0.01306	-0.01247	-0.05522
xy,xz(011)	-0.01480	0.01460	-0.01240
xy,d2(110)	0.02800	0.01139	0.06447
d2,d2(110)	-0.01981	0.02132	0.05424
d1,d1(110)	0.02155	0.02500	-0.02958
SECOND NEIGHBOR Ba-Ba			
s,s(200)	0.06023	0.02632	-0.02257
s,x(200)	0.06931	0.01308	-0.01395
s,d2(002)	0.03662	-0.00779	0.03404
x,x(200)	-0.03365	0.02641	0.04562
y,y(200)	0.02555	-0.01315	-0.01353
x,xy(020)	0.02346	-0.00650	-0.00981
z,d2(002)	-0.02159	-0.09908	-0.16653
xy,xy(200)	-0.02047	-0.01059	-0.01779
xy,xy(002)	0.00224	-0.00145	-0.00029
d2,d2(002)	0.01615	-0.02626	-0.02873
d1,d1(002)	0.00267	-0.01953	-0.02227
ON SITE H-H			
s,s(000)	0.40104	0.24579	
FIRST NEIGHBOR H-H			
s,s(110)	-0.02649	0.01959	0.02319
s,x(110)	0.00000	0.00000	0.00000
x,x(110)	0.00000	0.00000	0.00000
x,x(011)	0.00000	0.00000	0.00000
x,y(110)	0.00000	0.00000	0.00000
SECOND NEIGHBOR H-H			
s,s(200)	0.00771	-0.00964	-0.01685
s,x(200)	0.00000	0.00000	0.00000
x,x(200)	0.00000	0.00000	0.00000
y,y(200)	0.00000	0.00000	0.00000
FIRST NEIGHBOR Ba-H			
s,s(100)	-0.04890	0.07957	0.09956
s,x(100)	0.00000	0.00000	0.00000
x,s(100)	0.05411	0.06699	0.18762
x,x(100)	0.00000	0.00000	0.00000
y,y(100)	0.00000	0.00000	0.00000
xy,x(010)	0.00000	0.00000	0.00000
d2,s(001)	-0.10063	-0.02479	-0.20774
d2,z(001)	0.00000	0.00000	0.00000
SECOND NEIGHBOR Ba-H			
s,s(111)	0.01248	0.00306	0.00499
x,s(111)	-0.01003	0.00853	0.02162
s,x(111)	0.00000	0.00000	0.00000
x,x(111)	0.00000	0.00000	0.00000
x,y(111)	0.00000	0.00000	0.00000
xy,s(111)	0.01389	0.00172	-0.00066
xy,x(111)	0.00000	0.00000	0.00000
yz,x(111)	0.00000	0.00000	0.00000
d1,x(111)	0.00000	0.00000	0.00000

BaH

BAND	ORTHOGONAL RMS ERROR mRy	MAXIMUM DEVIATION k	mRy	NON-ORTHOGONAL RMS ERROR mRy	MAXIMUM DEVIATION k	mRy
1	5.1	(004)	15.6	0.9	(354)	1.9
2	9.8	(055)	22.0	1.9	(444)	4.1
3	5.9	(044)	13.3	2.1	(008)	4.9
4	6.3	(048)	14.6	2.2	(008)	4.3
5	10.7	(224)	24.1	3.2	(055)	8.1
6	11.4	(066)	31.0	4.5	(118)	15.6
7	11.9	(044)	36.8	5.3	(442)	20.6
1-7	9.1			3.2		

ENERGY VALUES IN Ry AT SELECTED k-POINTS

	ORTHOGONAL	APW	NON-ORTHOGONAL
GAMMA 1	0.07279	0.07355	0.07374
GAMMA 1	0.79085	0.79965	0.80076
GAMMA 12	0.77099	0.78575	0.78585
GAMMA 15	1.31754	1.33880	1.33750
GAMMA 25'	0.61099	0.61791	0.61425
X1 (008)	0.10712	0.10990	0.11056
X1 (008)	0.97508	0.97742	0.97243
X2 (008)	0.92247	0.92486	0.92145
X3 (008)	0.40201	0.41053	0.41172
X5 (008)	0.96349	0.97575	0.97048
X4' (008)	0.71306	0.71919	0.71434
X5' (008)	0.86049	0.87121	0.86691
L1 (444)	0.36137	0.36326	0.36734
L1 (444)	1.01830	1.11457	1.11352
L3 (444)	0.57424	0.57900	0.58014
L3 (444)	0.90398	0.92254	0.91683
L2' (444)	0.27130	0.26830	0.26947
L2' (444)	0.77572	0.78434	0.78247
W1 (048)	0.62091	0.63551	0.63369
W1 (048)	1.12766	0.99424	0.99505
W2' (048)	0.16943	0.16509	0.16348
W2' (048)	0.94822	0.95462	0.95304
W1' (048)	0.95452	0.97721	0.97820
W3 (048)	0.53939	0.53497	0.53343
W3 (048)	0.90982	0.90971	0.90504
ODD (224)	0.63085	0.62976	0.63086
ODD (224)	0.85920	0.86028	0.86033
EVEN(224)	0.19985	0.20130	0.20019
EVEN(224)	0.43811	0.44511	0.44366
EVEN(224)	0.55850	0.56575	0.56435
EVEN(224)	0.75932	0.73520	0.73602
EVEN(224)	0.81259	0.81602	0.82311

FERMI LEVEL QUANTITIES (Non-orthogonal fit)

ENERGY Ry	DENSITIES OF STATES (states/Ry/cell) Total	s-Ba	p-Ba	t2g-Ba	eg-Ba	s-H	p-H
0.4601	31.83	5.22	1.51	22.81	0.59	1.69	0.00

	INTEGRATED DENSITIES OF STATES (electrons) Total	s-Ba	p-Ba	t2g-Ba	eg-Ba	s-H	p-H
	3.00	0.41	0.06	0.67	0.05	1.80	0.00

VELOCITY cm/s	PLASMON ENERGY eV	EIGENVALUE SUM Ry
0.43x10E8	5.10	-0.5850

Table 4.25 BaH (NaCl) a = 10.61 Bohr Slater–Koster 2-center parameters

	ORTHOGONAL	NON-ORTHOGONAL	
	ENERGY INTEGRALS Ry	ENERGY INTEGRALS Ry	OVERLAP INTEGRALS
ON SITE Ba-Ba			
s	0.69759	0.51152	
p	0.90031	0.78805	
t2g	0.74946	0.62354	
eg	0.73073	0.60469	
FIRST NEIGHBOR Ba-Ba			
(sss)	-0.01525	0.01227	0.09375
(pps)	0.05998	-0.13502	-0.24938
(ppp)	0.02058	0.01027	0.02172
(dds)	-0.07746	0.04571	0.16541
(ddp)	0.01853	0.00614	-0.06535
(ddd)	0.00506	0.00029	0.00516
(sps)	0.03580	-0.05350	-0.14399
(sds)	-0.04599	0.03541	0.13622
(pds)	-0.07602	0.08917	0.20361
(pdp)	-0.00264	-0.01604	-0.05334
SECOND NEIGHBOR Ba-Ba			
(sss)	0.01899	0.02109	0.01161
(pps)	-0.00237	0.12130	0.12385
(ppp)	0.01405	0.01734	0.01392
(dds)	0.01031	-0.05570	-0.06121
(ddp)	-0.00419	-0.01272	-0.01269
(ddd)	0.00202	0.00962	0.01101
(sps)	-0.01089	-0.00012	0.00574
(sds)	0.02240	-0.01336	-0.01171
(pds)	0.00803	-0.09854	-0.10520
(pdp)	-0.01087	-0.01111	-0.00582
ON SITE H-H			
s	0.42941	0.25239	
FIRST NEIGHBOR H-H			
(sss)	-0.02229	0.00129	-0.04515
(sps)	0.00000	0.00000	0.00000
(pps)	0.00000	0.00000	0.00000
(ppp)	0.00000	0.00000	0.00000
SECOND NEIGHBOR H-H			
(sss)	0.00881	-0.01939	-0.01994
(sps)	0.00000	0.00000	0.00000
(pps)	0.00000	0.00000	0.00000
(ppp)	0.00000	0.00000	0.00000
FIRST NEIGHBOR Ba-H			
(sss)	0.06018	0.04259	-0.04817
(sps)	0.00000	0.00000	0.00000
(pss)	0.05595	0.02443	-0.12067
(pps)	0.00000	0.00000	0.00000
(ppp)	0.00000	0.00000	0.00000
(dpp)	0.00000	0.00000	0.00000
(dss)	0.09544	-0.02185	-0.24939
(dps)	0.00000	0.00000	0.00000
SECOND NEIGHBOR Ba-H			
(sss)	-0.01070	0.00296	0.02646
(pss)	-0.02132	0.00744	0.02985
(sps)	0.00000	0.00000	0.00000
(pps)	0.00000	0.00000	0.00000
(ppp)	0.00000	0.00000	0.00000
(dss)	-0.03049	0.01985	0.04575
(dps)	0.00000	0.00000	0.00000
(dpp)	0.00000	0.00000	0.00000

↑BaH

BAND	ORTHOGONAL RMS ERROR mRy	ORTHOGONAL MAXIMUM DEVIATION k	ORTHOGONAL MAXIMUM DEVIATION mRy	NON-ORTHOGONAL RMS ERROR mRy	NON-ORTHOGONAL MAXIMUM DEVIATION k	NON-ORTHOGONAL MAXIMUM DEVIATION mRy
1	8.3	(005)	19.9	2.4	(044)	6.5
2	7.7	(004)	20.2	4.1	(022)	11.4
3	10.6	(003)	20.3	3.7	(055)	8.0
4	11.0	(048)	24.5	3.4	(226)	8.6
5	18.7	(022)	36.7	3.8	(001)	8.8
6	16.4	(066)	43.9	7.1	(118)	21.3
7	15.4	(062)	48.1	6.5	(442)	16.7
1-7	13.2			4.7		

ENERGY VALUES IN Ry AT SELECTED k-POINTS

	ORTHOGONAL	APW	NON-ORTHOGONAL
GAMMA 1	0.07720	0.07355	0.06955
GAMMA 1	0.76621	0.79965	0.80112
GAMMA 12	0.78553	0.78575	0.78665
GAMMA 15	1.35632	1.33880	1.33861
GAMMA 25'	0.60379	0.61791	0.61293
X1 (008)	0.11505	0.10990	0.10697
X1 (008)	0.97039	0.97742	0.97458
X2 (008)	0.89821	0.92486	0.91390
X3 (008)	0.41500	0.41053	0.41064
X5 (008)	0.96405	0.97575	0.96726
X4' (008)	0.71185	0.71919	0.71279
X5' (008)	0.86947	0.87121	0.87044
L1 (444)	0.36307	0.36326	0.36382
L1 (444)	0.92888	1.11457	0.98725
L3 (444)	0.55998	0.57900	0.58431
L3 (444)	0.92289	0.92254	0.91848
L2' (444)	0.28423	0.26830	0.26923
L2' (444)	0.78353	0.78434	0.78128
W1 (048)	0.61104	0.63551	0.63314
W1 (048)	0.97427	0.99424	0.99534
W2' (048)	0.16522	0.16509	0.16369
W2' (048)	0.94823	0.95462	0.95008
W1' (048)	0.95597	0.97721	0.97690
W3 (048)	0.54108	0.53497	0.53482
W3 (048)	0.89967	0.90971	0.90426
ODD (224)	0.62407	0.62976	0.62486
ODD (224)	0.85730	0.86028	0.86926
EVEN(224)	0.19842	0.20130	0.20139
EVEN(224)	0.44226	0.44511	0.44321
EVEN(224)	0.55996	0.56575	0.56726
EVEN(224)	0.77162	0.73520	0.73715
EVEN(224)	0.81293	0.81602	0.81400

4.6 Radium Hydride (RaH)

See Fig. 4.26.

See Tables 4.26, 4.27 and 4.28.

See Figs. 4.27, 4.28, 4.29 and 4.30.

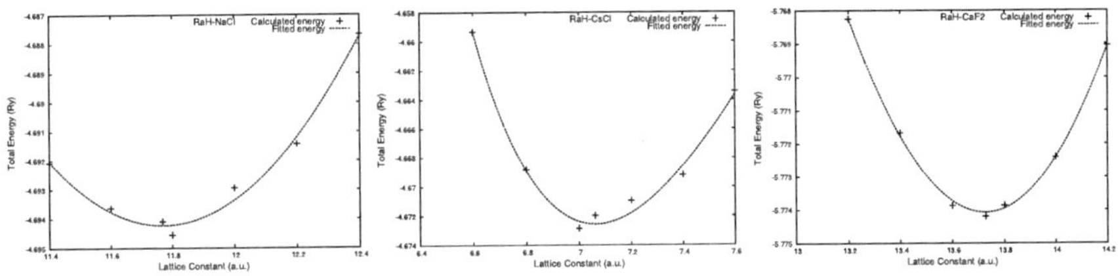

Fig. 4.26 Total energy versus lattice constant of RaH in the NaCl, CsCl and CaF2 structures

Table 4.26 Lattice Constant, Bulk Modulus, Gap, Total Energy

	a (Bohr)	B (MBar)	Gap (Ry)	Total Energy (Ry)
NaCl	11.77	0.18		-50074.69409
CsCl	7.06	0.20		-50074.67256
CaF2 (Insulator)	13.73	0.21		-50075.77407

Table 4.27 Birch Fit Coefficients

	A1	A2	A3	A4
NaCl	-3.163123E+00	-2.213606E+02	1.045684E+04	-1.602480E+05
CsCl	-4.658080E+00	2.536456E+01	-2.637808E+03	6.666952E+04
CaF2	-2.822406E+00	-5.842268E+02	3.786151E+04	-7.985679E+05

Table 4.28 DOS at Ef, Hopfield parameter, Stoner Criterion

RaH a=11.77 Bohr NaCl

Fermi Energy (Ry)	Total DOS (States/Ry)	Decomposed DOS inside the muffin tins s	p	eg	t2g	f
Ra 0.418	30.850	0.958	0.767	0.283	6.594	0.217
H 0.418	30.850	0.722	2.031	0.031	0.838	0.084

```
Fermi-Velocity (cm/s): 0.40421 x10E8
Plasmon Energy (eV)   : 3.99776
Electron-ion interaction (Hopfield parameter) (eV/A^2) Ra: 1.498 H: 0.172
-----------------------------------------------
  Ra     MUFFIN-TIN RADIUS and CHARGE =    3.0894   85.8150
   H     MUFFIN-TIN RADIUS and CHARGE =    2.7952    1.6879
          Ra        STONER I = 0.002
           H        STONER I = 0.002
         STONER PARAMETER (Ry) I =  0.004
         STONER CRITERION    N*I =  0.104
-----------------------------------------------
```

RaH a=7.06 Bohr CsCl

Fermi Energy (Ry)	Total DOS (States/Ry)	Decomposed DOS inside the muffin tins s	p	eg	t2g	f
Ra 0.510	24.442	1.859	0.512	4.407	3.113	0.475
H 0.510	24.442	0.610	3.926	0.511	0.181	0.054

```
Fermi-Velocity (cm/s): 0.65285 x10E8
Plasmon Energy (eV)   : 6.19232
Electron-ion interaction (Hopfield parameter) (eV/A^2) Ra: 1.374 H: 0.328
-----------------------------------------------
  Ra     MUFFIN-TIN RADIUS and CHARGE =    3.2107   86.1336
   H     MUFFIN-TIN RADIUS and CHARGE =    2.9049    1.7947
                  Ra        STONER I = 0.004
                   H        STONER I = 0.004
         STONER PARAMETER (Ry) I = 0.008
         STONER CRITERION    N*I = 0.201
-----------------------------------------------
```

RaH2 a=13.73 Bohr CaF2

Fermi Energy (Ry)	Total DOS (States/Ry)	Decomposed DOS inside the muffin tins s	p	eg	t2g	f
Ra 0.118	0.0	Insulator				
H 0.118	0.0					

```
        Ra     MUFFIN-TIN RADIUS and CHARGE =    3.1203   85.7225
         H     MUFFIN-TIN RADIUS and CHARGE =    2.8231    1.5595
```

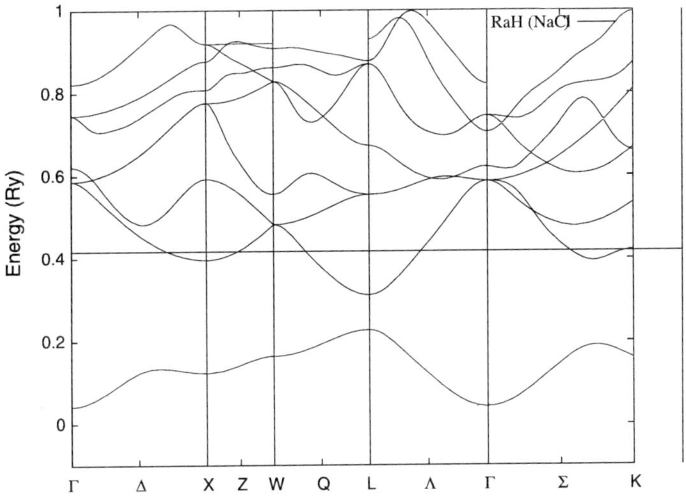

Fig. 4.27 Energy bands of RaH in the NaCl structure

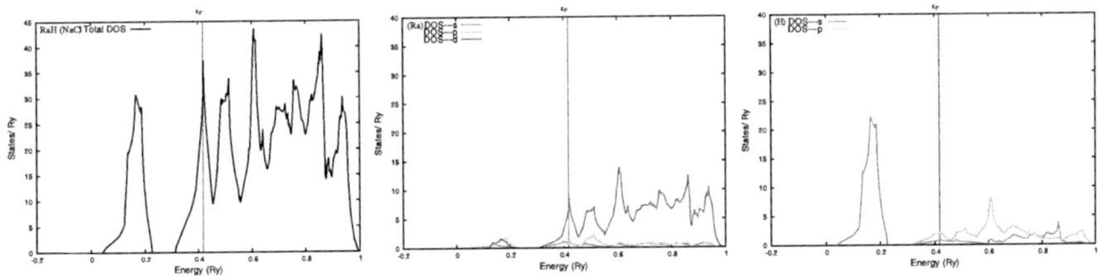

Fig. 4.28 Total, angular momentum and site-decomposed densities of states of RaH in the NaCl structure

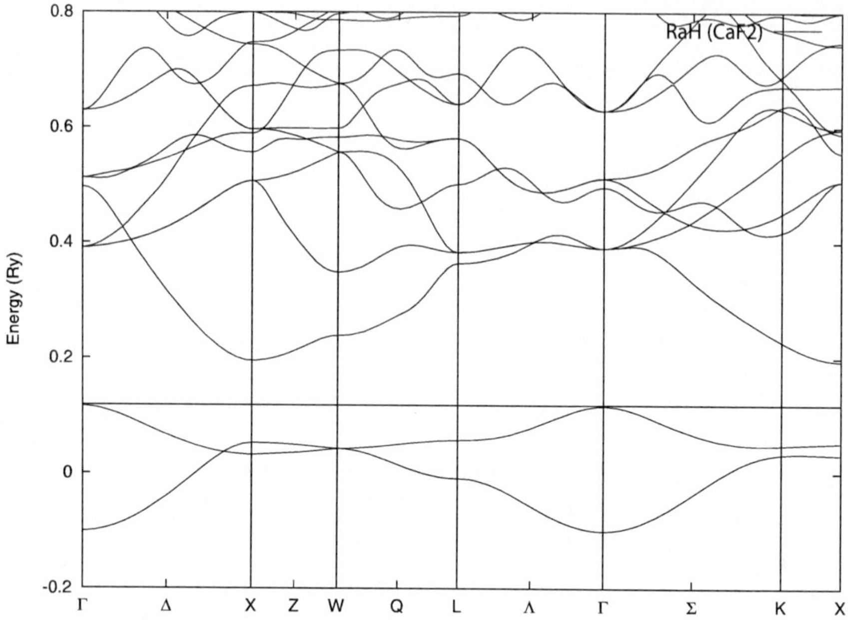

Fig. 4.29 Energy bands of RaH2 in the CaF2 structure

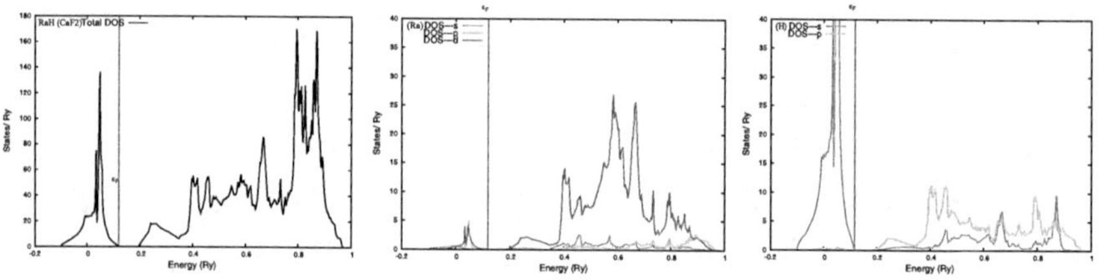

Fig. 4.30 Total, angular momentum and site decomposed densities of states of RaH2 in the CaF2 structure

References

1. G.G. Libowitz, J. Nucl. Mater. **2**, 1 (1960)
2. H. Smithson, C.A. Marianetti, D. Morgan, A. Van der Ven, A. Predith, G. Ceder, First-Principles study of the stability and electronic structure of metal hydrides. Phys. Rev. B **66**, 144107 (2002)
3. Z. Wang, Y. Yao, L. Zhu, H. Liu, T. Litaka, H. Wang, Metallization and superconductivity of BeH2 under high pressure. J. Chem. Phys. **140**, 124707 (2014)

Chapter 5
Rare Earth Hydrides

This chapter covers the rare earth hydrides ScH, YH, and LaH [1-2].

Results are presented for the crystal structures NaCl (B1), CaF2 (C1) and Im3m. These systems form as dihydrides in CaF2 structure. Due to the interest on superconductivity under high pressures in hydrides, results are also presented in the Im3m structure. Our results show that, unlike the light-element hydrides, ScH3 and YH3 show strong electron–ion interaction from the non-hydrogen sites. On the other hand the lanthanide shows high value of the Hopfield parameter due to the hydrogen sites, in a LaH10 Clathrate structure which leads to record high superconducting transition temperature as has been established by recent publications [3].

Examining the densities of states figures we observe that the lower occupied states have predominantly s-H character and near the Fermi level the d-2 g metal contribution is the dominant one. Comparing the energy bands of the fluorite structure to the bands of the NaCl structure it is noted that an additional band (second band) which is due to the second hydrogen in the CaF2 structure. Also Tight-binding parameters are given in the NaCl structure based on both orthogonal and non-orthogonal Hamiltonians using three- and two-center integrals.

D. A. Papaconstantopoulos, *Band Structure of Cubic Hydrides*,
https://doi.org/10.1007/978-3-031-06878-2_5

5.1 Scandium Hydride (ScH)

See Figs. 5.1 and Tables 5.1, 5.2, 5.3.

See Figs. 5.2, 5.3 and Tables 5.4, 5.5.

See Figs. 5.4, 5.5 and 5.6.

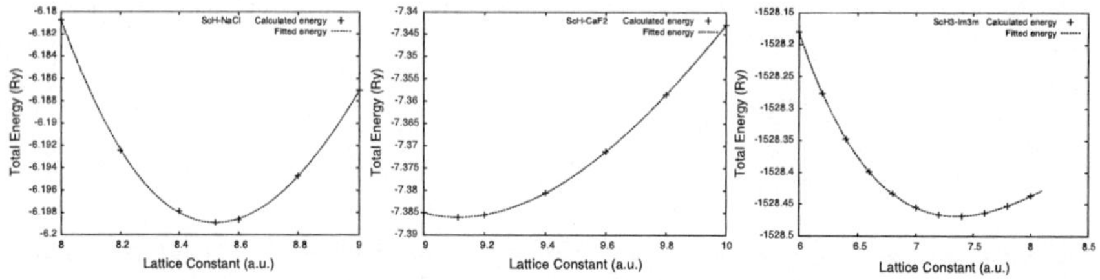

Fig. 5.1 Total energy versus lattice constant of ScH in the NaCl, CaF2 and Im3m structures

Table 5.1 Lattice constant, bulk modulus, gap, total energy

Stru	a (Bohr)	B (MBar)	Gap	Total Energy
NaCl	8.52	0.90	-	-1526.19894
CaF2	9.11	1.06	-	-1527.38604
Im3m (P=0)	7.35	0.83	-	-1528.46866
Im3m (P=1.49 MBar)	6.00	5.00	-	-1528.18028
exp (CaF2)	9.04			

Table 5.2 Birch fit coefficients

	A1	A2	A3	A4
NaCl	-4.668350E+00	-1.017000E+02	2.047976E+03	-1.118971E+04
CaF2	-6.014988E+00	-8.524237E+01	1.150976E+03	5.575163E+03
Im3m	-1.526981E+03	-1.090188E+02	2.249081E+03	-8.928303E+03

Table 5.3 DOS at Ef, Hopfield parameter, Stoner criterion

```
                    ScH    a=8.52 Bohr        NaCl
----------------------------------------------------------------------------
 Fermi Energy    Total DOS         Decomposed DOS inside the muffin tins
   (Ry)        (States/Ry)       s          p         eg         t2g        f
----------------------------------------------------------------------------
Sc  0.693       24.044          0.149     1.110     0.839       9.354     0.050
 H  0.693       24.044          0.166     1.590     0.003       0.337     0.055
----------------------------------------------------------------------------
Fermi-Velocity (cm/s): 0.39050 x10E8
Plasmon Energy (eV)   : 5.54205
Electron-ion interaction (Hopfield parameter) (eV/A^2) Sc: 4.435  H:0.164
-----------------------------------------------
  Sc      MUFFIN-TIN RADIUS and CHARGE =     2.237   18.959
  H       MUFFIN-TIN RADIUS and CHARGE =     2.024    1.385
           Sc        STONER I = 0.0076
            H        STONER I = 0.0013
          STONER PARAMETER (Ry) I = 0.009
          STONER CRITERION     N*I = 0.223
-----------------------------------------------

                    ScH2      a=9.11 Bohr      CaF2
----------------------------------------------------------------------------
 Fermi Energy    Total DOS         Decomposed DOS inside the muffin tins
   (Ry)         (States/Ry)       s         p          eg         t2g       f
----------------------------------------------------------------------------
Sc 0.514          9.080          0.007     0.106     3.420      0.951     0.033
 H 0.514          9.080          0.314     0.376     0.020      0.023     0.003
----------------------------------------------------------------------------
Fermi-Velocity (cm/s): 0.65436 x10E8
Plasmon Energy (eV)   : 5.14061
Electron-ion interaction (Hopfield parameter) (eV/A^2) Sc: 0.889  H: 0.779


-----------------------------------------------
          Sc      MUFFIN-TIN RADIUS and CHARGE =    2.3677   19.0794
           H      MUFFIN-TIN RADIUS and CHARGE =    1.5784    0.9845
                     Sc       STONER I = 0.0051
                      H       STONER I = 0.0004
          STONER PARAMETER (Ry)       I = 0.0060
          STONER CRITERION          N*I = 0.0544

-----------------------------------------------
                    ScH3    a=6.0 Bohr         Im3m
----------------------------------------------------------------------------
 Fermi Energy    Total DOS         Decomposed DOS inside the muffin tins
   (Ry)        (States/Ry)       s          p          d          f
----------------------------------------------------------------------------
Sc 1.242          9.123          0.077     0.160     3.266      0.049
 H 1.242          9.123          0.370     0.059     0.006      0.000
----------------------------------------------------------------------------
Fermi-Velocity (cm/s): 0.92001 x10E8
Plasmon Energy (eV)   : 9.62712
Electron-ion interaction (Hopfield parameter) (eV/A^2) Sc: 5.892  H3: 2.764
----------------------------------------------------------------------------
         Sc      MUFFIN-TIN RADIUS and CHARGE =    1.8000   18.4433
         H       MUFFIN-TIN RADIUS and CHARGE =    1.0000    0.5387
----------------------------------------------------------------------------
```

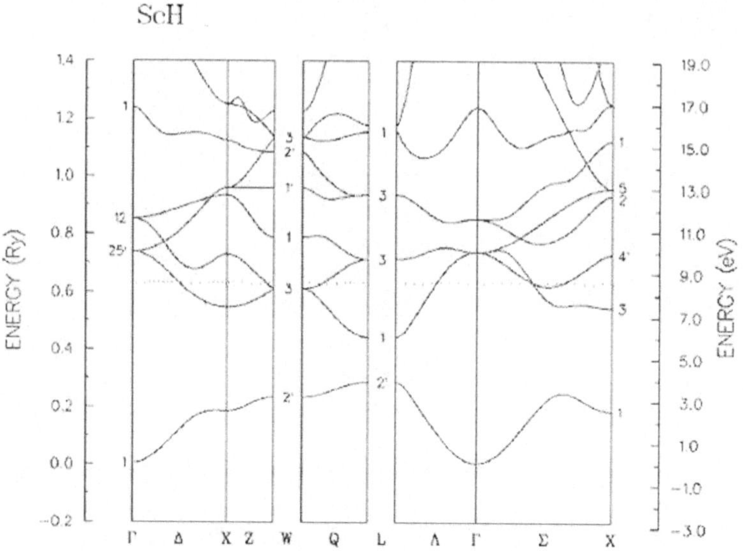

Fig. 5.2 Energy bands of ScH in the NaCl structure (tight-binding)

Fig. 5.3 Total, angular momentum and site decomposed densities of states of ScH in the NaCl structure (tight-binding)

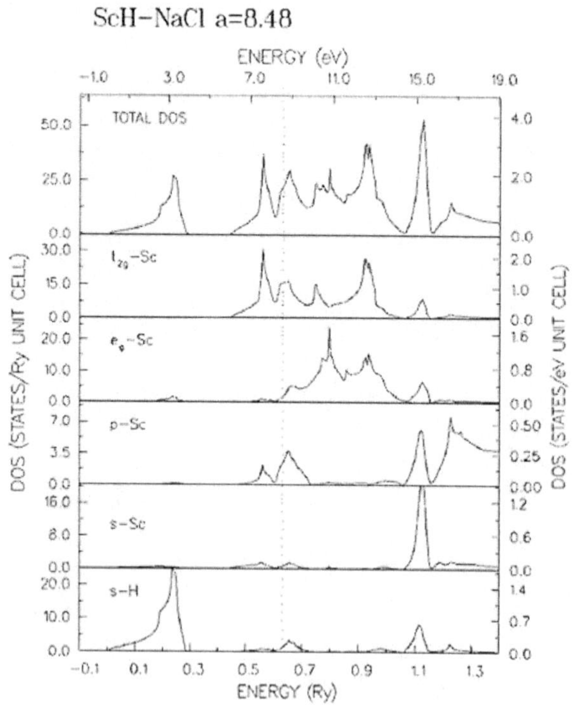

Table 5.4 ScH (NaCl) a = 8.48 Bohr Slater–Koster 3-center parameters

	ORTHOGONAL	NON-ORTHOGONAL	
	ENERGY INTEGRALS Ry	ENERGY INTEGRALS Ry	OVERLAP INTEGRALS
ON SITE Sc-Sc			
s,s(000)	1.03941	0.93857	
x,x(000)	1.39849	1.11712	
xy,xy(000)	0.83082	0.75615	
d2,d2(000)	0.87683	0.86893	
FIRST NEIGHBOR Sc-Sc			
s,s(110)	-0.01254	0.07461	0.09114
s,x(110)	-0.04089	0.02758	0.09808
s,xy(110)	-0.05147	-0.02546	0.02838
s,d2(110)	-0.00167	0.06274	0.04523
x,x(110)	0.05874	0.04148	-0.01891
x,x(011)	-0.00752	0.01704	0.02602
x,y(110)	0.05530	-0.08793	-0.15279
x,xy(110)	0.05731	-0.00230	-0.04983
x,xy(011)	-0.00480	-0.02147	0.00121
z,d2(011)	0.01699	0.00211	0.03753
z,d1(011)	0.00245	-0.05416	-0.03625
xy,xy(110)	-0.03991	-0.00650	0.04080
xy,xy(011)	0.01190	0.00893	-0.00717
xy,xz(011)	0.00375	0.02146	-0.00213
xy,d2(110)	0.02072	-0.00937	0.01722
d2,d2(110)	0.00122	0.04598	0.05154
d1,d1(110)	-0.02271	-0.03444	-0.01957
SECOND NEIGHBOR Sc-Sc			
s,s(200)	0.04569	-0.01719	-0.02178
s,x(200)	0.06086	-0.02466	0.01458
s,d2(002)	0.03696	-0.01180	0.03532
x,x(200)	-0.10488	0.09229	0.05530
y,y(200)	0.00984	-0.03461	-0.03281
x,xy(020)	0.00937	-0.00933	-0.01113
z,d2(002)	-0.04234	-0.07670	-0.13214
xy,xy(200)	-0.00815	-0.00573	-0.01119
xy,xy(002)	0.00145	0.00834	0.00885
d2,d2(002)	0.03470	0.02454	-0.00509
d1,d1(002)	0.00031	-0.02046	-0.02047
ON SITE H-H			
s,s(000)	0.30229	0.28023	
FIRST NEIGHBOR H-H			
s,s(110)	-0.01633	-0.00972	-0.00033
s,x(110)	0.00000	0.00000	0.00000
x,x(110)	0.00000	0.00000	0.00000
x,x(011)	0.00000	0.00000	0.00000
x,y(110)	0.00000	0.00000	0.00000
SECOND NEIGHBOR H-H			
s,s(200)	-0.00557	-0.00412	0.00371
s,x(200)	0.00000	0.00000	0.00000
x,x(200)	0.00000	0.00000	0.00000
y,y(200)	0.00000	0.00000	0.00000
FIRST NEIGHBOR Sc-H			
s,s(100)	-0.08256	0.07280	-0.06103
s,x(100)	0.00000	0.00000	0.00000
x,s(100)	0.06260	0.01553	0.22345
x,x(100)	0.00000	0.00000	0.00000
y,y(100)	0.00000	0.00000	0.00000
xy,x(010)	0.00000	0.00000	0.00000
d2,s(001)	-0.02855	-0.11013	-0.16549
d2,z(001)	0.00000	0.00000	0.00000
SECOND NEIGHBOR Sc-H			
s,s(111)	0.02756	0.00514	0.00428
x,s(111)	-0.03501	0.00328	0.01323
s,x(111)	0.00000	0.00000	0.00000
x,x(111)	0.00000	0.00000	0.00000
x,y(111)	0.00000	0.00000	0.00000
xy,s(111)	0.00897	0.00854	0.01576
xy,x(111)	0.00000	0.00000	0.00000
yz,x(111)	0.00000	0.00000	0.00000
d1,x(111)	0.00000	0.00000	0.00000

ScH

BAND	ORTHOGONAL RMS ERROR mRy	ORTHOGONAL MAXIMUM DEVIATION k	ORTHOGONAL MAXIMUM DEVIATION mRy	NON-ORTHOGONAL RMS ERROR mRy	NON-ORTHOGONAL MAXIMUM DEVIATION k	NON-ORTHOGONAL MAXIMUM DEVIATION mRy
1	4.9	(005)	11.9	1.5	(044)	4.5
2	3.6	(222)	11.0	0.9	(002)	2.0
3	3.1	(022)	5.8	0.8	(005)	2.0
4	2.3	(006)	4.6	0.8	(008)	1.7
5	2.8	(044)	5.5	0.9	(055)	2.3
6	3.3	(174)	7.1	1.4	(066)	4.9
7	4.9	(044)	13.5	2.3	(222)	6.0
1-7	3.7			1.3		

ENERGY VALUES IN Ry AT SELECTED k-POINTS

	ORTHOGONAL	APW	NON-ORTHOGONAL
GAMMA 1	0.00750	0.00531	0.00508
GAMMA 1	1.22851	1.23837	1.24037
GAMMA 12	0.85297	0.85343	0.85226
GAMMA 15	1.66795	1.67666	1.67707
GAMMA 25'	0.73666	0.73809	0.73728
X1 (008)	0.19410	0.18737	0.18528
X1 (008)	1.12585	1.12436	1.12567
X1 (008)	1.49857	1.49537	1.49498
X2 (008)	0.92910	0.93122	0.93297
X3 (008)	0.54629	0.54621	0.54666
X5 (008)	0.96078	0.95807	0.95825
X4' (008)	0.72807	0.72833	0.72856
X5' (008)	1.25818	1.25138	1.25113
L1 (444)	0.43980	0.44260	0.44288
L1 (444)	1.15596	1.15426	1.15531
L3 (444)	0.70966	0.71111	0.71141
L3 (444)	0.93763	0.93350	0.93480
L2' (444)	0.28013	0.28694	0.28642
L2' (444)	1.18201	1.17888	1.17899
L3' (444)	1.79010	1.75637	1.75646
W1 (048)	0.78753	0.78872	0.78788
W1 (048)	1.23159	1.22726	1.22590
W2' (048)	0.23243	0.23533	0.23430
W2' (048)	1.07815	1.08283	1.08351
W2' (048)	1.79503	1.85745	1.85724
W1' (048)	0.95497	0.95832	0.95898
W3 (048)	0.60773	0.61015	0.61040
W3 (048)	1.13217	1.13546	1.13531
ODD (224)	0.75659	0.75338	0.75475
ODD (224)	0.92110	0.91677	0.91592
EVEN(224)	0.19142	0.18543	0.18605
EVEN(224)	0.56706	0.56396	0.56432
EVEN(224)	0.69231	0.69485	0.69449
EVEN(224)	0.86921	0.87034	0.87191
EVEN(224)	1.08958	1.08999	1.09147

FERMI LEVEL QUANTITIES (Non-orthogonal fit)

ENERGY Ry	Total	s-Sc	p-Sc	t2g-Sc	eg-Sc	s-H	p-H
	DENSITIES OF STATES (states/Ry/cell)						
0.6299	21.31	0.53	2.34	15.62	1.59	1.22	0.00
	INTEGRATED DENSITIES OF STATES (electrons)						
	4.00	0.19	0.13	1.62	0.15	1.92	0.00

VELOCITY cm/s	PLASMON ENERGY eV	EIGENVALUE SUM Ry
0.42x10E8	5.67	-0.9841

Table 5.5 ScH (NaCl) a = 8.48 Bohr Slater–Koster 2-center parameters

	ORTHOGONAL	NON-ORTHOGONAL	
	ENERGY INTEGRALS Ry	ENERGY INTEGRALS Ry	OVERLAP INTEGRALS
ON SITE Sc-Sc			
s	0.91870	0.84216	
p	1.34973	0.99947	
t2g	0.80606	0.73954	
eg	0.79018	0.73310	
FIRST NEIGHBOR Sc-Sc			
(sss)	-0.06125	-0.04747	0.00651
(pps)	0.12727	-0.06886	-0.21785
(ppp)	-0.00960	-0.00906	0.03347
(dds)	-0.05329	0.01693	0.08675
(ddp)	0.02311	-0.00217	-0.05806
(ddd)	-0.00024	0.00358	0.01344
(sps)	0.09401	0.00694	-0.08632
(sds)	-0.05875	-0.01876	0.05315
(pds)	-0.08401	-0.00049	0.07621
(pdp)	0.01191	-0.00198	-0.04646
SECOND NEIGHBOR Sc-Sc			
(sss)	0.00734	0.02118	0.01864
(pps)	-0.06798	0.10481	0.02976
(ppp)	0.00435	-0.01641	0.00461
(dds)	0.00200	-0.00836	0.00698
(ddp)	-0.00156	0.00049	-0.00220
(ddd)	0.00063	-0.00566	-0.00723
(sps)	-0.02442	0.00890	-0.00669
(sds)	0.00820	-0.03439	-0.03699
(pds)	0.00835	-0.04134	-0.01614
(pdp)	-0.00528	0.00402	-0.00351
ON SITE H-H			
s	0.77198	0.44474	
FIRST NEIGHBOR H-H			
(sss)	0.00238	0.01055	0.01231
(sps)	0.00000	0.00000	0.00000
(pps)	0.00000	0.00000	0.00000
(ppp)	0.00000	0.00000	0.00000
SECOND NEIGHBOR H-H			
(sss)	0.03445	0.02285	-0.00624
(sps)	0.00000	0.00000	0.00000
(pps)	0.00000	0.00000	0.00000
(ppp)	0.00000	0.00000	0.00000
FIRST NEIGHBOR Sc-H			
(sss)	0.10917	0.09635	-0.02699
(sps)	0.00000	0.00000	0.00000
(pss)	0.12090	-0.06960	-0.31331
(pps)	0.00000	0.00000	0.00000
(ppp)	0.00000	0.00000	0.00000
(dpp)	0.00000	0.00000	0.00000
(dss)	0.10083	0.05788	-0.13638
(dps)	0.00000	0.00000	0.00000
SECOND NEIGHBOR Sc-H			
(sss)	-0.02239	-0.00689	-0.01207
(pss)	-0.04563	0.00759	-0.00361
(sps)	0.00000	0.00000	0.00000
(pps)	0.00000	0.00000	0.00000
(ppp)	0.00000	0.00000	0.00000
(dss)	-0.00875	-0.00104	-0.00855
(dps)	0.00000	0.00000	0.00000
(dpp)	0.00000	0.00000	0.00000

ScH

BAND	ORTHOGONAL RMS ERROR mRy	MAXIMUM DEVIATION k	MAXIMUM DEVIATION mRy	NON-ORTHOGONAL RMS ERROR mRy	MAXIMUM DEVIATION k	MAXIMUM DEVIATION mRy
1	3.3	(004)	8.4	2.3	(004)	4.9
2	6.8	(333)	16.9	1.8	(264)	3.8
3	6.9	(033)	17.7	1.6	(003)	3.8
4	4.7	(444)	9.8	2.0	(066)	4.9
5	4.0	(002)	9.1	2.0	(048)	4.1
6	6.9	(055)	16.0	3.3	(066)	11.8
7	7.6	(062)	17.1	7.0	(174)	15.6
1-7	6.0			3.3		

ENERGY VALUES IN Ry AT SELECTED k-POINTS

	ORTHOGONAL	APW	NON-ORTHOGONAL
GAMMA 1	0.00238	0.00531	0.00613
GAMMA 1	1.23259	1.23837	1.23518
GAMMA 12	0.85570	0.85343	0.85503
GAMMA 15	1.66340	1.67666	1.67709
GAMMA 25'	0.73245	0.73809	0.73714
X1 (008)	0.18751	0.18737	0.19048
X1 (008)	1.11389	1.12436	1.13965
X1 (008)	1.50785	1.49537	1.49208
X2 (008)	0.92529	0.93122	0.93170
X3 (008)	0.54950	0.54621	0.54513
X5 (008)	0.96118	0.95807	0.96128
X4' (008)	0.72211	0.72833	0.72786
X5' (008)	1.26960	1.25138	1.24955
L1 (444)	0.43501	0.44260	0.43943
L1 (444)	1.15729	1.15426	1.15656
L3 (444)	0.70134	0.71111	0.71274
L3 (444)	0.93870	0.93350	0.93477
L2' (444)	0.28808	0.28694	0.28755
L2' (444)	1.19802	1.17888	1.17947
L3' (444)	1.74202	1.75637	1.75474
W1 (048)	0.78642	0.78872	0.78392
W1 (048)	1.23603	1.22726	1.19411
W2' (048)	0.23465	0.23533	0.23470
W2' (048)	1.07536	1.08283	1.07915
W2' (048)	1.76593	1.85745	1.85787
W1' (048)	0.95867	0.95832	0.95421
W3 (048)	0.60714	0.61015	0.60802
W3 (048)	1.13289	1.13546	1.13565
ODD (224)	0.74939	0.75338	0.75383
ODD (224)	0.92040	0.91677	0.91445
EVEN(224)	0.18477	0.18543	0.18824
EVEN(224)	0.57392	0.56396	0.56441
EVEN(224)	0.69791	0.69485	0.69464
EVEN(224)	0.87624	0.87034	0.87442
EVEN(224)	1.08728	1.08999	1.08620

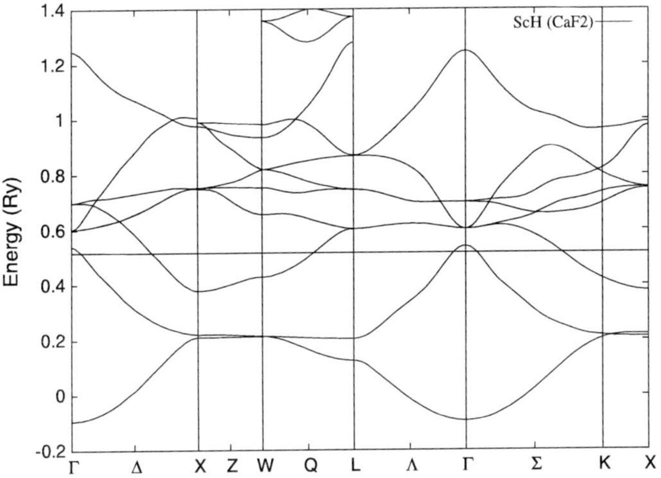

Fig. 5.4 Energy bands of ScH2 in the CaF2 structure

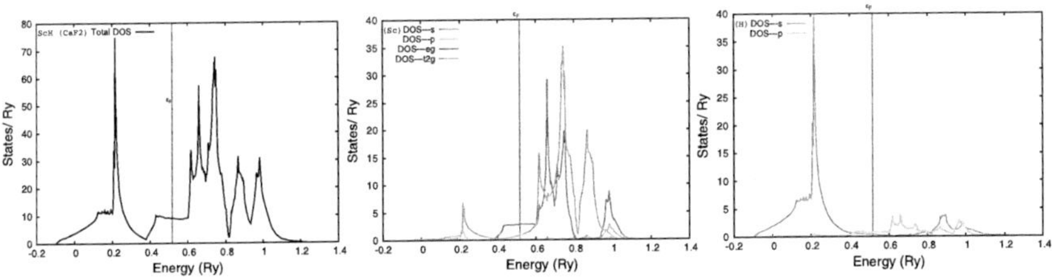

Fig. 5.5 Total, angular momentum and site decomposed densities of states of ScH2 in the CaF2 structure

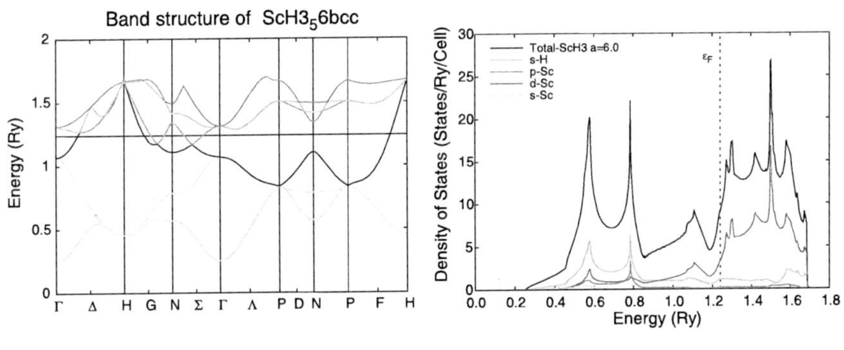

Fig. 5.6 Energy bands and DOS of ScH3

5.2 Yttrium Hydride (YH)

See Figs. 5.7 and Tables 5.6, 5.7, 5.8.

See Figs. 5.8, 5.9 and Tables 5.9, 5.10.

See Figs. 5.10, 5.11 and 5.12.

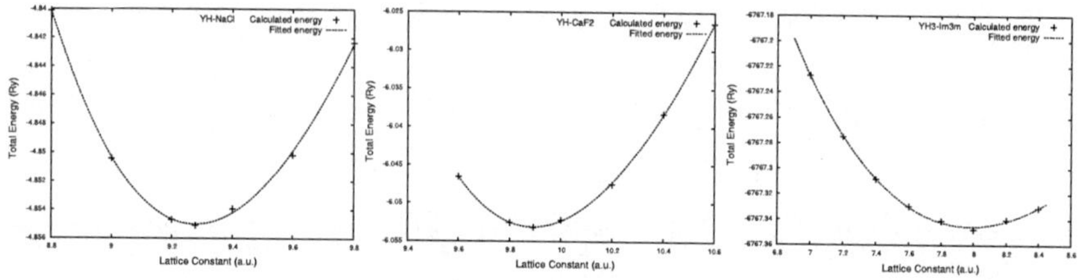

Fig. 5.7 Total energy versus lattice constant of YH in the NaCl, CaF2 and Im3m structures

Table 5.6 Lattice constant, bulk modulus, gap, total energy

Stru	a (Bohr)	B (MBar)	Gap	Total Energy
NaCl	9.28	0.78	–	-6764.85510
CaF2	9.89	0.90	–	-6766.05318
Im3m (P=0)	7.98	0.77	–	-6767.34546
Im3m (P=1.98 MBar)	6.00	3.71	–	-6766.73462
exp (CaF2)	9.84			

Table 5.7 Birch fit coefficients

	A1	A2	A3	A4
NaCl	-3.786025E+00	-6.911169E+01	9.795928E+02	4.572542E+03
CaF2	-4.948188E+00	-6.368586E+01	-5.241764E+01	3.336882E+04
Im3m	-6.765251E+03	-1.922004E+02	5.305912E+03	-3.878851E+04

Table 5.8 DOS at Ef, Hopfield parameter, Stoner criterion

```
                    YH     a=9.28 Bohr       NaCl
---------------------------------------------------------------------------
  Fermi Energy    Total DOS        Decomposed DOS inside the muffin tins
     (Ry)        (States/Ry)        s         p         eg          t2g         f
---------------------------------------------------------------------------
 Y 0.650          19.950        0.210     0.977     0.532       6.408      0.093
 H 0.650          19.950        0.192     1.236     0.004       0.302      0.053
---------------------------------------------------------------------------
 Fermi-Velocity (cm/s): 0.47910 x10E8
 Plasmon Energy (eV)  : 5.45288
 Electron-ion interaction (Hopfield parameter) (eV/A^2) Y: 4.194  H:
0.135
-------------------------------------------------
   Y      MUFFIN-TIN RADIUS and CHARGE =    2.5050   36.7239
    H      MUFFIN-TIN RADIUS and CHARGE =   2.1339    1.4672
            Y               STONER I = 0.0040
            H               STONER I = 0.0012
           STONER PARAMETER (Ry) I =  0.0055
           STONER CRITERION    N*I =  0.1105
-------------------------------------------------
                    YH2     a=9.89 Bohr       CaF2
---------------------------------------------------------------------------
  Fermi Energy    Total DOS        Decomposed DOS inside the muffin tins
     (Ry)        (States/Ry)        s         p          eg         t2g         f
---------------------------------------------------------------------------

 Y  0.476          8.481        0.008     0.126      2.295      0.527
0.049
 H  0.476          8.481        0.291     0.434      0.021      0.034
0.004


---------------------------------------------------------------------------
Fermi-Velocity (cm/s): 0.71474 x10E8
Plasmon Energy (eV)   : 4.81754
Electron-ion interaction (Hopfield parameter) (eV/A^2) Y: 0.738   H:
0.596
-------------------------------------------------
          Y      MUFFIN-TIN RADIUS and CHARGE =    2.5700   36.7149
          H      MUFFIN-TIN RADIUS and CHARGE =    1.7133    1.1039
                    Y       STONER I =  0.0014
                    H       STONER I =  0.0004
          STONER PARAMETER (Ry)      I = 0.0025
          STONER CRITERION         N*I = 0.0210
---------------------------------------------------------------------------
                    YH3      a=6.0 Bohr        Im3m
 --------------------------------------------------------------------------
  Fermi Energy    Total DOS        Decomposed DOS inside the muffin tins
     (Ry)        (States/Ry)        s         p          d          f
---------------------------------------------------------------------------

 Y  1.361          7.007        0.077     0.258      1.903      0.145
 H  1.361          7.007        0.358     0.056      0.006      0.000
---------------------------------------------------------------------------
Fermi-Velocity (cm/s): 1.20392 x10E8
Plasmon Energy (eV)   : 11.04028
Electron-ion interaction (Hopfield parameter) (eV/A^2) Y: 7.331  H3:
3.250
---------------------------------------------------------------------------
Y      MUFFIN-TIN RADIUS and CHARGE =    1.8000   36.2729
H      MUFFIN-TIN RADIUS and CHARGE =    1.0000    0.5938
```

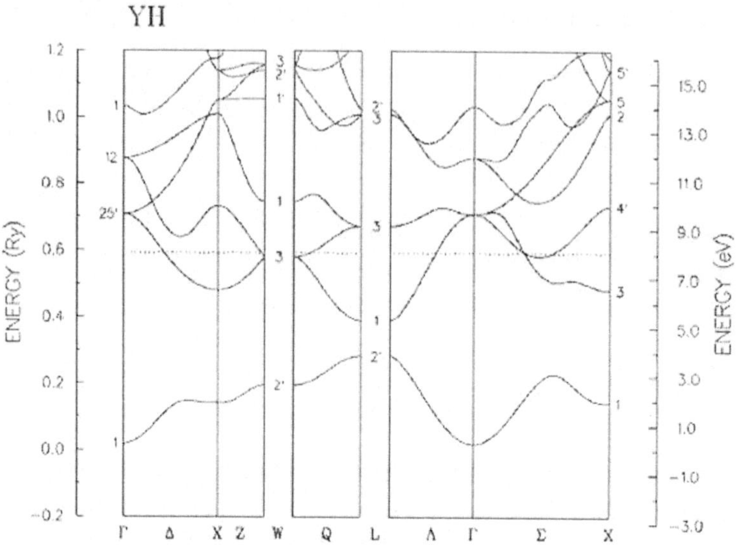

Fig. 5.8 Energy bands of YH in the NaCl structure (tight-binding)

Fig. 5.9 Total, angular momentum and site decomposed densities of states of YH in the NaCl structure (tight-binding)

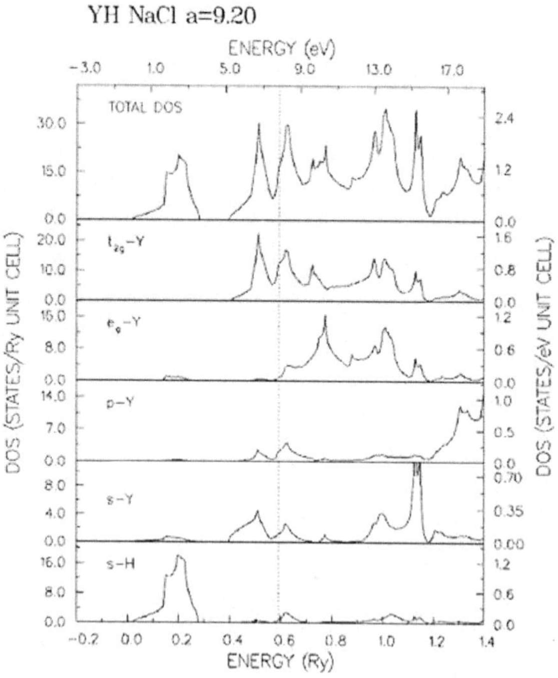

Table 5.9 YH (NaCl) a = 9.20 Bohr Slater–Koster 3-center parameters

	ORTHOGONAL	NON-ORTHOGONAL	
	ENERGY INTEGRALS Ry	ENERGY INTEGRALS Ry	OVERLAP INTEGRALS
ON SITE Y- Y			
s,s(000)	0.88977	0.85236	
x,x(000)	1.24785	1.09475	
xy,xy(000)	0.90075	0.74376	
d2,d2(000)	0.87920	0.76463	
FIRST NEIGHBOR Y- Y			
s,s(110)	-0.02142	0.05729	0.06351
s,x(110)	-0.03247	-0.00088	0.04026
s,xy(110)	-0.04811	0.00922	0.09505
s,d2(110)	0.00741	0.03955	0.03755
x,x(110)	0.04150	0.02265	-0.03860
x,x(011)	-0.00979	0.09810	0.10933
x,y(110)	0.05156	-0.02232	-0.07083
x,xy(110)	0.06868	0.00059	-0.06097
x,xy(011)	-0.00267	0.02002	0.05390
z,d2(011)	0.01355	0.03756	0.09151
z,d1(011)	-0.00329	0.00838	0.03190
xy,xy(110)	-0.05810	-0.01733	0.04541
xy,xy(011)	0.01451	0.00119	-0.02394
xy,xz(011)	-0.00923	-0.00115	-0.03488
xy,d2(110)	0.02614	0.00730	0.05155
d2,d2(110)	-0.00410	0.01054	0.03346
d1,d1(110)	-0.02013	0.04410	0.00912
SECOND NEIGHBOR Y- Y			
s,s(200)	0.05549	0.05816	0.01232
s,x(200)	0.07218	0.02034	-0.00939
s,d2(002)	0.04487	-0.00667	0.04225
x,x(200)	-0.09871	0.02673	0.04707
y,y(200)	0.01400	-0.00529	-0.00310
x,xy(020)	0.01281	0.01449	0.01999
z,d2(002)	-0.05356	-0.10191	-0.16693
xy,xy(200)	-0.02083	-0.01440	-0.02539
xy,xy(002)	0.00308	0.00542	0.00574
d2,d2(002)	0.04488	-0.04548	-0.05767
d1,d1(002)	0.00283	-0.00888	-0.00786
ON SITE H-H			
s,s(000)	0.27746	0.23415	
FIRST NEIGHBOR H-H			
s,s(110)	-0.01183	0.00220	0.02390
s,x(110)	0.00000	0.00000	0.00000
x,x(110)	0.00000	0.00000	0.00000
x,x(011)	0.00000	0.00000	0.00000
x,y(110)	0.00000	0.00000	0.00000
SECOND NEIGHBOR H-H			
s,s(200)	-0.00875	-0.01178	-0.01225
s,x(200)	0.00000	0.00000	0.00000
x,x(200)	0.00000	0.00000	0.00000
y,y(200)	0.00000	0.00000	0.00000
FIRST NEIGHBOR Y-H			
s,s(100)	-0.06639	0.07886	0.08890
s,x(100)	0.00000	0.00000	0.00000
x,s(100)	0.05191	0.06364	0.19480
x,x(100)	0.00000	0.00000	0.00000
y,y(100)	0.00000	0.00000	0.00000
xy,x(010)	0.00000	0.00000	0.00000
d2,s(001)	-0.05425	-0.00867	-0.16092
d2,z(001)	0.00000	0.00000	0.00000
SECOND NEIGHBOR Y-H			
s,s(111)	0.01969	0.01200	0.01318
x,s(111)	-0.02805	-0.00091	0.00699
s,x(111)	0.00000	0.00000	0.00000
x,x(111)	0.00000	0.00000	0.00000
x,y(111)	0.00000	0.00000	0.00000
xy,s(111)	0.01113	0.00180	0.01028
xy,x(111)	0.00000	0.00000	0.00000
yz,x(111)	0.00000	0.00000	0.00000
d1,x(111)	0.00000	0.00000	0.00000

YH

BAND	ORTHOGONAL RMS ERROR mRy	ORTHOGONAL MAXIMUM DEVIATION k	ORTHOGONAL MAXIMUM DEVIATION mRy	NON-ORTHOGONAL RMS ERROR mRy	NON-ORTHOGONAL MAXIMUM DEVIATION k	NON-ORTHOGONAL MAXIMUM DEVIATION mRy
1	4.4	(004)	9.1	1.5	(444)	3.4
2	4.6	(003)	13.3	1.5	(333)	3.3
3	4.0	(055)	11.0	1.0	(004)	2.9
4	5.4	(048)	14.1	1.0	(048)	2.5
5	4.4	(048)	9.6	1.6	(066)	6.4
6	4.6	(048)	11.7	1.4	(118)	3.0
7	5.9	(033)	11.4	3.0	(354)	11.2
1-7	4.8			1.7		

ENERGY VALUES IN Ry AT SELECTED k-POINTS

	ORTHOGONAL	APW	NON-ORTHOGONAL
GAMMA 1	0.02155	0.01837	0.01846
GAMMA 1	1.02710	1.03541	1.03295
GAMMA 12	0.87694	0.87783	0.87768
GAMMA 15	1.39924	1.40565	1.40523
GAMMA 25'	0.70725	0.70924	0.70796
X1 (008)	0.14536	0.13819	0.14136
X1 (008)	1.18601	1.17600	1.17647
X1 (008)	1.38422	1.35878	1.35798
X2 (008)	1.00664	1.00676	1.00909
X3 (008)	0.47515	0.48028	0.48177
X5 (008)	1.05599	1.05254	1.05226
X4' (008)	0.73529	0.73238	0.73212
X5' (008)	1.14556	1.13970	1.14006
L1 (444)	0.38915	0.38955	0.39081
L1 (444)	1.21940	1.21809	1.22238
L3 (444)	0.67455	0.67223	0.67267
L3 (444)	1.00252	1.00765	1.00776
L2' (444)	0.28334	0.28633	0.28295
L2' (444)	1.02344	1.02244	1.02327
L3' (444)	1.59553	1.53185	1.53261
W1 (048)	0.73480	0.74895	0.74641
W1 (048)	1.21083	1.20486	1.20186
W2' (048)	0.18602	0.19357	0.19479
W2' (048)	1.13009	1.14179	1.14019
W2' (048)	1.62311	1.73501	1.73444
W1' (048)	1.04367	1.05330	1.05578
W3 (048)	0.58074	0.57999	0.58037
W3 (048)	1.15259	1.15798	1.15770
ODD (224)	0.73324	0.72665	0.72705
ODD (224)	0.97897	0.97518	0.97552
EVEN(224)	0.19383	0.18770	0.18802
EVEN(224)	0.50641	0.50626	0.50596
EVEN(224)	0.64841	0.65019	0.65002
EVEN(224)	0.88067	0.88010	0.87961
EVEN(224)	1.01300	1.01054	1.01373

FERMI LEVEL QUANTITIES (Non-orthogonal fit)

DENSITIES OF STATES (states/Ry/cell)

ENERGY Ry	Total	s- Y	p- Y	t2g- Y	eg- Y	s-H	p-H
0.5911	17.50	1.16	2.23	12.51	0.88	0.71	0.00

INTEGRATED DENSITIES OF STATES (electrons)

Total	s- Y	p- Y	t2g- Y	eg- Y	s-H	p-H
4.00	0.47	0.17	1.38	0.13	1.86	0.00

VELOCITY cm/s	PLASMON ENERGY eV	EIGENVALUE SUM Ry
0.52x10E8	5.63	-0.9634

Table 5.10 YH (NaCl) a = 9.20 Bohr Slater–Koster 2-center parameters

	ORTHOGONAL	NON-ORTHOGONAL	
	ENERGY INTEGRALS Ry	ENERGY INTEGRALS Ry	OVERLAP INTEGRALS
ON SITE Y- Y			
s	0.86503	0.64131	
p	1.20727	1.01162	
t2g	0.83572	0.71173	
eg	0.80669	0.69881	
FIRST NEIGHBOR Y- Y			
(sss)	-0.04030	-0.05144	0.02961
(pps)	0.09670	0.02007	-0.11127
(ppp)	-0.01216	0.02816	0.04532
(dds)	-0.07817	0.02740	0.12463
(ddp)	0.02364	-0.01005	-0.08721
(ddd)	0.00327	0.00884	0.02093
(sps)	0.06806	-0.00742	-0.11068
(sds)	-0.06781	0.00678	0.11015
(pds)	-0.10015	0.02361	0.12181
(pdp)	0.00739	-0.01406	-0.05802
SECOND NEIGHBOR Y- Y			
(sss)	0.00463	0.01184	0.00769
(pps)	-0.04711	0.17935	0.12230
(ppp)	0.00015	-0.01621	-0.00392
(dds)	0.00930	-0.02720	-0.01064
(ddp)	-0.00398	-0.00906	-0.01250
(ddd)	0.00208	0.00391	0.00403
(sps)	-0.02917	-0.01068	-0.02217
(sds)	0.01889	-0.02908	-0.02472
(pds)	0.00981	-0.09474	-0.07833
(pdp)	-0.00056	-0.00626	-0.01062
ON SITE H-H			
s	0.65472	0.30246	
FIRST NEIGHBOR H-H			
(sss)	-0.01975	0.01687	0.00282
(sps)	0.00000	0.00000	0.00000
(pps)	0.00000	0.00000	0.00000
(ppp)	0.00000	0.00000	0.00000
SECOND NEIGHBOR H-H			
(sss)	0.03523	0.00010	-0.00865
(sps)	0.00000	0.00000	0.00000
(pps)	0.00000	0.00000	0.00000
(ppp)	0.00000	0.00000	0.00000
FIRST NEIGHBOR Y-H			
(sss)	0.10424	0.05203	-0.09493
(sps)	0.00000	0.00000	0.00000
(pss)	0.09299	-0.07728	-0.28016
(pps)	0.00000	0.00000	0.00000
(ppp)	0.00000	0.00000	0.00000
(dpp)	0.00000	0.00000	0.00000
(dss)	0.11409	-0.00990	-0.22937
(dps)	0.00000	0.00000	0.00000
SECOND NEIGHBOR Y-H			
(sss)	-0.01596	-0.01753	-0.00845
(pss)	-0.04392	0.00623	0.00576
(sps)	0.00000	0.00000	0.00000
(pps)	0.00000	0.00000	0.00000
(ppp)	0.00000	0.00000	0.00000
(dss)	-0.01493	0.00432	-0.00114
(dps)	0.00000	0.00000	0.00000
(dpp)	0.00000	0.00000	0.00000

♠ YH

BAND	ORTHOGONAL RMS ERROR mRy	ORTHOGONAL MAXIMUM DEVIATION k	ORTHOGONAL MAXIMUM DEVIATION mRy	NON-ORTHOGONAL RMS ERROR mRy	NON-ORTHOGONAL MAXIMUM DEVIATION k	NON-ORTHOGONAL MAXIMUM DEVIATION mRy
1	9.0	(005)	20.1	1.4	(066)	3.7
2	9.1	(264)	19.9	1.7	(022)	4.4
3	10.8	(444)	24.0	1.1	(033)	2.1
4	11.5	(444)	24.0	1.4	(033)	3.1
5	13.6	(226)	42.4	2.0	(226)	5.2
6	10.3	(044)	33.3	2.3	(174)	8.1
7	15.0	(333)	31.6	9.1	(354)	49.4
1-7	11.5			3.8		

ENERGY VALUES IN Ry AT SELECTED k-POINTS

	ORTHOGONAL	APW	NON-ORTHOGONAL
GAMMA 1	0.00943	0.01837	0.01842
GAMMA 1	1.02890	1.03541	1.03385
GAMMA 12	0.88012	0.87783	0.88012
GAMMA 15	1.40319	1.40565	1.40533
GAMMA 25'	0.70034	0.70924	0.70745
X1 (008)	0.13527	0.13819	0.13830
X1 (008)	1.16314	1.17600	1.17084
X1 (008)	1.36555	1.35878	1.35814
X2 (008)	0.99067	1.00676	1.00644
X3 (008)	0.48511	0.48028	0.48158
X5 (008)	1.05520	1.05254	1.05161
X4' (008)	0.72687	0.73238	0.73191
X5' (008)	1.16229	1.13970	1.13861
L1 (444)	0.39318	0.38955	0.39074
L1 (444)	1.21004	1.21809	1.22335
L3 (444)	0.64820	0.67223	0.67400
L3 (444)	1.01255	1.00765	1.00638
L2' (444)	0.26930	0.28633	0.28668
L2' (444)	1.03951	1.02244	1.02243
L3' (444)	1.51860	1.53185	1.53214
W1 (048)	0.72730	0.74895	0.74782
W1 (048)	1.18021	1.20486	1.19522
W2' (048)	0.19086	0.19357	0.19142
W2' (048)	1.14428	1.14179	1.13894
W2' (048)	1.55770	1.73501	1.73450
W1' (048)	1.04687	1.05330	1.05394
W3 (048)	0.58734	0.57999	0.57949
W3 (048)	1.14272	1.15798	1.15814
ODD (224)	0.71954	0.72665	0.72514
ODD (224)	0.97581	0.97518	0.97276
EVEN(224)	0.19275	0.18770	0.18924
EVEN(224)	0.50540	0.50626	0.50607
EVEN(224)	0.65708	0.65019	0.65202
EVEN(224)	0.90462	0.88010	0.88022
EVEN(224)	1.01118	1.01054	1.01494

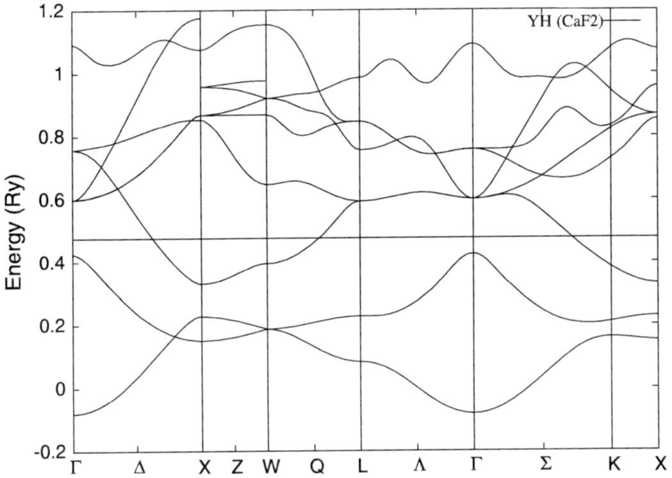

Fig. 5.10 Energy bands of YH2 in the CaF2 structure

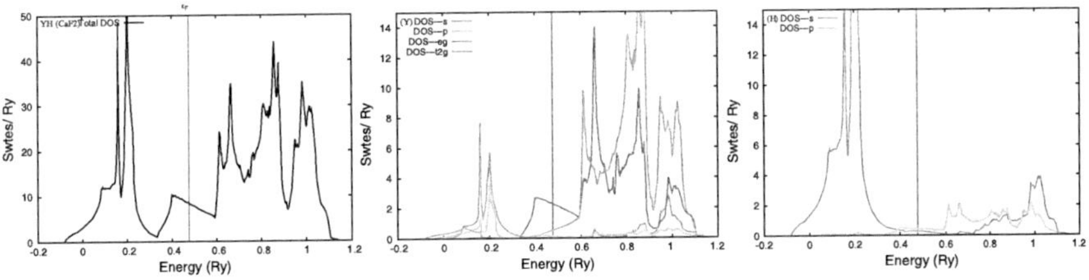

Fig. 5.11 Total, angular momentum and site decomposed densities of states of YH2 in the CaF2 structure

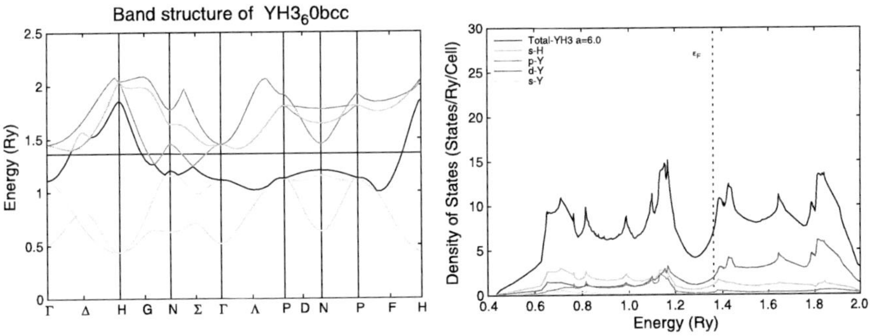

Fig. 5.12 Energy bands and DOS of YH3

5.3 Lanthanum Hydride (LaH)

See Figs. 5.13 and Tables 5.11, 5.12, 5.13.

See Figs. 5.14, 5.15 and Tables 5.14, 5.15.

See Figs. 5.16, 5.17, 5.18 and 5.19.

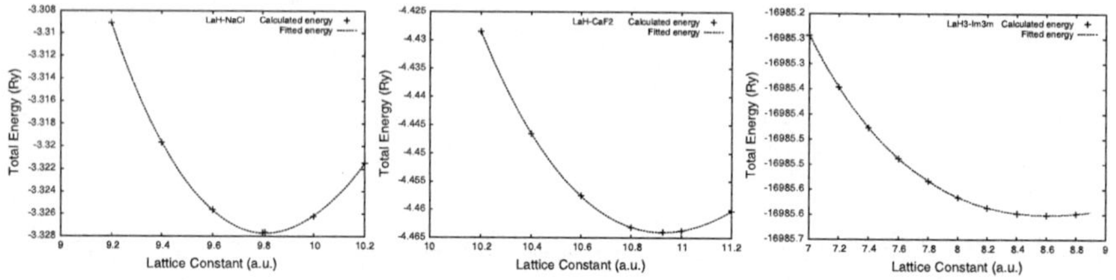

Fig. 5.13 Total energy versus lattice constant of LaH in the NaCl, CaF2 and Im3m (clathrate) structures

Table 5.11 Lattice constant, bulk modulus, gap, total energy

Stru	a (Bohr)	B (MBar)	Gap	Total Energy(Ry)
NaCl	9.81	0.59	-	-16983.32752
CaF2	10.93	0.65	-	-16984.46396
Im3m (P=0)	8.60	0.62	-	-16985.60125
Im3m (P=1.16 MBar)	7.00	3.91	-	-16985.24303
Clathr(Im3m) LaH10 P=0	11.43	1.07	-	-16993.48198
Clathr(Im3m) LaH10 P=2.83MBar	9.0	8.91	-	-16992.43261
exp (CaF2)	10.71			

Table 5.12 Birch fit coefficients

	A1	A2	A3	A4
NaCl	-1.749160E+00	-1.402005E+02	3.801865E+03	-2.863795E+04
CaF2	-2.953051E+00	-1.382962E+02	2.929150E+03	1.096628E+04
Im3m	-1.698385E+04	-1.740163E+02	4.839485E+03	-2.442460E+04
LaH10	-1.698998E+04	-3.864980E+02	1.180549E+04	-6.170307E+04

Table 5.13 DOS at Ef, Hopfield parameter, Stoner criterion

LaH a=9.81 Bohr NaCl

Fermi Energy (Ry)	Total DOS (States/Ry)	Decomposed DOS inside the muffin tins s	p	eg	t2g	f
La 0.594	10.827	0.259	0.387	0.318	3.425	1.551
H 0.594	10.827	0.072	0.346	0.002	0.078	0.015

Fermi-Velocity (cm/s): 0.47664 x10E8
Plasmon Energy (eV) : 3.67575
Electron-ion interaction (Hopfield parameter) (eV/A^2) La: 1.476 H: 0.035

La MUFFIN-TIN RADIUS and CHARGE = 2.8938 54.9205
H MUFFIN-TIN RADIUS and CHARGE = 2.0110 1.3117
La STONER I = 0.0022
H STONER I = 0.0004
STONER PARAMETER (Ry) I = 0.0028
STONER CRITERION N*I = 0.0305

LaH2 a=10.93 Bohr CaF2

Fermi Energy (Ry)	Total DOS (States/Ry)	Decomposed DOS inside the muffin tins s	p	eg	t2g	f
La 0.386	12.443	0.017	0.243	2.775	0.403	0.960
H 0.386	12.443	0.207	0.559	0.023	0.070	0.007

Fermi-Velocity (cm/s): 0.53916 x10E8
Plasmon Energy (eV) : 3.79158
Electron-ion interaction (Hopfield parameter) (eV/A^2) La: 0.564 H2: 0.246

La MUFFIN-TIN RADIUS and CHARGE = 2.8388 54.5150
H MUFFIN-TIN RADIUS and CHARGE = 1.8926 1.1948
La STONER I = 0.0008
H STONER I = 0.0002
STONER PARAMETER (Ry) I = 0.0015
STONER CRITERION N*I = 0.0184

LaH3 a=7.0 Bohr Im3m

Fermi Energy (Ry)	Total DOS (States/Ry)	Decomposed DOS inside the muffin tins s	p	eg	f
La 1.099	10.466	0.083	0.493	0.900	1.393
H 1.099	10.466	0.506	0.039	0.002	0.000

Fermi-Velocity (cm/s): 0.81703 x10E8
Plasmon Energy (eV) : 7.26656
Electron-ion interaction (Hopfield parameter) (eV/A^2) La: 3.248 H3: 2.109

La MUFFIN-TIN RADIUS and CHARGE = 2.0 52.3402
H MUFFIN-TIN RADIUS and CHARGE = 1.0 0.5187

LaH10 a=9.0 Bohr Clathrate

Fermi Energy (Ry)	Total DOS (States/Ry)	Decomposed DOS inside the muffin tins s	p	d	f
La 1.501	11.454	0.016	0.418	0.342	1.784
H1 1.501	11.454	0.170	0.095	0.007	0.0
H2 1.501	11.454	0.360	0.053	0.005	0.0

Fermi-Velocity (cm/s): 1.32995 x10E8
Plasmon Energy (eV) : 12.00345
Electron-ion interaction(Hopfield parameter) (eV/A^2) La:0.683 H1:1.196 H2:1.328

La MUFFIN-TIN RADIUS and CHARGE = 2.0 52.6534
H1 MUFFIN-TIN RADIUS and CHARGE = 1.0 0.5684
H2 MUFFIN-TIN RADIUS and CHARGE = 1.0 0.5838

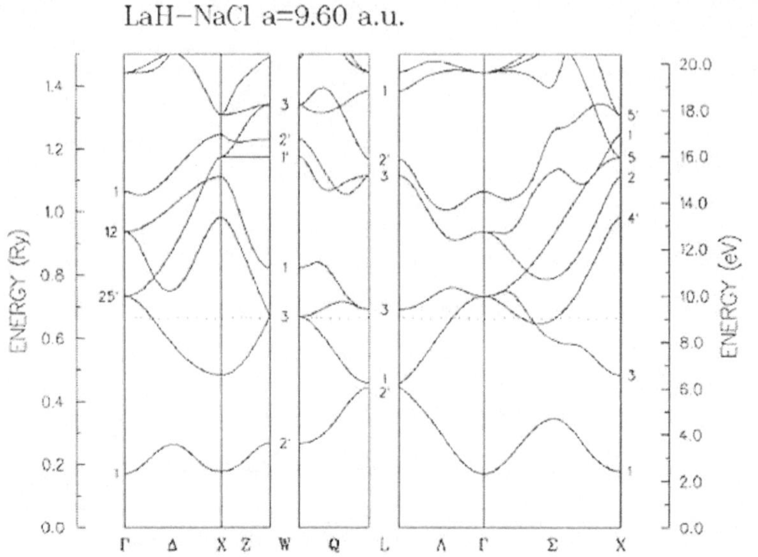

Fig. 5.14 Energy bands of LaH in the NaCl structure (tight-binding with f-states removed)

Fig. 5.15 Total, angular momentum and site decomposed densities of states of LaH in the NaCl structure (tight-binding with f-states removed)

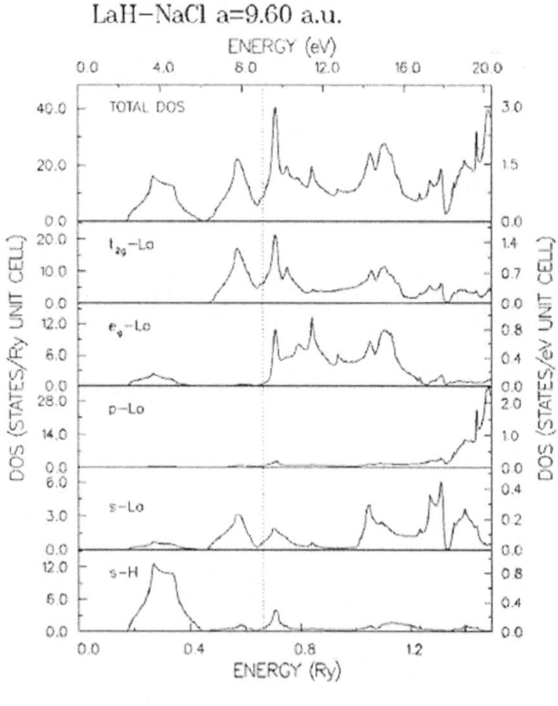

Table 5.14 LaH (NaCl) a = 9.60 Bohr Slater–Koster 3-center parameters

	ORTHOGONAL	NON-ORTHOGONAL	
	ENERGY INTEGRALS Ry	ENERGY INTEGRALS Ry	OVERLAP INTEGRALS
ON SITE La-La			
s,s(000)	1.05245	0.95537	
x,x(000)	1.41387	1.24481	
xy,xy(000)	0.98459	0.83278	
d2,d2(000)	0.91590	0.84072	
FIRST NEIGHBOR La-La			
s,s(110)	-0.04022	0.02363	0.04144
s,x(110)	-0.03963	-0.00971	0.02405
s,xy(110)	-0.05602	-0.00907	0.07886
s,d2(110)	0.01967	0.04550	0.03723
x,x(110)	0.02934	0.03272	-0.00214
x,x(011)	-0.01068	0.04249	0.04420
x,y(110)	0.05927	-0.02856	-0.05358
x,xy(110)	0.07638	-0.00795	-0.06863
x,xy(011)	-0.00519	0.01349	0.02813
z,d2(011)	0.00919	0.00962	0.03898
z,d1(011)	-0.01321	0.00280	0.01379
xy,xy(110)	-0.07117	-0.03673	0.04040
xy,xy(011)	0.01598	-0.00218	-0.03265
xy,xz(011)	-0.01326	-0.00445	-0.02959
xy,d2(110)	0.03287	-0.00924	0.03775
d2,d2(110)	-0.01183	0.00021	0.01892
d1,d1(110)	-0.00817	0.02777	0.00475
SECOND NEIGHBOR La-La			
s,s(200)	0.06502	0.01771	-0.03026
s,x(200)	0.07879	-0.04336	-0.07348
s,d2(002)	0.04434	0.00633	0.06290
x,x(200)	-0.10476	-0.05476	-0.02568
y,y(200)	0.01375	0.04422	0.02520
x,xy(020)	0.01365	0.01973	0.01172
z,d2(002)	-0.05337	-0.09923	-0.14191
xy,xy(200)	-0.02536	-0.00640	-0.00634
xy,xy(002)	0.00327	-0.00081	-0.00064
d2,d2(002)	0.04317	-0.04092	-0.06757
d1,d1(002)	0.00469	-0.00683	-0.00533
ON SITE H-H			
s,s(000)	0.45117	0.40878	
FIRST NEIGHBOR H-H			
s,s(110)	-0.00816	-0.00157	0.01322
s,x(110)	0.00000	0.00000	0.00000
x,x(110)	0.00000	0.00000	0.00000
x,x(011)	0.00000	0.00000	0.00000
x,y(110)	0.00000	0.00000	0.00000
SECOND NEIGHBOR H-H			
s,s(200)	-0.01217	-0.01179	-0.00759
s,x(200)	0.00000	0.00000	0.00000
x,x(200)	0.00000	0.00000	0.00000
y,y(200)	0.00000	0.00000	0.00000
FIRST NEIGHBOR La-H			
s,s(100)	-0.06410	0.05429	0.01546
s,x(100)	0.00000	0.00000	0.00000
x,s(100)	0.06202	0.12007	0.19860
x,x(100)	0.00000	0.00000	0.00000
y,y(100)	0.00000	0.00000	0.00000
xy,x(010)	0.00000	0.00000	0.00000
d2,s(001)	-0.09703	0.03572	-0.09609
d2,z(001)	0.00000	0.00000	0.00000
SECOND NEIGHBOR La-H			
s,s(111)	0.01241	0.01077	0.01127
x,s(111)	-0.01656	-0.00948	-0.01046
s,x(111)	0.00000	0.00000	0.00000
x,x(111)	0.00000	0.00000	0.00000
x,y(111)	0.00000	0.00000	0.00000
xy,s(111)	0.01084	0.00060	0.00411
xy,x(111)	0.00000	0.00000	0.00000
yz,x(111)	0.00000	0.00000	0.00000
d1,x(111)	0.00000	0.00000	0.00000

LaH

BAND	ORTHOGONAL RMS ERROR mRy	ORTHOGONAL MAXIMUM DEVIATION k	ORTHOGONAL MAXIMUM DEVIATION mRy	NON-ORTHOGONAL RMS ERROR mRy	NON-ORTHOGONAL MAXIMUM DEVIATION k	NON-ORTHOGONAL MAXIMUM DEVIATION mRy
1	3.7	(333)	8.0	3.1	(333)	8.7
2	4.5	(003)	11.3	2.8	(333)	10.0
3	3.6	(055)	8.0	1.8	(042)	5.9
4	5.1	(048)	12.9	1.0	(224)	2.8
5	5.1	(048)	13.2	1.3	(224)	3.7
6	5.1	(048)	10.5	1.2	(118)	3.3
7	6.1	(333)	15.1	1.8	(224)	4.0
1-7	4.8			2.0		

ENERGY VALUES IN Ry AT SELECTED k-POINTS

	ORTHOGONAL	APW	NON-ORTHOGONAL
GAMMA 1	0.17638	0.17057	0.17066
GAMMA 1	1.06386	1.06452	1.06396
GAMMA 12	0.93945	0.93887	0.93725
GAMMA 15	1.45129	1.44172	1.44221
GAMMA 25'	0.73285	0.73419	0.73380
X1 (008)	0.18158	0.17762	0.17919
X1 (008)	1.25621	1.24778	1.24597
X2 (008)	1.11414	1.11017	1.11000
X3 (008)	0.47715	0.48366	0.48426
X5 (008)	1.17440	1.17251	1.17333
X4' (008)	0.98193	0.98067	0.98150
X5' (008)	1.30206	1.30925	1.30891
L1 (444)	0.45587	0.45937	0.45762
L1 (444)	1.39202	1.41044	1.38429
L3 (444)	0.69426	0.69178	0.69205
L3 (444)	1.10450	1.11286	1.11350
L2' (444)	0.45228	0.44691	0.44389
L2' (444)	1.16614	1.16415	1.16543
W1 (048)	0.81100	0.82389	0.82269
W1 (048)	1.46651	1.40389	1.48826
W2' (048)	0.25862	0.26366	0.26713
W2' (048)	1.22159	1.23212	1.23132
W1' (048)	1.16130	1.17454	1.17363
W3 (048)	0.67120	0.67031	0.67018
W3 (048)	1.34303	1.33945	1.33998
ODD (224)	0.77178	0.76653	0.76935
ODD (224)	1.08091	1.07365	1.07187
EVEN(224)	0.33771	0.33542	0.33589
EVEN(224)	0.56827	0.56717	0.57317
EVEN(224)	0.69737	0.70200	0.70043
EVEN(224)	0.95385	0.95111	0.94740
EVEN(224)	1.12678	1.12165	1.12570

FERMI LEVEL QUANTITIES (Non-orthogonal fit)

ENERGY Ry		Total	s-La	p-La	t2g-La	eg-La	s-H	p-H
	DENSITIES OF STATES (states/Ry/cell)							
0.6631		8.56	0.65	0.75	6.16	0.43	0.57	0.00
	INTEGRATED DENSITIES OF STATES (electrons)							
		4.00	0.36	0.13	1.52	0.29	1.70	0.00

VELOCITY cm/s	PLASMON ENERGY eV	EIGENVALUE SUM Ry
0.63x10E8	4.49	-0.9057

Table 5.15 LaH (NaCl) a = 9.60 Bohr Slater–Koster 2-center parameters

	ORTHOGONAL	NON-ORTHOGONAL	
	ENERGY INTEGRALS Ry	ENERGY INTEGRALS Ry	OVERLAP INTEGRALS
ON SITE La-La			
s	1.03293	0.84252	
p	1.28107	1.10782	
t2g	0.89909	0.79094	
eg	0.86265	0.76380	
FIRST NEIGHBOR La-La			
(sss)	-0.03986	-0.02531	0.04429
(pps)	0.06885	-0.17264	-0.24316
(ppp)	-0.01240	0.03793	0.05948
(dds)	-0.09464	-0.00232	0.10404
(ddp)	0.02769	0.00972	-0.05597
(ddd)	0.00183	0.00165	0.00815
(sps)	0.06248	-0.05661	-0.15680
(sds)	-0.07780	0.00844	0.10526
(pds)	-0.10237	0.07212	0.16586
(pdp)	0.01388	-0.01927	-0.05801
SECOND NEIGHBOR La-La			
(sss)	-0.00874	0.00681	-0.01258
(pps)	-0.00010	0.13534	0.08208
(ppp)	-0.00525	0.03298	0.02790
(dds)	0.01342	-0.03430	-0.02890
(ddp)	-0.00354	-0.00529	-0.00482
(ddd)	0.00087	0.00478	0.00469
(sps)	0.00850	0.03505	0.03183
(sds)	0.01489	-0.01855	-0.01483
(pds)	0.01199	-0.07165	-0.05388
(pdp)	0.00061	-0.01904	-0.01942
ON SITE H-H			
s	0.75978	0.55653	
FIRST NEIGHBOR H-H			
(sss)	-0.01146	0.02704	-0.01308
(sps)	0.00000	0.00000	0.00000
(pps)	0.00000	0.00000	0.00000
(ppp)	0.00000	0.00000	0.00000
SECOND NEIGHBOR H-H			
(sss)	0.01878	-0.01201	-0.00672
(sps)	0.00000	0.00000	0.00000
(pps)	0.00000	0.00000	0.00000
(ppp)	0.00000	0.00000	0.00000
FIRST NEIGHBOR La-H			
(sss)	0.07600	0.09648	0.01566
(sps)	0.00000	0.00000	0.00000
(pss)	0.09427	0.04183	-0.12285
(pps)	0.00000	0.00000	0.00000
(ppp)	0.00000	0.00000	0.00000
(dpp)	0.00000	0.00000	0.00000
(dss)	0.12000	0.01394	-0.18085
(dps)	0.00000	0.00000	0.00000
SECOND NEIGHBOR La-H			
(sss)	-0.00383	-0.00722	0.02422
(pss)	-0.01363	0.01084	0.03553
(sps)	0.00000	0.00000	0.00000
(pps)	0.00000	0.00000	0.00000
(ppp)	0.00000	0.00000	0.00000
(dss)	-0.01981	0.01000	0.03350
(dps)	0.00000	0.00000	0.00000
(dpp)	0.00000	0.00000	0.00000

♠ LaH

BAND	ORTHOGONAL RMS ERROR mRy	ORTHOGONAL MAXIMUM DEVIATION k	ORTHOGONAL MAXIMUM DEVIATION mRy	NON-ORTHOGONAL RMS ERROR mRy	NON-ORTHOGONAL MAXIMUM DEVIATION k	NON-ORTHOGONAL MAXIMUM DEVIATION mRy
1	8.0	(005)	23.6	0.9	(048)	2.0
2	7.0	(004)	17.9	1.7	(022)	4.2
3	7.8	(003)	18.2	1.6	(003)	3.3
4	8.4	(444)	17.1	1.3	(022)	3.0
5	13.9	(226)	31.6	1.7	(055)	4.1
6	13.3	(066)	41.6	1.4	(118)	3.1
7	16.3	(226)	38.8	1.7	(222)	5.0
1-7	11.2			1.5		

ENERGY VALUES IN Ry AT SELECTED k-POINTS

	ORTHOGONAL	APW	NON-ORTHOGONAL
GAMMA 1	0.17757	0.17057	0.17041
GAMMA 1	1.05962	1.06452	1.06659
GAMMA 12	0.93796	0.93887	0.94033
GAMMA 15	1.43609	1.44172	1.44204
GAMMA 25'	0.72267	0.73419	0.73229
X1 (008)	0.18035	0.17762	0.17845
X1 (008)	1.23720	1.24778	1.24585
X2 (008)	1.09461	1.11017	1.10927
X3 (008)	0.48646	0.48366	0.48518
X5 (008)	1.16875	1.17251	1.17146
X4' (008)	0.98446	0.98067	0.97892
X5' (008)	1.30944	1.30925	1.30916
L1 (444)	0.45975	0.45937	0.46010
L1 (444)	1.43368	1.41044	1.38773
L3 (444)	0.67465	0.69178	0.69384
L3 (444)	1.10837	1.11286	1.11198
L2' (444)	0.44627	0.44691	0.44650
L2' (444)	1.17813	1.16415	1.16370
W1 (048)	0.81305	0.82389	0.82312
W1 (048)	1.30281	1.40389	1.41901
W2' (048)	0.26047	0.26366	0.26162
W2' (048)	1.23281	1.23212	1.23347
W1' (048)	1.16527	1.17454	1.17584
W3 (048)	0.67212	0.67031	0.67051
W3 (048)	1.33761	1.33945	1.33999
ODD (224)	0.76199	0.76653	0.76532
ODD (224)	1.07175	1.07365	1.07246
EVEN(224)	0.33694	0.33542	0.33442
EVEN(224)	0.56350	0.56717	0.56679
EVEN(224)	0.70106	0.70200	0.70151
EVEN(224)	0.97031	0.95111	0.95297
EVEN(224)	1.13078	1.12165	1.12282

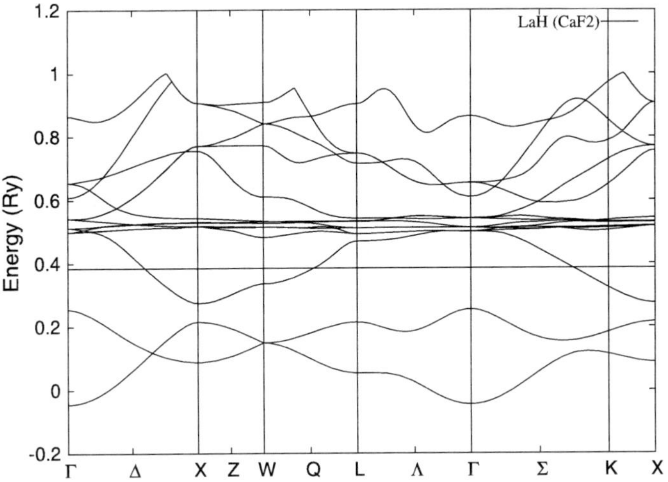

Fig. 5.16 Energy bands of LaH2 in the CaF2 structure

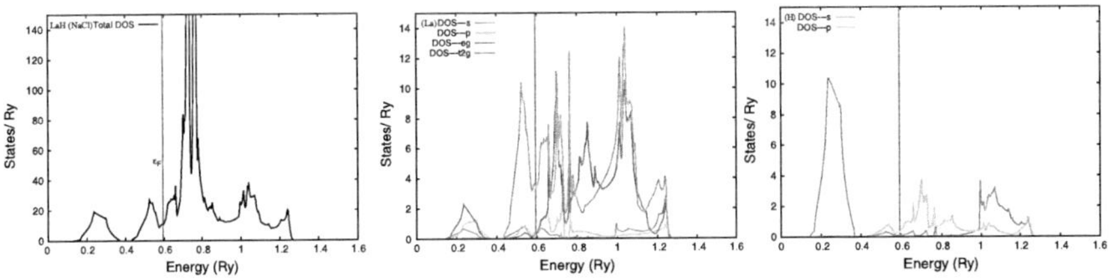

Fig. 5.17 Total, angular momentum and site decomposed densities of states of LaH2 in the CaF2 structure

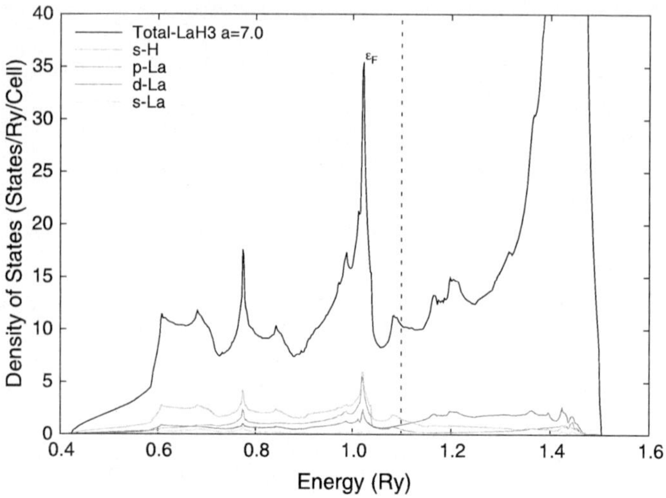

Fig. 5.18 Energy bands and DOS of LaH3

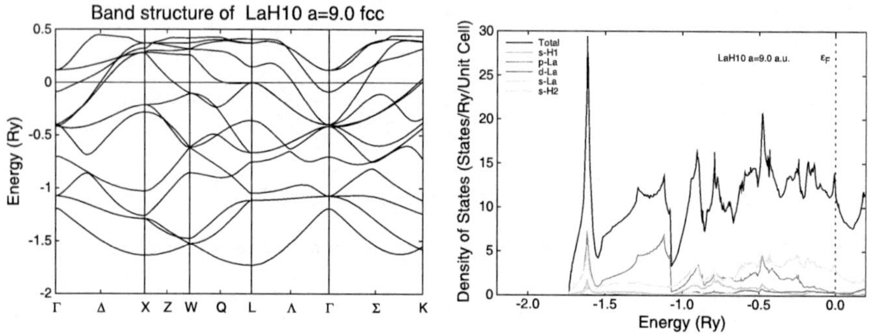

Fig. 5.19 Energy bands and DOS of LaH10

References

1. G.G. Libowitz, The nature and properties of transition metal hydrides. J. Nucl. Mater. **2**, 1 (1960)
2. H. Smithson, C.A. Marianetti, D. Morgan, A. Van der Ven, A. Predith, G. Ceder, First-principles study of the stability and electronic structure of metal hydrides. Phys. Rev. B **66**, 144107 (2002)
3. D.A. Papaconstantopoulos, M.J. Mehl, P.-H. Chang, High-temperature superconductivity in LaH10. Phys. Rev. B **101**, 060506(R) (2020)

Chapter 6
3D Transition-Metal Hydrides

This chapter covers the transition-metal hydrides from TiH to ZnH [1-3]. Results are presented for the crystal structures NaCl (B1) and CaF2 (C1). Most of these systems form as dihydrides in the CaF2 structure. However, in a few of them such as VH a body-centered tetragonal phase or/and hydrogen vacancies have been observed. The lattice parameters of these hydrides follow the trend of the corresponding elements that is they decrease from TiH to FeH and then increase from FeH to ZnH as expected because of the expansion introduced by hydrogen.

Comparing the energy bands of the CaF2 structure to the bands of the NaCl structure we note that a gap that appears between first and second band in the NaCl structure closes in the CaF2 structure and an additional antibonding band emerges above the Fermi level which is due to the second hydrogen in the CaF2 structure. Examining the densities of states figures we observe that for both crystal structures the lower occupied states have predominantly s-H hydrogen character and just below the Fermi level the d-t2g metal contribution dominates. However, the DOS details are significantly different for the two crystal structures. Moving along the 3d series the qualitative difference from TiH to CuH is the position of the Fermi level which is moving up and therefore changes the Fermi surface and the values of the DOS at the Fermi level. There is a crystal field splitting demonstrated by the positions of the points.

Gam25′ and Gam12 in the energy bands figures. It is important to mention that in ZnH the d-bands are positioned much deeper than in the other compounds. It is also noted that CuH in the NaCl structure is predicted to be a semi-metal while CoH is shown to be ferromagnetic [4]. Tight-binding parameters are given in the NaCl structure based on both orthogonal and non-orthogonal Hamiltonians using three- and two-center integrals.

D. A. Papaconstantopoulos, *Band Structure of Cubic Hydrides*,
https://doi.org/10.1007/978-3-031-06878-2_6

6.1 Titanium Hydride (TiH)

See Fig. 6.1 and Tables 6.1, 6.2, 6.3.

See Figs. 6.2, 6.3 and Tables 6.4, 6.5.

See Figs. 6.4 and 6.5.

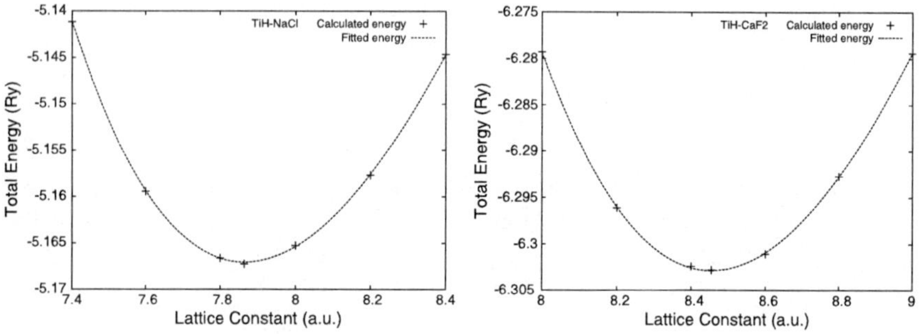

Fig. 6.1 Total energy versus lattice constant of TiH in the NaCl and CaF2 structures

Table 6.1 Lattice constant, bulk modulus, gap, total energy

Stru	a (Bohr)	B (MBar)	Gap	Total Energy
NaCl	7.86	1.62		-1705.16727
CaF2	8.45	1.48		-1706.30278
exp (CaF2)	8.43			

Table 6.2 Birch fit coefficients

	A1	A2	A3	A4
NaCl	-3.941896E+00	-5.324934E+01	3.995118E+02	4.157512E+03
CaF2	-4.704524E+00	-8.749349E+01	1.105489E+03	2.561289E+03

Table 6.3 DOS at Ef, Hopfield parameter, Stoner criterion

TiH a=7.86 Bohr NaCl

Fermi Energy (Ry)	Total DOS (States/Ry)	Decomposed DOS inside the muffin tins s	p	eg	t2g	f
Ti 0.810	21.878	0.209	0.715	4.025	7.527	0.047
H 0.810	21.878	0.142	1.678	0.014	0.149	0.041

Fermi-Velocity (cm/s): 0.38877 x10E8
Plasmon Energy (eV) : 5.93666
Electron-ion interaction (Hopfield parameter) (eV/A^2) Ti:5.738 H:0.207

Ti	MUFFIN-TIN RADIUS and CHARGE =	2.0645	19.6860
H	MUFFIN-TIN RADIUS and CHARGE =	1.8679	1.3696
Ti	STONER I = 0.0115		
H	STONER I = 0.0015		
	STONER PARAMETER (Ry) I = 0.0132		
	STONER CRITERION N*I = 0.2896		

TiH2 a=8.45 Bohr CaF2

Fermi Energy (Ry)	Total DOS (States/Ry)	Decomposed DOS inside the muffin tins s	p	eg	t2g	f
Ti 0.653	31.018	0.001	0.058	6.864	16.463	0.019
H 0.653	31.018	0.024	1.440	0.065	0.024	0.007

Fermi-Velocity (cm/s): 0.30510 x10E8
Plasmon Energy (eV) : 4.94429
Electron-ion interaction (Hopfield parameter) (eV/A^2) Ti: 1.215 H:0.176

Ti	MUFFIN-TIN RADIUS and CHARGE =	2.1965	19.9373
H	MUFFIN-TIN RADIUS and CHARGE =	1.4644	0.9292
Ti	STONER I = 0.015		
H	STONER I = 0.0002		
	STONER PARAMETER (Ry) I = 0.0152		
	STONER CRITERION N*I = 0.4706		

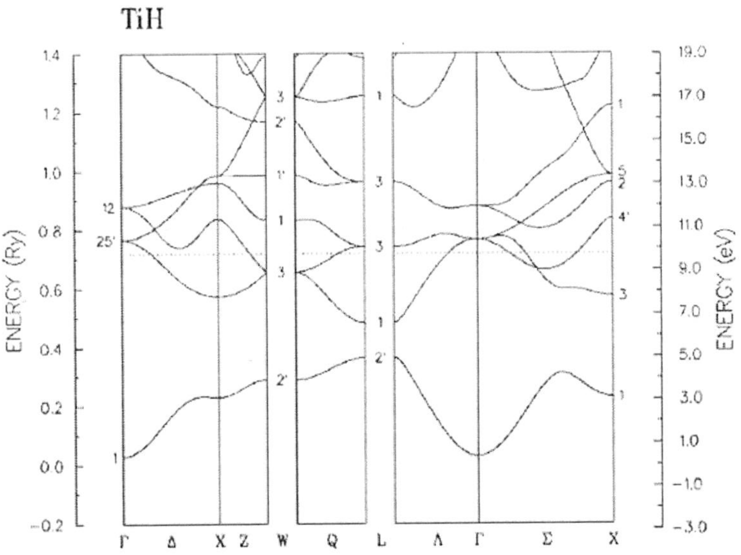

Fig. 6.2 Energy bands of TiH in the NaCl structure (tight-binding)

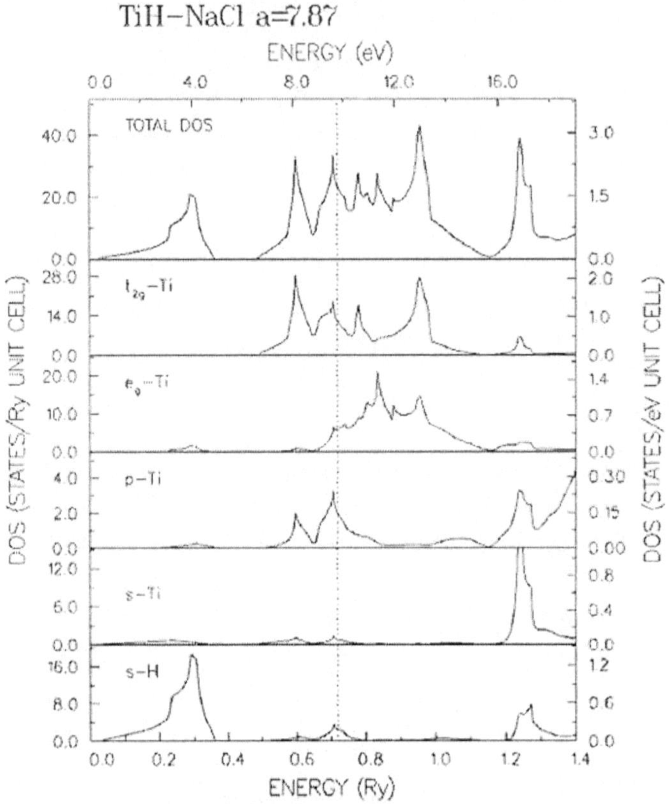

Fig. 6.3 Total, angular momentum and site decomposed densities of states of TiH in the NaCl structure (tight-binding)

Table 6.4 TiH (NaCl) a = 7.87 Bohr Slater–Koster 3-center parameters

	ORTHOGONAL	NON-ORTHOGONAL	
	ENERGY INTEGRALS Ry	ENERGY INTEGRALS Ry	OVERLAP INTEGRALS
ON SITE Ti-Ti			
s,s(000)	1.20648	1.01309	
x,x(000)	1.61866	1.25001	
xy,xy(000)	0.85600	0.80141	
d2,d2(000)	0.91468	0.93408	
FIRST NEIGHBOR Ti-Ti			
s,s(110)	-0.01135	0.04130	0.07014
s,x(110)	-0.04826	0.00988	0.09665
s,xy(110)	-0.05614	-0.03779	0.01349
s,d2(110)	-0.00468	0.07064	0.05147
x,x(110)	0.06850	0.06667	-0.00944
x,x(011)	-0.01108	0.04540	0.04042
x,y(110)	0.06201	-0.09438	-0.16023
x,xy(110)	0.06080	0.01448	-0.03228
x,xy(011)	-0.00515	-0.01285	0.01013
z,d2(011)	0.01767	-0.00313	0.02858
z,d1(011)	0.00428	-0.05540	-0.03880
xy,xy(110)	-0.03990	-0.02400	0.01745
xy,xy(011)	0.01203	0.01290	0.00058
xy,xz(011)	0.00434	0.02398	0.00508
xy,d2(110)	0.02109	-0.01443	0.01045
d2,d2(110)	0.00261	0.05470	0.05358
d1,d1(110)	-0.02659	-0.05672	-0.02528
SECOND NEIGHBOR Ti-Ti			
s,s(200)	0.05335	-0.04066	-0.03132
s,x(200)	0.06972	-0.04677	0.01576
s,d2(002)	0.03950	-0.00919	0.03600
x,x(200)	-0.12368	0.12413	0.05484
y,y(200)	0.01436	-0.02888	-0.02885
x,xy(020)	0.01028	-0.00376	-0.00650
z,d2(002)	-0.04168	-0.08730	-0.13157
xy,xy(200)	-0.00732	-0.00644	-0.01154
xy,xy(002)	0.00097	0.00993	0.01016
d2,d2(002)	0.03634	0.04170	0.00334
d1,d1(002)	0.00038	-0.01992	-0.01901
ON SITE H-H			
s,s(000)	0.39700	0.39877	
FIRST NEIGHBOR H-H			
s,s(110)	-0.01962	-0.00260	0.00836
s,x(110)	0.00000	0.00000	0.00000
x,x(110)	0.00000	0.00000	0.00000
x,x(011)	0.00000	0.00000	0.00000
x,y(110)	0.00000	0.00000	0.00000
SECOND NEIGHBOR H-H			
s,s(200)	-0.00589	-0.00513	-0.00367
s,x(200)	0.00000	0.00000	0.00000
x,x(200)	0.00000	0.00000	0.00000
y,y(200)	0.00000	0.00000	0.00000
FIRST NEIGHBOR Ti-H			
s,s(100)	-0.10060	0.09373	-0.07350
s,x(100)	0.00000	0.00000	0.00000
x,s(100)	0.07741	0.02086	0.25083
x,x(100)	0.00000	0.00000	0.00000
y,y(100)	0.00000	0.00000	0.00000
xy,x(010)	0.00000	0.00000	0.00000
d2,s(001)	-0.06029	-0.14266	-0.14809
d2,z(001)	0.00000	0.00000	0.00000

SECOND NEIGHBOR Ti-H

s,s(111)	0.03040	0.00436	0.00063
x,s(111)	-0.03926	0.00859	0.02054
s,x(111)	0.00000	0.00000	0.00000
x,x(111)	0.00000	0.00000	0.00000
x,y(111)	0.00000	0.00000	0.00000
xy,s(111)	0.00438	0.00669	0.01255
xy,x(111)	0.00000	0.00000	0.00000
yz,x(111)	0.00000	0.00000	0.00000
d1,x(111)	0.00000	0.00000	0.00000

↑TiH

	ORTHOGONAL			NON-ORTHOGONAL		
BAND	RMS ERROR	MAXIMUM DEVIATION		RMS ERROR	MAXIMUM DEVIATION	
	mRy	k	mRy	mRy	k	mRy
1	4.7	(005)	11.2	1.4	(044)	3.8
2	4.6	(222)	14.8	0.8	(003)	2.1
3	3.7	(264)	8.5	0.7	(044)	1.6
4	2.5	(006)	5.3	0.8	(008)	1.6
5	3.0	(044)	6.9	0.7	(066)	1.8
6	3.3	(055)	8.1	1.2	(066)	3.6
7	6.2	(000)	15.0	2.5	(062)	6.2
1-7	4.2			1.3		

ENERGY VALUES IN Ry AT SELECTED k-POINTS

	ORTHOGONAL	APW	NON-ORTHOGONAL
GAMMA 1	0.03070	0.02997	0.02888
GAMMA 1	1.48586	1.50090	1.50347
GAMMA 12	0.88093	0.88067	0.87933
GAMMA 15	1.93245	1.94103	1.94155
GAMMA 25'	0.76532	0.76642	0.76551
X1 (008)	0.23722	0.23053	0.22904
X1 (008)	1.21423	1.21734	1.22038
X1 (008)	1.75025	1.75616	1.75584
X2 (008)	0.95600	0.95843	0.96008
X3 (008)	0.57283	0.57291	0.57373
X5 (008)	0.98826	0.98575	0.98576
X4' (008)	0.83642	0.83752	0.83719
X5' (008)	1.47308	1.44453	1.44264
L1 (444)	0.47809	0.48434	0.48436
L1 (444)	1.25692	1.25289	1.25339
L3 (444)	0.73866	0.74137	0.74141
L3 (444)	0.96658	0.96142	0.96240
L2' (444)	0.35706	0.36430	0.36418
L2' (444)	1.38775	1.38608	1.38614
L3' (444)	2.05661	2.02770	2.02783
W1 (048)	0.83219	0.83312	0.83313
W1 (048)	1.39543	1.39154	1.39074
W2' (048)	0.28588	0.28861	0.28832
W2' (048)	1.16477	1.16636	1.16758
W2' (048)	2.11051	2.16429	2.16422
W1' (048)	0.98437	0.98597	0.98618
W3 (048)	0.65311	0.65601	0.65616
W3 (048)	1.25401	1.25216	1.25201
ODD (224)	0.78750	0.78549	0.78706
ODD (224)	0.94955	0.94531	0.94423
EVEN(224)	0.24421	0.23833	0.23849
EVEN(224)	0.61558	0.60995	0.60988
EVEN(224)	0.72937	0.73197	0.73069
EVEN(224)	0.90592	0.90626	0.90741
EVEN(224)	1.24138	1.24145	1.24473

FERMI LEVEL QUANTITIES (Non-orthogonal fit)

ENERGY		DENSITIES OF STATES					
	Total	s-Ti	p-Ti	t2g-Ti	eg-Ti	s-H	p-H
Ry		states/Ry/cell					
0.7173	23.77	0.89	2.04	12.44	5.86	2.54	0.00
		INTEGRATED DENSITIES OF STATES					
	Total	s-Ti	p-Ti	t2g-Ti	eg-Ti	s-H	p-H
		electrons					
	5.00	0.25	0.22	2.32	0.33	1.88	0.00

VELOCITY	PLASMON ENERGY	EIGENVALUE SUM
cm/s	eV	Ry
0.38x10E8	6.08	-1.1625

Table 6.5 TiH (NaCl) a = 7.87 Bohr Slater–Koster 2-center parameters

	ORTHOGONAL	NON-ORTHOGONAL	
	ENERGY INTEGRALS Ry	ENERGY INTEGRALS Ry	OVERLAP INTEGRALS
ON SITE Ti-Ti			
s	1.05057	0.58712	
p	1.57184	1.11317	
t2g	0.83102	0.79344	
eg	0.81763	0.78732	
FIRST NEIGHBOR Ti-Ti			
(sss)	-0.07237	0.06335	0.19235
(pps)	0.15030	-0.14491	-0.26432
(ppp)	-0.01355	-0.03765	0.02488
(dds)	-0.05307	-0.01171	0.04979
(ddp)	0.02415	0.02132	-0.01775
(ddd)	-0.00078	-0.00605	-0.00075
(sps)	0.11232	-0.10252	-0.22654
(sds)	-0.06137	0.01025	0.08491
(pds)	-0.08634	-0.02821	0.04797
(pdp)	0.01356	0.02796	-0.01107
SECOND NEIGHBOR Ti-Ti			
(sss)	0.00814	-0.02228	0.00378
(pps)	-0.08661	0.11164	0.02371
(ppp)	0.00920	-0.05163	-0.01520
(dds)	0.00089	-0.00819	-0.00154
(ddp)	-0.00078	0.00253	0.00108
(ddd)	0.00024	0.00475	0.00540
(sps)	-0.02897	0.05713	-0.00303
(sds)	0.00496	-0.01597	0.00115
(pds)	0.00533	-0.00949	0.01258
(pdp)	-0.00560	0.01253	0.00376
ON SITE H-H			
s	0.94672	0.57470	

FIRST NEIGHBOR H-H			
(sss)	0.00830	0.02893	-0.03673
(sps)	0.00000	0.00000	0.00000
(pps)	0.00000	0.00000	0.00000
(ppp)	0.00000	0.00000	0.00000
SECOND NEIGHBOR H-H			
(sss)	0.03993	-0.01845	0.01383
(sps)	0.00000	0.00000	0.00000
(pps)	0.00000	0.00000	0.00000
(ppp)	0.00000	0.00000	0.00000
FIRST NEIGHBOR Ti-H			
(sss)	0.12093	0.12122	-0.02236
(sps)	0.00000	0.00000	0.00000
(pss)	0.13896	0.10296	-0.12005
(pps)	0.00000	0.00000	0.00000
(ppp)	0.00000	0.00000	0.00000
(dpp)	0.00000	0.00000	0.00000
(dss)	0.10483	-0.06124	-0.24925
(dps)	0.00000	0.00000	0.00000
SECOND NEIGHBOR Ti-H			
(sss)	-0.02720	0.02686	0.03194
(pss)	-0.05657	0.03314	0.04729
(sps)	0.00000	0.00000	0.00000
(pps)	0.00000	0.00000	0.00000
(ppp)	0.00000	0.00000	0.00000
(dss)	-0.00779	0.00949	0.03108
(dps)	0.00000	0.00000	0.00000
(dpp)	0.00000	0.00000	0.00000

♠TiH

	ORTHOGONAL			NON-ORTHOGONAL		
BAND	RMS ERROR	MAXIMUM DEVIATION		RMS ERROR	MAXIMUM DEVIATION	
	mRy	k	mRy	mRy	k	mRy
1	3.5	(004)	8.3	1.5	(044)	4.0
2	7.1	(333)	21.7	2.1	(444)	6.5
3	5.9	(033)	15.4	1.9	(003)	4.5
4	4.1	(444)	8.7	1.6	(066)	4.2
5	3.2	(003)	6.8	1.1	(002)	3.0
6	5.7	(055)	14.4	1.7	(048)	5.4
7	8.6	(004)	18.7	4.1	(280)	11.3
1-7	5.8			2.2		

ENERGY VALUES IN Ry AT SELECTED k-POINTS

	ORTHOGONAL	APW	NON-ORTHOGONAL
GAMMA 1	0.02603	0.02997	0.03032
GAMMA 1	1.49072	1.50090	1.50077
GAMMA 12	0.88279	0.88067	0.88091
GAMMA 15	1.92824	1.94103	1.94097
GAMMA 25'	0.76186	0.76642	0.76435
X1 (008)	0.23358	0.23053	0.22962
X1 (008)	1.20684	1.21734	1.21462
X1 (008)	1.74994	1.75616	1.75476
X2 (008)	0.95242	0.95843	0.95547
X3 (008)	0.57489	0.57291	0.57429
X5 (008)	0.98838	0.98575	0.98591
X4' (008)	0.83422	0.83752	0.83644

X5' (008)	1.48960	1.44453	1.44351
L1 (444)	0.47606	0.48434	0.49083
L1 (444)	1.25963	1.25289	1.25775
L3 (444)	0.73267	0.74137	0.73964
L3 (444)	0.96510	0.96142	0.96224
L2' (444)	0.36855	0.36430	0.36427
L2' (444)	1.39149	1.38608	1.38731
L3' (444)	2.03595	2.02770	2.02733
W1 (048)	0.83397	0.83312	0.83318
W1 (048)	1.39524	1.39154	1.38823
W2' (048)	0.28983	0.28861	0.28718
W2' (048)	1.16059	1.16636	1.16093
W2' (048)	2.10245	2.16429	2.16360
W1' (048)	0.98742	0.98597	0.98860
W3 (048)	0.65265	0.65601	0.65510
W3 (048)	1.25722	1.25216	1.25087
ODD (224)	0.78237	0.78549	0.78645
ODD (224)	0.94787	0.94531	0.94476
EVEN(224)	0.23738	0.23833	0.23852
EVEN(224)	0.62084	0.60995	0.60896
EVEN(224)	0.73407	0.73197	0.73308
EVEN(224)	0.91152	0.90626	0.90449
EVEN(224)	1.24092	1.24145	1.24020

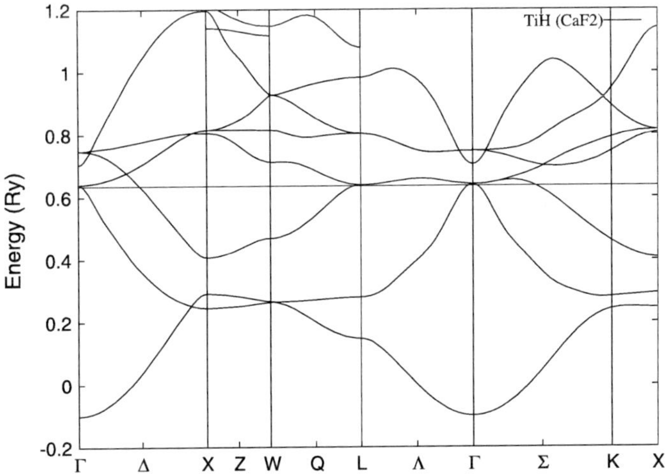

Fig. 6.4 Energy bands of TiH2 in the CaF2 structure

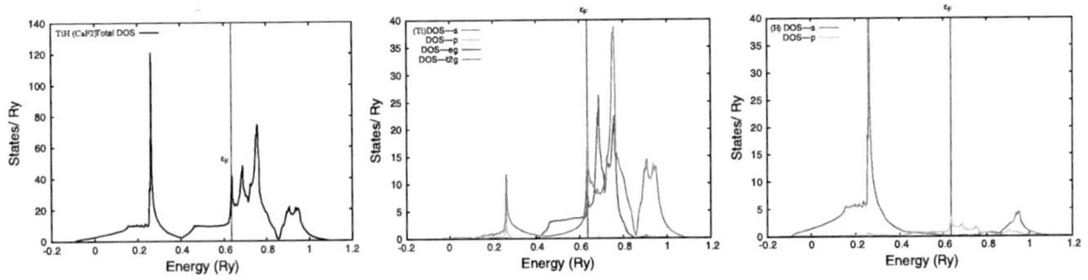

Fig. 6.5 Total, angular momentum and site decomposed densities of states of TiH2 in the CaF2 structure

6.2 Vanadium Hydride (VH)

See Figs. 6.6 and Tables 6.6, 6.7, 6.8.

See Figs. 6.7, 6.8 and Tables 6.9, 6.10.

See Figs. 6.9 and 6.10.

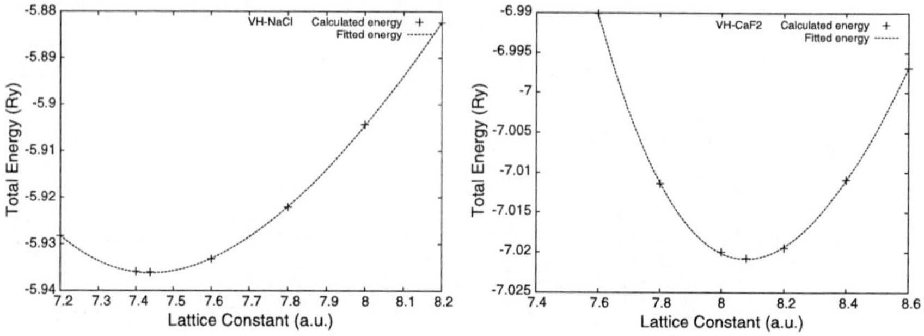

Fig. 6.6 Total energy versus lattice constant of VH in the NaCl and CaF2 structures

Table 6.6 Lattice constant, bulk modulus, gap, total energy

Stru	a (Bohr)	B (MBar)	Gap	Total Energy
NaCl	7.44	2.23		-1895.93606
CaF2	8.08	1.76		-1897.02073
exp (CaF2)	8.07			

Table 6.7 Birch fit coefficients

	A1	A2	A3	A4
NaCl	-4.387866E+00	-6.344222E+01	5.465853E+02	2.194289E+03
CaF2	-5.619660E+00	-6.291506E+01	4.393116E+02	6.481977E+03

Table 6.8 DOS at Ef, Hopfield parameter, Stoner criterion

VH a=7.44 Bohr NaCl

Fermi Energy (Ry)	Total DOS (States/Ry)	Decomposed DOS inside the muffin tins s	p	eg	t2g	f
V 0.905	20.981	0.042	0.233	6.188	8.897	0.019
H 0.905	20.981	0.042	1.049	0.048	0.132	0.044

Fermi-Velocity (cm/s): 0.31586 x10E8
Plasmon Energy (eV) : 5.13363
Electron-ion interaction (Hopfield parameter) (eV/A^2) V:3.147 H:0.056

V MUFFIN-TIN RADIUS and CHARGE = 1.9525 20.5524
H MUFFIN-TIN RADIUS and CHARGE = 1.7666 1.3637
V STONER I = 0.0197
H STONER I = 0.0007
STONER PARAMETER (Ry) I = 0.0206
STONER CRITERION N*I = 0.4322

VH2 a=8.08 Bohr CaF2

Fermi Energy (Ry)	Total DOS (States/Ry)	Decomposed DOS inside the muffin tins s	p	eg	t2g	f
V 0.711	25.738	0.014	0.111	14.647	6.020	0.029
H 0.711	25.738	0.041	0.593	0.054	0.029	0.018

Fermi-Velocity (cm/s): 0.26723 x10E8
Plasmon Energy (eV) : 4.25144
Electron-ion interaction (Hopfield parameter) (eV/A^2) V:2.565 H:0.147

V MUFFIN-TIN RADIUS and CHARGE = 2.0989 20.9226
H MUFFIN-TIN RADIUS and CHARGE = 1.3992 0.8897
V STONER I = 0.0188
H STONER I = 0.0001
STONER PARAMETER (Ry) I = 0.0188
STONER CRITERION N*I = 0.4842

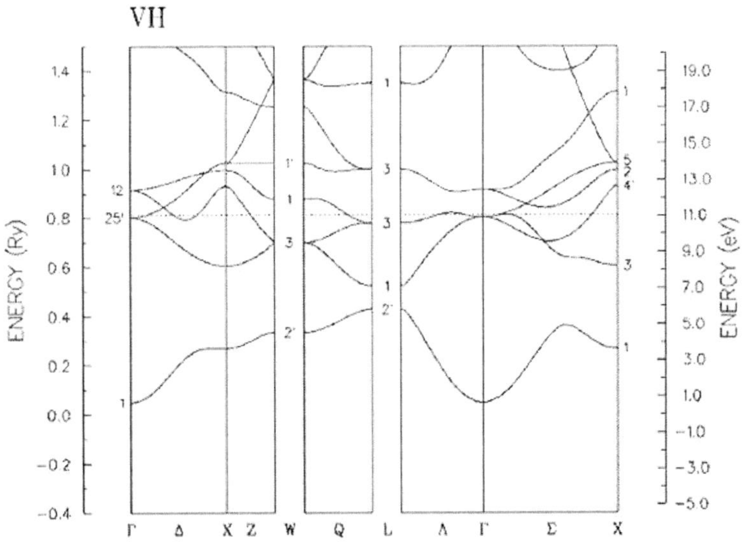

Fig. 6.7 Energy bands of VH in the NaCl structure (tight-binding)

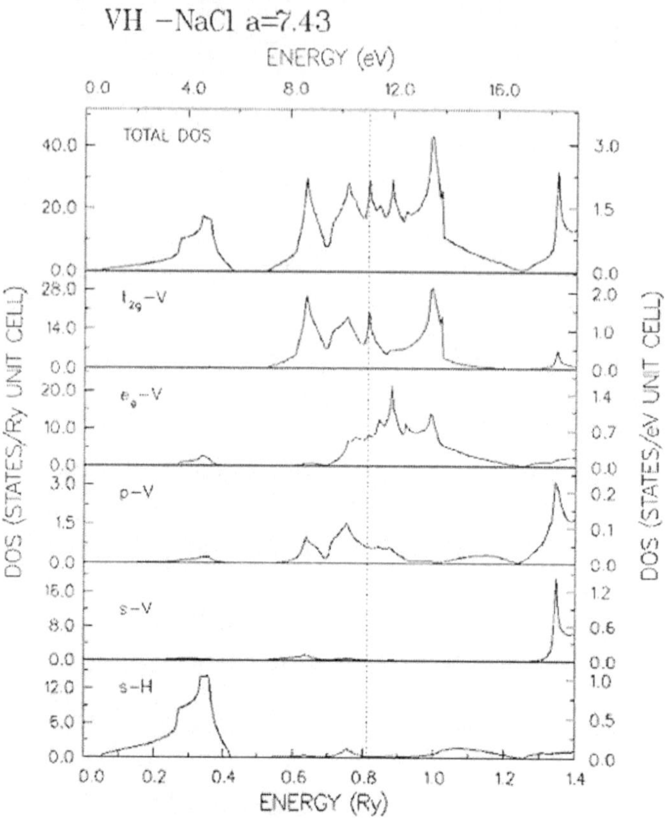

Fig. 6.8 Total, angular momentum and site decomposed densities of states of VH in the NaCl structure (tight-binding)

Table 6.9 VH (NaCl) a = 7.43 a.u. Slater–Koster 3-center parameters

	ORTHOGONAL	NON-ORTHOGONA	
	ENERGY INTEGRALS Ry	ENERGY INTEGRALS Ry	OVERLAP INTEGRALS
ON SITE V- V			
s,s(000)	1.35670	1.25479	
x,x(000)	1.81397	1.54406	
xy,xy(000)	0.89132	0.83171	
d2,d2(000)	0.96342	0.82207	
FIRST NEIGHBOR V- V			
s,s(110)	-0.00983	0.08997	0.08439
s,x(110)	-0.05512	0.05058	0.05377
s,xy(110)	-0.05991	-0.04128	0.01570
s,d2(110)	-0.00865	0.01032	0.02333
x,x(110)	0.07770	0.02833	-0.04789
x,x(011)	-0.01098	0.07692	0.07227
x,y(110)	0.06800	-0.12120	-0.12214
x,xy(110)	0.06422	0.03251	-0.00860
x,xy(011)	-0.00546	0.00974	0.03945
z,d2(011)	0.01940	0.04598	0.08837
z,d1(011)	0.00777	0.01088	0.02713
xy,xy(110)	-0.04045	-0.02581	0.01594
xy,xy(011)	0.01226	0.00647	-0.00880
xy,xz(011)	0.00478	0.00885	-0.01368
xy,d2(110)	0.02155	-0.00933	0.01558
d2,d2(110)	0.00444	0.00931	0.03041
d1,d1(110)	-0.03201	0.05144	0.01136
SECOND NEIGHBOR V- V			
s,s(200)	0.06127	-0.00083	-0.01912
s,x(200)	0.07778	-0.01868	-0.05507
s,d2(002)	0.04285	0.00814	0.02169
x,x(200)	-0.14310	0.01538	0.01558
y,y(200)	0.01703	-0.01841	-0.00119
x,xy(020)	0.01101	-0.00234	0.00476
z,d2(002)	-0.04183	-0.05358	-0.08972
xy,xy(200)	-0.00701	-0.00464	-0.01036
xy,xy(002)	0.00086	0.00381	0.00369
d2,d2(002)	0.03979	-0.04026	-0.03160
d1,d1(002)	0.00024	0.00053	0.00253
ON SITE H-H			
s,s(000)	0.46810	0.37839	
FIRST NEIGHBOR H-H			
s,s(110)	-0.02198	-0.02102	-0.00096
s,x(110)	0.00000	0.00000	0.00000
x,x(110)	0.00000	0.00000	0.00000
x,x(011)	0.00000	0.00000	0.00000
x,y(110)	0.00000	0.00000	0.00000
SECOND NEIGHBOR H-H			
s,s(200)	-0.00698	-0.00952	0.00315
s,x(200)	0.00000	0.00000	0.00000
x,x(200)	0.00000	0.00000	0.00000
y,y(200)	0.00000	0.00000	0.00000
FIRST NEIGHBOR V-H			
s,s(100)	-0.11536	-0.05407	0.06706
s,x(100)	0.00000	0.00000	0.00000
x,s(100)	0.08811	0.14522	0.23076
x,x(100)	0.00000	0.00000	0.00000
y,y(100)	0.00000	0.00000	0.00000
xy,x(010)	0.00000	0.00000	0.00000
d2,s(001)	-0.07122	-0.02067	-0.17917
d2,z(001)	0.00000	0.00000	0.00000

SECOND NEIGHBOR V-H

s,s(111)	0.03392	0.01834	0.02869
x,s(111)	-0.04445	0.00147	-0.00425
s,x(111)	0.00000	0.00000	0.00000
x,x(111)	0.00000	0.00000	0.00000
x,y(111)	0.00000	0.00000	0.00000
xy,s(111)	0.00378	-0.00873	-0.00166
xy,x(111)	0.00000	0.00000	0.00000
yz,x(111)	0.00000	0.00000	0.00000
d1,x(111)	0.00000	0.00000	0.00000

♠ VH

	ORTHOGONAL			NON-ORTHOGONAL		
BAND	RMS ERROR	MAXIMUM DEVIATION		RMS ERROR	MAXIMUM DEVIATION	
	mRy	k	mRy	mRy	k	mRy
1	5.2	(005)	12.9	0.8	(044)	2.5
2	5.1	(222)	16.1	0.5	(005)	1.0
3	4.4	(264)	10.3	0.6	(007)	1.6
4	2.7	(006)	5.0	0.4	(048)	0.8
5	3.2	(044)	7.9	0.4	(444)	1.0
6	3.4	(055)	9.3	0.6	(174)	1.9
7	7.5	(000)	17.4	1.1	(224)	2.8
1-7	4.8			0.7		

ENERGY VALUES IN Ry AT SELECTED k-POINTS

	ORTHOGONAL	APW	NON-ORTHOGONAL
GAMMA 1	0.04880	0.04835	0.04912
GAMMA 1	1.72000	1.73744	1.73725
GAMMA 12	0.91814	0.91732	0.91744
GAMMA 15	2.17353	2.18326	2.18329
GAMMA 25'	0.80126	0.80216	0.80192
X1 (008)	0.27657	0.26858	0.26891
X1 (008)	1.30645	1.31032	1.31255
X1 (008)	1.97922	1.98702	1.98783
X2 (008)	0.99285	0.99604	0.99577
X3 (008)	0.60513	0.60460	0.60513
X5 (008)	1.02679	1.02395	1.02418
X4' (008)	0.93036	0.93193	0.93148
X5' (008)	1.63982	1.61726	1.61867
L1 (444)	0.51563	0.52302	0.52305
L1 (444)	1.35291	1.34868	1.34829
L3 (444)	0.77457	0.77794	0.77815
L3 (444)	1.00548	0.99949	1.00050
L2' (444)	0.42252	0.42949	0.42867
L2' (444)	1.57546	1.57430	1.57434
L3' (444)	2.30408	2.27230	2.27225
W1 (048)	0.87739	0.87767	0.87848
W1 (048)	1.54774	1.54416	1.54427
W2' (048)	0.33177	0.33454	0.33459
W2' (048)	1.25147	1.25278	1.25231
W2' (048)	2.38162	2.43941	2.43959
W1' (048)	1.02336	1.02417	1.02405
W3 (048)	0.69513	0.69930	0.69970
W3 (048)	1.36585	1.36277	1.36299
ODD (224)	0.82616	0.82357	0.82297
ODD (224)	0.98736	0.98351	0.98313
EVEN(224)	0.28777	0.28173	0.28213
EVEN(224)	0.65980	0.65330	0.65352
EVEN(224)	0.76862	0.77202	0.77328
EVEN(224)	0.94866	0.94885	0.94818
EVEN(224)	1.38193	1.38043	1.38321

FERMI LEVEL QUANTITIES (Non-orthogonal fit)

ENERGY	DENSITIES OF STATES						
	Total	s- V	p- V	t2g- V	eg- V	s-H	p-H
Ry	states/Ry/cell						
0.8117	24.43	0.13	0.63	15.62	7.76	0.28	0.00
	INTEGRATED DENSITIES OF STATES						
	Total	s- V	p- V	t2g- V	eg- V	s-H	p-H
	electrons						
	6.00	0.18	0.19	3.01	0.77	1.85	0.00

VELOCITY	PLASMON ENERGY	EIGENVALUE SUM
cm/s	eV	Ry
0.31x10E8	5.45	-1.4476

Table 6.10 VH (NaCl) a = 7.43 Bohr Slater–Koster 2-center parameters

	ORTHOGONAL	NON-ORTHOGONAL	
	ENERGY INTEGRALS Ry	ENERGY INTEGRALS Ry	OVERLAP INTEGRALS
ON SITE V- V			
s	1.16996	0.81866	
p	1.76571	1.33350	
t2g	0.86614	0.82322	
eg	0.85382	0.81681	
FIRST NEIGHBOR V- V			
(sss)	-0.08108	-0.02112	0.09892
(pps)	0.16736	-0.05818	-0.19847
(ppp)	-0.01242	0.03344	0.05443
(dds)	-0.05369	-0.00792	0.05280
(ddp)	0.02508	0.01409	-0.02895
(ddd)	-0.00114	-0.00276	0.00401
(sps)	0.12629	-0.01613	-0.14591
(sds)	-0.06395	-0.01642	0.05958
(pds)	-0.08994	-0.00294	0.07371
(pdp)	0.01448	0.00810	-0.03279
SECOND NEIGHBOR V- V			
(sss)	0.01018	-0.05744	-0.03097
(pps)	-0.10554	0.17634	0.08686
(ppp)	0.01095	-0.04837	-0.02892
(dds)	0.00015	-0.01577	-0.00649
(ddp)	-0.00041	0.00341	0.00140
(ddd)	0.00016	0.00287	0.00294
(sps)	-0.03370	0.10298	0.05417
(sds)	0.00392	-0.01534	-0.00292
(pds)	0.00438	-0.02708	-0.00587
(pdp)	-0.00632	0.01433	0.01054
ON SITE H-H			
s	1.08454	0.45831	

FIRST NEIGHBOR H-H			
(sss)	0.01172	-0.01061	0.01205
(sps)	0.00000	0.00000	0.00000
(pps)	0.00000	0.00000	0.00000
(ppp)	0.00000	0.00000	0.00000
SECOND NEIGHBOR H-H			
(sss)	0.04779	-0.00798	-0.00528
(sps)	0.00000	0.00000	0.00000
(pps)	0.00000	0.00000	0.00000
(ppp)	0.00000	0.00000	0.00000
FIRST NEIGHBOR V-H			
(sss)	0.13556	0.01489	-0.19685
(sps)	0.00000	0.00000	0.00000
(pss)	0.15285	-0.05856	-0.27739
(pps)	0.00000	0.00000	0.00000
(ppp)	0.00000	0.00000	0.00000
(dpp)	0.00000	0.00000	0.00000
(dss)	0.10911	0.01625	-0.14478
(dps)	0.00000	0.00000	0.00000
SECOND NEIGHBOR V-H			
(sss)	-0.03167	0.00233	-0.01137
(pss)	-0.06620	0.00266	-0.02015
(sps)	0.00000	0.00000	0.00000
(pps)	0.00000	0.00000	0.00000
(ppp)	0.00000	0.00000	0.00000
(dss)	-0.00733	0.00868	-0.00541
(dps)	0.00000	0.00000	0.00000
(dpp)	0.00000	0.00000	0.00000

♀ VH

	ORTHOGONAL			NON-ORTHOGONAL		
BAND	RMS ERROR	MAXIMUM DEVIATION		RMS ERROR	MAXIMUM DEVIATION	
	mRy	k	mRy	mRy	k	mRy
1	4.3	(226)	9.5	0.4	(044)	1.6
2	7.6	(333)	26.5	0.5	(055)	1.3
3	6.0	(033)	13.8	0.4	(033)	1.3
4	3.8	(444)	7.9	0.4	(033)	1.0
5	3.1	(003)	7.0	0.6	(055)	1.6
6	5.5	(055)	14.8	0.5	(224)	1.2
7	10.2	(004)	22.7	0.8	(333)	2.6
1-7	6.2			0.5		

ENERGY VALUES IN Ry AT SELECTED k-POINTS

	ORTHOGONAL	APW	NON-ORTHOGONAL
GAMMA 1	0.04435	0.04835	0.04801
GAMMA 1	1.72567	1.73744	1.73595
GAMMA 12	0.91954	0.91732	0.91830
GAMMA 15	2.16854	2.18326	2.18330
GAMMA 25'	0.79832	0.80216	0.80195
X1 (008)	0.27417	0.26858	0.26832
X1 (008)	1.29933	1.31032	1.31075
X1 (008)	1.98201	1.98702	1.98685
X2 (008)	0.99060	0.99604	0.99561
X3 (008)	0.60684	0.60460	0.60516
X5 (008)	1.02703	1.02395	1.02361
X4' (008)	0.92898	0.93193	0.93168

X5' (008)	1.64811	1.61726	1.61720
L1 (444)	0.51478	0.52302	0.52326
L1 (444)	1.35670	1.34868	1.34812
L3 (444)	0.77000	0.77794	0.77763
L3 (444)	1.00281	0.99949	0.99952
L2' (444)	0.43692	0.42949	0.42968
L2' (444)	1.57470	1.57430	1.57426
L3' (444)	2.29255	2.27230	2.27229
W1 (048)	0.87885	0.87767	0.87704
W1 (048)	1.54708	1.54416	1.54435
W2' (048)	0.33518	0.33454	0.33448
W2' (048)	1.24703	1.25278	1.25315
W2' (048)	2.37509	2.43941	2.43940
W1' (048)	1.02640	1.02417	1.02427
W3 (048)	0.69452	0.69930	0.69887
W3 (048)	1.36882	1.36277	1.36262
ODD (224)	0.82055	0.82357	0.82314
ODD (224)	0.98574	0.98351	0.98233
EVEN(224)	0.28006	0.28173	0.28164
EVEN(224)	0.66469	0.65330	0.65271
EVEN(224)	0.77426	0.77202	0.77267
EVEN(224)	0.95412	0.94885	0.94905
EVEN(224)	1.38128	1.38043	1.38032

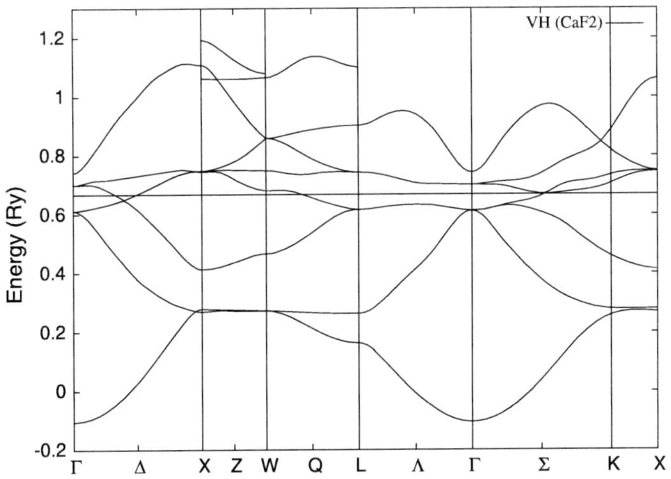

Fig. 6.9 Energy bands of VH2 in the CaF2 structure

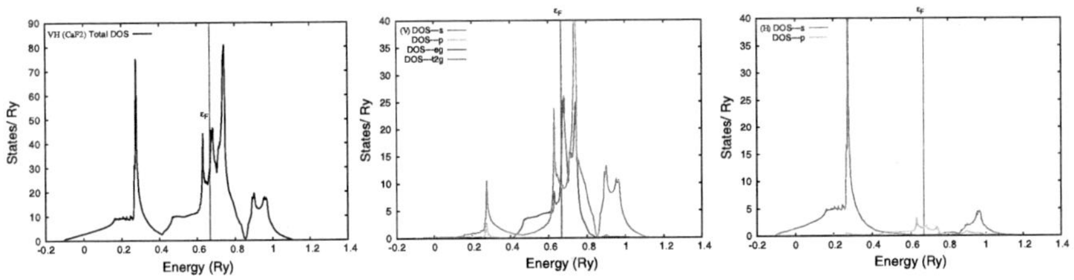

Fig. 6.10 Total, angular momentum and site decomposed densities of states of VH2 in the CaF2 structure

6.3 Chromium Hydride (CrH)

See Fig. 6.11 and Tables 6.11, 6.12, 6.13.

See Figs. 6.12, 6.13 and Tables 6.14, 6.15.

See Figs. 6.14 and 6.15.

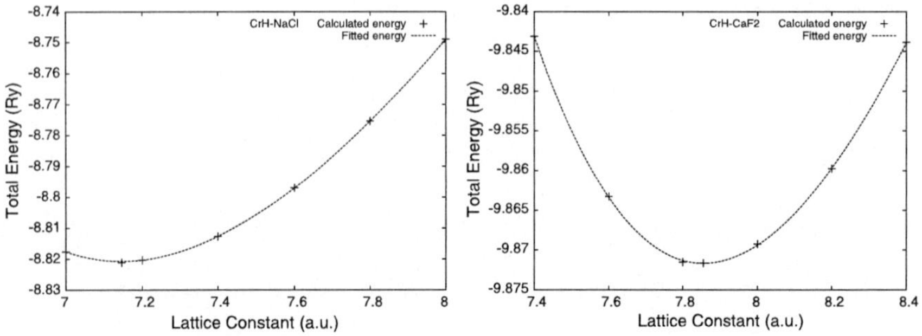

Fig. 6.11 Total energy versus lattice constant of CrH in the NaCl and CaF2 structures

Table 6.11 Lattice constant, bulk modulus, gap, total energy

Stru	a (Bohr)	B (MBar)	Gap	Total Energy (Ry)
NaCl	7.15	2.52	-	-2098.82114
CaF2	7.86	1.91	-	-2099.87165
exp (CaF2)	7.30			

Table 6.12 Birch fit coefficients

	A1	A2	A3	A4
NaCl	-7.084073E+00	-6.990776E+01	6.938871E+02	1.945547E+02
CaF2	-8.155074E+00	-8.277601E+01	9.660022E+02	7.750433E+04

Table 6.13 DOS at Ef, Hopfield parameter, Stoner criterion

CrH a=7.15 Bohr NaCl

Fermi Energy (Ry)	Total DOS (States/Ry)	Decomposed DOS inside the muffin tins s	p	eg	t2g	f
Cr 0.971	18.812	0.014	0.182	10.581	4.143	0.023
H 0.971	18.812	0.024	1.052	0.073	0.029	0.044

Fermi-Velocity (cm/s): 0.34113 x10E8
Plasmon Energy (eV) : 5.57764
Electron-ion interaction (Hopfield parameter) (eV/A^2) Cr: 4.240 H: 0.042

Cr	MUFFIN-TIN RADIUS and CHARGE =	1.8757	21.5155
H	MUFFIN-TIN RADIUS and CHARGE =	1.6971	1.3511
Cr	STONER	I = 0.0237	
H	STONER	I = 0.0008	
STONER PARAMETER (Ry)		I = 0.0246	
STONER CRITERION		N*I = 0.4630	

CrH2 a=7.86 Bohr CaF2

Fermi Energy (Ry)	Total DOS (States/Ry)	Decomposed DOS inside the muffin tins s	p	eg	t2g	f
Cr 0.728	28.585	0.035	0.096	14.372	9.622	0.028
H 0.662	28.585	0.039	0.496	0.047	0.039	0.018

Fermi-Velocity (cm/s): 0.31953 x10E8
Plasmon Energy (eV) : 5.58727
Electron-ion interaction (Hopfield parameter) (eV/A^2) Cr: 2.891 H: 0.125

Cr	MUFFIN-TIN RADIUS and CHARGE =	2.0408	21.9666
H	MUFFIN-TIN RADIUS and CHARGE =	1.3606	0.8550
Cr	STONER	I = 0.0217	
H	STONER	I = 0.0000	
STONER PARAMETER (Ry)		I = 0.0219	
STONER CRITERION		N*I = 0.6247	

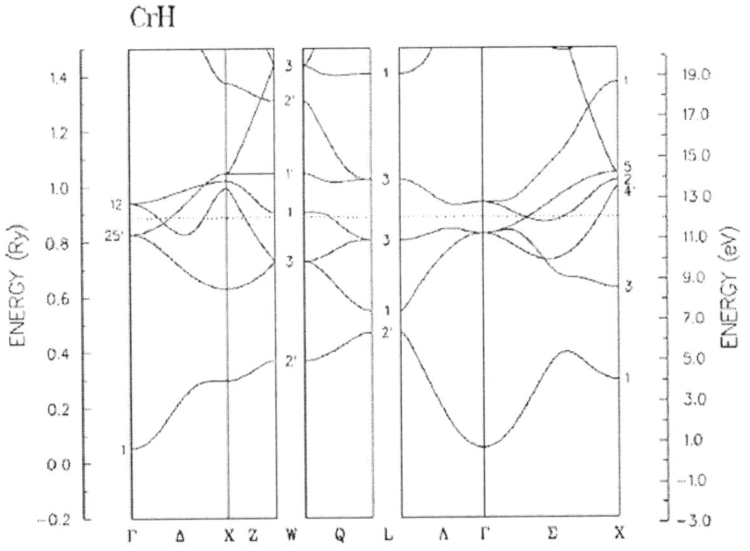

Fig. 6.12 Energy bands of CrH in the NaCl structure (tight-binding)

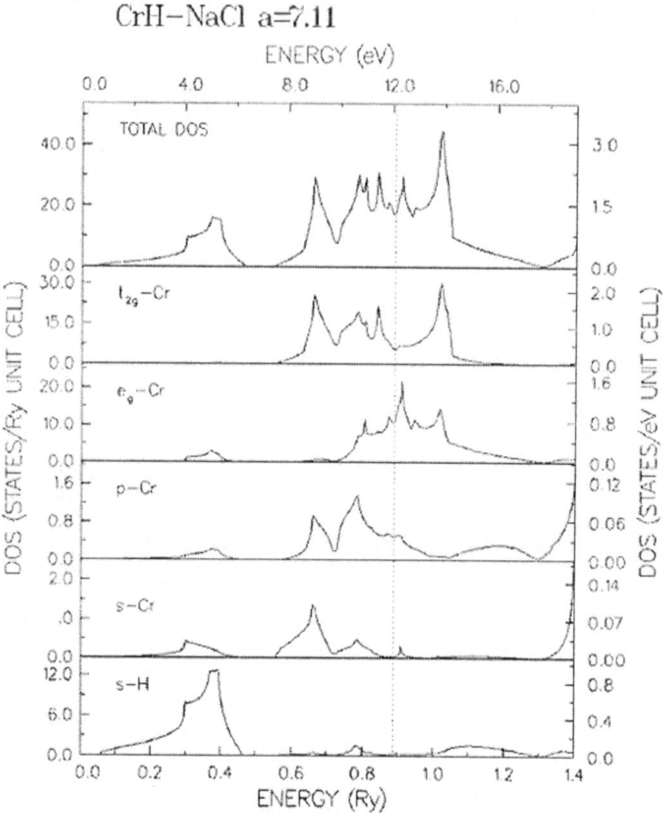

Fig. 6.13 Total, angular momentum and site decomposed densities of states of CrH in the NaCl structure (tight-binding)

Table 6.14 CrH (NaCl) a = 7.11 Bohr Slater–Koster 3-center parameters

	ORTHOGONAL	NON-ORTHOGONAL	
	ENERGY INTEGRALS Ry	ENERGY INTEGRALS Ry	OVERLAP INTEGRALS
ON SITE Cr-Cr			
s,s(000)	1.46467	1.34663	
x,x(000)	1.96385	1.68761	
xy,xy(000)	0.91365	0.86118	
d2,d2(000)	0.99764	0.84496	
FIRST NEIGHBOR Cr-Cr			
s,s(110)	-0.00755	0.08491	0.08377
s,x(110)	-0.06023	0.04600	0.04620
s,xy(110)	-0.06172	-0.05490	-0.00160
s,d2(110)	-0.01220	0.00028	0.01717
x,x(110)	0.08610	0.03235	-0.04927
x,x(011)	-0.01014	0.09892	0.08202
x,y(110)	0.07168	-0.09894	-0.10655
x,xy(110)	0.06581	0.04109	-0.00125
x,xy(011)	-0.00588	0.01863	0.04719
z,d2(011)	0.02077	0.04717	0.08807
z,d1(011)	0.01089	0.03117	0.04261
xy,xy(110)	-0.04014	-0.03401	0.00626
xy,xy(011)	0.01225	0.00584	-0.00886
xy,xz(011)	0.00539	0.00593	-0.01524
xy,d2(110)	0.02149	-0.01085	0.01295
d2,d2(110)	0.00602	0.00723	0.02823
d1,d1(110)	-0.03635	0.05373	0.01129
SECOND NEIGHBOR Cr-Cr			
s,s(200)	0.06742	0.00299	-0.01019
s,x(200)	0.08417	0.00963	-0.03975
s,d2(002)	0.04578	0.00357	0.01321
x,x(200)	-0.15968	-0.03170	-0.00667
y,y(200)	0.01910	-0.01171	0.00436
x,xy(020)	0.01125	-0.00210	0.00615
z,d2(002)	-0.04239	-0.05515	-0.08551
xy,xy(200)	-0.00642	-0.00243	-0.00800
xy,xy(002)	0.00077	0.00230	0.00211
d2,d2(002)	0.04223	-0.04051	-0.02861
d1,d1(002)	0.00008	0.00095	0.00273
ON SITE H-H			
s,s(000)	0.51207	0.40619	
FIRST NEIGHBOR H-H			
s,s(110)	-0.02413	-0.02362	-0.00257
s,x(110)	0.00000	0.00000	0.00000
x,x(110)	0.00000	0.00000	0.00000
x,x(011)	0.00000	0.00000	0.00000
x,y(110)	0.00000	0.00000	0.00000
SECOND NEIGHBOR H-H			
s,s(200)	-0.00726	-0.00941	0.00663
s,x(200)	0.00000	0.00000	0.00000
x,x(200)	0.00000	0.00000	0.00000
y,y(200)	0.00000	0.00000	0.00000
FIRST NEIGHBOR Cr-H			
s,s(100)	-0.12633	-0.04794	0.08275
s,x(100)	0.00000	0.00000	0.00000
x,s(100)	0.09453	0.14640	0.22631
x,x(100)	0.00000	0.00000	0.00000
y,y(100)	0.00000	0.00000	0.00000
xy,x(010)	0.00000	0.00000	0.00000
d2,s(001)	-0.07740	-0.03514	-0.19438
d2,z(001)	0.00000	0.00000	0.00000

SECOND NEIGHBOR Cr-H

s,s(111)	0.03634	0.02003	0.03331
x,s(111)	-0.04828	0.00605	-0.00548
s,x(111)	0.00000	0.00000	0.00000
x,x(111)	0.00000	0.00000	0.00000
x,y(111)	0.00000	0.00000	0.00000
xy,s(111)	0.00318	-0.01085	-0.00355
xy,x(111)	0.00000	0.00000	0.00000
yz,x(111)	0.00000	0.00000	0.00000
d1,x(111)	0.00000	0.00000	0.00000

♀ CrH

	ORTHOGONAL			NON-ORTHOGONAL		
BAND	RMS ERROR	MAXIMUM DEVIATION		RMS ERROR	MAXIMUM DEVIATION	
	mRy	k	mRy	mRy	k	mRy
1	5.5	(005)	13.8	0.9	(044)	2.7
2	5.4	(222)	16.6	0.5	(264)	1.0
3	5.0	(264)	11.9	0.7	(007)	1.7
4	2.8	(006)	4.7	0.3	(042)	0.9
5	3.3	(044)	8.4	0.4	(444)	0.9
6	3.5	(055)	10.2	0.7	(354)	1.7
7	8.4	(000)	19.0	1.2	(224)	3.4
1-7	5.2			0.7		

ENERGY VALUES IN Ry AT SELECTED k-POINTS

	ORTHOGONAL	APW	NON-ORTHOGONAL
GAMMA 1	0.05249	0.05235	0.05337
GAMMA 1	1.90511	1.92412	1.92423
GAMMA 12	0.94262	0.94147	0.94181
GAMMA 15	2.36913	2.37915	2.37917
GAMMA 25'	0.82698	0.82774	0.82773
X1 (008)	0.30638	0.29748	0.29796
X1 (008)	1.36900	1.37345	1.37387
X1 (008)	2.14368	2.15112	2.15130
X2 (008)	1.01572	1.01930	1.01906
X3 (008)	0.63094	0.62999	0.63037
X5 (008)	1.05010	1.04706	1.04741
X4' (008)	0.99154	0.99354	0.99301
X5' (008)	1.76146	1.74649	1.74802
L1 (444)	0.54187	0.54956	0.54894
L1 (444)	1.41295	1.40927	1.40872
L3 (444)	0.80082	0.80442	0.80446
L3 (444)	1.02923	1.02299	1.02388
L2' (444)	0.46396	0.47082	0.47018
L2' (444)	1.72508	1.72461	1.72450
L3' (444)	2.49351	2.46114	2.46115
W1 (048)	0.90661	0.90686	0.90744
W1 (048)	1.65465	1.65165	1.65159
W2' (048)	0.36553	0.36803	0.36779
W2' (048)	1.31097	1.31224	1.31091
W2' (048)	2.58736	2.64795	2.64794
W1' (048)	1.04701	1.04726	1.04700
W3 (048)	0.72358	0.72863	0.72901
W3 (048)	1.44270	1.43834	1.43847
ODD (224)	0.85293	0.85019	0.84963
ODD (224)	1.01097	1.00740	1.00699
EVEN(224)	0.31208	0.30596	0.30631
EVEN(224)	0.69097	0.68410	0.68422
EVEN(224)	0.79638	0.80028	0.80153
EVEN(224)	0.97654	0.97650	0.97600
EVEN(224)	1.48435	1.48056	1.48395

FERMI LEVEL QUANTITIES (Non-orthogonal fit)

ENERGY							
	DENSITIES OF STATES						
	Total	s-Cr	p-Cr	t2g-Cr	eg-Cr	s-H	p-H
Ry	states/Ry/cell						
0.8871	17.21	0.04	0.50	5.47	10.89	0.31	0.00
	INTEGRATED DENSITIES OF STATES						
	Total	s-Cr	p-Cr	t2g-Cr	eg-Cr	s-H	p-H
	electrons						
	7.00	0.17	0.20	3.54	1.25	1.83	0.00

VELOCITY	PLASMON ENERGY	EIGENVALUE SUM
cm/s	eV	Ry
0.40x10E8	6.31	-1.7581

Table 6.15 CrH (NaCl) a = 7.11 Bohr Slater–Koster 2-center parameters

	ORTHOGONAL	NON-ORTHOGONAL	
	ENERGY INTEGRALS Ry	ENERGY INTEGRALS Ry	OVERLAP INTEGRALS
ON SITE Cr-Cr			
s	1.24918	0.88214	
p	1.91392	1.42194	
t2g	0.88959	0.84916	
eg	0.87851	0.84330	
FIRST NEIGHBOR Cr-Cr			
(sss)	-0.08795	-0.02060	0.10144
(pps)	0.18200	-0.08961	-0.21573
(ppp)	-0.01069	0.02290	0.05098
(dds)	-0.05317	-0.00906	0.04943
(ddp)	0.02544	0.01321	-0.02880
(ddd)	-0.00144	-0.00197	0.00488
(sps)	0.13813	-0.03522	-0.16094
(sds)	-0.06516	-0.02042	0.05452
(pds)	-0.09170	-0.01631	0.05604
(pdp)	0.01538	0.01177	-0.02937
SECOND NEIGHBOR Cr-Cr			
(sss)	0.01196	-0.06230	-0.03133
(pps)	-0.12237	0.19043	0.08606
(ppp)	0.01245	-0.07435	-0.03949
(dds)	-0.00043	-0.01329	-0.00332
(ddp)	-0.00009	0.00365	0.00149
(ddd)	0.00005	0.00199	0.00194
(sps)	-0.03734	0.11471	0.05586
(sds)	0.00335	-0.01055	0.00263
(pds)	0.00357	-0.01475	0.00728
(pdp)	-0.00691	0.02043	0.01420
ON SITE H-H			
s	1.18626	0.48304	

FIRST NEIGHBOR H-H			
(sss)	0.01538	-0.01217	0.01292
(sps)	0.00000	0.00000	0.00000
(pps)	0.00000	0.00000	0.00000
(ppp)	0.00000	0.00000	0.00000
SECOND NEIGHBOR H-H			
(sss)	0.05391	-0.00946	-0.00382
(sps)	0.00000	0.00000	0.00000
(pps)	0.00000	0.00000	0.00000
(ppp)	0.00000	0.00000	0.00000
FIRST NEIGHBOR Cr-H			
(sss)	0.14618	0.01572	-0.20068
(sps)	0.00000	0.00000	0.00000
(pss)	0.16487	-0.07204	-0.28689
(pps)	0.00000	0.00000	0.00000
(ppp)	0.00000	0.00000	0.00000
(dpp)	0.00000	0.00000	0.00000
(dss)	0.11140	0.00687	-0.15334
(dps)	0.00000	0.00000	0.00000
SECOND NEIGHBOR Cr-H			
(sss)	-0.03498	0.00299	-0.01263
(pss)	-0.07341	0.00458	-0.02383
(sps)	0.00000	0.00000	0.00000
(pps)	0.00000	0.00000	0.00000
(ppp)	0.00000	0.00000	0.00000
(dss)	-0.00730	0.01101	-0.00436
(dps)	0.00000	0.00000	0.00000
(dpp)	0.00000	0.00000	0.00000

♀CrH

	ORTHOGONAL			NON-ORTHOGONAL		
BAND	RMS ERROR	MAXIMUM DEVIATION		RMS ERROR	MAXIMUM DEVIATION	
	mRy	k	mRy	mRy	k	mRy
1	5.3	(226)	11.4	0.4	(044)	1.5
2	7.9	(333)	29.9	0.5	(055)	1.7
3	6.1	(264)	14.0	0.5	(033)	1.7
4	3.6	(444)	7.1	0.6	(033)	1.3
5	3.0	(003)	7.0	0.7	(055)	2.0
6	5.5	(055)	15.6	0.5	(224)	1.3
7	11.7	(006)	30.0	0.8	(333)	2.5
1-7	6.7			0.6		

ENERGY VALUES IN Ry AT SELECTED k-POINTS

	ORTHOGONAL	APW	NON-ORTHOGONAL
GAMMA 1	0.04872	0.05235	0.05183
GAMMA 1	1.91104	1.92412	1.92290
GAMMA 12	0.94378	0.94147	0.94241
GAMMA 15	2.36143	2.37915	2.37924
GAMMA 25'	0.82440	0.82774	0.82782
X1 (008)	0.30562	0.29748	0.29727
X1 (008)	1.36235	1.37345	1.37383
X1 (008)	2.14902	2.15112	2.15087
X2 (008)	1.01446	1.01930	1.01918
X3 (008)	0.63239	0.62999	0.63052
X5 (008)	1.05028	1.04706	1.04696
X4' (008)	0.99098	0.99354	0.99295

X5' (008)	1.76173	1.74649	1.74490
L1 (444)	0.54183	0.54956	0.54963
L1 (444)	1.41794	1.40927	1.40871
L3 (444)	0.79732	0.80442	0.80416
L3 (444)	1.02593	1.02299	1.02307
L2' (444)	0.48184	0.47082	0.47094
L2' (444)	1.71905	1.72461	1.72455
L3' (444)	2.49424	2.46114	2.46115
W1 (048)	0.90776	0.90686	0.90576
W1 (048)	1.65371	1.65165	1.65169
W2' (048)	0.36804	0.36803	0.36796
W2' (048)	1.30579	1.31224	1.31242
W2' (048)	2.58539	2.64795	2.64796
W1' (048)	1.05009	1.04726	1.04714
W3 (048)	0.72263	0.72863	0.72819
W3 (048)	1.44562	1.43834	1.43805
ODD (224)	0.84738	0.85019	0.84936
ODD (224)	1.00941	1.00740	1.00614
EVEN(224)	0.30385	0.30596	0.30592
EVEN(224)	0.69581	0.68410	0.68369
EVEN(224)	0.80252	0.80028	0.80097
EVEN(224)	0.98149	0.97650	0.97713
EVEN(224)	1.48328	1.48056	1.48035

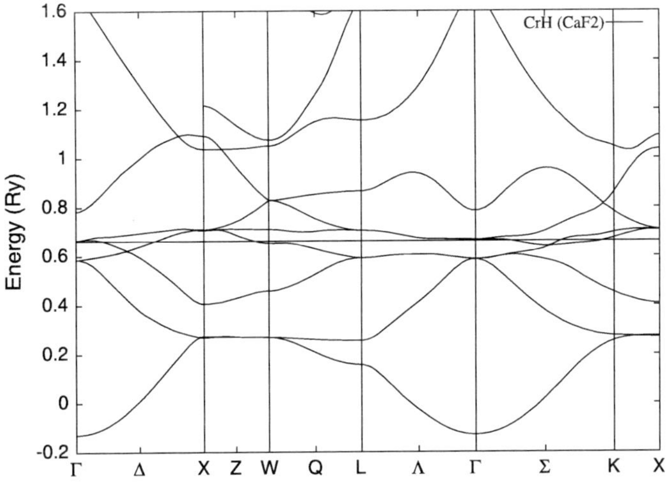

Fig. 6.14 Energy bands of CrH2 in the CaF2 structure

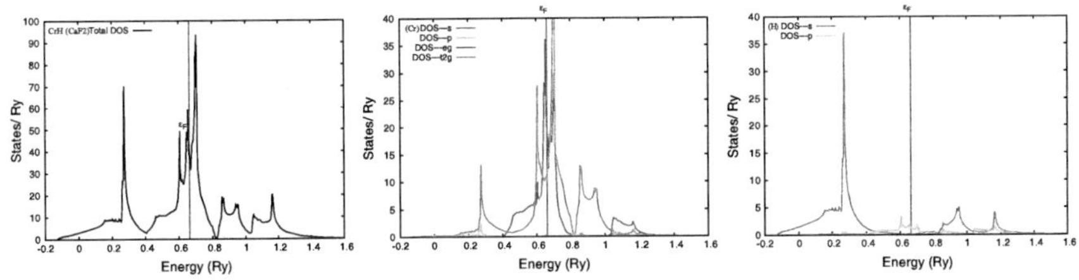

Fig. 6.15 Total, angular momentum and site decomposed densities of states of CrH2 in the CaF2 structure

6.4 Manganese Hydride (MnH)

See Fig. 6.16 and Tables 6.16, 6.17, 6.18.

See Figs. 6.17, 6.18 and Tables 6.19, 6.20.

See Figs. 6.19 and 6.20.

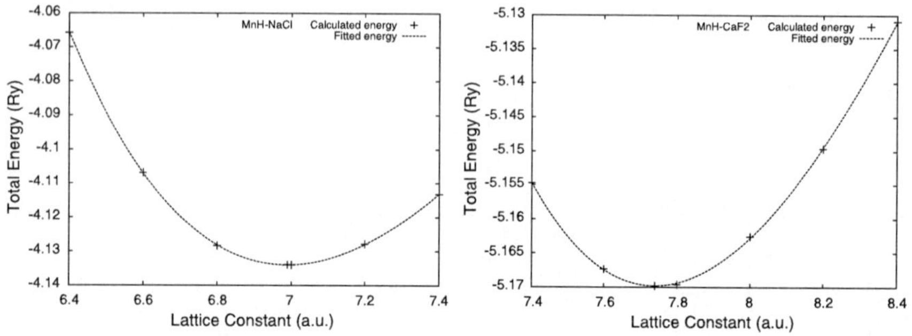

Fig. 6.16 Total energy versus lattice constant of MnH in the NaCl and CaF2 structures

Table 6.16 Lattice constant, bulk modulus, gap, total energy

Stru	a (Bohr)	B (MBar)	Gap	Total Energy (Ry)
NaCl	6.99	2.77	-	-2314.13392
CaF2	7.74	1.93		-2315.16980
exp (NiAs)				

Table 6.17 Birch fit coefficients

	A1	A2	A3	A4
NaCl	-2.497072E+00	-6.015660E+01	4.868684E+02	1.243706E+03
CaF2	-3.499548E+00	-7.854891E+01	9.029912E+02	4.850850E+02

Table 6.18 DOS at Ef, Hopfield parameter, Stoner criterion

MnH a=6.99 Bohr NaCl

	Fermi Energy (Ry)	Total DOS (States/Ry)	Decomposed DOS inside the muffin tins s	p	eg	t2g	f
Mn	0.988	18.558	0.016	0.122	8.649	6.533	0.019
H	0.988	18.558	0.034	0.489	0.116	0.042	0.033

```
Fermi-Velocity (cm/s): 0.41178 x10E8
Plasmon Energy (eV)  : 6.91270
Electron-ion interaction (Hopfield parameter) (eV/A^2) Mn: 3.839  H: 0.033
-------------------------------------------------
  Mn     MUFFIN-TIN RADIUS and CHARGE =    1.8347   22.5484
   H     MUFFIN-TIN RADIUS and CHARGE =    1.6600    1.3336
          Mn       STONER       I = 0.0262
           H       STONER       I = 0.0003
         STONER PARAMETER (Ry) I = 0.0266
         STONER CRITERION    N*I = 0.4929
-------------------------------------------------
```

MnH2 a=7.74 CaF2

	Fermi Energy (Ry)	Total DOS (States/Ry)	Decomposed DOS inside the muffin tins s	p	eg	t2g	f
Mn	0.726	36.858	0.008	0.076	14.743	18.239	0.020
H	0.726	36.858	0.053	0.407	0.037	0.079	0.025

```
Fermi-Velocity (cm/s):  0.34432 x10E8
Plasmon Energy (eV)  :  6.99087
Electron-ion interaction (Hopfield parameter) (eV/A^2) Mn: 2.313  H: 0.118
-------------------------------------------------
         Mn     MUFFIN-TIN RADIUS and CHARGE =     2.0108   23.0580
          H     MUFFIN-TIN RADIUS and CHARGE =     1.3405    0.8249
                 Mn       STONER I = 0.0259
                  H       STONER I = 0.0000
         STONER PARAMETER (Ry)      I = 0.0260
         STONER CRITERION         N*I = 0.9584
```

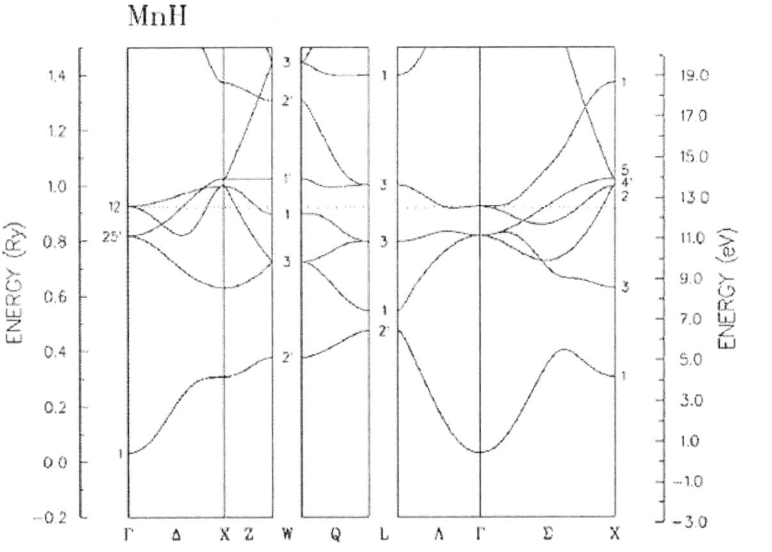

Fig. 6.17 Energy bands of MnH in the NaCl structure (tight-binding)

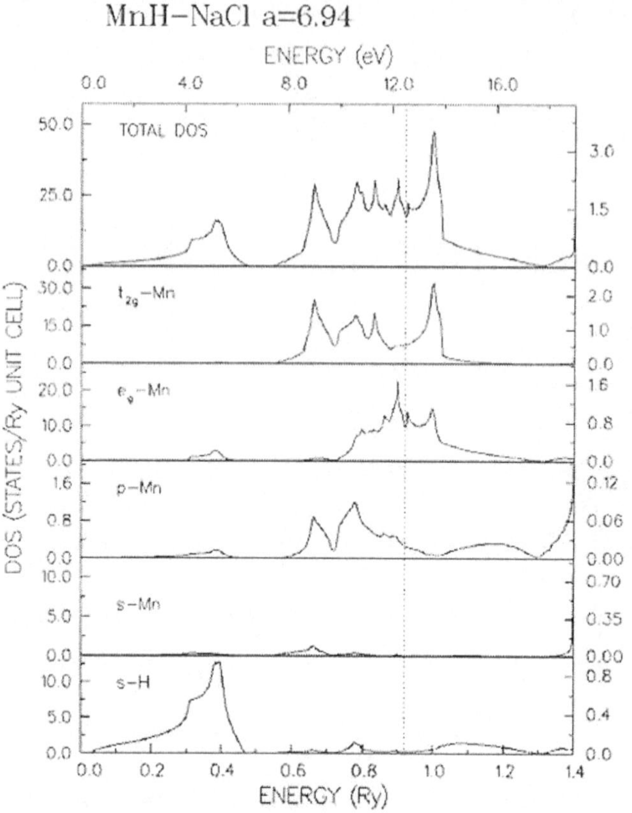

Fig. 6.18 Total, angular momentum and site decomposed densities of states of MnH in the NaCl structure (tight-binding)

Table 6.19 MnH (NaCl) a = 6.94 Bohr Slater–Koster 3-center parameters

	ORTHOGONAL	NON-ORTHOGONAL	
	ENERGY INTEGRALS Ry	ENERGY INTEGRALS Ry	OVERLAP INTEGRALS
ON SITE Mn-Mn			
s,s(000)	1.49698	1.35830	
x,x(000)	2.03227	1.73225	
xy,xy(000)	0.89785	0.85058	
d2,d2(000)	0.99142	0.83402	
FIRST NEIGHBOR Mn-Mn			
s,s(110)	-0.00514	0.08104	0.08013
s,x(110)	-0.06238	0.03732	0.03921
s,xy(110)	-0.06064	-0.04893	0.00296
s,d2(110)	-0.01782	-0.00277	0.01622
x,x(110)	0.09254	0.04389	-0.04635
x,x(011)	-0.00775	0.09168	0.07518
x,y(110)	0.07126	-0.09079	-0.10221
x,xy(110)	0.06482	0.03949	-0.00029
x,xy(011)	-0.00595	0.01671	0.04686
z,d2(011)	0.01963	0.04732	0.08978
z,d1(011)	0.01657	0.03244	0.04252
xy,xy(110)	-0.03792	-0.02680	0.01155
xy,xy(011)	0.01167	0.00824	-0.00508
xy,xz(011)	0.00557	0.00984	-0.01008
xy,d2(110)	0.02024	-0.01626	0.00596
d2,d2(110)	0.00790	0.00651	0.02712
d1,d1(110)	-0.04060	0.05149	0.01062
SECOND NEIGHBOR Mn-Mn			
s,s(200)	0.07143	0.01653	-0.00425
s,x(200)	0.08841	0.02012	-0.03631
s,d2(002)	0.04793	0.00999	0.01834
x,x(200)	-0.17202	-0.02473	-0.00243
y,y(200)	0.02032	0.00368	0.01130
x,xy(020)	0.01116	0.00625	0.01279
z,d2(002)	-0.04296	-0.05346	-0.08398
xy,xy(200)	-0.00578	-0.00490	-0.00961
xy,xy(002)	0.00070	0.00365	0.00355
d2,d2(002)	0.04408	-0.04170	-0.02982
d1,d1(002)	-0.00027	0.00056	0.00224
ON SITE H-H			
s,s(000)	0.50604	0.39716	
FIRST NEIGHBOR H-H			
s,s(110)	-0.02392	-0.02584	-0.00012
s,x(110)	0.00000	0.00000	0.00000
x,x(110)	0.00000	0.00000	0.00000
x,x(011)	0.00000	0.00000	0.00000
x,y(110)	0.00000	0.00000	0.00000
SECOND NEIGHBOR H-H			
s,s(200)	-0.00756	-0.00801	0.01250
s,x(200)	0.00000	0.00000	0.00000
x,x(200)	0.00000	0.00000	0.00000
y,y(200)	0.00000	0.00000	0.00000
FIRST NEIGHBOR Mn-H			
s,s(100)	-0.13177	-0.04053	0.09389
s,x(100)	0.00000	0.00000	0.00000
x,s(100)	0.09305	0.13834	0.22655
x,x(100)	0.00000	0.00000	0.00000
y,y(100)	0.00000	0.00000	0.00000
xy,x(010)	0.00000	0.00000	0.00000
d2,s(001)	-0.07483	-0.04062	-0.20077
d2,z(001)	0.00000	0.00000	0.00000

SECOND NEIGHBOR Mn-H

s,s(111)	0.03496	0.02069	0.03494
x,s(111)	-0.04876	0.00736	-0.00513
s,x(111)	0.00000	0.00000	0.00000
x,x(111)	0.00000	0.00000	0.00000
x,y(111)	0.00000	0.00000	0.00000
xy,s(111)	0.00315	-0.01192	-0.00435
xy,x(111)	0.00000	0.00000	0.00000
yz,x(111)	0.00000	0.00000	0.00000
d1,x(111)	0.00000	0.00000	0.00000

MnH

	ORTHOGONAL			NON-ORTHOGONAL		
BAND	RMS ERROR	MAXIMUM DEVIATION		RMS ERROR	MAXIMUM DEVIATION	
	mRy	k	mRy	mRy	k	mRy
1	5.3	(005)	13.3	0.9	(044)	2.7
2	5.3	(222)	15.8	0.5	(044)	1.2
3	5.7	(264)	13.4	0.7	(007)	1.6
4	2.8	(062)	5.1	0.5	(042)	1.5
5	3.2	(044)	8.3	0.4	(224)	1.1
6	3.5	(055)	10.6	0.9	(174)	2.7
7	8.9	(000)	19.6	1.5	(224)	4.5
1-7	5.3			0.9		

ENERGY VALUES IN Ry AT SELECTED k-POINTS

	ORTHOGONAL	APW	NON-ORTHOGONAL
GAMMA 1	0.03123	0.03090	0.03198
GAMMA 1	2.00624	2.02586	2.02563
GAMMA 12	0.92669	0.92519	0.92562
GAMMA 15	2.47876	2.48968	2.48955
GAMMA 25'	0.81784	0.81832	0.81812
X1 (008)	0.31522	0.30602	0.30609
X1 (008)	1.36780	1.37196	1.37148
X1 (008)	2.20165	2.20406	2.20424
X2 (008)	0.99423	0.99841	0.99844
X3 (008)	0.63108	0.62972	0.63018
X5 (008)	1.02785	1.02473	1.02496
X4' (008)	0.99820	1.00134	1.00070
X5' (008)	1.80051	1.79908	1.79924
L1 (444)	0.54058	0.54753	0.54720
L1 (444)	1.40286	1.39979	1.39875
L3 (444)	0.79364	0.79683	0.79707
L3 (444)	1.00815	1.00221	1.00332
L2' (444)	0.46871	0.47452	0.47380
L2' (444)	1.80764	1.80756	1.80749
L3' (444)	2.58010	2.54852	2.54862
W1 (048)	0.89592	0.89559	0.89647
W1 (048)	1.67939	1.67769	1.67754
W2' (048)	0.37445	0.37750	0.37694
W2' (048)	1.30912	1.31110	1.30857
W2' (048)	2.67794	2.73891	2.73877
W1' (048)	1.02504	1.02493	1.02489
W3 (048)	0.71887	0.72454	0.72526
W3 (048)	1.45178	1.44641	1.44635
ODD (224)	0.84326	0.84041	0.83950
ODD (224)	0.99108	0.98771	0.98767
EVEN(224)	0.30448	0.29855	0.29899
EVEN(224)	0.69031	0.68353	0.68404
EVEN(224)	0.78897	0.79366	0.79475
EVEN(224)	0.96099	0.96107	0.95993
EVEN(224)	1.51475	1.50910	1.51364

FERMI LEVEL QUANTITIES (Non-orthogonal fit)

ENERGY	DENSITIES OF STATES						
	Total	s-Mn	p-Mn	t2g-Mn	eg-Mn	s-H	p-H
Ry	states/Ry/cell						
0.9201	17.86	0.05	0.27	7.44	9.79	0.31	0.00
	INTEGRATED DENSITIES OF STATES						
	Total	s-Mn	p-Mn	t2g-Mn	eg-Mn	s-H	p-H
	electrons						
	8.00	0.16	0.20	3.85	1.94	1.85	0.00

VELOCITY	PLASMON ENERGY	EIGENVALUE SUM
cm/s	eV	Ry
0.40x10E8	6.63	-2.0236

Table 6.20 MnH (NaCl) a = 6.94 Bohr Slater–Koster 2-center parameters

	ORTHOGONAL	NON-ORTHOGONAL	
	ENERGY INTEGRALS Ry	ENERGY INTEGRALS Ry	OVERLAP INTEGRALS
ON SITE Mn-Mn			
s	1.26003	0.88281	
p	1.98245	1.44224	
t2g	0.87497	0.84156	
eg	0.86594	0.83584	
FIRST NEIGHBOR Mn-Mn			
(sss)	-0.09159	-0.02294	0.10372
(pps)	0.19170	-0.08654	-0.22027
(ppp)	-0.00777	0.03353	0.05705
(dds)	-0.05012	-0.01232	0.04313
(ddp)	0.02454	0.01487	-0.02398
(ddd)	-0.00165	-0.00310	0.00311
(sps)	0.14521	-0.03172	-0.16318
(sds)	-0.06373	-0.02260	0.05030
(pds)	-0.08944	-0.01786	0.05517
(pdp)	0.01588	0.01232	-0.02831
SECOND NEIGHBOR Mn-Mn			
(sss)	0.01342	-0.06031	-0.02889
(pps)	-0.13525	0.18509	0.07627
(ppp)	0.01363	-0.06453	-0.03372
(dds)	-0.00093	-0.01284	-0.00405
(ddp)	0.00027	0.00415	0.00225
(ddd)	-0.00011	0.00155	0.00150
(sps)	-0.03963	0.10862	0.04993
(sds)	0.00324	-0.00867	0.00445
(pds)	0.00320	-0.01421	0.00772
(pdp)	-0.00724	0.01835	0.01278
ON SITE H-H			
s	1.22400	0.49017	

FIRST NEIGHBOR H-H			
(sss)	0.01924	-0.01523	0.01097
(sps)	0.00000	0.00000	0.00000
(pps)	0.00000	0.00000	0.00000
(ppp)	0.00000	0.00000	0.00000
SECOND NEIGHBOR H-H			
(sss)	0.05706	-0.00888	-0.00153
(sps)	0.00000	0.00000	0.00000
(pps)	0.00000	0.00000	0.00000
(ppp)	0.00000	0.00000	0.00000
FIRST NEIGHBOR Mn-H			
(sss)	0.15100	0.02068	-0.20266
(sps)	0.00000	0.00000	0.00000
(pss)	0.17271	-0.07318	-0.29417
(pps)	0.00000	0.00000	0.00000
(ppp)	0.00000	0.00000	0.00000
(dpp)	0.00000	0.00000	0.00000
(dss)	0.10954	0.00489	-0.15314
(dps)	0.00000	0.00000	0.00000
SECOND NEIGHBOR Mn-H			
(sss)	-0.03636	0.00442	-0.01245
(pss)	-0.07661	0.00648	-0.02302
(sps)	0.00000	0.00000	0.00000
(pps)	0.00000	0.00000	0.00000
(ppp)	0.00000	0.00000	0.00000
(dss)	-0.00697	0.01227	-0.00279
(dps)	0.00000	0.00000	0.00000
(dpp)	0.00000	0.00000	0.00000

MnH

	ORTHOGONAL			NON-ORTHOGONAL		
BAND	RMS ERROR	MAXIMUM DEVIATION		RMS ERROR	MAXIMUM DEVIATION	
	mRy	k	mRy	mRy	k	mRy
1	6.2	(444)	14.5	0.5	(226)	1.2
2	7.9	(333)	31.4	0.5	(055)	1.4
3	6.0	(226)	14.0	0.4	(033)	1.2
4	3.1	(003)	6.7	0.4	(033)	0.9
5	2.8	(003)	6.7	0.6	(055)	1.7
6	5.5	(055)	16.1	0.5	(224)	1.1
7	13.1	(006)	38.0	0.8	(333)	1.9
1-7	7.1			0.5		

ENERGY VALUES IN Ry AT SELECTED k-POINTS

	ORTHOGONAL	APW	NON-ORTHOGONAL
GAMMA 1	0.02765	0.03090	0.03076
GAMMA 1	2.01106	2.02586	2.02451
GAMMA 12	0.92750	0.92519	0.92623
GAMMA 15	2.47105	2.48968	2.48968
GAMMA 25'	0.81540	0.81832	0.81838
X1 (008)	0.31636	0.30602	0.30536
X1 (008)	1.36162	1.37196	1.37221
X1 (008)	2.20637	2.20406	2.20413
X2 (008)	0.99453	0.99841	0.99813
X3 (008)	0.63224	0.62972	0.63041
X5 (008)	1.02785	1.02473	1.02479
X4' (008)	0.99965	1.00134	1.00090

X5' (008)	1.79753	1.79908	1.79895
L1 (444)	0.54055	0.54753	0.54717
L1 (444)	1.40832	1.39979	1.39915
L3 (444)	0.79099	0.79683	0.79677
L3 (444)	1.00453	1.00221	1.00222
L2' (444)	0.48897	0.47452	0.47445
L2' (444)	1.79324	1.80756	1.80753
L3' (444)	2.59740	2.54852	2.54862
W1 (048)	0.89559	0.89559	0.89547
W1 (048)	1.67925	1.67769	1.67744
W2' (048)	0.37658	0.37750	0.37748
W2' (048)	1.30298	1.31110	1.31156
W2' (048)	2.68525	2.73891	2.73885
W1' (048)	1.02827	1.02493	1.02475
W3 (048)	0.71771	0.72454	0.72393
W3 (048)	1.45535	1.44641	1.44610
ODD (224)	0.83784	0.84041	0.83981
ODD (224)	0.98951	0.98771	0.98661
EVEN(224)	0.29635	0.29855	0.29855
EVEN(224)	0.69532	0.68353	0.68331
EVEN(224)	0.79566	0.79366	0.79419
EVEN(224)	0.96547	0.96107	0.96146
EVEN(224)	1.51481	1.50910	1.50906

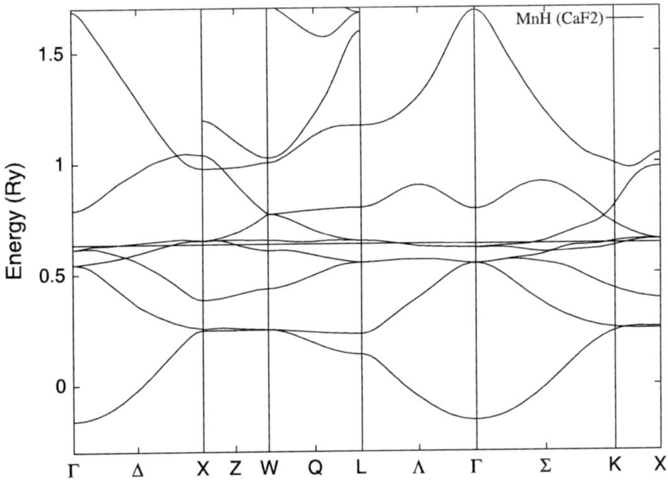

Fig. 6.19 Energy bands of MnH2 in the CaF2 structure

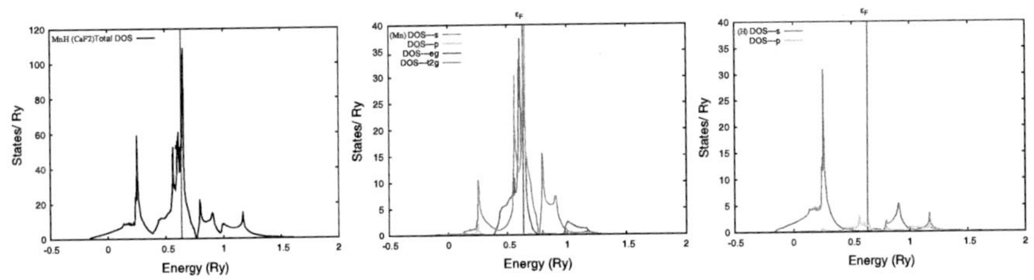

Fig. 6.20 Total, angular momentum and site decomposed densities of states of MnH2 in the CaF2 structure

6.5 Iron Hydride (FeH)

See Fig. 6.21 and Tables 6.21, 6.22, 6.23.

See Figs. 6.22, 6.23 and Tables 6.24, 6.25.

See Figs. 6.24 and 6.25.

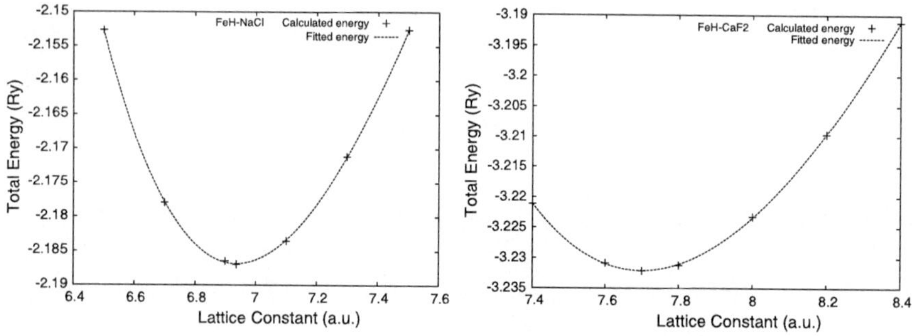

Fig. 6.21 Total energy versus lattice constant of FeH in the NaCl and CaF2 structures

Table 6.21 Lattice constant, bulk modulus, gap, total energy

Stru	a (Bohr)	B (MBar)	Gap	Total Energy (Ry)
NaCl	6.94	2.67	-	-2542.18700
CaF2	7.70	1.84		-2543.23214
exp (dhcp)				

Table 6.22 Birch fit coefficients

	A1	A2	A3	A4
NaCl	-9.218818E-01	-3.987614E+01	1.389906E+02	3.078250E+03
CaF2	-1.657238E+00	-7.336166E+01	8.370306E+02	4.057617E+02

Table 6.23 DOS at Ef, Hopfield parameter, Stoner criterion

FeH a=6.94 Bohr NaCl

Fermi Energy (Ry)		Total DOS (States/Ry)	Decomposed DOS inside the muffin tins s	p	eg	t2g	f
Fe	0.969	28.342	0.010	0.085	10.681	14.683	0.016
H	0.969	28.342	0.054	0.210	0.164	0.041	0.053

Fermi-Velocity (cm/s): 0.46252 x10E8
Plasmon Energy (eV) : 9.70353
Electron-ion interaction (Hopfield parameter) (eV/A^2) Fe: 3.368 H: 0.016

Fe	MUFFIN-TIN RADIUS and CHARGE =	1.8210	23.6537
H	MUFFIN-TIN RADIUS and CHARGE =	1.6476	1.2965
Fe	STONER	I = 0.0313	
H	STONER	I = 0.0001	
STONER PARAMETER (Ry)		I = 0.0314	
STONER CRITERION		N*I = 0.8894	

FeH2 a=7.70 Bohr CaF2

Fermi Energy (Ry)		Total DOS (States/Ry)	Decomposed DOS inside the muffin tins s	p	eg	t2g	f
Fe	0.694	88.810	0.031	0.090	27.470	54.503	0.017
H	0.694	88.810	0.084	0.658	0.026	0.238	0.065

Fermi-Velocity (cm/s): 0.21574 x10E8
Plasmon Energy (eV) : 6.85260
Electron-ion interaction (Hopfield parameter) (eV/A^2) Fe: 2.024 H: 0.155

Fe	MUFFIN-TIN RADIUS and CHARGE =	2.0003	24.1847
H	MUFFIN-TIN RADIUS and CHARGE =	1.3335	0.7999
Fe	STONER	I = 0.0285	
H	STONER	I = 0.0000	
STONER PARAMETER (Ry)		I = 0.0285	
STONER CRITERION		N*I = 2.5316	

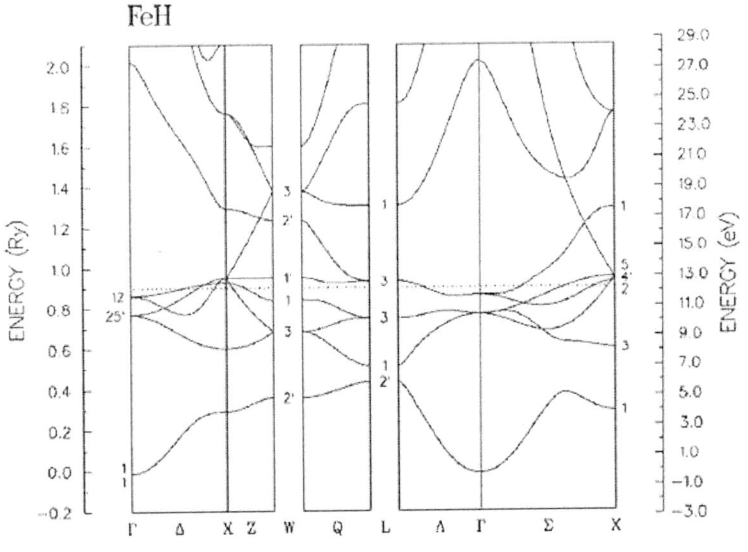

Fig. 6.22 Energy bands of FeH in the NaCl structure (tight-binding)

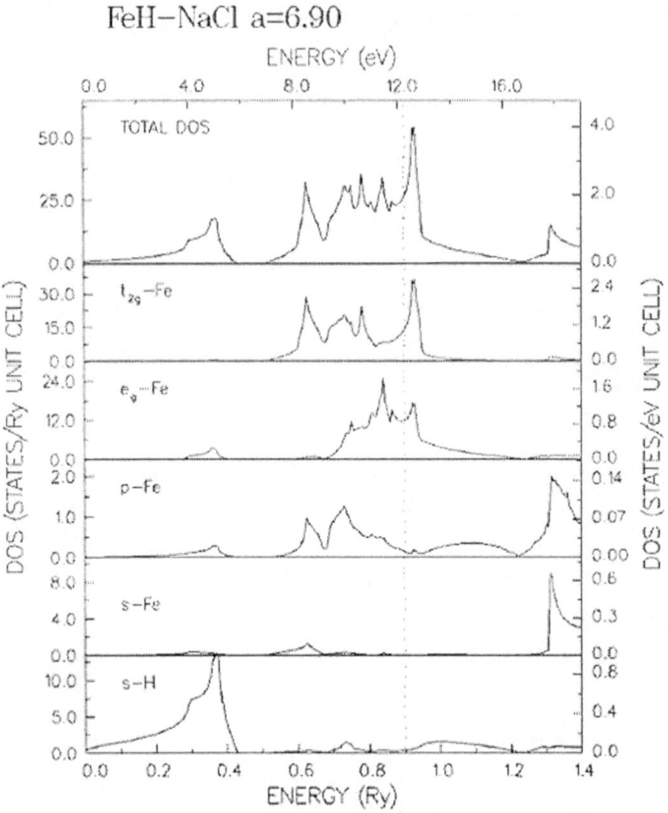

Fig. 6.23 Total, angular momentum and site decomposed densities of states of FeH in the NaCl structure (tight-binding)

Table 6.24 FeH (NaCl) a = 6.90 Bohr Slater–Koster 3-center parameters

	ORTHOGONAL	NON-ORTHOGONAL	
	ENERGY INTEGRALS Ry	ENERGY INTEGRALS Ry	OVERLAP INTEGRALS
ON SITE Fe-Fe			
s,s(000)	1.43711	1.30605	
x,x(000)	2.00497	1.71689	
xy,xy(000)	0.83598	0.80043	
d2,d2(000)	0.93231	0.77995	
FIRST NEIGHBOR Fe-Fe			
s,s(110)	-0.00144	0.08511	0.08486
s,x(110)	-0.05998	0.02128	0.03123
s,xy(110)	-0.05597	-0.05404	-0.00898
s,d2(110)	-0.02105	-0.01103	0.00925
x,x(110)	0.09621	0.05739	-0.04364
x,x(011)	-0.00549	0.08558	0.07175
x,y(110)	0.06629	-0.06165	-0.08555
x,xy(110)	0.05996	0.04631	0.01023
x,xy(011)	-0.00578	0.01768	0.05035
z,d2(011)	0.01735	0.04711	0.09277
z,d1(011)	0.02004	0.04552	0.05148
xy,xy(110)	-0.03356	-0.03001	0.00286
xy,xy(011)	0.01045	0.00944	-0.00127
xy,xz(011)	0.00576	0.01076	-0.00641
xy,d2(110)	0.01772	-0.01620	0.00334
d2,d2(110)	0.00878	0.00526	0.02600
d1,d1(110)	-0.04114	0.04564	0.00789
SECOND NEIGHBOR Fe-Fe			
s,s(200)	0.07124	0.01428	-0.00510
s,x(200)	0.08944	0.01315	-0.04010
s,d2(002)	0.04867	0.01099	0.01932
x,x(200)	-0.17677	-0.01323	0.00879
y,y(200)	0.02119	0.01456	0.01692
x,xy(020)	0.01001	0.01117	0.01909
z,d2(002)	-0.04448	-0.05209	-0.08138
xy,xy(200)	-0.00456	-0.00462	-0.00983
xy,xy(002)	0.00058	0.00442	0.00467
d2,d2(002)	0.04233	-0.04185	-0.03144
d1,d1(002)	-0.00039	0.00734	0.00964
ON SITE H-H			
s,s(000)	0.45445	0.36297	
FIRST NEIGHBOR H-H			
s,s(110)	-0.02374	-0.02843	0.00112
s,x(110)	0.00000	0.00000	0.00000
x,x(110)	0.00000	0.00000	0.00000
x,x(011)	0.00000	0.00000	0.00000
x,y(110)	0.00000	0.00000	0.00000
SECOND NEIGHBOR H-H			
s,s(200)	-0.00620	-0.00651	0.01642
s,x(200)	0.00000	0.00000	0.00000
x,x(200)	0.00000	0.00000	0.00000
y,y(200)	0.00000	0.00000	0.00000
FIRST NEIGHBOR Fe-H			
s,s(100)	-0.12943	-0.03404	0.10132
s,x(100)	0.00000	0.00000	0.00000
x,s(100)	0.08268	0.13317	0.22292
x,x(100)	0.00000	0.00000	0.00000
y,y(100)	0.00000	0.00000	0.00000
xy,x(010)	0.00000	0.00000	0.00000
d2,s(001)	-0.06928	-0.04547	-0.21006
d2,z(001)	0.00000	0.00000	0.00000

SECOND NEIGHBOR Fe-H

s,s(111)	0.03111	0.02286	0.04036
x,s(111)	-0.04583	0.01042	-0.00513
s,x(111)	0.00000	0.00000	0.00000
x,x(111)	0.00000	0.00000	0.00000
x,y(111)	0.00000	0.00000	0.00000
xy,s(111)	0.00151	-0.01314	-0.00622
xy,x(111)	0.00000	0.00000	0.00000
yz,x(111)	0.00000	0.00000	0.00000
d1,x(111)	0.00000	0.00000	0.00000

♀FeH

	ORTHOGONAL			NON-ORTHOGONAL		
BAND	RMS ERROR	MAXIMUM DEVIATION		RMS ERROR	MAXIMUM DEVIATION	
	mRy	k	mRy	mRy	k	mRy
1	4.4	(033)	10.7	0.8	(333)	2.4
2	5.0	(222)	14.6	0.6	(044)	1.8
3	6.0	(264)	14.0	0.9	(042)	1.7
4	2.5	(062)	5.2	0.6	(042)	1.7
5	2.7	(044)	7.4	0.4	(224)	1.5
6	3.5	(055)	10.8	1.0	(174)	2.8
7	9.2	(006)	20.3	1.8	(224)	5.1
1-7	5.2			1.0		

ENERGY VALUES IN Ry AT SELECTED k-POINTS

	ORTHOGONAL	APW	NON-ORTHOGONAL
GAMMA 1	-0.01699	-0.01699	-0.01558
GAMMA 1	1.99655	2.01665	2.01611
GAMMA 12	0.86394	0.86228	0.86280
GAMMA 15	2.48393	2.49258	2.49248
GAMMA 25'	0.76826	0.76845	0.76832
X1 (008)	0.29883	0.29106	0.29143
X1 (008)	1.28781	1.29198	1.29122
X1 (008)	2.11842	2.11657	2.11660
X2 (008)	0.92316	0.92669	0.92671
X3 (008)	0.60102	0.59987	0.60027
X5 (008)	0.95314	0.95005	0.95020
X4' (008)	0.94452	0.94889	0.94801
X5' (008)	1.75814	1.76119	1.76176
L1 (444)	0.50750	0.51352	0.51281
L1 (444)	1.30913	1.30637	1.30549
L3 (444)	0.74779	0.74977	0.75017
L3 (444)	0.93482	0.93031	0.93134
L2' (444)	0.42923	0.43459	0.43411
L2' (444)	1.80585	1.80672	1.80670
L3' (444)	2.53894	2.51258	2.51260
W1 (048)	0.83737	0.83792	0.83881
W1 (048)	1.60167	1.60137	1.60136
W2' (048)	0.35628	0.35926	0.35824
W2' (048)	1.23479	1.23659	1.23400
W2' (048)	2.63543	2.68894	2.68883
W1' (048)	0.95081	0.95024	0.95027
W3 (048)	0.67788	0.68289	0.68376
W3 (048)	1.38403	1.37649	1.37646
ODD (224)	0.79029	0.78852	0.78724
ODD (224)	0.92091	0.91769	0.91801
EVEN(224)	0.26092	0.25544	0.25587
EVEN(224)	0.65377	0.64734	0.64818
EVEN(224)	0.74244	0.74701	0.74840
EVEN(224)	0.89600	0.89585	0.89432
EVEN(224)	1.45592	1.44898	1.45412

FERMI LEVEL QUANTITIES (Non-orthogonal fit)

ENERGY	DENSITIES OF STATES						
	Total	s-Fe	p-Fe	t2g-Fe	eg-Fe	s-H	p-H
Ry	states/Ry/cell						
0.9011	26.74	0.01	0.14	14.27	11.92	0.40	0.00
	INTEGRATED DENSITIES OF STATES						
	Total	s-Fe	p-Fe	t2g-Fe	eg-Fe	s-H	p-H
	electrons						
	9.00	0.16	0.21	4.31	2.48	1.83	0.00

VELOCITY	PLASMON ENERGY	EIGENVALUE SUM
cm/s	eV	Ry
0.32x10E8	6.52	-2.2440

Table 6.25 FeH (NaCl) a = 6.90 Bohr Slater–Koster 2-center parameters

	ORTHOGONAL	NON-ORTHOGONAL	
	ENERGY INTEGRALS Ry	ENERGY INTEGRALS Ry	OVERLAP INTEGRALS
ON SITE Fe-Fe			
s	1.18960	0.83851	
p	1.95391	1.38359	
t2g	0.81676	0.79112	
eg	0.81001	0.78611	
FIRST NEIGHBOR Fe-Fe			
(sss)	-0.09038	-0.03496	0.09560
(pps)	0.19318	-0.08446	-0.22818
(ppp)	-0.00251	0.02438	0.05520
(dds)	-0.04417	-0.01286	0.03829
(ddp)	0.02222	0.01601	-0.01840
(ddd)	-0.00179	-0.00365	0.00175
(sps)	0.14501	-0.01106	-0.15392
(sds)	-0.05912	-0.02185	0.04838
(pds)	-0.08254	-0.01091	0.06254
(pdp)	0.01584	0.01604	-0.02383
SECOND NEIGHBOR Fe-Fe			
(sss)	0.01450	-0.04865	-0.02512
(pps)	-0.14355	0.14563	0.05703
(ppp)	0.01340	-0.04697	-0.02592
(dds)	-0.00129	-0.00999	-0.00288
(ddp)	0.00056	0.00410	0.00272
(ddd)	-0.00021	0.00181	0.00199
(sps)	-0.04013	0.08178	0.03751
(sds)	0.00388	-0.00681	0.00482
(pds)	0.00330	-0.01192	0.00838
(pdp)	-0.00727	0.01435	0.01034
ON SITE H-H			
s	1.17662	0.45741	

FIRST NEIGHBOR H-H			
(sss)	0.02248	-0.01894	0.01553
(sps)	0.00000	0.00000	0.00000
(pps)	0.00000	0.00000	0.00000
(ppp)	0.00000	0.00000	0.00000
SECOND NEIGHBOR H-H			
(sss)	0.05731	-0.00713	-0.00047
(sps)	0.00000	0.00000	0.00000
(pps)	0.00000	0.00000	0.00000
(ppp)	0.00000	0.00000	0.00000
FIRST NEIGHBOR Fe-H			
(sss)	0.14995	0.02786	-0.20703
(sps)	0.00000	0.00000	0.00000
(pss)	0.17393	-0.06594	-0.30294
(pps)	0.00000	0.00000	0.00000
(ppp)	0.00000	0.00000	0.00000
(dpp)	0.00000	0.00000	0.00000
(dss)	0.10283	0.01206	-0.14123
(dps)	0.00000	0.00000	0.00000
SECOND NEIGHBOR Fe-H			
(sss)	-0.03512	0.00487	-0.01348
(pss)	-0.07418	0.00653	-0.02360
(sps)	0.00000	0.00000	0.00000
(pps)	0.00000	0.00000	0.00000
(ppp)	0.00000	0.00000	0.00000
(dss)	-0.00637	0.00939	-0.00656
(dps)	0.00000	0.00000	0.00000
(dpp)	0.00000	0.00000	0.00000

♀FeH

	ORTHOGONA			NON-ORTHOGONAL		
BAND	RMS ERROR	MAXIMUM DEVIATION		RMS ERROR	MAXIMUM DEVIATION	
	mRy	k	mRy	mRy	k	mRy
1	7.0	(444)	17.2	0.5	(226)	1.2
2	7.4	(333)	30.6	0.4	(033)	1.1
3	5.9	(226)	14.2	0.3	(044)	0.8
4	2.6	(003)	6.0	0.2	(007)	0.5
5	2.5	(003)	6.0	0.4	(055)	1.2
6	5.6	(055)	16.2	0.5	(222)	1.1
7	14.3	(006)	45.5	0.7	(022)	1.7
1-7	7.4			0.5		

ENERGY VALUES IN Ry AT SELECTED k-POINTS

	ORTHOGONAL	APW	NON-ORTHOGONAL
GAMMA 1	-0.01953	-0.01699	-0.01712
GAMMA 1	2.00187	2.01665	2.01627
GAMMA 12	0.86450	0.86228	0.86316
GAMMA 15	2.47309	2.49258	2.49249
GAMMA 25'	0.76596	0.76845	0.76851
X1 (008)	0.30317	0.29106	0.29007
X1 (008)	1.28323	1.29198	1.29239
X1 (008)	2.12965	2.11657	2.11714
X2 (008)	0.92425	0.92669	0.92640
X3 (008)	0.60252	0.59987	0.60054
X5 (008)	0.95286	0.95005	0.95005
X4' (008)	0.94772	0.94889	0.94890

X5' (008)	1.73047	1.76119	1.75973
L1 (444)	0.50746	0.51352	0.51332
L1 (444)	1.31405	1.30637	1.30579
L3 (444)	0.74565	0.74977	0.74961
L3 (444)	0.93187	0.93031	0.93042
L2' (444)	0.45176	0.43459	0.43404
L2' (444)	1.78562	1.80672	1.80668
L3' (444)	2.57875	2.51258	2.51267
W1 (048)	0.83616	0.83792	0.83808
W1 (048)	1.60418	1.60137	1.60127
W2' (048)	0.35737	0.35926	0.35936
W2' (048)	1.22670	1.23659	1.23706
W2' (048)	2.63934	2.68894	2.68957
W1' (048)	0.95372	0.95024	0.95038
W3 (048)	0.67545	0.68289	0.68251
W3 (048)	1.38549	1.37649	1.37645
ODD (224)	0.78644	0.78852	0.78835
ODD (224)	0.91924	0.91769	0.91668
EVEN(224)	0.25329	0.25544	0.25569
EVEN(224)	0.65866	0.64734	0.64706
EVEN(224)	0.74868	0.74701	0.74743
EVEN(224)	0.89922	0.89585	0.89586
EVEN(224)	1.45816	1.44898	1.44944

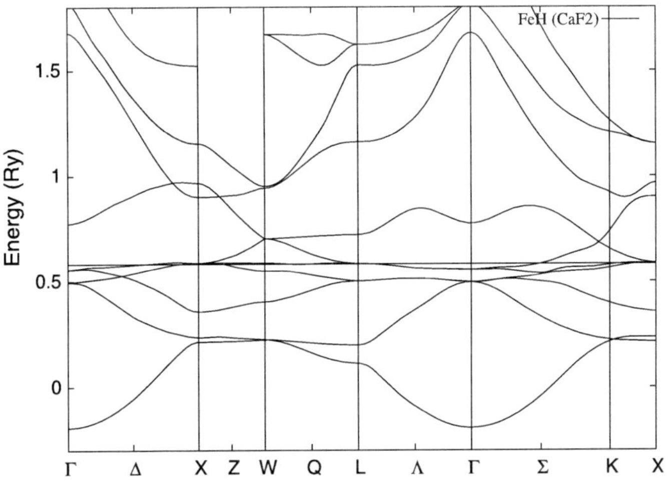

Fig. 6.24 Energy bands of FeH2 in the CaF2 structure

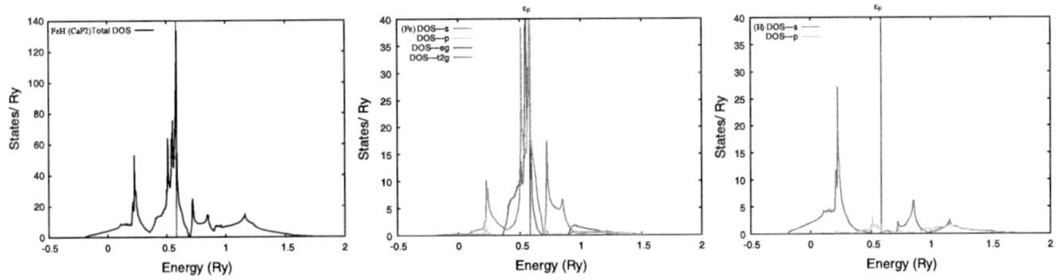

Fig. 6.25 Total, angular momentum and site decomposed densities of states of FeH2 in the CaF2 structure

6.6 Cobalt Hydride (CoH)

See Fig. 6.26 and Tables 6.26, 6.27, 6.28.

See Figs. 6.27, 6.28 and Tables 6.29, 6.30.

See Figs. 6.29 and 6.30.

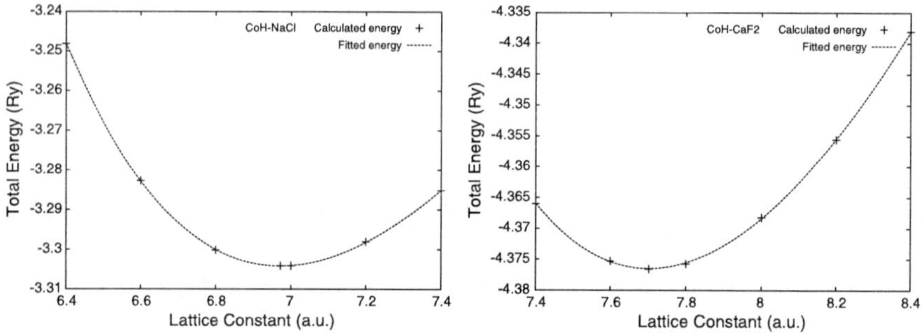

Fig. 6.26 Total energy versus lattice constant of CoH in the NaCl and CaF2 structures

Table 6.26 Lattice constant, bulk modulus, gap, total energy

Stru	a (Bohr)	B (MBar)	Gap	Total Energy (Ry)
NaCl	6.97	2.39	-	-2783.30412
CaF2	7.70	1.73		-2784.37657
exp (CaF2)	?			

Table 6.27 Birch fit coefficients

	A1	A2	A3	A4
NaCl	-2.060163E+00	-4.210759E+01	2.357279E+02	2.191846E+03
CaF2	-2.870207E+00	-7.074747E+01	8.267883E+02	9.275451E+01

Table 6.28 DOS at Ef, Hopfield parameter, Stoner criterion

CoH a=6.97 Bohr NaCl

Fermi Energy (Ry)		Total DOS (States/Ry)	Decomposed DOS inside the muffin tins s	p	eg	t2g	f
Co	0.899	67.448	0.032	0.108	17.162	46.074	0.013
H	0.899	67.448	0.111	0.299	0.123	0.026	0.141

Fermi-Velocity (cm/s): 0.29526 x10E8
Plasmon Energy (eV) : 9.48319
Electron-ion interaction (Hopfield parameter) (eV/A^2) Co: 2.819 H: 0.017

Co MUFFIN-TIN RADIUS and CHARGE = 1.8302 24.8156
H MUFFIN-TIN RADIUS and CHARGE = 1.6559 1.2513
Co STONER I = 0.03409
H STONER I = 0.00002
STONER PARAMETER (Ry) I = 0.03412
STONER CRITERION N*I = 2.3016

CoH2 a=7.70 Bohr CaF2

Fermi Energy (Ry)		Total DOS (States/Ry)	Decomposed DOS inside the muffin tins s	p	eg	t2g	f
Co	0.658	19.410	0.063	0.169	5.134	12.530	0.004
H	0.658	19.410	0.228	0.161	0.002	0.046	0.013

Fermi-Velocity (cm/s): 0.37158 x10E
Plasmon Energy (eV) : 5.51484
Electron-ion interaction (Hopfield parameter) (eV/A^2) Co: 0.534 H: 0.257

Co MUFFIN-TIN RADIUS and CHARGE = 2.0010 25.3216
H MUFFIN-TIN RADIUS and CHARGE = 1.3340 0.7756
Sc STONER I = 0.0287
H STONER I = 0.0000
STONER PARAMETER (Ry) I = 0.0288
STONER CRITERION N*I = 0.5589

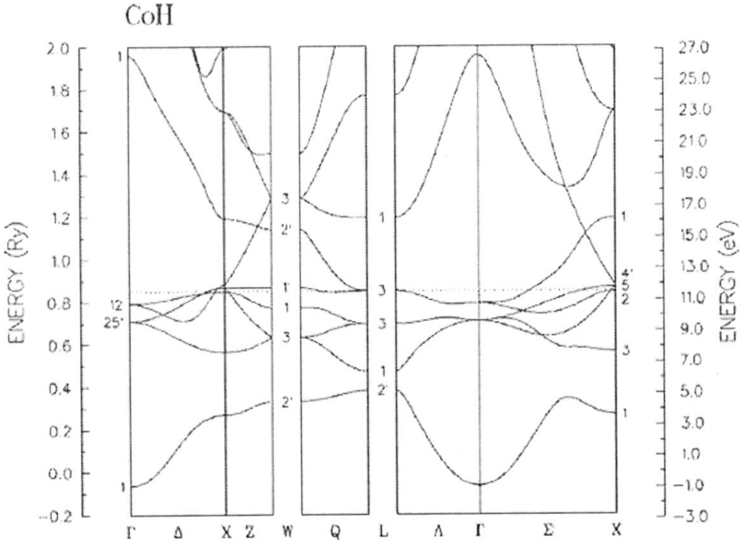

Fig. 6.27 Energy bands of CoH in the NaCl structure (tight-binding)

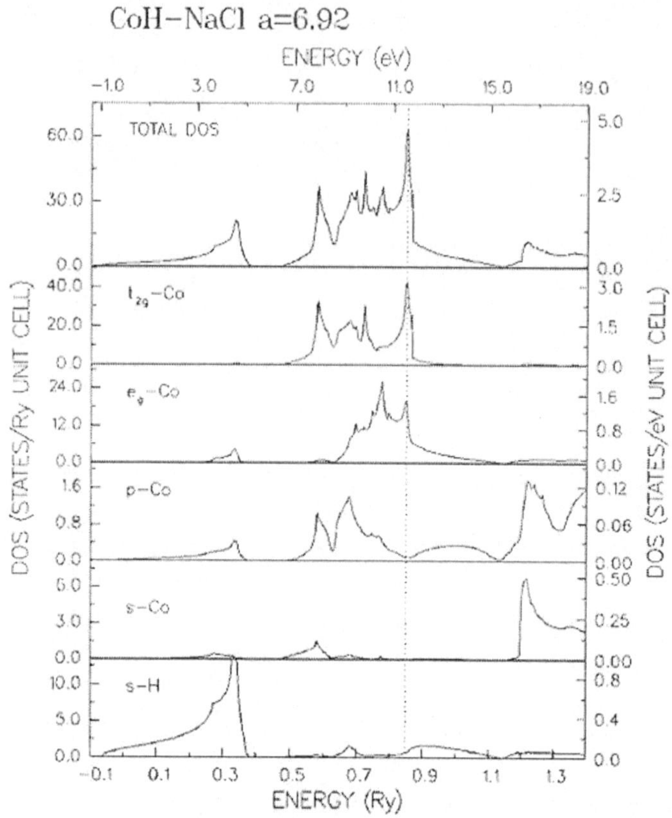

Fig. 6.28 Total, angular momentum and site decomposed densities of states of CoH in the NaCl structure (tight-binding)

Table 6.29 CoH (NaCl) a = 6.92 Bohr Slater–Koster 3-center parameters

	ORTHOGONAL	NON-ORTHOGONAL	
	ENERGY INTEGRALS Ry	ENERGY INTEGRALS Ry	OVERLAP INTEGRALS
ON SITE Co-Co			
s,s(000)	1.34942	1.23407	
x,x(000)	1.93950	1.65144	
xy,xy(000)	0.76675	0.74023	
d2,d2(000)	0.86121	0.71828	
FIRST NEIGHBOR Co-Co			
s,s(110)	0.00223	0.08684	0.08708
s,x(110)	-0.05521	-0.00418	0.01282
s,xy(110)	-0.05086	-0.05824	-0.02215
s,d2(110)	-0.02222	-0.01280	0.00698
x,x(110)	0.09778	0.06405	-0.04247
x,x(011)	-0.00310	0.08328	0.07181
x,y(110)	0.06025	-0.05539	-0.08460
x,xy(110)	0.05458	0.04901	0.01633
x,xy(011)	-0.00544	0.01986	0.05366
z,d2(011)	0.01489	0.04542	0.09646
z,d1(011)	0.02053	0.04105	0.04653
xy,xy(110)	-0.02916	-0.03183	-0.00472
xy,xy(011)	0.00919	0.00839	-0.00072
xy,xz(011)	0.00560	0.00920	-0.00553
xy,d2(110)	0.01516	-0.01738	-0.00113
d2,d2(110)	0.00895	0.00707	0.02827
d1,d1(110)	-0.03979	0.04118	0.00738
SECOND NEIGHBOR Co-Co			
s,s(200)	0.06996	0.01477	-0.00441
s,x(200)	0.08967	0.00239	-0.04656
s,d2(002)	0.04927	0.01555	0.02405
x,x(200)	-0.17720	-0.01174	0.01128
y,y(200)	0.02148	0.01573	0.01656
x,xy(020)	0.00886	0.01130	0.02090
z,d2(002)	-0.04756	-0.05220	-0.08455
xy,xy(200)	-0.00356	-0.00230	-0.00728
xy,xy(002)	0.00046	0.00365	0.00426
d2,d2(002)	0.03920	-0.04228	-0.03499
d1,d1(002)	-0.00025	0.00507	0.00762
ON SITE H-H			
s,s(000)	0.39246	0.32455	
FIRST NEIGHBOR H-H			
s,s(110)	-0.02400	-0.03017	-0.00223
s,x(110)	0.00000	0.00000	0.00000
x,x(110)	0.00000	0.00000	0.00000
x,x(011)	0.00000	0.00000	0.00000
x,y(110)	0.00000	0.00000	0.00000
SECOND NEIGHBOR H-H			
s,s(200)	-0.00464	-0.00538	0.01966
s,x(200)	0.00000	0.00000	0.00000
x,x(200)	0.00000	0.00000	0.00000
y,y(200)	0.00000	0.00000	0.00000
FIRST NEIGHBOR Co-H			
s,s(100)	-0.12394	-0.02758	0.10476
s,x(100)	0.00000	0.00000	0.00000
x,s(100)	0.06953	0.13699	0.23300
x,x(100)	0.00000	0.00000	0.00000
y,y(100)	0.00000	0.00000	0.00000
xy,x(010)	0.00000	0.00000	0.00000
d2,s(001)	-0.06026	-0.04506	-0.21532
d2,z(001)	0.00000	0.00000	0.00000

SECOND NEIGHBOR Co-H

s,s(111)	0.02808	0.02232	0.04185
x,s(111)	-0.04321	0.01267	-0.00044
s,x(111)	0.00000	0.00000	0.00000
x,x(111)	0.00000	0.00000	0.00000
x,y(111)	0.00000	0.00000	0.00000
xy,s(111)	-0.00029	-0.01398	-0.00977
xy,x(111)	0.00000	0.00000	0.00000
yz,x(111)	0.00000	0.00000	0.00000
d1,x(111)	0.00000	0.00000	0.00000

♀CoH

	ORTHOGONAL			NON-ORTHOGONAL		
BAND	RMS ERROR	MAXIMUM DEVIATION		RMS ERROR	MAXIMUM DEVIATION	
	mRy	k	mRy	mRy	k	mRy
1	3.9	(033)	10.2	1.0	(333)	2.3
2	4.8	(333)	14.7	0.7	(044)	2.3
3	6.1	(006)	14.4	0.9	(033)	2.1
4	2.0	(062)	4.8	0.8	(048)	1.8
5	2.4	(044)	6.4	0.5	(224)	2.0
6	3.6	(055)	11.0	1.0	(048)	2.7
7	9.6	(006)	22.6	2.2	(224)	6.6
1-7	5.2			1.1		

ENERGY VALUES IN Ry AT SELECTED k-POINTS

	ORTHOGONAL	APW	NON-ORTHOGONAL
GAMMA 1	-0.06788	-0.06798	-0.06613
GAMMA 1	1.94042	1.96065	1.96047
GAMMA 12	0.79303	0.79121	0.79183
GAMMA 15	2.44083	2.44766	2.44760
GAMMA 25'	0.71033	0.71025	0.71033
X1 (008)	0.27416	0.26796	0.26877
X1 (008)	1.18747	1.19156	1.19076
X1 (008)	1.99395	1.98838	1.98875
X2 (008)	0.84477	0.84685	0.84736
X3 (008)	0.56325	0.56242	0.56285
X5 (008)	0.87007	0.86713	0.86717
X4' (008)	0.87639	0.88195	0.88099
X5' (008)	1.68341	1.69164	1.69389
L1 (444)	0.46693	0.47262	0.47207
L1 (444)	1.19802	1.19551	1.19431
L3 (444)	0.69368	0.69428	0.69506
L3 (444)	0.85315	0.85014	0.85117
L2' (444)	0.37726	0.38310	0.38311
L2' (444)	1.76901	1.77021	1.77017
L3' (444)	2.44898	2.42738	2.42748
W1 (048)	0.76901	0.77114	0.77295
W1 (048)	1.49689	1.49658	1.49653
W2' (048)	0.32894	0.33203	0.33108
W2' (048)	1.14139	1.14300	1.14028
W2' (048)	2.53928	2.58660	2.58655
W1' (048)	0.86825	0.86728	0.86732
W3 (048)	0.62890	0.63329	0.63421
W3 (048)	1.29597	1.28666	1.28646
ODD (224)	0.72791	0.72802	0.72667
ODD (224)	0.84266	0.83931	0.83968
EVEN(224)	0.21006	0.20481	0.20508
EVEN(224)	0.60867	0.60262	0.60391
EVEN(224)	0.68796	0.69171	0.69299
EVEN(224)	0.82178	0.82139	0.81941
EVEN(224)	1.36843	1.36117	1.36782

FERMI LEVEL QUANTITIES (Non-orthogonal fit)

ENERGY	DENSITIES OF STATES						
	Total	s-Co	p-Co	t2g-Co	eg-Co	s-H	p-H
Ry	states/Ry/cell						
0.8483	62.57	0.05	0.10	41.65	20.02	0.75	0.00
	INTEGRATED DENSITIES OF STATES						
	Total	s-Co	p-Co	t2g-Co	eg-Co	s-H	p-H
	electrons						
	10.00	0.16	0.23	4.95	2.85	1.82	0.00

VELOCITY	PLASMON ENERGY	EIGENVALUE SUM
cm/s	eV	Ry
0.19x10E8	6.08	-2.2644

Table 6.30 CoH (NaCl) a = 6.92 Bohr Slater–Koster 2-center parameters

	ORTHOGONAL	NON-ORTHOGONAL	
	ENERGY INTEGRALS Ry	ENERGY INTEGRALS Ry	OVERLAP INTEGRALS
ON SITE Co-Co			
s	1.09583	0.76866	
p	1.88763	1.30686	
t2g	0.75095	0.73095	
eg	0.74633	0.72650	
FIRST NEIGHBOR Co-Co			
(sss)	-0.08808	-0.04285	0.09045
(pps)	0.19156	-0.07638	-0.23300
(ppp)	0.00166	0.01827	0.05448
(dds)	-0.03822	-0.01211	0.03467
(ddp)	0.01956	0.01444	-0.01631
(ddd)	-0.00175	-0.00314	0.00170
(sps)	0.14231	0.00047	-0.15103
(sds)	-0.05419	-0.02027	0.04835
(pds)	-0.07492	-0.01086	0.06217
(pdp)	0.01555	0.01687	-0.02097
SECOND NEIGHBOR Co-Co			
(sss)	0.01503	-0.03914	-0.02249
(pps)	-0.14718	0.12423	0.04677
(ppp)	0.01359	-0.03841	-0.02246
(dds)	-0.00141	-0.00712	-0.00083
(ddp)	0.00068	0.00366	0.00273
(ddd)	-0.00029	0.00111	0.00126
(sps)	-0.03957	0.06851	0.03292
(sds)	0.00446	-0.00650	0.00435
(pds)	0.00338	-0.00911	0.01020
(pdp)	-0.00717	0.01134	0.00825
ON SITE H-H			
s	1.10616	0.42400	

FIRST NEIGHBOR H-H			
(sss)	0.02596	-0.02127	0.01502
(sps)	0.00000	0.00000	0.00000
(pps)	0.00000	0.00000	0.00000
(ppp)	0.00000	0.00000	0.00000
SECOND NEIGHBOR H-H			
(sss)	0.05476	-0.00517	0.00193
(sps)	0.00000	0.00000	0.00000
(pps)	0.00000	0.00000	0.00000
(ppp)	0.00000	0.00000	0.00000
FIRST NEIGHBOR Co-H			
(sss)	0.14472	0.03726	-0.20582
(sps)	0.00000	0.00000	0.00000
(pss)	0.17369	-0.04507	-0.30122
(pps)	0.00000	0.00000	0.00000
(ppp)	0.00000	0.00000	0.00000
(dpp)	0.00000	0.00000	0.00000
(dss)	0.09522	0.01813	-0.13291
(dps)	0.00000	0.00000	0.00000
SECOND NEIGHBOR Co-H			
(sss)	-0.03336	0.00627	-0.01258
(pss)	-0.06969	0.01431	-0.01953
(sps)	0.00000	0.00000	0.00000
(pps)	0.00000	0.00000	0.00000
(ppp)	0.00000	0.00000	0.00000
(dss)	-0.00598	0.00948	-0.00583
(dps)	0.00000	0.00000	0.00000
(dpp)	0.00000	0.00000	0.00000

♀CoH

	ORTHOGONAL			NON-ORTHOGONAL		
BAND	RMS ERROR	MAXIMUM DEVIATION		RMS ERROR	MAXIMUM DEVIATION	
	mRy	k	mRy	mRy	k	mRy
1	7.7	(444)	19.5	0.4	(008)	0.9
2	6.8	(333)	28.7	0.4	(033)	0.9
3	5.5	(226)	13.5	0.3	(044)	0.8
4	2.4	(003)	5.3	0.2	(007)	0.4
5	2.2	(003)	5.3	0.4	(055)	1.1
6	5.6	(055)	15.8	0.5	(066)	1.3
7	15.8	(006)	52.5	0.6	(022)	1.7
1-7	7.8			0.4		

ENERGY VALUES IN Ry AT SELECTED k-POINTS

	ORTHOGONAL	APW	NON-ORTHOGONAL
GAMMA 1	-0.07013	-0.06798	-0.06814
GAMMA 1	1.94536	1.96065	1.95970
GAMMA 12	0.79336	0.79121	0.79210
GAMMA 15	2.42720	2.44766	2.44762
GAMMA 25'	0.70791	0.71025	0.71043
X1 (008)	0.28162	0.26796	0.26703
X1 (008)	1.18449	1.19156	1.19175
X1 (008)	2.00525	1.98838	1.98832
X2 (008)	0.84554	0.84685	0.84668
X3 (008)	0.56545	0.56242	0.56310
X5 (008)	0.86950	0.86713	0.86731
X4' (008)	0.88141	0.88195	0.88195

X5' (008)	1.64099	1.69164	1.69085
L1 (444)	0.46723	0.47262	0.47273
L1 (444)	1.20196	1.19551	1.19518
L3 (444)	0.69152	0.69428	0.69425
L3 (444)	0.85134	0.85014	0.85022
L2' (444)	0.40263	0.38310	0.38262
L2' (444)	1.74303	1.77021	1.77015
L3' (444)	2.50741	2.42738	2.42741
W1 (048)	0.76793	0.77114	0.77139
W1 (048)	1.50065	1.49658	1.49636
W2' (048)	0.32946	0.33203	0.33254
W2' (048)	1.13104	1.14300	1.14335
W2' (048)	2.54519	2.58660	2.58658
W1' (048)	0.87065	0.86728	0.86723
W3 (048)	0.62570	0.63329	0.63292
W3 (048)	1.29588	1.28666	1.28654
ODD (224)	0.72636	0.72802	0.72803
ODD (224)	0.84084	0.83931	0.83831
EVEN(224)	0.20303	0.20481	0.20480
EVEN(224)	0.61331	0.60262	0.60225
EVEN(224)	0.69290	0.69171	0.69197
EVEN(224)	0.82406	0.82139	0.82145
EVEN(224)	1.37354	1.36117	1.36174

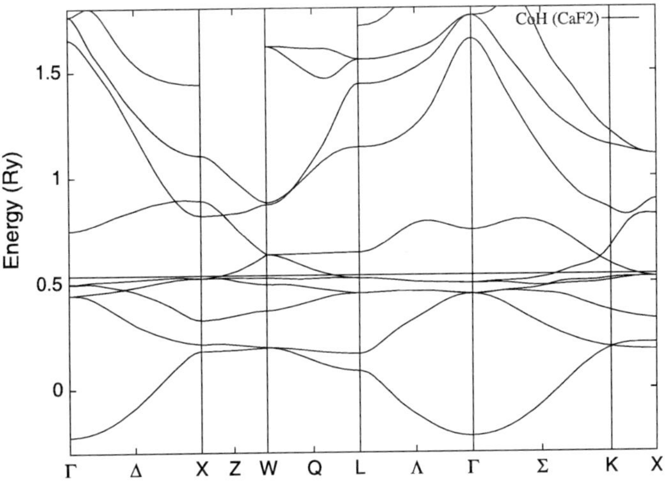

Fig. 6.29 Energy bands of CoH2 in the CaF2 structure

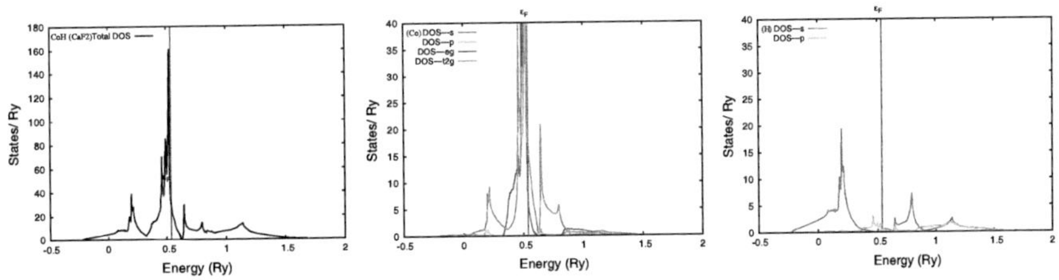

Fig. 6.30 Total, angular momentum and site decomposed densities of states of CoH2 in the CaF2 structure

6.7 Nickel Hydride (NiH)

See Fig. 6.31 and Tables 6.31, 6.32, 6.33.

See Figs. 6.32, 6.33 and Tables 6.34, 6.35.

See Figs. 6.34 and 6.35.

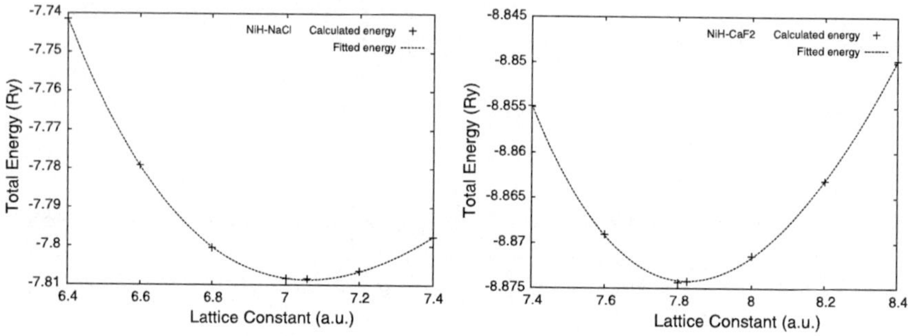

Fig. 6.31 Total energy versus lattice constant of NiH in the NaCl and CaF2 structures

Table 6.31 Lattice constant, bulk modulus, gap, total energy

Stru	a (Bohr)	B (MBar)	Gap	Total Energy (Ry)
NaCl	7.06	2.06		-3037.80840
CaF2	7.82	1.52		-3038.87423
exp (NaCl)	7.03			

Table 6.32 Birch fit coefficients

	A1	A2	A3	A4
NaCl	-6.685060E+00	-3.926600E+01	2.352680E+02	2.014019E+03
CaF2	-7.455102E+00	-6.964245E+01	8.708343E+02	-4.005462E+03

Table 6.33 DOS at Ef, Hopfield parameter, Stoner criterion

```
                 NiH       a=7.06 Bohr      NaCl
-------------------------------------------------------------------------------
 Fermi Energy   Total DOS           Decomposed DOS inside the muffin tins
    (Ry)        (States/Ry)       s          p         eg         t2g        f
-------------------------------------------------------------------------------
Ni  0.825       11.499          0.072      0.181      4.669      5.189      0.003
H   0.825       11.499          0.331      0.187      0.018      0.003      0.021
-------------------------------------------------------------------------------
Fermi-Velocity (cm/s): 0.60637 x10E8
Plasmon Energy (eV)  : 7.89626

Electron-ion interaction (Hopfield parameter) (eV/A^2) Ni: 0.730 H: 0.168
------------------------------------------------
  Ni      MUFFIN-TIN RADIUS and CHARGE =    1.8526    25.9837
   H      MUFFIN-TIN RADIUS and CHARGE =    1.6762     1.2083
           Ni        STONER       I = 0.0293
            H        STONER       I = 0.0004
          STONER PARAMETER (Ry) I = 0.0298
          STONER CRITERION    N*I = 0.3422
------------------------------------------------

                 NiH2      a=7.82 Bohr          CaF2
-------------------------------------------------------------------------------
 Fermi Energy   Total DOS           Decomposed DOS inside the muffin tins
    (Ry)        (States/Ry)       s          p         eg         t2g        f
-------------------------------------------------------------------------------
Ni 0.661         0.638          0.005      0.032      0.007      0.435     0.001
 H 0.661         0.638          0.070      0.010      0.000      0.001     0.000
-------------------------------------------------------------------------------
Fermi-Velocity (cm/s): 0.57010 x10E8
Plasmon Energy (eV)  : 1.49777
Electron-ion interaction (Hopfield parameter) (eV/A^2) Ni: 0.066  H: 0.131
------------------------------------------------
         Ni     MUFFIN-TIN RADIUS and CHARGE =    2.0330    26.4653
          H     MUFFIN-TIN RADIUS and CHARGE =    1.3553     0.7618
                  Ni        STONER I = 0.0191
                   H        STONER I = 0.0012
         STONER PARAMETER (Ry)      I = 0.0216
         STONER CRITERION         N*I = 0.0137
```

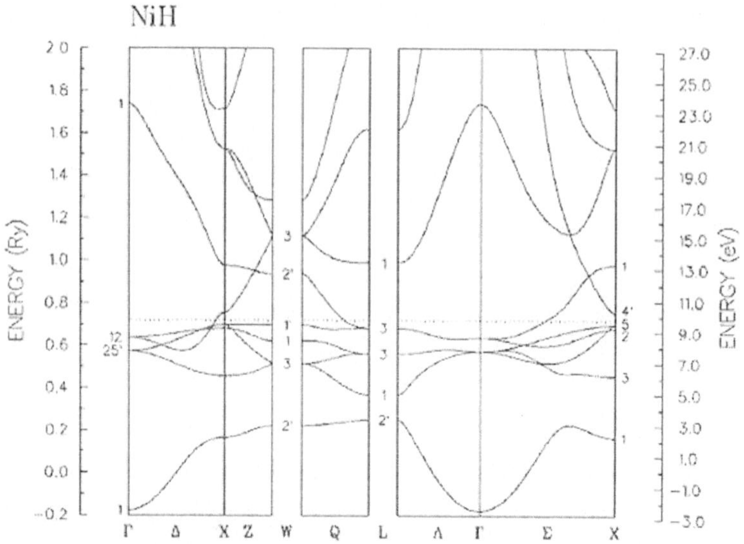

Fig. 6.32 Energy bands of NiH in the NaCl structure (tight-binding)

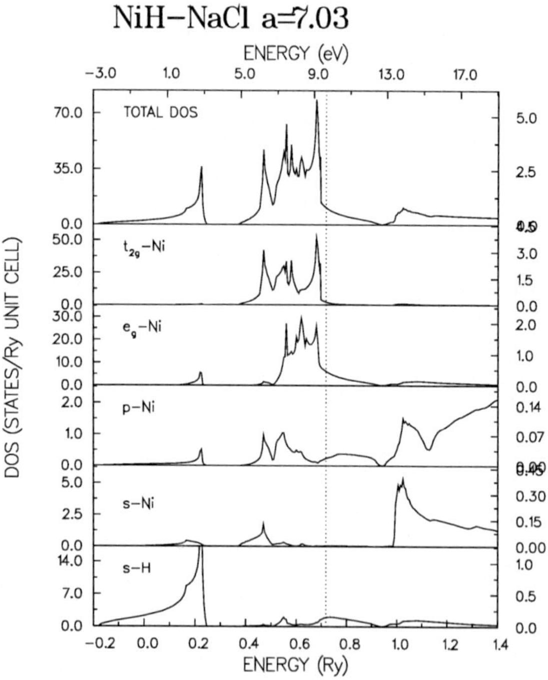

Fig. 6.33 Total, angular momentum and site decomposed densities of states of NiH in the NaCl structure (tight-binding)

Table 6.34 NiH (NaCl) a = 7.03 Bohr Slater–Koster 3-center parameters

	ORTHOGONAL	NON-ORTHOGONAL	
	ENERGY INTEGRALS Ry	ENERGY INTEGRALS Ry	OVERLAP INTEGRALS
ON SITE Ni-Ni			
s,s(000)	1.16080	0.99267	
x,x(000)	1.76867	1.48256	
xy,xy(000)	0.61357	0.59788	
d2,d2(000)	0.69896	0.56601	
FIRST NEIGHBOR Ni-Ni			
s,s(110)	0.00420	0.06602	0.09310
s,x(110)	-0.04728	0.00443	0.03319
s,xy(110)	-0.04271	-0.04923	-0.02059
s,d2(110)	-0.02124	-0.01164	0.01529
x,x(110)	0.09461	0.01568	-0.08719
x,x(011)	-0.00209	0.07066	0.09133
x,y(110)	0.05297	-0.00334	-0.04336
x,xy(110)	0.04445	0.05305	0.03683
x,xy(011)	-0.00546	0.01196	0.05898
z,d2(011)	0.01130	0.05788	0.12329
z,d1(011)	0.01869	0.05093	0.06002
xy,xy(110)	-0.02322	-0.02812	-0.00993
xy,xy(011)	0.00734	0.00853	0.00337
xy,xz(011)	0.00568	0.01185	0.00171
xy,d2(110)	0.01203	-0.01103	0.00287
d2,d2(110)	0.00800	0.00312	0.02771
d1,d1(110)	-0.03443	0.03562	0.00199
SECOND NEIGHBOR Ni-Ni			
s,s(200)	0.06344	-0.00809	-0.02271
s,x(200)	0.08596	-0.03436	-0.07757
s,d2(002)	0.04589	0.00925	0.01154
x,x(200)	-0.16859	0.04397	0.05879
y,y(200)	0.02120	-0.00557	0.01920
x,xy(020)	0.00553	0.00408	0.01958
z,d2(002)	-0.04730	-0.02895	-0.05050
xy,xy(200)	-0.00175	-0.00262	-0.00908
xy,xy(002)	0.00018	0.00363	0.00509
d2,d2(002)	0.03282	-0.02563	-0.01575
d1,d1(002)	-0.00027	0.01078	0.01769
ON SITE H-H			
s,s(000)	0.24217	0.16702	
FIRST NEIGHBOR H-H			
s,s(110)	-0.02257	-0.02885	0.01103
s,x(110)	0.00000	0.00000	0.00000
x,x(110)	0.00000	0.00000	0.00000
x,x(011)	0.00000	0.00000	0.00000
x,y(110)	0.00000	0.00000	0.00000
SECOND NEIGHBOR H-H			
s,s(200)	-0.00355	-0.00720	0.02936
s,x(200)	0.00000	0.00000	0.00000
x,x(200)	0.00000	0.00000	0.00000
y,y(200)	0.00000	0.00000	0.00000
FIRST NEIGHBOR Ni-H			
s,s(100)	-0.11025	-0.02555	0.14611
s,x(100)	0.00000	0.00000	0.00000
x,s(100)	0.04823	0.06178	0.16105
x,x(100)	0.00000	0.00000	0.00000
y,y(100)	0.00000	0.00000	0.00000
xy,x(010)	0.00000	0.00000	0.00000
d2,s(001)	-0.04713	-0.03639	-0.24048
d2,z(001)	0.00000	0.00000	0.00000

SECOND NEIGHBOR Ni-H			
s,s(111)	0.02283	0.01275	0.03516
x,s(111)	-0.03802	0.01385	-0.00249
s,x(111)	0.00000	0.00000	0.00000
x,x(111)	0.00000	0.00000	0.00000
x,y(111)	0.00000	0.00000	0.00000
xy,s(111)	-0.00289	-0.01051	-0.00495
xy,x(111)	0.00000	0.00000	0.00000
yz,x(111)	0.00000	0.00000	0.00000
d1,x(111)	0.00000	0.00000	0.00000

♀NiH

	ORTHOGONAL			NON-ORTHOGONAL		
BAND	RMS ERROR	MAXIMUM DEVIATION		RMS ERROR	MAXIMUM DEVIATION	
	mRy	k	mRy	mRy	k	mRy
1	3.2	(033)	8.7	0.7	(280)	1.2
2	4.1	(333)	12.5	0.7	(044)	1.5
3	5.1	(006)	13.7	0.8	(033)	1.9
4	1.7	(222)	4.0	0.4	(042)	1.0
5	1.7	(044)	4.3	0.4	(224)	1.8
6	3.5	(055)	9.1	0.8	(055)	2.2
7	8.8	(006)	22.3	1.6	(224)	4.1
1-7	4.6			0.9		

ENERGY VALUES IN Ry AT SELECTED k-POINTS

	ORTHOGONAL	APW	NON-ORTHOGONAL
GAMMA 1	-0.17945	-0.17986	-0.17885
GAMMA 1	1.72126	1.73738	1.73781
GAMMA 12	0.63802	0.63634	0.63676
GAMMA 15	2.26482	2.26913	2.26903
GAMMA 25'	0.57279	0.57273	0.57262
X1 (008)	0.16835	0.16463	0.16365
X1 (008)	0.97138	0.97618	0.97627
X1 (008)	1.71525	1.71565	1.71580
X2 (008)	0.67979	0.68075	0.68043
X3 (008)	0.45535	0.45564	0.45630
X5 (008)	0.69984	0.69746	0.69765
X4' (008)	0.75103	0.75631	0.75555
X5' (008)	1.52466	1.52658	1.52654
L1 (444)	0.36425	0.36922	0.36811
L1 (444)	0.99070	0.98812	0.98778
L3 (444)	0.55961	0.55942	0.55952
L3 (444)	0.68458	0.68367	0.68413
L2' (444)	0.24287	0.24885	0.24928
L2' (444)	1.61792	1.61964	1.61958
L3' (444)	2.23293	2.22034	2.22015
W1 (048)	0.61846	0.62117	0.62105
W1 (048)	1.28579	1.28391	1.28389
W2' (048)	0.21921	0.22189	0.22078
W2' (048)	0.93589	0.93654	0.93568
W2' (048)	2.30469	2.33755	2.33789
W1' (048)	0.69911	0.69755	0.69734
W3 (048)	0.51104	0.51382	0.51427
W3 (048)	1.12658	1.11535	1.11560
ODD (224)	0.58478	0.58682	0.58643
ODD (224)	0.67776	0.67512	0.67546
EVEN(224)	0.08766	0.08305	0.08319
EVEN(224)	0.49022	0.48543	0.48661
EVEN(224)	0.55612	0.55812	0.55922
EVEN(224)	0.66251	0.66173	0.65997
EVEN(224)	1.16739	1.16218	1.16627

FERMI LEVEL QUANTITIES (Non-orthogonal fit)

ENERGY	DENSITIES OF STATES						
	Total	s-Ni	p-Ni	t2g-Ni	eg-Ni	s-H	p-H
Ry	states/Ry/cell						
0.7185	10.72	0.05	0.23	2.18	6.29	1.97	0.00
	INTEGRATED DENSITIES OF STATES						
	Total	s-Ni	p-Ni	t2g-Ni	eg-Ni	s-H	p-H
	electrons						
	11.00	0.14	0.15	5.53	3.20	1.99	0.00

VELOCITY	PLASMON ENERGY	EIGENVALUE SUM
cm/s	eV	Ry
0.45x10E8	5.63	-2.3547

Table 6.35 NiH (NaCl) a = 7.03 Slater–Koster 2-center parameters

	ORTHOGONAL	NON-ORTHOGONAL	
	ENERGY INTEGRALS Ry	ENERGY INTEGRALS Ry	OVERLAP INTEGRALS
ON SITE Ni-Ni			
s	0.89031	0.52629	
p	1.69324	1.21402	
t2g	0.60637	0.58983	
eg	0.60409	0.58830	
FIRST NEIGHBOR Ni-Ni			
(sss)	-0.08849	0.01182	0.16654
(pps)	0.18926	0.01491	-0.19612
(ppp)	-0.00203	0.10423	0.09490
(dds)	-0.03038	-0.01306	0.02930
(ddp)	0.01441	0.01881	-0.00382
(ddd)	-0.00071	-0.00722	-0.00559
(sps)	0.14126	-0.02902	-0.19106
(sds)	-0.04784	-0.02072	0.03842
(pds)	-0.06391	-0.03226	0.02161
(pdp)	0.01359	-0.00387	-0.05051
SECOND NEIGHBOR Ni-Ni			
(sss)	0.01340	-0.05206	-0.01763
(pps)	-0.12838	0.18395	0.06527
(ppp)	0.01858	-0.03107	-0.01729
(dds)	-0.00039	0.00517	0.01576
(ddp)	0.00034	0.00768	0.00961
(ddd)	-0.00051	-0.00022	-0.00040
(sps)	-0.03249	0.09436	0.03669
(sds)	0.00237	-0.00301	0.01094
(pds)	-0.00164	-0.01178	-0.00359
(pdp)	-0.00577	0.01937	0.01862
ON SITE H-H			
s	0.97008	0.37086	

FIRST NEIGHBOR H-H			
(sss)	0.03677	0.01111	-0.04059
(sps)	0.00000	0.00000	0.00000
(pps)	0.00000	0.00000	0.00000
(ppp)	0.00000	0.00000	0.00000
SECOND NEIGHBOR H-H			
(sss)	0.03594	-0.01277	0.02150
(sps)	0.00000	0.00000	0.00000
(pps)	0.00000	0.00000	0.00000
(ppp)	0.00000	0.00000	0.00000
FIRST NEIGHBOR Ni-H			
(sss)	0.11380	0.09471	-0.11962
(sps)	0.00000	0.00000	0.00000
(pss)	0.18263	0.07281	-0.20721
(pps)	0.00000	0.00000	0.00000
(ppp)	0.00000	0.00000	0.00000
(dpp)	0.00000	0.00000	0.00000
(dss)	0.08515	-0.01991	-0.21754
(dps)	0.00000	0.00000	0.00000
SECOND NEIGHBOR Ni-H			
(sss)	-0.03470	0.02852	0.01818
(pss)	-0.06147	0.06642	0.03697
(sps)	0.00000	0.00000	0.00000
(pps)	0.00000	0.00000	0.00000
(ppp)	0.00000	0.00000	0.00000
(dss)	-0.00880	0.02613	0.04273
(dps)	0.00000	0.00000	0.00000
(dpp)	0.00000	0.00000	0.00000

♀ NiH

BAND	ORTHOGONAL RMS ERROR mRy	MAXIMUM DEVIATION k	mRy	NON-ORTHOGONAL RMS ERROR mRy	MAXIMUM DEVIATION k	mRy
1	9.9	(444)	26.4	0.9	(044)	2.8
2	5.7	(333)	22.1	0.5	(333)	1.2
3	3.6	(264)	9.1	0.6	(004)	1.3
4	2.3	(002)	4.3	0.6	(224)	1.2
5	2.1	(003)	4.2	0.6	(444)	1.1
6	5.7	(048)	14.2	1.0	(226)	2.0
7	16.5	(006)	57.0	1.2	(380)	2.5
1-7	8.1			0.8		

ENERGY VALUES IN Ry AT SELECTED k-POINTS

	ORTHOGONAL	APW	NON-ORTHOGONAL
GAMMA 1	-0.18198	-0.17986	-0.18164
GAMMA 1	1.71775	1.73738	1.73550
GAMMA 12	0.63908	0.63634	0.63718
GAMMA 15	2.25160	2.26913	2.26925
GAMMA 25'	0.56965	0.57273	0.57290
X1 (008)	0.18151	0.16463	0.16387
X1 (008)	0.97119	0.97618	0.97549
X1 (008)	1.70931	1.71565	1.71547
X2 (008)	0.67901	0.68075	0.67954
X3 (008)	0.46007	0.45564	0.45558
X5 (008)	0.69857	0.69746	0.69772
X4' (008)	0.75378	0.75631	0.75738

X5' (008)	1.51891	1.52658	1.52655
L1 (444)	0.36170	0.36922	0.36975
L1 (444)	0.99373	0.98812	0.98789
L3 (444)	0.55689	0.55942	0.55893
L3 (444)	0.68616	0.68367	0.68480
L2' (444)	0.27524	0.24885	0.24884
L2' (444)	1.58971	1.61964	1.61962
L3' (444)	2.25824	2.22034	2.22035
W1 (048)	0.61855	0.62117	0.62235
W1 (048)	1.28802	1.28391	1.28305
W2' (048)	0.22216	0.22189	0.22203
W2' (048)	0.92236	0.93654	0.93639
W2' (048)	2.32857	2.33755	2.33748
W1' (048)	0.70059	0.69755	0.69748
W3 (048)	0.50920	0.51382	0.51378
W3 (048)	1.13078	1.11535	1.11611
ODD (224)	0.58575	0.58682	0.58801
ODD (224)	0.67742	0.67512	0.67449
EVEN(224)	0.08360	0.08305	0.08300
EVEN(224)	0.49394	0.48543	0.48543
EVEN(224)	0.55742	0.55812	0.55809
EVEN(224)	0.66481	0.66173	0.66105
EVEN(224)	1.17103	1.16218	1.16368

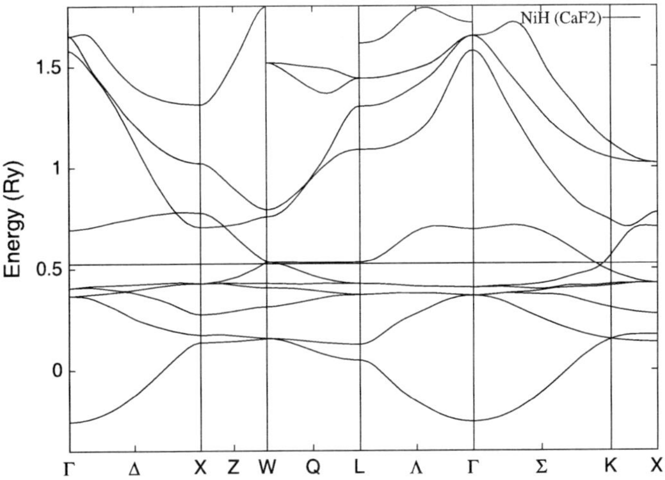

Fig. 6.34 Energy bands of NiH2 in the CaF2 structure

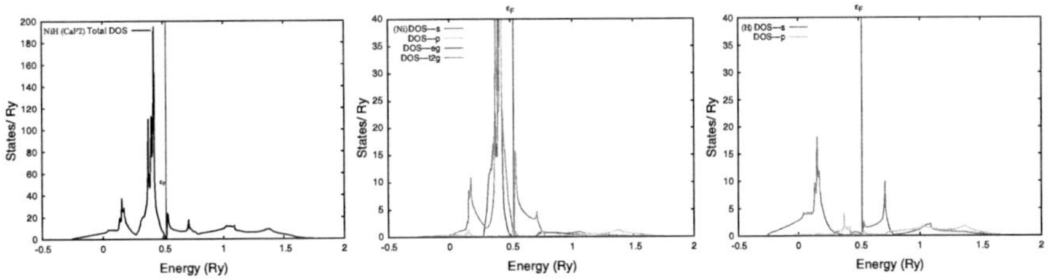

Fig. 6.35 Total, angular momentum and site decomposed densities of states of NiH2 in the CaF2 structure

6.8 Copper Hydride (CuH)

See Fig. 6.36 and Tables 6.36, 6.37, 6.38.

See Figs. 6.37, 6.38 and Tables 6.39, 6.40.

See Figs. 6.39, 6.40, 6.41 and 6.42.

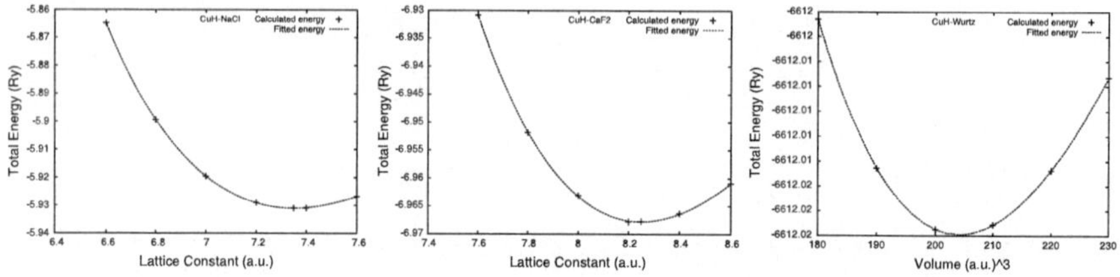

Fig. 6.36 Total energy versus lattice constant of CuH in the NaCl, wurtzite, and CaF2 structures

Table 6.36 Lattice constant, bulk modulus, gap, total energy

Stru	a (Bohr)	c (Bohr)	B (MBar)	Gap (Ry)	Total Energy (Ry)
NaCl (semimetal)	7.35		1.37		-3305.93100
CsCl	4.64		1.40		-3305.93038
CaF2	8.25		1.03		-3306.96771
wurtzite semicond	5.28	8.45			
exp (wurtz)	5.46	8.71			

Table 6.37 Birch fit coefficients

	A1	A2	A3	A4
NaCl	-5.315807E+00	-1.720324E+01	-1.107757E+02	4.219858E+03
CsCl	-5.075025E+00	-3.216777E+01	1.985228E+02	2.111685E+03
CaF2	-6.177498E+00	-3.418333E+01	1.169988E+02	6.205575E+03
Wurz	-6.610479E+03	-8.373897E+01	2.513990E+02	2.773389E+04

Table 6.38 DOS at Ef, Hopfield parameter, Stoner criterion

```
                    CuH          a=7.35 Bohr          NaCl
 Fermi Energy    Total DOS          Decomposed DOS inside the muffin tins
   (Ry)         (States/Ry)       s           p          eg          t2g          f
-------------------------------------------------------------------------------------
Cu  0.835       0.405          0.067     0.004      0.033      0.126       0.000
H   0.835       0.405          0.016     0.029      0.000      0.000       0.000
-------------------------------------------------------------------------------------
Fermi-Velocity (cm/s): 1.33484 x10E8
Plasmon Energy (eV)   : 2.54043
Electron-ion interaction (Hopfield parameter) (eV/A^2) Cu: 0.010 H: 0.034
-----------------------------------------------
  Cu      MUFFIN-TIN RADIUS and CHARGE =     1.9295   27.1631
   H      MUFFIN-TIN RADIUS and CHARGE =     1.7457    1.2098
           Cu        STONER I = 0.0104
            H        STONER I = 0.0021
    STONER PARAMETER (Ry) I = 0.0129
    STONER CRITERION    N*I = 0.0052
---------------------------------------------
                  CuH       a=4.64 Bohr          CsCl
---------------------------------------------
 Fermi Energy    Total DOS          Decomposed DOS inside the muffin tins
   (Ry)         (States/Ry)       s           p          eg          t2g          f
---------------------------------------------
Cu  0.780       8.532          0.848     0.201      2.393      1.327      0.053
H   0.780       8.532          2.394     0.423      0.001      0.022      0.005
--------------------------------------------
Fermi-Velocity (cm/s): 0.60184 x10E8
Plasmon Energy (eV)   : 6.33844
Electron-ion interaction (Hopfield parameter) (eV/A^2) Cu: 1.599 H: 1.908
-------------------------------------------------------------------------------------
  Cu      MUFFIN-TIN RADIUS and CHARGE =     2.1083   27.6376
   H      MUFFIN-TIN RADIUS and CHARGE =     1.9075    1.2913
           Cu        STONER I = 0.0102
            H        STONER I = 0.0198
    STONER PARAMETER (Ry) I = 0.0300
    STONER CRITERION    N*I = 0.2561
-------------------------------------------------------------------------------------
                  CuH2      a=8.25 Bohr      CaF2
          ---------------------------------------------------------------------------
-----  Fermi Energy   Total DOS          Decomposed DOS inside the muffin tins
    (Ry)           (States/Ry)       s          p          eg          t2g          f
-------------------------------------------------------------------------------
 Cu  0.548        6.193          0.190     0.376      0.019      2.776       0.018
  H  0.548        6.193          1.589     0.099      0.001      0.009       0.002
-------------------------------------------------------------------------------
Fermi-Velocity (cm/s): 1.01581 x10E8
Plasmon Energy (eV)   : 7.68194
Electron-ion interaction (Hopfield parameter) (eV/A^2) Cu: 0.649  H: 2.517
-----------------------------------------------
         Cu      MUFFIN-TIN RADIUS and CHARGE =     2.1434   27.6330
          H      MUFFIN-TIN RADIUS and CHARGE =     1.4289    0.7736
                  Cu         STONER I = 0.0093
                   H         STONER I = 0.0055
         STONER PARAMETER (Ry)      I = 0.0204
         STONER CRITERION         N*I = 0.1262
```

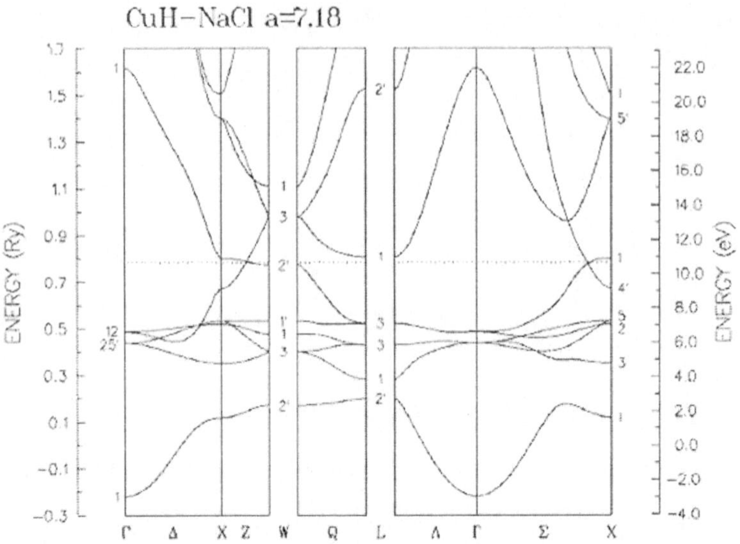

Fig. 6.37 Energy bands of CuH in the NaCl structure (tight-binding)

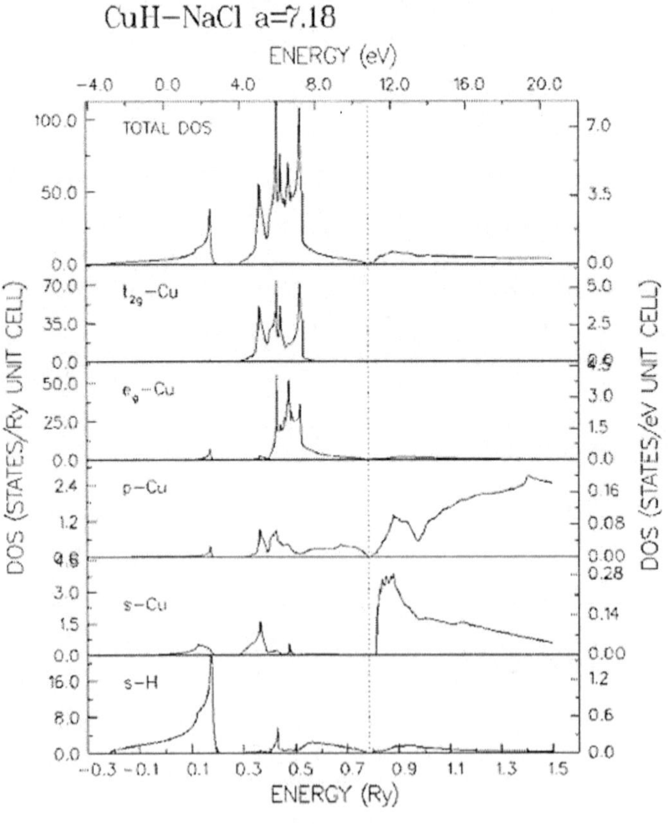

Fig. 6.38 Total, angular momentum and site decomposed densities of states of CuH in the NaCl structure (tight-binding)

Table 6.39 CuH (NaCl) a = 7.18 Bohr Slater–Koster 3-center parameters

	ORTHOGONAL	NON-ORTHOGONAL	
	ENERGY INTEGRALS Ry	ENERGY INTEGRALS Ry	OVERLAP INTEGRALS
ON SITE Cu-Cu			
s,s(000)	1.01334	0.80219	
x,x(000)	1.62997	1.37037	
xy,xy(000)	0.46691	0.45991	
d2,d2(000)	0.55233	0.42508	
FIRST NEIGHBOR Cu-Cu			
s,s(110)	0.00710	0.04347	0.09264
s,x(110)	-0.03904	-0.00315	0.02396
s,xy(110)	-0.03478	-0.03935	-0.02197
s,d2(110)	-0.02505	-0.01151	0.02747
x,x(110)	0.09060	0.02818	-0.07828
x,x(011)	-0.00008	0.07997	0.09826
x,y(110)	0.04423	0.00810	-0.04341
x,xy(110)	0.03523	0.04350	0.03958
x,xy(011)	-0.00501	0.00532	0.05384
z,d2(011)	0.00631	0.04254	0.12072
z,d1(011)	0.01931	0.04476	0.06065
xy,xy(110)	-0.01746	-0.02417	-0.01651
xy,xy(011)	0.00548	0.00823	0.00834
xy,xz(011)	0.00460	0.01210	0.00943
xy,d2(110)	0.00905	-0.01199	-0.00519
d2,d2(110)	0.00905	0.00404	0.03371
d1,d1(110)	-0.03488	0.02347	-0.01383
SECOND NEIGHBOR Cu-Cu			
s,s(200)	0.05931	-0.01285	-0.02590
s,x(200)	0.08298	-0.03150	-0.08385
s,d2(002)	0.04678	0.01193	0.00649
x,x(200)	-0.16235	0.01349	0.04159
y,y(200)	0.02108	-0.01410	0.01000
x,xy(020)	0.00278	-0.00251	0.00823
z,d2(002)	-0.04721	-0.02619	-0.05163
xy,xy(200)	-0.00078	-0.00144	-0.00749
xy,xy(002)	0.00007	0.00351	0.00631
d2,d2(002)	0.03030	-0.01634	-0.00210
d1,d1(002)	-0.00035	0.00750	0.01608
ON SITE H-H			
s,s(000)	0.17113	0.11166	
FIRST NEIGHBOR H-H			
s,s(110)	-0.01776	-0.02843	0.00824
s,x(110)	0.00000	0.00000	0.00000
x,x(110)	0.00000	0.00000	0.00000
x,x(011)	0.00000	0.00000	0.00000
x,y(110)	0.00000	0.00000	0.00000
SECOND NEIGHBOR H-H			
s,s(200)	-0.00503	-0.00716	0.03863
s,x(200)	0.00000	0.00000	0.00000
x,x(200)	0.00000	0.00000	0.00000
y,y(200)	0.00000	0.00000	0.00000
FIRST NEIGHBOR Cu-H			
s,s(100)	-0.10162	-0.02658	0.17746
s,x(100)	0.00000	0.00000	0.00000
x,s(100)	0.02952	0.04136	0.16643
x,x(100)	0.00000	0.00000	0.00000
y,y(100)	0.00000	0.00000	0.00000
xy,x(010)	0.00000	0.00000	0.00000
d2,s(001)	-0.03130	-0.03375	-0.27240
d2,z(001)	0.00000	0.00000	0.00000

SECOND NEIGHBOR Cu-H

s,s(111)	0.01419	0.00710	0.02559
x,s(111)	-0.03144	0.01513	0.00374
s,x(111)	0.00000	0.00000	0.00000
x,x(111)	0.00000	0.00000	0.00000
x,y(111)	0.00000	0.00000	0.00000
xy,s(111)	-0.00301	-0.00886	-0.00565
xy,x(111)	0.00000	0.00000	0.00000
yz,x(111)	0.00000	0.00000	0.00000
d1,x(111)	0.00000	0.00000	0.00000

CuH

	ORTHOGONAL			NON-ORTHOGONAL		
BAND	RMS ERROR	MAXIMUM DEVIATION		RMS ERROR	MAXIMUM DEVIATION	
	mRy	k	mRy	mRy	k	mRy
1	2.8	(033)	7.0	0.8	(280)	1.5
2	3.5	(333)	9.6	0.7	(066)	1.8
3	4.8	(006)	14.1	0.7	(055)	1.6
4	2.0	(222)	4.8	0.4	(033)	1.0
5	1.7	(044)	4.0	0.4	(224)	1.4
6	3.9	(055)	9.4	0.9	(062)	2.5
7	9.2	(006)	25.3	1.4	(333)	4.4
1-7	4.6			0.8		

ENERGY VALUES IN Ry AT SELECTED k-POINTS

	ORTHOGONAL	APW	NON-ORTHOGONAL
GAMMA 1	-0.21929	-0.21936	-0.21845
GAMMA 1	1.60150	1.61613	1.61689
GAMMA 12	0.48718	0.48493	0.48537
GAMMA 15	2.11407	2.11951	2.11952
GAMMA 25'	0.43793	0.43776	0.43767
X1 (008)	0.12221	0.11979	0.11906
X1 (008)	0.79597	0.80217	0.80185
X1 (008)	1.50418	1.50582	1.50587
X2 (008)	0.51810	0.51857	0.51879
X3 (008)	0.35026	0.35077	0.35142
X5 (008)	0.53373	0.53155	0.53195
X4' (008)	0.66449	0.66983	0.66996
X5' (008)	1.38992	1.39943	1.39847
L1 (444)	0.27950	0.28398	0.28292
L1 (444)	0.81113	0.80928	0.80927
L3 (444)	0.42921	0.42821	0.42815
L3 (444)	0.52157	0.52112	0.52144
L2' (444)	0.19339	0.19952	0.20032
L2' (444)	1.52439	1.52601	1.52598
L3' (444)	2.04727	2.03703	2.03675
W1 (048)	0.47279	0.47611	0.47588
W1 (048)	1.11554	1.11086	1.11090
W2' (048)	0.16915	0.17219	0.17104
W2' (048)	0.77235	0.77352	0.77306
W2' (048)	2.11307	2.14456	2.14480
W1' (048)	0.53347	0.53164	0.53134
W3 (048)	0.39759	0.39988	0.39988
W3 (048)	0.99040	0.97876	0.97936
ODD (224)	0.44566	0.44915	0.44879
ODD (224)	0.51729	0.51476	0.51486
EVEN(224)	0.03874	0.03413	0.03449
EVEN(224)	0.38178	0.37765	0.37891
EVEN(224)	0.42779	0.42868	0.42959
EVEN(224)	0.50756	0.50753	0.50614
EVEN(224)	1.01950	1.01513	1.01764

FERMI LEVEL QUANTITIES (Non-orthogonal fit)

ENERGY		DENSITIES OF STATES					
	Total	s-Cu	p-Cu	t2g-Cu	eg-Cu	s-H	p-H
Ry		states/Ry/cell					
0.7810	0.05	0.00	0.02	0.00	0.01	0.01	0.00
		INTEGRATED DENSITIES OF STATES					
	Total	s-Cu	p-Cu	t2g-Cu	eg-Cu	s-H	p-H
		electrons					
	12.00	0.12	0.17	5.53	3.84	2.35	0.00
VELOCITY	PLASMON ENERGY		EIGENVALUE SUM				
cm/s	eV		Ry				
1.22x10E8	0.99		-4.4853				

Table 6.40 CuH (NaCl) a = 7.18 Bohr Slater–Koster 2-center parameters

	ORTHOGONAL	NON-ORTHOGONAL	
	ENERGY INTEGRALS	ENERGY INTEGRALS	OVERLAP INTEGRALS
	Ry	Ry	
ON SITE Cu-Cu			
s	0.74003	0.45006	
p	1.50600	1.10500	
t2g	0.46528	0.45369	
eg	0.46262	0.45266	
FIRST NEIGHBOR Cu-Cu			
(sss)	-0.08626	-0.00572	0.15204
(pps)	0.18878	0.02347	-0.19574
(ppp)	-0.01024	0.09342	0.09368
(dds)	-0.02281	-0.01317	0.02115
(ddp)	0.00998	0.01580	0.00195
(ddd)	-0.00015	-0.00522	-0.00539
(sps)	0.14059	-0.01349	-0.18277
(sds)	-0.03981	-0.01918	0.03519
(pds)	-0.05237	-0.03196	0.01219
(pdp)	0.01194	-0.00146	-0.04876
SECOND NEIGHBOR Cu-Cu			
(sss)	0.01254	-0.03868	-0.01494
(pps)	-0.09926	0.16359	0.06243
(ppp)	0.02790	-0.02623	-0.01710
(dds)	0.00003	0.00760	0.02215
(ddp)	-0.00021	0.00580	0.00981
(ddd)	-0.00017	-0.00069	-0.00154
(sps)	-0.02424	0.07530	0.03474
(sds)	-0.00339	-0.00086	0.01146
(pds)	-0.00472	-0.00848	-0.00865
(pdp)	-0.00217	0.01423	0.01664
ON SITE H-H			
s	0.97187	0.31906	

FIRST NEIGHBOR H-H			
(sss)	0.04965	0.01025	-0.04466
(sps)	0.00000	0.00000	0.00000
(pps)	0.00000	0.00000	0.00000
(ppp)	0.00000	0.00000	0.00000
SECOND NEIGHBOR H-H			
(sss)	0.00457	-0.01024	0.02772
(sps)	0.00000	0.00000	0.00000
(pps)	0.00000	0.00000	0.00000
(ppp)	0.00000	0.00000	0.00000
FIRST NEIGHBOR Cu-H			
(sss)	0.05879	0.09451	-0.11647
(sps)	0.00000	0.00000	0.00000
(pss)	0.18699	0.07371	-0.20707
(pps)	0.00000	0.00000	0.00000
(ppp)	0.00000	0.00000	0.00000
(dpp)	0.00000	0.00000	0.00000
(dss)	0.07138	-0.00880	-0.21504
(dps)	0.00000	0.00000	0.00000
SECOND NEIGHBOR Cu-H			
(sss)	-0.03599	0.02270	0.01761
(pss)	-0.04571	0.05944	0.03578
(sps)	0.00000	0.00000	0.00000
(pps)	0.00000	0.00000	0.00000
(ppp)	0.00000	0.00000	0.00000
(dss)	-0.01405	0.02132	0.04441
(dps)	0.00000	0.00000	0.00000
(dpp)	0.00000	0.00000	0.00000

♀ CuH

	ORTHOGONAL			NON-ORTHOGONAL		
BAND	RMS ERROR	MAXIMUM DEVIATION		RMS ERROR	MAXIMUM DEVIATION	
	mRy	k	mRy	mRy	k	mRy
1	8.2	(226)	22.7	0.7	(044)	2.2
2	4.8	(333)	12.3	0.4	(044)	1.2
3	3.4	(066)	7.3	0.5	(004)	1.2
4	1.8	(022)	3.2	0.5	(048)	1.1
5	1.8	(224)	3.9	0.5	(444)	1.0
6	4.8	(007)	12.1	0.7	(280)	1.8
7	15.4	(006)	44.0	1.1	(066)	3.0
1-7	7.3			0.7		

ENERGY VALUES IN Ry AT SELECTED k-POINTS

	ORTHOGONAL	APW	NON-ORTHOGONAL
GAMMA 1	-0.22212	-0.21936	-0.22067
GAMMA 1	1.59735	1.61613	1.61528
GAMMA 12	0.48719	0.48493	0.48557
GAMMA 15	2.09230	2.11951	2.11980
GAMMA 25'	0.43483	0.43776	0.43784
X1 (008)	0.12933	0.11979	0.11950
X1 (008)	0.79774	0.80217	0.80110
X1 (008)	1.42464	1.50582	1.50589
X2 (008)	0.51705	0.51857	0.51805
X3 (008)	0.35620	0.35077	0.35061
X5 (008)	0.53268	0.53155	0.53191

X4' (008)	0.66397	0.66983	0.67052
X5' (008)	1.46003	1.39943	1.39875
L1 (444)	0.27833	0.28398	0.28413
L1 (444)	0.81240	0.80928	0.80973
L3 (444)	0.42537	0.42821	0.42796
L3 (444)	0.52439	0.52112	0.52210
L2' (444)	0.21870	0.19952	0.19978
L2' (444)	1.52261	1.52601	1.52593
L3' (444)	1.99096	2.03703	2.03695
W1 (048)	0.47842	0.47611	0.47721
W1 (048)	1.11705	1.11086	1.11062
W2' (048)	0.17743	0.17219	0.17311
W2' (048)	0.76945	0.77352	0.77378
W2' (048)	2.11561	2.14456	2.14461
W1' (048)	0.53337	0.53164	0.53139
W3 (048)	0.39776	0.39988	0.39975
W3 (048)	0.99792	0.97876	0.97928
ODD (224)	0.44803	0.44915	0.45020
ODD (224)	0.51715	0.51476	0.51411
EVEN(224)	0.03479	0.03413	0.03421
EVEN(224)	0.38173	0.37765	0.37756
EVEN(224)	0.42663	0.42868	0.42859
EVEN(224)	0.51138	0.50753	0.50731
EVEN(224)	1.01816	1.01513	1.01678

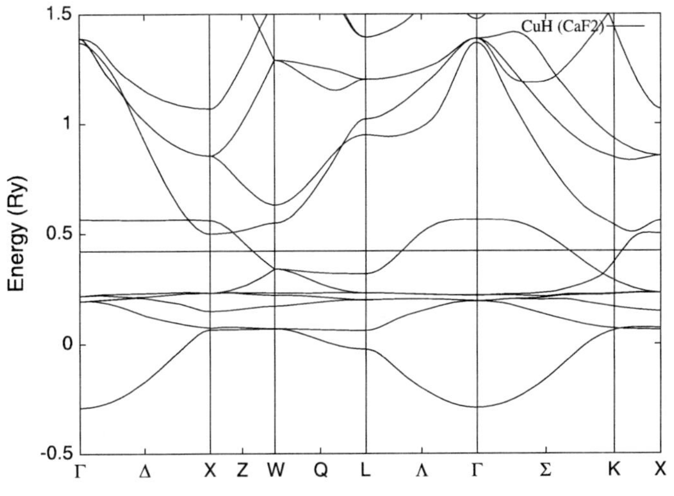

Fig. 6.39 Energy bands of CuH2 in the CaF2 structure

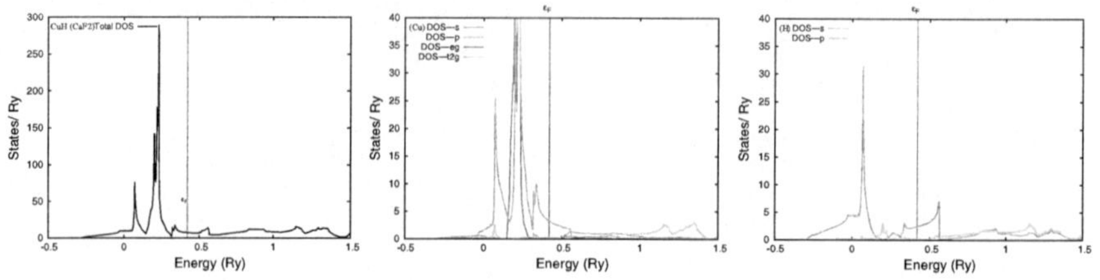

Fig. 6.40 Total, angular momentum and site decomposed densities of states of CuH2 in the CaF2 structure

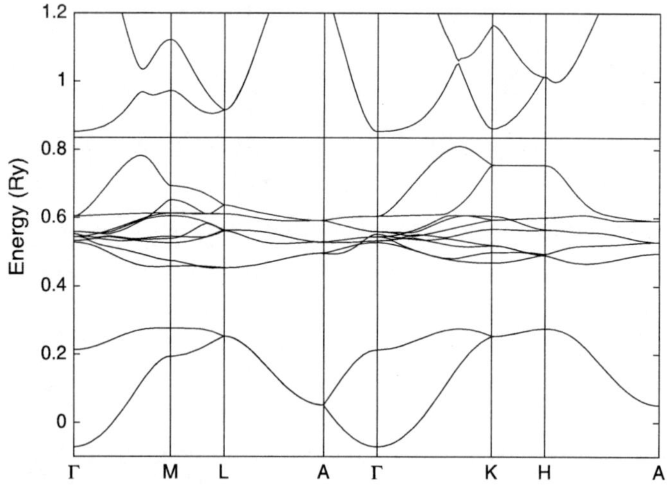

Fig. 6.41 Energy bands of CuH in the wurtzite structure

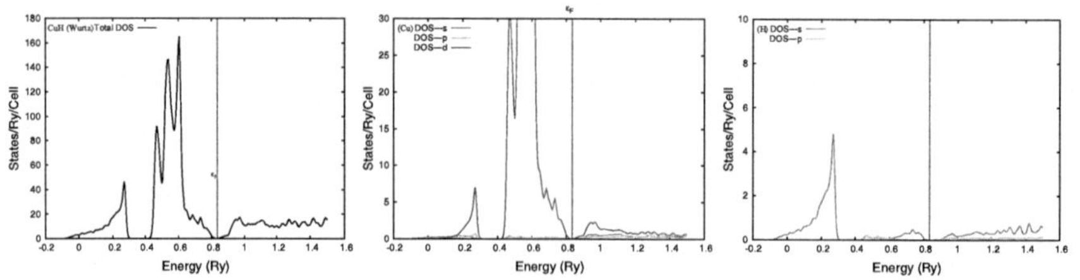

Fig. 6.42 Total, angular momentum and site decomposed densities of states of CuH in the wurtzite structure

6.9 Zinc Hydride (ZnH)

See Fig. 6.43 and Tables 6.41, 6.42, 6.43.

See Figs. 6.44, 6.45 and Tables 6.44, 6.45.

See Figs. 6.46 and 6.47.

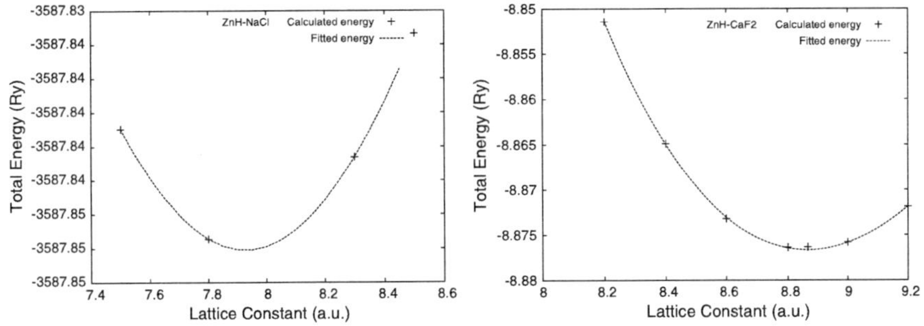

Fig. 6.43 Total energy versus lattice constant of ZnH in the NaCl and CaF2 structures

Table 6.41 Lattice constant, bulk modulus, gap, total energy

Stru	a (Bohr)	B (MBar)	Gap	Total Energy (Ry)
NaCl	7.89	0.66	-	-3587.81396
CaF2	8.87	0.69	-	-3588.87634
exp	-	-		

Table 6.42 Birch fit coefficients

	A1	A2	A3	A4
NaCl	-3.586281E+03	-1.015776E+02	2.141587E+03	-1.454721E+04
CaF2	-7.716639E+00	-7.970502E+01	1.586474E+03	-7.140945E+03

Table 6.43 DOS at Ef, Hopfield parameter, Stoner criterion

ZnH a=7.89 Bohr NaCl

	Fermi Energy (Ry)	Total DOS (States/Ry)	Decomposed DOS inside the muffin tins s	p	eg	t2g	f
Zn	0.671	6.828	1.249	0.557	0.266	0.320	0.025
H	0.671	6.828	1.150	0.702	0.034	0.012	0.005

```
Fermi-Velocity (cm/s): 1.10722 x10E8
Plasmon Energy (eV)  :  9.39948
Electron-ion interaction (Hopfield parameter) (eV/A^2) Zn: 0.261 H: 2.553

  Zn     MUFFIN-TIN RADIUS and CHARGE =     2.0711   28.2519
   H     MUFFIN-TIN RADIUS and CHARGE =     1.8739    1.2297
          Zn        STONER         I = 0.0051
           H        STONER         I = 0.0120
         STONER PARAMETER (Ry)     I = 0.0174
         STONER CRITERION        N*I = 0.1189
```

ZnH2 a=8.87 Bohr CaF2

	Fermi Energy (Ry)	Total DOS (States/Ry)	Decomposed DOS inside the muffin tins s	p	eg	t2g	f
Zn	0.431	8.620	1.740	0.381	0.114	0.636	0.053
H	0.431	8.620	2.890	0.322	0.005	0.016	0.001

```
Fermi-Velocity (cm/s): 1.04205 x10E8
Plasmon Energy (eV)  : 8.34583
Electron-ion interaction (Hopfield parameter) (eV/A^2) Zn: 0.274   H:  7.735

          Zn     MUFFIN-TIN RADIUS and CHARGE =    2.3033   28.6446
           H     MUFFIN-TIN RADIUS and CHARGE =    1.5356    0.8405
                   Zn        STONER I = 0.0013
                    H        STONER I = 0.0089
         STONER PARAMETER (Ry)      I = 0.0192
         STONER CRITERION         N*I = 0.1653
```

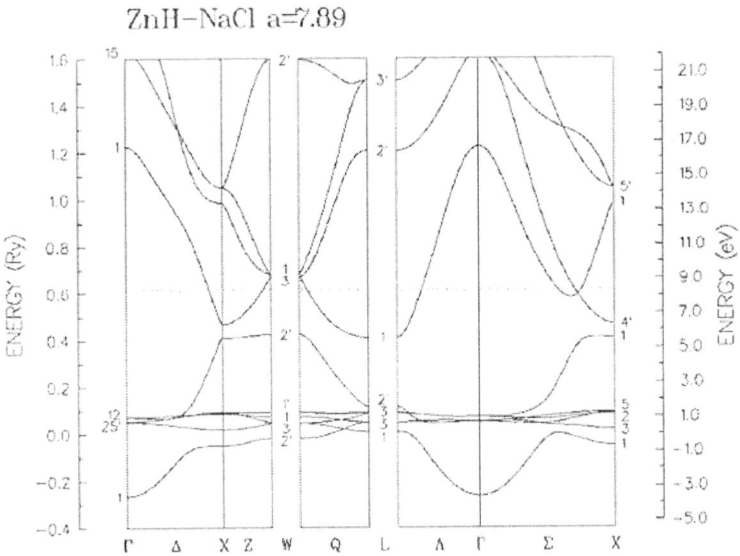

Fig. 6.44 Energy bands of ZnH in the NaCl structure (tight-binding)

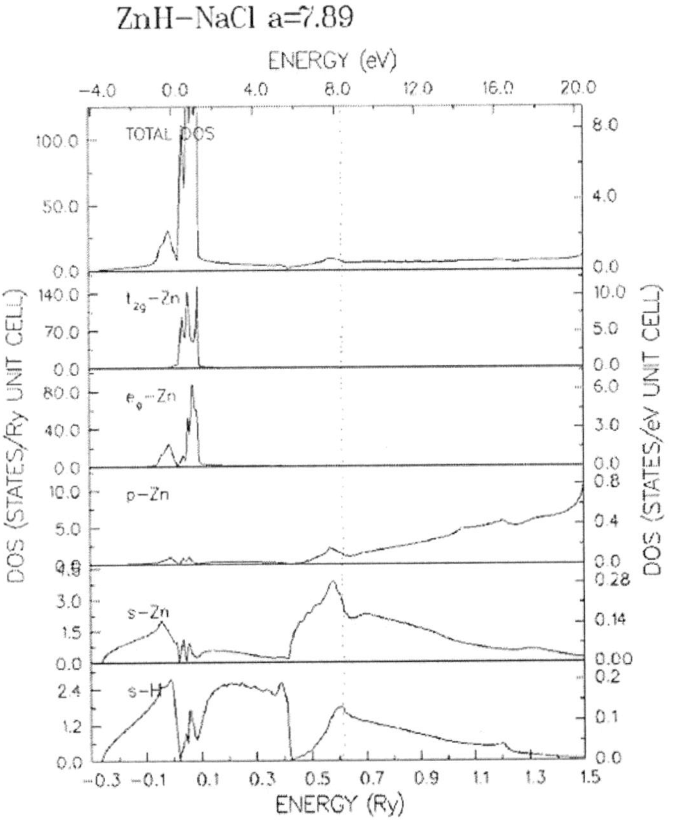

Fig. 6.45 Total, angular momentum and site decomposed densities of states of ZnH in the NaCl structure (tight-binding)

Table 6.44 ZnH (NaCl) a = 7.89 Bohr Slater–Koster 3-center parameters

	ORTHOGONAL	NON-ORTHOGONAL	
	ENERGY INTEGRALS Ry	ENERGY INTEGRALS Ry	OVERLAP INTEGRALS
ON SITE Zn-Zn			
s,s(000)	0.60337	0.45836	
x,x(000)	1.20209	0.99269	
xy,xy(000)	0.07151	0.06530	
d2,d2(000)	0.05372	0.05220	
FIRST NEIGHBOR Zn-Zn			
s,s(110)	-0.00073	-0.01248	0.01527
s,x(110)	-0.03005	-0.00147	0.05228
s,xy(110)	0.02094	-0.01562	0.05572
s,d2(110)	0.02025	-0.01353	0.00636
x,x(110)	0.07240	-0.01108	-0.10626
x,x(011)	-0.00047	-0.03378	0.02666
x,y(110)	0.03743	-0.05648	-0.07039
x,xy(110)	-0.02489	0.01274	-0.05882
x,xy(011)	-0.00047	-0.00757	0.00115
z,d2(011)	0.00416	0.00575	0.00264
z,d1(011)	0.01837	0.01598	0.00245
xy,xy(110)	-0.00749	-0.00935	-0.03028
xy,xy(011)	0.00216	0.00168	-0.00580
xy,xz(011)	0.00101	0.00152	-0.01369
xy,d2(110)	0.00361	-0.00247	0.01724
d2,d2(110)	-0.00433	-0.00429	0.00589
d1,d1(110)	0.00918	0.01060	0.00321
SECOND NEIGHBOR Zn-Zn			
s,s(200)	0.04190	0.02442	-0.03151
s,x(200)	0.06397	0.02349	-0.05018
s,d2(002)	-0.03699	0.00909	0.04480
x,x(200)	-0.12137	-0.03633	0.07538
y,y(200)	0.02152	0.04292	0.03122
x,xy(020)	0.00019	0.00104	0.03722
z,d2(002)	-0.01031	-0.01269	-0.05727
xy,xy(200)	-0.00153	0.00107	0.00411
xy,xy(002)	0.00022	-0.00039	-0.00476
d2,d2(002)	-0.00361	-0.00354	0.00977
d1,d1(002)	0.00083	0.00020	0.00296
ON SITE H-H			
s,s(000)	0.32390	0.41729	
FIRST NEIGHBOR H-H			
s,s(110)	-0.02378	0.00157	-0.00370
s,x(110)	0.00000	0.00000	0.00000
x,x(110)	0.00000	0.00000	0.00000
x,x(011)	0.00000	0.00000	0.00000
x,y(110)	0.00000	0.00000	0.00000
SECOND NEIGHBOR H-H			
s,s(200)	0.01078	0.00103	-0.00323
s,x(200)	0.00000	0.00000	0.00000
x,x(200)	0.00000	0.00000	0.00000
y,y(200)	0.00000	0.00000	0.00000
FIRST NEIGHBOR Zn-H			
s,s(100)	-0.10196	-0.12211	-0.00359
s,x(100)	0.00000	0.00000	0.00000
x,s(100)	0.10713	0.17610	0.04676
x,x(100)	0.00000	0.00000	0.00000
y,y(100)	0.00000	0.00000	0.00000
xy,x(010)	0.00000	0.00000	0.00000
d2,s(001)	0.01384	0.00566	-0.10660
d2,z(001)	0.00000	0.00000	0.00000

SECOND NEIGHBOR Zn-H

s,s(111)	-0.00444	0.00292	0.00542
x,s(111)	-0.00008	-0.00079	-0.00770
s,x(111)	0.00000	0.00000	0.00000
x,x(111)	0.00000	0.00000	0.00000
x,y(111)	0.00000	0.00000	0.00000
xy,s(111)	0.00419	0.00124	0.00472
xy,x(111)	0.00000	0.00000	0.00000
yz,x(111)	0.00000	0.00000	0.00000
d1,x(111)	0.00000	0.00000	0.00000

ZnH

	ORTHOGONAL			NON-ORTHOGONAL		
BAND	RMS ERROR	MAXIMUM DEVIATION		RMS ERROR	MAXIMUM DEVIATION	
	mRy	k	mRy	mRy	k	mRy
1	3.9	(044)	8.6	0.2	(000)	0.4
2	2.3	(222)	8.1	0.1	(442)	0.3
3	3.0	(004)	9.5	0.1	(042)	0.3
4	3.3	(333)	18.3	0.1	(044)	0.5
5	1.0	(444)	2.2	0.1	(003)	0.3
6	4.3	(048)	11.4	0.2	(442)	0.3
7	8.7	(006)	21.1	0.4	(380)	1.4
1-7	4.4			0.2		

ENERGY VALUES IN Ry AT SELECTED k-POINTS

	ORTHOGONAL	APW	NON-ORTHOGONAL
GAMMA 1	-0.27174	-0.26372	-0.26417
GAMMA 1	1.22089	1.22842	1.22822
GAMMA 12	0.07450	0.07319	0.07327
GAMMA 15	1.62278	1.63081	1.63080
GAMMA 25'	0.05312	0.05329	0.05339
X1 (008)	-0.04860	-0.04983	-0.04969
X1 (008)	0.41915	0.41052	0.41069
X1 (008)	0.95246	0.98562	0.98565
X2 (008)	0.08974	0.08822	0.08826
X3 (008)	0.01863	0.01922	0.01929
X5 (008)	0.09582	0.09407	0.09417
X4' (008)	0.46432	0.46802	0.46821
X5' (008)	1.04730	1.04952	1.04947
L1 (444)	0.00790	0.00774	0.00789
L1 (444)	0.41313	0.41012	0.41067
L3 (444)	0.05153	0.05067	0.05085
L3 (444)	0.09175	0.08951	0.08956
L2' (444)	0.11361	0.11619	0.11651
L2' (444)	1.20496	1.20774	1.20774
L3' (444)	1.50848	1.50805	1.50800
W1 (048)	0.06993	0.07309	0.07309
W1 (048)	0.69732	0.68394	0.68386
W2' (048)	-0.01944	-0.01908	-0.01889
W2' (048)	0.41293	0.42429	0.42428
W2' (048)	1.58522	1.59494	1.59509
W1' (048)	0.09492	0.09411	0.09398
W3 (048)	0.04798	0.04758	0.04751
W3 (048)	0.68508	0.67139	0.67179
ODD (224)	0.06015	0.05963	0.05958
ODD (224)	0.08714	0.08678	0.08696
EVEN(224)	-0.05114	-0.05852	-0.05877
EVEN(224)	0.04386	0.03993	0.03976
EVEN(224)	0.05077	0.05432	0.05429
EVEN(224)	0.09372	0.09821	0.09802
EVEN(224)	0.67958	0.66797	0.66822

FERMI LEVEL QUANTITIES (Non-orthogonal fit)

ENERGY				DENSITIES OF STATES			
	Total	s-Zn	p-Zn	t2g-Zn	eg-Zn	s-H	p-H
Ry		states/Ry/cell					
0.6131	6.79	2.78	1.40	0.57	0.28	1.77	0.00
		INTEGRATED DENSITIES OF STATES					
	Total	s-Zn	p-Zn	t2g-Zn	eg-Zn	s-H	p-H
		electrons					
	13.00	0.97	0.39	6.11	4.16	1.37	0.00

VELOCITY	PLASMON ENERGY	EIGENVALUE SUM
cm/s	eV	Ry
1.18x10E8	9.97	-6.6430

Table 6.45 ZnH (NaCl) a = 7.89 Bohr Slater–Koster 2-center parameters

	ORTHOGONAL	NON-ORTHOGONAL	
	ENERGY INTEGRALS Ry	ENERGY INTEGRALS Ry	OVERLAP INTEGRALS
ON SITE Zn-Zn			
s	0.42080	0.21774	
p	1.10533	0.77410	
t2g	0.06687	0.06351	
eg	0.06797	0.06230	
FIRST NEIGHBOR Zn-Zn			
(sss)	-0.06281	-0.03673	0.13203
(pps)	0.14529	-0.01093	-0.22682
(ppp)	-0.00426	0.00989	0.05655
(dds)	-0.00956	-0.00916	0.00707
(ddp)	0.00334	0.00507	-0.00220
(ddd)	0.00025	-0.00091	-0.00360
(sps)	0.10125	0.02128	-0.16534
(sds)	-0.02184	-0.01669	0.02032
(pds)	-0.03120	-0.01977	0.01495
(pdp)	0.00495	0.00756	-0.02790
SECOND NEIGHBOR Zn-Zn			
(sss)	0.01065	-0.02244	-0.00145
(pps)	-0.07488	0.05423	0.01418
(ppp)	0.02162	-0.00867	-0.00797
(dds)	0.00010	-0.00173	-0.01156
(ddp)	-0.00008	0.00062	0.00482
(ddd)	-0.00028	-0.00020	-0.00231
(sps)	-0.01811	0.03377	0.00419
(sds)	-0.00244	-0.00382	0.00792
(pds)	-0.00819	-0.00899	0.00330
(pdp)	0.00008	0.00254	-0.00181
ON SITE H-H			
s	0.73316	0.33980	

FIRST NEIGHBOR H-H			
(sss)	0.04283	0.00268	-0.03638
(sps)	0.00000	0.00000	0.00000
(pps)	0.00000	0.00000	0.00000
(ppp)	0.00000	0.00000	0.00000
SECOND NEIGHBOR H-H			
(sss)	-0.00456	-0.01042	0.01649
(sps)	0.00000	0.00000	0.00000
(pps)	0.00000	0.00000	0.00000
(ppp)	0.00000	0.00000	0.00000
FIRST NEIGHBOR Zn-H			
(sss)	0.03889	0.08380	-0.11173
(sps)	0.00000	0.00000	0.00000
(pss)	0.15426	0.09327	-0.18818
(pps)	0.00000	0.00000	0.00000
(ppp)	0.00000	0.00000	0.00000
(dpp)	0.00000	0.00000	0.00000
(dss)	0.03946	0.03328	-0.16500
(dps)	0.00000	0.00000	0.00000
SECOND NEIGHBOR Zn-H			
(sss)	-0.02697	0.00728	0.00528
(pss)	-0.03003	0.00810	0.01181
(sps)	0.00000	0.00000	0.00000
(pps)	0.00000	0.00000	0.00000
(ppp)	0.00000	0.00000	0.00000
(dss)	-0.00802	0.00344	0.02984
(dps)	0.00000	0.00000	0.00000
(dpp)	0.00000	0.00000	0.00000

♀ZnH

	ORTHOGONAL			NON-ORTHOGONAL		
BAND	RMS ERROR	MAXIMUM DEVIATION		RMS ERROR	MAXIMUM DEVIATION	
	mRy	k	mRy	mRy	k	mRy
1	5.5	(226)	14.3	0.2	(000)	0.3
2	2.7	(354)	7.5	0.1	(226)	0.3
3	1.5	(380)	3.2	0.2	(226)	0.3
4	2.1	(333)	9.0	0.2	(354)	0.5
5	1.2	(444)	2.7	0.2	(044)	0.6
6	5.2	(007)	16.5	0.2	(007)	0.5
7	14.6	(007)	42.0	0.5	(380)	1.1
1-7	6.4			0.2		

ENERGY VALUES IN Ry AT SELECTED k-POINTS

	ORTHOGONAL	APW	NON-ORTHOGONAL
GAMMA 1	-0.26922	-0.26372	-0.26407
GAMMA 1	1.21997	1.22842	1.22858
GAMMA 12	0.07424	0.07319	0.07353
GAMMA 15	1.58914	1.63081	1.63084
GAMMA 25'	0.05191	0.05329	0.05335
X1 (008)	-0.04442	-0.04983	-0.04968
X1 (008)	0.41182	0.41052	0.41077
X1 (008)	0.94601	0.98562	0.98480
X2 (008)	0.08733	0.08822	0.08802
X3 (008)	0.02322	0.01922	0.01947
X5 (008)	0.09444	0.09407	0.09426

X4' (008)	0.46088	0.46802	0.46834
X5' (008)	1.05909	1.04952	1.04943
L1 (444)	0.00638	0.00774	0.00803
L1 (444)	0.40592	0.41012	0.41067
L3 (444)	0.05022	0.05067	0.05088
L3 (444)	0.09220	0.08951	0.08958
L2' (444)	0.12270	0.11619	0.11634
L2' (444)	1.20820	1.20774	1.20776
L3' (444)	1.46772	1.50805	1.50801
W1 (048)	0.07398	0.07309	0.07347
W1 (048)	0.69519	0.68394	0.68389
W2' (048)	-0.01596	-0.01908	-0.01926
W2' (048)	0.42439	0.42429	0.42436
W2' (048)	1.55813	1.59494	1.59400
W1' (048)	0.09556	0.09411	0.09418
W3 (048)	0.04610	0.04758	0.04755
W3 (048)	0.68654	0.67139	0.67177
ODD (224)	0.05921	0.05963	0.05957
ODD (224)	0.08808	0.08678	0.08673
EVEN(224)	-0.06061	-0.05852	-0.05860
EVEN(224)	0.04098	0.03993	0.03984
EVEN(224)	0.05398	0.05432	0.05446
EVEN(224)	0.10381	0.09821	0.09803
EVEN(224)	0.67556	0.66797	0.66777

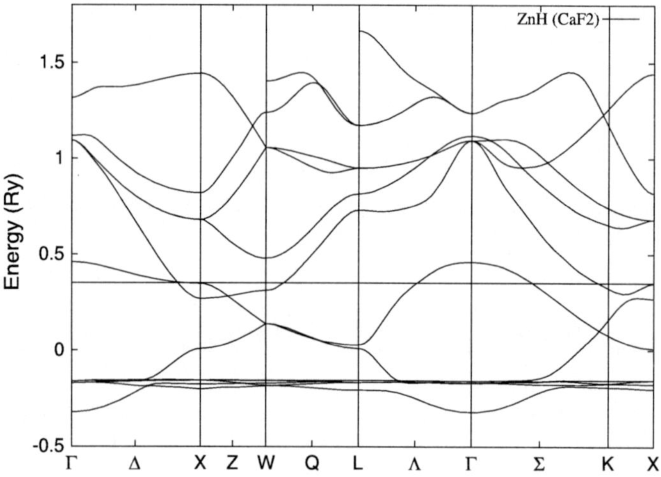

Fig. 6.46 Energy bands of ZnH2 in the CaF2 structure

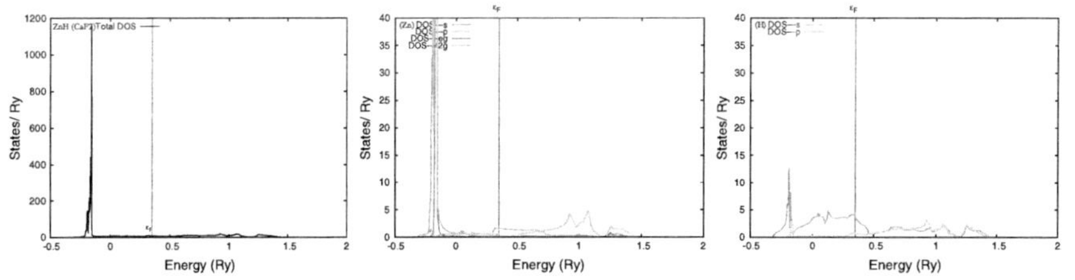

Fig. 6.47 Total, angular momentum and site decomposed densities of states of ZnH2 in the CaF2 structure

References

1. G.G. Libowitz, The nature and properties of transition metal hydrides. J. Nucl. Mater. **2**, 1 (1960)
2. W.M. Mueller, J.P. Blackledge, G.G. Libowitz (eds.), *Metal Hydrides* (Academic Press, N.Y., 1968)
3. D.A. Papaconstantopoulos, Magnetism in transition metal hydrides. Europhys. Lett. **15**, 621 (1991)
4. H. Smithson, C.A. Marianetti, D. Morgan, A. Van der Ven, A. Predith, G. Ceder, First-principles study of the stability and electronic structure of metal hydrides. Phys. Rev. B **66**, 144107 (2002)

Chapter 7
4D Transition-Metal Hydrides

This chapter covers the transition-metal hydrides from ZrH to CdH [1-3]. Results are presented for the crystal structures NaCl(B1) and CaF2 (C1). The first two compounds (ZrH and NbH) form as dihydrides in the CaF2 structure while PdH is found in the NaCl structure. However, in a few of them such as ZrH a body-centered tetragonal phase or/and hydrogen vacancies have been observed. The lattice constants, as expected, are larger than in the single elements because of the expansion of the lattice upon hydrogenation. Exactly, as in the elements, the lattice parameter decreases from ZrH up to the middle of the 4d series where it reaches a minimum and then increases up to CdH. Comparing the energy bands of the CaF2 structure to the bands of the NaCl structure we note that a gap that appears between first and second band in the NaCl structure closes in the CaF2 structure and an additional antibonding band emerges above the Fermi level which is due to the second hydrogen in the CaF2 structure. Examining the densities of states figures we observe that for both crystal structures the lower occupied states have predominantly s–H hydrogen character and just below the Fermi level the d-t2g metal contribution dominates. However, the DOS details are significantly different for the two crystal structures. Moving along the 4d series the qualitative difference from ZrH to AgH is the position of the Fermi level which is moving up and therefore changes the Fermi surface and the values of the DOS at the Fermi level. There is a small crystal field splitting that one can see in the separation of the t2g and eg states in the DOS figures, and in the positioning of the Gam25' and Gam12 points in the energy bands diagrams. It is important to mention that in CdH the d-bands are positioned much deeper than in the other compounds. It is also noted that AgH in the NaCl structure is predicted to be a semi-metal, PdH is a superconductor [4] and RhH is shown to be near a ferromagnetic [5] instability. Tight-binding parameters are given in the NaCl structure based on both orthogonal and non-orthogonal Hamiltonians using three- and two-center integrals.

D. A. Papaconstantopoulos, *Band Structure of Cubic Hydrides*,
https://doi.org/10.1007/978-3-031-06878-2_7

7.1 Zirconium Hydride (ZrH)

See Fig. 7.1 and Tables 7.1, 7.2 and 7.3.

See Figs. 7.2, 7.3, Tables 7.4, and 7.5.

See Figs. 7.4 and 7.5.

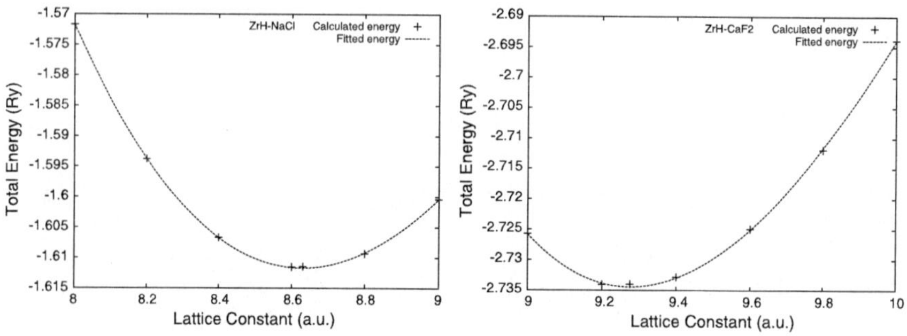

Fig. 7.1 Total energy versus lattice constant of ZrH in the NaCl and CaF2 structures

Table 7.1 Lattice constant, bulk modulus, gap, total energy

Stru	a (Bohr)	B (MBar)	Gap	Total Energy
NaCl	8.63	1.35	-	-7191.61145
CaF2	9.28	1.43	-	-7192.73396
exp (CaF2)	9.04			

Table 7.2 Birch fit coefficients

	A1	A2	A3	A4
NaCl	1.088237E+00	-1.903641E+02	4.176857E+03	-2.686620E+04
CaF2	-1.304219E+00	-7.207980E+01	-8.012101E+01	2.983715E+04

Table 7.3 DOS at Ef, Hopfield parameter, Stoner criterion

ZrH a=8.63 Bohr NaCl

Fermi Energy (Ry)	Total DOS (States/Ry)	Decomposed DOS inside the muffin tins s	p	eg	t2g	f
Zr 0.751	21.086	0.331	0.959	2.768	6.048	0.117
H 0.751	21.086	0.184	1.629	0.014	0.150	0.039

Fermi-Velocity (cm/s): 0.45817 x10E8
Plasmon Energy (eV) : 5.97465
Electron-ion interaction (Hopfield parameter) (eV/A^2) Zr: 6.416 H: 0.234

Zr	MUFFIN-TIN RADIUS and CHARGE =	2.3734	37.4282
H	MUFFIN-TIN RADIUS and CHARGE =	1.9419	1.3622
Zr	STONER I = 0.0051		
H	STONER I = 0.0015		
STONER PARAMETER (Ry)	I = 0.0070		
STONER CRITERION	N*I = 0.1471		

ZrH2 a=9.28 Bohr CaF2

Fermi Energy (Ry)	Total DOS (States/Ry)	Decomposed DOS inside the muffin tins s	p	eg	t2g	f
Zr 0.624	20.933	0.002	0.052	3.953	8.799	0.029
H 0.624	20.933	0.034	1.542	0.071	0.021	0.007

Fermi-Velocity (cm/s): 0.41429 x10E8
Plasmon Energy (eV) : 4.82372
Electron-ion interaction (Hopfield parameter) (eV/A^2) Zr: 1.035 H: 0.214

Zr	MUFFIN-TIN RADIUS and CHARGE =	2.4099	37.4180
H	MUFFIN-TIN RADIUS and CHARGE =	1.6066	1.0437
Zr	STONER I = 0.0054		
H	STONER I = 0.0004		
STONER PARAMETER (Ry)	I = 0.0063		
STONER CRITERION	N*I = 0.1321		

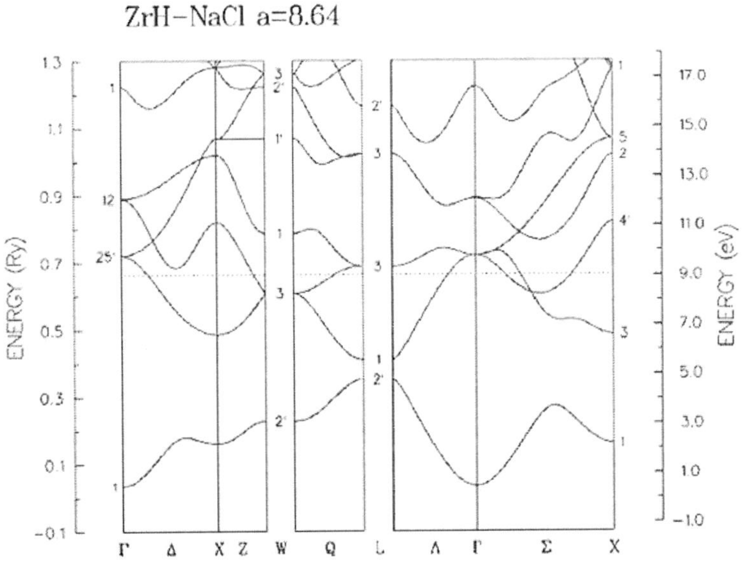

Fig. 7.2 Energy bands of ZrH in the NaCl structure (tight-binding)

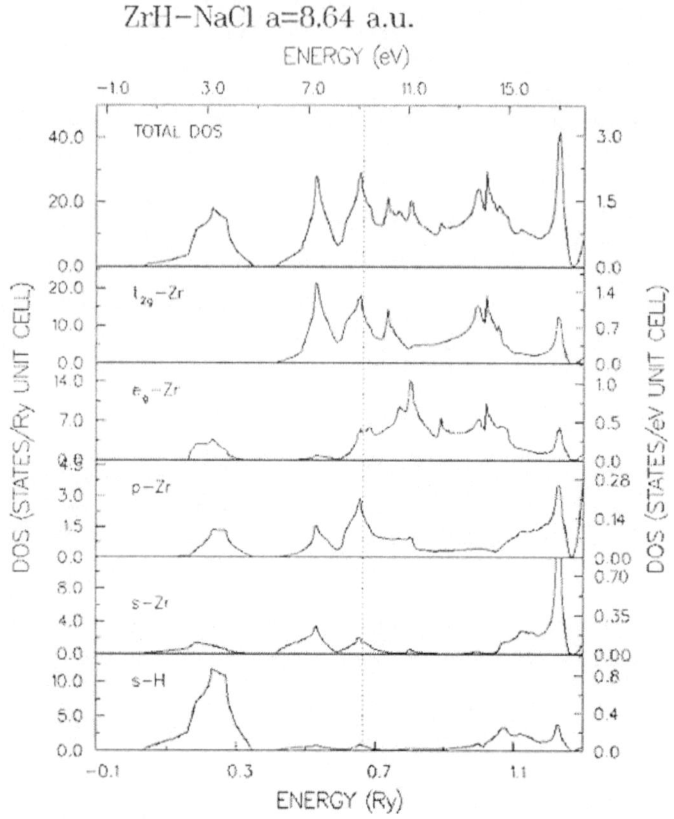

Fig. 7.3 Total, angular momentum and site decomposed densities of states of ZrH in the NaCl structure (tight-binding)

Table 7.4 ZrH (NaCl) a = 8.64 Bohr Slater–Koster 3-center parameters

	ORTHOGONAL	NON-ORTHOGONAL	
	ENERGY INTEGRALS Ry	ENERGY INTEGRALS Ry	OVERLAP INTEGRALS
ON SITE Zr-Zr			
s,s(000)	1.01190	0.85780	
x,x(000)	1.40095	1.26527	
xy,xy(000)	0.91294	0.75418	
d2,d2(000)	0.90995	0.76103	
FIRST NEIGHBOR Zr-Zr			
s,s(110)	-0.02035	0.01281	0.04973
s,x(110)	-0.03675	0.03010	0.07382
s,xy(110)	-0.05793	0.00427	0.09999
s,d2(110)	0.00385	-0.01181	0.00266
x,x(110)	0.04689	0.01984	-0.04881
x,x(011)	-0.01176	0.13371	0.13184
x,y(110)	0.05756	-0.05860	-0.07457
x,xy(110)	0.07603	0.00436	-0.05575
x,xy(011)	-0.00280	0.04021	0.07408
z,d2(011)	0.01446	0.04133	0.07923
z,d1(011)	0.00060	-0.00261	0.00411
xy,xy(110)	-0.05851	-0.01712	0.05021
xy,xy(011)	0.01475	-0.00427	-0.03498
xy,xz(011)	-0.00867	-0.01303	-0.04931
xy,d2(110)	0.02730	0.00943	0.04964
d2,d2(110)	-0.00155	-0.00810	0.01673
d1,d1(110)	-0.02741	0.05083	0.00709
SECOND NEIGHBOR Zr-Zr			
s,s(200)	0.06003	0.03095	-0.01688
s,x(200)	0.07580	0.01935	-0.02556
s,d2(002)	0.04865	0.02488	0.05496
x,x(200)	-0.10651	-0.03609	-0.00033
y,y(200)	0.01389	0.02395	0.02914
x,xy(020)	0.01352	0.01060	0.01826
z,d2(002)	-0.05513	-0.04446	-0.08085
xy,xy(200)	-0.02068	-0.00472	-0.01145
xy,xy(002)	0.00282	-0.00264	-0.00236
d2,d2(002)	0.04875	-0.04503	-0.05119
d1,d1(002)	0.00273	-0.00018	0.00043
ON SITE H-H			
s,s(000)	0.34760	0.46613	
FIRST NEIGHBOR H-H			
s,s(110)	-0.01398	-0.01221	-0.01255
s,x(110)	0.00000	0.00000	0.00000
x,x(110)	0.00000	0.00000	0.00000
x,x(011)	0.00000	0.00000	0.00000
x,y(110)	0.00000	0.00000	0.00000
SECOND NEIGHBOR H-H			
s,s(200)	-0.01045	-0.00589	-0.00650
s,x(200)	0.00000	0.00000	0.00000
x,x(200)	0.00000	0.00000	0.00000
y,y(200)	0.00000	0.00000	0.00000
FIRST NEIGHBOR Zr-H			
s,s(100)	-0.08084	-0.09792	-0.00455
s,x(100)	0.00000	0.00000	0.00000
x,s(100)	0.06483	0.17825	0.17649
x,x(100)	0.00000	0.00000	0.00000
y,y(100)	0.00000	0.00000	0.00000
xy,x(010)	0.00000	0.00000	0.00000
d2,s(001)	-0.06764	0.02148	-0.14122
d2,z(001)	0.00000	0.00000	0.00000
SECOND NEIGHBOR Zr-H			
s,s(111)	0.02390	0.00623	0.01667
x,s(111)	-0.03377	-0.00894	-0.01321
s,x(111)	0.00000	0.00000	0.00000
x,x(111)	0.00000	0.00000	0.00000
x,y(111)	0.00000	0.00000	0.00000
xy,s(111)	0.00818	-0.00078	0.00273
xy,x(111)	0.00000	0.00000	0.00000
yz,x(111)	0.00000	0.00000	0.00000
d1,x(111)	0.00000	0.00000	0.00000

ZrH

BAND	ORTHOGONAL RMS ERROR mRy	MAXIMUM DEVIATION k	mRy	NON-ORTHOGONAL RMS ERROR mRy	MAXIMUM DEVIATION k	mRy
1	5.4	(004)	11.3	0.7	(044)	1.8
2	4.9	(222)	15.2	0.6	(333)	1.5
3	4.4	(055)	10.0	0.6	(226)	1.7
4	5.3	(048)	12.9	0.7	(118)	1.9
5	4.2	(226)	9.3	0.4	(007)	1.1
6	5.1	(048)	12.3	0.6	(033)	1.6
7	7.0	(000)	16.0	1.3	(174)	3.0
1-7	5.3			0.8		

ENERGY VALUES IN Ry AT SELECTED k-POINTS

	ORTHOGONAL	APW	NON-ORTHOGONAL
GAMMA 1	0.03792	0.03513	0.03493
GAMMA 1	1.20707	1.22304	1.22205
GAMMA 12	0.89061	0.89036	0.89008
GAMMA 15	1.57153	1.58247	1.58243
GAMMA 25'	0.71978	0.72107	0.72072
X1 (008)	0.17002	0.16199	0.16190
X1 (008)	1.28661	1.27882	1.27999
X1 (008)	1.56343	1.53188	1.53125
X2 (008)	1.01886	1.01973	1.01857
X3 (008)	0.48384	0.48652	0.48642
X5 (008)	1.06992	1.06637	1.06656
X4' (008)	0.82137	0.81916	0.81869
X5' (008)	1.29051	1.28168	1.28118
L1 (444)	0.40543	0.40907	0.40926
L1 (444)	1.30567	1.30385	1.30215
L3 (444)	0.68852	0.68648	0.68733
L3 (444)	1.02235	1.02217	1.02194
L2' (444)	0.34343	0.35136	0.35157
L2' (444)	1.16479	1.16223	1.16364
L3' (444)	1.78865	1.71937	1.71940
W1 (048)	0.77500	0.78794	0.78650
W1 (048)	1.31397	1.31050	1.31099
W2' (048)	0.21843	0.22795	0.22862
W2' (048)	1.20958	1.22192	1.22159
W2' (048)	1.83426	1.95899	1.95933
W1' (048)	1.05862	1.06699	1.06723
W3 (048)	0.60693	0.60694	0.60749
W3 (048)	1.25307	1.26085	1.26147
ODD (224)	0.75116	0.74460	0.74446
ODD (224)	0.99606	0.99166	0.99092
EVEN(224)	0.23584	0.22764	0.22789
EVEN(224)	0.53717	0.53442	0.53376
EVEN(224)	0.66977	0.67283	0.67245
EVEN(224)	0.91042	0.91057	0.91048
EVEN(224)	1.12466	1.12249	1.12375

FERMI LEVEL QUANTITIES (Non-orthogonal fit)

DENSITIES OF STATES (states/Ry/cell)

ENERGY Ry	Total	s-Zr	p-Zr	t2g-Zr	eg-Zr	s-H	p-H
0.6637	22.64	1.49	2.10	13.28	5.03	0.74	0.00

INTEGRATED DENSITIES OF STATES (electrons)

Total	s-Zr	p-Zr	t2g-Zr	eg-Zr	s-H	p-H
5.00	0.49	0.33	2.15	0.52	1.50	0.00

VELOCITY cm/s	PLASMON ENERGY eV	EIGENVALUE SUM Ry
0.50x10E8	6.69	-1.1538

Table 7.5 ZrH (NaCl) a = 8.64 Bohr Slater–Koster 2-center

	ORTHOGONAL	NON-ORTHOGONAL	
	ENERGY INTEGRALS Ry	ENERGY INTEGRALS Ry	OVERLAP INTEGRALS
ON SITE Zr-Zr			
s	0.96213	0.69745	
p	1.36343	1.09097	
t2g	0.84846	0.72716	
eg	0.81924	0.71670	
FIRST NEIGHBOR Zr-Zr			
(sss)	-0.04971	-0.05734	0.03673
(pps)	0.10846	-0.02735	-0.15822
(ppp)	-0.01439	0.03577	0.05438
(dds)	-0.07857	0.02268	0.11778
(ddp)	0.02485	-0.00735	-0.08286
(ddd)	0.00262	0.00816	0.02044
(sps)	0.08231	-0.00684	-0.11433
(sds)	-0.07276	0.00589	0.10684
(pds)	-0.10688	0.03441	0.12600
(pdp)	0.00950	-0.00935	-0.05661
SECOND NEIGHBOR Zr-Zr			
(sss)	0.00315	0.01577	0.01437
(pps)	-0.05868	0.21681	0.12611
(ppp)	0.00129	-0.03794	-0.01866
(dds)	0.00796	-0.01943	-0.00112
(ddp)	-0.00423	-0.00809	-0.01155
(ddd)	0.00242	0.00271	0.00267
(sps)	-0.03238	0.00142	-0.01814
(sds)	0.01712	-0.03102	-0.02376
(pds)	0.00877	-0.09069	-0.06455
(pdp)	-0.00178	-0.00434	-0.00838
ON SITE H-H			
s	0.79024	0.37003	
FIRST NEIGHBOR H-H			
(sss)	-0.01596	0.01989	0.01481
(sps)	0.00000	0.00000	0.00000
(pps)	0.00000	0.00000	0.00000
(ppp)	0.00000	0.00000	0.00000
SECOND NEIGHBOR H-H			
(sss)	0.04302	-0.00697	-0.01609
(sps)	0.00000	0.00000	0.00000
(pps)	0.00000	0.00000	0.00000
(ppp)	0.00000	0.00000	0.00000
FIRST NEIGHBOR Zr-H			
(sss)	0.11484	0.04560	-0.11946
(sps)	0.00000	0.00000	0.00000
(pss)	0.10403	-0.06420	-0.27167
(pps)	0.00000	0.00000	0.00000
(ppp)	0.00000	0.00000	0.00000
(dpp)	0.00000	0.00000	0.00000
(dss)	0.12033	-0.00345	-0.21715
(dps)	0.00000	0.00000	0.00000
SECOND NEIGHBOR Zr-H			
(sss)	-0.01788	-0.01737	-0.01442
(pss)	-0.04599	0.00767	-0.00146
(sps)	0.00000	0.00000	0.00000
(pps)	0.00000	0.00000	0.00000
(ppp)	0.00000	0.00000	0.00000
(dss)	-0.01247	0.00005	-0.00911
(dps)	0.00000	0.00000	0.00000
(dpp)	0.00000	0.00000	0.00000

♀ ZrH

BAND	ORTHOGONAL RMS ERROR	MAXIMUM DEVIATION		NON-ORTHOGONAL RMS ERROR	MAXIMUM DEVIATION	
	mRy	k	mRy	mRy	k	mRy
1	8.3	(444)	18.6	1.5	(066)	4.5
2	9.4	(264)	22.6	1.4	(222)	3.4
3	10.8	(264)	25.6	1.5	(354)	4.1
4	9.7	(444)	21.8	1.3	(226)	3.2
5	10.1	(226)	27.3	1.8	(442)	4.0
6	10.7	(066)	35.0	2.0	(174)	5.5
7	18.2	(118)	38.4	4.4	(174)	15.2
1.7	11.5			2.2		

ENERGY VALUES IN Ry AT SELECTED k-POINTS

	ORTHOGONAL	APW	NON-ORTHOGONAL
GAMMA 1	0.02579	0.03513	0.03490
GAMMA 1	1.21550	1.22304	1.21917
GAMMA 12	0.89341	0.89036	0.89305
GAMMA 15	1.56993	1.58247	1.58157
GAMMA 25'	0.71316	0.72107	0.71971
X1 (008)	0.15977	0.16199	0.16115
X1 (008)	1.25311	1.27882	1.27963
X1 (008)	1.54511	1.53188	1.52932
X2 (008)	1.00614	1.01973	1.01897
X3 (008)	0.49340	0.48652	0.48857
X5 (008)	1.06945	1.06637	1.06598
X4' (008)	0.81740	0.81916	0.81868
X5' (008)	1.30880	1.28168	1.28089
L1 (444)	0.41252	0.40907	0.40812
L1 (444)	1.30230	1.30385	1.31022
L3 (444)	0.66468	0.68648	0.68748
L3 (444)	1.02843	1.02217	1.02218
L2' (444)	0.33278	0.35136	0.35249
L2' (444)	1.18359	1.16223	1.15960
L3' (444)	1.72134	1.71937	1.71929
W1 (048)	0.77307	0.78794	0.78739
W1 (048)	1.28347	1.31050	1.29593
W2' (048)	0.22863	0.22795	0.22520
W2' (048)	1.23128	1.22192	1.21820
W2' (048)	1.76498	1.95899	1.95860
W1' (048)	1.05977	1.06699	1.06695
W3 (048)	0.61164	0.60694	0.60458
W3 (048)	1.24467	1.26085	1.26045
ODD (224)	0.73803	0.74460	0.74225
ODD (224)	0.99363	0.99166	0.98866
EVEN(224)	0.23063	0.22764	0.22840
EVEN(224)	0.53564	0.53442	0.53452
EVEN(224)	0.67834	0.67283	0.67491
EVEN(224)	0.93150	0.91057	0.91236
EVEN(224)	1.12584	1.12249	1.12439

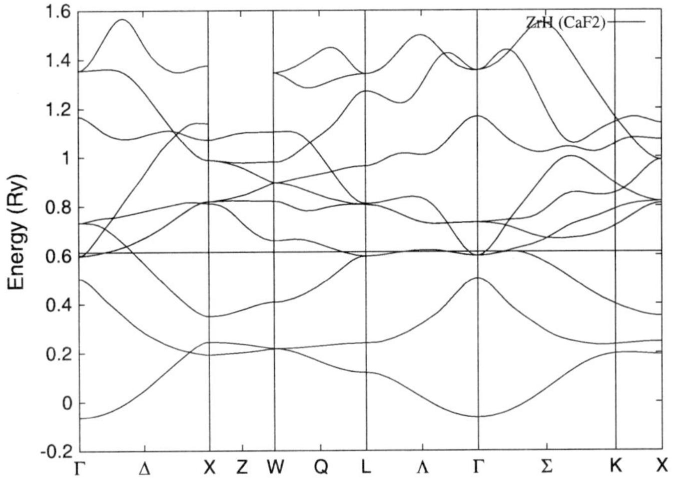

Fig. 7.4 Energy bands of ZrH2 in the CaF2 structure

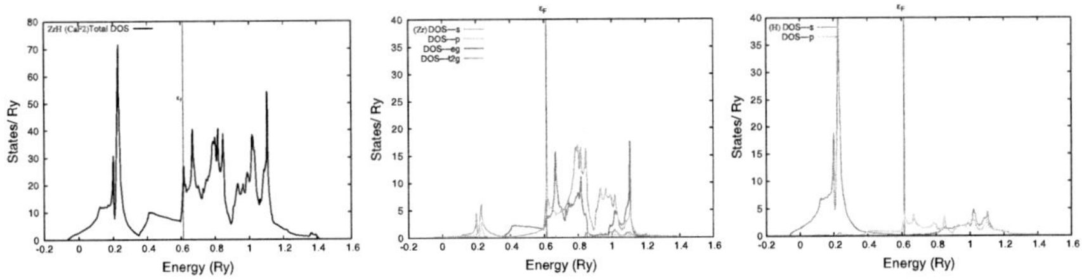

Fig. 7.5 Total, angular momentum and site decomposed densities of states of ZrH2 in the CaF2 structure

7.2 Niobium Hydride (NbH)

See Fig. 7.6, Tables 7.6, 7.7, and 7.8.

See Figs. 7.7, 7.8, Tables 7.9 and 7.10.

See Figs. 7.9 and 7.10.

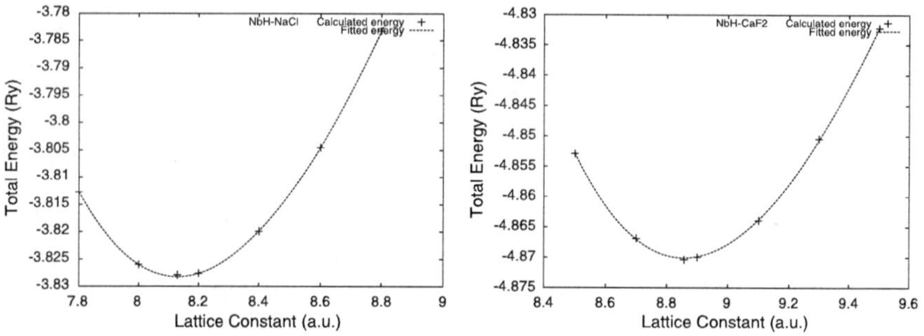

Fig. 7.6 Total energy versus lattice constant of NbH in the NaCl and CaF2 structures

Table 7.6 Lattice constant, bulk modulus, gap, total energy

Stru	a (Bohr)	B (MBar)	Gap	Total Energy
NaCl	8.13	2.02	-	-7633.82788
CaF2	8.86	1.73	-	-7634.87036
exp(CaF2)	8.63			

Table 7.7 Birch fit coefficients

	A1	A2	A3	A4
NaCl	-1.622310E+00	-1.190471E+02	1.691914E+03	-2.282129E+03
CaF2	-3.225335E+00	-8.209760E+01	3.301445E+02	1.966373E+04

Table 7.8 DOS at Ef, Hopfield parameter, Stoner criterion

NbH a=8.13 Bohr NaCl

	Fermi Energy (Ry)	Total DOS (States/Ry)	Decomposed DOS inside the muffin tins s	p	eg	t2g	f
Nb	0.863	17.578	0.072	0.312	3.802	6.765	0.052
H	0.863	17.578	0.058	1.037	0.041	0.153	0.034

Fermi-Velocity (cm/s): 0.44916 x10E8
Plasmon Energy (eV) : 5.84417
Electron-ion interaction (Hopfield parameter) (eV/A^2) Nb: 3.971 H: 0.074

Nb	MUFFIN-TIN RADIUS and CHARGE =	2.2358	38.1675
H	MUFFIN-TIN RADIUS and CHARGE =	1.8293	1.3690

Nb STONER I = 0.0097
H STONER I = 0.0010
STONER PARAMETER (Ry) I = 0.0110
STONER CRITERION N*I = 0.1927

NbH2 a=8.86 Bohr CaF2

	Fermi Energy (Ry)	Total DOS (States/Ry)	Decomposed DOS inside the muffin tins s	p	eg	t2g	f
Nb	0.678	19.056	0.021	0.126	8.674	3.915	0.057
H	0.678	19.056	0.086	0.788	0.065	0.037	0.021

Fermi-Velocity (cm/s): 0.40973 x10E8
Plasmon Energy (eV) : 4.88640
Electron-ion interaction (Hopfield parameter) (eV/A^2) Nb: 2.914 H: 0.296

Nb	MUFFIN-TIN RADIUS and CHARGE =	2.3010	38.2975
H	MUFFIN-TIN RADIUS and CHARGE =	1.5340	1.0095

Nb STONER I = 0.0072
H STONER I = 0.0002
STONER PARAMETER (Ry) I = 0.0077
STONER CRITERION N*I = 0.1465

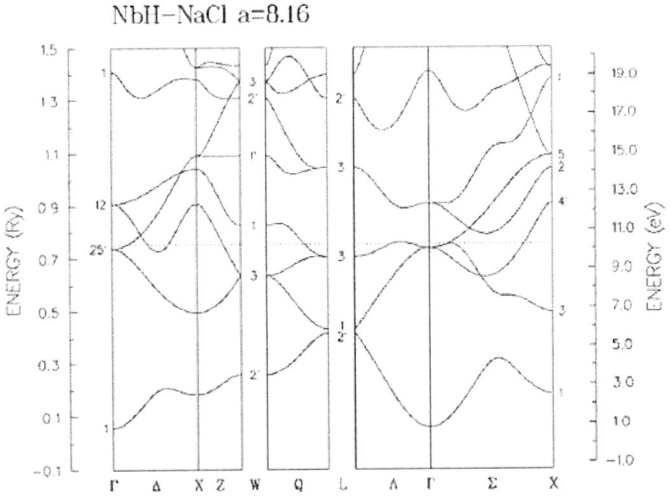

Fig. 7.7 Energy bands of NbH in the NaCl structure (tight-binding)

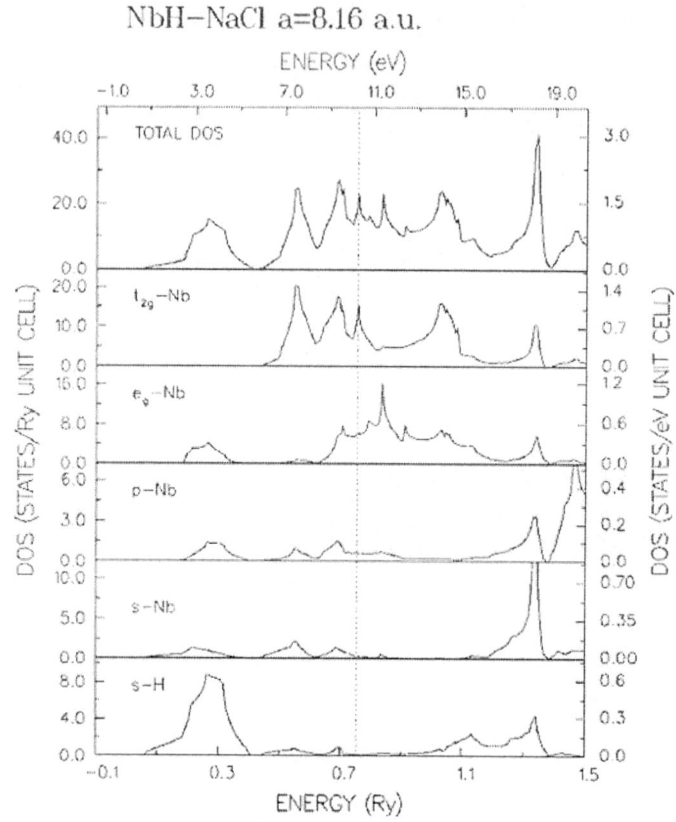

Fig. 7.8 Total, angular momentum and site decomposed densities of states of NbH in the NaCl structure (tight-binding)

Table 7.9 NbH (NaCl) a = 8.16 Bohr Slater–Koster 3-center parameters

	ORTHOGONAL	NON-ORTHOGONAL	
	ENERGY INTEGRALS Ry	ENERGY INTEGRALS Ry	OVERLAP INTEGRALS
ON SITE Nb-Nb			
s,s(000)	1.14174	0.96847	
x,x(000)	1.55861	1.41349	
xy,xy(000)	0.93343	0.76850	
d2,d2(000)	0.94300	0.78623	
FIRST NEIGHBOR Nb-Nb			
s,s(110)	-0.01979	0.01197	0.04687
s,x(110)	-0.04124	0.03632	0.07846
s,xy(110)	-0.06637	0.00317	0.09876
s,d2(110)	0.00108	-0.01207	0.00099
x,x(110)	0.05211	0.00296	-0.06233
x,x(011)	-0.01387	0.14377	0.13222
x,y(110)	0.06373	-0.05126	-0.06452
x,xy(110)	0.08295	-0.00262	-0.05855
x,xy(011)	-0.00299	0.03570	0.07029
z,d2(011)	0.01560	0.06088	0.09127
z,d1(011)	0.00299	0.00453	0.00680
xy,xy(110)	-0.05967	-0.01233	0.05608
xy,xy(011)	0.01513	-0.00059	-0.02988
xy,xz(011)	-0.00864	-0.00707	-0.04411
xy,d2(110)	0.02838	0.01057	0.05035
d2,d2(110)	0.00012	-0.01550	0.00827
d1,d1(110)	-0.03285	0.05882	0.01866
SECOND NEIGHBOR Nb-Nb			
s,s(200)	0.06570	0.02708	-0.02332
s,x(200)	0.08072	0.01283	-0.03089
s,d2(002)	0.05235	0.02322	0.05338
x,x(200)	-0.11641	-0.04584	-0.00106
y,y(200)	0.01479	0.01401	0.02330
x,xy(020)	0.01467	0.00763	0.01557
z,d2(002)	-0.05692	-0.03707	-0.06543
xy,xy(200)	-0.02094	-0.00431	-0.01033
xy,xy(002)	0.00259	-0.00090	-0.00070
d2,d2(002)	0.05205	-0.04677	-0.05535
d1,d1(002)	0.00285	0.00343	0.00368
ON SITE H-H			
s,s(000)	0.42037	0.58306	
FIRST NEIGHBOR H-H			
s,s(110)	-0.01666	-0.01700	-0.01982
s,x(110)	0.00000	0.00000	0.00000
x,x(110)	0.00000	0.00000	0.00000
x,x(011)	0.00000	0.00000	0.00000
x,y(110)	0.00000	0.00000	0.00000
SECOND NEIGHBOR H-H			
s,s(200)	-0.01172	-0.00327	-0.00113
s,x(200)	0.00000	0.00000	0.00000
x,x(200)	0.00000	0.00000	0.00000
y,y(200)	0.00000	0.00000	0.00000
FIRST NEIGHBOR Nb-H			
s,s(100)	-0.09262	-0.11752	-0.01539
s,x(100)	0.00000	0.00000	0.00000
x,s(100)	0.07655	0.21043	0.17069
x,x(100)	0.00000	0.00000	0.00000
y,y(100)	0.00000	0.00000	0.00000
xy,x(010)	0.00000	0.00000	0.00000
d2,s(001)	-0.08685	0.01979	-0.13335
d2,z(001)	0.00000	0.00000	0.00000
SECOND NEIGHBOR Nb-H			
s,s(111)	0.02749	0.01206	0.02141
x,s(111)	-0.03816	-0.01101	-0.01564
s,x(111)	0.00000	0.00000	0.00000
x,x(111)	0.00000	0.00000	0.00000
x,y(111)	0.00000	0.00000	0.00000
xy,s(111)	0.00519	-0.00095	0.00213
xy,x(111)	0.00000	0.00000	0.00000
yz,x(111)	0.00000	0.00000	0.00000
d1,x(111)	0.00000	0.00000	0.00000

NbH

BAND	ORTHOGONAL RMS ERROR mRy	MAXIMUM DEVIATION k	mRy	NON-ORTHOGONAL RMS ERROR mRy	MAXIMUM DEVIATION k	mRy
1	6.1	(004)	11.8	0.6	(066)	1.4
2	5.7	(222)	19.8	0.6	(022)	1.3
3	4.9	(044)	10.6	0.5	(003)	1.1
4	5.2	(048)	11.7	0.6	(118)	1.5
5	4.2	(003)	9.9	0.4	(055)	1.0
6	6.0	(044)	12.7	0.7	(033)	1.8
7	8.2	(000)	23.1	1.4	(174)	3.4
1-7	5.9			0.7		

ENERGY VALUES IN Ry AT SELECTED k-POINTS

	ORTHOGONAL	APW	NON-ORTHOGONAL
GAMMA 1	0.05915	0.05803	0.05816
GAMMA 1	1.38941	1.41251	1.41176
GAMMA 12	0.91132	0.91029	0.91012
GAMMA 15	1.74635	1.76267	1.76276
GAMMA 25'	0.73719	0.73807	0.73786
X1 (008)	0.19148	0.18297	0.18359
X1 (008)	1.38325	1.37832	1.38069
X1 (008)	1.76215	1.72547	1.72453
X2 (008)	1.04127	1.04194	1.04103
X3 (008)	0.49512	0.49617	0.49619
X5 (008)	1.09351	1.08987	1.08982
X4' (008)	0.91260	0.91078	0.91060
X5' (008)	1.44041	1.42823	1.42788
L1 (444)	0.42522	0.43229	0.43267
L1 (444)	1.40352	1.40049	1.39947
L3 (444)	0.70686	0.70483	0.70570
L3 (444)	1.04887	1.04551	1.04494
L2' (444)	0.40469	0.41604	0.41555
L2' (444)	1.30845	1.30518	1.30650
L3' (444)	1.98720	1.91184	1.91160
W1 (048)	0.81414	0.82586	0.82486
W1 (048)	1.43717	1.43348	1.43441
W2' (048)	0.24840	0.25844	0.25873
W2' (048)	1.29489	1.30757	1.30777
W2' (048)	2.05406	2.19221	2.19236
W1' (048)	1.08317	1.09041	1.09052
W3 (048)	0.63373	0.63499	0.63539
W3 (048)	1.36171	1.37106	1.37171
ODD (224)	0.77315	0.76644	0.76527
ODD (224)	1.02046	1.01567	1.01553
EVEN(224)	0.27859	0.26895	0.26949
EVEN(224)	0.56743	0.56273	0.56213
EVEN(224)	0.69409	0.69777	0.69782
EVEN(224)	0.94401	0.94373	0.94372
EVEN(224)	1.24498	1.24416	1.24539

FERMI LEVEL QUANTITIES (Non-orthogonal fit)

ENERGY	DENSITIES OF STATES						
	Total	s-Nb	p-Nb	t2g-Nb	eg-Nb	s-H	p-H
Ry		states/Ry/cell					
0.7526	21.52	0.35	0.60	14.33	6.04	0.20	0.00
	INTEGRATED DENSITIES OF STATES						
	Total	s-Nb	p-Nb	t2g-Nb	eg-Nb	s-H	p-H
		electrons					
	6.00	0.44	0.33	2.87	0.98	1.39	0.00

VELOCITY	PLASMON ENERGY	EIGENVALUE SUM
cm/s	eV	Ry
0.45x10E8	6.41	-1.4766

Table 7.10 NbH (NaCl) a = 8.16 Bohr Slater–Koster 2-center parameters

	ORTHOGONAL	NON-ORTHOGONAL	
	ENERGY INTEGRALS Ry	ENERGY INTEGRALS Ry	OVERLAP INTEGRALS
ON SITE Nb-Nb			
s	1.07770	0.72557	
p	1.50921	1.14703	
t2g	0.86430	0.75140	
eg	0.83659	0.74130	
FIRST NEIGHBOR Nb-Nb			
(sss)	-0.06999	-0.05753	0.05590
(pps)	0.12158	-0.13266	-0.23443
(ppp)	-0.01783	-0.03303	0.01825
(dds)	-0.07955	0.01572	0.10943
(ddp)	0.02701	0.00081	-0.07102
(ddd)	0.00137	0.00448	0.01605
(sps)	0.10155	-0.02795	-0.13499
(sds)	-0.07788	0.00643	0.10307
(pds)	-0.11077	0.05826	0.14494
(pdp)	0.01331	0.01092	-0.03886
SECOND NEIGHBOR Nb-Nb			
(sss)	0.00589	0.01149	0.01807
(pps)	-0.06303	0.16987	0.07588
(ppp)	0.00373	-0.04363	-0.01664
(dds)	0.00773	-0.01520	0.00218
(ddp)	-0.00334	-0.00542	-0.00838
(ddd)	0.00088	0.00075	0.00039
(sps)	-0.02067	0.00705	-0.01610
(sds)	0.00947	-0.03226	-0.02302
(pds)	0.01516	-0.07427	-0.04595
(pdp)	-0.00302	0.00116	-0.00388
ON SITE H-H			
s	0.95448	0.42281	
FIRST NEIGHBOR H-H			
(sss)	-0.00264	0.02061	0.03503
(sps)	0.00000	0.00000	0.00000
(pps)	0.00000	0.00000	0.00000
(ppp)	0.00000	0.00000	0.00000
SECOND NEIGHBOR H-H			
(sss)	0.04455	-0.00970	-0.01893
(sps)	0.00000	0.00000	0.00000
(pps)	0.00000	0.00000	0.00000
(ppp)	0.00000	0.00000	0.00000
FIRST NEIGHBOR Nb-H			
(sss)	0.11550	0.02563	-0.16957
(sps)	0.00000	0.00000	0.00000
(pss)	0.12314	-0.09888	-0.30749
(pps)	0.00000	0.00000	0.00000
(ppp)	0.00000	0.00000	0.00000
(dpp)	0.00000	0.00000	0.00000
(dss)	0.12254	0.00467	-0.19622
(dps)	0.00000	0.00000	0.00000
SECOND NEIGHBOR Nb-H			
(sss)	-0.02455	-0.02264	-0.02764
(pss)	-0.05128	0.00752	-0.00641
(sps)	0.00000	0.00000	0.00000
(pps)	0.00000	0.00000	0.00000
(ppp)	0.00000	0.00000	0.00000
(dss)	-0.01366	-0.01161	-0.02961
(dps)	0.00000	0.00000	0.00000
(dpp)	0.00000	0.00000	0.00000

♀ NbH

BAND	ORTHOGONAL RMS ERROR	MAXIMUM DEVIATION		NON-ORTHOGONAL RMS ERROR	MAXIMUM DEVIATION	
	mRy	k	mRy	mRy	k	mRy
1	6.7	(005)	17.5	2.0	(066)	5.8
2	9.0	(333)	20.0	1.5	(048)	3.2
3	8.8	(444)	17.1	1.5	(048)	3.2
4	8.2	(444)	17.1	1.2	(226)	2.9
5	8.1	(002)	15.5	1.6	(224)	3.5
6	12.8	(066)	51.8	2.4	(174)	5.5
7	18.7	(118)	48.5	5.7	(174)	14.8
1.7	11.0			2.7		

ENERGY VALUES IN Ry AT SELECTED k-POINTS

	ORTHOGONAL	APW	NON-ORTHOGONAL
GAMMA 1	0.05575	0.05803	0.05664
GAMMA 1	1.40746	1.41251	1.40636
GAMMA 12	0.91130	0.91029	0.91274
GAMMA 15	1.74177	1.76267	1.76280
GAMMA 25'	0.72893	0.73807	0.73652
X1 (008)	0.18204	0.18297	0.18177
X1 (008)	1.35082	1.37832	1.38723
X1 (008)	1.75509	1.72547	1.72281
X2 (008)	1.02963	1.04194	1.04215
X3 (008)	0.50191	0.49617	0.49736
X5 (008)	1.09002	1.08987	1.09102
X4' (008)	0.91176	0.91078	0.91100
X5' (008)	1.46942	1.42823	1.42697
L1 (444)	0.42158	0.43229	0.43106
L1 (444)	1.39410	1.40049	1.40673
L3 (444)	0.68771	0.70483	0.70539
L3 (444)	1.05023	1.04551	1.04497
L2' (444)	0.40886	0.41604	0.41620
L2' (444)	1.34099	1.30518	1.30096
L3' (444)	1.89916	1.91184	1.91129
W1 (048)	0.83658	0.82586	0.82524
W1 (048)	1.43982	1.43348	1.41376
W2' (048)	0.25391	0.25844	0.25586
W2' (048)	1.29176	1.30757	1.30255
W2' (048)	1.98085	2.19221	2.19199
W1' (048)	1.08649	1.09041	1.08894
W3 (048)	0.63329	0.63499	0.63184
W3 (048)	1.35167	1.37106	1.37119
ODD (224)	0.75881	0.76644	0.76475
ODD (224)	1.02120	1.01567	1.01295
EVEN(224)	0.26761	0.26895	0.26955
EVEN(224)	0.57146	0.56273	0.56290
EVEN(224)	0.69916	0.69777	0.70071
EVEN(224)	0.95848	0.94373	0.94728
EVEN(224)	1.24506	1.24416	1.24329

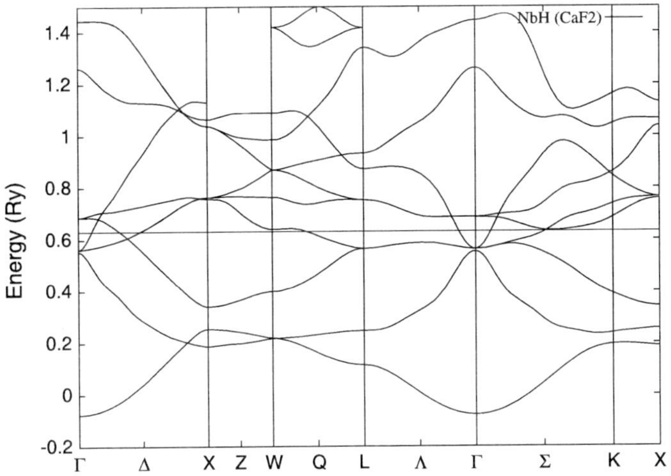

Fig. 7.9 Energy bands of NbH2 in the CaF2 structure

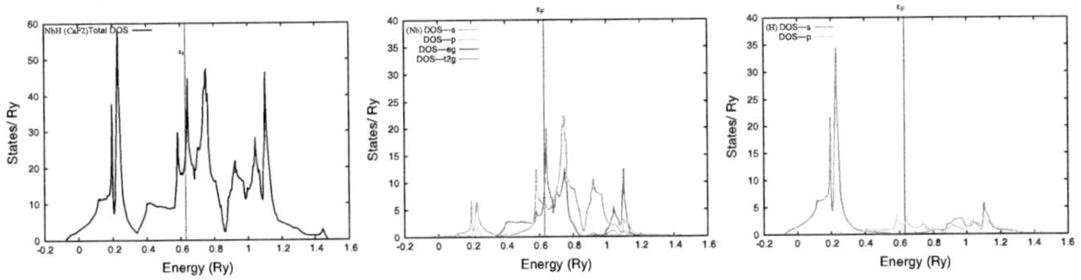

Fig. 7.10 Total, angular momentum and site decomposed densities of states of NbH2 in the CaF2 structure

7.3 Molybdenum Hydride (MoH)

See Fig. 7.11, Tables 7.11, 7.12, and 7.13.

See Figs. 7.12, 7.13, and Tables 7.14, and 7.15.

See Figs. 7.14 and 7.15.

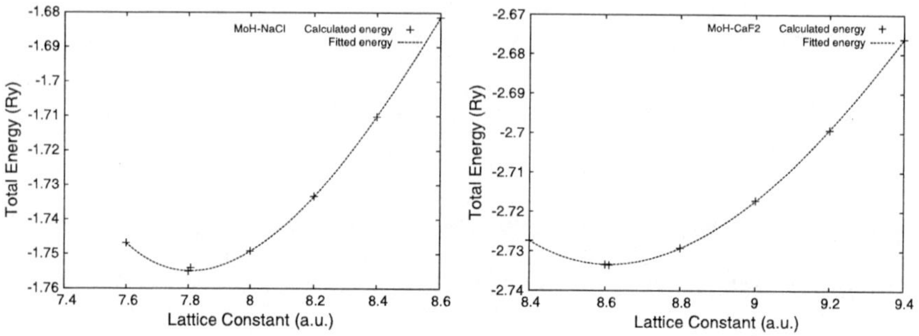

Fig. 7.11 Total energy versus lattice constant of MoH in the NaCl and CaF2 structures

Table 7.11 Lattice constant, bulk modulus, gap, total energy

Stru	a (Bohr)	B (MBar)	Gap	Total Energy
NaCl	7.81	2.82	-	-8091.75403
CaF2	8.61	1.88	-	-8092.73355
exp				

Table 7.12 Birch fit coefficients

	A1	A2	A3	A4
NaCl	-2.547956E-01	-4.679343E+01	-3.697887E+02	1.509280E+04
CaF2	-9.530273E-01	-8.980309E+01	6.604076E+02	1.296390E+04

Table 7.13 DOS at Ef, Hopfield parameter, Stoner criterion

MoH a=7.81 Bohr NaCl

Fermi Energy (Ry)		Total DOS (States/Ry)	Decomposed DOS inside the muffin tins s	p	eg	t2g	f
Mo	0.935	13.018	0.027	0.258	6.107	2.406	0.059
H	0.935	13.018	0.034	1.043	0.061	0.024	0.030

Fermi-Velocity (cm/s): 0.58796 x10E8
Plasmon Energy (eV) : 7.00076
Electron-ion interaction (Hopfield parameter) (eV/A^2) Mo: 5.044 H: 0.068

Mo	MUFFIN-TIN RADIUS and CHARGE =	2.1471	39.0514
H	MUFFIN-TIN RADIUS and CHARGE =	1.7567	1.3651
Mo	STONER I = 0.0135		
H	STONER I = 0.0015		
STONER PARAMETER (Ry)	I = 0.0135		
STONER CRITERION	N*I = 0.1754		

MoH2 a=8.61 Bohr CaF2

Fermi Energy (Ry)		Total DOS (States/Ry)	Decomposed DOS inside the muffin tins s	p	eg	t2g	f
Mo	0.686	18.532	0.034	0.097	7.722	5.687	0.056
H	0.686	18.532	0.153	0.600	0.054	0.045	0.020

Fermi-Velocity (cm/s): 0.54480 x10E8
Plasmon Energy (eV) : 6.68341
Electron-ion interaction (Hopfield parameter) (eV/A^2) Mo: 3.642 H: 0.427

Mo	MUFFIN-TIN RADIUS and CHARGE =	2.2372	39.2922
H	MUFFIN-TIN RADIUS and CHARGE =	1.4915	0.9697
Mo	STONER I = 0.0095		
H	STONER I = 0.0001		
STONER PARAMETER (Ry)	I = 0.0099		
STONER CRITERION	N*I = 0.1831		

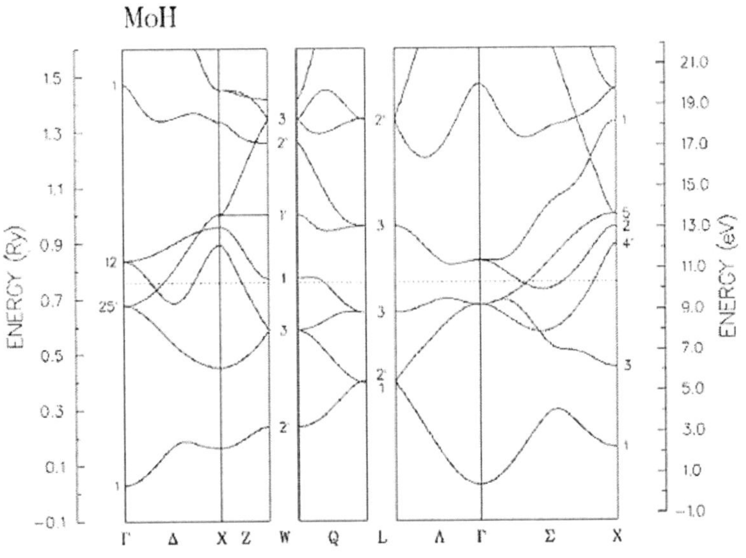

Fig. 7.12 Energy bands of MoH in the NaCl structure (tight-binding)

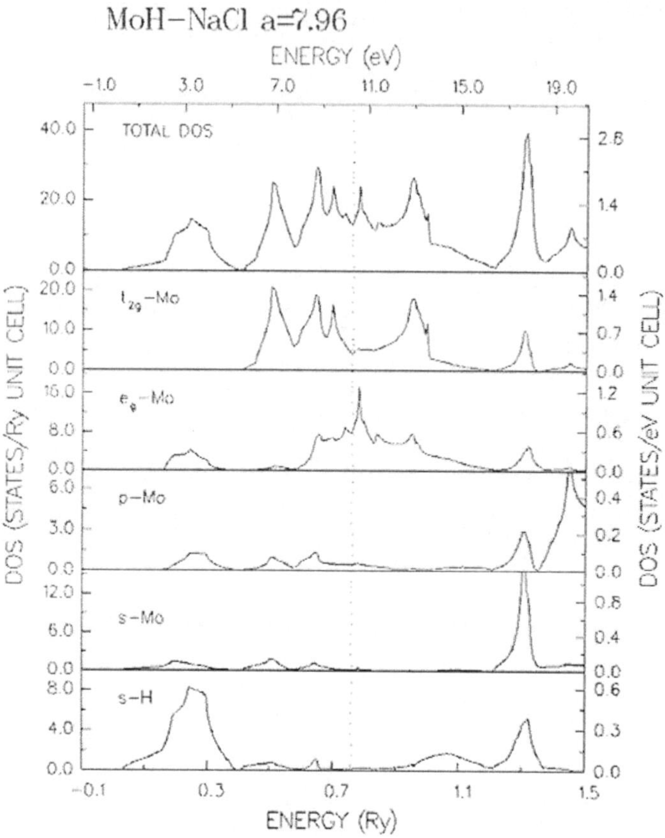

Fig. 7.13 Total, angular momentum and site decomposed densities of states of MoH in the NaCl structure (tight-binding)

Table 7.14 MoH (NaCl) a = 7.96 Bohr Slater–Koster 3-center parameters

	ORTHOGONAL	NON-ORTHOGONAL	
	ENERGY INTEGRALS Ry	ENERGY INTEGRALS Ry	OVERLAP INTEGRALS
ON SITE Mo-Mo			
s,s(000)	1.15719	0.98245	
x,x(000)	1.59270	1.43354	
xy,xy(000)	0.85116	0.71962	
d2,d2(000)	0.88317	0.72513	
FIRST NEIGHBOR Mo-Mo			
s,s(110)	-0.01713	0.01868	0.05026
s,x(110)	-0.04445	0.03746	0.08088
s,xy(110)	-0.06833	-0.00737	0.08808
s,d2(110)	-0.00185	-0.01278	-0.00340
x,x(110)	0.05749	-0.00082	-0.06712
x,x(011)	-0.01388	0.12003	0.11619
x,y(110)	0.06529	-0.06829	-0.07499
x,xy(110)	0.08185	0.01261	-0.04558
x,xy(011)	-0.00324	0.02929	0.06540
z,d2(011)	0.01739	0.05406	0.08437
z,d1(011)	0.00446	0.00252	0.00221
xy,xy(110)	-0.05504	-0.02236	0.04243
xy,xy(011)	0.01409	0.00137	-0.02602
xy,xz(011)	-0.00597	-0.00503	-0.03922
xy,d2(110)	0.02736	0.00443	0.04386
d2,d2(110)	0.00217	-0.01216	0.01094
d1,d1(110)	-0.03625	0.05438	0.01810
SECOND NEIGHBOR Mo-Mo			
s,s(200)	0.06376	0.02952	-0.02298
s,x(200)	0.07717	0.01884	-0.02545
s,d2(002)	0.05074	0.02306	0.05469
x,x(200)	-0.11968	-0.05590	-0.00693
y,y(200)	0.01418	0.01169	0.02267
x,xy(020)	0.01433	0.00573	0.01389
z,d2(002)	-0.05318	-0.03990	-0.06379
xy,xy(200)	-0.01765	-0.00301	-0.00869
xy,xy(002)	0.00210	-0.00128	-0.00120
d2,d2(002)	0.05071	-0.04310	-0.05344
d1,d1(002)	0.00223	0.00283	0.00370
ON SITE H-H			
s,s(000)	0.41885	0.58897	
FIRST NEIGHBOR H-H			
s,s(110)	-0.01703	-0.01943	-0.02272
s,x(110)	0.00000	0.00000	0.00000
x,x(110)	0.00000	0.00000	0.00000
x,x(011)	0.00000	0.00000	0.00000
x,y(110)	0.00000	0.00000	0.00000
SECOND NEIGHBOR H-H			
s,s(200)	-0.01198	-0.00178	0.00109
s,x(200)	0.00000	0.00000	0.00000
x,x(200)	0.00000	0.00000	0.00000
y,y(200)	0.00000	0.00000	0.00000
FIRST NEIGHBOR Mo-H			
s,s(100)	-0.10160	-0.13224	-0.02711
s,x(100)	0.00000	0.00000	0.00000
x,s(100)	0.08408	0.21816	0.15887
x,x(100)	0.00000	0.00000	0.00000
y,y(100)	0.00000	0.00000	0.00000
xy,x(010)	0.00000	0.00000	0.00000
d2,s(001)	-0.09124	0.02058	-0.13305
d2,z(001)	0.00000	0.00000	0.00000
SECOND NEIGHBOR Mo-H			
s,s(111)	0.02793	0.01699	0.02843
x,s(111)	-0.03932	-0.01769	-0.02285
s,x(111)	0.00000	0.00000	0.00000
x,x(111)	0.00000	0.00000	0.00000
x,y(111)	0.00000	0.00000	0.00000
xy,s(111)	0.00433	-0.00158	0.00106
xy,x(111)	0.00000	0.00000	0.00000
yz,x(111)	0.00000	0.00000	0.00000
d1,x(111)	0.00000	0.00000	0.00000

MoH

BAND	ORTHOGONAL RMS ERROR mRy	MAXIMUM DEVIATION k	mRy	NON-ORTHOGONAL RMS ERROR mRy	MAXIMUM DEVIATION k	mRy
1	5.6	(033)	11.5	0.7	(066)	1.5
2	6.1	(222)	20.2	0.5	(044)	1.3
3	4.8	(044)	10.2	0.5	(005)	1.4
4	4.3	(226)	8.7	0.5	(224)	1.3
5	4.1	(003)	8.6	0.3	(055)	0.8
6	5.6	(174)	12.2	0.7	(174)	2.1
7	8.2	(000)	23.5	1.2	(226)	3.3
1-7	5.7			0.7		

ENERGY VALUES IN Ry AT SELECTED k-POINTS

	ORTHOGONAL	APW	NON-ORTHOGONAL
GAMMA 1	0.02838	0.02896	0.02874
GAMMA 1	1.44845	1.47198	1.47095
GAMMA 12	0.83757	0.83671	0.83677
GAMMA 15	1.81442	1.83360	1.83357
GAMMA 25'	0.67728	0.67795	0.67773
X1 (008)	0.16974	0.16161	0.16261
X1 (008)	1.33644	1.33532	1.33702
X1 (008)	1.78102	1.74444	1.74328
X2 (008)	0.95647	0.95758	0.95703
X3 (008)	0.45191	0.45091	0.45101
X5 (008)	1.00495	1.00164	1.00167
X4' (008)	0.89466	0.89299	0.89255
X5' (008)	1.46556	1.45179	1.45152
L1 (444)	0.38705	0.39512	0.39536
L1 (444)	1.35283	1.34934	1.34879
L3 (444)	0.65046	0.64977	0.65042
L3 (444)	0.96757	0.96173	0.96122
L2' (444)	0.39218	0.40400	0.40310
L2' (444)	1.35156	1.34283	1.34286
L3' (444)	2.03651	1.96634	1.96621
W1 (048)	0.76537	0.77379	0.77335
W1 (048)	1.41830	1.41624	1.41709
W2' (048)	0.22853	0.23772	0.23793
W2' (048)	1.25159	1.26153	1.26185
W2' (048)	2.10525	2.23371	2.23373
W1' (048)	0.99656	1.00206	1.00209
W3 (048)	0.58419	0.58744	0.58769
W3 (048)	1.33823	1.34697	1.34739
ODD (224)	0.71395	0.70796	0.70668
ODD (224)	0.94021	0.93537	0.93494
EVEN(224)	0.25427	0.24508	0.24529
EVEN(224)	0.52893	0.52294	0.52229
EVEN(224)	0.64276	0.64673	0.64686
EVEN(224)	0.87910	0.87837	0.87835
EVEN(224)	1.25160	1.25160	1.25273

FERMI LEVEL QUANTITIES (Non-orthogonal fit)

DENSITIES OF STATES (states/Ry/cell)

ENERGY Ry	Total	s-Mo	p-Mo	t2g-Mo	eg-Mo	s-H	p-H
0.7594	13.57	0.13	0.49	4.47	8.26	0.24	0.00

INTEGRATED DENSITIES OF STATES (electrons)

Total	s-Mo	p-Mo	t2g-Mo	eg-Mo	s-H	p-H
7.00	0.42	0.34	3.38	1.49	1.37	0.00

VELOCITY cm/s	PLASMON ENERGY eV	EIGENVALUE SUM Ry
0.61x10E8	7.17	-1.7750

Table 7.15 MoH (NaCl) a = 7.96 Bohr Slater–Koster 2-center parameters

	ORTHOGONAL	NON-ORTHOGONAL	
	ENERGY INTEGRALS Ry	ENERGY INTEGRALS Ry	OVERLAP INTEGRALS
ON SITE Mo-Mo			
s	1.06436	0.58114	
p	1.54307	1.22746	
t2g	0.79196	0.70428	
eg	0.76865	0.69278	
FIRST NEIGHBOR Mo-Mo			
(sss)	-0.07304	-0.00959	0.13166
(pps)	0.13077	-0.08222	-0.19431
(ppp)	-0.01676	-0.03118	0.01581
(dds)	-0.07333	-0.00420	0.08958
(ddp)	0.02614	0.01717	-0.04649
(ddd)	0.00068	-0.00673	0.00150
(sps)	0.10954	-0.02757	-0.13871
(sds)	-0.07529	-0.00425	0.09103
(pds)	-0.10754	0.06803	0.16727
(pdp)	0.01424	0.04275	0.00273
SECOND NEIGHBOR Mo-Mo			
(sss)	0.00493	-0.06513	-0.02436
(pps)	-0.06962	0.07925	0.03545
(ppp)	0.00347	-0.02279	-0.00737
(dds)	0.00584	-0.02534	-0.01541
(ddp)	-0.00266	0.00387	0.00195
(ddd)	0.00059	0.00074	0.00034
(sps)	-0.02151	0.11441	0.06490
(sds)	0.00566	-0.03957	-0.02474
(pds)	0.01121	-0.02257	0.00518
(pdp)	-0.00403	0.01540	0.01259
ON SITE H-H			
s	0.99388	0.60761	
FIRST NEIGHBOR H-H			
(sss)	-0.00034	0.00946	0.04181
(sps)	0.00000	0.00000	0.00000
(pps)	0.00000	0.00000	0.00000
(ppp)	0.00000	0.00000	0.00000
SECOND NEIGHBOR H-H			
(sss)	0.04730	-0.00362	-0.02205
(sps)	0.00000	0.00000	0.00000
(pps)	0.00000	0.00000	0.00000
(ppp)	0.00000	0.00000	0.00000
FIRST NEIGHBOR Mo-H			
(sss)	0.11560	-0.01461	-0.22587
(sps)	0.00000	0.00000	0.00000
(pss)	0.13203	0.01914	-0.19342
(pps)	0.00000	0.00000	0.00000
(ppp)	0.00000	0.00000	0.00000
(dpp)	0.00000	0.00000	0.00000
(dss)	0.11884	0.06513	-0.10686
(dps)	0.00000	0.00000	0.00000
SECOND NEIGHBOR Mo-H			
(sss)	-0.02566	-0.00162	-0.02151
(pss)	-0.05543	0.00826	0.00913
(sps)	0.00000	0.00000	0.00000
(pps)	0.00000	0.00000	0.00000
(ppp)	0.00000	0.00000	0.00000
(dss)	-0.01312	-0.01653	-0.03438
(dps)	0.00000	0.00000	0.00000
(dpp)	0.00000	0.00000	0.00000

♀MoH

BAND	ORTHOGONAL RMS ERROR	ORTHOGONAL MAXIMUM DEVIATION		NON-ORTHOGONAL RMS ERROR	NON-ORTHOGONAL MAXIMUM DEVIATION	
	mRy	k	mRy	mRy	k	mRy
1	5.7	(005)	13.2	2.0	(444)	6.1
2	8.1	(333)	21.0	1.7	(226)	4.7
3	7.5	(264)	15.6	1.7	(007)	5.1
4	6.8	(444)	14.2	1.3	(004)	2.6
5	6.2	(224)	11.8	1.2	(004)	2.6
6	9.8	(066)	39.3	2.6	(044)	6.0
7	14.2	(004)	34.1	4.8	(226)	15.3
1.7	8.7			2.5		

ENERGY VALUES IN Ry AT SELECTED k-POINTS

	ORTHOGONAL	APW	NON-ORTHOGONAL
GAMMA 1	0.02633	0.02896	0.02896
GAMMA 1	1.46473	1.47198	1.46895
GAMMA 12	0.83786	0.83671	0.83900
GAMMA 15	1.80674	1.83360	1.83387
GAMMA 25'	0.67050	0.67795	0.67686
X1 (008)	0.16589	0.16161	0.15906
X1 (008)	1.31287	1.33532	1.33948
X1 (008)	1.78187	1.74444	1.74396
X2 (008)	0.94714	0.95758	0.95726
X3 (008)	0.45590	0.45091	0.45256
X5 (008)	1.00183	1.00164	1.00316
X4' (008)	0.89466	0.89299	0.89279
X5' (008)	1.48476	1.45179	1.45117
L1 (444)	0.38533	0.39512	0.40121
L1 (444)	1.34901	1.34934	1.35305
L3 (444)	0.63559	0.64977	0.64953
L3 (444)	0.96607	0.96173	0.96275
L2' (444)	0.40115	0.40400	0.40132
L2' (444)	1.38722	1.34283	1.34024
L3' (444)	1.96346	1.96634	1.96540
W1 (048)	0.78431	0.77379	0.77583
W1 (048)	1.41794	1.41624	1.40948
W2' (048)	0.23184	0.23772	0.23521
W2' (048)	1.25195	1.26153	1.25589
W2' (048)	2.05031	2.23371	2.23432
W1' (048)	0.99947	1.00206	1.00099
W3 (048)	0.58371	0.58744	0.58549
W3 (048)	1.33159	1.34697	1.34907
ODD (224)	0.70266	0.70796	0.70560
ODD (224)	0.93965	0.93537	0.93341
EVEN(224)	0.24446	0.24508	0.24501
EVEN(224)	0.53077	0.52294	0.52061
EVEN(224)	0.64902	0.64673	0.64922
EVEN(224)	0.89013	0.87837	0.87887
EVEN(224)	1.25189	1.25160	1.25218

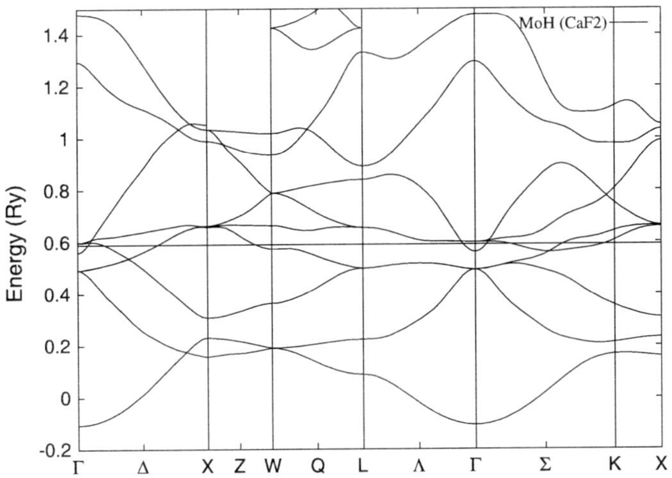

Fig. 7.14 Energy bands of MoH2 in the CaF2 structure

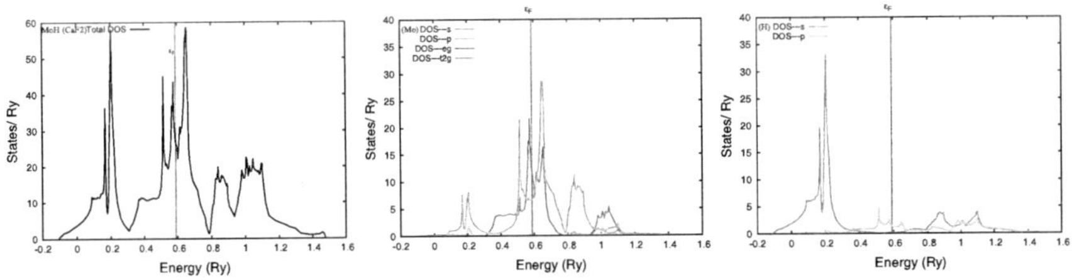

Fig. 7.15 Total, angular momentum and site decomposed densities of states of MoH2 in the CaF2 structure

7.4 Technetium Hydride (TcH)

See Fig. 7.16, Table 7.16, 7.17, and 7.18.

See Figs. 7.17, 7.18, Tables 7.19 and 7.20.

See Figs. 7.19 and 7.20.

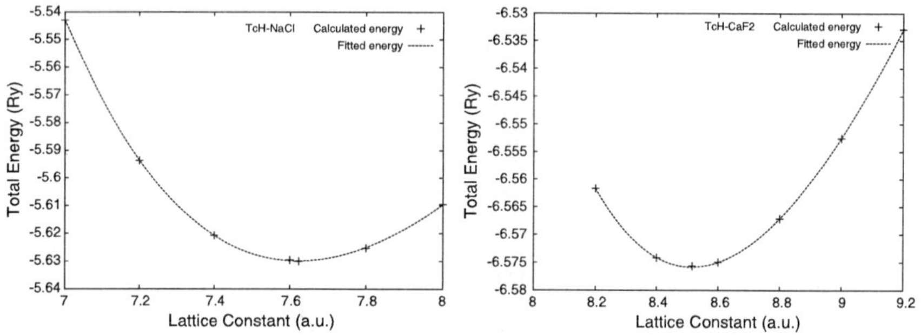

Fig. 7.16 Total energy versus lattice constant of TcH in the NaCl and CaF2 structures

Table 7.16 Lattice constant, bulk modulus, gap, total energy

Stru	a (Bohr)	B (MBar)	Gap	Total Energy
NaCl	7.62	2.88	-	-8565.63013
CaF2	8.52	1.88	-	-8566.57567

Table 7.17 Birch fit coefficients

	A1	A2	A3	A4
NaCl	-3.612176E+00	-8.339360E+01	6.270984E+02	5.142289E+03
CaF2	-5.263080E+00	-4.936377E+01	-4.206735E+02	2.170037E+04

Table 7.18 DOS at Ef, Hopfield parameter, Stoner criterion

TcH a=7.62 Bohr NaCl

	Fermi Energy (Ry)	Total DOS (States/Ry)	Decomposed DOS inside the muffin tins s	p	eg	t2g	f
Tc	0.963	11.594	0.019	0.143	5.141	3.303	0.043
H	0.963	11.594	0.030	0.407	0.131	0.031	0.021

```
Fermi-Velocity (cm/s): 0.89930 x10E8
Plasmon Energy (eV)  : 10.47040
Electron-ion interaction (Hopfield parameter) (eV/A^2) Tc: 4.267 H:  0.033
-------------------------------------------------
 Tc     MUFFIN-TIN RADIUS and CHARGE =     2.0963   40.0457
  H     MUFFIN-TIN RADIUS and CHARGE =     1.7152    1.3527
         Tc             STONER I = 0.0147
          H             STONER I = 0.0005
        STONER PARAMETER (Ry) I = 0.0154
        STONER CRITERION    N*I = 0.1781
```

TcH2 a=8.52 Bohr CaF2

	Fermi Energy (Ry)	Total DOS (States/Ry)	Decomposed DOS inside the muffin tins s	p	eg	t2g	f
Tc	0.660	24.002	0.009	0.074	8.735	10.894	0.035
H	0.660	24.002	0.071	0.498	0.046	0.094	0.029

```
Fermi-Velocity (cm/s): 0.56692 x10E8
Plasmon Energy (eV)  : 8.04775
Electron-ion interaction (Hopfield parameter) (eV/A^2) Tc:  2.800  H: 0.181
-------------------------------------------------
         Tc     MUFFIN-TIN RADIUS and CHARGE =     2.2125   40.4231
          H     MUFFIN-TIN RADIUS and CHARGE =     1.4750    0.9305
                   Tc        STONER I = 0.0132
                    H        STONER I = 0.0001
         STONER PARAMETER (Ry)      I = 0.0134
         STONER CRITERION         N*I = 0.3216
```

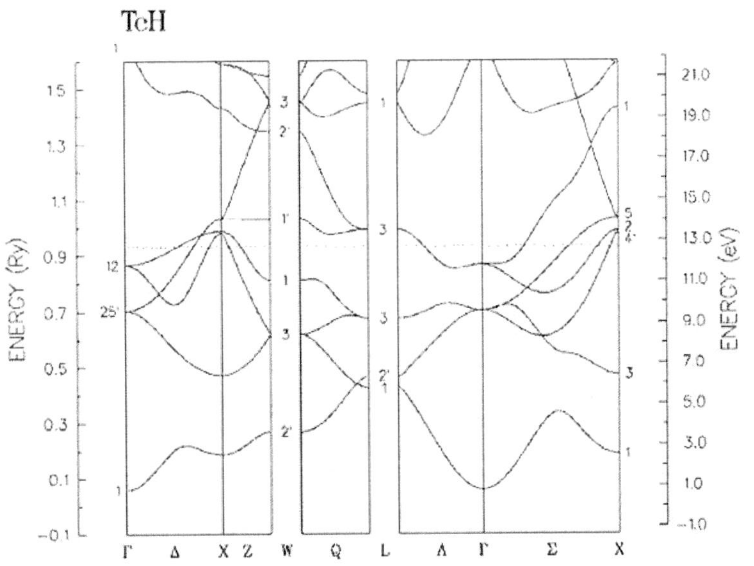

Fig. 7.17 Energy bands of TcH in the NaCl structure (tight-binding)

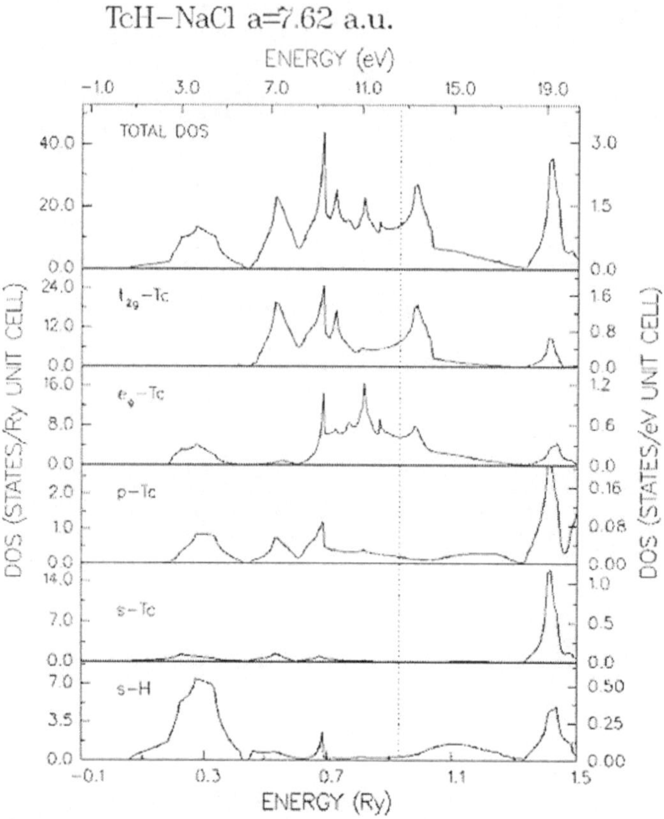

Fig. 7.18 Total, angular momentum and site decomposed densities of states of TcH in the NaCl structure (tight-binding)

Table 7.19 TcH (NaCl) a = 7.62 Bohr Slater–Koster 3-center parameters

	ORTHOGONAL	NON-ORTHOGONAL	
	ENERGY INTEGRALS Ry	ENERGY INTEGRALS Ry	OVERLAP INTEGRALS
ON SITE Tc-Tc			
s,s(000)	1.28792	1.11663	
x,x(000)	1.74459	1.56481	
xy,xy(000)	0.87947	0.74810	
d2,d2(000)	0.92350	0.74722	
FIRST NEIGHBOR Tc-Tc			
s,s(110)	-0.01653	0.02457	0.04647
s,x(110)	-0.04967	0.04887	0.08459
s,xy(110)	-0.07422	-0.00814	0.08474
s,d2(110)	-0.00473	-0.01068	-0.00201
x,x(110)	0.06249	-0.01984	-0.07669
x,x(011)	-0.01537	0.11377	0.10761
x,y(110)	0.07094	-0.08503	-0.08109
x,xy(110)	0.08660	0.00674	-0.04932
x,xy(011)	-0.00359	0.02001	0.05582
z,d2(011)	0.01890	0.05147	0.07653
z,d1(011)	0.00621	-0.00014	-0.00047
xy,xy(110)	-0.05608	-0.02231	0.04169
xy,xy(011)	0.01443	0.00430	-0.02064
xy,xz(011)	-0.00559	0.00060	-0.03237
xy,d2(110)	0.02825	0.00247	0.04088
d2,d2(110)	0.00391	-0.00984	0.01496
d1,d1(110)	-0.04163	0.05399	0.01193
SECOND NEIGHBOR Tc-Tc			
s,s(200)	0.06857	0.02710	-0.02781
s,x(200)	0.08117	0.01536	-0.02951
s,d2(002)	0.05401	0.02528	0.05476
x,x(200)	-0.13129	-0.06569	-0.00560
y,y(200)	0.01559	0.00029	0.01438
x,xy(020)	0.01521	0.00453	0.01143
z,d2(002)	-0.05362	-0.04405	-0.06226
xy,xy(200)	-0.01762	-0.00373	-0.00948
xy,xy(002)	0.00197	0.00062	0.00055
d2,d2(002)	0.05436	-0.04203	-0.04667
d1,d1(002)	0.00232	0.00392	0.00521
ON SITE H-H			
s,s(000)	0.48468	0.63539	
FIRST NEIGHBOR H-H			
s,s(110)	-0.01847	-0.01637	-0.02208
s,x(110)	0.00000	0.00000	0.00000
x,x(110)	0.00000	0.00000	0.00000
x,x(011)	0.00000	0.00000	0.00000
x,y(110)	0.00000	0.00000	0.00000
SECOND NEIGHBOR H-H			
s,s(200)	-0.01355	-0.00182	0.00065
s,x(200)	0.00000	0.00000	0.00000
x,x(200)	0.00000	0.00000	0.00000
y,y(200)	0.00000	0.00000	0.00000
FIRST NEIGHBOR Tc-H			
s,s(100)	-0.11239	-0.14866	-0.03523
s,x(100)	0.00000	0.00000	0.00000
x,s(100)	0.09480	0.22597	0.14542
x,x(100)	0.00000	0.00000	0.00000
y,y(100)	0.00000	0.00000	0.00000
xy,x(010)	0.00000	0.00000	0.00000
d2,s(001)	-0.10361	0.01926	-0.14204
d2,z(001)	0.00000	0.00000	0.00000
SECOND NEIGHBOR Tc-H			
s,s(111)	0.03125	0.01676	0.02642
x,s(111)	-0.04341	-0.01553	-0.02092
s,x(111)	0.00000	0.00000	0.00000
x,x(111)	0.00000	0.00000	0.00000
x,y(111)	0.00000	0.00000	0.00000
xy,s(111)	0.00379	-0.00233	-0.00038
xy,x(111)	0.00000	0.00000	0.00000
yz,x(111)	0.00000	0.00000	0.00000
d1,x(111)	0.00000	0.00000	0.00000

TcH

BAND	ORTHOGONAL RMS ERROR mRy	MAXIMUM DEVIATION k	MAXIMUM DEVIATION mRy	NON-ORTHOGONAL RMS ERROR mRy	MAXIMUM DEVIATION k	MAXIMUM DEVIATION mRy
1	6.0	(033)	12.6	0.7	(003)	1.6
2	6.9	(222)	22.5	0.5	(022)	1.3
3	5.3	(044)	11.1	0.5	(006)	1.3
4	4.4	(006)	8.9	0.5	(033)	1.1
5	4.4	(006)	8.9	0.4	(055)	0.9
6	5.6	(174)	11.8	0.9	(174)	3.0
7	9.6	(000)	28.1	1.3	(226)	3.5
1-7	6.3			0.7		

ENERGY VALUES IN Ry AT SELECTED k-POINTS

	ORTHOGONAL	APW	NON-ORTHOGONAL
GAMMA 1	0.05702	0.05855	0.05851
GAMMA 1	1.62574	1.65384	1.65247
GAMMA 12	0.86720	0.86575	0.86571
GAMMA 15	1.98279	2.00527	2.00513
GAMMA 25'	0.70406	0.70465	0.70431
X1 (008)	0.19347	0.18480	0.18579
X1 (008)	1.42760	1.42828	1.42982
X1 (008)	1.97282	1.93663	1.93551
X2 (008)	0.98683	0.98852	0.98790
X3 (008)	0.47316	0.47122	0.47164
X5 (008)	1.03725	1.03391	1.03419
X4' (008)	0.98298	0.98051	0.97996
X5' (008)	1.60584	1.59073	1.59035
L1 (444)	0.41625	0.42535	0.42574
L1 (444)	1.45094	1.44679	1.44668
L3 (444)	0.67610	0.67608	0.67634
L3 (444)	1.00102	0.99349	0.99305
L2' (444)	0.44972	0.46393	0.46326
L2' (444)	1.49355	1.48336	1.48329
L3' (444)	2.22858	2.15326	2.15309
W1 (048)	0.80394	0.81085	0.81067
W1 (048)	1.54771	1.54653	1.54743
W2' (048)	0.25780	0.26731	0.26806
W2' (048)	1.33810	1.34646	1.34713
W2' (048)	2.31734	2.45456	2.45454
W1' (048)	1.02939	1.03429	1.03394
W3 (048)	0.61483	0.61944	0.61965
W3 (048)	1.44290	1.45259	1.45301
ODD (224)	0.74348	0.73684	0.73586
ODD (224)	0.97185	0.96724	0.96669
EVEN(224)	0.29659	0.28672	0.28702
EVEN(224)	0.56181	0.55483	0.55391
EVEN(224)	0.67285	0.67756	0.67781
EVEN(224)	0.91727	0.91601	0.91625
EVEN(224)	1.37399	1.37377	1.37533

FERMI LEVEL QUANTITIES (Non-orthogonal fit)

ENERGY Ry		Total	s-Tc	p-Tc	t2g-Tc	eg-Tc	s-H	p-H
	DENSITIES OF STATES (states/Ry/cell)							
0.8566		11.91	0.08	0.28	5.12	6.16	0.26	0.00
	INTEGRATED DENSITIES OF STATES (electrons)							
		8.00	0.37	0.28	3.80	2.16	1.39	0.00

VELOCITY cm/s	PLASMON ENERGY eV	EIGENVALUE SUM Ry
0.69x10E8	8.15	-2.2670

Table 7.20 TcH (NaCl) a = 7.62 Bohr Slater–Koster 2-center parameters

	ORTHOGONAL	NON-ORTHOGONAL	
	ENERGY INTEGRALS Ry	ENERGY INTEGRALS Ry	OVERLAP INTEGRALS
ON SITE Tc-Tc			
s	1.17253	0.64991	
p	1.69028	1.44048	
t2g	0.81947	0.73439	
eg	0.79596	0.72277	
FIRST NEIGHBOR Tc-Tc			
(sss)	-0.07992	-0.00318	0.13626
(pps)	0.14218	-0.06779	-0.16613
(ppp)	-0.01801	0.00297	0.03170
(dds)	-0.07473	-0.00635	0.08546
(ddp)	0.02732	0.01852	-0.04401
(ddd)	0.00044	-0.00670	0.00188
(sps)	0.12014	-0.02589	-0.12297
(sds)	-0.07861	-0.00552	0.08793
(pds)	-0.11222	0.09042	0.18560
(pdp)	0.01506	0.05382	0.01723
SECOND NEIGHBOR Tc-Tc			
(sss)	0.00486	-0.07729	-0.02850
(pps)	-0.07739	0.01163	0.02185
(ppp)	0.00388	-0.04265	-0.02783
(dds)	0.00530	-0.01968	-0.00791
(ddp)	-0.00243	0.00454	0.00272
(ddd)	0.00048	-0.00112	-0.00204
(sps)	-0.02272	0.12951	0.07865
(sds)	0.00350	-0.04073	-0.02262
(pds)	0.01011	0.01965	0.04690
(pdp)	-0.00466	0.02037	0.02047
ON SITE H-H			
s	1.12486	0.71107	
FIRST NEIGHBOR H-H			
(sss)	0.00083	0.01378	0.04046
(sps)	0.00000	0.00000	0.00000
(pps)	0.00000	0.00000	0.00000
(ppp)	0.00000	0.00000	0.00000
SECOND NEIGHBOR H-H			
(sss)	0.05373	-0.01347	-0.02655
(sps)	0.00000	0.00000	0.00000
(pps)	0.00000	0.00000	0.00000
(ppp)	0.00000	0.00000	0.00000
FIRST NEIGHBOR Tc-H			
(sss)	0.12455	-0.01281	-0.21768
(sps)	0.00000	0.00000	0.00000
(pss)	0.14382	0.09673	-0.11097
(pps)	0.00000	0.00000	0.00000
(ppp)	0.00000	0.00000	0.00000
(dpp)	0.00000	0.00000	0.00000
(dss)	0.12412	0.04637	-0.11860
(dps)	0.00000	0.00000	0.00000
SECOND NEIGHBOR Tc-H			
(sss)	-0.02940	-0.00743	-0.02408
(pss)	-0.06359	-0.02912	-0.00646
(sps)	0.00000	0.00000	0.00000
(pps)	0.00000	0.00000	0.00000
(ppp)	0.00000	0.00000	0.00000
(dss)	-0.01356	-0.02757	-0.04190
(dps)	0.00000	0.00000	0.00000
(dpp)	0.00000	0.00000	0.00000

♀ TcH

BAND	ORTHOGONAL RMS ERROR mRy	ORTHOGONAL MAXIMUM DEVIATION k	ORTHOGONAL MAXIMUM DEVIATION mRy	NON-ORTHOGONAL RMS ERROR mRy	NON-ORTHOGONAL MAXIMUM DEVIATION k	NON-ORTHOGONAL MAXIMUM DEVIATION mRy
1	6.0	(005)	12.3	1.7	(444)	5.1
2	8.3	(333)	23.7	2.1	(264)	5.8
3	7.3	(264)	16.3	2.7	(007)	8.1
4	6.3	(444)	13.4	1.5	(048)	4.3
5	5.8	(224)	11.4	1.7	(264)	2.9
6	8.8	(066)	35.2	3.0	(044)	6.9
7	14.6	(004)	38.5	4.8	(226)	15.3
1.7	8.6			2.7		

ENERGY VALUES IN Ry AT SELECTED k-POINTS

	ORTHOGONAL	APW	NON-ORTHOGONAL
GAMMA 1	0.05552	0.05855	0.05872
GAMMA 1	1.64423	1.65384	1.65253
GAMMA 12	0.86708	0.86575	0.86842
GAMMA 15	1.97567	2.00527	2.00532
GAMMA 25'	0.69798	0.70465	0.70390
X1 (008)	0.19146	0.18480	0.18388
X1 (008)	1.40713	1.42828	1.43256
X1 (008)	1.97940	1.93663	1.93757
X2 (008)	0.97804	0.98852	0.98875
X3 (008)	0.47593	0.47122	0.47240
X5 (008)	1.03444	1.03391	1.03675
X4' (008)	0.98228	0.98051	0.97957
X5' (008)	1.62305	1.59073	1.59046
L1 (444)	0.41524	0.42535	0.43044
L1 (444)	1.44888	1.44679	1.45075
L3 (444)	0.66270	0.67608	0.67605
L3 (444)	0.99794	0.99349	0.99573
L2' (444)	0.46128	0.46393	0.46029
L2' (444)	1.52998	1.48336	1.48191
L3' (444)	2.14994	2.15326	2.15246
W1 (048)	0.82236	0.81085	0.81516
W1 (048)	1.54508	1.54653	1.53980
W2' (048)	0.26072	0.26731	0.26506
W2' (048)	1.33953	1.34646	1.33977
W2' (048)	2.26348	2.45456	2.45517
W1' (048)	1.03253	1.03429	1.03151
W3 (048)	0.61496	0.61944	0.61709
W3 (048)	1.43712	1.45259	1.45372
ODD (224)	0.73201	0.73684	0.73458
ODD (224)	0.97108	0.96724	0.96498
EVEN(224)	0.28602	0.28672	0.28562
EVEN(224)	0.56263	0.55483	0.55462
EVEN(224)	0.68051	0.67756	0.68063
EVEN(224)	0.92743	0.91601	0.91503
EVEN(224)	1.37387	1.37377	1.37214

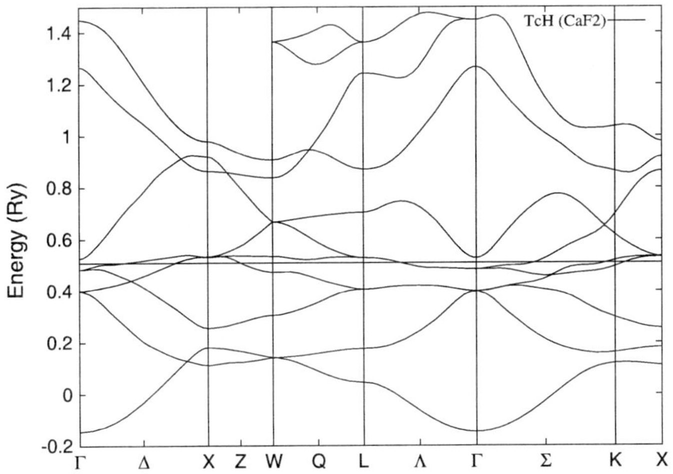

Fig. 7.19 Energy bands of TcH2 in the CaF2 structure

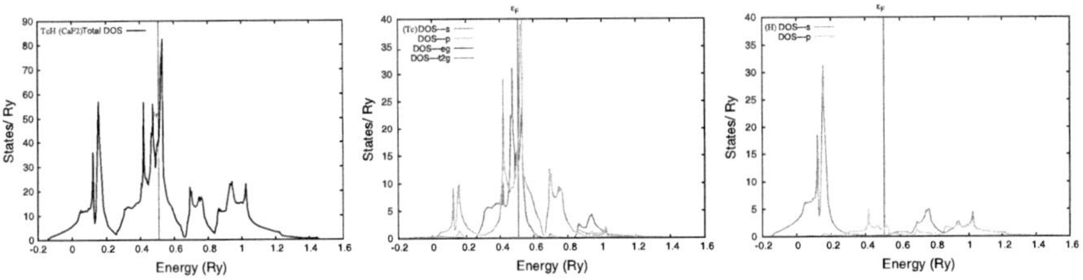

Fig. 7.20 Total, angular momentum and site decomposed densities of states of TcH2 in the CaF2 structure

7.5 Ruthenium Hydride (RuH)

See Fig. 7.21, Tables 7.21, 7.22, and 7.23.

See Figs. 7.22, 7.23, Tables 7.24 and 7.25.

See Figs. 7.24 and 7.25.

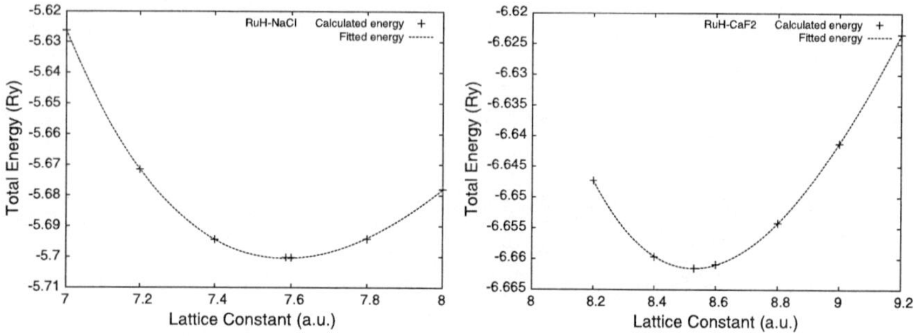

Fig. 7.21 Total energy versus lattice constant of RuH in the NaCl and CaF2 structures

Table 7.21 Lattice constant, bulk modulus, gap, total energy

Stru	a (Bohr)	B (MBar)	Gap	Total Energy
NaCl	7.59	2.75	-	-9055.70015
CaF2	8.53	1.73	-	-9056.66149

Table 7.22 Birch fit coefficients

	A1	A2	A3	A4
NaCl	-4.209612E+00	-4.979040E+01	-5.884740E+01	9.554965E+03
CaF2	-5.539931E+00	-3.778024E+01	-6.224021E+02	2.246927E+04

Table 7.23 DOS at Ef, Hopfield parameter, Stoner criterion

RuH a=7.59 Bohr NaCl

	Fermi Energy (Ry)	Total DOS (States/Ry)	Decomposed DOS inside the muffin tins s	p	eg	t2g	f
Ru	0.919	17.563	0.015	0.114	6.392	8.238	0.041
H	0.919	17.563	0.072	0.224	0.149	0.037	0.040

Fermi-Velocity (cm/s): 0.78359 x10E8
Plasmon Energy (eV) : 11.31603
Electron-ion interaction (Hopfield parameter) (eV/A^2) Ru: 4.616 H: 0.029

Ru	MUFFIN-TIN RADIUS and CHARGE =	2.0863	41.2029
H	MUFFIN-TIN RADIUS and CHARGE =	1.7070	1.3040

Ru STONER I =	0.0190
H STONER I =	0.0002
STONER PARAMETER (Ry) I =	0.0192
STONER CRITERION N*I =	0.3378

RuH2 a=8.53 Bohr CaF2

	Fermi Energy (Ry)	Total DOS (States/Ry)	Decomposed DOS inside the muffin tins s	p	eg	t2g	f
Ru	0.587	57.635	0.032	0.085	16.817	33.309	0.028
H	0.587	57.635	0.121	0.792	0.031	0.283	0.077

Fermi-Velocity (cm/s): 0.35587 x10E8
Plasmon Energy (eV) : 7.80995
Electron-ion interaction (Hopfield parameter) (eV/A^2) Ru: 2.345 H: 0.236

Ru	MUFFIN-TIN RADIUS and CHARGE =	2.2158	41.6505
H	MUFFIN-TIN RADIUS and CHARGE =	1.4772	0.8897

Ru STONER I =	0.0156
H STONER I =	0.00004
STONER PARAMETER (Ry) I =	0.0157
STONER CRITERION N*I =	0.9044

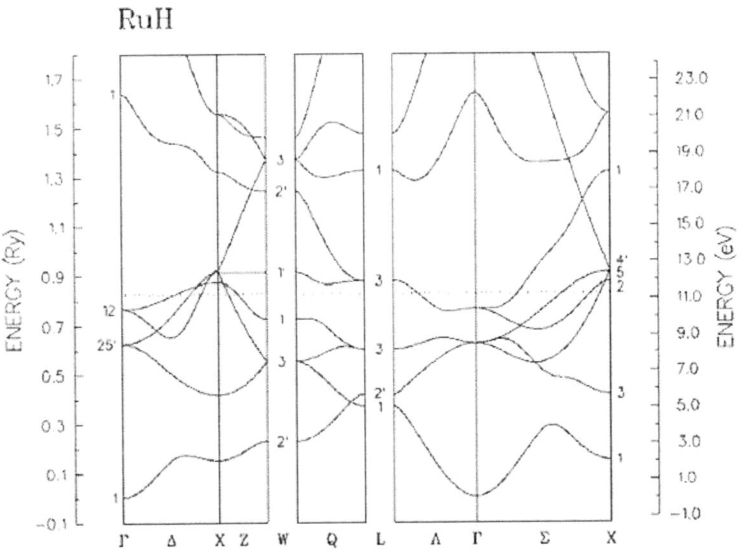

Fig. 7.22 Energy bands of RuH in the NaCl structure (tight-binding)

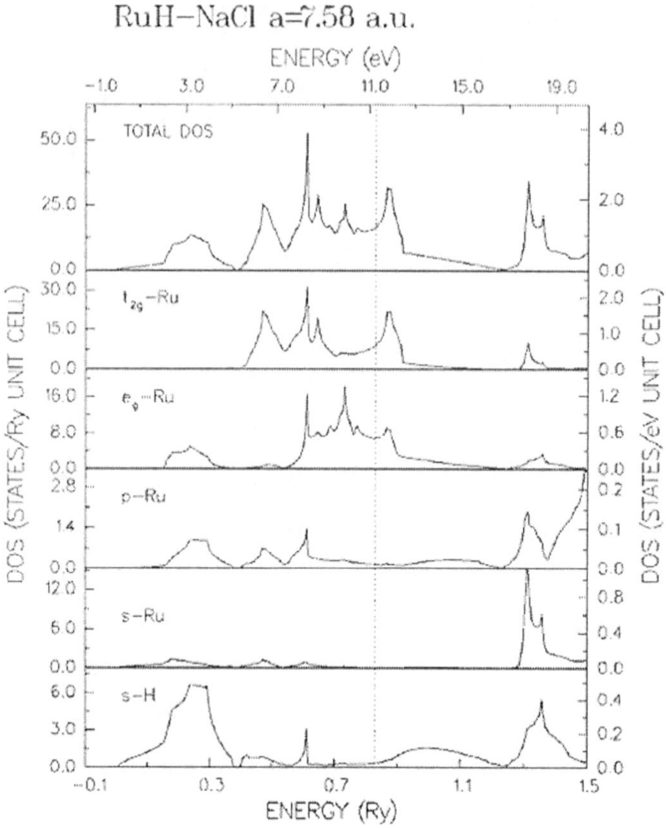

Fig. 7.23 Total, angular momentum and site decomposed densities of states of RuH in the NaCl structure (tight-binding)

Table 7.24 RuH (NaCl) a = 7.58 Bohr Slater–Koster 3-center parameters

	ORTHOGONAL	NON-ORTHOGONAL	
	ENERGY INTEGRALS Ry	ENERGY INTEGRALS Ry	OVERLAP INTEGRALS
ON SITE Ru-Ru			
s,s(000)	1.23725	1.02481	
x,x(000)	1.72249	1.50170	
xy,xy(000)	0.76999	0.66751	
d2,d2(000)	0.83421	0.65042	
FIRST NEIGHBOR Ru-Ru			
s,s(110)	-0.01299	0.02505	0.05578
s,x(110)	-0.05112	0.05619	0.09377
s,xy(110)	-0.06970	-0.01088	0.08199
s,d2(110)	-0.00682	-0.01556	-0.00635
x,x(110)	0.06694	-0.04716	-0.10070
x,x(011)	-0.01362	0.09133	0.10439
x,y(110)	0.07007	-0.10877	-0.09502
x,xy(110)	0.07969	0.00988	-0.04470
x,xy(011)	-0.00396	0.01023	0.05262
z,d2(011)	0.02087	0.04777	0.07590
z,d1(011)	0.00655	0.00160	0.00250
xy,xy(110)	-0.04912	-0.02524	0.03301
xy,xy(011)	0.01278	0.00580	-0.01618
xy,xz(011)	-0.00233	0.00519	-0.02562
xy,d2(110)	0.02579	0.00188	0.03978
d2,d2(110)	0.00568	-0.00688	0.02015
d1,d1(110)	-0.04293	0.04774	0.00236
SECOND NEIGHBOR Ru-Ru			
s,s(200)	0.06365	0.01517	-0.03627
s,x(200)	0.07574	-0.00306	-0.04653
s,d2(002)	0.05051	0.02680	0.05318
x,x(200)	-0.13333	-0.04364	0.00984
y,y(200)	0.01496	-0.01386	0.00895
x,xy(020)	0.01368	-0.00028	0.00879
z,d2(002)	-0.04733	-0.04264	-0.05840
xy,xy(200)	-0.01330	-0.00215	-0.00903
xy,xy(002)	0.00153	0.00112	0.00126
d2,d2(002)	0.05090	-0.03733	-0.03862
d1,d1(002)	0.00152	0.00659	0.00936
ON SITE H-H			
s,s(000)	0.44239	0.65471	
FIRST NEIGHBOR H-H			
s,s(110)	-0.01728	-0.01250	-0.02295
s,x(110)	0.00000	0.00000	0.00000
x,x(110)	0.00000	0.00000	0.00000
x,x(011)	0.00000	0.00000	0.00000
x,y(110)	0.00000	0.00000	0.00000
SECOND NEIGHBOR H-H			
s,s(200)	-0.01336	-0.00131	-0.00240
s,x(200)	0.00000	0.00000	0.00000
x,x(200)	0.00000	0.00000	0.00000
y,y(200)	0.00000	0.00000	0.00000
FIRST NEIGHBOR Ru-H			
s,s(100)	-0.11776	-0.15293	-0.03327
s,x(100)	0.00000	0.00000	0.00000
x,s(100)	0.09642	0.25850	0.16611
x,x(100)	0.00000	0.00000	0.00000
y,y(100)	0.00000	0.00000	0.00000
xy,x(010)	0.00000	0.00000	0.00000
d2,s(001)	-0.09693	0.01644	-0.14743
d2,z(001)	0.00000	0.00000	0.00000
SECOND NEIGHBOR Ru-H			
s,s(111)	0.02927	0.01027	0.02233
x,s(111)	-0.04280	-0.00889	-0.01690
s,x(111)	0.00000	0.00000	0.00000
x,x(111)	0.00000	0.00000	0.00000
x,y(111)	0.00000	0.00000	0.00000
xy,s(111)	0.00379	-0.00137	-0.00087
xy,x(111)	0.00000	0.00000	0.00000
yz,x(111)	0.00000	0.00000	0.00000
d1,x(111)	0.00000	0.00000	0.00000

RuH

BAND	ORTHOGONAL RMS ERROR mRy	MAXIMUM DEVIATION k	mRy	NON-ORTHOGONAL RMS ERROR mRy	MAXIMUM DEVIATION k	mRy
1	5.4	(033)	10.9	0.6	(264)	1.3
2	6.3	(222)	19.6	0.5	(444)	1.0
3	4.9	(044)	9.9	0.5	(006)	1.3
4	3.7	(007)	7.0	0.6	(224)	1.5
5	4.1	(044)	8.8	0.3	(226)	0.7
6	4.9	(064)	10.9	0.8	(174)	3.0
7	8.7	(000)	23.1	1.2	(226)	2.9
1-7	5.6			0.7		

ENERGY VALUES IN Ry AT SELECTED k-POINTS

	ORTHOGONAL	APW	NON-ORTHOGONAL
GAMMA 1	0.00210	0.00405	0.00345
GAMMA 1	1.61601	1.63915	1.63870
GAMMA 12	0.76798	0.76679	0.76713
GAMMA 15	1.99669	2.01837	2.01823
GAMMA 25'	0.62563	0.62626	0.62602
X1 (008)	0.16016	0.15186	0.15285
X1 (008)	1.31775	1.31916	1.32082
X1 (008)	1.88495	1.85505	1.85430
X2 (008)	0.87148	0.87353	0.87302
X3 (008)	0.42113	0.41802	0.41845
X5 (008)	0.91634	0.91332	0.91343
X4' (008)	0.92565	0.92289	0.92230
X5' (008)	1.57014	1.55763	1.55754
L1 (444)	0.36405	0.37174	0.37225
L1 (444)	1.33003	1.32539	1.32509
L3 (444)	0.60091	0.60221	0.60231
L3 (444)	0.88687	0.87859	0.87841
L2' (444)	0.40661	0.41966	0.41862
L2' (444)	1.48469	1.47652	1.47657
L3' (444)	2.20963	2.14268	2.14267
W1 (048)	0.72120	0.72616	0.72639
W1 (048)	1.46044	1.46045	1.46111
W2' (048)	0.22222	0.23094	0.23111
W2' (048)	1.23872	1.24413	1.24467
W2' (048)	2.28442	2.40291	2.40287
W1' (048)	0.91024	0.91362	0.91360
W3 (048)	0.54721	0.55304	0.55323
W3 (048)	1.36277	1.37063	1.37089
ODD (224)	0.66222	0.65659	0.65505
ODD (224)	0.86076	0.85638	0.85603
EVEN(224)	0.24482	0.23632	0.23609
EVEN(224)	0.50301	0.49657	0.49593
EVEN(224)	0.60092	0.60505	0.60588
EVEN(224)	0.81954	0.81844	0.81857
EVEN(224)	1.32136	1.31979	1.32190

FERMI LEVEL QUANTITIES (Non-orthogonal fit)

ENERGY Ry		Total	s-Ru	p-Ru	t2g-Ru	eg-Ru	s-H	p-H
	DENSITIES OF STATES (states/Ry/cell)							
0.8289		16.56	0.04	0.15	9.14	6.87	0.37	0.00
	INTEGRATED DENSITIES OF STATES (electrons)							
		9.00	0.35	0.27	4.31	2.76	1.30	0.00

VELOCITY cm/s	PLASMON ENERGY eV	EIGENVALUE SUM Ry
0.59x10E8	8.25	-2.5838

Table 7.25 RuH (NaCl) a = 7.58 Bohr Slater–Koster 2-center parameters

	ORTHOGONAL	NON-ORTHOGONAL	
	ENERGY INTEGRALS Ry	ENERGY INTEGRALS Ry	OVERLAP INTEGRALS
ON SITE Ru-Ru			
s	1.10117	0.59225	
p	1.67094	1.46612	
t2g	0.72396	0.65322	
eg	0.70541	0.64267	
FIRST NEIGHBOR Ru-Ru			
(sss)	-0.07806	-0.00826	0.13944
(pps)	0.14878	-0.04919	-0.15516
(ppp)	-0.01564	0.04721	0.05527
(dds)	-0.06540	-0.00866	0.08035
(ddp)	0.02507	0.01609	-0.04448
(ddd)	-0.00018	-0.00581	0.00226
(sps)	0.12104	-0.01156	-0.11358
(sds)	-0.07189	-0.01239	0.08128
(pds)	-0.10355	0.08004	0.18153
(pdp)	0.01451	0.04599	0.01667
SECOND NEIGHBOR Ru-Ru			
(sss)	0.00568	-0.08051	-0.03244
(pps)	-0.08578	-0.07295	-0.01845
(ppp)	0.00506	-0.04360	-0.03189
(dds)	0.00348	-0.01621	-0.00501
(ddp)	-0.00162	0.00508	0.00392
(ddd)	0.00028	-0.00331	-0.00518
(sps)	-0.02477	0.12345	0.08077
(sds)	0.00299	-0.03826	-0.02099
(pds)	0.00856	0.04907	0.08990
(pdp)	-0.00515	0.02108	0.02390
ON SITE H-H			
s	1.07707	0.73852	
FIRST NEIGHBOR H-H			
(sss)	0.00343	0.00602	0.03697
(sps)	0.00000	0.00000	0.00000
(pps)	0.00000	0.00000	0.00000
(ppp)	0.00000	0.00000	0.00000
SECOND NEIGHBOR H-H			
(sss)	0.05193	-0.01584	-0.02913
(sps)	0.00000	0.00000	0.00000
(pps)	0.00000	0.00000	0.00000
(ppp)	0.00000	0.00000	0.00000
FIRST NEIGHBOR Ru-H			
(sss)	0.12523	-0.00263	-0.21118
(sps)	0.00000	0.00000	0.00000
(pss)	0.14639	0.17965	-0.04233
(pps)	0.00000	0.00000	0.00000
(ppp)	0.00000	0.00000	0.00000
(dpp)	0.00000	0.00000	0.00000
(dss)	0.11593	0.04717	-0.10735
(dps)	0.00000	0.00000	0.00000
SECOND NEIGHBOR Ru-H			
(sss)	-0.02751	-0.00241	-0.02361
(pss)	-0.06238	-0.03285	-0.00860
(sps)	0.00000	0.00000	0.00000
(pps)	0.00000	0.00000	0.00000
(ppp)	0.00000	0.00000	0.00000
(dss)	-0.01112	-0.02435	-0.04044
(dps)	0.00000	0.00000	0.00000
(dpp)	0.00000	0.00000	0.00000

♀ RuH

BAND	ORTHOGONAL RMS ERROR mRy	ORTHOGONAL MAXIMUM DEVIATION k	ORTHOGONAL MAXIMUM DEVIATION mRy	NON-ORTHOGONAL RMS ERROR mRy	NON-ORTHOGONAL MAXIMUM DEVIATION k	NON-ORTHOGONAL MAXIMUM DEVIATION mRy
1	5.5	(004)	10.9	2.4	(264)	5.2
2	7.3	(333)	23.6	2.4	(264)	6.5
3	6.3	(264)	14.3	2.7	(006)	7.5
4	4.5	(444)	9.9	2.3	(007)	7.8
5	4.2	(224)	8.5	2.0	(048)	4.0
6	6.5	(066)	24.2	3.4	(044)	7.9
7	12.9	(004)	35.5	5.7	(226)	18.0
1-7	7.3			3.2		

ENERGY VALUES IN Ry AT SELECTED k-POINTS

	ORTHOGONAL	APW	NON-ORTHOGONAL
GAMMA 1	0.00103	0.00405	0.00359
GAMMA 1	1.62731	1.63915	1.64005
GAMMA 12	0.76818	0.76679	0.76990
GAMMA 15	1.98960	2.01837	2.01915
GAMMA 25'	0.62119	0.62626	0.62630
X1 (008)	0.15924	0.15186	0.15279
X1 (008)	1.30313	1.31916	1.32381
X1 (008)	1.89338	1.85505	1.85738
X2 (008)	0.86574	0.87353	0.87413
X3 (008)	0.42209	0.41802	0.41903
X5 (008)	0.91442	0.91332	0.91691
X4' (008)	0.92450	0.92289	0.92198
X5' (008)	1.58215	1.55763	1.55744
L1 (444)	0.36419	0.37174	0.37690
L1 (444)	1.33176	1.32539	1.32899
L3 (444)	0.59234	0.60221	0.60277
L3 (444)	0.88218	0.87859	0.88194
L2' (444)	0.42009	0.41966	0.41368
L2' (444)	1.51004	1.47652	1.47317
L3' (444)	2.15111	2.14268	2.14315
W1 (048)	0.73362	0.72616	0.73202
W1 (048)	1.45985	1.46045	1.45009
W2' (048)	0.22505	0.23094	0.22769
W2' (048)	1.23973	1.24413	1.23642
W2' (048)	2.24304	2.40291	2.40348
W1' (048)	0.91331	0.91362	0.90957
W3 (048)	0.54799	0.55304	0.55096
W3 (048)	1.35985	1.37063	1.37181
ODD (224)	0.65290	0.65659	0.65378
ODD (224)	0.85943	0.85638	0.85459
EVEN(224)	0.23539	0.23632	0.23426
EVEN(224)	0.50392	0.49657	0.49831
EVEN(224)	0.60831	0.60505	0.60845
EVEN(224)	0.82698	0.81844	0.81620
EVEN(224)	1.32026	1.31979	1.31943

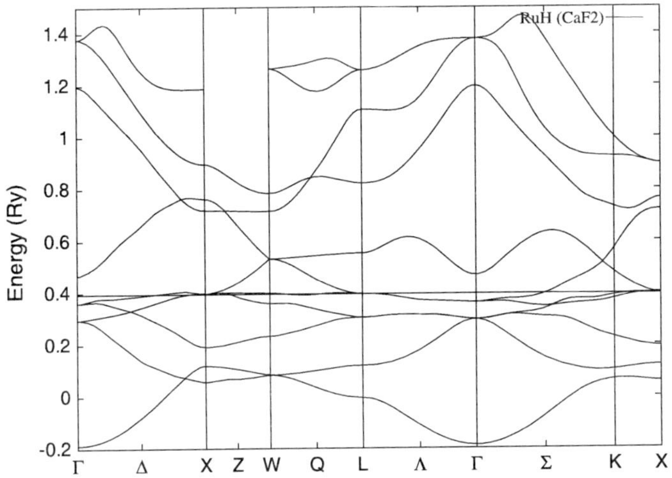

Fig. 7.24 Energy bands of RuH2 in the CaF2 structure

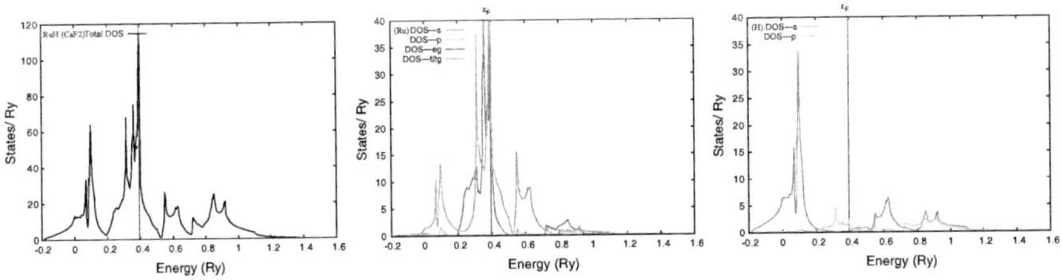

Fig. 7.25 Total, angular momentum and site decomposed densities of states of RuH2 in the CaF2 structure

7.6 Rhodium Hydride (RhH)

See Fig. 7.26, Tables 7.26, 7.27, and 7.28.

See Figs. 7.27, 7.28, Tables 7.29 and 7.30.

See Figs. 7.29 and 7.30.

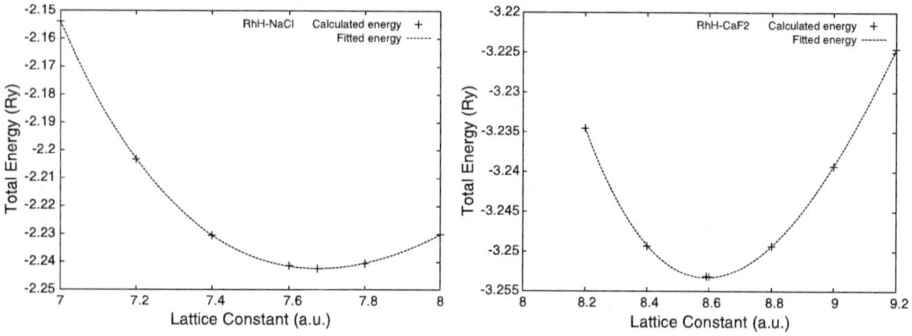

Fig. 7.26 Total energy versus lattice constant of RhH in the NaCl and CaF2 structures

Table 7.26 Lattice constant, bulk modulus, gap, total energy

Stru	a (Bohr)	B (MBar)	Gap	Total Energy (Ry)
NaCl	7.68	2.30	-	-9562.24230
CsCl	4.88	2.28	-	-9562.23993
CaF2	8.59	1.54	-	-9563.25314

Table 7.27 Birch fit coefficients

	A1	A2	A3	A4
NaCl	-1.107853E+00	-3.301419E+01	-3.163231E+02	1.094348E+04
CsCl	-9.954373E-01	-4.077440E+01	-1.711043E+02	1.039032E+04
CaF2	-2.281627E+00	-3.061956E+01	-7.076354E+02	2.257928E+04

Table 7.28 DOS at Ef, Hopfield parameter, Stoner criterion

```
                  RhH       a=7.68 Bohr        NaCl
--------------------------------------------------------------------------------
 Fermi Energy   Total DOS         Decomposed DOS inside the muffin tins
   (Ry)        (States/Ry)        s         p         eg         t2g         f
--------------------------------------------------------------------------------
Rh  0.799       40.004          0.033     0.093      9.947      26.198     0.030
H   0.799       40.004          0.148     0.267      0.105       0.021     0.105
--------------------------------------------------------------------------------
Fermi-Velocity (cm/s): 0.49853 x10E8
Plasmon Energy (eV)   : 10.67792
Electron-ion interaction (Hopfield parameter) (eV/A^2) Rh: 3.452 H: 0.029
-------------------------------------------------
  Rh      MUFFIN-TIN RADIUS and CHARGE =    2.1106   42.4818
   H      MUFFIN-TIN RADIUS and CHARGE =    1.7269    1.2335
           Rh              STONER I = 0.0220
            H              STONER I = 0.0000
          STONER PARAMETER (Ry) I = 0.0220
          STONER CRITERION    N*I = 0.8786
-------------------------------------------------
                  RhH       a=4.88 Bohr         CsCl
--------------------------------------------------------------------------------
 Fermi Energy   Total DOS         Decomposed DOS inside the muffin tins
   (Ry)        (States/Ry)        s         p         eg         t2g         f
Rh  0.654        30.900         0.380     1.027      4.308      22.604     0.226
H   0.654        30.900         1.015     0.264      0.005       0.111     0.035
Fermi-Velocity (cm/s): 0.56403 x10E8
Plasmon Energy (eV)   : 10.48567
Electron-ion interaction (Hopfield parameter) (eV/A^2) Rh: 7.757 H: 0.284
--------------------------------------------------------------------------------
  Rh      MUFFIN-TIN RADIUS and CHARGE =    2.5333   44.0742
   H      MUFFIN-TIN RADIUS and CHARGE =    1.6889    0.9603
           Rh              STONER I = 0.0177
            H              STONER I = 0.0004
          STONER PARAMETER (Ry) I = 0.0182
          STONER CRITERION    N*I = 0.5617
--------------------------------------------------------------------------------
                  RhH2      a=8.59 Bohr        CaF2
--------------------------------------------------------------------------------
 Fermi Energy   Total DOS         Decomposed DOS inside the muffin tins
   (Ry)        (States/Ry)        s         p         eg         t2g         f
--------------------------------------------------------------------------------
Rh  0.515       12.411          0.069     0.163      2.769      7.606      0.007
```

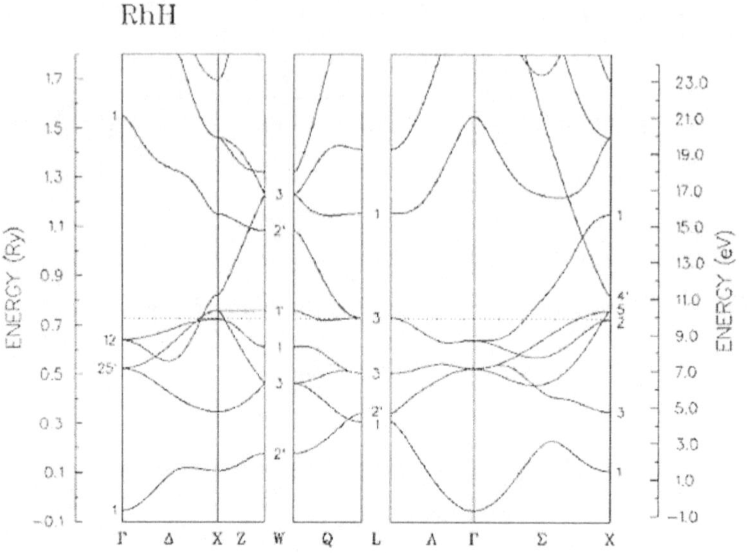

Fig. 7.27 Energy bands of RhH in the NaCl structure (tight-binding)

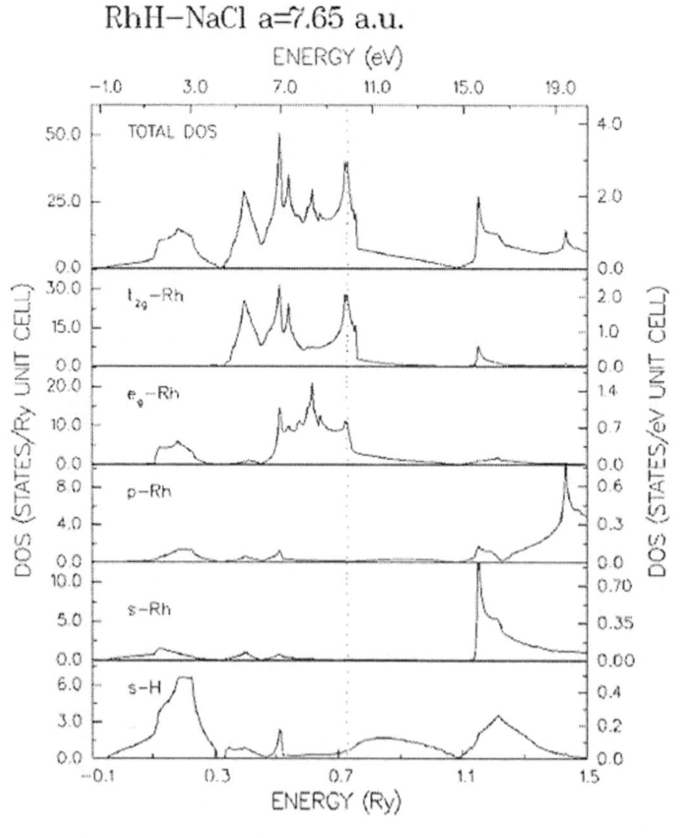

Fig. 7.28 Total, angular momentum and site decomposed densities of states of RhH in the NaCl structure (tight-binding)

Table 7.29 RhH (NaCl) a = 7.65 Bohr Slater–Koster 3-center parameters

	ORTHOGONAL	NON-ORTHOGONAL	
	ENERGY INTEGRALS Ry	ENERGY INTEGRALS Ry	OVERLAP INTEGRALS
ON SITE Rh-Rh			
s,s(000)	1.13727	0.91483	
x,x(000)	1.63545	1.40446	
xy,xy(000)	0.63309	0.55809	
d2,d2(000)	0.70908	0.54188	
FIRST NEIGHBOR Rh-Rh			
s,s(110)	-0.00954	-0.00043	0.03906
s,x(110)	-0.04984	0.02856	0.07659
s,xy(110)	-0.06120	-0.01933	0.06979
s,d2(110)	-0.00920	-0.01517	-0.00155
x,x(110)	0.07006	-0.05574	-0.11420
x,x(011)	-0.01061	0.05686	0.09190
x,y(110)	0.06541	-0.09133	-0.08670
x,xy(110)	0.06893	0.00940	-0.04202
x,xy(011)	-0.00415	0.00411	0.05038
z,d2(011)	0.02069	0.04498	0.07955
z,d1(011)	0.00799	0.00642	0.00194
xy,xy(110)	-0.04048	-0.02126	0.03262
xy,xy(011)	0.01069	0.00613	-0.01379
xy,xz(011)	0.00060	0.00746	-0.02069
xy,d2(110)	0.02184	-0.00224	0.03295
d2,d2(110)	0.00677	-0.00824	0.01729
d1,d1(110)	-0.04128	0.04314	0.00496
SECOND NEIGHBOR Rh-Rh			
s,s(200)	0.05697	-0.01530	-0.05838
s,x(200)	0.06963	-0.02975	-0.06867
s,d2(002)	0.04543	0.02831	0.05406
x,x(200)	-0.13205	-0.05258	0.01036
y,y(200)	0.01424	-0.00663	0.01546
x,xy(020)	0.01153	-0.00198	0.00822
z,d2(002)	-0.04109	-0.02795	-0.04118
xy,xy(200)	-0.00897	0.00022	-0.00519
xy,xy(002)	0.00108	0.00054	0.00085
d2,d2(002)	0.04524	-0.02920	-0.03261
d1,d1(002)	0.00068	0.00292	0.00556
ON SITE H-H			
s,s(000)	0.36262	0.63870	
FIRST NEIGHBOR H-H			
s,s(110)	-0.01513	-0.00484	-0.01318
s,x(110)	0.00000	0.00000	0.00000
x,x(110)	0.00000	0.00000	0.00000
x,x(011)	0.00000	0.00000	0.00000
x,y(110)	0.00000	0.00000	0.00000
SECOND NEIGHBOR H-H			
s,s(200)	-0.01168	-0.00086	-0.00087
s,x(200)	0.00000	0.00000	0.00000
x,x(200)	0.00000	0.00000	0.00000
y,y(200)	0.00000	0.00000	0.00000
FIRST NEIGHBOR Rh-H			
s,s(100)	-0.11463	-0.13233	-0.01560
s,x(100)	0.00000	0.00000	0.00000
x,s(100)	0.09034	0.26483	0.16904
x,x(100)	0.00000	0.00000	0.00000
y,y(100)	0.00000	0.00000	0.00000
xy,x(010)	0.00000	0.00000	0.00000
d2,s(001)	-0.08716	0.01701	-0.13682
d2,z(001)	0.00000	0.00000	0.00000
SECOND NEIGHBOR Rh-H			
s,s(111)	0.02520	0.00671	0.01739
x,s(111)	-0.03914	-0.01063	-0.01904
s,x(111)	0.00000	0.00000	0.00000
x,x(111)	0.00000	0.00000	0.00000
x,y(111)	0.00000	0.00000	0.00000
xy,s(111)	0.00275	0.00103	0.00163
xy,x(111)	0.00000	0.00000	0.00000
yz,x(111)	0.00000	0.00000	0.00000
d1,x(111)	0.00000	0.00000	0.00000

RhH

BAND	ORTHOGONAL RMS ERROR mRy	MAXIMUM DEVIATION k	MAXIMUM DEVIATION mRy	NON-ORTHOGONAL RMS ERROR mRy	MAXIMUM DEVIATION k	MAXIMUM DEVIATION mRy
1	4.4	(033)	8.9	0.4	(333)	1.1
2	5.2	(222)	15.7	0.4	(222)	1.1
3	4.6	(006)	9.4	0.3	(062)	0.7
4	2.7	(226)	4.7	0.4	(055)	1.0
5	3.4	(044)	7.8	0.3	(005)	0.6
6	4.1	(055)	11.4	0.5	(174)	2.1
7	7.1	(000)	17.1	0.8	(224)	1.9
1-7	4.7			0.5		

ENERGY VALUES IN Ry AT SELECTED k-POINTS

	ORTHOGONAL	APW	NON-ORTHOGONAL
GAMMA 1	-0.05552	-0.05349	-0.05357
GAMMA 1	1.53109	1.54818	1.54805
GAMMA 12	0.63973	0.63871	0.63893
GAMMA 15	1.94631	1.96486	1.96484
GAMMA 25'	0.52299	0.52354	0.52351
X1 (008)	0.11353	0.10620	0.10665
X1 (008)	1.14689	1.14878	1.14918
X1 (008)	1.71799	1.69567	1.69536
X2 (008)	0.72365	0.72594	0.72567
X3 (008)	0.35190	0.34855	0.34866
X5 (008)	0.76129	0.75876	0.75859
X4' (008)	0.82541	0.82352	0.82331
X5' (008)	1.47074	1.46204	1.46204
L1 (444)	0.29951	0.30544	0.30604
L1 (444)	1.15789	1.15332	1.15270
L3 (444)	0.50261	0.50434	0.50457
L3 (444)	0.73795	0.73066	0.73065
L2' (444)	0.33337	0.34309	0.34244
L2' (444)	1.41862	1.41348	1.41349
L3' (444)	2.10424	2.05177	2.05179
W1 (048)	0.60798	0.61140	0.61186
W1 (048)	1.32020	1.32035	1.32064
W2' (048)	0.16971	0.17694	0.17701
W2' (048)	1.08045	1.08350	1.08401
W2' (048)	2.15780	2.25076	2.25076
W1' (048)	0.75695	0.75899	0.75904
W3 (048)	0.45761	0.46355	0.46393
W3 (048)	1.22601	1.23038	1.23049
ODD (224)	0.55424	0.55003	0.54915
ODD (224)	0.71673	0.71299	0.71292
EVEN(224)	0.17550	0.16903	0.16842
EVEN(224)	0.42197	0.41663	0.41647
EVEN(224)	0.50473	0.50801	0.50848
EVEN(224)	0.68801	0.68703	0.68723
EVEN(224)	1.20753	1.20494	1.20689

FERMI LEVEL QUANTITIES (Non-orthogonal fit)

ENERGY Ry		s-Rh	p-Rh	t2g-Rh	eg-Rh	s-H	p-H
	DENSITIES OF STATES (states/Ry/cell) Total						
0.7277	39.32	0.08	0.13	27.62	10.81	0.68	0.00
	INTEGRATED DENSITIES OF STATES (electrons) Total						
	10.00	0.34	0.30	5.02	3.15	1.20	0.00

VELOCITY cm/s	PLASMON ENERGY eV	EIGENVALUE SUM Ry
0.34x10E8	7.29	-2.5324

Table 7.30 RhH (NaCl) a = 7.65 Bohr Slater–Koster 2-center parameters

	ORTHOGONAL	NON-ORTHOGONAL	
	ENERGY INTEGRALS Ry	ENERGY INTEGRALS Ry	OVERLAP INTEGRALS
ON SITE Rh-Rh			
s	0.98673	0.51640	
p	1.58881	1.50637	
t2g	0.60137	0.54792	
eg	0.58788	0.53848	
FIRST NEIGHBOR Rh-Rh			
(sss)	-0.07337	-0.01361	0.13775
(pps)	0.15129	0.01975	-0.09959
(ppp)	-0.01170	0.06160	0.05626
(dds)	-0.05378	-0.01131	0.07302
(ddp)	0.02162	0.01517	-0.04107
(ddd)	-0.00065	-0.00631	0.00004
(sps)	0.11712	0.01147	-0.08568
(sds)	-0.06212	-0.01601	0.07544
(pds)	-0.09062	0.05303	0.15972
(pdp)	0.01339	0.03841	0.02347
SECOND NEIGHBOR Rh-Rh			
(sss)	0.00677	-0.08027	-0.03877
(pps)	-0.09281	-0.15237	-0.04781
(ppp)	0.00678	-0.04092	-0.03441
(dds)	0.00193	-0.01068	0.00000
(ddp)	-0.00088	0.00629	0.00701
(ddd)	0.00011	-0.00651	-0.01124
(sps)	-0.02581	0.10913	0.08054
(sds)	0.00284	-0.03818	-0.02619
(pds)	0.00715	0.07643	0.14363
(pdp)	-0.00557	0.02227	0.03006
ON SITE H-H			
s	0.96526	0.71292	
FIRST NEIGHBOR H-H			
(sss)	0.00712	0.01891	0.04691
(sps)	0.00000	0.00000	0.00000
(pps)	0.00000	0.00000	0.00000
(ppp)	0.00000	0.00000	0.00000
SECOND NEIGHBOR H-H			
(sss)	0.04707	-0.02801	-0.03904
(sps)	0.00000	0.00000	0.00000
(pps)	0.00000	0.00000	0.00000
(ppp)	0.00000	0.00000	0.00000
FIRST NEIGHBOR Rh-H			
(sss)	0.12203	0.00218	-0.20635
(sps)	0.00000	0.00000	0.00000
(pss)	0.14469	0.25978	0.05411
(pps)	0.00000	0.00000	0.00000
(ppp)	0.00000	0.00000	0.00000
(dpp)	0.00000	0.00000	0.00000
(dss)	0.10361	0.04016	-0.10093
(dps)	0.00000	0.00000	0.00000
SECOND NEIGHBOR Rh-H			
(sss)	-0.02513	0.00243	-0.02292
(pss)	-0.05783	-0.01479	0.00680
(sps)	0.00000	0.00000	0.00000
(pps)	0.00000	0.00000	0.00000
(ppp)	0.00000	0.00000	0.00000
(dss)	-0.00832	-0.02092	-0.04018
(dps)	0.00000	0.00000	0.00000
(dpp)	0.00000	0.00000	0.00000

♀RhH

BAND	ORTHOGONAL RMS ERROR mRy	MAXIMUM DEVIATION k	MAXIMUM DEVIATION mRy	NON-ORTHOGONAL RMS ERROR mRy	MAXIMUM DEVIATION k	MAXIMUM DEVIATION mRy
1	4.8	(004)	9.7	3.4	(062)	8.9
2	6.2	(333)	21.9	2.6	(264)	5.5
3	5.1	(264)	11.1	3.1	(006)	11.6
4	3.2	(444)	6.5	2.2	(048)	6.8
5	2.9	(224)	5.8	2.4	(048)	5.6
6	4.6	(066)	13.9	4.1	(007)	9.1
7	10.9	(004)	30.0	7.5	(226)	23.6
1.7	5.9			4.0		

ENERGY VALUES IN Ry AT SELECTED k-POINTS

	ORTHOGONAL	APW	NON-ORTHOGONAL
GAMMA 1	-0.05612	-0.05349	-0.05405
GAMMA 1	1.53611	1.54818	1.55172
GAMMA 12	0.64012	0.63871	0.64199
GAMMA 15	1.94189	1.96486	1.96675
GAMMA 25'	0.51995	0.52354	0.52355
X1 (008)	0.11336	0.10620	0.10423
X1 (008)	1.13718	1.14878	1.15148
X1 (008)	1.72590	1.69567	1.69866
X2 (008)	0.72085	0.72594	0.72653
X3 (008)	0.35219	0.34855	0.34879
X5 (008)	0.76006	0.75876	0.76333
X4' (008)	0.82515	0.82352	0.82306
X5' (008)	1.47713	1.46204	1.46130
L1 (444)	0.30042	0.30544	0.30992
L1 (444)	1.16126	1.15332	1.15956
L3 (444)	0.49785	0.50434	0.50569
L3 (444)	0.73314	0.73066	0.73454
L2' (444)	0.34709	0.34309	0.33911
L2' (444)	1.43109	1.41348	1.40801
L3' (444)	2.07329	2.05177	2.05208
W1 (048)	0.61492	0.61140	0.61824
W1 (048)	1.32085	1.32035	1.30624
W2' (048)	0.17254	0.17694	0.17609
W2' (048)	1.08047	1.08350	1.07535
W2' (048)	2.13336	2.25076	2.25163
W1' (048)	0.75962	0.75899	0.75336
W3 (048)	0.45824	0.46355	0.46129
W3 (048)	1.22543	1.23038	1.23217
ODD (224)	0.54748	0.55003	0.54764
ODD (224)	0.71517	0.71299	0.71136
EVEN(224)	0.16764	0.16903	0.16836
EVEN(224)	0.42318	0.41663	0.41948
EVEN(224)	0.51096	0.50801	0.51162
EVEN(224)	0.69287	0.68703	0.68387
EVEN(224)	1.20675	1.20494	1.20533

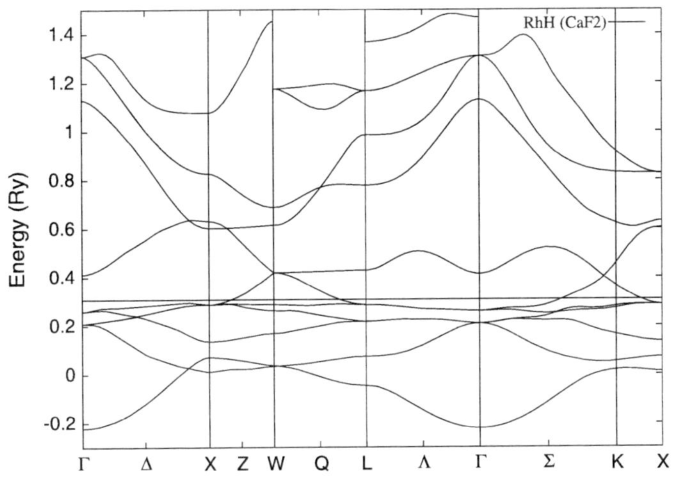

Fig. 7.29 Energy bands of RhH2 in the CaF2 structure

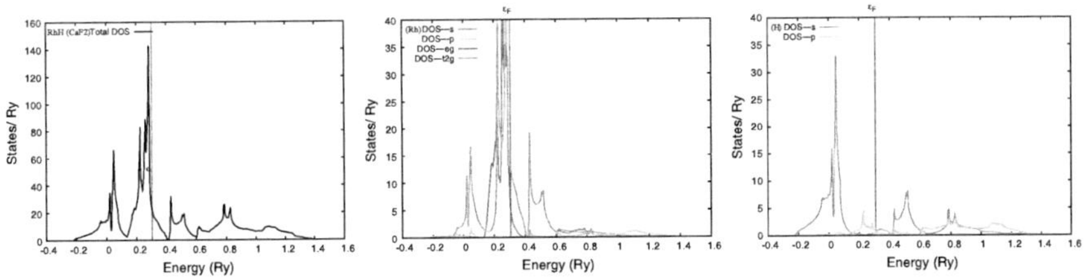

Fig. 7.30 Total, angular momentum and site decomposed densities of states of RhH2 in the CaF2 structure

7.7 Palladium Hydride (PdH)

See Figs. 7.31, Tables 7.31, 7.32, and 7.33.

See Figs. 7.32, 7.33, Tables 7.34 and 7.35.

See Figs. 7.34 and 7.35.

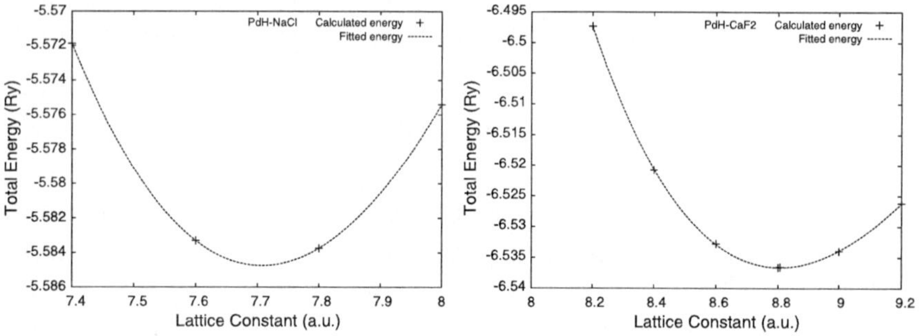

Fig. 7.31 Total energy versus lattice constant of PdH in the NaCl and CaF2 structures

Table 7.31 Lattice constant, bulk modulus, gap, total energy

Stru	a (Bohr)	B (MBar)	Gap	Total Energy (Ry)
NaCl	7.71	2.06	-	-10085.58463
CaF2	8.81	1.19	-	-10086.53655
exp (NaCl)	7.73			

Table 7.32 Birch fit coefficients

	A1	A2	A3	A4
NaCl	-3.659245E+00	-9.373190E+01	1.209109E+03	-1.639386E+03
CaF2	-5.778095E+00	-2.237799E+01	-7.779558E+02	2.302705E+04

Table 7.33 DOS at Ef, Hopfield parameter, Stoner criterion

```
                  PdH        a=7.73          NaCl
 -------------------------------------------------------------------------------
  Fermi Energy   Total DOS        Decomposed DOS inside the muffin tins
    (Ry)        (States/Ry)       s          p          eg         t2g        f
 -------------------------------------------------------------------------------
 Pd  0.691       6.645          0.119      0.302      2.650      2.378      0.014
 H   0.691       6.645          0.424      0.107      0.007      0.001      0.005
 -------------------------------------------------------------------------------
 Fermi-Velocity (cm/s): 0.97672 x10E8
 Plasmon Energy (eV)   : 8.47343
 Electron-ion interaction (Hopfield parameter) (eV/A^2) Pd: 0.913 H: 0.372
 -------------------------------------------------
   Pd      MUFFIN-TIN RADIUS and CHARGE =     2.3122    44.3576
    H      MUFFIN-TIN RADIUS and CHARGE =     1.5414     0.9086
                 Pd        STONER I = 0.0142
                  H        STONER I = 0.0016
           STONER PARAMETER (Ry) I = 0.0158
           STONER CRITERION    N*I = 0.1051

 -------------------------------------------------
                  PdH2     a=8.81 Bohr            CaF2
 -------------------------------------------------------------------------------
  Fermi Energy   Total DOS        Decomposed DOS inside the muffin tins
    (Ry)        (States/Ry)       s          p          eg         t2g        f
 -------------------------------------------------------------------------------
 Pd   0.509         0.537         0.012     0.029      0.010      0.285      0.001
  H   0.509         0.537         0.089     0.014      0.000      0.001      0.000
 -------------------------------------------------------------------------------
 Fermi-Velocity (cm/s): 0.75991 x10E8
 Plasmon Energy (eV)   : 1.53416
 Electron-ion interaction (Hopfield parameter) (eV/A^2) Pd:  0.057  H: 0.171
 -------------------------------------------------
           Pd     MUFFIN-TIN RADIUS and CHARGE =     2.2880   44.1813
            H     MUFFIN-TIN RADIUS and CHARGE =     1.5253    0.8272
                     Pd        STONER I = 0.0081
                      H        STONER I = 0.0025
           STONER PARAMETER (Ry)       I = 0.0131
           STONER CRITERION          N*I = 0.0070
```

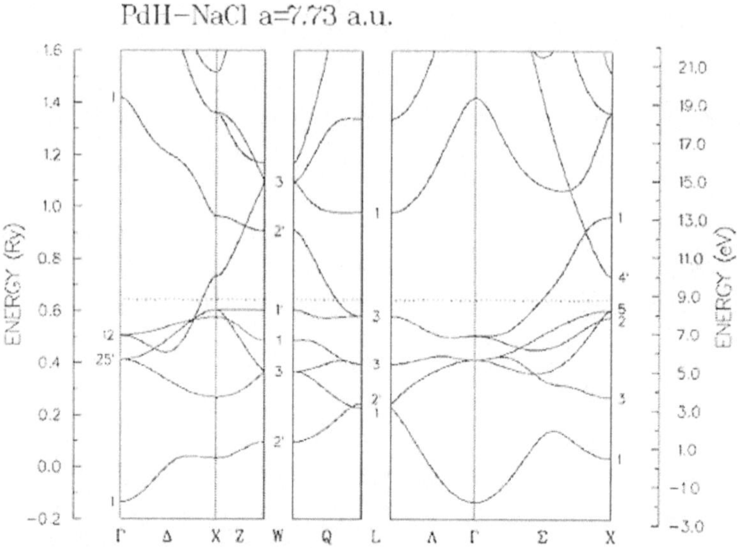

Fig. 7.32 Energy bands of PdH in the NaCl structure (tight-binding)

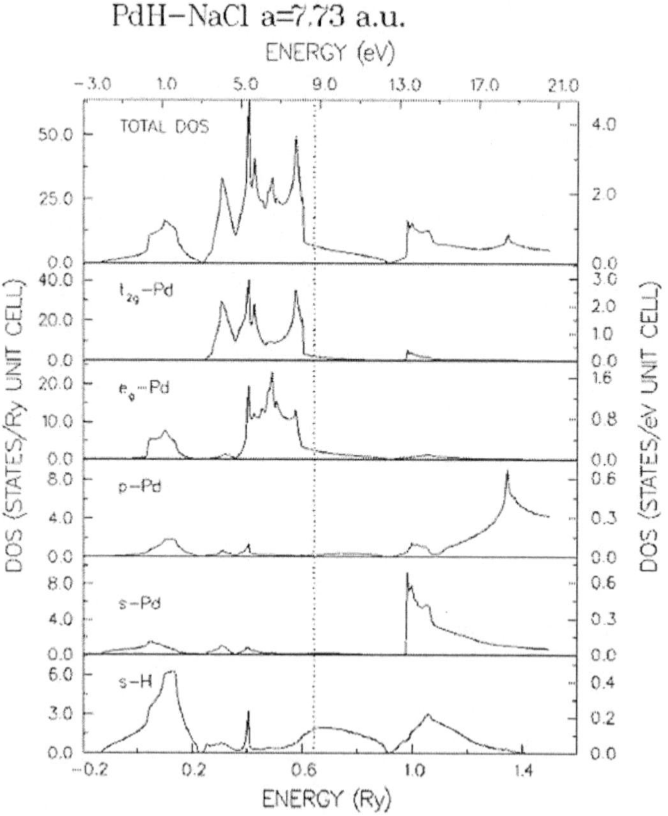

Fig. 7.33 Total, angular momentum and site decomposed densities of states of PdH in the NaCl structure (tight-binding)

Table 7.34 PdH (NaCl) a = 7.73 Bohr Slater–Koster 3-center parameters

	ORTHOGONAL	NON-ORTHOGONAL	
	ENERGY INTEGRALS Ry	ENERGY INTEGRALS Ry	OVERLAP INTEGRALS
ON SITE Pd-Pd			
s,s(000)	1.01771	0.80899	
x,x(000)	1.54460	1.30456	
xy,xy(000)	0.49542	0.44337	
d2,d2(000)	0.57246	0.42578	
FIRST NEIGHBOR Pd-Pd			
s,s(110)	-0.00695	-0.00445	0.02770
s,x(110)	-0.04719	0.01521	0.06702
s,xy(110)	-0.05261	-0.02363	0.06209
s,d2(110)	-0.01090	-0.01317	0.00403
x,x(110)	0.07107	-0.04010	-0.11117
x,x(011)	-0.00859	0.03607	0.08050
x,y(110)	0.06053	-0.08835	-0.08711
x,xy(110)	0.05857	0.00634	-0.04872
x,xy(011)	-0.00419	-0.00175	0.04373
z,d2(011)	0.01875	0.04140	0.08184
z,d1(011)	0.00889	0.01537	-0.00168
xy,xy(110)	-0.03293	-0.01840	0.03125
xy,xy(011)	0.00878	0.00591	-0.01190
xy,xz(011)	0.00242	0.00759	-0.01760
xy,d2(110)	0.01801	-0.00561	0.02652
d2,d2(110)	0.00699	-0.01256	0.00758
d1,d1(110)	-0.03797	0.04139	0.01555
SECOND NEIGHBOR Pd-Pd			
s,s(200)	0.05149	-0.01645	-0.06560
s,x(200)	0.06541	-0.03079	-0.07267
s,d2(002)	0.04145	0.02759	0.06222
x,x(200)	-0.13195	-0.05116	0.01854
y,y(200)	0.01438	0.01527	0.02663
x,xy(020)	0.00920	0.00226	0.01220
z,d2(002)	-0.03661	-0.02870	-0.03946
xy,xy(200)	-0.00580	0.00035	-0.00398
xy,xy(002)	0.00070	0.00005	0.00021
d2,d2(002)	0.03954	-0.02600	-0.03661
d1,d1(002)	0.00020	0.00171	0.00392
ON SITE H-H			
s,s(000)	0.25182	0.60125	
FIRST NEIGHBOR H-H			
s,s(110)	-0.01349	-0.00592	-0.01140
s,x(110)	0.00000	0.00000	0.00000
x,x(110)	0.00000	0.00000	0.00000
x,x(011)	0.00000	0.00000	0.00000
x,y(110)	0.00000	0.00000	0.00000
SECOND NEIGHBOR H-H			
s,s(200)	-0.01030	-0.00121	-0.00156
s,x(200)	0.00000	0.00000	0.00000
x,x(200)	0.00000	0.00000	0.00000
y,y(200)	0.00000	0.00000	0.00000
FIRST NEIGHBOR Pd-H			
s,s(100)	-0.10749	-0.12602	-0.02160
s,x(100)	0.00000	0.00000	0.00000
x,s(100)	0.08053	0.26530	0.16714
x,x(100)	0.00000	0.00000	0.00000
y,y(100)	0.00000	0.00000	0.00000
xy,x(010)	0.00000	0.00000	0.00000
d2,s(001)	-0.07783	0.01651	-0.11987
d2,z(001)	0.00000	0.00000	0.00000
SECOND NEIGHBOR Pd-H			
s,s(111)	0.02103	0.00646	0.01434
x,s(111)	-0.03517	-0.00957	-0.01768
s,x(111)	0.00000	0.00000	0.00000
x,x(111)	0.00000	0.00000	0.00000
x,y(111)	0.00000	0.00000	0.00000
xy,s(111)	0.00103	0.00155	0.00283
xy,x(111)	0.00000	0.00000	0.00000
yz,x(111)	0.00000	0.00000	0.00000
d1,x(111)	0.00000	0.00000	0.00000

PdH

BAND	ORTHOGONAL RMS ERROR mRy	MAXIMUM DEVIATION k	MAXIMUM DEVIATION mRy	NON-ORTHOGONAL RMS ERROR mRy	MAXIMUM DEVIATION k	MAXIMUM DEVIATION mRy
1	3.6	(033)	7.5	0.3	(333)	0.6
2	4.3	(222)	12.8	0.3	(222)	0.7
3	4.1	(006)	9.7	0.2	(062)	0.7
4	2.0	(062)	4.5	0.3	(224)	0.8
5	2.5	(044)	6.4	0.2	(042)	0.4
6	3.4	(055)	10.6	0.3	(174)	1.3
7	6.0	(000)	12.8	0.6	(224)	1.4
1-7	3.9			0.3		

ENERGY VALUES IN Ry AT SELECTED k-POINTS

	ORTHOGONAL	APW	NON-ORTHOGONAL
GAMMA 1	-0.13656	-0.13434	-0.13417
GAMMA 1	1.40790	1.42071	1.42051
GAMMA 12	0.50581	0.50493	0.50529
GAMMA 15	1.87248	1.88607	1.88605
GAMMA 25'	0.41215	0.41255	0.41264
X1 (008)	0.04016	0.03414	0.03419
X1 (008)	0.96108	0.96385	0.96397
X1 (008)	1.53066	1.51826	1.51819
X2 (008)	0.57381	0.57556	0.57574
X3 (008)	0.27162	0.26872	0.26893
X5 (008)	0.60536	0.60323	0.60322
X4' (008)	0.73528	0.73385	0.73391
X5' (008)	1.37259	1.36207	1.36201
L1 (444)	0.22389	0.22855	0.22873
L1 (444)	0.98276	0.97844	0.97819
L3 (444)	0.39445	0.39584	0.39613
L3 (444)	0.58570	0.58030	0.58044
L2' (444)	0.23797	0.24558	0.24495
L2' (444)	1.34241	1.33935	1.33935
L3' (444)	1.99307	1.95372	1.95370
W1 (048)	0.48314	0.48611	0.48642
W1 (048)	1.17092	1.17032	1.17026
W2' (048)	0.09194	0.09787	0.09781
W2' (048)	0.90682	0.90858	0.90858
W2' (048)	2.02957	2.09861	2.09864
W1' (048)	0.60257	0.60348	0.60356
W3 (048)	0.36231	0.36749	0.36757
W3 (048)	1.09578	1.09705	1.09701
ODD (224)	0.43711	0.43445	0.43363
ODD (224)	0.56936	0.56633	0.56633
EVEN(224)	0.08691	0.08185	0.08131
EVEN(224)	0.33000	0.32568	0.32585
EVEN(224)	0.39898	0.40123	0.40146
EVEN(224)	0.55039	0.54916	0.54910
EVEN(224)	1.06906	1.06611	1.06755

FERMI LEVEL QUANTITIES (Non-orthogonal fit)

DENSITIES OF STATES (states/Ry/cell)

ENERGY Ry	Total	s-Pd	p-Pd	t2g-Pd	eg-Pd	s-H	p-H
0.6431	6.75	0.15	0.22	2.15	2.41	1.83	0.00

INTEGRATED DENSITIES OF STATES (electrons)

Total	s-Pd	p-Pd	t2g-Pd	eg-Pd	s-H	p-H
11.00	0.33	0.31	5.73	3.48	1.13	0.00

VELOCITY cm/s	PLASMON ENERGY eV	EIGENVALUE SUM Ry
0.75x10E8	6.56	-2.8183

Table 7.35 PdH (NaCl) a = 7.73 Bohr Slater–Koster 2-center parameters

	ORTHOGONAL	NON-ORTHOGONAL	
	ENERGY INTEGRALS Ry	ENERGY INTEGRALS Ry	OVERLAP INTEGRALS
ON SITE Pd-Pd			
s	0.85702	0.40706	
p	1.50168	1.52398	
t2g	0.47428	0.43622	
eg	0.46336	0.42683	
FIRST NEIGHBOR Pd-Pd			
(sss)	-0.06952	-0.02312	0.13937
(pps)	0.15073	0.07801	-0.04610
(ppp)	-0.00925	0.07001	0.05398
(dds)	-0.04366	-0.01374	0.06596
(ddp)	0.01812	0.01464	-0.03585
(ddd)	-0.00084	-0.00594	-0.00186
(sps)	0.11223	0.01870	-0.05793
(sds)	-0.05337	-0.01894	0.07152
(pds)	-0.07780	0.03033	0.14380
(pdp)	0.01230	0.03547	0.04220
SECOND NEIGHBOR Pd-Pd			
(sss)	0.00773	-0.07529	-0.04170
(pps)	-0.09933	-0.19991	-0.06756
(ppp)	0.00897	-0.03276	-0.03118
(dds)	0.00119	-0.00570	0.00526
(ddp)	-0.00037	0.00643	0.01025
(ddd)	-0.00003	-0.00760	-0.01660
(sps)	-0.02663	0.09783	0.08469
(sds)	0.00285	-0.03484	-0.02839
(pds)	0.00569	0.09231	0.19436
(pdp)	-0.00560	0.02263	0.03928
ON SITE H-H			
s	0.81677	0.59861	
FIRST NEIGHBOR H-H			
(sss)	0.01178	0.02769	0.06081
(sps)	0.00000	0.00000	0.00000
(pps)	0.00000	0.00000	0.00000
(ppp)	0.00000	0.00000	0.00000
SECOND NEIGHBOR H-H			
(sss)	0.04080	-0.02913	-0.04430
(sps)	0.00000	0.00000	0.00000
(pps)	0.00000	0.00000	0.00000
(ppp)	0.00000	0.00000	0.00000
FIRST NEIGHBOR Pd-H			
(sss)	0.11855	0.02022	-0.20855
(sps)	0.00000	0.00000	0.00000
(pss)	0.14356	0.27878	0.11790
(pps)	0.00000	0.00000	0.00000
(ppp)	0.00000	0.00000	0.00000
(dpp)	0.00000	0.00000	0.00000
(dss)	0.09416	0.04103	-0.09743
(dps)	0.00000	0.00000	0.00000
SECOND NEIGHBOR Pd-H			
(sss)	-0.02320	0.00748	-0.02426
(pss)	-0.05256	0.00115	0.02971
(sps)	0.00000	0.00000	0.00000
(pps)	0.00000	0.00000	0.00000
(ppp)	0.00000	0.00000	0.00000
(dss)	-0.00630	-0.01540	-0.03773
(dps)	0.00000	0.00000	0.00000
(dpp)	0.00000	0.00000	0.00000

♀ PdH

BAND	ORTHOGONAL RMS ERROR mRy	MAXIMUM DEVIATION k	mRy	NON-ORTHOGONAL RMS ERROR mRy	MAXIMUM DEVIATION k	mRy
1	4.3	(004)	8.9	4.5	(003)	11.3
2	5.2	(333)	19.8	2.8	(333)	6.9
3	4.0	(264)	8.0	4.0	(006)	18.0
4	2.3	(354)	5.0	2.6	(048)	8.5
5	2.0	(224)	4.2	2.7	(048)	6.4
6	3.4	(055)	8.2	4.5	(007)	10.7
7	9.7	(004)	25.1	8.9	(226)	28.5
1.7	5.0			4.8		

ENERGY VALUES IN Ry AT SELECTED k-POINTS

	ORTHOGONAL	APW	NON-ORTHOGONAL
GAMMA 1	-0.13703	-0.13434	-0.13303
GAMMA 1	1.40919	1.42071	1.42610
GAMMA 12	0.50630	0.50493	0.50800
GAMMA 15	1.86782	1.88607	1.88862
GAMMA 25'	0.41003	0.41255	0.41253
X1 (008)	0.04082	0.03414	0.03027
X1 (008)	0.95572	0.96385	0.96404
X1 (008)	1.53440	1.51826	1.52280
X2 (008)	0.57235	0.57556	0.57684
X3 (008)	0.27180	0.26872	0.26928
X5 (008)	0.60453	0.60323	0.60847
X4' (008)	0.73598	0.73385	0.73662
X5' (008)	1.37588	1.36207	1.36022
L1 (444)	0.22490	0.22855	0.23302
L1 (444)	0.98571	0.97844	0.98661
L3 (444)	0.39208	0.39584	0.39810
L3 (444)	0.58155	0.58030	0.58467
L2' (444)	0.25192	0.24558	0.24466
L2' (444)	1.34465	1.33935	1.33476
L3' (444)	1.98441	1.95372	1.95427
W1 (048)	0.48721	0.48611	0.49463
W1 (048)	1.17161	1.17032	1.15368
W2' (048)	0.09573	0.09787	0.09977
W2' (048)	0.90540	0.90858	0.90073
W2' (048)	2.01777	2.09861	2.10006
W1' (048)	0.60465	0.60348	0.59707
W3 (048)	0.36244	0.36749	0.36451
W3 (048)	1.09727	1.09705	1.10159
ODD (224)	0.43304	0.43445	0.43232
ODD (224)	0.56759	0.56633	0.56525
EVEN(224)	0.08029	0.08185	0.08309
EVEN(224)	0.33137	0.32568	0.32945
EVEN(224)	0.40359	0.40123	0.40441
EVEN(224)	0.55332	0.54916	0.54509
EVEN(224)	1.06965	1.06611	1.06859

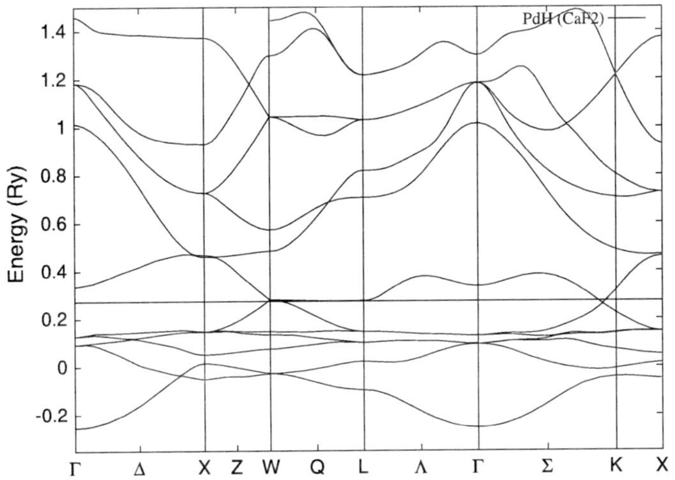

Fig. 7.34 Energy bands of PdH2 in the CaF2 structure

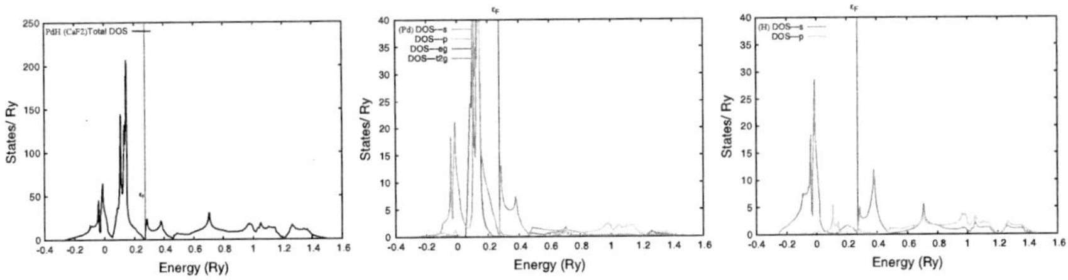

Fig. 7.35 Total, angular momentum and site decomposed densities of states of PdH2 in the CaF2 structure

7.8 Silver Hydride (AgH)

See Fig. 7.36, Tables 7.36, 7.37, and 7.38.

See Figs. 7.37, 7.38, and Tables 7.39 and 7.40.

See Figs. 7.39 and 7.40.

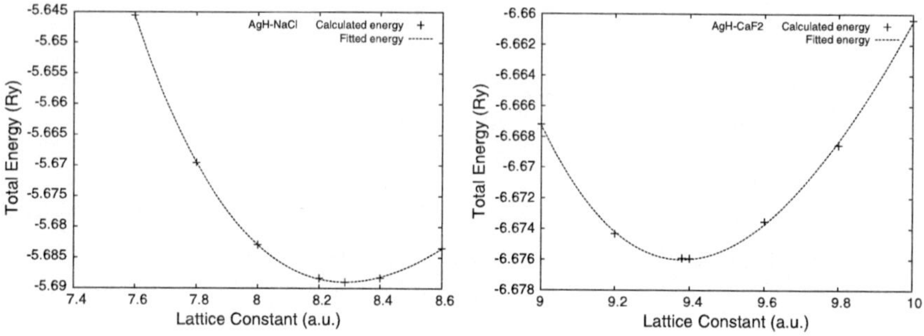

Fig. 7.36 Total energy versus lattice constant of AgH in the NaCl and CaF2 structures

Table 7.36 Lattice constant, bulk modulus, gap, total energy

Stru	a (Bohr)	B (MBar)	Gap	Total Energy (Ry)
NaCl (semimetal)	8.29	1.01		-10625.68900
CaF2	9.38	0.72		-10626.67592
exp				

Table 7.37 Birch fit coefficients

	A1	A2	A3	A4
NaCl	-5.214661E+00	-8.722886E+00	-5.808258E+02	1.270705E+04
CaF2	-6.058041E+00	-2.500315E+01	-5.139451E+02	2.212182E+04

Table 7.38 DOS at Ef, Hopfield parameter, Stoner criterion

AgH2 CaF2

	Fermi Energy (Ry)	Total DOS (States/Ry)	Decomposed DOS inside the muffin tins s	p	eg	t2g	f
Ag	0.374	7.308	0.213	0.381	0.018	2.473	0.038
H	0.374	7.308	2.715	0.108	0.001	0.012	0.003

Fermi-Velocity (cm/s): 1.01307 x10E8
Plasmon Energy (eV) : 6.86498
Electron-ion interaction (Hopfield parameter) (eV/A^2) Ag: 0.682 H: 2.300

Ag	MUFFIN-TIN RADIUS and CHARGE =	2.4369	45.4796
H	MUFFIN-TIN RADIUS and CHARGE =	1.6246	0.8433

Ag STONER I = 0.0036
H STONER I = 0.0100
STONER PARAMETER (Ry) I = 0.0223
STONER CRITERION N*I = 0.1665

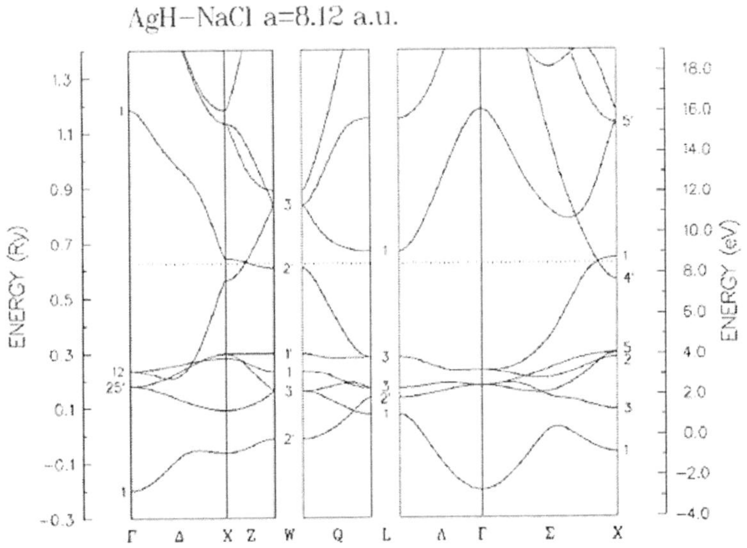

Fig. 7.37 Energy bands of AgH in the NaCl structure (tight-binding)

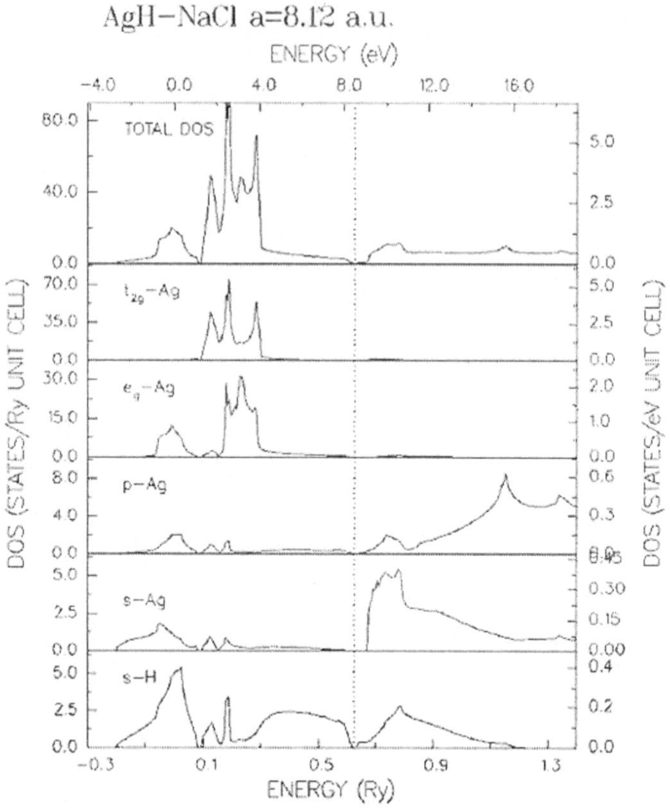

Fig. 7.38 Total, angular momentum and site decomposed densities of states of AgH in the NaCl structure (tight-binding)

Table 7.39 AgH lattice constant = 8.11900 a.u. Slater–Koster 3-center parameters

AgH lattice constant= 8.11900 a.u.
SLATER-KOSTER 3-CENTER PARAMETERS

	ORTHOGONAL	NON-ORTHOGONAL	
	ENERGY INTEGRALS Ry	ENERGY INTEGRALS Ry	OVERLAP INTEGRALS
ON SITE Ag-Ag			
s,s(000)	0.78692	0.63538	
x,x(000)	1.30519	1.11519	
xy,xy(000)	0.22720	0.20648	
d2,d2(000)	0.30016	0.19142	
FIRST NEIGHBOR Ag-Ag			
s,s(110)	-0.00223	-0.00951	0.01442
s,x(110)	-0.03863	-0.00349	0.04953
s,xy(110)	-0.03624	-0.02751	0.04165
s,d2(110)	-0.01291	-0.01447	-0.00390
x,x(110)	0.06837	-0.00065	-0.08976
x,x(011)	-0.00355	0.02948	0.07288
x,y(110)	0.04782	-0.07287	-0.08466
x,xy(110)	0.04003	0.01542	-0.03783
x,xy(011)	-0.00336	-0.00347	0.03849
z,d2(011)	0.01396	0.01826	0.05349
z,d1(011)	0.00863	0.01997	0.00540
xy,xy(110)	-0.02040	-0.01640	0.01946
xy,xy(011)	0.00545	0.00463	-0.00974
xy,xz(011)	0.00315	0.00557	-0.01363
xy,d2(110)	0.01117	-0.00799	0.01520
d2,d2(110)	0.00727	-0.00951	0.00282
d1,d1(110)	-0.03263	0.02490	0.01839
SECOND NEIGHBOR Ag-Ag			
s,s(200)	0.04214	-0.01048	-0.06980
s,x(200)	0.05578	-0.01804	-0.07280
s,d2(002)	0.03515	0.01788	0.06525
x,x(200)	-0.11907	-0.04232	0.03372
y,y(200)	0.01428	0.01565	0.02089
x,xy(020)	0.00568	0.00125	0.01155
z,d2(002)	-0.02892	-0.02803	-0.04764
xy,xy(200)	-0.00232	0.00063	-0.00097
xy,xy(002)	0.00034	-0.00049	-0.00148
d2,d2(002)	0.02996	-0.01373	-0.03613
d1,d1(002)	-0.00004	-0.00015	0.00005
ON SITE H-H			
s,s(000)	0.12971	0.53746	
FIRST NEIGHBOR H-H			
s,s(110)	-0.00975	0.00146	-0.00140
s,x(110)	0.00000	0.00000	0.00000
x,x(110)	0.00000	0.00000	0.00000
x,x(011)	0.00000	0.00000	0.00000
x,y(110)	0.00000	0.00000	0.00000
SECOND NEIGHBOR H-H			
s,s(200)	-0.00846	0.00033	0.00155
s,x(200)	0.00000	0.00000	0.00000
x,x(200)	0.00000	0.00000	0.00000
y,y(200)	0.00000	0.00000	0.00000
FIRST NEIGHBOR Ag-H			
s,s(100)	-0.09151	-0.10930	-0.03011
s,x(100)	0.00000	0.00000	0.00000
x,s(100)	0.06332	0.24849	0.17776
x,x(100)	0.00000	0.00000	0.00000
y,y(100)	0.00000	0.00000	0.00000
xy,x(010)	0.00000	0.00000	0.00000
d2,s(001)	-0.07131	0.01147	-0.08956
d2,z(001)	0.00000	0.00000	0.00000
SECOND NEIGHBOR Ag-H			
s,s(111)	0.01281	0.00014	0.00822
x,s(111)	-0.02506	-0.00248	-0.01364
s,x(111)	0.00000	0.00000	0.00000
x,x(111)	0.00000	0.00000	0.00000
x,y(111)	0.00000	0.00000	0.00000
xy,s(111)	-0.00141	0.00142	0.00385
xy,x(111)	0.00000	0.00000	0.00000
yz,x(111)	0.00000	0.00000	0.00000
d1,x(111)	0.00000	0.00000	0.00000

AgH

BAND	ORTHOGONAL RMS ERROR mRy	MAXIMUM DEVIATION k	MAXIMUM DEVIATION mRy	NON-ORTHOGONAL RMS ERROR mRy	MAXIMUM DEVIATION k	MAXIMUM DEVIATION mRy
1	2.3	(033)	4.9	0.2	(224)	0.6
2	3.0	(333)	10.0	0.2	(222)	0.4
3	2.4	(004)	5.8	0.2	(062)	0.5
4	1.4	(062)	3.8	0.2	(224)	0.6
5	1.5	(044)	3.6	0.2	(003)	0.5
6	3.2	(006)	10.0	0.2	(055)	0.6
7	5.1	(006)	12.7	0.6	(111)	1.5
1-7	2.9			0.3		

ENERGY VALUES IN Ry AT SELECTED k-POINTS

	ORTHOGONAL	APW	NON-ORTHOGONAL
GAMMA 1	-0.20212	-0.19881	-0.19899
GAMMA 1	1.17713	1.18482	1.18439
GAMMA 12	0.23776	0.23719	0.23762
GAMMA 15	1.65691	1.66455	1.66455
GAMMA 25'	0.18056	0.18058	0.18067
X1 (008)	-0.05871	-0.06163	-0.06166
X1 (008)	0.63772	0.64258	0.64254
X1 (008)	1.18778	1.18334	1.18308
X2 (008)	0.28105	0.28126	0.28152
X3 (008)	0.09338	0.09152	0.09174
X5 (008)	0.30019	0.29882	0.29890
X4' (008)	0.56303	0.56282	0.56294
X5' (008)	1.13833	1.13474	1.13465
L1 (444)	0.07363	0.07512	0.07540
L1 (444)	0.67103	0.66890	0.66913
L3 (444)	0.17148	0.17163	0.17193
L3 (444)	0.28735	0.28473	0.28491
L2' (444)	0.13101	0.13800	0.13806
L2' (444)	1.15313	1.15186	1.15185
L3' (444)	1.67752	1.65730	1.65724
W1 (048)	0.22887	0.23238	0.23285
W1 (048)	0.89225	0.89092	0.89087
W2' (048)	-0.01579	-0.01205	-0.01202
W2' (048)	0.60590	0.60613	0.60624
W2' (048)	1.69685	1.73697	1.73704
W1' (048)	0.29884	0.29895	0.29911
W3 (048)	0.15862	0.16172	0.16173
W3 (048)	0.84103	0.83717	0.83725
ODD (224)	0.19693	0.19627	0.19568
ODD (224)	0.27825	0.27636	0.27621
EVEN(224)	-0.01201	-0.01486	-0.01544
EVEN(224)	0.13791	0.13605	0.13613
EVEN(224)	0.17906	0.17963	0.17959
EVEN(224)	0.27948	0.27773	0.27797
EVEN(224)	0.81799	0.81356	0.81428

FERMI LEVEL QUANTITIES (Non-orthogonal fit)

DENSITIES OF STATES (states/Ry/cell)

ENERGY Ry	Total	s-Ag	p-Ag	t2g-Ag	eg-Ag	s-H	p-H
0.6230	0.04	0.00	0.02	0.00	0.01	0.02	0.00

INTEGRATED DENSITIES OF STATES (electrons)

Total	s-Ag	p-Ag	t2g-Ag	eg-Ag	s-H	p-H
12.00	0.36	0.40	5.94	3.84	1.46	0.00

VELOCITY cm/s	PLASMON ENERGY eV	EIGENVALUE SUM Ry
1.35x10E8	0.84	-5.1279

Table 7.40 AgH lattice constant = 8.11900 a.u. Slater–Koster 2-center parameters

AgH lattice constant= 8.11900 a.u.
SLATER-KOSTER 2-CENTER PARAMETERS

	ORTHOGONAL	NON-ORTHOGONAL	
	ENERGY INTEGRALS Ry	ENERGY INTEGRALS Ry	OVERLAP INTEGRALS
ON SITE Ag-Ag			
s	0.62595	0.51701	
p	1.26465	1.01598	
t2g	0.21783	0.20645	
eg	0.21347	0.20330	
FIRST NEIGHBOR Ag-Ag			
(sss)	-0.06033	-0.03787	0.05177
(pps)	0.13916	0.03062	-0.13980
(ppp)	-0.00314	-0.00513	0.02571
(dds)	-0.02681	-0.02145	0.02733
(ddp)	0.01143	0.01427	-0.00563
(ddd)	-0.00080	-0.00323	-0.00380
(sps)	0.09782	0.04437	-0.07592
(sds)	-0.03633	-0.02670	0.03262
(pds)	-0.05259	-0.02315	0.05553
(pdp)	0.01008	0.01571	-0.00480
SECOND NEIGHBOR Ag-Ag			
(sss)	0.00812	-0.02588	-0.04078
(pps)	-0.09401	0.02820	0.04152
(ppp)	0.01093	-0.01731	-0.01950
(dds)	0.00021	-0.00170	0.00275
(ddp)	0.00016	0.00250	0.00708
(ddd)	-0.00027	-0.00003	-0.00021
(sps)	-0.02300	0.03436	0.04675
(sds)	0.00134	-0.00153	0.00025
(pds)	0.00092	0.00161	0.01318
(pdp)	-0.00533	0.00694	0.01175
ON SITE H-H			
s	0.62556	0.14784	
FIRST NEIGHBOR H-H			
(sss)	0.01865	-0.00385	0.01015
(sps)	0.00000	0.00000	0.00000
(pps)	0.00000	0.00000	0.00000
(ppp)	0.00000	0.00000	0.00000
SECOND NEIGHBOR H-H			
(sss)	0.02897	-0.00593	0.00891
(sps)	0.00000	0.00000	0.00000
(pps)	0.00000	0.00000	0.00000
(ppp)	0.00000	0.00000	0.00000
FIRST NEIGHBOR Ag-H			
(sss)	0.09737	0.06106	-0.14760
(sps)	0.00000	0.00000	0.00000
(pss)	0.13230	0.02301	-0.22139
(pps)	0.00000	0.00000	0.00000
(ppp)	0.00000	0.00000	0.00000
(dpp)	0.00000	0.00000	0.00000
(dss)	0.06841	0.04700	-0.14324
(dps)	0.00000	0.00000	0.00000
SECOND NEIGHBOR Ag-H			
(sss)	-0.01917	0.00628	-0.00486
(pss)	-0.04059	0.02100	0.00232
(sps)	0.00000	0.00000	0.00000
(pps)	0.00000	0.00000	0.00000
(ppp)	0.00000	0.00000	0.00000
(dss)	-0.00453	0.00820	0.01191
(dps)	0.00000	0.00000	0.00000
(dpp)	0.00000	0.00000	0.00000

♀AgH

BAND	ORTHOGONAL RMS ERROR mRy	MAXIMUM DEVIATION k	MAXIMUM DEVIATION mRy	NON-ORTHOGONAL RMS ERROR mRy	MAXIMUM DEVIATION k	MAXIMUM DEVIATION mRy
1	4.2	(226)	9.7	0.1	(044)	0.3
2	3.7	(333)	13.7	0.2	(004)	0.7
3	1.8	(226)	4.4	0.2	(055)	0.5
4	1.7	(354)	6.4	0.2	(048)	0.4
5	1.2	(005)	5.2	0.2	(044)	0.4
6	2.8	(174)	6.4	0.2	(044)	0.6
7	9.7	(006)	27.6	0.3	(174)	0.7
1-7	4.5			0.2		

ENERGY VALUES IN Ry AT SELECTED k-POINTS

	ORTHOGONAL	APW	NON-ORTHOGONAL
GAMMA 1	-0.20103	-0.19881	-0.19877
GAMMA 1	1.17477	1.18482	1.18451
GAMMA 12	0.23809	0.23719	0.23754
GAMMA 15	1.65183	1.66455	1.66459
GAMMA 25'	0.17926	0.18058	0.18058
X1 (008)	-0.05397	-0.06163	-0.06165
X1 (008)	0.64075	0.64258	0.64265
X1 (008)	1.18410	1.18334	1.18316
X2 (008)	0.27997	0.28126	0.28113
X3 (008)	0.09418	0.09152	0.09147
X5 (008)	0.29918	0.29882	0.29901
X4' (008)	0.56369	0.56282	0.56300
X5' (008)	1.13292	1.13474	1.13463
L1 (444)	0.07434	0.07512	0.07518
L1 (444)	0.67172	0.66890	0.66891
L3 (444)	0.17090	0.17163	0.17158
L3 (444)	0.28490	0.28473	0.28497
L2' (444)	0.14808	0.13800	0.13829
L2' (444)	1.14340	1.15186	1.15183
L3' (444)	1.69356	1.65730	1.65722
W1 (048)	0.23232	0.23238	0.23198
W1 (048)	0.89286	0.89092	0.89104
W2' (048)	-0.01278	-0.01205	-0.01185
W2' (048)	0.60045	0.60613	0.60649
W2' (048)	1.69885	1.73697	1.73707
W1' (048)	0.30024	0.29895	0.29890
W3 (048)	0.15776	0.16172	0.16190
W3 (048)	0.84285	0.83717	0.83746
ODD (224)	0.19630	0.19627	0.19634
ODD (224)	0.27669	0.27636	0.27619
EVEN(224)	-0.01630	-0.01486	-0.01479
EVEN(224)	0.13877	0.13605	0.13578
EVEN(224)	0.18109	0.17963	0.17973
EVEN(224)	0.28047	0.27773	0.27760
EVEN(224)	0.82023	0.81356	0.81350

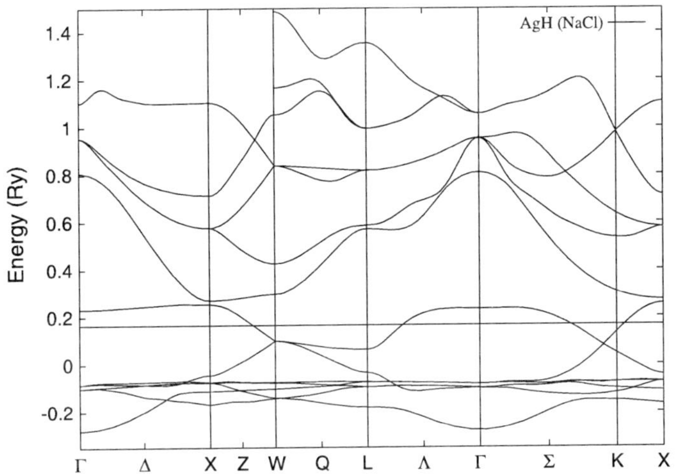

Fig. 7.39 Energy bands of AgH2 in the CaF2 structure

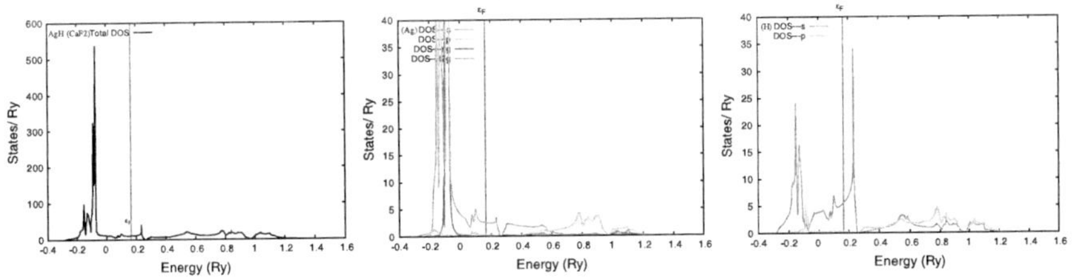

Fig. 7.40 Total, angular momentum and site decomposed densities of states of AgH2 in the CaF2 structure

7.9 Cadmium Hydride (CdH)

See Figs. 7.41, Tables 7.41, 7.42, and 7.43.

See Figs. 7.42, 7.43, Tables 7.44 and 7.45.

See Figs. 7.44 and 7.45.

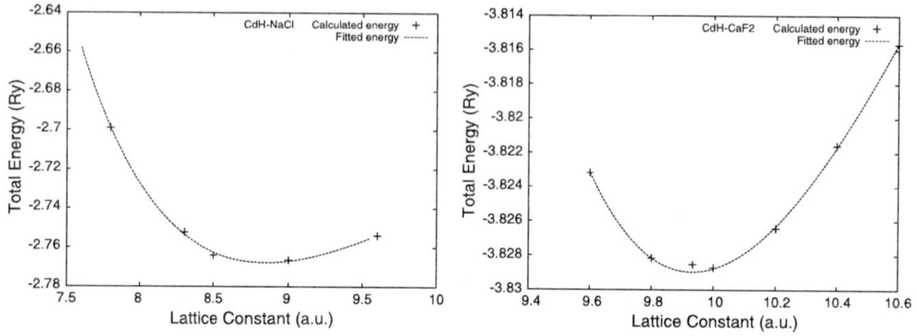

Fig. 7.41 Total energy versus lattice constant of CdH in the NaCl and CaF2 structures

Table 7.41 Lattice constant, bulk modulus, gap, total energy

Stru	a (Bohr)	B (MBar)	Gap	Total Energy (Ry)
NaCl	8.89	0.53		-11182.76999
CaF2	9.93	0.57		-11183.82853
exp				

Table 7.42 Birch fit coefficients

	A1	A2	A3	A4
NaCl	-2.550033E+00	1.328306E+00	-7.136776E+02	1.436115E+04
CaF2	-4.169565E+00	8.172043E+01	-4.832381E+03	8.437124E+04

Table 7.43 DOS at Ef, Hopfield parameter, Stoner criterion

CdH a=8.89 Bohr NaCl

Fermi Energy (Ry)	Total DOS (States/Ry)	Decomposed DOS inside the muffin tins s	p	eg	t2g	f
Cd 0.592	5.372	0.978	0.374	0.139	0.194	0.034
H 0.592	5.372	1.117	0.544	0.030	0.016	0.004

Fermi-Velocity (cm/s): 3.28687 x10E8
Plasmon Energy (eV) : 20.71011
Electron-ion interaction (Hopfield parameter) (eV/A^2) Cd: 0.148 H: 1.525

Cd	MUFFIN-TIN RADIUS and CHARGE =	2.3326	46.0319
H	MUFFIN-TIN RADIUS and CHARGE =	2.1104	1.3331
Cd	STONER I =	0.0031	
H	STONER I =	0.0141	
STONER PARAMETER (Ry)	I =	0.0174	
STONER CRITERION	N*I =	0.0934	

CdH2 a=9.93 Bohr CaF2

Fermi Energy (Ry)	Total DOS (States/Ry)	Decomposed DOS inside the muffin tins s	p	eg	t2g	f
Cd 0.310	12.085	2.208	0.316	0.096	0.761	0.117
H 0.310	12.085	5.095	0.375	0.004	0.021	0.002

Fermi-Velocity (cm/s): 0.77431 x10E8
Plasmon Energy (eV) : 6.19227
Electron-ion interaction (Hopfield parameter) (eV/A^2) Cd: 0.275 H: 6.815

Cd	MUFFIN-TIN RADIUS and CHARGE =	2.5804	46.4589
H	MUFFIN-TIN RADIUS and CHARGE =	1.7203	0.9388
Cd	STONER I =	0.0001	
H	STONER I =	0.0113	
STONER PARAMETER (Ry)	I =	0.0232	
STONER CRITERION	N*I =	0.2802	

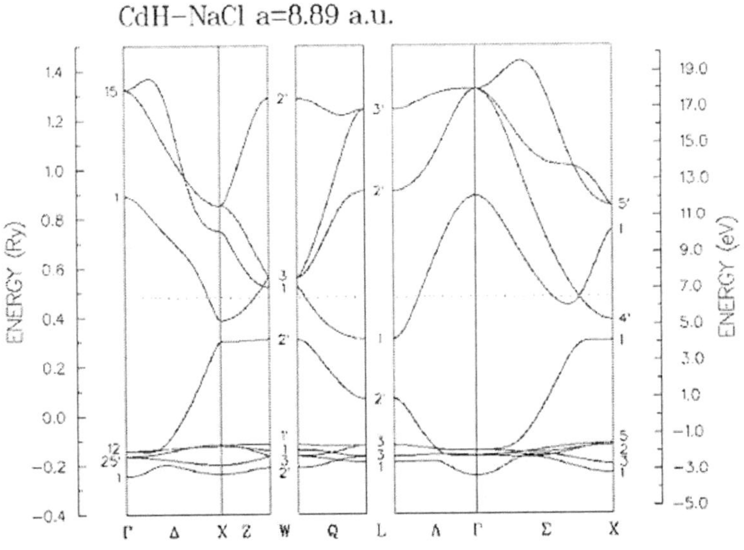

Fig. 7.42 Energy bands of CdH in the NaCl structure (tight-binding)

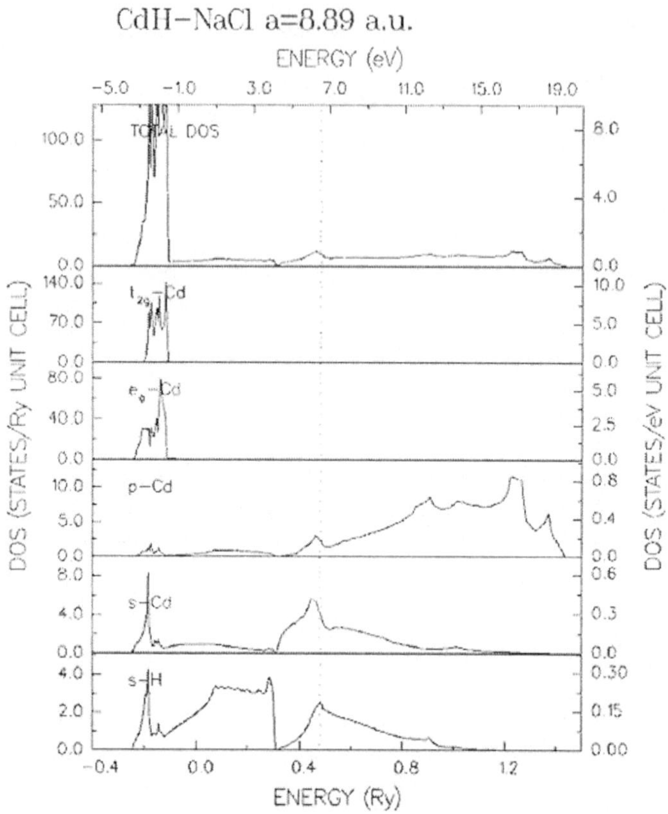

Fig. 7.43 Total, angular momentum and site decomposed densities of states of CdH in the NaCl structure (tight-binding)

Table 7.44 CdH (NaCl) a = 8.89 Bohr Slater–Koster 3-center parameters

	ORTHOGONAL	NON-ORTHOGONAL	
	ENERGY INTEGRALS Ry	ENERGY INTEGRALS Ry	OVERLAP INTEGRALS
ON SITE Cd-Cd			
s,s(000)	0.40486	0.35966	
x,x(000)	0.98030	0.86168	
xy,xy(000)	-0.14996	-0.14383	
d2,d2(000)	-0.12194	-0.15064	
FIRST NEIGHBOR Cd-Cd			
s,s(110)	-0.02194	-0.02168	0.00186
s,x(110)	-0.03005	-0.01057	0.02799
s,xy(110)	-0.00542	-0.02571	0.03592
s,d2(110)	0.02640	-0.01186	-0.01359
x,x(110)	0.05818	0.01429	-0.06529
x,x(011)	-0.00050	-0.02325	0.00827
x,y(110)	0.03582	-0.02780	-0.05941
x,xy(110)	-0.00151	0.02743	-0.03919
x,xy(011)	0.00720	-0.00275	-0.00295
z,d2(011)	0.01052	0.00334	0.01333
z,d1(011)	-0.01432	0.01714	0.01224
xy,xy(110)	-0.00835	-0.00570	-0.01883
xy,xy(011)	0.00226	0.00345	-0.00804
xy,xz(011)	0.00395	0.00309	-0.01065
xy,d2(110)	0.00463	-0.00352	-0.00599
d2,d2(110)	0.00156	-0.00221	-0.00981
d1,d1(110)	-0.00845	0.00104	0.03278
SECOND NEIGHBOR Cd-Cd			
s,s(200)	0.01615	0.02329	-0.02143
s,x(200)	0.03681	0.03334	-0.03415
s,d2(002)	-0.02553	0.00547	-0.01303
x,x(200)	-0.09130	-0.03848	0.06563
y,y(200)	0.01340	0.02148	0.00830
x,xy(020)	0.01537	-0.00282	0.01755
z,d2(002)	0.01146	-0.00819	-0.00621
xy,xy(200)	0.00077	-0.00087	0.00383
xy,xy(002)	0.00004	0.00036	-0.00333
d2,d2(002)	0.00795	0.00087	-0.01127
d1,d1(002)	0.00039	-0.00046	0.00319
ON SITE H-H			
s,s(000)	0.21797	0.39373	
FIRST NEIGHBOR H-H			
s,s(110)	0.00934	-0.00431	0.00462
s,x(110)	0.00000	0.00000	0.00000
x,x(110)	0.00000	0.00000	0.00000
x,x(011)	0.00000	0.00000	0.00000
x,y(110)	0.00000	0.00000	0.00000
SECOND NEIGHBOR H-H			
s,s(200)	0.01135	-0.00168	0.00087
s,x(200)	0.00000	0.00000	0.00000
x,x(200)	0.00000	0.00000	0.00000
y,y(200)	0.00000	0.00000	0.00000
FIRST NEIGHBOR Cd-H			
s,s(100)	-0.08389	-0.09663	-0.00905
s,x(100)	0.00000	0.00000	0.00000
x,s(100)	0.07483	0.17399	0.11348
x,x(100)	0.00000	0.00000	0.00000
y,y(100)	0.00000	0.00000	0.00000
xy,x(010)	0.00000	0.00000	0.00000
d2,s(001)	0.07071	0.00203	-0.02865
d2,z(001)	0.00000	0.00000	0.00000
SECOND NEIGHBOR Cd-H			
s,s(111)	-0.00762	0.00553	0.01362
x,s(111)	-0.00170	-0.00772	-0.00969
s,x(111)	0.00000	0.00000	0.00000
x,x(111)	0.00000	0.00000	0.00000
x,y(111)	0.00000	0.00000	0.00000
xy,s(111)	0.00473	0.00011	0.00867
xy,x(111)	0.00000	0.00000	0.00000
yz,x(111)	0.00000	0.00000	0.00000
d1,x(111)	0.00000	0.00000	0.00000

CdH

BAND	ORTHOGONAL RMS ERROR mRy	MAXIMUM DEVIATION k	mRy	NON-ORTHOGONAL RMS ERROR mRy	MAXIMUM DEVIATION k	mRy
1	2.3	(022)	5.7	0.1	(444)	0.2
2	1.7	(033)	5.9	0.1	(062)	0.2
3	2.4	(033)	9.6	0.1	(226)	0.3
4	2.0	(022)	4.8	0.1	(044)	0.4
5	1.5	(222)	3.7	0.1	(044)	0.5
6	4.4	(044)	9.6	0.1	(174)	0.3
7	4.3	(006)	12.7	0.2	(380)	1.0
1-7	2.9			0.1		

ENERGY VALUES IN Ry AT SELECTED k-POINTS

	ORTHOGONAL	APW	NON-ORTHOGONAL
GAMMA 1	-0.25161	-0.24649	-0.24640
GAMMA 1	0.88824	0.89102	0.89077
GAMMA 12	-0.13828	-0.14059	-0.14051
GAMMA 15	1.31475	1.32113	1.32114
GAMMA 25'	-0.16219	-0.16332	-0.16318
X1 (008)	-0.23581	-0.23608	-0.23612
X1 (008)	0.29612	0.29950	0.29955
X1 (008)	0.73485	0.74898	0.74890
X2 (008)	-0.12316	-0.12190	-0.12186
X3 (008)	-0.19829	-0.19754	-0.19744
X5 (008)	-0.11340	-0.11433	-0.11421
X4' (008)	0.38388	0.38554	0.38567
X5' (008)	0.85329	0.85182	0.85178
L1 (444)	-0.18755	-0.19072	-0.19048
L1 (444)	0.31079	0.31019	0.31043
L3 (444)	-0.16533	-0.16544	-0.16525
L3 (444)	-0.11896	-0.12009	-0.11998
L2' (444)	0.06162	0.06777	0.06770
L2' (444)	0.91103	0.91102	0.91100
L3' (444)	1.25258	1.24021	1.24016
W1 (048)	-0.13498	-0.13831	-0.13831
W1 (048)	0.52493	0.52059	0.52043
W2' (048)	-0.21004	-0.21062	-0.21057
W2' (048)	0.30526	0.30970	0.30971
W2' (048)	1.26192	1.28667	1.28677
W1' (048)	-0.11356	-0.11427	-0.11414
W3 (048)	-0.16058	-0.16267	-0.16265
W3 (048)	0.56666	0.56227	0.56254
ODD (224)	-0.15552	-0.15514	-0.15529
ODD (224)	-0.12429	-0.12345	-0.12349
EVEN(224)	-0.18639	-0.18530	-0.18537
EVEN(224)	-0.16716	-0.16812	-0.16816
EVEN(224)	-0.14818	-0.14820	-0.14813
EVEN(224)	-0.02379	-0.02418	-0.02420
EVEN(224)	0.51823	0.51496	0.51512

FERMI LEVEL QUANTITIES (Non-orthogonal fit)

DENSITIES OF STATES (states/Ry/cell)

ENERGY Ry	Total	s-Cd	p-Cd	t2g-Cd	eg-Cd	s-H	p-H
0.4791	9.71	4.28	2.45	0.47	0.02	2.48	0.00

INTEGRATED DENSITIES OF STATES (electrons)

Total	s-Cd	p-Cd	t2g-Cd	eg-Cd	s-H	p-H
13.00	1.08	0.50	6.10	3.91	1.40	0.00

VELOCITY cm/s	PLASMON ENERGY eV	EIGENVALUE SUM Ry
0.96x10E8	8.10	-7.1925

Table 7.45 CdH (NaCl) a = 8.89 Bohr Slater–Koster 2-center parameters

	ORTHOGONAL	NON-ORTHOGONAL	
	ENERGY INTEGRALS Ry	ENERGY INTEGRALS Ry	OVERLAP INTEGRALS
ON SITE Cd-Cd			
s	0.31885	0.28422	
p	0.92603	0.78400	
t2g	-0.14438	-0.14745	
eg	-0.14520	-0.14905	
FIRST NEIGHBOR Cd-Cd			
(sss)	-0.04816	-0.03541	0.03577
(pps)	0.11665	0.05765	-0.10649
(ppp)	-0.00146	-0.00421	0.01885
(dds)	-0.01098	-0.01092	-0.00032
(ddp)	0.00337	0.00404	0.00457
(ddd)	0.00036	-0.00009	-0.00278
(sps)	0.07532	0.05053	-0.04437
(sds)	-0.02129	-0.01870	0.00319
(pds)	-0.03314	-0.02701	0.02140
(pdp)	0.00416	0.00563	0.00792
SECOND NEIGHBOR Cd-Cd			
(sss)	0.00791	-0.01103	-0.03781
(pps)	-0.06847	0.00553	0.03621
(ppp)	0.01497	-0.00159	-0.01282
(dds)	0.00030	0.00051	-0.00550
(ddp)	-0.00053	-0.00029	0.00261
(ddd)	-0.00009	0.00001	-0.00003
(sps)	-0.01658	0.01490	0.03864
(sds)	-0.00128	-0.00121	-0.00405
(pds)	-0.01231	-0.00291	0.00331
(pdp)	-0.00144	0.00175	-0.00607
ON SITE H-H			
s	0.48408	0.11648	
FIRST NEIGHBOR H-H			
(sss)	0.02707	-0.00591	0.01352
(sps)	0.00000	0.00000	0.00000
(pps)	0.00000	0.00000	0.00000
(ppp)	0.00000	0.00000	0.00000
SECOND NEIGHBOR H-H			
(sss)	0.00610	-0.00446	0.02271
(sps)	0.00000	0.00000	0.00000
(pps)	0.00000	0.00000	0.00000
(ppp)	0.00000	0.00000	0.00000
FIRST NEIGHBOR Cd-H			
(sss)	0.05440	0.05748	-0.14497
(sps)	0.00000	0.00000	0.00000
(pss)	0.12037	0.04270	-0.18286
(pps)	0.00000	0.00000	0.00000
(ppp)	0.00000	0.00000	0.00000
(dpp)	0.00000	0.00000	0.00000
(dss)	0.03696	0.05348	-0.10287
(dps)	0.00000	0.00000	0.00000
SECOND NEIGHBOR Cd-H			
(sss)	-0.01545	0.00660	-0.00469
(pss)	-0.02097	0.01233	0.00079
(sps)	0.00000	0.00000	0.00000
(pps)	0.00000	0.00000	0.00000
(ppp)	0.00000	0.00000	0.00000
(dss)	-0.00458	-0.00021	0.01485
(dps)	0.00000	0.00000	0.00000
(dpp)	0.00000	0.00000	0.00000

♀ CdH

	ORTHOGONAL			NON-ORTHOGONAL		
BAND	RMS ERROR	MAXIMUM DEVIATION		RMS ERROR	MAXIMUM DEVIATION	
	mRy	k	mRy	mRy	k	mRy
1	3.4	(226)	9.3	0.1	(044)	0.2
2	1.6	(008)	3.4	0.1	(004)	0.4
3	1.3	(033)	5.1	0.1	(044)	0.4
4	1.0	(444)	3.7	0.1	(264)	0.4
5	1.2	(444)	3.7	0.1	(044)	0.5
6	3.5	(007)	9.4	0.2	(007)	0.3
7	9.4	(007)	31.2	0.3	(380)	1.0
1.7	4.1			0.2		

ENERGY VALUES IN Ry AT SELECTED k-POINTS

	ORTHOGONAL	APW	NON-ORTHOGONAL
GAMMA 1	-0.24918	-0.24649	-0.24643
GAMMA 1	0.88314	0.89102	0.89092
GAMMA 12	-0.13919	-0.14059	-0.14034
GAMMA 15	1.30392	1.32113	1.32115
GAMMA 25'	-0.16431	-0.16332	-0.16336
X1 (008)	-0.23093	-0.23608	-0.23611
X1 (008)	0.30274	0.29950	0.29952
X1 (008)	0.72975	0.74898	0.74880
X2 (008)	-0.12301	-0.12190	-0.12195
X3 (008)	-0.19417	-0.19754	-0.19733
X5 (008)	-0.11411	-0.11433	-0.11413
X4' (008)	0.38239	0.38554	0.38566
X5' (008)	0.85483	0.85182	0.85178
L1 (444)	-0.18953	-0.19072	-0.19061
L1 (444)	0.30681	0.31019	0.31030
L3 (444)	-0.16548	-0.16544	-0.16531
L3 (444)	-0.11640	-0.12009	-0.12005
L2' (444)	0.07004	0.06777	0.06793
L2' (444)	0.90809	0.91102	0.91099
L3' (444)	1.23930	1.24021	1.24014
W1 (048)	-0.13799	-0.13831	-0.13808
W1 (048)	0.52940	0.52059	0.52057
W2' (048)	-0.20900	-0.21062	-0.21056
W2' (048)	0.30734	0.30970	0.30990
W2' (048)	1.26066	1.28667	1.28671
W1' (048)	-0.11374	-0.11427	-0.11415
W3 (048)	-0.16399	-0.16267	-0.16261
W3 (048)	0.57066	0.56227	0.56253
ODD (224)	-0.15472	-0.15514	-0.15513
ODD (224)	-0.12243	-0.12345	-0.12353
EVEN(224)	-0.18582	-0.18530	-0.18541
EVEN(224)	-0.16982	-0.16812	-0.16807
EVEN(224)	-0.14910	-0.14820	-0.14797
EVEN(224)	-0.01983	-0.02418	-0.02431
EVEN(224)	0.52102	0.51496	0.51482

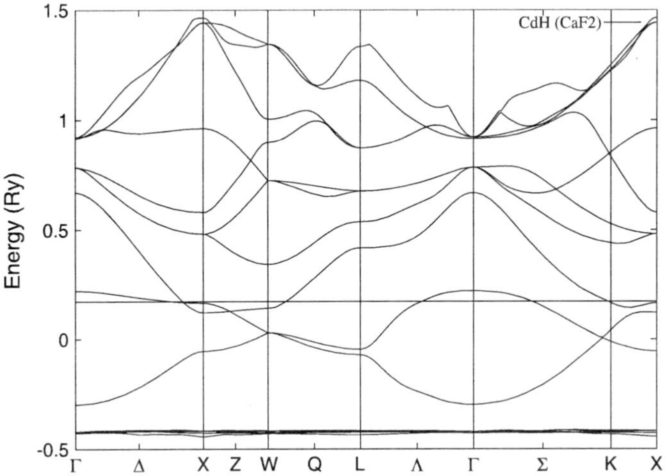

Fig. 7.44 Energy bands of CdH2 in the CaF2 structure

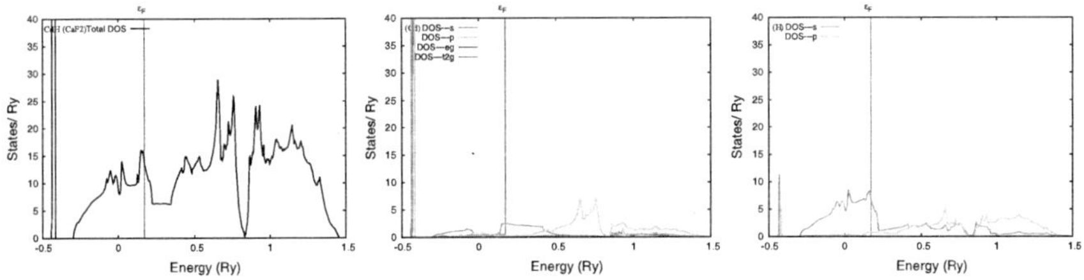

Fig. 7.45 Total, angular momentum and site decomposed densities of states of CdH2 in the CaF2 structure

References

1. G.G. Libowitz, "The nature and properties of transition metal hydrides, J. Nucl. Mater. **2**, 1 (1960)
2. W.M. Mueller, J.P. Blackledge, G.G. Libowitz (ed.), *Metal Hydrides* (Academic Press, N.Y., 1968)
3. D.A. Papaconstantopoulos, B.M. Klein, E.N. Economou, L.L. Boyer, Band structure and superconductivity of PdDx and PdHx. Phys. Rev. B **17**, 141 (1978)
4. D.A. Papaconstantopoulos, Magnetism in transition metal hydrides, Europhys. Lett.,**15**, 621 (1991)
5. H. Smithson, C.A. Marianetti, D. Morgan, A. Van der Ven, A. Predith, G. Ceder, First-Principles study of the stability and electronic structure of metal hydrides. Phys. Rev. B **66**, 144107 (2002)

Chapter 8
5d Transition-Metal Hydrides

This chapter covers the 5d row of the transition-metal hydrides from HfH to HgH [1-3]. Results are presented for the crystal structures NaCl (B1) and CaF2 (C1). The first two compounds (HfH and TaH) form as dihydrides in the CaF2 structure while PtH was predicted to be a superconductor [3, 4] in the NaCl structure but was found to form a hexagonal closed-packed structure under pressure. A few of the 5d hydrides can be found in a body-centered tetragonal phase and as substoichiometric materials with hydrogen vacancies. The lattice constants, as expected, are larger than in the single elements because of the expansion of the lattice upon hydrogenation. Exactly, as in the elements and in the 3d and 4d hydrides, the lattice parameter decreases from HfH up to the middle of the 5d series where it reaches a minimum and then increases up to HgH. Comparing the energy bands of the CaF2 structure to the bands of the NaCl structure we note that a gap that appears between first and second band in the NaCl structure closes in the CaF2 structure and an additional antibonding band emerges above the Fermi level which is due to the second hydrogen in the CaF2 structure. Examining the densities of states figures we observe that for both crystal structures the lower occupied states have predominantly s–H hydrogen character and just below the Fermi level the d-t2g metal contribution dominates. However, the DOS details are significantly different for the two crystal structures. Moving along the 5d series the qualitative difference from HfH to AuH is the position of the Fermi level which is moving up and therefore changes the Fermi surface and the values of the DOS at the Fermi level. There is a crystal field splitting that one can see in the separation of the t2g and eg states in the DOS figures, and in the positioning of the Gam25' and Gam12 points in the energy bands diagrams. It is important to mention that, unlike the other noble-metal hydrides CuH and AgH, AuH is not predicted to be a semimetal. Also, in HgH the d-bands are positioned much deeper than in the other compounds. It should be noted that spin–orbit interaction, which has not be considered in this work, would cause minor modifications of the band structure in the 5d series. Tight-binding parameters are given in the NaCl structure based on both orthogonal and non-orthogonal Hamiltonians using three- and two-center integrals.

D. A. Papaconstantopoulos, *Band Structure of Cubic Hydrides*,
https://doi.org/10.1007/978-3-031-06878-2_8

8.1 Hafnium Hydride (HfH)

See Fig. 8.1 and Tables 8.1, 8.2, 8.3.

See Figs. 8.2, 8.3 and Tables 8.4, 8.5.

See Figs. 8.4 and 8.5.

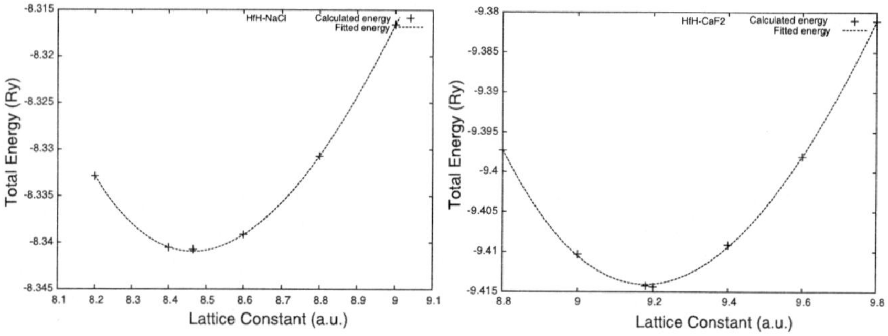

Fig. 8.1 Total energy versus lattice constant of HfH in the NaCl and CaF2 structures

Table 8.1 Lattice constant, bulk modulus, gap, total energy

Stru	a (Bohr)	B (MBar)	Gap	Total Energy
NaCl	8.47	1.60	-	-30178.34075
CaF2	9.18	1.47	-	-30179.41423
exp (CaF2)	8.90			

Table 8.2 Birch fit coefficients

	A1	A2	A3	A4
NaCl	-6.704369E+00	-8.689470E+01	9.716962E+02	5.001373E+03
CaF2	-7.076896E+00	-1.617556E+02	2.977378E+03	-6.077885E+03

Table 8.3 DOS at Ef, Hopfield parameter, Stoner criterion

HfH a=8.47 Bohr NaCl

Fermi Energy (Ry)	Total DOS (States/Ry)	Decomposed DOS inside the muffin tins s	p	eg	t2g	f
Hf 0.756	18.230	0.457	1.202	2.376	5.227	0.188
H 0.756	18.230	0.258	1.041	0.007	0.082	0.014

Fermi-Velocity (cm/s): 0.54780 x10E8
Plasmon Energy (eV) : 6.83798
Electron-ion interaction (Hopfield parameter) (eV/A^2) Hf: 6.393 H: 0.417

Hf	MUFFIN-TIN RADIUS and CHARGE =	2.4972	69.7606
H	MUFFIN-TIN RADIUS and CHARGE =	1.7353	1.1728
Hf	STONER I = 0.0024		
H	STONER I = 0.0011		
STONER PARAMETER (Ry)	I = 0.0039		
STONER CRITERION	N*I = 0.0714		

HfH2 a=9.18 Bohr CaF2

Fermi Energy (Ry)	Total DOS (States/Ry)	Decomposed DOS inside the muffin tins s	p	eg	t2g	f
Hf 0.673	18.540	0.004	0.053	3.261	7.311	0.030
H 0.673	18.540	0.037	1.513	0.067	0.020	0.007

Fermi-Velocity (cm/s): 0.44225 x10E8
Plasmon Energy (eV) : 4.91585
Electron-ion interaction (Hopfield parameter) (eV/A^2) Hf: 1.131 H: 0.255

Hf	MUFFIN-TIN RADIUS and CHARGE =	2.3850	69.2987
H	MUFFIN-TIN RADIUS and CHARGE =	1.5900	1.0484
Hf	STONER I = 0.0034		
H	STONER I = 0.0005		
STONER PARAMETER (Ry)	I = 0.0045		
STONER CRITERION	N*I = 0.0839		

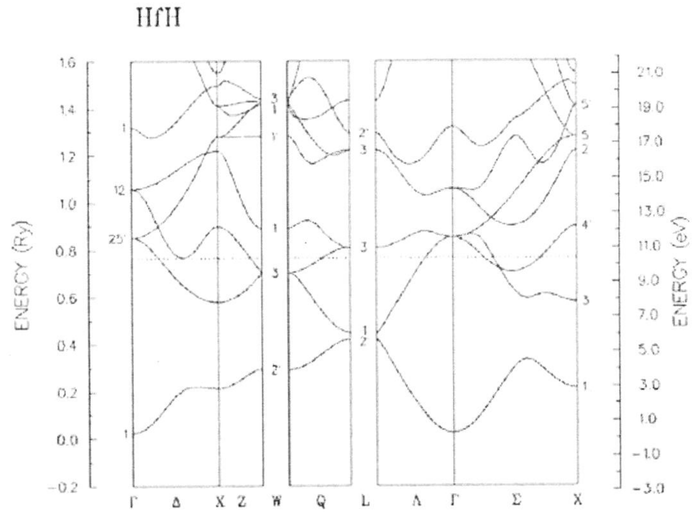

Fig. 8.2 Energy bands of HfH in the NaCl structure (tight-binding)

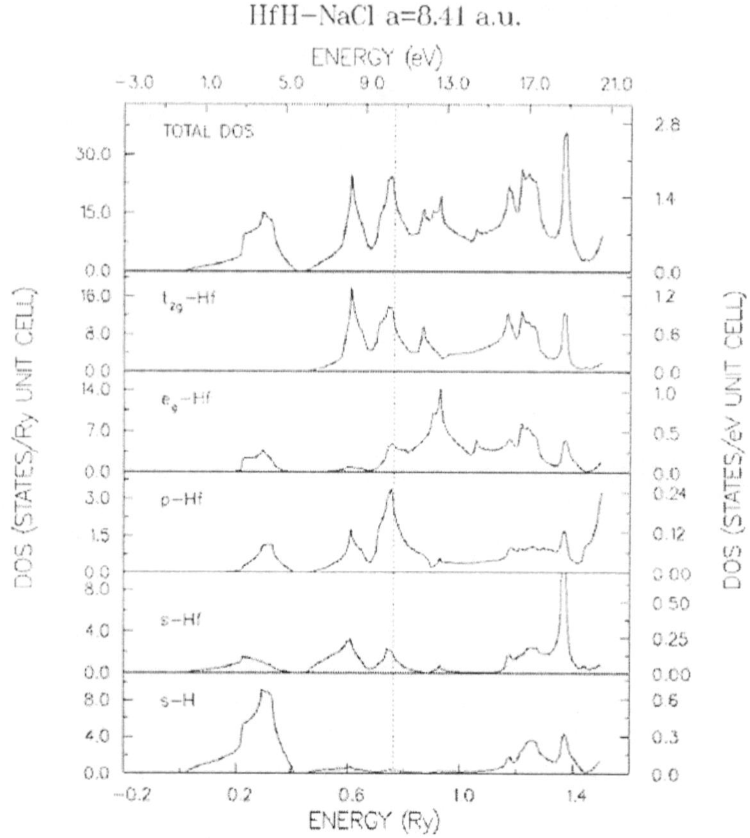

Fig. 8.3 Total, angular momentum and site decomposed densities of states of HfH in the NaCl structure (tight-binding)

Table 8.4 HfH (NaCl) a = 8.41 a.u. Slater–Koster 3-center parameters

	ORTHOGONAL	NON-ORTHOGONAL	
	ENERGY INTEGRALS Ry	ENERGY INTEGRALS Ry	OVERLAP INTEGRALS
ON SITE Hf-Hf			
s,s(000)	1.02796	0.90073	
x,x(000)	1.55355	1.33396	
xy,xy(000)	1.08334	0.89432	
d2,d2(000)	1.07466	0.87408	
FIRST NEIGHBOR Hf-Hf			
s,s(110)	-0.02371	-0.00161	0.04081
s,x(110)	-0.03673	0.02128	0.07589
s,xy(110)	-0.05645	0.01339	0.10316
s,d2(110)	0.00891	-0.02053	-0.00264
x,x(110)	0.05244	-0.07235	-0.11440
x,x(011)	-0.01215	0.06941	0.09506
x,y(110)	0.06265	-0.06602	-0.07911
x,xy(110)	0.08510	-0.04033	-0.10675

```
          x,xy(011)        -0.00371            0.02122            0.05734
          z,d2(011)         0.02001            0.05441            0.06934
          z,d1(011)        -0.00300            0.00283           -0.00162
          xy,xy(110)       -0.07123           -0.01252            0.06168
          xy,xy(011)        0.01734           -0.00565           -0.03726
          xy,xz(011)       -0.00945           -0.00520           -0.03982
          xy,d2(110)        0.03366            0.01520            0.05399
          d2,d2(110)       -0.00370           -0.00811            0.02070
          d1,d1(110)       -0.02921            0.04983           -0.01115
      SECOND NEIGHBOR Hf-Hf
          s,s(200)          0.06440            0.04622           -0.00934
          s,x(200)          0.08752            0.04995           -0.01723
          s,d2(002)         0.05365            0.01684            0.04501
          x,x(200)         -0.12279           -0.06703            0.02557
          y,y(200)          0.01548            0.02556            0.01385
          x,xy(020)         0.01473            0.00869            0.00833
          z,d2(002)        -0.06558           -0.05272           -0.07566
          xy,xy(200)       -0.02318           -0.00084           -0.00436
          xy,xy(002)        0.00348           -0.00209           -0.00130
          d2,d2(002)        0.05713           -0.05718           -0.05238
          d1,d1(002)        0.00248            0.00581            0.00553
      ON SITE H-H
          s,s(000)          0.41351            0.55326
      FIRST NEIGHBOR H-H
          s,s(110)         -0.01189           -0.01853           -0.01168
          s,x(110)          0.00000            0.00000            0.00000
          x,x(110)          0.00000            0.00000            0.00000
          x,x(011)          0.00000            0.00000            0.00000
          x,y(110)          0.00000            0.00000            0.00000
      SECOND NEIGHBOR H-H
          s,s(200)         -0.01090           -0.00536           -0.00444
          s,x(200)          0.00000            0.00000            0.00000
          x,x(200)          0.00000            0.00000            0.00000
          y,y(200)          0.00000            0.00000            0.00000
      FIRST NEIGHBOR Hf-H
          s,s(100)         -0.09636           -0.10998            0.00217
          s,x(100)          0.00000            0.00000            0.00000
          x,s(100)          0.06455            0.15505            0.06559
          x,x(100)          0.00000            0.00000            0.00000

          y,y(100)          0.00000            0.00000            0.00000
          xy,x(010)         0.00000            0.00000            0.00000
          d2,s(001)        -0.07818            0.01085           -0.15775
          d2,z(001)         0.00000            0.00000            0.00000
      SECOND NEIGHBOR Hf-H
          s,s(111)          0.01668            0.01281            0.01813
          x,s(111)         -0.03032           -0.01909           -0.01693
          s,x(111)          0.00000            0.00000            0.00000
          x,x(111)          0.00000            0.00000            0.00000
          x,y(111)          0.00000            0.00000            0.00000
          xy,s(111)         0.01290            0.00488            0.00801
          xy,x(111)         0.00000            0.00000            0.00000
          yz,x(111)         0.00000            0.00000            0.00000
          d1,x(111)         0.00000            0.00000            0.00000
HfH
```

BAND	ORTHOGONAL RMS ERROR mRy	MAXIMUM DEVIATION k	mRy	NON-ORTHOGONAL RMS ERROR mRy	MAXIMUM DEVIATION k	mRy
1	4.4	(048)	9.4	0.8	(044)	2.8
2	5.0	(003)	14.5	0.7	(333)	1.7
3	4.8	(055)	12.4	0.6	(033)	1.9
4	6.7	(048)	18.8	0.7	(055)	1.3
5	5.4	(226)	10.8	0.8	(224)	2.0
6	5.0	(044)	12.4	1.1	(022)	2.7
7	7.4	(033)	11.7	1.1	(066)	3.7
1-7	5.6			0.8		

ENERGY VALUES IN Ry AT SELECTED k-POINTS

	ORTHOGONAL	APW	NON-ORTHOGONAL
GAMMA 1	0.02625	0.02255	0.02302
GAMMA 1	1.30904	1.32062	1.31939
GAMMA 12	1.05604	1.05815	1.05832
GAMMA 15	1.74077	1.75148	1.75142
GAMMA 25'	0.85140	0.85305	0.85236
X1 (008)	0.22156	0.21422	0.21431
X1 (008)	1.50524	1.49434	1.49427
X1 (008)	1.59946	1.55202	1.55023
X2 (008)	1.21725	1.21891	1.21781
X3 (008)	0.57396	0.57811	0.57836
X5 (008)	1.28250	1.27785	1.27770
X4' (008)	0.90175	0.89837	0.89867
X5' (008)	1.41850	1.41058	1.41058
L1 (444)	0.44924	0.44772	0.44794
L1 (444)	1.43697	1.43499	1.43304
L3 (444)	0.81093	0.80852	0.80904
L3 (444)	1.21618	1.22089	1.22055
L2' (444)	0.41779	0.41868	0.41858
L2' (444)	1.29716	1.29283	1.29311
L3' (444)	1.98779	1.90918	1.90922
W1 (048)	0.87396	0.89271	0.89163
W1 (048)	1.41739	1.41295	1.41342
W2' (048)	0.28342	0.29279	0.29287
W2' (048)	1.42506	1.43862	1.43888
W2' (048)	2.00086	2.14051	2.14049
W1' (048)	1.26857	1.27883	1.28000
W3 (048)	0.70224	0.70225	0.70242
W3 (048)	1.41692	1.42304	1.42347
ODD (224)	0.88616	0.87675	0.87664
ODD (224)	1.18657	1.18204	1.18156
EVEN(224)	0.26463	0.25752	0.25789
EVEN(224)	0.61613	0.61672	0.61661
EVEN(224)	0.78167	0.78451	0.78403
EVEN(224)	1.07236	1.07315	1.07518
EVEN(224)	1.25893	1.25324	1.25439

FERMI LEVEL QUANTITIES (Non-orthogonal fit)

ENERGY	DENSITIES OF STATES						
	Total	s-Hf	p-Hf	t2g-Hf	eg-Hf	s-H	p-H
Ry		states/Ry/cell					
0.7643	19.21	1.78	2.49	9.96	4.63	0.36	0.00
	INTEGRATED DENSITIES OF STATES						
	Total	s-Hf	p-Hf	t2g-Hf	eg-Hf	s-H	p-H
		electrons					
	5.00	0.68	0.40	1.90	0.60	1.41	0.00

VELOCITY	PLASMON ENERGY	EIGENVALUE SUM
cm/s	eV	Ry
0.57x10E8	7.40	-1.3011

Table 8.5 HfH (NaCl) a = 8.41 Bohr Slater–Koster 2-center parameters

	ORTHOGONAL	NON-ORTHOGONAL	
	ENERGY INTEGRALS Ry	ENERGY INTEGRALS Ry	OVERLAP INTEGRALS
ON SITE Hf-Hf			
s	1.01176	0.66001	
p	1.50916	1.02662	
t2g	1.01231	0.87049	
eg	0.97417	0.85423	
FIRST NEIGHBOR Hf-Hf			
(sss)	-0.04995	-0.03097	0.07789
(pps)	0.12103	-0.20570	-0.30966
(ppp)	-0.01666	0.04513	0.07760
(dds)	-0.09514	0.01734	0.11302
(ddp)	0.02833	0.00493	-0.06732
(ddd)	0.00341	-0.00034	0.00961
(sps)	0.08987	-0.03991	-0.18115
(sds)	-0.08096	0.00572	0.10803
(pds)	-0.12290	0.08074	0.17790
(pdp)	0.01156	-0.01676	-0.07052
SECOND NEIGHBOR Hf-Hf			
(sss)	0.01177	-0.01054	-0.00698
(pps)	-0.07060	0.11386	0.03122
(ppp)	0.00289	-0.00271	0.00042
(dds)	0.01282	-0.03572	-0.01733
(ddp)	-0.00486	0.00030	-0.00246
(ddd)	0.00159	0.00364	0.00324
(sps)	-0.03409	0.04358	0.01685
(sds)	0.02200	-0.02619	-0.01326
(pds)	0.01943	-0.05473	-0.01785
(pdp)	-0.00149	-0.00454	-0.00707
ON SITE H-H			
s	0.85960	0.51795	
FIRST NEIGHBOR H-H			
(sss)	-0.01882	0.02134	-0.01147
(sps)	0.00000	0.00000	0.00000
(pps)	0.00000	0.00000	0.00000
(ppp)	0.00000	0.00000	0.00000
SECOND NEIGHBOR H-H			
(sss)	0.03538	-0.00788	-0.00598
(sps)	0.00000	0.00000	0.00000
(pps)	0.00000	0.00000	0.00000
(ppp)	0.00000	0.00000	0.00000
FIRST NEIGHBOR Hf-H			
(sss)	0.11388	0.06263	-0.08653
(sps)	0.00000	0.00000	0.00000
(pss)	0.11529	-0.00749	-0.23848
(pps)	0.00000	0.00000	0.00000
(ppp)	0.00000	0.00000	0.00000
(dpp)	0.00000	0.00000	0.00000
(dss)	0.13996	-0.00183	-0.21162
(dps)	0.00000	0.00000	0.00000
SECOND NEIGHBOR Hf-H			
(sss)	-0.00793	-0.00331	0.01173
(pss)	-0.03174	0.02679	0.02432
(sps)	0.00000	0.00000	0.00000
(pps)	0.00000	0.00000	0.00000
(ppp)	0.00000	0.00000	0.00000
(dss)	-0.01702	0.01909	0.02226
(dps)	0.00000	0.00000	0.00000
(dpp)	0.00000	0.00000	0.00000

HfH

BAND	ORTHOGONAL RMS ERROR mRy	MAXIMUM DEVIATION k	MAXIMUM DEVIATION mRy	NON-ORTHOGONAL RMS ERROR mRy	MAXIMUM DEVIATION k	MAXIMUM DEVIATION mRy
1	8.2	(005)	22.5	0.9	(044)	2.6
2	10.8	(004)	21.3	1.5	(111)	3.7
3	12.6	(264)	27.8	1.2	(044)	2.5
4	11.7	(444)	25.5	1.1	(022)	2.9
5	16.4	(226)	47.6	1.4	(055)	3.5
6	12.5	(044)	39.5	1.3	(066)	3.3
7	18.6	(008)	35.0	1.2	(222)	2.7
1-7	13.4			1.2		

ENERGY VALUES IN Ry AT SELECTED k-POINTS

	ORTHOGONAL	APW	NON-ORTHOGONAL
GAMMA 1	0.01864	0.02255	0.02279
GAMMA 1	1.31044	1.32062	1.32077
GAMMA 12	1.05998	1.05815	1.05936
GAMMA 15	1.73033	1.75148	1.75150
GAMMA 25'	0.84097	0.85305	0.85156
X1 (008)	0.21445	0.21422	0.21419
X1 (008)	1.44555	1.49434	1.49528
X1 (008)	1.57431	1.55202	1.55150
X2 (008)	1.20143	1.21891	1.21803
X3 (008)	0.58710	0.57811	0.57881
X5 (008)	1.27806	1.27785	1.27677
X4' (008)	0.89539	0.89837	0.89732
X5' (008)	1.44616	1.41058	1.41029
L1 (444)	0.44922	0.44772	0.44743
L1 (444)	1.42078	1.43499	1.43448
L3 (444)	0.78301	0.80852	0.80982
L3 (444)	1.22634	1.22089	1.22099
L2' (444)	0.41155	0.41868	0.41825
L2' (444)	1.32382	1.29283	1.29452
L3' (444)	1.91419	1.90918	1.90929
W1 (048)	0.87674	0.89271	0.89188
W1 (048)	1.40941	1.41295	1.41322
W2' (048)	0.28959	0.29279	0.29112
W2' (048)	1.43330	1.43862	1.43824
W2' (048)	1.90901	2.14051	2.14061
W1' (048)	1.27170	1.27883	1.27984
W3 (048)	0.70617	0.70225	0.70256
W3 (048)	1.40508	1.42304	1.42254
ODD (224)	0.86656	0.87675	0.87546
ODD (224)	1.18828	1.18204	1.18032
EVEN(224)	0.25924	0.25752	0.25773
EVEN(224)	0.61640	0.61672	0.61712
EVEN(224)	0.78986	0.78451	0.78381
EVEN(224)	1.09751	1.07315	1.07295
EVEN(224)	1.25839	1.25324	1.25379

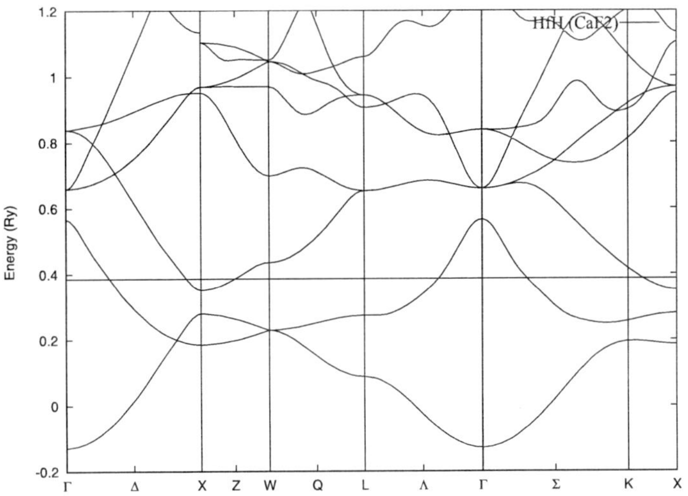

Fig. 8.4 Energy bands of HfH2 in the CaF2 structure

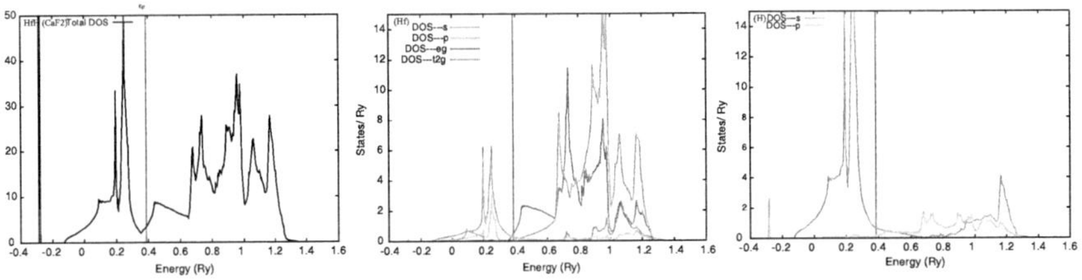

Fig. 8.5 Total, angular momentum and site decomposed densities of states of HfH2 in the CaF2 structure

8.2 Tantalum Hydride (TaH)

See Fig. 8.6 and Tables 8.6, 8.7, 8.8.

See Figs. 8.7, 8.8 and Tables 8.9, 8.10.

See Figs. 8.9 and 8.10.

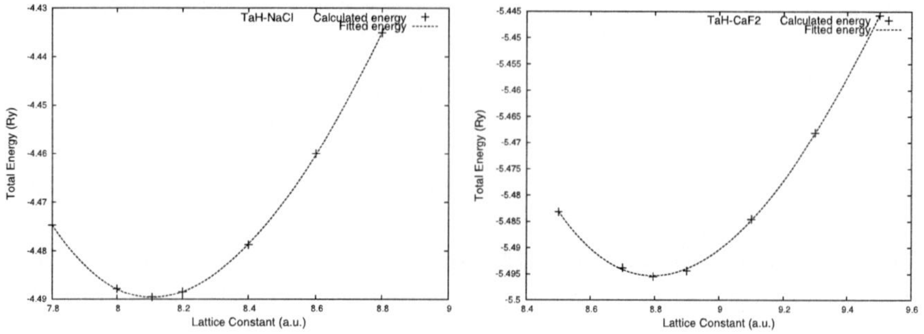

Fig. 8.6 Total energy versus lattice constant of TaH in the NaCl and CaF2 structures

Table 8.7 Birch fit coefficients

	A1	A2	A3	A4
NaCl	-1.796820E+00	-1.498577E+02	2.321288E+03	-6.377481E+03
CaF2	-3.142559E+00	-1.417625E+02	2.052543E+03	2.521549E+03

Table 8.6 Lattice constant, bulk modulus, gap, total energy

Stru	a (Bohr)	B (MBar)	Gap	Total Energy (Ry)
NaCl	8.11	2.29	-	-31234.48954
CaF2	8.79	1.88	-	-31235.49546

Table 8.8 DOS at Ef, Hopfield parameter, Stoner criterion

```
                TaH      a=8.11 Bohr        NaCl
-----------------------------------------------------------------------------
 Fermi Energy   Total DOS         Decomposed DOS inside the muffin tins
   (Ry)        (States/Ry)       s          p          eg          t2g          f
-----------------------------------------------------------------------------
Ta 0.918        11.353        0.046     0.240     2.821      3.336     0.039
 H 0.918        11.353        0.069     0.922     0.031      0.077     0.028
-----------------------------------------------------------------------------
Fermi-Velocity (cm/s): 0.49624 x10E8
Plasmon Energy (eV)  : 5.21525
Electron-ion interaction (Hopfield parameter) (eV/A^2) Ta: 3.342  H: 0.116
------------------------------------------------
  Ta     MUFFIN-TIN RADIUS and CHARGE =   2.2296   69.9407
   H     MUFFIN-TIN RADIUS and CHARGE =   1.8242    1.4026
          Ta        STONER       I = 0.0068
           H        STONER       I = 0.0016
         STONER PARAMETER (Ry) I = 0.0088
         STONER CRITERION    N*I = 0.0997
------------------------------------------------

                TaH2       a=8.79 Bohr       CaF2
-----------------------------------------------------------------------------
 Fermi Energy   Total DOS         Decomposed DOS inside the muffin tins
   (Ry)        (States/Ry)       s          p          eg          t2g          f
-----------------------------------------------------------------------------
Ta  0.745       24.454        0.375     0.185    10.199      3.589      0.107
H   0.745       24.454        0.118     1.292     0.079      0.045      0.032
-----------------------------------------------------------------------------
Fermi-Velocity (cm/s): 0.38831 x10E8
Plasmon Energy (eV)  : 5.29279
Electron-ion interaction (Hopfield parameter) (eV/A^2) Ta: 5.596   H: 0.519
------------------------------------------------
          Ta    MUFFIN-TIN RADIUS and CHARGE =    2.2850   70.0757
          H     MUFFIN-TIN RADIUS and CHARGE =    1.5233    1.0276
                   Ta        STONER   I = 0.0037
                    H        STONER   I = 0.0003
          STONER PARAMETER (Ry)       I = 0.0045
          STONER CRITERION          N*I = 0.1089
```

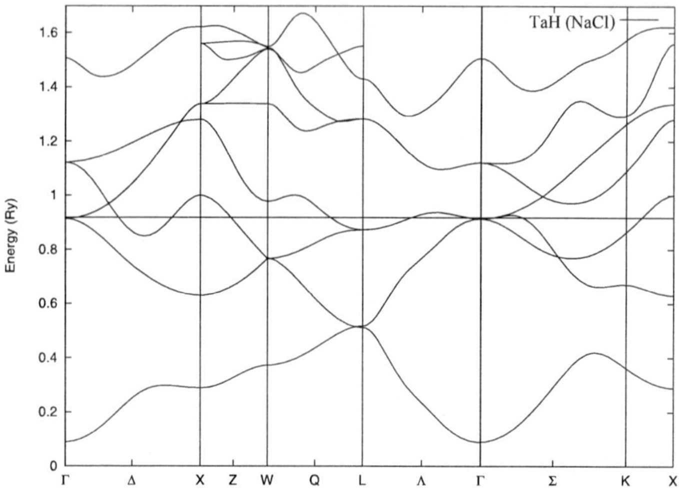

Fig. 8.7 Energy bands of TaH in the NaCl structure

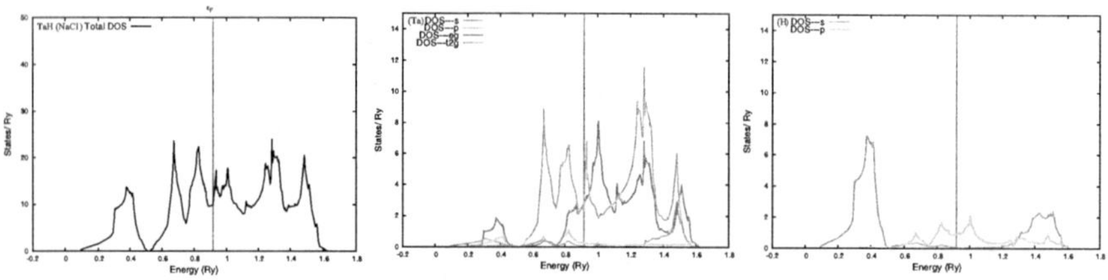

Fig. 8.8 Total, angular momentum and site decomposed densities of states of TaH in the NaCl structure

Table 8.9 TaH (NaCl) a = 8.11 Bohr Slater–Koster 3-center parameters

	ORTHOGONAL	NON-ORTHOGONAL	
	ENERGY INTEGRALS Ry	ENERGY INTEGRALS Ry	OVERLAP INTEGRALS
ON SITE Ta-Ta			
s,s(000)	1.15556	0.82686	
x,x(000)	1.70853	1.19825	
xy,xy(000)	1.14404	0.92464	
d2,d2(000)	1.13515	0.93727	
FIRST NEIGHBOR Ta-Ta			
s,s(110)	-0.03003	-0.03729	0.04984
s,x(110)	-0.04320	0.01428	0.10222
s,xy(110)	-0.06463	0.02456	0.11562
s,d2(110)	0.00887	-0.01615	0.01422
x,x(110)	0.07515	-0.17286	-0.17713
x,x(011)	0.00546	0.03507	0.08595
x,y(110)	0.06178	-0.03041	-0.11369
x,xy(110)	0.09543	-0.07758	-0.13090
x,xy(011)	-0.00040	0.01811	0.05962
z,d2(011)	0.01659	0.07544	0.05807
z,d1(011)	-0.00141	0.01577	-0.03068
xy,xy(110)	-0.07171	0.01049	0.08147
xy,xy(011)	0.01770	-0.00650	-0.03848
xy,xz(011)	-0.00809	-0.00305	-0.04338
xy,d2(110)	0.03356	0.01479	0.05032
d2,d2(110)	-0.00456	-0.01314	0.01629
d1,d1(110)	-0.02637	0.06014	-0.00407
SECOND NEIGHBOR Ta-Ta			
s,s(200)	0.06454	-0.00231	-0.03005
s,x(200)	0.08069	-0.03801	-0.04370
s,d2(002)	0.05293	0.03790	0.04552
x,x(200)	-0.08693	0.06727	0.04671
y,y(200)	0.01075	0.00916	0.00956
x,xy(020)	0.01723	-0.00153	0.00353
z,d2(002)	-0.06664	-0.08642	-0.05369
xy,xy(200)	-0.02391	0.00296	-0.00150
xy,xy(002)	0.00437	-0.00234	-0.00169
d2,d2(002)	0.05486	-0.06177	-0.05083
d1,d1(002)	0.00245	-0.00095	-0.00048
ON SITE H -H			
s,s(000)	0.54925	0.69831	
FIRST NEIGHBOR H -H			
s,s(110)	-0.00845	-0.00639	0.00112
s,x(110)	0.00000	0.00000	0.00000
x,x(110)	0.00000	0.00000	0.00000
x,x(011)	0.00000	0.00000	0.00000
x,y(110)	0.00000	0.00000	0.00000
SECOND NEIGHBOR H -H			
s,s(200)	-0.00942	0.00806	0.00688
s,x(200)	0.00000	0.00000	0.00000
x,x(200)	0.00000	0.00000	0.00000
y,y(200)	0.00000	0.00000	0.00000
FIRST NEIGHBOR Ta-H			
s,s(100)	-0.10545	-0.05732	0.09080
s,x(100)	0.00000	0.00000	0.00000
x,s(100)	0.07916	-0.05083	-0.22887
x,x(100)	0.00000	0.00000	0.00000
y,y(100)	0.00000	0.00000	0.00000
xy,x(010)	0.00000	0.00000	0.00000
d2,s(001)	-0.11266	-0.01334	-0.16258
d2,z(001)	0.00000	0.00000	0.00000

SECOND NEIGHBOR Ta-H

s,s(111)	0.00702	0.00361	0.01156
x,s(111)	-0.01176	-0.01476	-0.01562
s,x(111)	0.00000	0.00000	0.00000
x,x(111)	0.00000	0.00000	0.00000
x,y(111)	0.00000	0.00000	0.00000
xy,s(111)	0.00943	0.00703	0.01196
xy,x(111)	0.00000	0.00000	0.00000
yz,x(111)	0.00000	0.00000	0.00000
d1,x(111)	0.00000	0.00000	0.00000

TaH

	ORTHOGONAL			NON-ORTHOGONAL		
BAND	RMS ERROR	MAXIMUM DEVIATION		RMS ERROR	MAXIMUM DEVIATION	
	mRy	k	mRy	mRy	k	mRy
1	1.6	(044)	3.3	0.5	(066)	1.3
2	4.1	(003)	11.4	0.5	(008)	1.2
3	2.7	(444)	5.6	0.3	(444)	0.8
4	4.4	(048)	11.1	0.4	(224)	0.9
5	4.1	(048)	12.4	0.6	(044)	1.9
6	3.4	(055)	8.2	0.5	(022)	1.5
7	7.7	(000)	20.0	0.6	(033)	1.3
1-7	4.4			0.5		

ENERGY VALUES IN Ry AT SELECTED k-POINTS

	ORTHOGONAL	APW	NON-ORTHOGONAL
GAMMA 1	0.08769	0.08879	0.08879
GAMMA 1	1.48609	1.50613	1.50585
GAMMA 12	1.12149	1.12093	1.12159
GAMMA 25'	0.91185	0.91379	0.91381
X1 (008)	0.28918	0.28902	0.28985
X1 (008)	1.62623	1.62140	1.62162
X2 (008)	1.28172	1.28089	1.28107
X3 (008)	0.62868	0.63027	0.63151
X5 (008)	1.34398	1.33777	1.33859
X4' (008)	0.99828	0.99989	1.00003
X5' (008)	1.55584	1.56005	1.56018
L1 (444)	0.51238	0.51256	0.51290
L1 (444)	1.55162	1.55087	1.55044
L3 (444)	0.87874	0.87311	0.87394
L3 (444)	1.28311	1.28349	1.28334
L2' (444)	0.51519	0.51722	0.51651
L2' (444)	1.43579	1.42971	1.42994
W1 (048)	0.96647	0.97756	0.97740
W1 (048)	1.54417	1.53988	1.54004
W2' (048)	0.37320	0.37294	0.37309
W2' (048)	1.54006	1.54919	1.54984
W1' (048)	1.32650	1.33887	1.33902
W3 (048)	0.76725	0.76717	0.76706
W3 (048)	1.54881	1.54462	1.54474
ODD (224)	0.94680	0.94342	0.94248
ODD (224)	1.24819	1.24651	1.24628
EVEN(224)	0.34047	0.33742	0.33729
EVEN(224)	0.68835	0.68742	0.68754
EVEN(224)	0.84817	0.85191	0.85210
EVEN(224)	1.14496	1.14711	1.14818
EVEN(224)	1.38330	1.38257	1.38381

Table 8.10 TaH (NaCl) a = 8.11 Bohr Slater–Koster 2-center parameters

	ORTHOGONAL	NON-ORTHOGONAL	
	ENERGY INTEGRALS Ry	ENERGY INTEGRALS Ry	OVERLAP INTEGRALS
ON SITE Ta-Ta			
s	1.13370	0.77986	
p	1.59018	1.18042	
t2g	1.06262	0.93283	
eg	1.03225	0.92020	
FIRST NEIGHBOR Ta-Ta			
(sss)	-0.06210	-0.00776	0.09427
(pps)	0.12386	-0.19452	-0.28953
(ppp)	-0.01660	0.06283	0.08202
(dds)	-0.09499	0.01779	0.10768
(ddp)	0.03275	0.00763	-0.06163
(ddd)	0.00092	-0.00275	0.00737
(sps)	0.09736	-0.06607	-0.18855
(sds)	-0.08514	0.01492	0.10757
(pds)	-0.12391	0.08935	0.17700
(pdp)	0.01638	-0.01328	-0.06362
SECOND NEIGHBOR Ta-Ta			
(sss)	0.00664	0.00359	0.00412
(pps)	-0.02851	0.19022	0.07855
(ppp)	-0.00824	-0.01016	-0.00381
(dds)	0.00931	-0.03607	-0.01584
(ddp)	-0.00351	0.00127	-0.00171
(ddd)	0.00116	0.00352	0.00284
(sps)	-0.02021	0.03543	0.01058
(sds)	0.01729	-0.01427	-0.00158
(pds)	0.01613	-0.06845	-0.02758
(pdp)	0.00158	0.00115	-0.00199
ON SITE H-H			
s	1.05527	0.67780	
p	10.00000	10.00000	
FIRST NEIGHBOR H-H			
(sss)	-0.01090	0.01617	-0.02206
(sps)	0.00000	0.00000	0.00000
(pps)	0.00000	0.00000	0.00000
(ppp)	0.00000	0.00000	0.00000
SECOND NEIGHBOR H-H			
(sss)	0.03628	-0.02011	-0.01205
(sps)	0.00000	0.00000	0.00000
(pps)	0.00000	0.00000	0.00000
(ppp)	0.00000	0.00000	0.00000
FIRST NEIGHBOR Ta-H			
(sss)	0.10439	0.08678	-0.05307
(sps)	0.00000	0.00000	0.00000
(pss)	0.12906	0.01039	-0.20322
(pps)	0.00000	0.00000	0.00000
(ppp)	0.00000	0.00000	0.00000
(dpp)	0.00000	0.00000	0.00000
(dss)	0.14299	-0.00725	-0.20639
(dps)	0.00000	0.00000	0.00000
SECOND NEIGHBOR Ta-H			
(sss)	-0.00383	0.00814	0.02271
(pss)	-0.01561	0.04071	0.04015
(sps)	0.00000	0.00000	0.00000
(pps)	0.00000	0.00000	0.00000
(ppp)	0.00000	0.00000	0.00000
(dss)	-0.01558	0.02185	0.02837
(dps)	0.00000	0.00000	0.00000
(dpp)	0.00000	0.00000	0.00000

TaH

BAND	ORTHOGONAL RMS ERROR mRy	MAXIMUM DEVIATION k	mRy	NON-ORTHOGONAL RMS ERROR mRy	MAXIMUM DEVIATION k	mRy
1	6.7	(005)	19.4	0.8	(044)	2.9
2	8.7	(004)	19.3	1.5	(222)	3.6
3	10.9	(003)	24.3	1.2	(044)	3.0
4	8.5	(444)	18.4	1.0	(022)	2.2
5	11.8	(226)	32.1	1.4	(055)	3.6
6	12.9	(066)	51.4	1.0	(000)	2.1
7	17.5	(222)	47.4	1.4	(022)	3.4
1-7	11.5			1.2		

ENERGY VALUES IN Ry AT SELECTED k-POINTS

	ORTHOGONAL	APW	NON-ORTHOGONAL
GAMMA 1	0.09084	0.08879	0.08891
GAMMA 1	1.47969	1.50613	1.50564
GAMMA 12	1.12179	1.12093	1.12307
GAMMA 25'	0.90155	0.91379	0.91266
X1 (008)	0.29225	0.28902	0.28936
X1 (008)	1.57388	1.62140	1.62237
X2 (008)	1.26748	1.28089	1.28070
X3 (008)	0.63218	0.63027	0.63124
X5 (008)	1.33498	1.33777	1.33796
X4' (008)	1.00478	0.99989	0.99926
X5' (008)	1.56663	1.56005	1.55924
L1 (444)	0.50436	0.51256	0.51233
L1 (444)	1.53652	1.55087	1.55070
L3 (444)	0.85470	0.87311	0.87474
L3 (444)	1.28412	1.28349	1.28394
L2' (444)	0.50935	0.51722	0.51734
L2' (444)	1.44651	1.42971	1.42881
W1 (048)	0.98049	0.97756	0.97711
W1 (048)	1.53353	1.53988	1.53964
W2' (048)	0.36501	0.37294	0.37183
W2' (048)	1.55430	1.54919	1.54918
W1' (048)	1.33034	1.33887	1.33966
W3 (048)	0.77318	0.76717	0.76780
W3 (048)	1.54306	1.54462	1.54464
ODD (224)	0.93616	0.94342	0.94212
ODD (224)	1.24926	1.24651	1.24454
EVEN(224)	0.33786	0.33742	0.33800
EVEN(224)	0.68541	0.68742	0.68722
EVEN(224)	0.85533	0.85191	0.85167
EVEN(224)	1.16284	1.14711	1.14688
EVEN(224)	1.39516	1.38257	1.38319

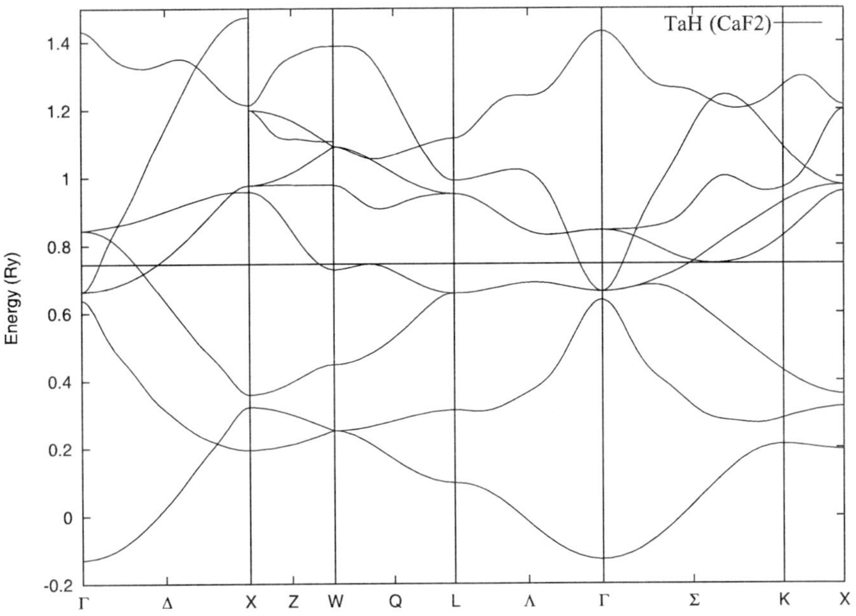

Fig. 8.9 Energy bands of TaH2 in the CaF2 structure

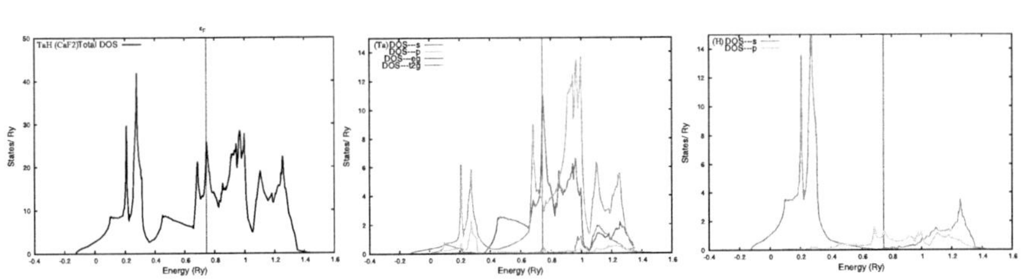

Fig. 8.10 Total, angular momentum and site decomposed densities of states of TaH2 in the CaF2 structure

8.3 Tungsten Hydride (WH)

See Fig. 8.11 and Tables 8.11, 8.12, 8.13.

See Figs. 8.12, 8.13 and Tables 8.14, 8.15.

See Figs. 8.14 and 8.15.

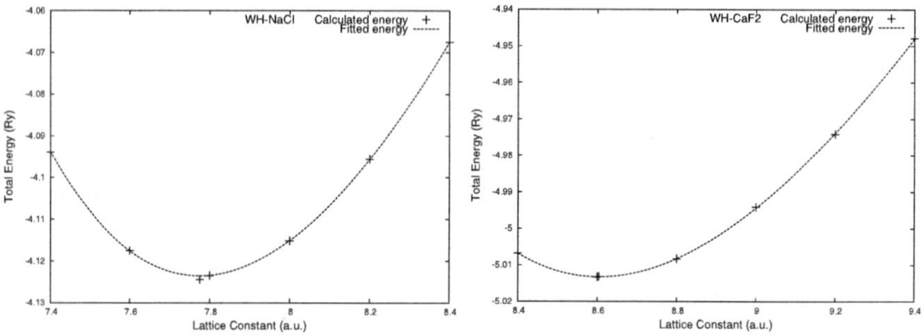

Fig. 8.11 Total energy versus lattice constant of WH in the NaCl and CaF2 structures

Table 8.11 Lattice constant, bulk modulus, gap, total energy

Stru	a (Bohr)	B (MBar)	Gap	Total Energy
NaCl	7.78	3.05	-	-32314.12443
CaF2	8.60	2.14	-	-32315.01315
exp (CaF2)	?			

Table 8.12 Birch fit coefficients

	A1	A2	A3	A4
NaCl	-1.450815E+00	-1.266224E+02	1.460679E+03	9.305390E+02
CaF2	-3.186650E+00	-8.434859E+01	2.262538E+02	1.983845E+04

Table 8.13 DOS at Ef, Hopfield parameter, Stoner criterion

```
              WH      a=7.78 Bohr       NaCl
--------------------------------------------------------------------------------
 Fermi Energy   Total DOS          Decomposed DOS inside the muffin tins
   (Ry)        (States/Ry)       s         p         eg         t2g         f
--------------------------------------------------------------------------------
 W 0.958        11.997         0.090     0.349     5.685      2.190      0.120
 H 0.958        11.997         0.029     0.748     0.026      0.013      0.012
--------------------------------------------------------------------------------
Fermi-Velocity (cm/s): 0.58351 x10E8
Plasmon Energy (eV)   : 6.71216
Electron-ion interaction (Hopfield parameter) (eV/A^2) W: 6.546  H: 0.088
-------------------------------------------------
   W     MUFFIN-TIN RADIUS and CHARGE =    2.3325    71.4691
   H     MUFFIN-TIN RADIUS and CHARGE =    1.5550     1.1004
           W               STONER I = 0.0057
           H               STONER I = 0.0011
         STONER PARAMETER (Ry)    I = 0.0070
         STONER CRITERION       N*I = 0.0839
-------------------------------------------------

              WH2      a=8.60 Bohr       CaF2
--------------------------------------------------------------------------------
 Fermi Energy   Total DOS          Decomposed DOS inside the muffin tins
   (Ry)        (States/Ry)       s         p         eg         t2g         f
--------------------------------------------------------------------------------
W   0.760       14.226         0.057     0.104     5.115      4.223      0.054
H   0.760       14.226         0.104     0.613     0.050      0.040      0.016
--------------------------------------------------------------------------------
Fermi-Velocity (cm/s): 0.69311 x10E8
Plasmon Energy (eV)   : 7.45828
Electron-ion interaction (Hopfield parameter) (eV/A^2) W: 3.855   H: 0.401
-------------------------------------------------
         W     MUFFIN-TIN RADIUS and CHARGE =    2.2355    70.9721
         H     MUFFIN-TIN RADIUS and CHARGE =    1.4903     1.0029
                   W        STONER I = 0.0056
                   H       STONER  I = 0.0002
         STONER PARAMETER (Ry)     I = 0.0062
         STONER CRITERION        N*I = 0.0875
```

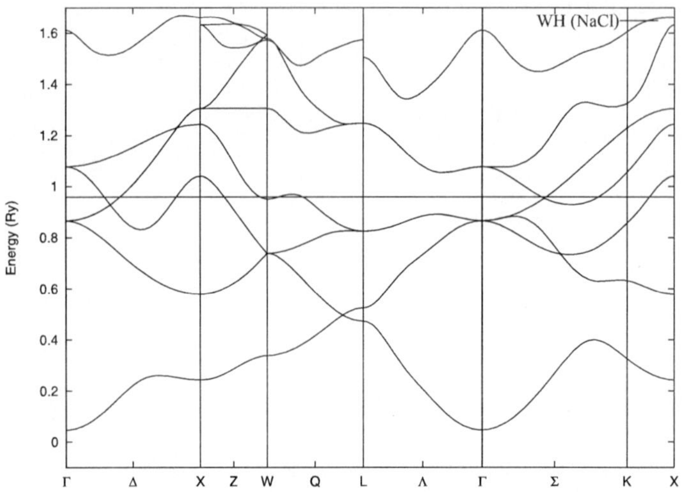

Fig. 8.12 Energy bands of WH in the NaCl structure

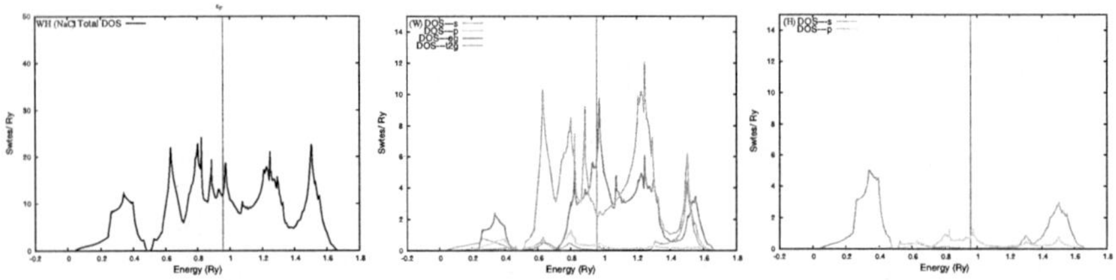

Fig. 8.13 Total, angular momentum and site decomposed densities of states of WH in the NaCl structure

Table 8.14 WH (NaCl) a = 7.78 Bohr Slater–Koster 3-center parameters

	ORTHOGONAL	NON-ORTHOGONAL	
	ENERGY INTEGRALS Ry	ENERGY INTEGRALS Ry	OVERLAP INTEGRALS
ON SITE W- W			
s,s(000)	1.19253	0.92327	
x,x(000)	1.76489	1.30294	
xy,xy(000)	1.10293	0.89607	
d2,d2(000)	1.10265	1.00327	
FIRST NEIGHBOR W- W			
s,s(110)	-0.02997	-0.05677	0.01324
s,x(110)	-0.04445	-0.00622	0.07326
s,xy(110)	-0.07210	0.01549	0.11627
s,d2(110)	0.00720	-0.01662	-0.00690
x,x(110)	0.07749	-0.16120	-0.16181
x,x(011)	0.00380	0.05851	0.10104
x,y(110)	0.06094	-0.00674	-0.08076
x,xy(110)	0.10115	-0.06153	-0.12003
x,xy(011)	-0.00078	0.01555	0.05726
z,d2(011)	0.01661	0.07461	0.07313
z,d1(011)	-0.00005	0.02484	-0.00389
xy,xy(110)	-0.07389	-0.00534	0.07043
xy,xy(011)	0.01798	-0.00142	-0.03337
xy,xz(011)	-0.00821	0.00129	-0.03814
xy,d2(110)	0.03497	0.00941	0.04991
d2,d2(110)	-0.00351	-0.01060	0.00387
d1,d1(110)	-0.03114	0.04094	0.02859
SECOND NEIGHBOR W- W			
s,s(200)	0.06715	0.00431	-0.04810
s,x(200)	0.08185	-0.03291	-0.06366
s,d2(002)	0.05692	0.02929	0.06586
x,x(200)	-0.08122	0.02199	0.03640
y,y(200)	0.01012	0.01795	0.01435
x,xy(020)	0.01652	0.00365	0.00737
z,d2(002)	-0.06861	-0.07687	-0.07054
xy,xy(200)	-0.02398	0.00136	-0.00289
xy,xy(002)	0.00406	-0.00113	-0.00080
d2,d2(002)	0.05808	-0.04226	-0.07068
d1,d1(002)	0.00301	0.00037	0.00027
ON SITE H -H			
s,s(000)	0.55402	0.68381	
FIRST NEIGHBOR H -H			
s,s(110)	-0.00767	0.03933	0.04793
s,x(110)	0.00000	0.00000	0.00000
x,x(110)	0.00000	0.00000	0.00000
x,x(011)	0.00000	0.00000	0.00000
x,y(110)	0.00000	0.00000	0.00000
SECOND NEIGHBOR H -H			
s,s(200)	-0.01177	0.01709	0.01573
s,x(200)	0.00000	0.00000	0.00000
x,x(200)	0.00000	0.00000	0.00000
y,y(200)	0.00000	0.00000	0.00000
FIRST NEIGHBOR W-H			
s,s(100)	-0.11264	-0.09060	0.04912
s,x(100)	0.00000	0.00000	0.00000
x,s(100)	0.08808	-0.04210	-0.20095
x,x(100)	0.00000	0.00000	0.00000
y,y(100)	0.00000	0.00000	0.00000
xy,x(010)	0.00000	0.00000	0.00000
d2,s(001)	-0.12896	0.08486	-0.04098
d2,z(001)	0.00000	0.00000	0.00000

SECOND NEIGHBOR W-H

s,s(111)	0.00412	0.00622	0.02369
x,s(111)	-0.00758	-0.00139	-0.01527
s,x(111)	0.00000	0.00000	0.00000
x,x(111)	0.00000	0.00000	0.00000
x,y(111)	0.00000	0.00000	0.00000
xy,s(111)	0.00568	0.00846	0.01565
xy,x(111)	0.00000	0.00000	0.00000
yz,x(111)	0.00000	0.00000	0.00000
d1,x(111)	0.00000	0.00000	0.00000

WH

	ORTHOGONAL			NON-ORTHOGONAL		
BAND	RMS ERROR	MAXIMUM DEVIATION		RMS ERROR	MAXIMUM DEVIATION	
	mRy	k	mRy	mRy	k	mRy
1	1.8	(033)	4.0	0.4	(174)	0.9
2	4.0	(003)	10.7	0.5	(111)	0.9
3	2.9	(062)	5.0	0.3	(222)	0.6
4	4.3	(048)	11.9	0.4	(062)	1.0
5	3.6	(048)	11.7	0.4	(002)	0.9
6	3.3	(044)	7.8	0.3	(442)	1.1
7	8.9	(000)	28.3	16.9	(000)	97.1
1-7	4.6			6.4		

ENERGY VALUES IN Ry AT SELECTED k-POINTS

	ORTHOGONAL	APW	NON-ORTHOGONAL
GAMMA 1	0.04443	0.04720	0.04744
GAMMA 1	1.58272	1.61102	1.61071
GAMMA 12	1.07802	1.07703	1.07742
GAMMA 25'	0.86344	0.86542	0.86517
X1 (008)	0.24234	0.24297	0.24307
X1 (008)	1.66553	1.66126	1.66129
X2 (008)	1.24469	1.24361	1.24353
X3 (008)	0.57573	0.57841	0.57915
X5 (008)	1.31071	1.30529	1.30561
X4' (008)	1.03821	1.04000	1.04012
X5' (008)	1.62774	1.63173	1.63169
L1 (444)	0.47162	0.47281	0.47272
L1 (444)	1.57436	1.57288	1.57261
L3 (444)	0.82988	0.82493	0.82539
L3 (444)	1.24735	1.24783	1.24742
L2' (444)	0.51889	0.52405	0.52371
L2' (444)	1.50505	1.50502	1.50506
W1 (048)	0.93827	0.95018	0.95007
W1 (048)	1.57608	1.57286	1.57297
W2' (048)	0.33819	0.33803	0.33820
W2' (048)	1.57111	1.57850	1.57886
W1' (048)	1.29445	1.30611	1.30607
W3 (048)	0.73885	0.73833	0.73809
W3 (048)	1.59276	1.59168	1.59180
ODD (224)	0.90080	0.89922	0.89848
ODD (224)	1.21083	1.20969	1.20992
EVEN(224)	0.31919	0.31588	0.31575
EVEN(224)	0.65627	0.65464	0.65437
EVEN(224)	0.80952	0.81128	0.81161
EVEN(224)	1.11569	1.11696	1.11731
EVEN(224)	1.42705	1.42794	1.42882

Table 8.15 WH (NaCl) a = 7.78 Bohr Slater–Koster 2-center parameters

	ORTHOGONAL	NON-ORTHOGONAL	
	ENERGY INTEGRALS Ry	ENERGY INTEGRALS Ry	OVERLAP INTEGRALS
ON SITE W- W			
s	1.15587	0.78560	
p	1.60950	1.23481	
t2g	1.01684	0.89435	
eg	0.98589	0.88046	
FIRST NEIGHBOR W- W			
(sss)	-0.07260	0.01697	0.12083
(pps)	0.11484	-0.23819	-0.32890
(ppp)	-0.03009	0.04010	0.06610
(dds)	-0.09752	0.00593	0.10337
(ddp)	0.03434	0.01512	-0.05612
(ddd)	0.00020	-0.00414	0.00605
(sps)	0.10143	-0.08667	-0.20624
(sds)	-0.08886	0.01446	0.10654
(pds)	-0.12291	0.09381	0.18869
(pdp)	0.02077	-0.00569	-0.05980
SECOND NEIGHBOR W- W			
(sss)	0.00440	0.02758	0.02375
(pps)	-0.03033	0.20982	0.08698
(ppp)	-0.01017	-0.00849	-0.00369
(dds)	0.00912	-0.03269	-0.01471
(ddp)	-0.00271	0.00209	-0.00073
(ddd)	0.00080	0.00274	0.00223
(sps)	-0.02344	0.00696	-0.00811
(sds)	0.01676	0.00043	0.01138
(pds)	0.01536	-0.07174	-0.03133
(pdp)	0.00428	0.00398	0.00083
ON SITE H-H			
s	1.12187	0.74887	
FIRST NEIGHBOR H-H			
(sss)	-0.00382	0.01217	-0.03031
(sps)	0.00000	0.00000	0.00000
(pps)	0.00000	0.00000	0.00000
(ppp)	0.00000	0.00000	0.00000
SECOND NEIGHBOR H-H			
(sss)	0.03936	-0.02610	-0.01338
(sps)	0.00000	0.00000	0.00000
(pps)	0.00000	0.00000	0.00000
(ppp)	0.00000	0.00000	0.00000
FIRST NEIGHBOR W-H			
(sss)	0.10314	0.13025	-0.01790
(sps)	0.00000	0.00000	0.00000
(pss)	0.13925	0.04509	-0.18111
(pps)	0.00000	0.00000	0.00000
(ppp)	0.00000	0.00000	0.00000
(dpp)	0.00000	0.00000	0.00000
(dss)	0.14921	0.01744	-0.19559
(dps)	0.00000	0.00000	0.00000
SECOND NEIGHBOR W-H			
(sss)	-0.00501	0.01015	0.02707
(pss)	-0.01092	0.03669	0.04809
(sps)	0.00000	0.00000	0.00000
(pps)	0.00000	0.00000	0.00000
(ppp)	0.00000	0.00000	0.00000
(dss)	-0.01242	0.01912	0.03266
(dps)	0.00000	0.00000	0.00000
(dpp)	0.00000	0.00000	0.00000

WH

BAND	ORTHOGONAL RMS ERROR mRy	ORTHOGONAL MAXIMUM DEVIATION k	ORTHOGONAL MAXIMUM DEVIATION mRy	NON-ORTHOGONAL RMS ERROR mRy	NON-ORTHOGONAL MAXIMUM DEVIATION k	NON-ORTHOGONAL MAXIMUM DEVIATION mRy
1	5.5	(005)	13.8	0.8	(044)	2.6
2	9.1	(264)	19.4	1.4	(003)	3.3
3	10.6	(006)	23.7	1.1	(044)	2.8
4	8.2	(444)	18.2	0.8	(022)	1.8
5	9.7	(002)	19.6	1.1	(055)	3.0
6	13.6	(066)	54.6	0.9	(174)	1.8
7	18.3	(222)	50.1	1.3	(022)	4.0
1-7	11.4			1.1		

ENERGY VALUES IN Ry AT SELECTED k-POINTS

	ORTHOGONAL	APW	NON-ORTHOGONAL
GAMMA 1	0.04645	0.04720	0.04707
GAMMA 1	1.57676	1.61102	1.60946
GAMMA 12	1.07630	1.07703	1.07850
GAMMA 25'	0.85338	0.86542	0.86427
X1 (008)	0.24466	0.24297	0.24363
X1 (008)	1.61977	1.66126	1.66147
X2 (008)	1.22975	1.24361	1.24333
X3 (008)	0.57709	0.57841	0.57835
X5 (008)	1.29998	1.30529	1.30520
X4' (008)	1.04876	1.04000	1.03978
X5' (008)	1.62850	1.63173	1.63170
L1 (444)	0.46301	0.47281	0.47208
L1 (444)	1.55598	1.57288	1.57283
L3 (444)	0.80672	0.82493	0.82625
L3 (444)	1.24376	1.24783	1.24817
L2' (444)	0.51069	0.52405	0.52414
L2' (444)	1.50617	1.50502	1.50495
W1 (048)	0.95606	0.95018	0.94958
W1 (048)	1.57510	1.57286	1.57298
W2' (048)	0.33517	0.33803	0.33766
W2' (048)	1.58167	1.57850	1.57897
W1' (048)	1.29677	1.30611	1.30618
W3 (048)	0.74522	0.73833	0.73918
W3 (048)	1.58348	1.59168	1.59165
ODD (224)	0.89276	0.89922	0.89826
ODD (224)	1.21197	1.20969	1.20816
EVEN(224)	0.31768	0.31588	0.31685
EVEN(224)	0.65420	0.65464	0.65492
EVEN(224)	0.81463	0.81128	0.81082
EVEN(224)	1.13313	1.11696	1.11658
EVEN(224)	1.43916	1.42794	1.42906

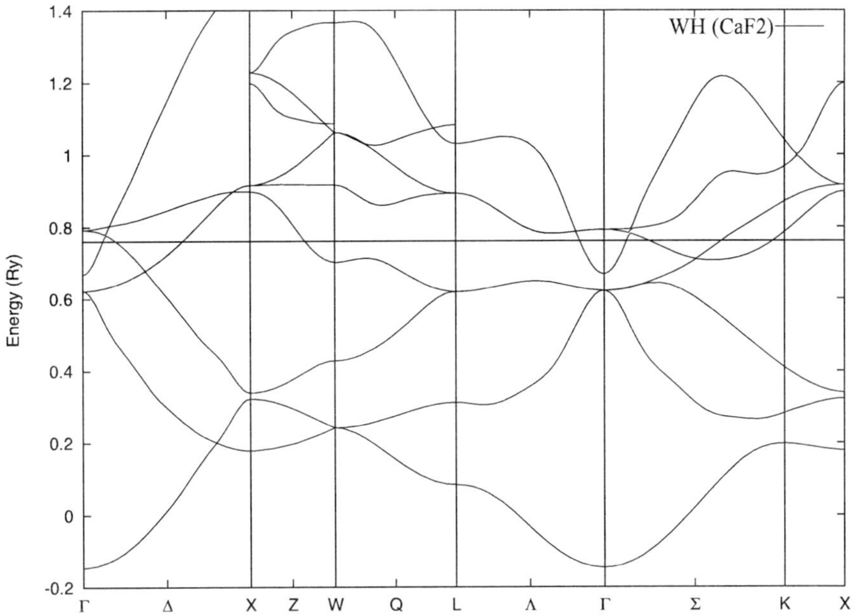

Fig. 8.14 Energy bands of WH2 in the CaF2 structure

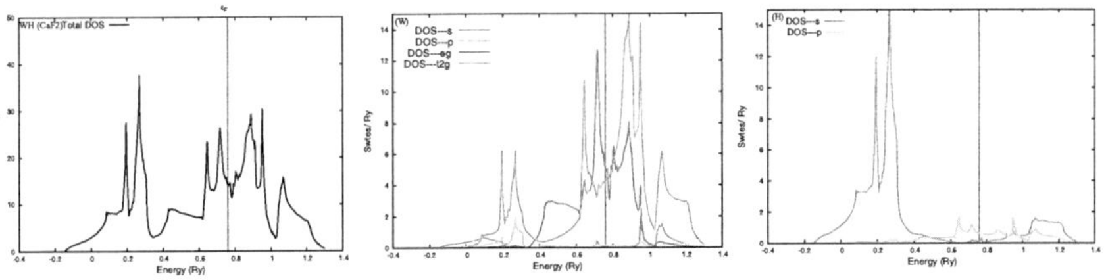

Fig. 8.15 Total, angular momentum and site decomposed densities of states of WH2 in the CaF2 structure

8.4 Rhenium Hydride (ReH)

See Fig. 8.16 and Tables 8.16, 8.17, 8.18.

See Figs. 8.17, 8.18 and Tables 8.19, 8.20.

See Figs. 8.19 and 8.20.

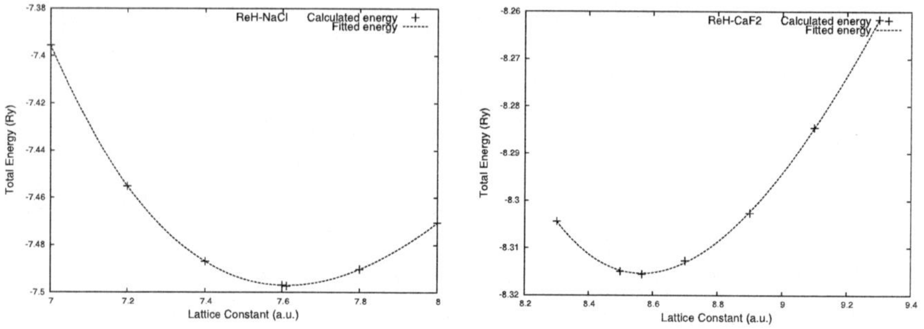

Fig. 8.16 Total energy versus lattice constant of ReH in the NaCl and CaF2 structures

Table 8.16 Lattice constant, bulk modulus, gap, total energy

Stru	a (Bohr)	B (MBar)	Gap	Total Energy
NaCl	7.61	3.53	-	-33417.49727
CaF2	8.57	2.07	-	-33418.31554

Table 8.17 Birch fit coefficients

	A1	A2	A3	A4
NaCl	-4.916929E+00	-1.094418E+02	9.410859E+02	4.858250E+03
CaF2	-6.790688E+00	-6.071390E+01	-3.435331E+02	2.383866E+04

Table 8.18 DOS at Ef, Hopfield parameter, Stoner criterion

```
                ReH      a=7.61 Bohr           NaCl
 ---------------------------------------------------------------------------
  Fermi Energy   Total DOS        Decomposed DOS inside the muffin tins
    (Ry)        (States/Ry)       s          p         eg         t2g         f
 ---------------------------------------------------------------------------
  Re 1.016        8.695        0.025     0.173     3.847      2.542      0.078
   H 1.016        8.695        0.023     0.248     0.059      0.016      0.007
 ---------------------------------------------------------------------------
 Fermi-Velocity (cm/s): 1.16181 x10E8
 Plasmon Energy (eV)  : 11.74759
 Electron-ion interaction (Hopfield parameter) (eV/A^2) Re: 5.101  H: 0.041
 ---------------------------------------------
    Re     MUFFIN-TIN RADIUS and CHARGE =   2.2834   72.4109
     H     MUFFIN-TIN RADIUS and CHARGE =   1.5223    1.0925
             Re             STONER I = 0.0080
             H              STONER I = 0.0004
           STONER PARAMETER (Ry)  I = 0.0085
           STONER CRITERION     N*I = 0.0739
 ---------------------------------------------

                ReH2     A=8.57 Bohr           CaF2
 ---------------------------------------------------------------------------
  Fermi Energy   Total DOS        Decomposed DOS inside the muffin tins
    (Ry)         (States/Ry)       s         p         eg         t2g         f
 ---------------------------------------------------------------------------
 Re  0.737        18.751        0.020     0.078      6.262      8.154       0.044
 H   0.737        18.751        0.115     0.527      0.049      0.087       0.025
 ---------------------------------------------------------------------------
 Fermi-Velocity (cm/s): 0.68000 x10E8
 Plasmon Energy (eV)  : 8.45563
 Electron-ion interaction (Hopfield parameter) (eV/A^2) Re: 3.792   H: 0.325
 ---------------------------------------------
         Re     MUFFIN-TIN RADIUS and CHARGE =    2.2258    72.0443
          H     MUFFIN-TIN RADIUS and CHARGE =    1.4838     0.9746
                    Re        STONER I = 0.0084
                    H         STONER I = 0.0001
           STONER PARAMETER (Ry)      I = 0.0087
           STONER CRITERION         N*I = 0.1629
```

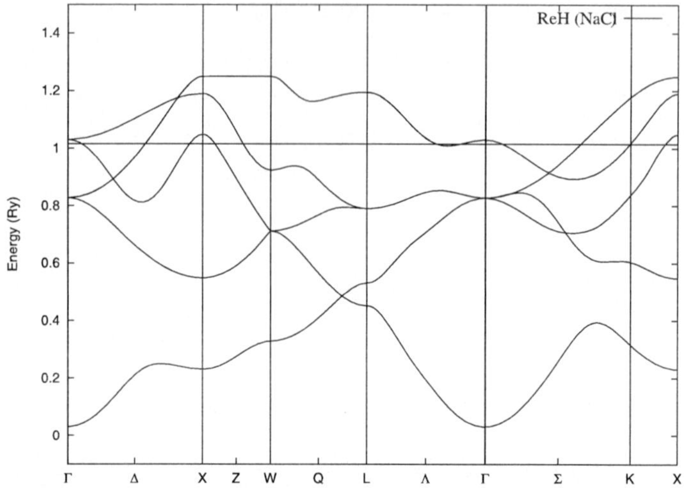

Fig. 8.17 Energy bands of ReH in the NaCl structure

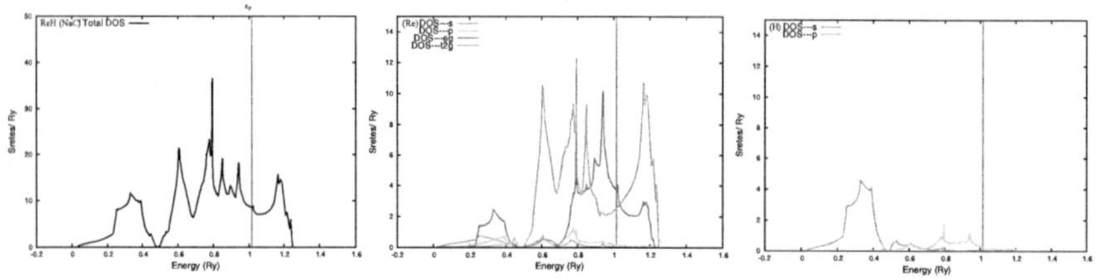

Fig. 8.18 Total, angular momentum and site decomposed densities of states of ReH in the NaCl structure

Table 8.19 ReH (NaCl) a = 7.61 Bohr Slater–Koster 3-center parameters

	ORTHOGONAL	NON-ORTHOGONAL	
	ENERGY INTEGRALS Ry	ENERGY INTEGRALS Ry	OVERLAP INTEGRALS
ON SITE Re-Re			
s,s(000)	1.23149	1.18845	
x,x(000)	1.76785	1.54610	
xy,xy(000)	1.05038	0.87899	
d2,d2(000)	1.08552	0.90589	
FIRST NEIGHBOR Re-Re			
s,s(110)	-0.02288	-0.05153	-0.01618
s,x(110)	-0.03930	-0.03671	0.03746
s,xy(110)	-0.07504	-0.00873	0.08477
s,d2(110)	0.00036	-0.05732	-0.04022
x,x(110)	0.07283	-0.09715	-0.11919
x,x(011)	-0.00464	0.04720	0.08288
x,y(110)	0.05085	0.00484	-0.05040
x,xy(110)	0.09971	-0.03328	-0.10105
x,xy(011)	-0.00250	0.00813	0.04507
z,d2(011)	0.01886	0.03960	0.05009
z,d1(011)	0.00590	0.04611	0.01295
xy,xy(110)	-0.07094	-0.03175	0.04452
xy,xy(011)	0.01749	-0.00061	-0.03077
xy,xz(011)	-0.00657	0.00513	-0.02942
xy,d2(110)	0.03436	0.00075	0.04174
d2,d2(110)	0.00141	-0.01163	0.01279
d1,d1(110)	-0.04401	0.05748	0.01399
SECOND NEIGHBOR Re-Re			
s,s(200)	0.06876	0.04106	-0.02693
s,x(200)	0.08491	0.02216	-0.04594
s,d2(002)	0.06104	0.02484	0.05965
x,x(200)	-0.07684	0.00067	0.07016
y,y(200)	0.00806	0.02295	0.01495
x,xy(020)	0.01493	-0.00276	0.00255
z,d2(002)	-0.07065	-0.06963	-0.06469
xy,xy(200)	-0.02190	0.00811	0.00225
xy,xy(002)	0.00341	-0.00208	-0.00154
d2,d2(002)	0.06433	-0.04996	-0.05392
d1,d1(002)	0.00271	-0.00019	0.00034
ON SITE H -H			
s,s(000)	0.50102	0.83657	
FIRST NEIGHBOR H -H			
s,s(110)	-0.00870	0.01646	0.00258
s,x(110)	0.00000	0.00000	0.00000
x,x(110)	0.00000	0.00000	0.00000
x,x(011)	0.00000	0.00000	0.00000
x,y(110)	0.00000	0.00000	0.00000
SECOND NEIGHBOR H -H			
s,s(200)	-0.01473	0.01363	0.01183
s,x(200)	0.00000	0.00000	0.00000
x,x(200)	0.00000	0.00000	0.00000
y,y(200)	0.00000	0.00000	0.00000
FIRST NEIGHBOR Re-H			
s,s(100)	-0.11322	-0.15688	-0.06330
s,x(100)	0.00000	0.00000	0.00000
x,s(100)	0.07373	0.11391	-0.03041
x,x(100)	0.00000	0.00000	0.00000
y,y(100)	0.00000	0.00000	0.00000
xy,x(010)	0.00000	0.00000	0.00000
d2,s(001)	-0.11566	0.01207	-0.12952
d2,z(001)	0.00000	0.00000	0.00000

SECOND NEIGHBOR Re-H

s,s(111)	0.00808	0.00048	0.01091
x,s(111)	-0.01412	-0.00897	-0.01055
s,x(111)	0.00000	0.00000	0.00000
x,x(111)	0.00000	0.00000	0.00000
x,y(111)	0.00000	0.00000	0.00000
xy,s(111)	0.00381	0.00998	0.00894
xy,x(111)	0.00000	0.00000	0.00000
yz,x(111)	0.00000	0.00000	0.00000
d1,x(111)	0.00000	0.00000	0.00000

ReH

	ORTHOGONAL			NON-ORTHOGONAL		
BAND	RMS ERROR	MAXIMUM DEVIATION		RMS ERROR	MAXIMUM DEVIATION	
	mRy	k	mRy	mRy	k	mRy
1	2.6	(033)	7.0	0.4	(333)	0.7
2	4.6	(003)	10.6	0.3	(333)	0.9
3	3.2	(264)	7.4	0.3	(222)	0.6
4	3.9	(048)	9.8	0.4	(055)	1.0
5	3.3	(048)	10.1	0.4	(000)	0.8
6	3.6	(044)	8.5	0.4	(226)	0.9
7	9.2	(000)	32.4	0.7	(033)	1.7
1-7	4.8			0.4		

ENERGY VALUES IN Ry AT SELECTED k-POINTS

	ORTHOGONAL	APW	NON-ORTHOGONAL
GAMMA 1	0.02681	0.03117	0.03072
GAMMA 1	1.65084	1.68327	1.68338
GAMMA 12	1.03102	1.02977	1.03057
GAMMA 25'	0.82583	0.82685	0.82678
X1 (008)	0.22915	0.23086	0.23145
X1 (008)	1.65333	1.64914	1.64844
X1 (008)	1.85396	1.85655	1.85651
X2 (008)	1.19019	1.18951	1.18933
X3 (008)	0.54592	0.54843	0.54901
X5 (008)	1.25337	1.24911	1.24936
X4' (008)	1.04523	1.04808	1.04830
X5' (008)	1.66497	1.67003	1.66991
L1 (444)	0.45037	0.45143	0.45171
L1 (444)	1.55226	1.55206	1.55156
L3 (444)	0.79343	0.78985	0.79002
L3 (444)	1.19584	1.19469	1.19420
L2' (444)	0.52154	0.53099	0.53080
L2' (444)	1.55033	1.55212	1.55220
W1 (048)	0.91342	0.92326	0.92341
W1 (048)	1.57080	1.56937	1.57016
W2' (048)	0.32746	0.32876	0.32876
W2' (048)	1.55943	1.56724	1.56779
W1' (048)	1.23971	1.24985	1.25009
W3 (048)	0.71312	0.71193	0.71192
W3 (048)	1.59696	1.59722	1.59723
ODD (224)	0.86286	0.86279	0.86255
ODD (224)	1.16049	1.15904	1.15949
EVEN(224)	0.31126	0.30674	0.30672
EVEN(224)	0.63494	0.63361	0.63329
EVEN(224)	0.78141	0.78047	0.78058
EVEN(224)	1.07872	1.07866	1.07920
EVEN(224)	1.44676	1.44985	1.45098

Table 8.20 ReH (NaCl) a = 7.61 Bohr Slater–Koster 2-center parameters

	ORTHOGONAL	NON-ORTHOGONAL	
	ENERGY INTEGRALS Ry	ENERGY INTEGRALS Ry	OVERLAP INTEGRALS
ON SITE Re-Re			
s	1.16510	0.77834	
p	1.64307	1.22488	
t2g	0.97025	0.86154	
eg	0.94085	0.84947	
FIRST NEIGHBOR Re-Re			
(sss)	-0.07472	-0.00960	0.10204
(pps)	0.11162	-0.24360	-0.32568
(ppp)	-0.04145	0.04635	0.07678
(dds)	-0.09395	-0.00280	0.09526
(ddp)	0.03419	0.01838	-0.05075
(ddd)	0.00003	-0.00501	0.00489
(sps)	0.10787	-0.06772	-0.19737
(sds)	-0.08855	0.00209	0.09960
(pds)	-0.11992	0.08160	0.17931
(pdp)	0.02148	-0.00088	-0.05628
SECOND NEIGHBOR Re-Re			
(sss)	0.00428	0.00218	0.00558
(pps)	-0.04901	0.21370	0.08702
(ppp)	-0.01075	-0.01356	-0.00525
(dds)	0.00785	-0.02867	-0.01269
(ddp)	-0.00195	0.00289	0.00004
(ddd)	0.00018	0.00227	0.00189
(sps)	-0.02698	0.03904	0.01054
(sds)	0.01182	-0.01101	0.00133
(pds)	0.01253	-0.06162	-0.02362
(pdp)	0.00400	0.00509	0.00123
ON SITE H-H			
s	1.18380	0.76801	
FIRST NEIGHBOR H-H			
(sss)	-0.00550	0.01712	-0.02430
(sps)	0.00000	0.00000	0.00000
(pps)	0.00000	0.00000	0.00000
(ppp)	0.00000	0.00000	0.00000
SECOND NEIGHBOR H-H			
(sss)	0.04720	-0.01945	-0.01081
(sps)	0.00000	0.00000	0.00000
(pps)	0.00000	0.00000	0.00000
(ppp)	0.00000	0.00000	0.00000
FIRST NEIGHBOR Re-H			
(sss)	0.11171	0.10865	-0.04370
(sps)	0.00000	0.00000	0.00000
(pss)	0.14356	0.03190	-0.19824
(pps)	0.00000	0.00000	0.00000
(ppp)	0.00000	0.00000	0.00000
(dpp)	0.00000	0.00000	0.00000
(dss)	0.14611	0.02535	-0.18616
(dps)	0.00000	0.00000	0.00000
SECOND NEIGHBOR Re-H			
(sss)	-0.00872	0.00716	0.02302
(pss)	-0.02908	0.03757	0.04121
(sps)	0.00000	0.00000	0.00000
(pps)	0.00000	0.00000	0.00000
(ppp)	0.00000	0.00000	0.00000
(dss)	-0.01268	0.01696	0.02688
(dps)	0.00000	0.00000	0.00000
(dpp)	0.00000	0.00000	0.00000

ReH

	ORTHOGONAL			NON-ORTHOGONAL		
BAND	RMS ERROR	MAXIMUM DEVIATION		RMS ERROR	MAXIMUM DEVIATION	
	mRy	k	mRy	mRy	k	mRy
1	6.8	(354)	18.2	0.7	(044)	2.4
2	8.4	(264)	21.1	1.2	(222)	3.4
3	10.3	(264)	24.6	1.0	(044)	2.5
4	8.4	(444)	18.1	0.7	(022)	1.4
5	8.3	(044)	17.0	1.0	(055)	2.8
6	11.8	(066)	46.9	0.7	(224)	1.5
7	21.4	(000)	66.3	1.4	(022)	3.3
1-7	11.7			1.0		

ENERGY VALUES IN Ry AT SELECTED k-POINTS

	ORTHOGONAL	APW	NON-ORTHOGONAL
GAMMA 1	0.03099	0.03117	0.03138
GAMMA 1	1.66427	1.68327	1.68285
GAMMA 12	1.02933	1.02977	1.03113
GAMMA 25'	0.81789	0.82685	0.82589
X1 (008)	0.23663	0.23086	0.23142
X1 (008)	1.61332	1.64914	1.64865
X1 (008)	1.84164	1.85655	1.85651
X2 (008)	1.17411	1.18951	1.18906
X3 (008)	0.54410	0.54843	0.54855
X5 (008)	1.24463	1.24911	1.24926
X4' (008)	1.05557	1.04808	1.04751
X5' (008)	1.66783	1.67003	1.66999
L1 (444)	0.43838	0.45143	0.45164
L1 (444)	1.54209	1.55206	1.55194
L3 (444)	0.77179	0.78985	0.79088
L3 (444)	1.19096	1.19469	1.19494
L2' (444)	0.52073	0.53099	0.53097
L2' (444)	1.55168	1.55212	1.55206
W1 (048)	0.93391	0.92326	0.92276
W1 (048)	1.56602	1.56937	1.56963
W2' (048)	0.32357	0.32876	0.32807
W2' (048)	1.57283	1.56724	1.56728
W1' (048)	1.24390	1.24985	1.24973
W3 (048)	0.71893	0.71193	0.71262
W3 (048)	1.58794	1.59722	1.59720
ODD (224)	0.85703	0.86279	0.86198
ODD (224)	1.16098	1.15904	1.15758
EVEN(224)	0.30723	0.30674	0.30719
EVEN(224)	0.63696	0.63361	0.63339
EVEN(224)	0.78442	0.78047	0.78037
EVEN(224)	1.09409	1.07866	1.07792
EVEN(224)	1.45812	1.44985	1.45048

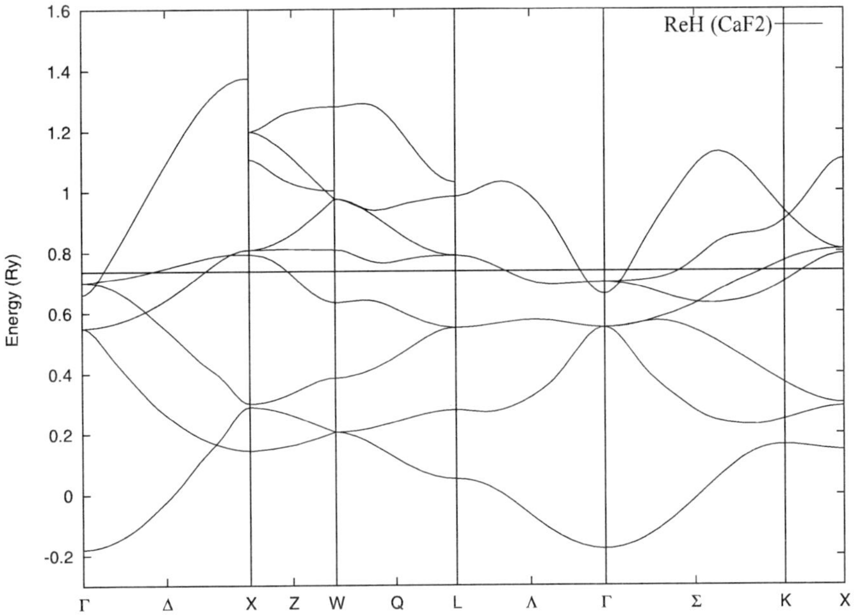

Fig. 8.19 Energy bands of ReH2 in the CaF2 structure

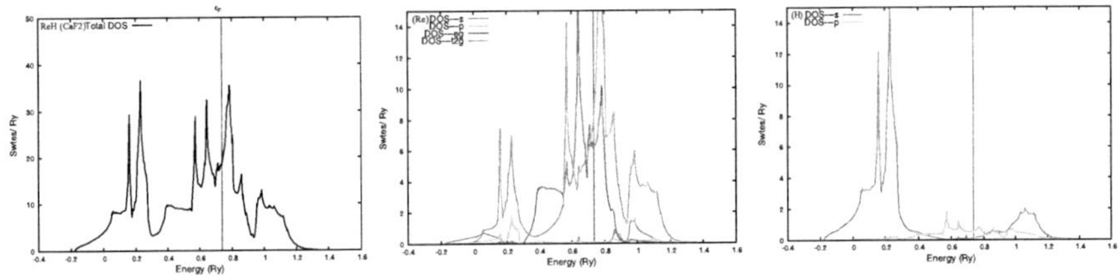

Fig. 8.20 Total, angular momentum and site decomposed densities of states of ReH2 in the CaF2 structure

8.5 Osmium Hydride (OsH)

See Fig. 8.21 and Tables 8.21, 8.22, 8.23.

See Figs. 8.22, 8.23 and Tables 8.24, 8.25.

See Figs. 8.24 and 8.25.

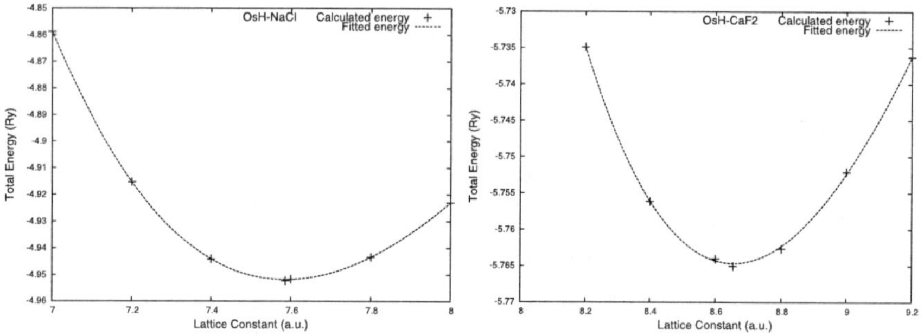

Fig. 8.21 Total energy versus lattice constant of OsH in the NaCl and CaF2 structures

Table 8.21 Lattice constant, bulk modulus, gap, total energy

Stru	a (Bohr)	B (MBar)	Gap	Total Energy
NaCl	7.59	3.49	-	-34544.95219
CaF2	8.65	1.79	-	-34545.76504
exp (CaF2)	?			

Table 8.22 Birch fit coefficients

	A1	A2	A3	A4
NaCl	-2.817075E+00	-7.967002E+01	2.985347E+02	9.307940E+03
CaF2	-4.159090E+00	-7.721377E+01	3.344582E+02	1.611380E+04

Table 8.23 DOS at Ef, Hopfield parameter, Stoner criterion

OsH a=7.59 Bohr NaCl

Fermi Energy (Ry)	Total DOS (States/Ry)	Decomposed DOS inside the muffin tins s	p	eg	t2g	f
Os 1.031	11.205	0.015	0.067	2.682	7.093	0.064
H 1.031	11.205	0.021	0.094	0.030	0.013	0.016

```
Fermi-Velocity (cm/s): 0.72543 x10E8
Plasmon Energy (eV)  : 8.36694
Electron-ion interaction (Hopfield parameter) (eV/A^2) Os: 4.790  H: 0.011
---------------------------------------------------
  Os     MUFFIN-TIN RADIUS and CHARGE =   2.2759   73.4986
  H      MUFFIN-TIN RADIUS and CHARGE =   1.5173    1.0574
          Os            STONER I = 0.0126
          H             STONER I = 0.0001
        STONER PARAMETER (Ry) I = 0.0128
        STONER CRITERION    N*I = 0.1428
---------------------------------------------------
```

OsH2 a=8.65 Bohr CaF2

Fermi Energy (Ry)	Total DOS (States/Ry)	Decomposed DOS inside the muffin tins s	p	eg	t2g	f
Os 0.646	42.982	0.068	0.080	12.297	23.680	0.036
H 0.646	42.982	0.138	0.781	0.034	0.261	0.071

```
Fermi-Velocity (cm/s): 0.42238 x10E8
Plasmon Energy (eV)  : 7.83149
Electron-ion interaction (Hopfield parameter) (eV/A^2) Os: 3.186  H: 0.303
---------------------------------------------------
    Os     MUFFIN-TIN RADIUS and CHARGE =    2.2485   73.2609
     H     MUFFIN-TIN RADIUS and CHARGE =    1.4990    0.9359
                Os        STONER I = 0.0108
                 H        STONER I = 0.0001
    STONER PARAMETER (Ry)        I = 0.0109
    STONER CRITERION           N*I = 0.4697
```

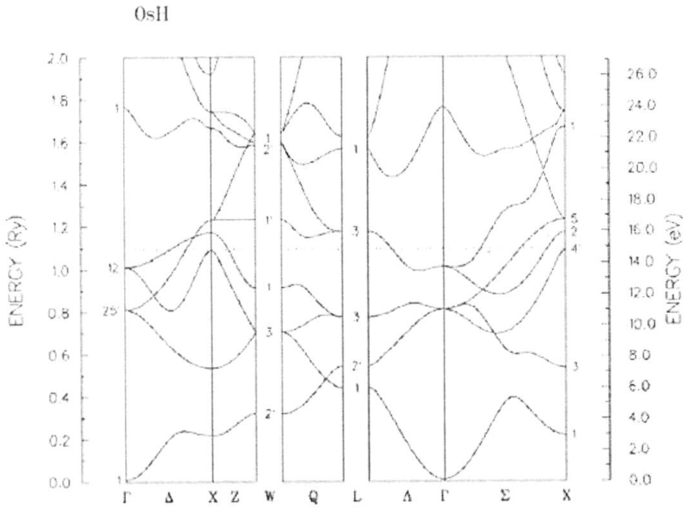

Fig. 8.22 Energy bands of OsH in the NaCl structure (tight-binding)

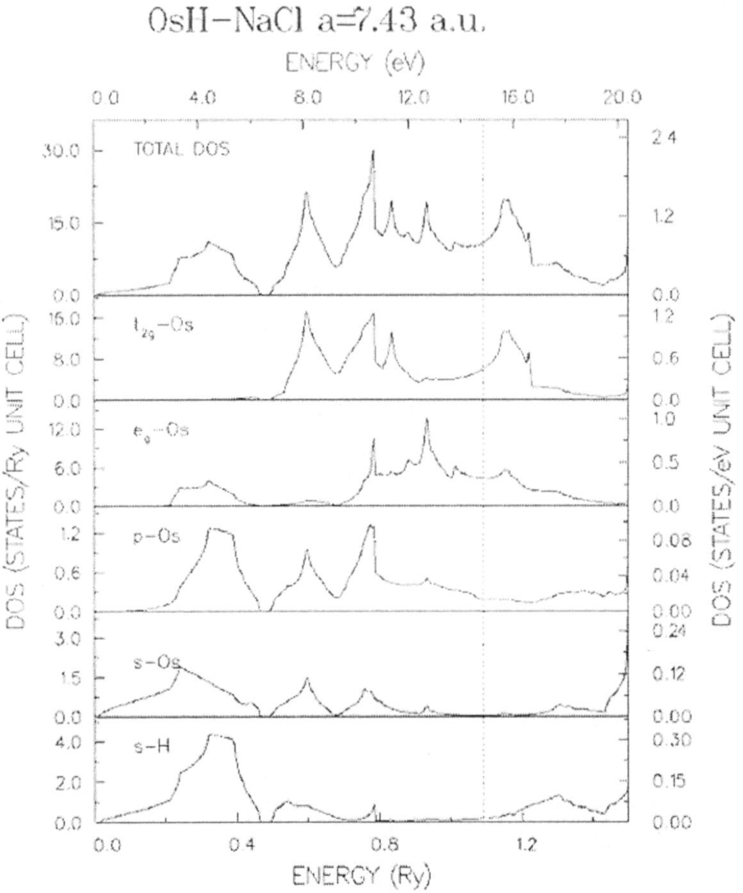

Fig. 8.23 Total, angular momentum and site decomposed densities of states of OsH in the NaCl structure (tight-binding)

Table 8.24 OsH lattice constant = 7.41160 a.u. Slater–Koster 3-center parameters

	ORTHOGONAL	NON-ORTHOGONAL	
	ENERGY INTEGRALS Ry	ENERGY INTEGRALS Ry	OVERLAP INTEGRALS
ON SITE Os-Os			
s,s(000)	1.26561	1.13914	
x,x(000)	1.91722	1.53929	
xy,xy(000)	1.03637	0.85698	
d2,d2(000)	1.08121	0.88319	
FIRST NEIGHBOR Os-Os			
s,s(110)	-0.01844	-0.05636	-0.00753
s,x(110)	-0.04708	-0.01925	0.05706
s,xy(110)	-0.07748	0.00176	0.09842
s,d2(110)	0.00000	-0.05131	-0.03234
x,x(110)	0.06653	-0.16040	-0.16128
x,x(011)	-0.01690	0.06199	0.09794
x,y(110)	0.07684	-0.04784	-0.07482

x,xy(110)	0.10136	-0.05873	-0.12280
x,xy(011)	-0.00422	0.00482	0.04848
z,d2(011)	0.02510	0.04947	0.05639
z,d1(011)	0.00579	0.03278	0.00324
xy,xy(110)	-0.07013	-0.01950	0.05539
xy,xy(011)	0.01714	0.00341	-0.02595
xy,xz(011)	-0.00828	0.00762	-0.02924
xy,d2(110)	0.03539	-0.00063	0.04066
d2,d2(110)	0.00360	-0.01618	0.00976
d1,d1(110)	-0.04987	0.06030	0.01441
SECOND NEIGHBOR Os-Os			
s,s(200)	0.07428	0.04823	-0.02592
s,x(200)	0.09548	0.04882	-0.03109
s,d2(002)	0.06295	0.02136	0.05277
x,x(200)	-0.14579	-0.07754	0.03258
y,y(200)	0.01606	0.03204	0.01304
x,xy(020)	0.01651	0.00682	0.00688
z,d2(002)	-0.06676	-0.06051	-0.05781
xy,xy(200)	-0.02219	0.00098	-0.00373
xy,xy(002)	0.00307	-0.00018	-0.00011
d2,d2(002)	0.06692	-0.04519	-0.04857
d1,d1(002)	0.00242	-0.00090	-0.00059
ON SITE H-H			
s,s(000)	0.53729	0.85697	
FIRST NEIGHBOR H-H			
s,s(110)	-0.01572	0.00381	-0.01277
s,x(110)	0.00000	0.00000	0.00000
x,x(110)	0.00000	0.00000	0.00000
x,x(011)	0.00000	0.00000	0.00000
x,y(110)	0.00000	0.00000	0.00000
SECOND NEIGHBOR H-H			
s,s(200)	-0.01625	0.01118	0.01297
s,x(200)	0.00000	0.00000	0.00000
x,x(200)	0.00000	0.00000	0.00000
y,y(200)	0.00000	0.00000	0.00000
FIRST NEIGHBOR Os-H			
s,s(100)	-0.13025	-0.16999	-0.05169
s,x(100)	0.00000	0.00000	0.00000
x,s(100)	0.09338	0.16662	0.00156
x,x(100)	0.00000	0.00000	0.00000
y,y(100)	0.00000	0.00000	0.00000
xy,x(010)	0.00000	0.00000	0.00000
d2,s(001)	-0.11349	-0.00361	-0.14843
d2,z(001)	0.00000	0.00000	0.00000
SECOND NEIGHBOR Os-H			
s,s(111)	0.02290	0.02148	0.02237
x,s(111)	-0.04318	-0.03343	-0.02220
s,x(111)	0.00000	0.00000	0.00000
x,x(111)	0.00000	0.00000	0.00000
x,y(111)	0.00000	0.00000	0.00000
xy,s(111)	0.00830	0.01483	0.01189
xy,x(111)	0.00000	0.00000	0.00000
yz,x(111)	0.00000	0.00000	0.00000
d1,x(111)	0.00000	0.00000	0.00000

OsH

BAND	ORTHOGONAL RMS ERROR mRy	MAXIMUM DEVIATION k	mRy	NON-ORTHOGONAL RMS ERROR mRy	MAXIMUM DEVIATION k	mRy
1	6.4	(048)	15.0	0.5	(333)	1.3
2	7.0	(222)	23.5	0.5	(333)	1.1
3	6.2	(044)	13.5	0.4	(226)	1.5
4	6.3	(048)	16.3	0.4	(033)	1.2
5	5.1	(003)	10.3	0.4	(033)	1.1
6	7.7	(174)	16.2	0.6	(174)	2.5
7	10.3	(000)	31.0	1.1	(033)	2.6
1-7	7.2			0.6		

ENERGY VALUES IN Ry AT SELECTED k-POINTS

	ORTHOGONAL	APW	NON-ORTHOGONAL
GAMMA 1	0.00937	0.00904	0.00844
GAMMA 1	1.73175	1.76279	1.76246
GAMMA 12	1.01161	1.01152	1.01215
GAMMA 15	2.15451	2.18024	2.18019
GAMMA 25'	0.81036	0.81103	0.81077
X1 (008)	0.22597	0.21651	0.21661
X1 (008)	1.66800	1.66595	1.66593
X1 (008)	1.98943	1.91923	1.91869
X2 (008)	1.16786	1.16963	1.16993
X3 (008)	0.53609	0.53373	0.53416
X5 (008)	1.23429	1.22916	1.22966
X4' (008)	1.09002	1.08728	1.08722
X5' (008)	1.75750	1.74254	1.74262
L1 (444)	0.43557	0.43957	0.44003
L1 (444)	1.56964	1.56352	1.56355
L3 (444)	0.77614	0.77520	0.77507
L3 (444)	1.18288	1.17555	1.17547
L2' (444)	0.53011	0.54377	0.54380
L2' (444)	1.63445	1.62399	1.62397
L3' (444)	2.45192	2.35657	2.35667
W1 (048)	0.89942	0.91570	0.91617
W1 (048)	1.59565	1.59651	1.59697
W2' (048)	0.30483	0.31981	0.31960
W2' (048)	1.57130	1.58452	1.58426
W2' (048)	2.51127	2.68748	2.68730
W1' (048)	1.22201	1.22987	1.22987
W3 (048)	0.70247	0.70554	0.70570
W3 (048)	1.63097	1.64214	1.64196
ODD (224)	0.85884	0.84896	0.84874
ODD (224)	1.14686	1.14100	1.14092
EVEN(224)	0.31149	0.29946	0.29902
EVEN(224)	0.63103	0.62675	0.62624
EVEN(224)	0.76509	0.76926	0.76901
EVEN(224)	1.06805	1.06803	1.06841
EVEN(224)	1.49299	1.49030	1.49215

FERMI LEVEL QUANTITIES (Non-orthogonal fit)

ENERGY Ry	DENSITIES OF STATES (states/Ry/cell) Total	s-Os	p-Os	t2g-Os	eg-Os	s-H	p-H
1.0917	10.76	0.02	0.18	6.02	4.34	0.19	0.00

	INTEGRATED DENSITIES OF STATES (electrons) Total	s-Os	p-Os	t2g-Os	eg-Os	s-H	p-H
	9.00	0.63	0.46	4.17	2.72	1.01	0.00

VELOCITY cm/s	PLASMON ENERGY eV	EIGENVALUE SUM Ry
0.81x10E8	9.40	-3.5740

Table 8.25 OsH (NaCl) a = 7.41 Bohr Slater–Koster 2-center parameters

	ORTHOGONAL	NON-ORTHOGONAL	
	ENERGY INTEGRALS Ry	ENERGY INTEGRALS Ry	OVERLAP INTEGRALS
ON SITE Os-Os			
s	1.18009	0.81517	
p	1.86656	1.27657	
t2g	0.96480	0.85433	
eg	0.93223	0.84271	
FIRST NEIGHBOR Os-Os			
(sss)	-0.07396	-0.02433	0.08712
(pps)	0.15167	-0.20998	-0.28849
(ppp)	-0.02155	0.07971	0.08840
(dds)	-0.09356	-0.01237	0.08600
(ddp)	0.03056	0.01958	-0.04701
(ddd)	0.00179	-0.00378	0.00533
(sps)	0.12423	-0.04321	-0.17726
(sds)	-0.09172	-0.00867	0.09202
(pds)	-0.13482	0.05762	0.15515
(pdp)	0.01615	-0.00928	-0.06050
SECOND NEIGHBOR Os-Os			
(sss)	0.00847	-0.00592	-0.00240
(pps)	-0.09292	0.17367	0.04676
(ppp)	0.00482	-0.00807	-0.00236
(dds)	0.00909	-0.02893	-0.01418
(ddp)	-0.00406	0.00277	0.00013
(ddd)	0.00105	0.00208	0.00168
(sps)	-0.03625	0.04767	0.01401
(sds)	0.01245	-0.01435	-0.00298
(pds)	0.01595	-0.05917	-0.02244
(pdp)	-0.00336	0.00211	-0.00087
ON SITE H-H			
s	1.19403	0.75692	
FIRST NEIGHBOR H-H			
(sss)	-0.00899	0.01565	-0.02415
(sps)	0.00000	0.00000	0.00000
(pps)	0.00000	0.00000	0.00000
(ppp)	0.00000	0.00000	0.00000
SECOND NEIGHBOR H-H			
(sss)	0.05333	-0.01563	-0.00832
(sps)	0.00000	0.00000	0.00000
(pps)	0.00000	0.00000	0.00000
(ppp)	0.00000	0.00000	0.00000
FIRST NEIGHBOR Os-H			
(sss)	0.13262	0.10058	-0.05899
(sps)	0.00000	0.00000	0.00000
(pss)	0.14943	0.02429	-0.20640
(pps)	0.00000	0.00000	0.00000
(ppp)	0.00000	0.00000	0.00000
(dpp)	0.00000	0.00000	0.00000
(dss)	0.14929	0.03007	-0.18326
(dps)	0.00000	0.00000	0.00000
SECOND NEIGHBOR Os-H			
(sss)	-0.01323	0.00312	0.01914
(pss)	-0.05089	0.03431	0.02973
(sps)	0.00000	0.00000	0.00000
(pps)	0.00000	0.00000	0.00000
(ppp)	0.00000	0.00000	0.00000
(dss)	-0.01536	0.01789	0.02375
(dps)	0.00000	0.00000	0.00000
(dpp)	0.00000	0.00000	0.00000

OsH

BAND	ORTHOGONAL RMS ERROR mRy	MAXIMUM DEVIATION k	MAXIMUM DEVIATION mRy	NON-ORTHOGONAL RMS ERROR mRy	MAXIMUM DEVIATION k	MAXIMUM DEVIATION mRy
1	9.1	(005)	19.8	0.6	(044)	2.3
2	10.4	(264)	24.9	1.1	(222)	2.7
3	11.2	(264)	29.6	1.1	(044)	2.8
4	8.2	(444)	19.2	0.8	(022)	1.5
5	8.6	(224)	16.9	0.9	(055)	2.7
6	11.6	(066)	41.4	0.8	(044)	1.6
7	20.9	(004)	49.8	1.5	(022)	3.1
1-7	12.1			1.0		

ENERGY VALUES IN Ry AT SELECTED k-POINTS

	ORTHOGONAL	APW	NON-ORTHOGONAL
GAMMA 1	0.00402	0.00904	0.00878
GAMMA 1	1.74557	1.76279	1.76163
GAMMA 12	1.01373	1.01152	1.01272
GAMMA 15	2.13430	2.18024	2.18015
GAMMA 25'	0.80117	0.81103	0.81040
X1 (008)	0.22689	0.21651	0.21698
X1 (008)	1.62820	1.66595	1.66525
X1 (008)	1.95680	1.91923	1.91867
X2 (008)	1.15608	1.16963	1.16965
X3 (008)	0.54236	0.53373	0.53384
X5 (008)	1.22957	1.22916	1.22942
X4' (008)	1.09336	1.08728	1.08647
X5' (008)	1.78621	1.74254	1.74241
L1 (444)	0.43047	0.43957	0.44005
L1 (444)	1.56262	1.56352	1.56358
L3 (444)	0.75596	0.77520	0.77637
L3 (444)	1.18234	1.17555	1.17580
L2' (444)	0.53943	0.54377	0.54416
L2' (444)	1.67484	1.62399	1.62378
L3' (444)	2.37952	2.35657	2.35661
W1 (048)	0.91959	0.91570	0.91463
W1 (048)	1.58774	1.59651	1.59701
W2' (048)	0.31052	0.31981	0.31959
W2' (048)	1.59031	1.58452	1.58603
W2' (048)	2.42990	2.68748	2.68755
W1' (048)	1.22539	1.22987	1.22937
W3 (048)	0.70475	0.70554	0.70629
W3 (048)	1.61793	1.64214	1.64195
ODD (224)	0.84125	0.84896	0.84773
ODD (224)	1.14741	1.14100	1.13964
EVEN(224)	0.29949	0.29946	0.29935
EVEN(224)	0.63138	0.62675	0.62650
EVEN(224)	0.77466	0.76926	0.76923
EVEN(224)	1.08491	1.06803	1.06770
EVEN(224)	1.49460	1.49030	1.49017

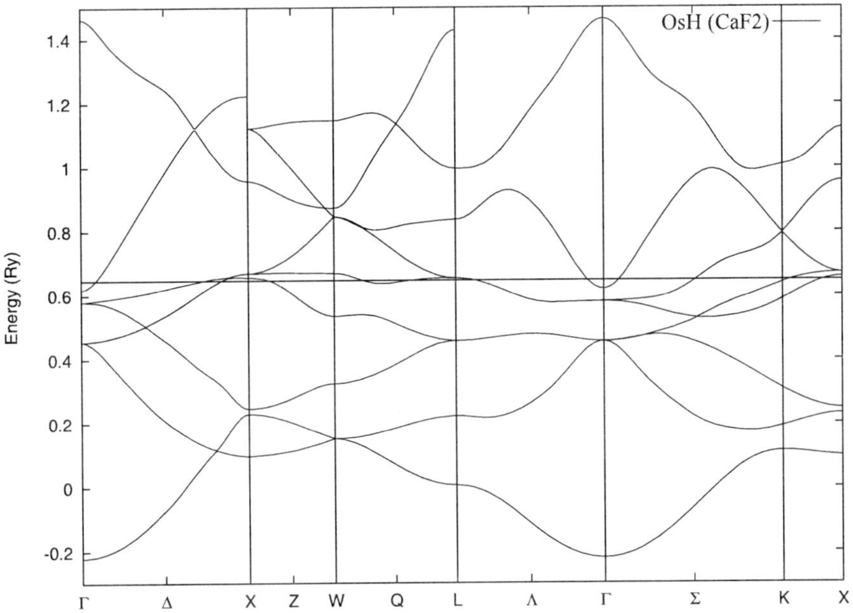

Fig. 8.24 Energy bands of OsH2 in the CaF2 structure

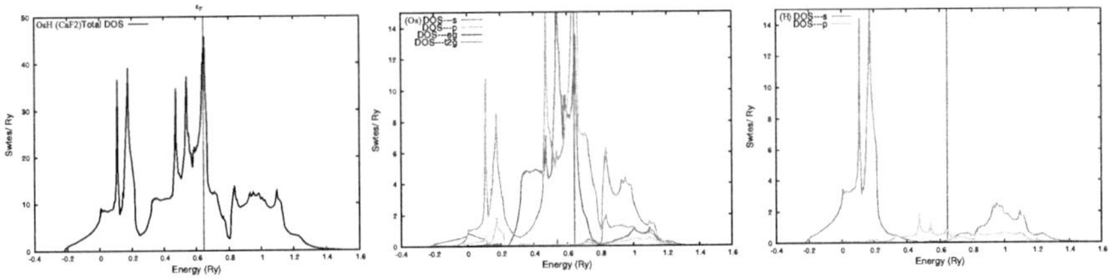

Fig. 8.25 Total, angular momentum and site decomposed densities of states of OsH2 in the CaF2 structure

8.6 Iridium Hydride (IrH)

See Fig. 8.26 and Tables 8.26, 8.27, 8.28.

See Figs. 8.27, 8.28 and Tables 8.29, 8.30.

See Figs. 8.29 and 8.30.

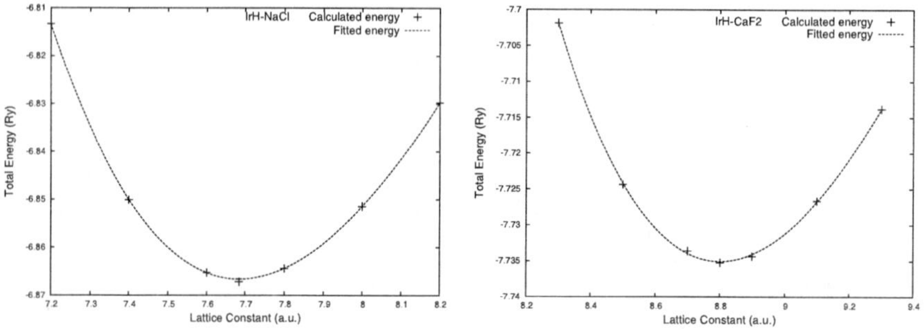

Fig. 8.26 Total energy versus lattice constant of IrH in the NaCl and CaF2 structures

Table 8.26 Lattice constant, bulk modulus, gap, total energy

Stru	a (Bohr)	B (MBar)	Gap	Total Energy
NaCl	7.68	3.01	-	-35696.86719
CaF2	8.80	1.56	-	-35697.73520
exp (CaF2)	?			

Table 8.27 Birch fit coefficients

	A1	A2	A3	A4
NaCl	-5.269325E+00	-5.109541E+01	-2.367832E+02	1.305267E+04
CaF2	-6.580987E+00	-4.383312E+01	-5.761828E+02	2.560929E+04

Table 8.28 DOS at Ef, Hopfield parameter, Stoner criterion

IrH a=7.68 Bohr NaCl

Fermi Energy (Ry)	Total DOS (States/Ry)	Decomposed DOS inside the muffin tins s	p	eg	t2g	f
Ir 0.889	26.683	0.064	0.136	6.978	17.348	0.068
H 0.889	26.683	0.110	0.170	0.051	0.012	0.037

Fermi-Velocity (cm/s): 0.63334 x10E8
Plasmon Energy (eV) : 11.05726
Electron-ion interaction (Hopfield parameter) (eV/A^2) Ir: 5.169 H: 0.038

Ir	MUFFIN-TIN RADIUS and CHARGE =	2.3053	74.7186
H	MUFFIN-TIN RADIUS and CHARGE =	1.5369	1.0047
Ir	STONER I = 0.0147		
H	STONER I = 0.0000		
STONER PARAMETER (Ry)	I = 0.0147		
STONER CRITERION	N*I = 0.3909		

IrH2 a=8.80 Bohr CaF2

Fermi Energy (Ry)	Total DOS (States/Ry)	Decomposed DOS inside the muffin tins s	p	eg	t2g	f
Ir 0.545	13.281	0.147	0.144	3.076	7.865	0.007
H 0.545	13.281	0.258	0.237	0.003	0.074	0.020

Fermi-Velocity (cm/s): 0.59365 x10E8
Plasmon Energy (eV) : 5.96686
Electron-ion interaction (Hopfield parameter) (eV/A^2) Ir: 0.683 H: 0.358

Ir	MUFFIN-TIN RADIUS and CHARGE =	2.2864	74.5502
H	MUFFIN-TIN RADIUS and CHARGE =	1.5243	0.8944
Ir	STONER I = 0.0113		
H	STONER I = 0.0001		
STONER PARAMETER (Ry)	I = 0.0116		
STONER CRITERION	N*I = 0.1535		

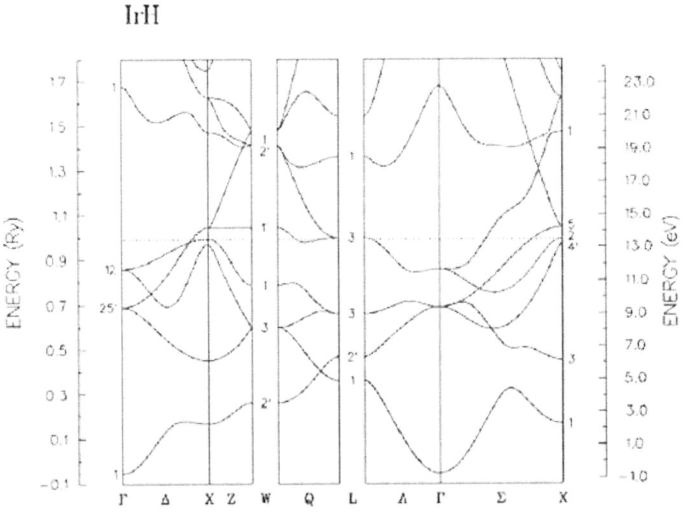

Fig. 8.27 Energy bands of IrH in the NaCl structure (tight-binding)

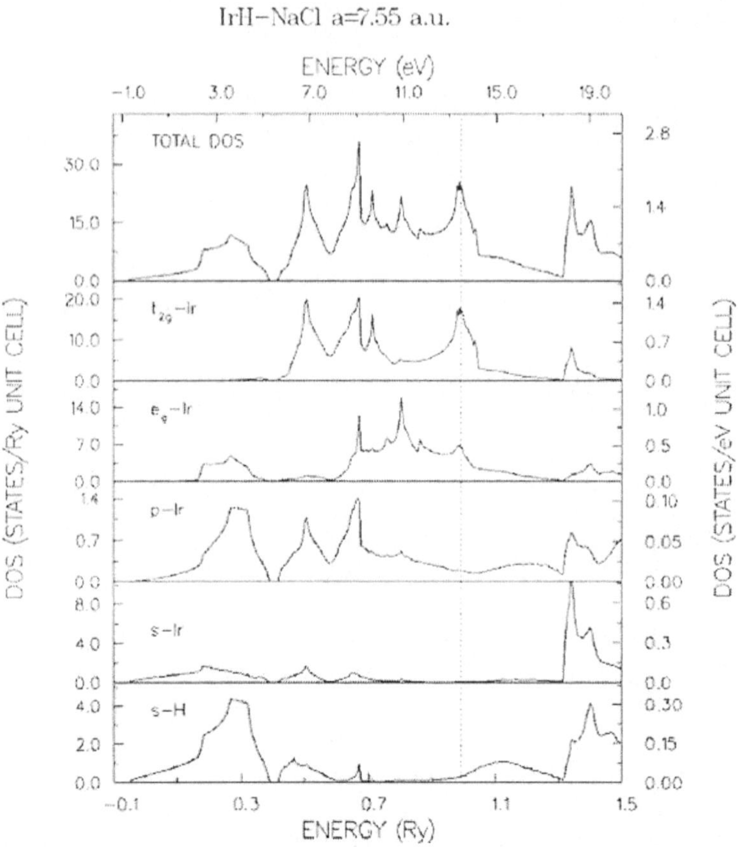

Fig. 8.28 Total, angular momentum and site decomposed densities of states of IrH in the NaCl structure (tight-binding)

Table 8.29 IrH (NaCl) a = 7.5 Bohr Slater–Koster 3-center parameters

	ORTHOGONAL	NON-ORTHOGONAL	
	ENERGY INTEGRALS Ry	ENERGY INTEGRALS Ry	OVERLAP INTEGRALS
ON SITE Ir-Ir			
s,s(000)	1.13185	1.03931	
x,x(000)	1.76304	1.43814	
xy,xy(000)	0.87178	0.74207	
d2,d2(000)	0.88961	0.76402	
FIRST NEIGHBOR Ir-Ir			
s,s(110)	-0.02393	-0.04688	-0.00785
s,x(110)	-0.04789	-0.02253	0.05333
s,xy(110)	-0.07143	-0.01622	0.08019
s,d2(110)	0.00165	-0.04519	-0.03541
x,x(110)	0.07108	-0.14173	-0.16145
x,x(011)	-0.01122	0.03845	0.09035
x,y(110)	0.07090	-0.03364	-0.06744
x,xy(110)	0.09099	-0.02854	-0.10482
x,xy(011)	-0.00336	0.00316	0.04663
z,d2(011)	0.01413	0.04206	0.05644
z,d1(011)	0.00502	0.03650	0.01145
xy,xy(110)	-0.05960	-0.02718	0.04349
xy,xy(011)	0.01509	0.00293	-0.02572
xy,xz(011)	-0.00322	0.00523	-0.02752
xy,d2(110)	0.02907	-0.00628	0.03326
d2,d2(110)	-0.00166	-0.01624	0.00535
d1,d1(110)	-0.02736	0.05169	0.02026
SECOND NEIGHBOR Ir-Ir			
s,s(200)	0.06259	0.03845	-0.02748
s,x(200)	0.08001	0.03749	-0.03480
s,d2(002)	0.04929	0.01661	0.05308
x,x(200)	-0.10637	-0.06367	0.03369
y,y(200)	0.00711	0.02649	0.01766
x,xy(020)	0.01283	-0.00008	0.00405
z,d2(002)	-0.05977	-0.05150	-0.05818
xy,xy(200)	-0.01710	0.00397	0.00103
xy,xy(002)	0.00209	-0.00207	-0.00187
d2,d2(002)	0.04660	-0.03869	-0.05138
d1,d1(002)	0.00254	-0.00112	-0.00118
ON SITE H-H			
s,s(000)	0.56505	0.76884	
FIRST NEIGHBOR H-H			
s,s(110)	-0.01210	0.01243	0.00495
s,x(110)	0.00000	0.00000	0.00000
x,x(110)	0.00000	0.00000	0.00000
x,x(011)	0.00000	0.00000	0.00000
x,y(110)	0.00000	0.00000	0.00000
SECOND NEIGHBOR H-H			
s,s(200)	-0.00988	0.00909	0.01295
s,x(200)	0.00000	0.00000	0.00000
x,x(200)	0.00000	0.00000	0.00000
y,y(200)	0.00000	0.00000	0.00000
FIRST NEIGHBOR Ir-H			
s,s(100)	-0.12391	-0.14869	-0.03776
s,x(100)	0.00000	0.00000	0.00000
x,s(100)	0.11270	0.15939	-0.00087
x,x(100)	0.00000	0.00000	0.00000
y,y(100)	0.00000	0.00000	0.00000
xy,x(010)	0.00000	0.00000	0.00000
d2,s(001)	-0.14808	0.02244	-0.11947
d2,z(001)	0.00000	0.00000	0.00000

SECOND NEIGHBOR Ir-H

s,s(111)	0.00100	0.00224	0.01582
x,s(111)	-0.00294	-0.01277	-0.01794
s,x(111)	0.00000	0.00000	0.00000
x,x(111)	0.00000	0.00000	0.00000
x,y(111)	0.00000	0.00000	0.00000
xy,s(111)	-0.00063	0.00872	0.01039
xy,x(111)	0.00000	0.00000	0.00000
yz,x(111)	0.00000	0.00000	0.00000
d1,x(111)	0.00000	0.00000	0.00000

IrH

	ORTHOGONAL			NON-ORTHOGONAL		
BAND	RMS ERROR	MAXIMUM DEVIATION		RMS ERROR	MAXIMUM DEVIATION	
	mRy	k	mRy	mRy	k	mRy
1	3.1	(222)	6.8	0.3	(066)	0.8
2	3.5	(003)	8.3	0.3	(222)	0.7
3	2.6	(062)	6.4	0.3	(226)	0.9
4	3.2	(226)	6.8	0.4	(033)	0.9
5	2.8	(048)	6.4	0.4	(000)	0.9
6	3.7	(226)	7.5	0.5	(174)	1.4
7	9.4	(000)	34.0	0.7	(333)	1.5
1-7	4.6			0.4		

ENERGY VALUES IN Ry AT SELECTED k-POINTS

	ORTHOGONAL	APW	NON-ORTHOGONAL
GAMMA 1	-0.06146	-0.05485	-0.05464
GAMMA 1	1.64223	1.67622	1.67635
GAMMA 12	0.86292	0.85983	0.86071
GAMMA 15	2.10248	2.10177	2.10175
GAMMA 25'	0.68990	0.69024	0.69021
X1 (008)	0.16593	0.16633	0.16700
X1 (008)	1.46864	1.47018	1.47072
X1 (008)	1.72064	1.74900	1.74879
X2 (008)	0.99223	0.99296	0.99370
X3 (008)	0.44842	0.45022	0.45042
X5 (008)	1.04596	1.04347	1.04366
X4' (008)	0.96524	0.96894	0.96895
X5' (008)	1.62361	1.62631	1.62625
L1 (444)	0.35362	0.35919	0.35954
L1 (444)	1.36446	1.36075	1.36073
L3 (444)	0.66393	0.66131	0.66140
L3 (444)	1.00138	0.99912	0.99879
L2' (444)	0.45887	0.46622	0.46572
L2' (444)	1.54559	1.54475	1.54478
L3' (444)	2.23094	2.23718	2.23725
W1 (048)	0.78421	0.78839	0.78917
W1 (048)	1.42579	1.41474	1.41520
W2' (048)	0.26370	0.26169	0.26176
W2' (048)	1.40608	1.41002	1.41029
W1' (048)	1.03760	1.04397	1.04424
W3 (048)	0.60128	0.60045	0.60049
W3 (048)	1.48030	1.48041	1.48023
ODD (224)	0.72761	0.72515	0.72499
ODD (224)	0.97351	0.97075	0.97106
EVEN(224)	0.22775	0.22444	0.22451
EVEN(224)	0.53956	0.53525	0.53487
EVEN(224)	0.65515	0.65735	0.65737
EVEN(224)	0.91527	0.91645	0.91627
EVEN(224)	1.35986	1.36519	1.36636

FERMI LEVEL QUANTITIES (Non-orthogonal fit)

ENERGY		DENSITIES OF STATES					
	Total	s-Ir	p-Ir	t2g-Ir	eg-Ir	s-H	p-H
Ry		states/Ry/cell					
0.9896	23.78	0.09	0.19	16.76	6.40	0.34	0.00
		INTEGRATED DENSITIES OF STATES					
	Total	s-Ir	p-Ir	t2g-Ir	eg-Ir	s-H	p-H
		electrons					
	10.00	0.58	0.47	4.84	3.15	0.96	0.00

VELOCITY	PLASMON ENERGY	EIGENVALUE SUM
cm/s	eV	Ry
0.52x10E8	8.79	-3.6335

Table 8.30 IrH (NaCl) a = 7.50 Bohr Slater–Koster 2-center parameters

	ORTHOGONAL	NON-ORTHOGONAL	
	ENERGY INTEGRALS Ry	ENERGY INTEGRALS Ry	OVERLAP INTEGRALS
ON SITE Ir-Ir			
s	1.05256	0.70611	
p	1.71560	1.13839	
t2g	0.81001	0.73042	
eg	0.78985	0.72172	
FIRST NEIGHBOR Ir-Ir			
(sss)	-0.07243	-0.03265	0.08365
(pps)	0.15382	-0.20333	-0.30839
(ppp)	-0.01230	0.06479	0.08809
(dds)	-0.07924	-0.01633	0.07861
(ddp)	0.02923	0.02019	-0.04196
(ddd)	-0.00028	-0.00413	0.00460
(sps)	0.11217	-0.01352	-0.16978
(sds)	-0.07996	-0.01827	0.08308
(pds)	-0.11588	0.04618	0.15361
(pdp)	0.01635	-0.00256	-0.05914
SECOND NEIGHBOR Ir-Ir			
(sss)	0.00749	-0.01817	-0.01012
(pps)	-0.05241	0.14367	0.03177
(ppp)	-0.00750	-0.00506	-0.00157
(dds)	0.00497	-0.02009	-0.00906
(ddp)	-0.00156	0.00322	0.00087
(ddd)	0.00034	0.00132	0.00132
(sps)	-0.02883	0.05919	0.02048
(sds)	0.01055	-0.01804	-0.00785
(pds)	0.01254	-0.04773	-0.01781
(pdp)	-0.00107	0.00336	0.00042
ON SITE H-H			
s	1.09270	0.73850	
FIRST NEIGHBOR H-H			
(sss)	-0.00016	0.00962	-0.01870
(sps)	0.00000	0.00000	0.00000
(pps)	0.00000	0.00000	0.00000
(ppp)	0.00000	0.00000	0.00000

SECOND NEIGHBOR H-H			
(sss)	0.04322	-0.00997	-0.00849
(sps)	0.00000	0.00000	0.00000
(pps)	0.00000	0.00000	0.00000
(ppp)	0.00000	0.00000	0.00000
FIRST NEIGHBOR Ir-H			
(sss)	0.11791	0.07857	-0.08166
(sps)	0.00000	0.00000	0.00000
(pss)	0.15054	0.04166	-0.21083
(pps)	0.00000	0.00000	0.00000
(ppp)	0.00000	0.00000	0.00000
(dpp)	0.00000	0.00000	0.00000
(dss)	0.13240	0.05623	-0.15117
(dps)	0.00000	0.00000	0.00000
SECOND NEIGHBOR Ir-H			
(sss)	-0.00879	0.00741	0.01640
(pss)	-0.02304	0.03747	0.02439
(sps)	0.00000	0.00000	0.00000
(pps)	0.00000	0.00000	0.00000
(ppp)	0.00000	0.00000	0.00000
(dss)	-0.00853	0.01508	0.01893
(dps)	0.00000	0.00000	0.00000
(dpp)	0.00000	0.00000	0.00000

IrH

	ORTHOGONAL			NON-ORTHOGONAL		
BAND	RMS ERROR	MAXIMUM DEVIATION		RMS ERROR	MAXIMUM DEVIATION	
	mRy	k	mRy	mRy	k	mRy
1	5.2	(354)	13.1	0.4	(044)	1.5
2	6.2	(264)	14.2	0.7	(222)	1.8
3	7.3	(264)	18.9	0.7	(044)	1.9
4	5.4	(444)	11.9	0.5	(022)	1.0
5	5.0	(002)	10.7	0.7	(055)	2.0
6	8.4	(066)	35.8	0.5	(226)	1.1
7	15.6	(000)	38.2	0.9	(042)	2.2
1-7	8.3			0.6		

ENERGY VALUES IN Ry AT SELECTED k-POINTS

	ORTHOGONAL	APW	NON-ORTHOGONAL
GAMMA 1	-0.05963	-0.05485	-0.05532
GAMMA 1	1.63803	1.67622	1.67543
GAMMA 12	0.86104	0.85983	0.86081
GAMMA 15	2.09769	2.10177	2.10161
GAMMA 25'	0.68223	0.69024	0.69001
X1 (008)	0.16984	0.16633	0.16671
X1 (008)	1.45251	1.47018	1.47006
X1 (008)	1.70792	1.74900	1.74858
X2 (008)	0.98439	0.99296	0.99288

X3 (008)	0.45065	0.45022	0.45028
X5 (008)	1.04244	1.04347	1.04355
X4' (008)	0.96551	0.96894	0.96855
X5' (008)	1.62997	1.62631	1.62623
L1 (444)	0.34838	0.35919	0.35996
L1 (444)	1.35675	1.36075	1.36075
L3 (444)	0.64936	0.66131	0.66197
L3 (444)	0.99915	0.99912	0.99936
L2' (444)	0.45917	0.46622	0.46657
L2' (444)	1.56015	1.54475	1.54478
L3' (444)	2.18264	2.23718	2.23719
W1 (048)	0.79354	0.78839	0.78774
W1 (048)	1.42972	1.41474	1.41514
W2' (048)	0.25719	0.26169	0.26122
W2' (048)	1.40621	1.41002	1.41026
W1' (048)	1.04107	1.04397	1.04390
W3 (048)	0.60254	0.60045	0.60111
W3 (048)	1.47799	1.48041	1.48028
ODD (224)	0.72082	0.72515	0.72475
ODD (224)	0.97291	0.97075	0.96991
EVEN(224)	0.22508	0.22444	0.22411
EVEN(224)	0.54145	0.53525	0.53493
EVEN(224)	0.66168	0.65735	0.65716
EVEN(224)	0.92439	0.91645	0.91589
EVEN(224)	1.36571	1.36519	1.36521

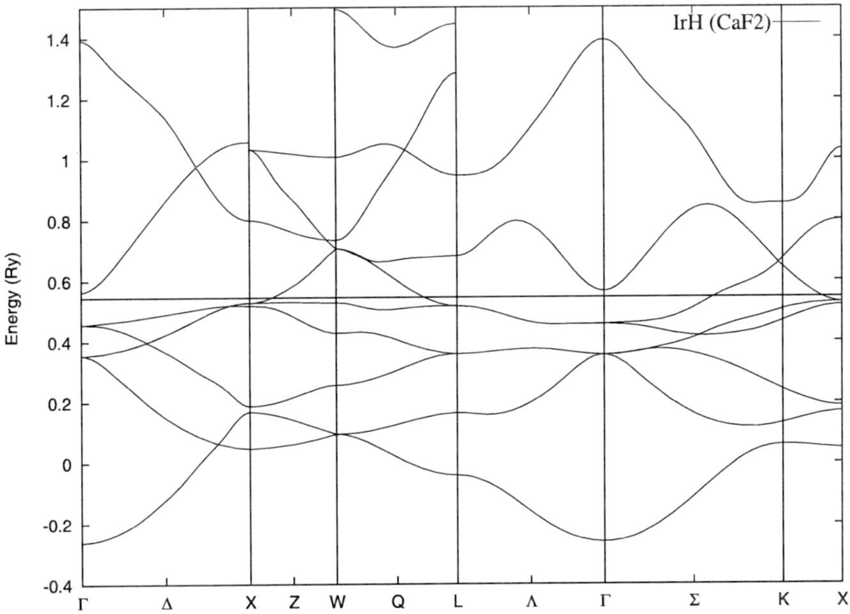

Fig. 8.29 Energy bands of IrH2 in the CaF2 structure

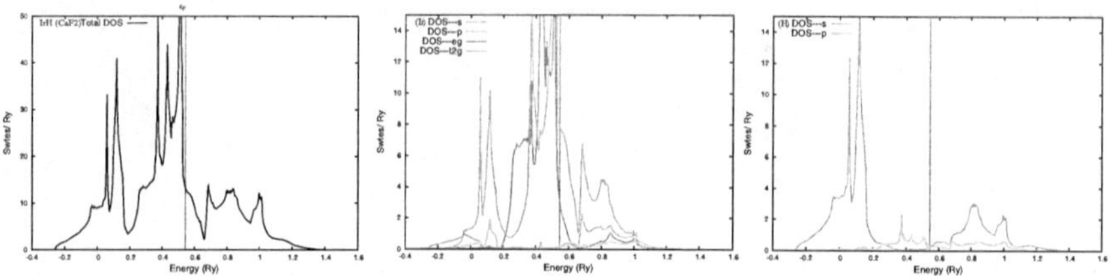

Fig. 8.30 Total, angular momentum and site decomposed densities of states of IrH2 in the CaF2 structure

8.7 Platinum Hydride (PtH)

See Fig. 8.31 and Tables 8.31, 8.32, 8.33.

See Figs. 8.32, 8.33 and Tables 8.34, 8.35.

See Figs. 8.34 and 8.35.

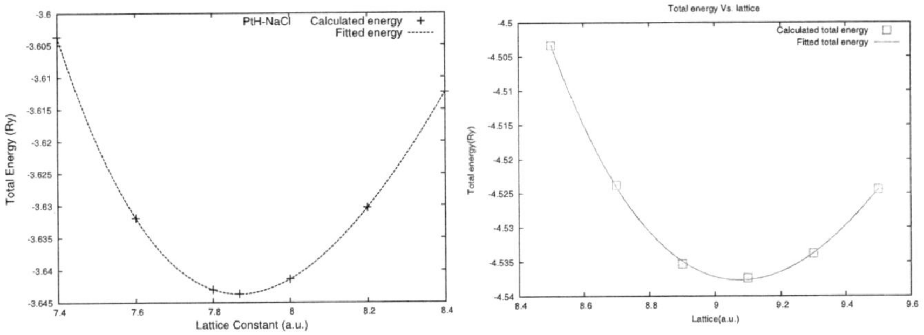

Fig. 8.31 Total energy versus lattice constant of PtH in the NaCl and CaF2 structures

Table 8.31 Lattice constant, bulk modulus, gap, total energy

Stru	a (Bohr)	B (MBar)	Gap	Total Energy
NaCl	7.87	2.37	-	-36873.64367
CaF2	9.08	1.22	-	-36874.53745
exp (CaF2)	?			

Table 8.32 Birch fit coefficients

	A1	A2	A3	A4
NaCl	-2.464223E+00	-3.259490E+01	-5.342665E+02	1.530929E+04
CaF2	-2.849885E+00	-1.084264E+02	1.678006E+03	2.051991E+03

Table 8.33 DOS at Ef, Hopfield parameter, Stoner criterion

PtH a=7.87 Bohr NaCl

Fermi Energy (Ry)	Total DOS (States/Ry)	Decomposed DOS inside the muffin tins s	p	eg	t2g	f
Pt 0.753	7.226	0.138	0.250	2.657	3.107	0.014
H 7.753	7.226	0.333	0.110	0.010	0.001	0.008

Fermi-Velocity (cm/s): 0.99196 x10E8
Plasmon Energy (eV) : 8.69931
Electron-ion interaction (Hopfield parameter) (eV/A^2) Pt: 1.080 H: 0.249

Pt	MUFFIN-TIN RADIUS and CHARGE =	2.3605	76.0053
H	MUFFIN-TIN RADIUS and CHARGE =	1.5737	0.9492
Pt	STONER I = 0.0124		
H	STONER I = 0.0009		
	STONER PARAMETER (Ry) I = 0.0133		
	STONER CRITERION N*I = 0.0961		

PtH2 a=9.08 Bohr CaF2

Fermi Energy (Ry)	Total DOS (States/Ry)	Decomposed DOS inside the muffin tins s	p	eg	t2g	f
Pt 0.485	6.423	1.138	0.092	0.072	3.330	0.010
H 0.485	6.423	0.327	0.231	0.002	0.037	0.070

Fermi-Velocity (cm/s): 0.53757 x10E8
Plasmon Energy (eV) : 3.58816
Electron-ion interaction (Hopfield parameter) (eV/A^2) Pt: 0.449 H: 0.793

Pt	MUFFIN-TIN RADIUS and CHARGE =	2.3580	75.8816
H	MUFFIN-TIN RADIUS and CHARGE =	1.5720	0.8586
Pt	STONER I = 0.0064		
H	STONER I = 0.0001		
	STONER PARAMETER (Ry) I = 0.0074		
	STONER CRITERION N*I = 0.0478		

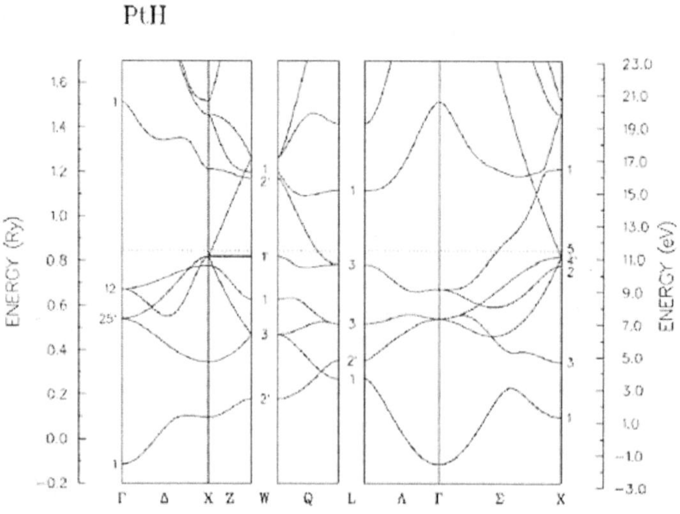

Fig. 8.32 Energy bands of PtH in the NaCl structure (tight-binding)

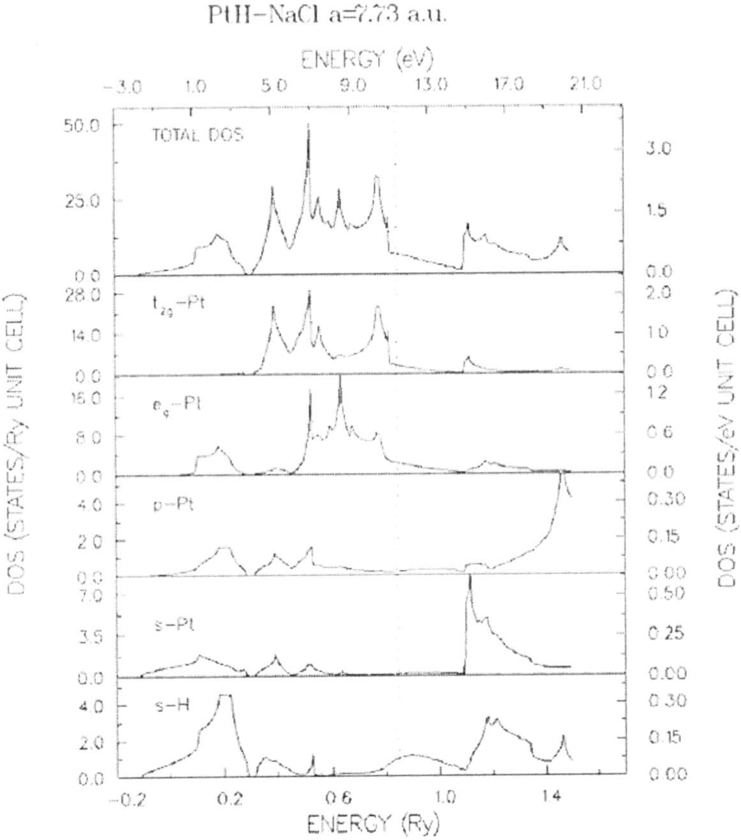

Fig. 8.33 Total, angular momentum and site decomposed densities of states of PtH in the NaCl structure (tight-binding)

Table 8.34 PtH (NaCl) a = 7.69 Bohr Slater–Koster 3-center parameters

	ORTHOGONAL	NON-ORTHOGONAL	
	ENERGY INTEGRALS Ry	ENERGY INTEGRALS Ry	OVERLAP INTEGRALS
ON SITE Pt-Pt			
s,s(000)	0.99780	0.88429	
x,x(000)	1.63693	1.28139	
xy,xy(000)	0.67044	0.58730	
d2,d2(000)	0.75651	0.59880	
FIRST NEIGHBOR Pt-Pt			
s,s(110)	-0.00694	-0.04285	-0.00994
s,x(110)	-0.04317	-0.02356	0.05206
s,xy(110)	-0.06178	-0.02233	0.07121
s,d2(110)	-0.00579	-0.03653	-0.03368
x,x(110)	0.07009	-0.12052	-0.16568
x,x(011)	-0.01085	0.03075	0.09505
x,y(110)	0.06538	-0.02573	-0.06388
x,xy(110)	0.07526	-0.01666	-0.10260
x,xy(011)	-0.00464	-0.00178	0.04351
z,d2(011)	0.02614	0.03556	0.05753
z,d1(011)	0.00462	0.03379	0.01231
xy,xy(110)	-0.04701	-0.02374	0.03941
xy,xy(011)	0.01181	0.00421	-0.02189
xy,xz(011)	0.00015	0.00432	-0.02598
xy,d2(110)	0.02556	-0.00992	0.02569
d2,d2(110)	0.00717	-0.01656	0.00085
d1,d1(110)	-0.04740	0.04476	0.02479
SECOND NEIGHBOR Pt-Pt			
s,s(200)	0.05297	0.03445	-0.02696
s,x(200)	0.06942	0.04008	-0.03221
s,d2(002)	0.05048	0.01195	0.04971
x,x(200)	-0.13157	-0.07657	0.03348
y,y(200)	0.01225	0.02889	0.02085
x,xy(020)	0.01142	0.00193	0.00537
z,d2(002)	-0.04915	-0.03547	-0.04786
xy,xy(200)	-0.01022	0.00098	-0.00081
xy,xy(002)	0.00157	-0.00204	-0.00245
d2,d2(002)	0.05165	-0.02995	-0.04905
d1,d1(002)	0.00103	-0.00048	-0.00069
ON SITE H-H			
s,s(000)	0.35578	0.69709	
FIRST NEIGHBOR H-H			
s,s(110)	-0.01124	0.00878	0.00211
s,x(110)	0.00000	0.00000	0.00000
x,x(110)	0.00000	0.00000	0.00000
x,x(011)	0.00000	0.00000	0.00000
x,y(110)	0.00000	0.00000	0.00000
SECOND NEIGHBOR H-H			
s,s(200)	-0.01297	0.00344	0.00865
s,x(200)	0.00000	0.00000	0.00000
x,x(200)	0.00000	0.00000	0.00000
y,y(200)	0.00000	0.00000	0.00000
FIRST NEIGHBOR Pt-H			
s,s(100)	-0.12365	-0.13689	-0.03740
s,x(100)	0.00000	0.00000	0.00000
x,s(100)	0.08732	0.18064	0.02247
x,x(100)	0.00000	0.00000	0.00000
y,y(100)	0.00000	0.00000	0.00000
xy,x(010)	0.00000	0.00000	0.00000
d2,s(001)	-0.08318	0.02167	-0.11381
d2,z(001)	0.00000	0.00000	0.00000
SECOND NEIGHBOR Pt-H			
s,s(111)	0.01877	0.00522	0.01663
x,s(111)	-0.03758	-0.01211	-0.01800
s,x(111)	0.00000	0.00000	0.00000
x,x(111)	0.00000	0.00000	0.00000
x,y(111)	0.00000	0.00000	0.00000
xy,s(111)	0.00532	0.00788	0.01097
xy,x(111)	0.00000	0.00000	0.00000
yz,x(111)	0.00000	0.00000	0.00000
d1,x(111)	0.00000	0.00000	0.00000

PtH

BAND	ORTHOGONAL RMS ERROR mRy	MAXIMUM DEVIATION k	MAXIMUM DEVIATION mRy	NON-ORTHOGONAL RMS ERROR mRy	MAXIMUM DEVIATION k	MAXIMUM DEVIATION mRy
1	4.6	(048)	10.9	0.4	(004)	0.8
2	5.3	(222)	16.0	0.3	(222)	0.9
3	4.9	(044)	9.3	0.4	(226)	1.1
4	3.5	(048)	10.1	0.3	(048)	0.9
5	3.7	(044)	8.6	0.3	(044)	0.9
6	5.0	(055)	12.4	0.5	(174)	1.8
7	7.3	(000)	20.2	0.6	(003)	1.3
1-7	5.0			0.4		

ENERGY VALUES IN Ry AT SELECTED k-POINTS

	ORTHOGONAL	APW	NON-ORTHOGONAL
GAMMA 1	-0.11653	-0.11519	-0.11479
GAMMA 1	1.49192	1.51216	1.51233
GAMMA 12	0.67315	0.67296	0.67367
GAMMA 15	1.94013	1.96281	1.96279
GAMMA 25'	0.53914	0.53979	0.53976
X1 (008)	0.10417	0.09718	0.09739
X1 (008)	1.21391	1.21180	1.21234
X1 (008)	1.56149	1.51947	1.51879
X2 (008)	0.77677	0.77720	0.77789
X3 (008)	0.35026	0.34613	0.34612
X5 (008)	0.82075	0.81726	0.81762
X4' (008)	0.81868	0.81618	0.81615
X5' (008)	1.46621	1.45532	1.45530
L1 (444)	0.26519	0.26868	0.26883
L1 (444)	1.12171	1.11569	1.11572
L3 (444)	0.51676	0.51782	0.51791
L3 (444)	0.79047	0.78300	0.78295
L2' (444)	0.34095	0.35260	0.35227
L2' (444)	1.42075	1.41459	1.41457
L3' (444)	2.11256	2.04692	2.04696
W1 (048)	0.61716	0.62727	0.62817
W1 (048)	1.19362	1.19646	1.19672
W2' (048)	0.16772	0.17864	0.17873
W2' (048)	1.16537	1.16922	1.16903
W2' (048)	2.13425	2.24587	2.24585
W1' (048)	0.81445	0.81760	0.81780
W3 (048)	0.46408	0.46961	0.46964
W3 (048)	1.25572	1.26356	1.26342
ODD (224)	0.57396	0.56957	0.56950
ODD (224)	0.76631	0.76145	0.76154
EVEN(224)	0.14397	0.13596	0.13568
EVEN(224)	0.42210	0.41767	0.41721
EVEN(224)	0.51475	0.51672	0.51682
EVEN(224)	0.72607	0.72502	0.72468
EVEN(224)	1.18303	1.18077	1.18182

FERMI LEVEL QUANTITIES (Non-orthogonal fit)

ENERGY Ry	DENSITIES OF STATES (states/Ry/cell) Total	s-Pt	p-Pt	t2g-Pt	eg-Pt	s-H	p-H
0.8435	6.68	0.15	0.20	3.00	2.36	0.97	0.00

INTEGRATED DENSITIES OF STATES (electrons) Total	s-Pt	p-Pt	t2g-Pt	eg-Pt	s-H	p-H
11.00	0.56	0.48	5.54	3.49	0.93	0.00

VELOCITY cm/s	PLASMON ENERGY eV	EIGENVALUE SUM Ry
0.81x10E8	7.01	-3.6418

Table 8.35 PtH (NaCl) a = 7.69 Bohr Slater–Koster 2-center parameters

	ORTHOGONAL	NON-ORTHOGONAL	
	ENERGY INTEGRALS Ry	ENERGY INTEGRALS Ry	OVERLAP INTEGRALS
ON SITE Pt-Pt			
s	0.87568	0.58320	
p	1.58575	0.98093	
t2g	0.63694	0.57654	
eg	0.61932	0.57015	
FIRST NEIGHBOR Pt-Pt			
(sss)	-0.06618	-0.03286	0.08658
(pps)	0.15141	-0.17040	-0.31715
(ppp)	-0.01278	0.05703	0.09155
(dds)	-0.06257	-0.02086	0.06697
(ddp)	0.02290	0.01935	-0.03538
(ddd)	-0.00004	-0.00418	0.00324
(sps)	0.11334	0.00015	-0.17129
(sds)	-0.06795	-0.02326	0.07322
(pds)	-0.10133	0.02409	0.14164
(pdp)	0.01376	0.00209	-0.05782
SECOND NEIGHBOR Pt-Pt			
(sss)	0.00726	-0.02045	-0.01194
(pps)	-0.09140	0.12720	0.02655
(ppp)	0.00574	-0.01061	-0.00599
(dds)	0.00384	-0.01363	-0.00632
(ddp)	-0.00169	0.00380	0.00242
(ddd)	0.00016	0.00045	0.00058
(sps)	-0.02923	0.05627	0.02010
(sds)	0.00603	-0.01479	-0.00761
(pds)	0.01143	-0.03627	-0.01405
(pdp)	-0.00423	0.00544	0.00271
ON SITE H-H			
s	0.96619	0.63487	
FIRST NEIGHBOR H-H			
(sss)	0.00209	0.00296	-0.01548
(sps)	0.00000	0.00000	0.00000
(pps)	0.00000	0.00000	0.00000
(ppp)	0.00000	0.00000	0.00000
SECOND NEIGHBOR H-H			
(sss)	0.04283	-0.00425	-0.00600
(sps)	0.00000	0.00000	0.00000
(pps)	0.00000	0.00000	0.00000
(ppp)	0.00000	0.00000	0.00000
FIRST NEIGHBOR Pt-H			
(sss)	0.11281	0.07086	-0.09934
(sps)	0.00000	0.00000	0.00000
(pss)	0.14700	0.06018	-0.21688
(pps)	0.00000	0.00000	0.00000
(ppp)	0.00000	0.00000	0.00000
(dpp)	0.00000	0.00000	0.00000
(dss)	0.11414	0.07064	-0.12425
(dps)	0.00000	0.00000	0.00000
SECOND NEIGHBOR Pt-H			
(sss)	-0.01234	0.01172	0.01462
(pss)	-0.04472	0.04074	0.01995
(sps)	0.00000	0.00000	0.00000
(pps)	0.00000	0.00000	0.00000
(ppp)	0.00000	0.00000	0.00000
(dss)	-0.00939	0.01398	0.01657
(dps)	0.00000	0.00000	0.00000
(dpp)	0.00000	0.00000	0.00000

PtH

BAND	ORTHOGONAL RMS ERROR mRy	MAXIMUM DEVIATION k	MAXIMUM DEVIATION mRy	NON-ORTHOGONAL RMS ERROR mRy	MAXIMUM DEVIATION k	MAXIMUM DEVIATION mRy
1	7.1	(005)	13.7	0.2	(044)	0.7
2	6.5	(333)	19.2	0.4	(222)	1.2
3	6.0	(264)	16.7	0.4	(044)	1.0
4	4.1	(444)	8.4	0.4	(264)	0.7
5	4.0	(224)	7.6	0.4	(055)	1.3
6	5.6	(066)	17.1	0.3	(000)	0.7
7	13.1	(004)	35.3	0.4	(022)	0.9
1-7	7.2			0.4		

ENERGY VALUES IN Ry AT SELECTED k-POINTS

	ORTHOGONAL	APW	NON-ORTHOGONAL
GAMMA 1	-0.11928	-0.11519	-0.11517
GAMMA 1	1.49272	1.51216	1.51175
GAMMA 12	0.67468	0.67296	0.67370
GAMMA 15	1.92927	1.96281	1.96278
GAMMA 25'	0.53420	0.53979	0.53989
X1 (008)	0.10697	0.09718	0.09731
X1 (008)	1.20130	1.21180	1.21161
X1 (008)	1.55303	1.51947	1.51916
X2 (008)	0.77113	0.77720	0.77716
X3 (008)	0.35127	0.34613	0.34625
X5 (008)	0.81825	0.81726	0.81744
X4' (008)	0.82026	0.81618	0.81580
X5' (008)	1.47702	1.45532	1.45531
L1 (444)	0.26072	0.26868	0.26881
L1 (444)	1.12300	1.11569	1.11578
L3 (444)	0.50947	0.51782	0.51828
L3 (444)	0.78715	0.78300	0.78318
L2' (444)	0.35555	0.35260	0.35265
L2' (444)	1.44249	1.41459	1.41453
L3' (444)	2.07399	2.04692	2.04694
W1 (048)	0.62987	0.62727	0.62694
W1 (048)	1.20369	1.19646	1.19656
W2' (048)	0.16977	0.17864	0.17840
W2' (048)	1.16629	1.16922	1.16914
W2' (048)	2.08911	2.24587	2.24588
W1' (048)	0.81763	0.81760	0.81754
W3 (048)	0.46646	0.46961	0.47009
W3 (048)	1.25074	1.26356	1.26337
ODD (224)	0.56582	0.56957	0.56939
ODD (224)	0.76545	0.76145	0.76096
EVEN(224)	0.13611	0.13596	0.13603
EVEN(224)	0.42365	0.41767	0.41750
EVEN(224)	0.52034	0.51672	0.51652
EVEN(224)	0.73262	0.72502	0.72473
EVEN(224)	1.18191	1.18077	1.18079

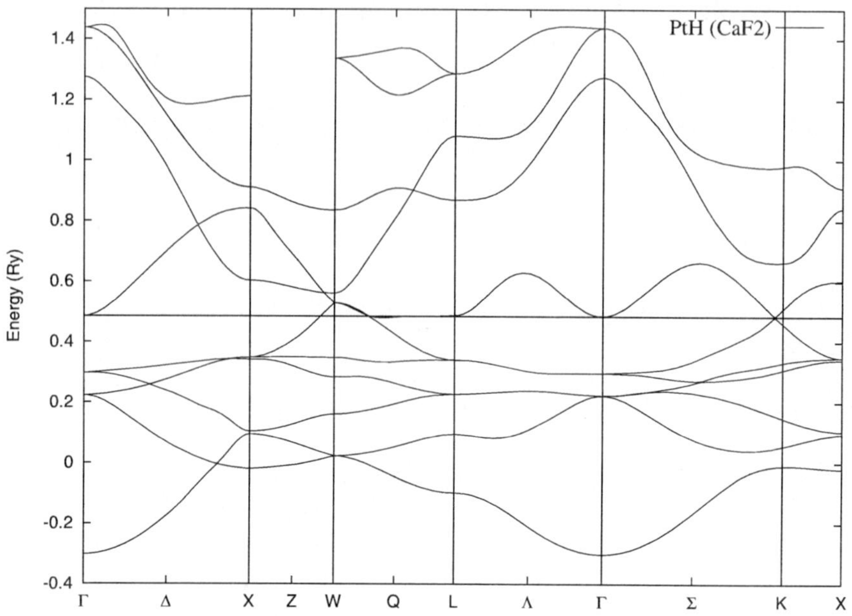

Fig. 8.34 Energy bands of PtH2 in the CaF2 structure

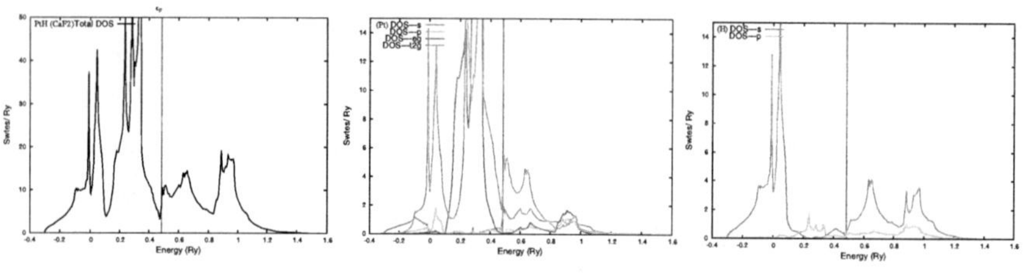

Fig. 8.35 Total, angular momentum and site decomposed densities of states of PtH2 in the CaF2 structure

8.8 Gold Hydride (AuH)

See Fig. 8.36 and Tables 8.36, 8.37, 8.38.

See Figs. 8.37, 8.38 and Tables 8.39, 8.40.

See Figs. 8.39 and 8.40.

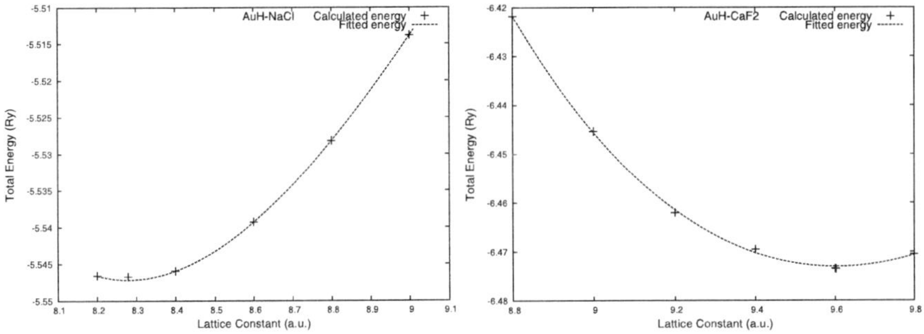

Fig. 8.36 Total energy versus lattice constant of AuH in the NaCl and CaF2 structures

Table 8.36 Lattice constant, bulk modulus, gap, total energy

Stru	a (Bohr)	B (MBar)	Gap	Total Energy
NaCl	8.28	1.41	-	-38075.54671
CaF2	9.60	0.88	-	-38076.47352
exp (CaF2)	?			

Table 8.37 Birch fit coefficients

	A1	A2	A3	A4
NaCl	-4.766314E+00	-2.224457E+01	-5.232314E+02	1.497577E+04
CaF2	-4.585337E+00	-1.527945E+02	3.599461E+03	-1.961216E+04

Table 8.38 DOS at Ef, Hopfield parameter, Stoner criterion

AuH a=8.28 Bohr NaCl

Fermi Energy (Ry)	Total DOS (States/Ry)	s	p	eg	t2g	f
Au 0.677	8.421	2.486	0.306	0.629	2.299	0.048
H 0.677	8.421	0.532	0.365	0.017	0.001	0.008

(Columns s, p, eg, t2g, f: Decomposed DOS inside the muffin tins)

```
Fermi-Velocity (cm/s): 0.71761 x10E8
Plasmon Energy (eV)  : 6.29393
Electron-ion interaction (Hopfield parameter) (eV/A^2) Au: 1.080  H: 0.862

   Au      MUFFIN-TIN RADIUS and CHARGE =   2.4424   77.1957
   H       MUFFIN-TIN RADIUS and CHARGE =   1.6973    0.9805
             Au           STONER I = 0.0050
             H            STONER I = 0.0023
         STONER PARAMETER (Ry) I = 0.0074
         STONER CRITERION    N*I = 0.0624
```

AuH2 a=9.60 Bohr CaF2

Fermi Energy (Ry)	Total DOS (States/Ry)	s	p	eg	t2g	f
Au 0.357	8.985	1.575	0.366	0.239	2.352	0.037
H 0.357	8.985	1.917	0.275	0.001	0.034	0.005

(Columns s, p, eg, t2g, f: Decomposed DOS inside the muffin tins)

```
Fermi-Velocity (cm/s): 1.02850 x10E8
Plasmon Energy (eV)  : 7.45892
Electron-ion interaction (Hopfield parameter) (eV/A^2) Au: 0.780  H: 3.128

     Au      MUFFIN-TIN RADIUS and CHARGE =   2.4948   77.2720
     H       MUFFIN-TIN RADIUS and CHARGE =   1.6632    0.8335
                  Au        STONER I = 0.0027
                  H         STONER I = 0.0037
         STONER PARAMETER (Ry)     I = 0.0102
         STONER CRITERION        N*I = 0.0916
```

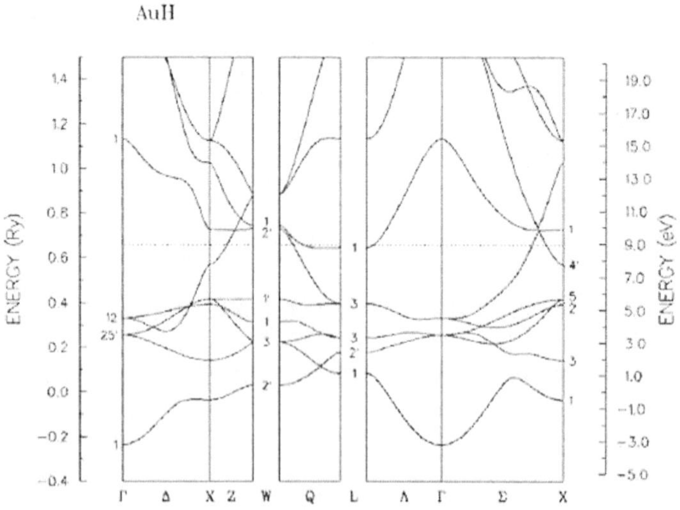

Fig. 8.37 Energy bands of AuH in the NaCl structure (tight-binding)

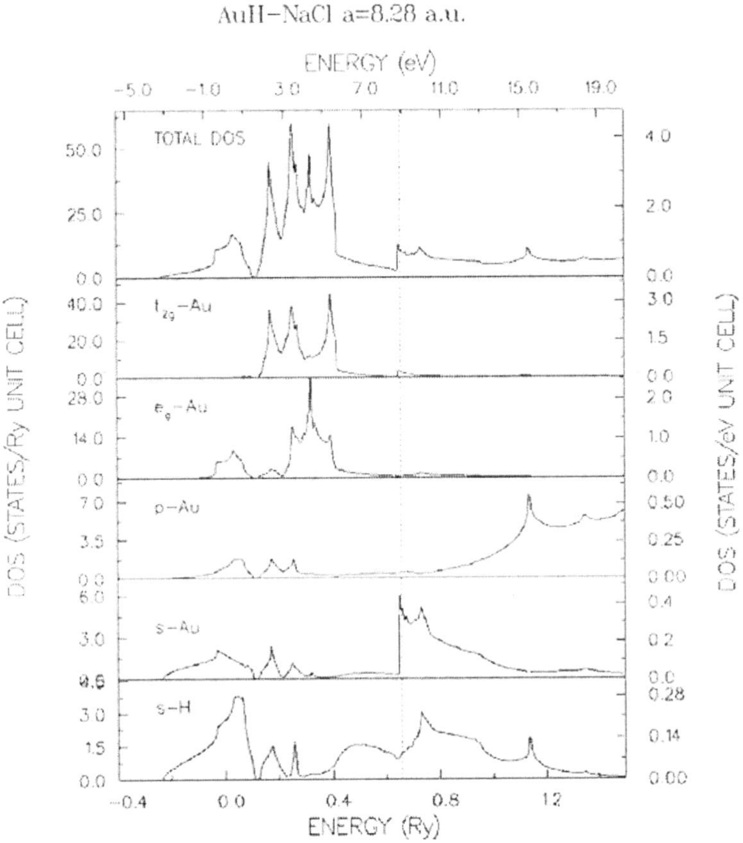

Fig. 8.38 Total, angular momentum and site decomposed densities of states of AuH in the NaCl structure (tight-binding)

Table 8.39 AuH (NaCl) a = 8.28 Bohr Slater–Koster 3-center parameters

	ORTHOGONAL	NON-ORTHOGONAL	
	ENERGY INTEGRALS Ry	ENERGY INTEGRALS Ry	OVERLAP INTEGRALS
ON SITE Au-Au			
s,s(000)	0.63845	0.57935	
x,x(000)	1.29080	1.02191	
xy,xy(000)	0.32217	0.29614	
d2,d2(000)	0.39929	0.28344	
FIRST NEIGHBOR Au-Au			
s,s(110)	-0.00009	-0.02244	-0.01428
s,x(110)	-0.03073	-0.02303	0.04528
s,xy(110)	-0.04026	-0.02854	0.04734
s,d2(110)	-0.00602	-0.02608	-0.04093
x,x(110)	0.06535	-0.06688	-0.14938
x,x(011)	-0.00464	0.00723	0.08085
x,y(110)	0.04891	-0.00452	-0.04892
x,xy(110)	0.04823	0.01130	-0.07744
x,xy(011)	-0.00421	-0.00677	0.03157
z,d2(011)	0.02223	0.01750	0.04776
z,d1(011)	0.00177	0.02967	0.01847
xy,xy(110)	-0.02733	-0.02108	0.02217
xy,xy(011)	0.00696	0.00465	-0.01560
xy,xz(011)	0.00301	0.00309	-0.02147
xy,d2(110)	0.01529	-0.01160	0.01101
d2,d2(110)	0.00619	-0.01188	-0.00030
d1,d1(110)	-0.03390	0.02634	0.01467
SECOND NEIGHBOR Au-Au			
s,s(200)	0.03660	0.03410	-0.01897
s,x(200)	0.05222	0.03977	-0.02451
s,d2(002)	0.03584	0.00498	0.04462
x,x(200)	-0.11127	-0.07084	0.03036
y,y(200)	0.01121	0.01923	0.02186
x,xy(020)	0.00691	0.00032	0.00496
z,d2(002)	-0.03345	-0.02201	-0.03725
xy,xy(200)	-0.00351	-0.00055	-0.00109
xy,xy(002)	0.00069	-0.00137	-0.00335
d2,d2(002)	0.03284	-0.01562	-0.03823
d1,d1(002)	0.00042	0.00086	0.00210
ON SITE H-H			
s,s(000)	0.17947	0.53058	
FIRST NEIGHBOR H-H			
s,s(110)	-0.00905	0.00455	0.00213
s,x(110)	0.00000	0.00000	0.00000
x,x(110)	0.00000	0.00000	0.00000
x,x(011)	0.00000	0.00000	0.00000
x,y(110)	0.00000	0.00000	0.00000
SECOND NEIGHBOR H-H			
s,s(200)	-0.00786	0.00037	0.00105
s,x(200)	0.00000	0.00000	0.00000
x,x(200)	0.00000	0.00000	0.00000
y,y(200)	0.00000	0.00000	0.00000
FIRST NEIGHBOR Au-H			
s,s(100)	-0.10371	-0.11678	-0.05667
s,x(100)	0.00000	0.00000	0.00000
x,s(100)	0.06673	0.17126	0.04605
x,x(100)	0.00000	0.00000	0.00000
y,y(100)	0.00000	0.00000	0.00000
xy,x(010)	0.00000	0.00000	0.00000
d2,s(001)	-0.07374	0.01955	-0.10420
d2,z(001)	0.00000	0.00000	0.00000

SECOND NEIGHBOR Au-H			
s,s(111)	0.01044	0.00064	0.01145
x,s(111)	-0.02502	-0.00404	-0.01633
s,x(111)	0.00000	0.00000	0.00000
x,x(111)	0.00000	0.00000	0.00000
x,y(111)	0.00000	0.00000	0.00000
xy,s(111)	0.00034	0.00338	0.00828
xy,x(111)	0.00000	0.00000	0.00000
yz,x(111)	0.00000	0.00000	0.00000
d1,x(111)	0.00000	0.00000	0.00000

AuH

BAND	ORTHOGONAL RMS ERROR mRy	MAXIMUM DEVIATION k	mRy	NON-ORTHOGONAL RMS ERROR mRy	MAXIMUM DEVIATION k	mRy
1	2.6	(033)	5.9	0.3	(111)	0.7
2	3.6	(333)	10.2	0.3	(222)	0.7
3	3.3	(006)	9.1	0.3	(226)	0.6
4	2.2	(048)	7.4	0.3	(066)	0.6
5	2.2	(044)	6.0	0.3	(044)	0.9
6	3.3	(055)	11.0	0.4	(174)	1.0
7	5.4	(007)	15.3	0.6	(111)	1.8
1-7	3.4			0.3		

ENERGY VALUES IN Ry AT SELECTED k-POINTS

	ORTHOGONAL	APW	NON-ORTHOGONAL
GAMMA 1	-0.24072	-0.23739	-0.23711
GAMMA 1	1.12146	1.13300	1.13271
GAMMA 12	0.33278	0.33257	0.33329
GAMMA 15	1.61737	1.63128	1.63129
GAMMA 25'	0.25586	0.25606	0.25616
X1 (008)	-0.03145	-0.03463	-0.03440
X1 (008)	0.72678	0.72711	0.72730
X1 (008)	1.04637	1.02587	1.02518
X2 (008)	0.39413	0.39295	0.39350
X3 (008)	0.14456	0.14109	0.14109
X5 (008)	0.41884	0.41675	0.41711
X4' (008)	0.57177	0.57037	0.57041
X5' (008)	1.13169	1.12548	1.12545
L1 (444)	0.08066	0.08278	0.08256
L1 (444)	0.64895	0.64515	0.64549
L3 (444)	0.24461	0.24468	0.24478
L3 (444)	0.40176	0.39732	0.39748
L2' (444)	0.16787	0.17642	0.17608
L2' (444)	1.13598	1.13260	1.13259
L3' (444)	1.66411	1.62852	1.62850
W1 (048)	0.30776	0.31514	0.31564
W1 (048)	0.74718	0.74419	0.74421
W2' (048)	0.02368	0.02906	0.02890
W2' (048)	0.72755	0.73107	0.73048
W2' (048)	1.65844	1.72293	1.72300
W1' (048)	0.41609	0.41691	0.41697
W3 (048)	0.22067	0.22497	0.22484
W3 (048)	0.88193	0.88402	0.88403
ODD (224)	0.27764	0.27635	0.27618
ODD (224)	0.38845	0.38548	0.38553
EVEN(224)	-0.01820	-0.02233	-0.02250
EVEN(224)	0.19430	0.19227	0.19194
EVEN(224)	0.24722	0.24698	0.24707
EVEN(224)	0.37594	0.37378	0.37353
EVEN(224)	0.79031	0.78802	0.78856

FERMI LEVEL QUANTITIES (Non-orthogonal fit)

ENERGY		DENSITIES OF STATES					
	Total	s-Au	p-Au	t2g-Au	eg-Au	s-H	p-H
Ry		states/Ry/cell					
0.6570	10.65	5.09	0.55	3.16	0.65	1.19	0.00
		INTEGRATED DENSITIES OF STATES					
	Total	s-Au	p-Au	t2g-Au	eg-Au	s-H	p-H
		electrons					
	12.00	0.68	0.48	5.92	3.86	1.06	0.00

VELOCITY	PLASMON ENERGY	EIGENVALUE SUM
cm/s	eV	Ry
0.76x10E8	7.45	-4.6860

Table 8.40 AuH (NaCl) a = 8.28 Bohr Slater–Koster 2-center parameters

	ORTHOGONAL	NON-ORTHOGONAL	
	ENERGY INTEGRALS Ry	ENERGY INTEGRALS Ry	OVERLAP INTEGRALS
ON SITE Au-Au			
s	0.50860	0.36052	
p	1.24801	0.73688	
t2g	0.31025	0.28306	
eg	0.30248	0.27880	
FIRST NEIGHBOR Au-Au			
(sss)	-0.05185	-0.03419	0.07174
(pps)	0.13480	-0.08618	-0.30546
(ppp)	-0.00533	0.07899	0.11833
(dds)	-0.03617	-0.02124	0.04957
(ddp)	0.01421	0.01316	-0.03211
(ddd)	-0.00063	-0.00189	0.00489
(sps)	0.08890	0.04080	-0.12445
(sds)	-0.04375	-0.02347	0.05742
(pds)	-0.06583	0.00211	0.13248
(pdp)	0.01080	0.00217	-0.06632
SECOND NEIGHBOR Au-Au			
(sss)	0.00554	0.00774	0.02035
(pps)	-0.08473	0.07132	-0.00767
(ppp)	0.00852	0.01490	0.01058
(dds)	0.00107	-0.00623	-0.00257
(ddp)	-0.00047	0.00215	0.00275
(ddd)	-0.00005	-0.00088	-0.00265
(sps)	-0.02131	0.00848	-0.02038
(sds)	0.00364	-0.00721	-0.00974
(pds)	0.00598	-0.03125	-0.04206
(pdp)	-0.00430	0.00047	-0.00277
ON SITE H-H			
s	0.66133	0.40689	
FIRST NEIGHBOR H-H			
(sss)	0.01095	-0.04104	0.00502
(sps)	0.00000	0.00000	0.00000
(pps)	0.00000	0.00000	0.00000
(ppp)	0.00000	0.00000	0.00000

SECOND NEIGHBOR H-H			
(sss)	0.02748	0.01139	-0.01107
(sps)	0.00000	0.00000	0.00000
(pps)	0.00000	0.00000	0.00000
(ppp)	0.00000	0.00000	0.00000
FIRST NEIGHBOR Au-H			
(sss)	0.08642	0.08096	-0.13961
(sps)	0.00000	0.00000	0.00000
(pss)	0.13023	0.04530	-0.28119
(pps)	0.00000	0.00000	0.00000
(ppp)	0.00000	0.00000	0.00000
(dpp)	0.00000	0.00000	0.00000
(dss)	0.07871	0.08590	-0.02558
(dps)	0.00000	0.00000	0.00000
SECOND NEIGHBOR Au-H			
(sss)	-0.00996	0.00701	-0.01248
(pss)	-0.02944	-0.00121	-0.04814
(sps)	0.00000	0.00000	0.00000
(pps)	0.00000	0.00000	0.00000
(ppp)	0.00000	0.00000	0.00000
(dss)	-0.00461	-0.00050	-0.03699
(dps)	0.00000	0.00000	0.00000
(dpp)	0.00000	0.00000	0.00000

AuH

BAND	ORTHOGONAL RMS ERROR mRy	MAXIMUM DEVIATION k	MAXIMUM DEVIATION mRy	NON-ORTHOGONAL RMS ERROR mRy	MAXIMUM DEVIATION k	MAXIMUM DEVIATION mRy
1	4.5	(004)	9.0	0.5	(444)	1.3
2	3.8	(333)	13.4	0.4	(354)	1.0
3	2.7	(264)	5.7	0.4	(033)	1.1
4	2.0	(354)	7.0	0.4	(280)	0.9
5	1.6	(224)	3.5	0.4	(055)	0.9
6	2.6	(066)	5.9	0.5	(044)	1.6
7	9.5	(007)	27.2	1.1	(048)	3.0
1-7	4.6			0.6		

ENERGY VALUES IN Ry AT SELECTED k-POINTS

	ORTHOGONAL	APW	NON-ORTHOGONAL
GAMMA 1	-0.24101	-0.23739	-0.23722
GAMMA 1	1.11827	1.13300	1.13333
GAMMA 12	0.33368	0.33257	0.33335
GAMMA 15	1.60920	1.63128	1.63132
GAMMA 25'	0.25343	0.25606	0.25643

X1 (008)	-0.02789	-0.03463	-0.03395
X1 (008)	0.72212	0.72711	0.72719
X1 (008)	1.03867	1.02587	1.02487
X2 (008)	0.39105	0.39295	0.39351
X3 (008)	0.14483	0.14109	0.14133
X5 (008)	0.41743	0.41675	0.41726
X4' (008)	0.57340	0.57037	0.57090
X5' (008)	1.13392	1.12548	1.12548
L1 (444)	0.07901	0.08278	0.08405
L1 (444)	0.64923	0.64515	0.64600
L3 (444)	0.24233	0.24468	0.24510
L3 (444)	0.39899	0.39732	0.39774
L2' (444)	0.17993	0.17642	0.17693
L2' (444)	1.13941	1.13260	1.13267
L3' (444)	1.66366	1.62852	1.62843
W1 (048)	0.31410	0.31514	0.31514
W1 (048)	0.75228	0.74419	0.74117
W2' (048)	0.02659	0.02906	0.02868
W2' (048)	0.72755	0.73107	0.73127
W2' (048)	1.64012	1.72293	1.72307
W1' (048)	0.41763	0.41691	0.41623
W3 (048)	0.22150	0.22497	0.22482
W3 (048)	0.88109	0.88402	0.88433
ODD (224)	0.27521	0.27635	0.27604
ODD (224)	0.38720	0.38548	0.38549
EVEN(224)	-0.02260	-0.02233	-0.02288
EVEN(224)	0.19624	0.19227	0.19237
EVEN(224)	0.24887	0.24698	0.24636
EVEN(224)	0.37723	0.37378	0.37376
EVEN(224)	0.79068	0.78802	0.78718

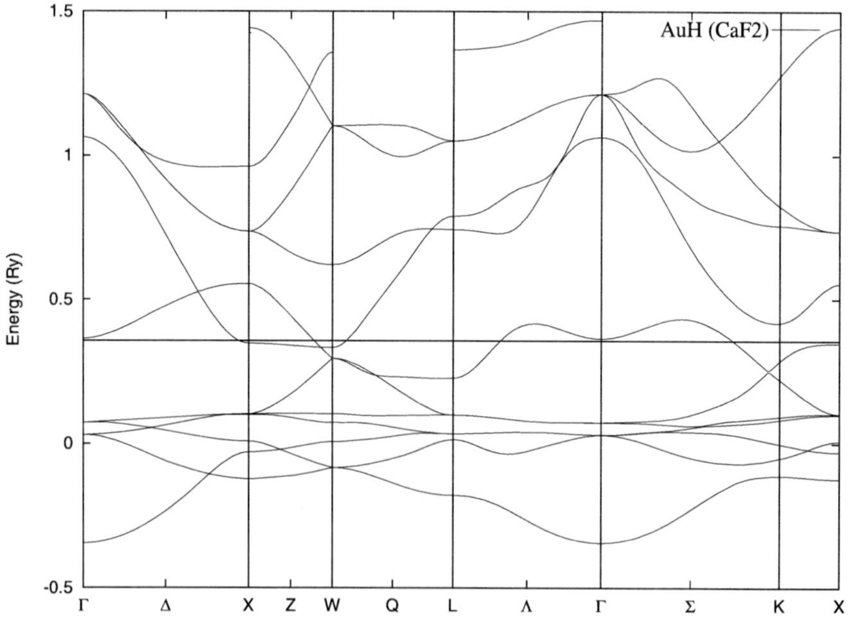

Fig. 8.39 Energy bands of AuH2 in the CaF2 structure

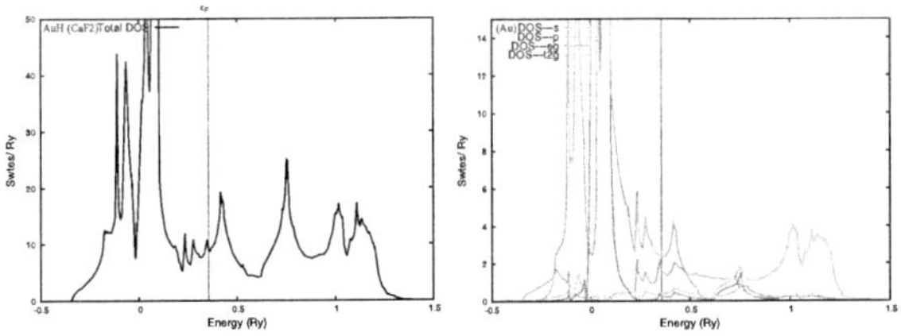

Fig. 8.40 Total, angular momentum and site decomposed densities of states of AuH2 in the CaF2 structure

8.9 Mercury Hydride (HgH)

See Fig. 8.41 and Tables 8.41, 8.42, 8.43.

See Figs. 8.42, 8.43 and Tables 8.44, 8.45.

See Figs. 8.44 and 8.45.

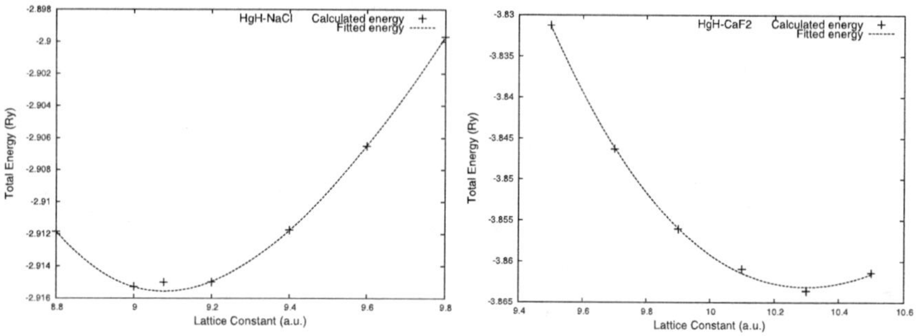

Fig. 8.41 Total energy versus lattice constant of HgH in the NaCl and CaF2 structures

Table 8.41 Lattice constant, bulk modulus, gap, total energy

Stru	a (Bohr)	B (MBar)	Gap	Total Energy
NaCl	9.08	0.61	-	-39302.91500
CaF2	10.30	0.51	-	-39303.86353
exp (CaF2)	?			

Table 8.42 Birch fit coefficients

	A1	A2	A3	A4
NaCl	-2.645172E+00	1.932507E+00	-9.936186E+02	2.097076E+04
CaF2	-2.557380E+00	-1.196144E+02	3.129388E+03	-1.717213E+04

Table 8.43 DOS at Ef, Hopfield parameter, Stoner criterion

HgH a=9.08 Bohr NaCl

Fermi Energy (Ry)	Total DOS (States/Ry)	Decomposed DOS inside the muffin tins: s	p	eg	t2g	f
Hg 0.411	7.092	2.011	0.767	0.337	0.272	0.085
H 0.411	7.092	1.548	0.251	0.021	0.001	0.002

Fermi-Velocity (cm/s): 1.13525 x10E8
Plasmon Energy (eV) : 7.96022
Electron-ion interaction (Hopfield parameter) (eV/A^2) Hg: 0.557 H: 1.646

Hg	MUFFIN-TIN RADIUS and CHARGE =	2.7231	78.7149
H	MUFFIN-TIN RADIUS and CHARGE =	1.8154	0.9445
Hg	STONER I = 0.0014		
H	STONER I = 0.0137		
STONER PARAMETER (Ry)	I = 0.0152		
STONER CRITERION	N*I = 0.1078		

HgH2 a=10.30 Bohr CaF2

Fermi Energy (Ry)	Total DOS (States/Ry)	Decomposed DOS inside the muffin tins: s	p	eg	t2g	f
Hg 0.232	13.676	2.094	0.801	0.087	1.387	0.103
H 0.232	13.676	5.938	0.318	0.005	0.030	0.004

Fermi-Velocity (cm/s): 0.91255 x10E8
Plasmon Energy (eV) : 7.35235
Electron-ion interaction (Hopfield parameter) (eV/A^2) Hg: 0.866 H: 5.406

Hg	MUFFIN-TIN RADIUS and CHARGE =	2.6757	78.5214
H	MUFFIN-TIN RADIUS and CHARGE =	1.7838	0.8764
Hg	STONER I = 0.0008		
H	STONER I = 0.0120		
STONER PARAMETER (Ry)	I = 0.0248		
STONER CRITERION	N*I = 0.3398		

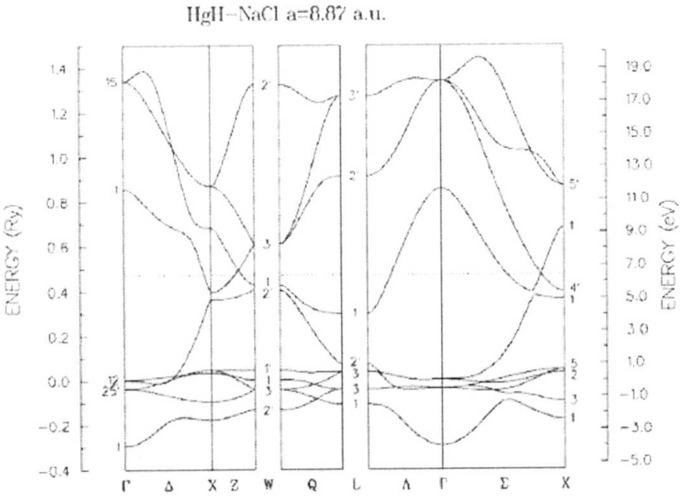

Fig. 8.42 Energy bands of HgH in the NaCl structure (tight-binding)

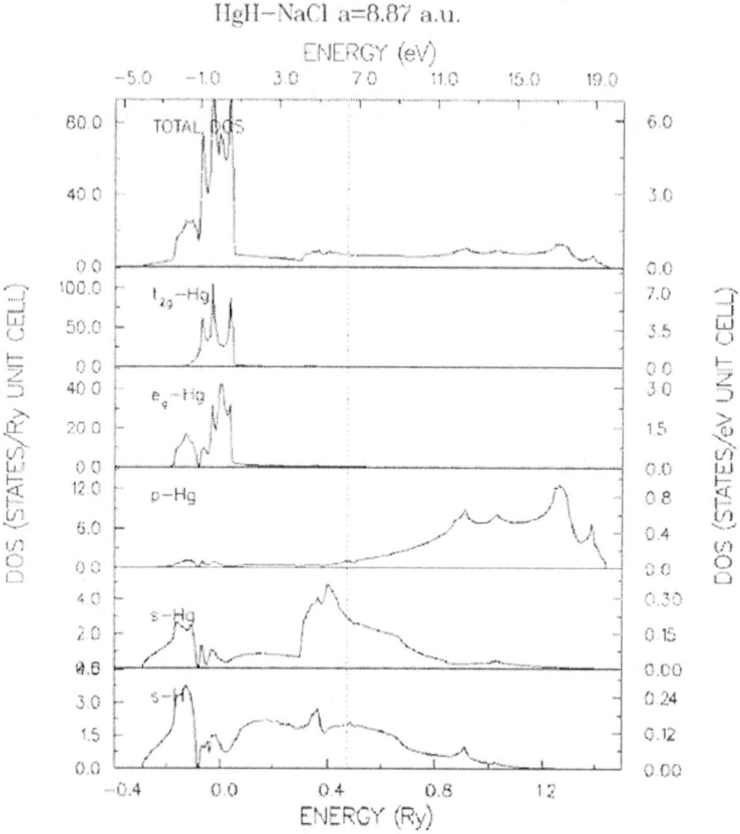

Fig. 8.43 Total, angular momentum and site decomposed densities of states of HgH in the NaCl structure (tight-binding)

Table 8.44 HgH (NaCl) a = 8.87 Bohr Slater–Koster 3-center parameters

	ORTHOGONAL	NON-ORTHOGONAL	
	ENERGY INTEGRALS Ry	ENERGY INTEGRALS Ry	OVERLAP INTEGRALS
ON SITE Hg-Hg			
s,s(000)	0.34693	0.33277	
x,x(000)	0.99236	0.84033	
xy,xy(000)	0.00452	-0.01564	
d2,d2(000)	0.03556	-0.02187	
FIRST NEIGHBOR Hg-Hg			
s,s(110)	-0.00462	-0.00464	-0.00639
s,x(110)	-0.02610	-0.01321	0.03049
s,xy(110)	-0.02757	-0.02332	0.04339
s,d2(110)	-0.00741	-0.00967	-0.01816
x,x(110)	0.05832	-0.00263	-0.09008
x,x(011)	-0.00109	-0.02721	0.01765
x,y(110)	0.04321	-0.02884	-0.06284
x,xy(110)	0.03682	0.01843	-0.07161
x,xy(011)	-0.00050	-0.00830	0.00977
z,d2(011)	0.00340	0.00174	0.01390
z,d1(011)	0.00561	0.01526	0.00949
xy,xy(110)	-0.01411	-0.01450	0.00822
xy,xy(011)	0.00354	0.00433	-0.00840
xy,xz(011)	-0.00035	0.00533	-0.00959
xy,d2(110)	0.00714	-0.00790	0.01129
d2,d2(110)	0.00345	-0.00635	0.00121
d1,d1(110)	-0.01810	0.01150	0.01554
SECOND NEIGHBOR Hg-Hg			
s,s(200)	0.02930	0.02631	-0.02319
s,x(200)	0.03990	0.02975	-0.03177
s,d2(002)	0.02287	-0.00162	0.03948
x,x(200)	-0.09162	-0.04039	0.06711
y,y(200)	0.01463	0.03064	0.01930
x,xy(020)	0.00564	-0.00024	0.01369
z,d2(002)	-0.01688	-0.00875	-0.05087
xy,xy(200)	-0.00354	0.00068	0.00249
xy,xy(002)	0.00010	-0.00010	-0.00217
d2,d2(002)	0.01759	-0.00244	-0.02566
d1,d1(002)	0.00083	-0.00008	0.00216
ON SITE H-H			
s,s(000)	0.16292	0.35933	
FIRST NEIGHBOR H-H			
s,s(110)	-0.00205	0.00446	0.00336
s,x(110)	0.00000	0.00000	0.00000
x,x(110)	0.00000	0.00000	0.00000
x,x(011)	0.00000	0.00000	0.00000
x,y(110)	0.00000	0.00000	0.00000
SECOND NEIGHBOR H-H			
s,s(200)	-0.00815	0.00009	-0.00268
s,x(200)	0.00000	0.00000	0.00000
x,x(200)	0.00000	0.00000	0.00000
y,y(200)	0.00000	0.00000	0.00000
FIRST NEIGHBOR Hg-H			
s,s(100)	-0.08988	-0.09851	-0.05154
s,x(100)	0.00000	0.00000	0.00000
x,s(100)	0.09198	0.15498	0.07718
x,x(100)	0.00000	0.00000	0.00000
y,y(100)	0.00000	0.00000	0.00000
xy,x(010)	0.00000	0.00000	0.00000
d2,s(001)	-0.08110	0.02850	-0.07199
d2,z(001)	0.00000	0.00000	0.00000
SECOND NEIGHBOR Hg-H			
s,s(111)	-0.00046	-0.00088	-0.00003
x,s(111)	-0.00238	0.00114	-0.00563
s,x(111)	0.00000	0.00000	0.00000
x,x(111)	0.00000	0.00000	0.00000
x,y(111)	0.00000	0.00000	0.00000
xy,s(111)	-0.00085	0.00078	0.00287
xy,x(111)	0.00000	0.00000	0.00000
yz,x(111)	0.00000	0.00000	0.00000
d1,x(111)	0.00000	0.00000	0.00000

HgH

BAND	ORTHOGONAL RMS ERROR mRy	MAXIMUM DEVIATION k	mRy	NON-ORTHOGONAL RMS ERROR mRy	MAXIMUM DEVIATION k	mRy
1	1.4	(004)	3.8	0.1	(055)	0.3
2	1.4	(222)	4.0	0.1	(044)	0.3
3	1.7	(226)	4.3	0.1	(000)	0.2
4	1.6	(333)	5.4	0.1	(224)	0.2
5	1.4	(044)	3.9	0.2	(044)	0.7
6	2.7	(006)	7.3	0.1	(044)	0.4
7	4.1	(007)	17.0	0.2	(000)	0.4
1-7	2.2			0.1		

ENERGY VALUES IN Ry AT SELECTED k-POINTS

	ORTHOGONAL	APW	NON-ORTHOGONAL
GAMMA 1	-0.29659	-0.29339	-0.29334
GAMMA 1	0.85326	0.85833	0.85788
GAMMA 12	0.00295	0.00119	0.00146
GAMMA 15	1.32985	1.33982	1.33982
GAMMA 25'	-0.03751	-0.03779	-0.03758
X1 (008)	-0.17563	-0.17573	-0.17581
X1 (008)	0.36045	0.36408	0.36398
X1 (008)	0.68492	0.68129	0.68159
X2 (008)	0.03393	0.03306	0.03322
X3 (008)	-0.09419	-0.09581	-0.09568
X5 (008)	0.04700	0.04593	0.04612
X4' (008)	0.39675	0.39667	0.39662
X5' (008)	0.87202	0.86865	0.86864
L1 (444)	-0.10901	-0.10882	-0.10869
L1 (444)	0.30142	0.29901	0.29918
L3 (444)	-0.04086	-0.04211	-0.04195
L3 (444)	0.03822	0.03591	0.03608
L2' (444)	0.06757	0.07152	0.07156
L2' (444)	0.91566	0.91354	0.91354
L3' (444)	1.28989	1.27177	1.27175
W1 (048)	-0.00231	0.00056	0.00065
W1 (048)	0.43297	0.42818	0.42824
W2' (048)	-0.13321	-0.13338	-0.13329
W2' (048)	0.40281	0.40532	0.40541
W2' (048)	1.28361	1.32279	1.32283
W1' (048)	0.04658	0.04603	0.04614
W3 (048)	-0.04532	-0.04243	-0.04234
W3 (048)	0.61173	0.60958	0.60969
ODD (224)	-0.02442	-0.02503	-0.02525
ODD (224)	0.03055	0.02999	0.03005
EVEN(224)	-0.13349	-0.13349	-0.13371
EVEN(224)	-0.05465	-0.05518	-0.05517
EVEN(224)	-0.03343	-0.03361	-0.03349
EVEN(224)	0.04825	0.04659	0.04644
EVEN(224)	0.50432	0.50323	0.50325

FERMI LEVEL QUANTITIES (Non-orthogonal fit)

ENERGY Ry	DENSITIES OF STATES (states/Ry/cell) Total	s-Hg	p-Hg	t2g-Hg	eg-Hg	s-H	p-H
0.4723	6.84	2.96	1.03	0.47	0.38	2.00	0.00

	INTEGRATED DENSITIES OF STATES (electrons) Total	s-Hg	p-Hg	t2g-Hg	eg-Hg	s-H	p-H
	13.00	1.19	0.35	6.11	4.00	1.35	0.00

VELOCITY cm/s	PLASMON ENERGY eV	EIGENVALUE SUM Ry
1.20x10E8	8.55	-5.8481

Table 8.45 HgH (NaCl) a = 8.87 Bohr Slater–Koster 2-center parameters

	ORTHOGONAL	NON-ORTHOGONAL	
	ENERGY INTEGRALS Ry	ENERGY INTEGRALS Ry	OVERLAP INTEGRALS
ON SITE Hg-Hg			
s	0.24827	0.15696	
p	0.95968	0.53206	
t2g	-0.00942	-0.01905	
eg	-0.01190	-0.02123	
FIRST NEIGHBOR Hg-Hg			
(sss)	-0.04226	-0.03347	0.09166
(pps)	0.11747	-0.08281	-0.33423
(ppp)	-0.00155	0.01319	0.08335
(dds)	-0.01850	-0.01925	0.03330
(ddp)	0.00728	0.01074	-0.01629
(ddd)	-0.00048	-0.00148	-0.00164
(sps)	0.07028	0.03660	-0.13692
(sds)	-0.02560	-0.02202	0.02339
(pds)	-0.03776	-0.01804	0.11966
(pdp)	0.00797	0.01435	-0.05001
SECOND NEIGHBOR Hg-Hg			
(sss)	0.00338	-0.01041	-0.00013
(pps)	-0.06763	-0.00422	-0.05743
(ppp)	0.01007	0.01053	0.00969
(dds)	0.00017	-0.00198	-0.00815
(ddp)	-0.00009	0.00058	0.00462
(ddd)	-0.00026	-0.00007	0.00114
(sps)	-0.01126	0.00347	-0.03643
(sds)	0.00065	-0.00370	0.02188
(pds)	-0.00297	-0.01249	0.02019
(pdp)	-0.00290	0.00211	-0.00396
ON SITE H-H			
s	0.49182	0.57796	
FIRST NEIGHBOR H-H			
(sss)	0.01761	0.01630	0.00993
(sps)	0.00000	0.00000	0.00000
(pps)	0.00000	0.00000	0.00000
(ppp)	0.00000	0.00000	0.00000
SECOND NEIGHBOR H-H			
(sss)	0.01454	0.00670	0.00944
(sps)	0.00000	0.00000	0.00000
(pps)	0.00000	0.00000	0.00000
(ppp)	0.00000	0.00000	0.00000
FIRST NEIGHBOR Hg-H			
(sss)	0.05641	0.08273	-0.02596
(sps)	0.00000	0.00000	0.00000
(pss)	0.11875	0.16033	-0.05396
(pps)	0.00000	0.00000	0.00000
(ppp)	0.00000	0.00000	0.00000
(dpp)	0.00000	0.00000	0.00000
(dss)	0.04933	0.05016	0.03597
(dps)	0.00000	0.00000	0.00000
SECOND NEIGHBOR Hg-H			
(sss)	-0.01084	0.01404	0.01475
(pss)	-0.02181	0.01935	0.01178
(sps)	0.00000	0.00000	0.00000
(pps)	0.00000	0.00000	0.00000
(ppp)	0.00000	0.00000	0.00000
(dss)	-0.00382	0.00064	0.01107
(dps)	0.00000	0.00000	0.00000
(dpp)	0.00000	0.00000	0.00000

HgH

BAND	ORTHOGONAL RMS ERROR mRy	MAXIMUM DEVIATION k	mRy	NON-ORTHOGONAL RMS ERROR mRy	MAXIMUM DEVIATION k	mRy
1	3.7	(226)	8.7	0.3	(444)	1.0
2	1.6	(354)	4.4	0.5	(055)	1.2
3	0.9	(048)	1.9	0.3	(033)	0.9
4	1.4	(333)	6.6	0.3	(333)	0.7
5	0.7	(042)	2.1	0.3	(044)	1.1
6	2.5	(007)	6.8	0.5	(066)	1.4
7	8.4	(007)	34.1	0.8	(007)	2.0
1-7	3.7			0.5		

ENERGY VALUES IN Ry AT SELECTED k-POINTS

	ORTHOGONAL	APW	NON-ORTHOGONAL
GAMMA 1	-0.29682	-0.29339	-0.29319
GAMMA 1	0.84863	0.85833	0.85797
GAMMA 12	0.00163	0.00119	0.00166
GAMMA 15	1.32216	1.33982	1.33977
GAMMA 25'	-0.03907	-0.03779	-0.03733
X1 (008)	-0.17008	-0.17573	-0.17578
X1 (008)	0.36346	0.36408	0.36458
X1 (008)	0.68699	0.68129	0.68271
X2 (008)	0.03231	0.03306	0.03284
X3 (008)	-0.09352	-0.09581	-0.09503
X5 (008)	0.04568	0.04593	0.04591
X4' (008)	0.39480	0.39667	0.39673
X5' (008)	0.87091	0.86865	0.86870
L1 (444)	-0.10950	-0.10882	-0.10784
L1 (444)	0.29791	0.29901	0.29884
L3 (444)	-0.04167	-0.04211	-0.04196
L3 (444)	0.03704	0.03591	0.03631
L2' (444)	0.07111	0.07152	0.07111
L2' (444)	0.91204	0.91354	0.91363
L3' (444)	1.29273	1.27177	1.27181
W1 (048)	0.00024	0.00056	0.00111
W1 (048)	0.43088	0.42818	0.42706
W2' (048)	-0.13284	-0.13338	-0.13337
W2' (048)	0.40369	0.40532	0.40562
W2' (048)	1.28081	1.32279	1.32261
W1' (048)	0.04674	0.04603	0.04644
W3 (048)	-0.04434	-0.04243	-0.04221
W3 (048)	0.61336	0.60958	0.60940
ODD (224)	-0.02418	-0.02503	-0.02548
ODD (224)	0.03058	0.02999	0.03017
EVEN(224)	-0.13514	-0.13349	-0.13411
EVEN(224)	-0.05458	-0.05518	-0.05435
EVEN(224)	-0.03236	-0.03361	-0.03329
EVEN(224)	0.05032	0.04659	0.04612
EVEN(224)	0.50807	0.50323	0.50262

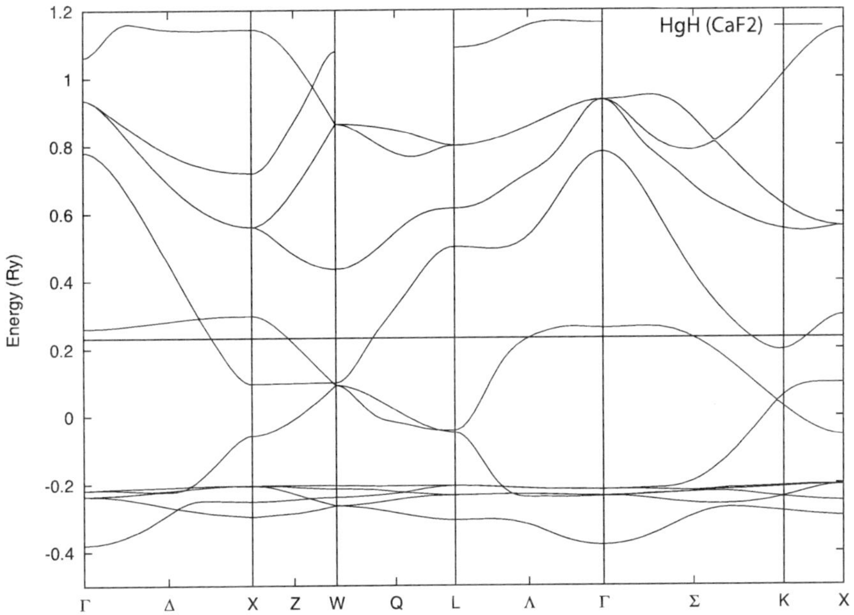

Fig. 8.44 Energy bands of HgH2 in the CaF2 structure

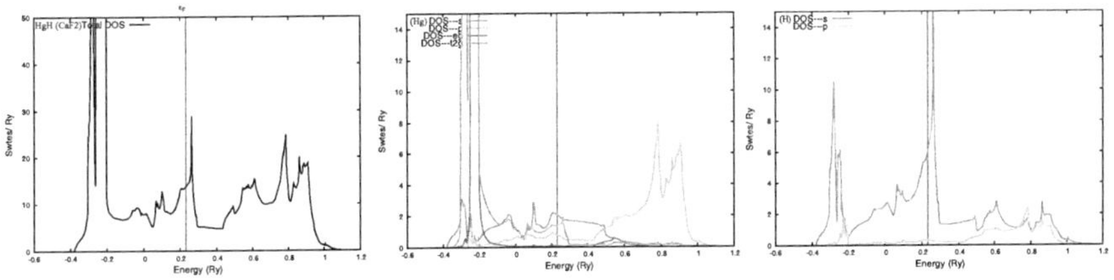

Fig. 8.45 Total, angular momentum and site decomposed densities of states of HgH2 in the CaF2 structure

References

1. G.G. Libowitz, The nature and properties of transition metal hydrides. J. Nucl. Mater. **2**, 1 (1960)
2. W.M. Mueller, J.P. Blackledge, G.G. Libowitz (eds.), *Metal Hydrides* (Academic Press, N.Y., 1968)
3. H. Smithson, C.A. Marianetti, D. Morgan, A. Van der Ven, A. Predith, G. Ceder, First-Principles study of the stability and electronic structure of metal hydrides. Phys. Rev. B **66**, 144107 (2002)
4. D.A. Papaconstantopoulos, Platinum Hydride: A possible high temperature superconductor. J. Less. Comm. Metals **73**, 305 (1980)
5. T. Scheler, O. Degtyareva, M. Marques, C.L. Guillaume, J.E. Proctor, S. Evans, E. Gregoryanz, Synthesis and Properties of platinum hydride. Phys. Rev. B **83**, 214106 (2011)

Chapter 9
Group 13 Hydrides

This chapter covers the Group 13 column of the periodic table hydrides from BH to TlH [1]. Results are presented for the crystal structures NaCl(B1),CaF2 (C1) and Im3m. These structures are not found experimentally and are presented here for the purpose of comparison through the periodic Table and to establish various trends. These compounds form non-metal hydrides. BH is found in a cluster form. AlH exists as a polymer (Alane) and GaH is a dimer. It is possible that under high pressure conditions a structure like the Im3m may emerge in the future in the search for high temperature superconductivity. The lattice constants, as expected, are larger than in the single elements because of the expansion of the lattice upon hydrogenation. Comparing the energy bands of the CaF2 structure to the bands of the NaCl structure we note the additional antibonding band above the Fermi level due to the second hydrogen in the CaF2 structure. Examining the densities of states figures we observe that for both crystal structures the lowest band has a mixture of s–H hydrogen and s-character of the other element. At the Fermi level the states are of s- and p- character. There are significant differences in the DOS among different crystal structures. In GaH, InH and TlH we show d-states appearing below the GAM1 point in the energy band diagrams. Tight-binding parameters are given in the NaCl structure based on both orthogonal and non-orthogonal Hamiltonians using three- and two-center integrals.

D. A. Papaconstantopoulos, *Band Structure of Cubic Hydrides*,
https://doi.org/10.1007/978-3-031-06878-2_9

9.1 Boron Hydride (BH)

See Fig. 9.1.

See Tables 9.1, 9.2 and 9.3.

See Figs. 9.2 and 9.3.

See Tables 9.4 and 9.5.

See Figs. 9.4, 9.5 and 9.6.

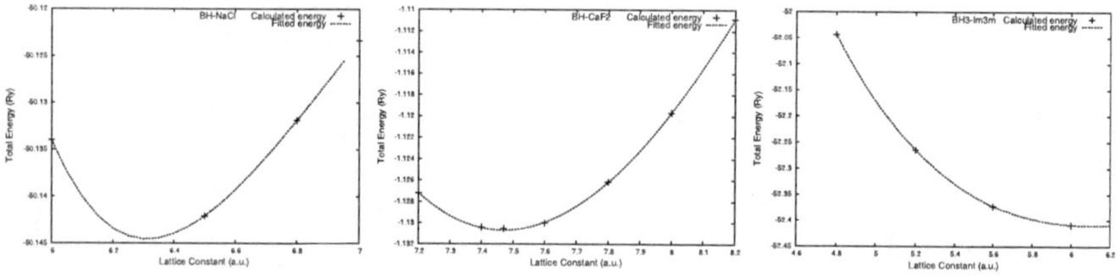

Fig. 9.1 Total energy versus lattice constant of BH in the NaCl, CaF2 and Im3m structures

Table 9.1 Lattice constant, bulk modulus, gap, total energy

Stru	a (Bohr)	B (MBar)	Gap	Total Energy (Ry)
NaCl	6.35	1.30	-	-50.14480
CaF2	7.47	0.78	-	-51.13052
Im3m (P=0)	6.10	1.29	-	-52.40984
Im3m(P=2.14 MBar)	5.00	7.09	-	-52.17134

Table 9.2 Birch fit coefficients

	A1	A2	A3	A4
NaCl	-5.043013E+01	2.635024E+01	-6.189931E+02	4.328484E+03
CaF2	-1.586179E-01	-5.074936E+01	8.176688E+02	-3.776197E+03
Im3m	-5.104443E+01	-6.978906E+01	1.020913E+03	-3.170487E+03

Table 9.3 DOS at Ef, Hopfield parameter, Stoner criterion

```
        BH        a=6.35 Bohr          NaCl
-------------------------------------------------------------------------------
 Fermi Energy    Total DOS        Decomposed DOS inside the muffin tins
    ( Ry )     (States/Ry)       s         p          eg          t2g      f
-------------------------------------------------------------------------------
 B  0.998         5.595       0.961      1.323      0.036       0.043   0.010
 H  0.998         5.595       0.519      0.573      0.058       0.034   0.012
-------------------------------------------------------------------------------
Fermi-Velocity (cm/s): 1.61152 x10E8
Plasmon Energy (eV)   : 17.15404
Electron-ion interaction (Hopfield parameter) (eV/A^2) B:3.496  H:2.168
-------------------------------------------------
   B      MUFFIN-TIN RADIUS and CHARGE =     1.5875    3.2806
   H      MUFFIN-TIN RADIUS and CHARGE =     1.5875    1.1863
                    B        STONER I =  0.0263
                    H        STONER I =  0.0081
          STONER PARAMETER (Ry)     I =  0.0352
          STONER CRITERION        N*I =  0.1971
-------------------------------------------------
        BH2       a=7.47 Bohr          CaF2
-------------------------------------------------
 Fermi Energy    Total DOS        Decomposed DOS inside the muffin tins
    ( Ry )     (States/Ry)       s         p          eg          t2g      f
 B  0.678         7.326       1.013     1.669      0.020       0.026  0.005
 H  0.678         7.326       0.945     0.928      0.032       0.094  0.013
-------------------------------------------------------------------------------
Fermi-Velocity (cm/s): 1.43692 x10E8
Plasmon Energy (eV)   : 13.718
Electron-ion interaction (Hopfield parameter) (eV/A^2) B:1.859  H2:5.744
-------------------------------------------------
   B      MUFFIN-TIN RADIUS and CHARGE =     1.6172    3.2795
   H      MUFFIN-TIN RADIUS and CHARGE =     1.6172    1.0473
                    B        STONER I =  0.0200
                    H        STONER I =  0.0036
          STONER PARAMETER (Ry)     I =  0.0276
          STONER CRITERION        N*I =  0.2019
-------------------------------------------------------------------------------
        BH3  a=5.0 Bohr                Im3m
-------------------------------------------------------------------------------
 Fermi Energy    Total DOS        Decomposed DOS inside the muffin tins
    ( Ry )     (States/Ry)       s       p        d       f
 B  1.379         6.950       0.488   1.675    0.045   0.012
 H  1.379         6.950       0.487   0.086    0.012   0.001
-------------------------------------------------------------------------------
Fermi-Velocity (cm/s): 1.60638 x10E8
Plasmon Energy (eV)   : 19.28549
Electron-ion interaction (Hopfield parameter) (eV/A^2) B:5.369  H3:6.292
-------------------------------------------------
   B      MUFFIN-TIN RADIUS and CHARGE =    1.40     3.2806
   H      MUFFIN-TIN RADIUS and CHARGE =    1.00     0.5676
```

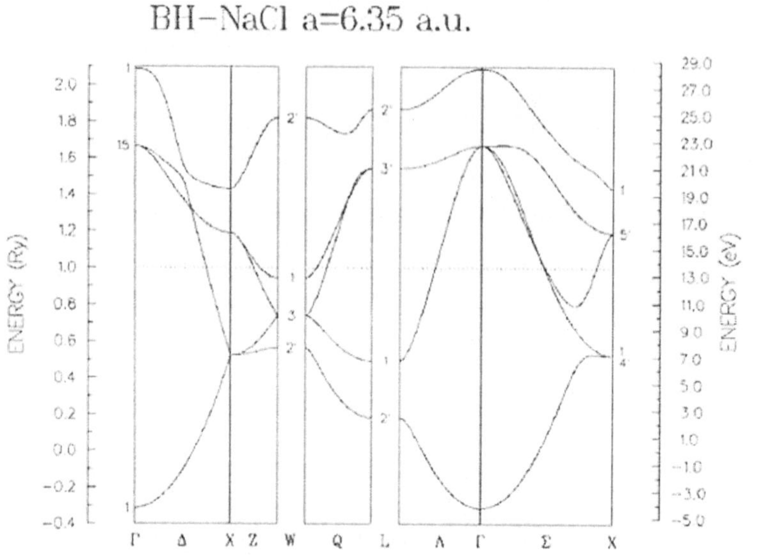

Fig. 9.2 Energy bands of BH in the NaCl structure (tight binding)

Fig. 9.3 Total, angular momentum and site decomposed densities of states of BH in the NaCl structure (tight binding)

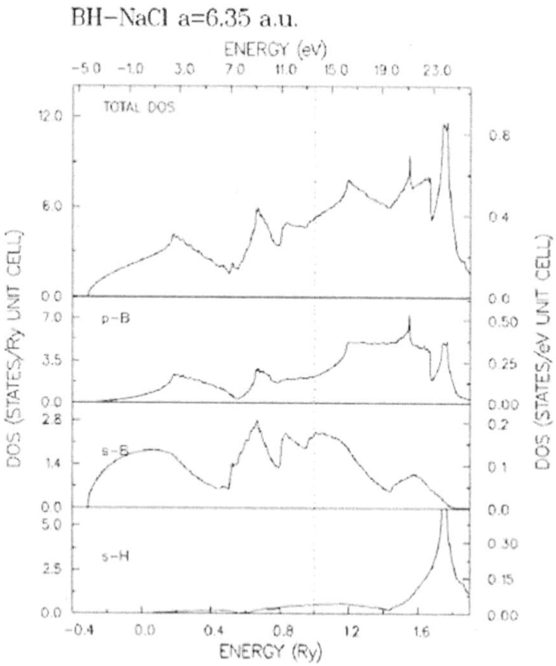

Table 9.4 BH (NaCl) a = 6.35 Bohr Slater–Koster 3-center parameters

	ORTHOGONAL	NON-ORTHOGONAL	
	ENERGY INTEGRALS Ry	ENERGY INTEGRALS Ry	OVERLAP INTEGRALS
ON SITE B- B			
s,s(000)	1.28275	1.05904	
x,x(000)	1.92092	1.77999	
FIRST NEIGHBOR B- B			
s,s(110)	-0.04287	0.00979	0.11281
s,x(110)	0.03533	0.07457	0.08901
s,xy(110)	0.00000	0.00000	0.00000
x,x(110)	-0.15656	-0.21121	-0.05211
x,x(011)	0.14557	0.10076	0.03856
x,y(110)	0.04532	-0.06768	-0.12128
SECOND NEIGHBOR B- B			
s,s(200)	-0.02284	-0.09039	-0.01664
s,x(200)	-0.04529	-0.11390	0.00472
x,x(200)	0.05825	0.16065	-0.02355
y,y(200)	0.11178	-0.02566	-0.06830
ON SITE H-H			
s,s(000)	0.90163	0.47925	
FIRST NEIGHBOR H-H			
s,s(110)	0.03231	0.00347	0.08713
s,x(110)	0.00336	0.02932	0.04188
x,x(110)	-0.00450	-0.00932	-0.02956
x,x(011)	0.06884	-0.04260	-0.01535
x,y(110)	-0.08190	0.05546	0.06010
SECOND NEIGHBOR H-H			
s,s(200)	-0.02342	-0.01258	-0.00416
s,x(200)	-0.01801	-0.04632	-0.00437
x,x(200)	0.08537	-0.06191	-0.11712
y,y(200)	-0.04692	0.03584	0.02050
FIRST NEIGHBOR B-H			
s,s(100)	-0.18575	-0.19830	0.16429
s,x(100)	0.63452	0.50134	-0.21710
x,s(100)	0.24276	0.19810	-0.15997
x,x(100)	0.83984	0.84724	-0.18396
y,y(100)	-0.62305	-0.37087	0.18191
SECOND NEIGHBOR B-H			
s,s(111)	-0.00764	-0.04586	0.01834
x,s(111)	0.00917	0.02991	-0.02635
s,x(111)	0.00001	0.03353	0.00751
x,x(111)	0.01056	0.01743	0.00026
x,y(111)	0.00701	0.01359	-0.01330

BH

BAND	ORTHOGONAL RMS ERROR mRy	MAXIMUM DEVIATION k	MAXIMUM DEVIATION mRy	NON-ORTHOGONAL RMS ERROR mRy	MAXIMUM DEVIATION k	MAXIMUM DEVIATION mRy
1	4.2	(066)	9.8	0.4	(033)	1.1
2	5.8	(004)	13.5	0.5	(226)	1.4
3	9.8	(022)	23.5	0.6	(180)	2.6
4	6.2	(008)	11.4	0.5	(008)	1.1
5	10.8	(066)	25.1	1.3	(118)	3.0
1-5	7.8			0.8		

ENERGY VALUES IN Ry AT SELECTED k-POINTS

	ORTHOGONAL	APW	NON-ORTHOGONAL
GAMMA 1	-0.31373	-0.31339	-0.31320
GAMMA 1	2.09391	2.08973	2.08936
GAMMA 15	1.66535	1.66708	1.66736
X1 (008)	0.51225	0.52137	0.52093
X1 (008)	1.43684	1.43382	1.43313
X4' (008)	0.52263	0.52063	0.52061
X5' (008)	1.18079	1.19216	1.19104
L1 (444)	0.48787	0.49514	0.49457
L2' (444)	0.17716	0.18215	0.18236
L2' (444)	1.85969	1.87027	1.86919
L3' (444)	1.53858	1.54509	1.54539
W1 (048)	0.93872	0.94779	0.94814
W2' (048)	0.57090	0.56207	0.56198
W2' (048)	1.82396	1.81739	1.81745
W3 (048)	0.73963	0.74222	0.74191
ODD (224)	1.44781	1.44755	1.44826

FERMI LEVEL QUANTITIES (Non-orthogonal fit)

ENERGY		DENSITIES OF STATES					
	Total	s- B	p- B	t2g- B	eg- B	s-H	p-H
Ry		states/Ry/cell					
0.9992	5.29	2.37	2.23	0.00	0.00	0.53	0.16
		INTEGRATED DENSITIES OF STATES					
	Total	s- B	p- B	t2g- B	eg- B	s-H	p-H
		electrons					
	4.00	2.05	1.70	0.00	0.00	0.20	0.06

VELOCITY	PLASMON ENERGY	EIGENVALUE SUM
cm/s	eV	Ry
1.72x10E8	17.81	-2.1220

Table 9.5 BH (NaCl) a = 6.35 Bohr Slater–Koster 2-center parameters

	ORTHOGONAL	NON-ORTHOGONAL	
	ENERGY INTEGRALS Ry	ENERGY INTEGRALS Ry	OVERLAP INTEGRALS
ON SITE B-B			
s	0.86939	0.31349	
p	1.38502	0.82880	
FIRST NEIGHBOR B-B			
(sss)	-0.02014	-0.06598	0.11477
(pps)	0.07188	-0.03549	-0.23657
(ppp)	0.00375	-0.01249	0.04644
(sps)	0.04214	0.03941	-0.15907
SECOND NEIGHBOR B-B			
(sss)	0.05513	-0.02973	0.00267
(pps)	-0.07850	0.03204	-0.01989
(ppp)	0.03586	-0.01358	-0.00688
(sps)	-0.06175	0.02858	-0.01357
(sds)	0.00000	0.00000	0.00000
ON SITE H-H			
s	0.80021	0.33978	

FIRST NEIGHBOR H-H			
(sss)	0.00981	-0.06210	0.01425
(sps)	-0.14494	0.00000	0.00000
(pps)	-1.18848	0.00000	0.00000
(ppp)	0.37050	0.00000	0.00000
SECOND NEIGHBOR H-H			
(sss)	-0.01748	0.01723	0.01132
(sps)	0.06034	0.00000	0.00000
(pps)	0.88164	0.00000	0.00000
(ppp)	-0.20975	0.00000	0.00000
FIRST NEIGHBOR B-H			
(sss)	0.19205	-0.06310	0.26245
(sps)	0.26587	0.00000	0.00000
(pss)	0.24199	0.00103	0.33909
(pps)	-0.60879	0.00000	0.00000
(ppp)	0.41826	0.00000	0.00000
SECOND NEIGHBOR B-H			
(sss)	0.00567	-0.02798	0.00906
(pss)	0.01599	-0.04009	0.01539
(sps)	-0.05214	0.00000	0.00000
(pps)	-0.01459	0.00000	0.00000
(ppp)	0.00625	0.00000	0.00000

BH

BAND	ORTHOGONAL RMS ERROR mRy	MAXIMUM DEVIATION k	mRy	NON-ORTHOGONAL RMS ERROR mRy	MAXIMUM DEVIATION k	mRy
1	5.8	(008)	13.0	3.6	(444)	9.1
2	8.1	(055)	22.0	3.5	(062)	10.4
3	13.9	(022)	30.3	4.1	(062)	9.2
4	6.8	(005)	16.2	3.0	(062)	7.4
5	11.5	(066)	33.1	5.7	(224)	21.5
1-5	9.7			4.1		

ENERGY VALUES IN Ry AT SELECTED k-POINTS

	ORTHOGONAL	APW	NON-ORTHOGONAL
GAMMA 1	-0.31411	-0.31339	-0.31237
GAMMA 1	2.08571	2.08973	2.09057
GAMMA 15	1.66186	1.66708	1.66724
X1 (008)	0.50765	0.52137	0.51757
X1 (008)	1.42914	1.43382	1.43659
X4' (008)	0.53020	0.52063	0.52184
X5' (008)	1.17869	1.19216	1.18952
L1 (444)	0.48140	0.49514	0.49989
L2' (444)	0.17009	0.18215	0.17300
L2' (444)	1.86111	1.87027	1.85753
L3' (444)	1.53486	1.54509	1.54211
W1 (048)	0.94025	0.94779	0.94817
W2' (048)	0.57278	0.56207	0.56190
W2' (048)	1.82369	1.81739	1.81623
W3 (048)	0.74337	0.74222	0.73766

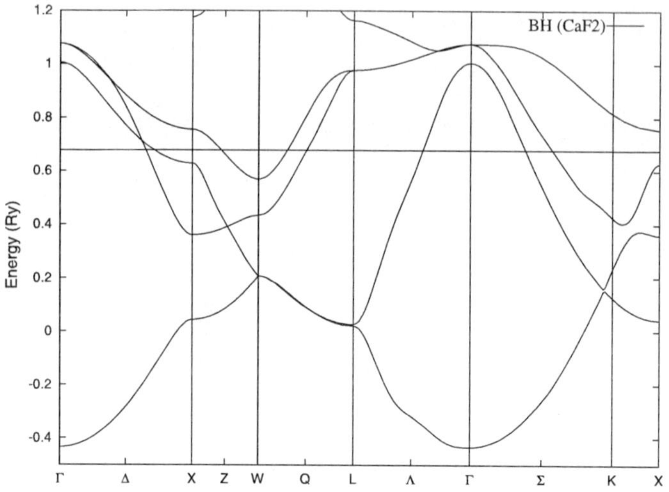

Fig. 9.4 Energy bands of BH2 in the CaF2 structure

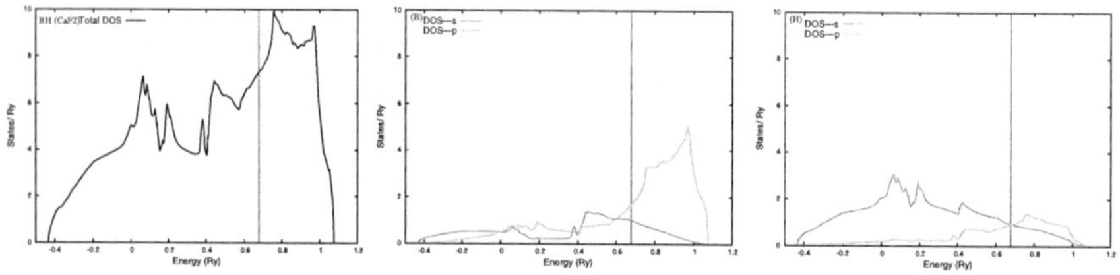

Fig. 9.5 Total, angular momentum and site decomposed densities of states of BH2 in the CaF2 structure

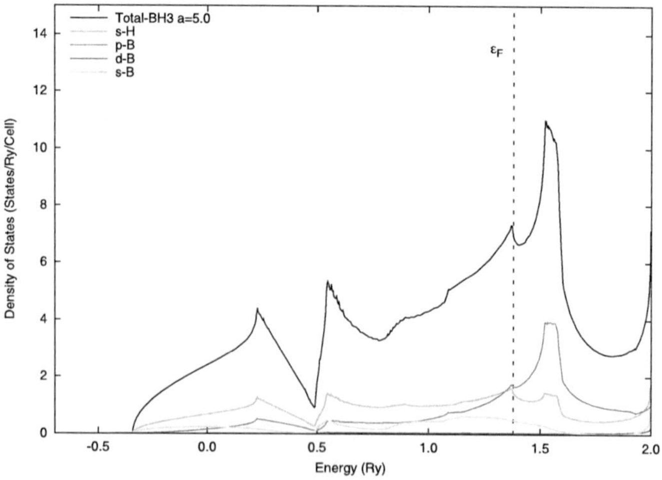

Fig. 9.6 Total, angular momentum and site decomposed densities of states of BH3 in the Im3m structure

9.2 Aluminum Hydride (AlH)

See Figs. 9.7 and 9.8.

See Tables 9.6, 9.7 and 9.8.

See Figs. 9.9 and 9.10.

See Tables 9.9 and 9.10.

See Figs. 9.11 and 9.12.

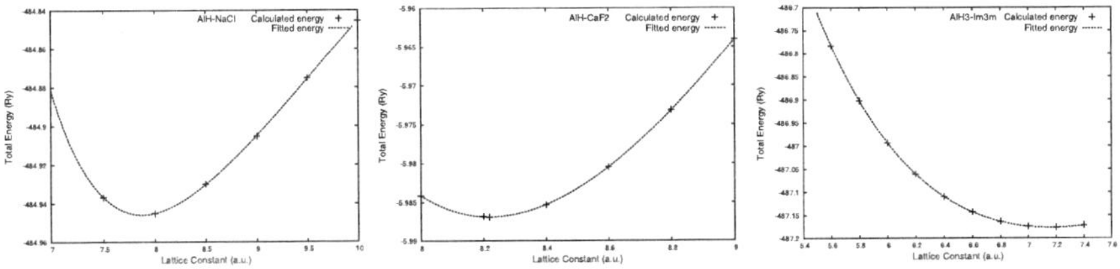

Fig. 9.7 Total energy versus lattice constant of AlH in the NaCl, CaF2 and Im3m structures

Table 9.6 Lattice constant, bulk modulus, gap, total energy

Stru	a (Bohr)	B (MBar)	Gap	Total Energy (Ry)
NaCl	7.87	0.88	-	-484.93298
CaF2	8.22	0.82	-	-485.98688
Im3m (P=0)	7.16	0.78	-	-487.17594
Im3m (P=2.12MBar)	5.60	6.44	-	-486.78306

Table 9.7 Birch fit coefficients

	A1	A2	A3	A4
NaCl	-4.844590E+02	-1.569051E+01	-1.122623E+02	5.009548E+03
CaF2	-5.379560E+00	-2.543763E+01	5.452137E+01	5.119546E+03
Im3m	-4.858903E+02	-8.922755E+01	1.739847E+03	-6.431134E+03

Table 9.8 DOS at Ef, Hopfield parameter, Stoner criterion

```
        AlH     a=7.87 Bohr             NaCl
 --------------------------------------------------------------------------------
 Fermi Energy   Total DOS        Decomposed DOS inside the muffin tins
     ( Ry )    (States/Ry)       s         p         eg         t2g      f
 -------------------------------------------------------------------------------
 Al  0.818          5.912       0.421     0.690     0.071      0.168    0.017
 H   0.818          5.912       0.744     0.852     0.016      0.088    0.006
 --------------------------------------------------------------------------------
Fermi-Velocity (cm/s): 1.50829 x10E8
Plasmon Energy (eV)   : 11.98336
Electron-ion interaction (Hopfield parameter) (eV/A^2) Al:0.203  H:1.669
-------------------------------------------------
  Al      MUFFIN-TIN RADIUS and CHARGE =     1.9650   10.8055
   H      MUFFIN-TIN RADIUS and CHARGE =     1.9650    1.4964
                    Al       STONER I = 0.0069
                     H       STONER I = 0.0104
         STONER PARAMETER (Ry)     I = 0.0179
         STONER CRITERION        N*I = 0.1058
-------------------------------------------------
        AlH2      a=8.22 Bohr            CaF2
-------------------------------------------------
 Fermi Energy   Total DOS        Decomposed DOS inside the muffin tins
     ( Ry )    (States/Ry)       s         p         eg         t2g      f
 --------------------------------------------------------------------------------
 Al  0.809          6.151       0.657     0.256     0.083      0.032    0.011
 H   0.809          6.151       0.736     1.150     0.034      0.074    0.006
 --------------------------------------------------------------------------------
Fermi-Velocity (cm/s): 1.28277 x10E8
Plasmon Energy (eV)   : 9.72250
Electron-ion interaction (Hopfield parameter) (eV/A^2) Al:0.138  H2:3.888
 --------------------------------------------------------------------------------
  Al      MUFFIN-TIN RADIUS and CHARGE =     1.7795   10.4800
   H      MUFFIN-TIN RADIUS and CHARGE =     1.7795    1.3470
                    Al       STONER I = 0.0046
                     H       STONER I = 0.0040
         STONER PARAMETER (Ry)     I = 0.0131
         STONER CRITERION        N*I = 0.0805
 --------------------------------------------------------------------------------
        AlH3    a=5.6 Bohr              Im3m
 Fermi Energy   Total DOS        Decomposed DOS inside the muffin tins
     ( Ry )    (States/Ry)       s         p         d        f
 --------------------------------------------------------------------------------
 Al  1.340          5.030       0.371     0.618     0.181    0.042
 H   1.340          5.030       0.363     0.063     0.004    0.000
Fermi-Velocity (cm/s): 1.23149 x10E8
Plasmon Energy (eV)   : 10.61212
Electron-ion interaction (Hopfield parameter) (eV/A^2) Al:1.261  H3:2.315
  Al      MUFFIN-TIN RADIUS and CHARGE =     1.80    11.0194
   H      MUFFIN-TIN RADIUS and CHARGE =     1.00     0.6064
```

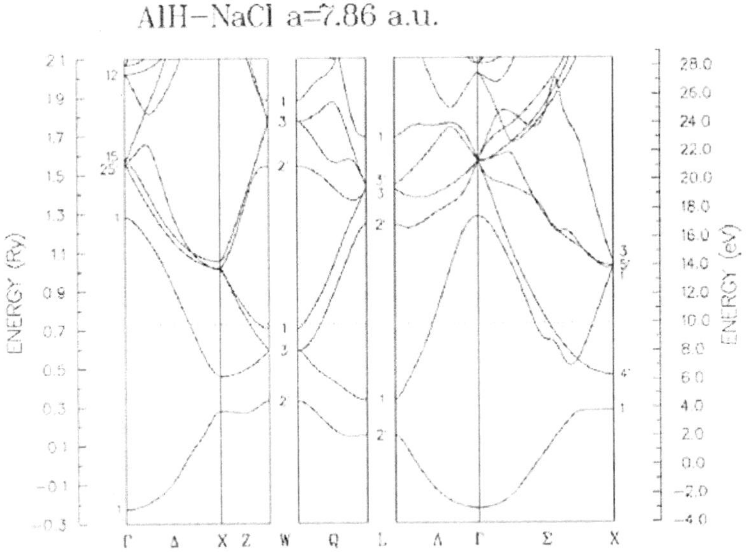

Fig. 9.8 Energy bands of AlH in the NaCl structure

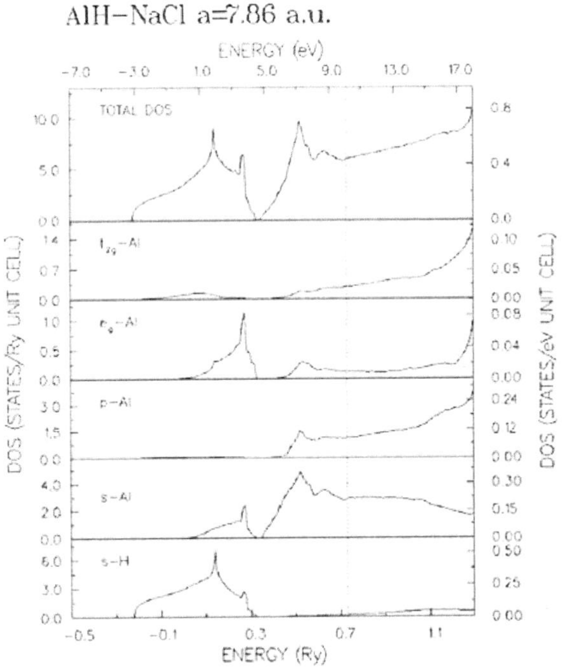

Fig. 9.9 Total, angular momentum and site decomposed densities of states of AlH in the NaCl structure (tight binding)

Table 9.9 AlH (NaCl) a=7.86 Bohr Slater–Koster 3-center parameters

	ORTHOGONAL	NON-ORTHOGONAL	
	ENERGY INTEGRALS Ry	ENERGY INTEGRALS Ry	OVERLAP INTEGRALS
ON SITE Al-Al			
s,s(000)	0.58122	0.68223	
x,x(000)	1.33873	1.57220	
xy,xy(000)	2.63126	1.85179	
d2,d2(000)	1.93933	1.47140	
FIRST NEIGHBOR Al-Al			
s,s(110)	-0.04780	0.16585	0.26474
s,x(110)	0.05505	-0.12179	-0.15607
s,xy(110)	0.10286	-0.04862	-0.06151
s,d2(110)	-0.03641	0.06651	0.12904
x,x(110)	-0.01393	0.08223	0.04017
x,x(011)	0.07127	0.02187	0.04473
x,y(110)	0.05849	-0.05233	-0.09164
x,xy(110)	-0.01204	0.05235	-0.01481
x,xy(011)	0.05672	0.16100	0.09738
z,d2(011)	-0.00458	0.02457	0.10247
z,d1(011)	-0.05065	0.06256	0.06904
xy,xy(110)	-0.15625	-0.12693	0.02131
xy,xy(011)	0.03266	-0.06027	-0.07158
xy,xz(011)	-0.18696	0.04012	0.02025
xy,d2(110)	-0.02198	-0.07798	0.01687
d2,d2(110)	-0.03129	0.20682	0.16069
d1,d1(110)	-0.00685	0.03408	-0.15811
SECOND NEIGHBOR Al-Al			
s,s(200)	-0.03031	0.08874	0.12449
s,x(200)	0.02653	-0.05515	-0.06548
s,d2(002)	0.07014	-0.03961	-0.11153
x,x(200)	-0.03891	0.05523	-0.05302
y,y(200)	0.04505	-0.01005	-0.03818
x,xy(020)	0.08163	-0.04897	-0.04246
z,d2(002)	-0.03835	0.11703	-0.03986
xy,xy(200)	-0.18497	-0.02001	-0.00456
xy,xy(002)	0.00990	-0.01622	-0.00754
d2,d2(002)	0.08238	-0.16092	0.05487
d1,d1(002)	0.01811	0.06484	0.04003
ON SITE H-H			
s,s(000)	0.76929	0.30802	
FIRST NEIGHBOR H-H			
s,s(110)	0.01914	-0.05121	-0.01762
s,x(110)	0.01519	0.05162	0.00849
x,x(110)	-0.07479	0.00226	-0.04381
x,x(011)	-0.03094	-0.05159	0.00577
x,y(110)	0.01774	0.03437	0.02853
SECOND NEIGHBOR H-H			
s,s(200)	0.03416	0.02348	-0.00259
s,x(200)	-0.00186	-0.07354	-0.02474
x,x(200)	-0.06113	-0.09193	-0.02306
y,y(200)	0.01419	-0.03310	-0.02340

FIRST NEIGHBOR Al-H			
s,s(100)	0.07571	-0.03798	-0.25844
s,x(100)	-0.07977	-0.10227	-0.06246
x,s(100)	0.15460	-0.07942	-0.24459
x,x(100)	0.35984	0.33991	0.00898
y,y(100)	-0.27679	0.03728	0.08403
xy,x(010)	-0.50678	-0.02291	0.04212
d2,s(001)	-0.13604	0.05510	0.33460
d2,z(001)	0.26126	0.43381	0.21992
SECOND NEIGHBOR Al-H			
s,s(111)	-0.01776	0.04449	0.02135
x,s(111)	-0.01862	-0.02953	-0.02577
s,x(111)	-0.00644	0.02466	0.01622
x,x(111)	0.02821	-0.02116	-0.01283
x,y(111)	0.02114	0.05829	0.06222
xy,s(111)	0.01951	0.05904	0.03335
xy,x(111)	0.04641	0.06562	0.02395
yz,x(111)	0.05397	0.13536	0.10836
d1,x(111)	-0.00387	-0.01368	-0.00449

AlH

	ORTHOGONAL			NON-ORTHOGONAL		
BAND	RMS ERROR	MAXIMUM DEVIATION		RMS ERROR	MAXIMUM DEVIATION	
	mRy	k	mRy	mRy	k	mRy
1	8.3	(222)	20.0	10.0	(048)	21.9
2	10.2	(222)	26.4	11.7	(022)	26.3
3	14.5	(022)	39.5	9.2	(066)	26.8
4	10.2	(000)	20.3	9.9	(066)	30.2
5	20.3	(003)	59.6	22.5	(042)	67.3
6	13.9	(044)	39.6	19.3	(064)	55.6
7	20.7	(226)	62.3	15.8	(264)	45.4
8	23.2	(264)	51.5	12.5	(042)	29.0
9	25.1	(033)	58.9	14.6	(224)	51.8
10	28.2	(048)	60.3	24.5	(033)	81.2
1-10	18.7			15.9		

ENERGY VALUES IN Ry AT SELECTED k-POINTS

	ORTHOGONAL	APW	NON-ORTHOGONAL
GAMMA 1	-0.24158	-0.22897	-0.22407
GAMMA 1	1.27140	1.29562	1.28351
GAMMA 12	2.01198	2.02191	2.02171
GAMMA 15	1.57931	1.57730	1.57057
GAMMA 25'	1.54749	1.56781	1.56381
X1 (008)	0.25952	0.26908	0.28195
X1 (008)	1.01611	1.02146	1.01329
X1 (008)	2.43196	2.43260	2.42817
X2 (008)	2.41482	2.41340	2.41184
X3 (008)	1.02486	1.05351	1.05487
X5 (008)	2.53615	2.54091	2.54928
X4' (008)	0.45907	0.45371	0.46318
X5' (008)	1.04107	1.02310	1.02050
L1 (444)	0.32745	0.34490	0.32392
L1 (444)	1.72078	1.71325	1.69060
L3 (444)	1.40595	1.42405	1.41737
L3 (444)	2.31556	2.36989	2.36885
L2' (444)	0.12118	0.13436	0.15144
L2' (444)	1.21157	1.23268	1.24093
L3' (444)	1.47034	1.44579	1.44881
W1 (048)	0.69712	0.71084	0.70867
W1 (048)	1.84526	1.87290	1.88111
W2' (048)	0.30412	0.30899	0.33093
W2' (048)	1.55620	1.54955	1.54377
W2' (048)	2.40811	2.43972	2.43140

W1' (048)	2.49654	2.55686	2.53607
W3 (048)	0.59418	0.59916	0.59375
W3 (048)	1.77181	1.76631	1.77521
ODD (224)	1.22027	1.22870	1.21450
ODD (224)	1.80655	1.77532	1.79271
ODD (224)	2.30271	2.29189	2.25736
EVEN(224)	0.00333	-0.00797	-0.01619
EVEN(224)	0.63986	0.62564	0.62249
EVEN(224)	1.11072	1.10010	1.10134
EVEN(224)	1.31325	1.32976	1.36507
EVEN(224)	1.65583	1.65984	1.62042
EVEN(224)	1.83544	1.81934	1.80922
EVEN(224)	2.01388	1.97671	2.02854

FERMI LEVEL QUANTITIES (Non-orthogonal fit)

ENERGY		DENSITIES OF STATES					
	Total	s-Al	p-Al	t2g-Al	eg-Al	s-H	p-H
Ry		states/Ry/cell					
0.7312	5.92	2.82	1.13	0.29	0.12	0.20	1.36
		INTEGRATED DENSITIES OF STATES					
	Total	s-Al	p-Al	t2g-Al	eg-Al	s-H	p-H
		electrons					
	4.00	1.36	0.32	0.09	0.15	1.47	0.61

VELOCITY	PLASMON ENERGY	EIGENVALUE SUM
cm/s	eV	Ry
1.56x10E8	12.43	-1.5936

Table 9.10 AlH (NaCl) a=7.86 Bohr Slater–Koster 2-center parameters

	ORTHOGONAL	NON-ORTHOGONAL	
	ENERGY INTEGRALS Ry	ENERGY INTEGRALS Ry	OVERLAP INTEGRALS
ON SITE Al-Al			
s	1.51237	1.25997	
p	1.14274	1.66613	
t2g	1.89594	1.67718	
eg	2.14936	2.24831	
FIRST NEIGHBOR Al-Al			
(sss)	-0.00691	-0.03203	0.06254
(pps)	0.12934	0.03839	-0.08079
(ppp)	0.01263	-0.03056	-0.00912
(dds)	-0.20939	0.07533	0.14216
(ddp)	0.05293	0.00726	-0.05518
(ddd)	0.00766	-0.00740	-0.00823
(sps)	-0.02474	-0.06116	0.01066
(sds)	0.01343	-0.10188	-0.02780
(pds)	-0.11758	0.01886	-0.10226
(pdp)	-0.01293	-0.03189	0.00562
SECOND NEIGHBOR Al-Al			
(sss)	-0.09076	-0.07124	0.03642
(pps)	-0.07377	0.17387	-0.06611
(ppp)	0.01705	0.05111	0.03147
(dds)	-0.05976	-0.58552	-0.17928
(ddp)	-0.00214	-0.07794	-0.03280
(ddd)	0.00979	0.02983	0.01310
(sps)	0.12575	-0.11797	0.04278
(sds)	-0.18093	-0.19361	-0.03772
(pds)	0.03300	0.37327	0.08206
(pdp)	-0.00285	0.08472	0.04529
ON SITE H-H			
s	0.82925	0.82309	

FIRST NEIGHBOR H-H			
(sss)	-0.06242	-0.07549	-0.01384
(sps)	0.10905	0.08381	-0.03362
(pps)	0.29205	0.23843	-0.12000
(ppp)	-0.21345	0.02002	0.05556
SECOND NEIGHBOR H-H			
(sss)	0.02400	0.03526	0.00641
(sps)	0.11724	-0.03243	-0.02496
(pps)	-0.07695	-0.05581	-0.02120
(ppp)	-0.11234	0.03726	0.01132
FIRST NEIGHBOR Al-H			
(sss)	0.12945	0.10036	-0.05767
(sps)	-0.60714	-0.20003	0.10925
(pss)	0.16965	-0.16241	0.02419
(pps)	0.12434	-0.30254	0.32982
(ppp)	0.00068	-0.10372	-0.10236
(dpp)	0.09617	-0.12837	-0.16941
(dss)	0.25891	0.04024	-0.12878
(dps)	-0.66204	-0.56308	0.05130
SECOND NEIGHBOR Al-H			
(sss)	-0.00838	0.00986	-0.03112
(pss)	-0.00525	0.03543	0.01673
(sps)	0.07743	0.05890	-0.04811
(pps)	-0.00326	0.01316	-0.00191
(ppp)	0.03373	-0.04100	-0.02247
(dss)	-0.05996	-0.05426	0.00010
(dps)	-0.08218	-0.07885	-0.00018
(dpp)	-0.00154	-0.00212	0.00010

AlH

BAND	ORTHOGONAL RMS ERROR mRy	MAXIMUM DEVIATION k	MAXIMUM DEVIATION mRy	NON-ORTHOGONAL RMS ERROR mRy	MAXIMUM DEVIATION k	MAXIMUM DEVIATION mRy
1	16.3	(226)	41.9	7.0	(226)	22.0
2	24.0	(062)	69.5	8.2	(062)	23.7
3	28.4	(444)	76.3	8.2	(066)	21.1
4	20.6	(044)	43.7	6.6	(000)	17.1
5	25.1	(003)	52.9	9.6	(002)	35.8
6	20.9	(002)	51.2	5.7	(224)	13.9
7	28.4	(042)	57.7	8.8	(011)	30.9
8	36.8	(044)	70.1	15.2	(022)	51.3
9	40.7	(224)	111.3	13.6	(224)	33.1
10	36.9	(066)	88.3	14.4	(442)	44.5
1-10	28.8			10.2		

ENERGY VALUES IN Ry AT SELECTED k-POINTS

	ORTHOGONAL	APW	NON-ORTHOGONAL
GAMMA 1	-0.22825	-0.22897	-0.22969
GAMMA 1	1.33732	1.29562	1.29654
GAMMA 12	2.03740	2.02191	2.03266
GAMMA 15	1.58398	1.57730	1.57762
GAMMA 15	3.50797	2.28616	6.00084
GAMMA 25'	1.52877	1.56781	1.55070
X1 (008)	0.28609	0.26908	0.27228
X1 (008)	1.02115	1.02146	1.01771
X1 (008)	2.49981	2.43260	2.43087
X2 (008)	2.38496	2.41340	2.40093
X3 (008)	1.04410	1.05351	1.04936
X5 (008)	2.52747	2.54091	2.53044

X4'	(008)	0.47073	0.45371	0.45518
X5'	(008)	1.01128	1.02310	1.03109
L1	(444)	0.35288	0.34490	0.34728
L1	(444)	1.68809	1.71325	1.70307
L3	(444)	1.39484	1.42405	1.42355
L3	(444)	2.28329	2.36989	2.33859
L2'	(444)	0.13155	0.13436	0.12788
L2'	(444)	1.30900	1.23268	1.23984
L3'	(444)	1.45549	1.44579	1.45056
W1	(048)	0.73429	0.71084	0.71526
W1	(048)	1.89532	1.87290	1.87472
W2'	(048)	0.29556	0.30899	0.30657
W2'	(048)	1.50086	1.54955	1.54707
W2'	(048)	2.38700	2.43972	2.43268
W1'	(048)	2.48829	2.55686	2.56177
W3	(048)	0.62463	0.59916	0.60224
W3	(048)	1.80137	1.76631	1.77096

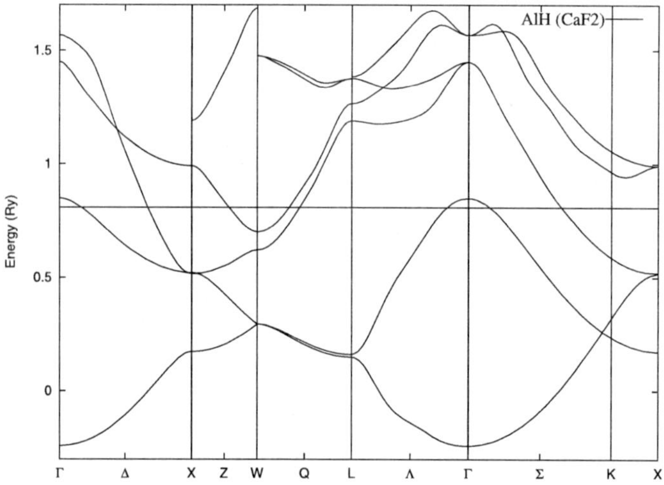

Fig. 9.10 Energy bands of AlH2 in the CaF2 structure

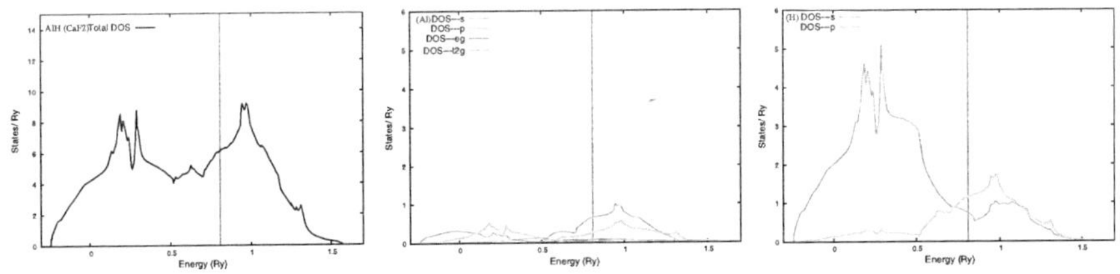

Fig. 9.11 Total, angular momentum and site decomposed densities of states of AlH2 in the CaF2 structure

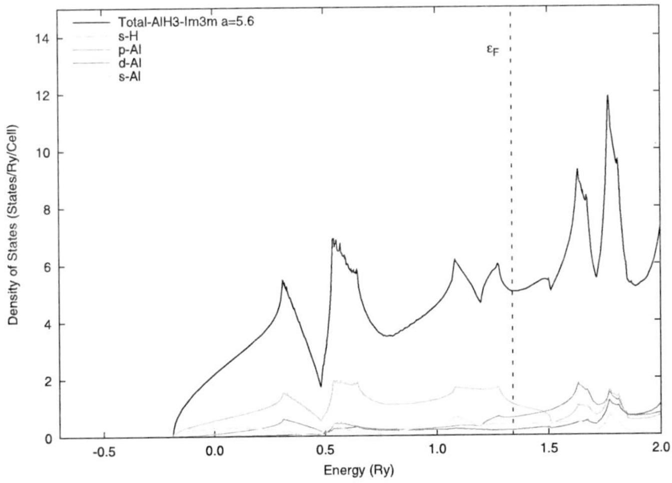

Fig. 9.12 Total, angular momentum and site decomposed densities of states of AlH3 in the Im3m structure

9.3 Gallium Hydride (GaH)

See Fig. 9.13.

See Tables 9.11, 9.12 and 9.13.

See Figs. 9.14 and 9.15.

See Tables 9.14 and 9.15.

See Figs. 9.16, 9.17 and 9.18.

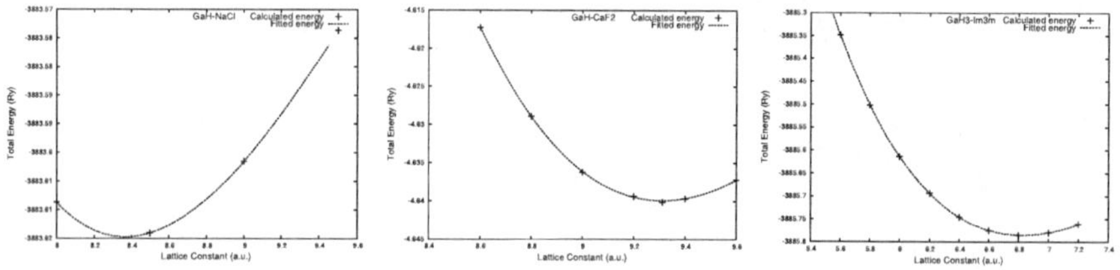

Fig. 9.13 Total energy versus lattice constant of GaH in the NaCl, CaF2 and Im3m structures

Table 9.11 Lattice constant, bulk modulus, gap, total energy

Stru	a (Bohr)	B (MBar)	Gap	Total Energy (Ry)
NaCl	8.36	0.63	-	-3883.58009
CaF2	9.31	0.50	-	-3884.64009
Im3m (P=0)	6.82	1.80		-3885.78518
Im3m (P=1.92 MBar)	5.80	6.81		-3885.50067

Table 9.12 Birch fit coefficients

	A1	A2	A3	A4
NaCl	-3.882860E+03	-4.332571E+01	6.601555E+02	-1.086027E+03
CaF2	-3.859076E+00	-5.427624E+01	9.617924E+02	-6.421615E+02
Im3m	-3.882927E+03	-1.875606E+02	3.623549E+03	-1.705506E+04

Table 9.13 DOS at Ef, Hopfield parameter, Stoner criterion

GaH a=8.36 NaCl

Fermi Energy (Ry)		Total DOS (States/Ry)	Decomposed DOS inside the muffin tins s	p	eg	t2g	f
Ga	0.601	6.763	0.991	1.155	0.070	0.096	0.023
H	0.601	6.763	1.124	0.625	0.044	0.034	0.007

Fermi-Velocity (cm/s): 1.46533 x10E8
Plasmon Energy (eV) : 11.35213
Electron-ion interaction (Hopfield parameter) (eV/A^2) Ga:0.412 H:1.709

Ga MUFFIN-TIN RADIUS and CHARGE = 2.1945 29.1632
H MUFFIN-TIN RADIUS and CHARGE = 1.9855 1.2997
Ga STONER I = 0.0061
H STONER I = 0.0104
STONER PARAMETER (Ry) I = 0.0168
STONER CRITERION N*I = 0.1134

GaH2 a=9.31 Bohr CaF2

Fermi Energy (Ry)		Total DOS (States/Ry)	Decomposed DOS inside the muffin tins s	p	eg	t2g	f
Ga	0.388	5.800	1.444	0.707	0.011	0.102	0.046
H	0.388	5.800	1.879	0.186	0.006	0.023	0.002

Velocity (cm/s): 1.39463 x10E8
Plasmon Energy (eV) : 8.51138
Electron-ion interaction (Hopfield parameter) (eV/A^2) Ga:0.370 H2:3.474

Ga MUFFIN-TIN RADIUS and CHARGE = 2.4192 29.5449
H MUFFIN-TIN RADIUS and CHARGE = 1.6128 0.8888

Ga STONER I = 0.0024
H STONER I = 0.0076
STONER PARAMETER (Ry) I = 0.0178
STONER CRITERION N*I = 0.1030

GaH3 a=5.8 Bohr Im3m

Fermi Energy (Ry)		Total DOS (States/Ry)	Decomposed DOS inside the muffin tins s	p	d	f
Ga	1.276	5.084	0.277	0.508	0.178	0.025
H	1.276	5.084	0.388	0.049	0.003	0.000

Fermi-Velocity (cm/s): 1.62925 x10E8
Plasmon Energy (eV) : 13.39103
Electron-ion interaction (Hopfield parameter) (eV/A^2) Ga:3.706 H3:1.785

Ga MUFFIN-TIN RADIUS and CHARGE = 1.70 28.3324
H MUFFIN-TIN RADIUS and CHARGE = 1.00 0.5602

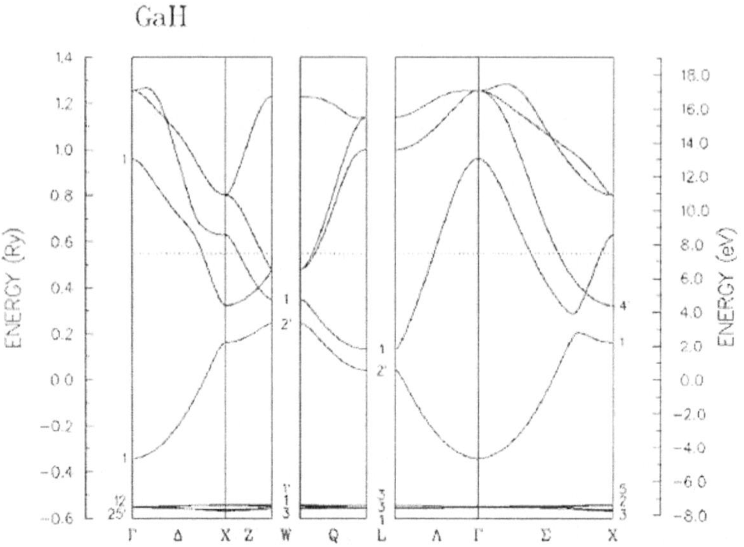

Fig. 9.14 Energy bands of GaH in the NaCl structure (tight binding)

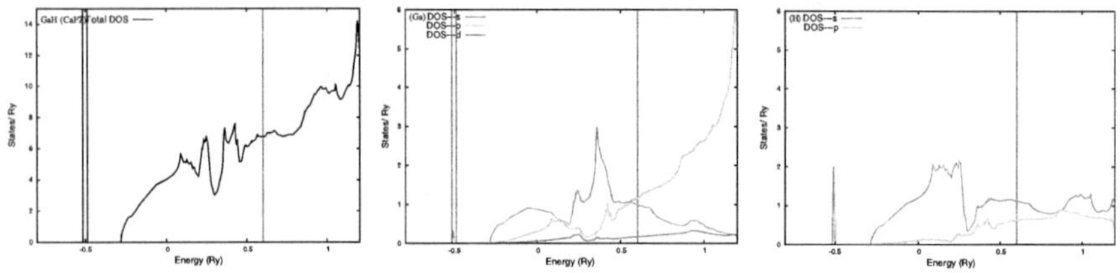

Fig. 9.15 Total, angular momentum and site decomposed densities of states of GaH in the NaCl structure

Table 9.14 GaH (NaCl) a = 8.36 Bohr Slater–Koster 3-center parameters

	ORTHOGONAL	NON-ORTHOGONAL	
	ENERGY INTEGRALS Ry	ENERGY INTEGRALS Ry	OVERLAP INTEGRALS
ON SITE Ga-Ga			
s,s(000)	0.30198	0.26023	
x,x(000)	0.91778	0.77631	
xy,xy(000)	-0.54923	-0.55048	
d2,d2(000)	-0.49207	-0.54752	
FIRST NEIGHBOR Ga-Ga			
s,s(110)	0.00082	-0.00144	0.01411
s,x(110)	-0.01534	0.00279	0.05792
s,xy(110)	-0.01037	-0.01554	0.03803
s,d2(110)	0.01481	-0.01810	0.01402
x,x(110)	0.05826	-0.01788	-0.12348
x,x(011)	-0.00017	-0.00848	0.06392
x,y(110)	0.02155	-0.04360	-0.06305
x,xy(110)	0.00913	0.01484	-0.03603
x,xy(011)	-0.00045	-0.00270	0.00838
z,d2(011)	0.01114	-0.00518	0.01666
z,d1(011)	0.00357	0.00522	0.01235
xy,xy(110)	-0.00240	0.00485	-0.01298
xy,xy(011)	0.00056	0.00307	-0.00454
xy,xz(011)	0.00031	0.00438	-0.00669
xy,d2(110)	0.00123	0.00264	-0.00268
d2,d2(110)	0.00891	-0.00321	0.00528
d1,d1(110)	-0.02783	0.00357	-0.00714
SECOND NEIGHBOR Ga-Ga			
s,s(200)	0.02925	0.02431	-0.01275
s,x(200)	0.05104	0.03893	-0.02407
s,d2(002)	-0.02535	0.00496	0.00246
x,x(200)	-0.10124	-0.07332	0.01357
y,y(200)	0.01779	0.00503	0.02045
x,xy(020)	-0.01376	-0.01184	0.02689
z,d2(002)	0.00776	-0.02529	0.02809
xy,xy(200)	-0.00016	-0.00403	0.00757
xy,xy(002)	-0.00004	0.00211	-0.00386
d2,d2(002)	0.01912	-0.00459	0.01001
d1,d1(002)	0.00027	-0.00086	0.00155
ON SITE H-H			
s,s(000)	0.09013	0.29950	
FIRST NEIGHBOR H-H			
s,s(110)	0.00853	-0.00072	0.00996
s,x(110)	0.00000	0.00000	0.00000
x,x(110)	0.00000	0.00000	0.00000
x,x(011)	0.00000	0.00000	0.00000
x,y(110)	0.00000	0.00000	0.00000
SECOND NEIGHBOR H-H			
s,s(200)	-0.01080	0.00067	-0.00622
s,x(200)	0.00000	0.00000	0.00000
x,x(200)	0.00000	0.00000	0.00000
y,y(200)	0.00000	0.00000	0.00000
FIRST NEIGHBOR Ga-H			
s,s(100)	-0.10143	-0.11901	-0.01337
s,x(100)	0.00000	0.00000	0.00000
x,s(100)	0.09231	0.16174	0.06780
x,x(100)	0.00000	0.00000	0.00000
y,y(100)	0.00000	0.00000	0.00000
xy,x(010)	0.00000	0.00000	0.00000
d2,s(001)	0.10563	0.00044	-0.06334
d2,z(001)	0.00000	0.00000	0.00000

SECOND NEIGHBOR Ga-H

s,s(111)	-0.00204	0.00083	0.00876
x,s(111)	-0.00197	0.00076	-0.01950
s,x(111)	0.00000	0.00000	0.00000
x,x(111)	0.00000	0.00000	0.00000
x,y(111)	0.00000	0.00000	0.00000
xy,s(111)	-0.00027	-0.00657	0.01146
xy,x(111)	0.00000	0.00000	0.00000
yz,x(111)	0.00000	0.00000	0.00000
d1,x(111)	0.00000	0.00000	0.00000

GaH

	ORTHOGONAL			NON-ORTHOGONAL		
BAND	RMS ERROR	MAXIMUM DEVIATION		RMS ERROR	MAXIMUM DEVIATION	
	mRy	k	mRy	mRy	k	mRy
1	0.8	(226)	1.7	0.0	(044)	0.1
2	1.0	(005)	3.4	0.0	(008)	0.1
3	1.0	(066)	3.6	0.0	(044)	0.1
4	0.9	(022)	2.3	0.0	(044)	0.1
5	0.8	(042)	1.5	0.0	(044)	0.2
6	3.9	(226)	7.0	0.1	(333)	0.3
7	5.6	(006)	14.9	0.2	(118)	0.5
1-7	2.7			0.1		

ENERGY VALUES IN Ry AT SELECTED k-POINTS

	ORTHOGONAL	APW	NON-ORTHOGONAL
GAMMA 1	-0.34277	-0.34093	-0.34088
GAMMA 1	0.95778	0.95693	0.95711
GAMMA 12	-0.54736	-0.54887	-0.54881
GAMMA 15	1.25182	1.25495	1.25495
GAMMA 25'	-0.55510	-0.55495	-0.55489
X1 (008)	-0.56851	-0.56714	-0.56717
X1 (008)	0.15736	0.16303	0.16314
X1 (008)	0.62749	0.62851	0.62856
X2 (008)	-0.54302	-0.54364	-0.54362
X3 (008)	-0.56403	-0.56364	-0.56354
X5 (008)	-0.54037	-0.54150	-0.54144
X4' (008)	0.31974	0.32283	0.32260
X5' (008)	0.78714	0.80134	0.80131
L1 (444)	-0.55852	-0.55855	-0.55850
L1 (444)	0.13399	0.13263	0.13283
L3 (444)	-0.55496	-0.55524	-0.55518
L3 (444)	-0.54255	-0.54305	-0.54301
L2' (444)	0.03364	0.04008	0.04019
L2' (444)	0.99801	0.99943	0.99942
L3' (444)	1.13530	1.14017	1.14016
W1 (048)	-0.54799	-0.54750	-0.54752
W1 (048)	0.35916	0.34974	0.34983
W2' (048)	-0.56182	-0.56113	-0.56104
W2' (048)	0.23813	0.24308	0.24322
W2' (048)	1.22653	1.22836	1.22837
W1' (048)	-0.54019	-0.54148	-0.54149
W3 (048)	-0.55298	-0.55328	-0.55325
W3 (048)	0.48371	0.47601	0.47612
ODD (224)	-0.55240	-0.55241	-0.55238
ODD (224)	-0.54508	-0.54397	-0.54396
EVEN(224)	-0.55691	-0.55758	-0.55759
EVEN(224)	-0.55615	-0.55465	-0.55466
EVEN(224)	-0.54690	-0.54804	-0.54809
EVEN(224)	-0.12664	-0.12862	-0.12877
EVEN(224)	0.42902	0.42703	0.42691

Table 9.15 GaH (NaCl) a=8.36 Bohr Slater–Koster 2-center parameters

	ORTHOGONAL	NON-ORTHOGONAL	
	ENERGY INTEGRALS Ry	ENERGY INTEGRALS Ry	OVERLAP INTEGRALS
ON SITE Ga-Ga			
s	0.15633	0.05316	
p	0.84201	0.59299	
t2g	-0.54658	-0.55061	
eg	-0.54492	-0.55101	
FIRST NEIGHBOR Ga-Ga			
(sss)	-0.04580	-0.04446	0.10389
(pps)	0.11980	0.01185	-0.20873
(ppp)	-0.00249	0.00644	0.05501
(dds)	-0.00324	-0.00775	0.00870
(ddp)	-0.00057	-0.00054	0.00337
(ddd)	0.00112	0.00532	-0.00997
(sps)	0.07811	0.03939	-0.13175
(sds)	-0.02049	-0.00092	0.01061
(pds)	-0.02663	-0.01085	0.01447
(pdp)	-0.00015	0.01579	-0.01767
SECOND NEIGHBOR Ga-Ga			
(sss)	0.00860	-0.01318	0.00043
(pps)	-0.06092	0.03035	0.00665
(ppp)	0.01813	-0.00505	-0.00871
(dds)	0.00071	0.00465	-0.00877
(ddp)	-0.00087	-0.00183	0.00338
(ddd)	0.00003	0.00327	-0.00592
(sps)	-0.01168	0.01982	-0.00023
(sds)	0.00227	-0.00102	0.00158
(pds)	-0.03305	-0.00171	-0.00150
(pdp)	0.00637	0.00282	-0.00305
ON SITE H-H			
s	0.57548	0.24440	
FIRST NEIGHBOR H-H			
(sss)	0.03564	0.00273	-0.03416
(sps)	0.00000	0.00000	0.00000
(pps)	0.00000	0.00000	0.00000
(ppp)	0.00000	0.00000	0.00000
SECOND NEIGHBOR H-H			
(sss)	-0.00964	-0.00202	0.02620
(sps)	0.00000	0.00000	0.00000
(pps)	0.00000	0.00000	0.00000
(ppp)	0.00000	0.00000	0.00000
FIRST NEIGHBOR Ga-H			
(sss)	0.02067	0.07485	-0.10991
(sps)	0.00000	0.00000	0.00000
(pss)	0.13534	0.08373	-0.19095
(pps)	0.00000	0.00000	0.00000
(ppp)	0.00000	0.00000	0.00000
(dpp)	0.00000	0.00000	0.00000
(dss)	0.02698	0.06588	-0.14429
(dps)	0.00000	0.00000	0.00000
SECOND NEIGHBOR Ga-H			
(sss)	-0.02535	0.01043	0.00768
(pss)	-0.02468	0.01207	0.01306
(sps)	0.00000	0.00000	0.00000
(pps)	0.00000	0.00000	0.00000
(ppp)	0.00000	0.00000	0.00000
(dss)	0.00897	-0.02225	0.04197
(dps)	0.00000	0.00000	0.00000
(dpp)	0.00000	0.00000	0.00000

GaH

BAND	ORTHOGONAL RMS ERROR mRy	ORTHOGONAL MAXIMUM DEVIATION k	ORTHOGONAL MAXIMUM DEVIATION mRy	NON-ORTHOGONAL RMS ERROR mRy	NON-ORTHOGONAL MAXIMUM DEVIATION k	NON-ORTHOGONAL MAXIMUM DEVIATION mRy
1	3.0	(226)	10.8	0.1	(044)	0.2
2	2.4	(004)	6.2	0.1	(444)	0.3
3	1.0	(222)	2.1	0.1	(444)	0.3
4	1.6	(444)	5.5	0.1	(000)	0.2
5	1.9	(444)	5.5	0.1	(044)	0.2
6	4.7	(007)	12.6	0.2	(264)	0.5
7	13.4	(007)	39.1	0.4	(062)	1.0
1-7	5.6			0.2		

ENERGY VALUES IN Ry AT SELECTED k-POINTS

	ORTHOGONAL	APW	NON-ORTHOGONAL
GAMMA 1	-0.34648	-0.34093	-0.34106
GAMMA 1	0.95019	0.95693	0.95694
GAMMA 12	-0.54596	-0.54887	-0.54866
GAMMA 15	1.25196	1.25495	1.25496
GAMMA 25'	-0.55643	-0.55495	-0.55491
X1 (008)	-0.56516	-0.56714	-0.56702
X1 (008)	0.16459	0.16303	0.16306
X1 (008)	0.62761	0.62851	0.62860
X2 (008)	-0.54403	-0.54364	-0.54366
X3 (008)	-0.56083	-0.56364	-0.56352
X5 (008)	-0.54142	-0.54150	-0.54130
X4' (008)	0.31348	0.32283	0.32320
X5' (008)	0.80265	0.80134	0.80130
L1 (444)	-0.55914	-0.55855	-0.55842
L1 (444)	0.12748	0.13263	0.13288
L3 (444)	-0.55608	-0.55524	-0.55491
L3 (444)	-0.53756	-0.54305	-0.54304
L2' (444)	0.03487	0.04008	0.03968
L2' (444)	1.00059	0.99943	0.99942
L3' (444)	1.13592	1.14017	1.14017
W1 (048)	-0.54844	-0.54750	-0.54760
W1 (048)	0.35952	0.34974	0.34968
W2' (048)	-0.55669	-0.56113	-0.56135
W2' (048)	0.24791	0.24308	0.24297
W2' (048)	1.22822	1.22836	1.22866
W1' (048)	-0.54152	-0.54148	-0.54157
W3 (048)	-0.55168	-0.55328	-0.55331
W3 (048)	0.48960	0.47601	0.47645
ODD (224)	-0.55321	-0.55241	-0.55234
ODD (224)	-0.54437	-0.54397	-0.54400
EVEN(224)	-0.55735	-0.55758	-0.55759
EVEN(224)	-0.55604	-0.55465	-0.55467
EVEN(224)	-0.54909	-0.54804	-0.54805
EVEN(224)	-0.12530	-0.12862	-0.12859
EVEN(224)	0.43555	0.42703	0.42722

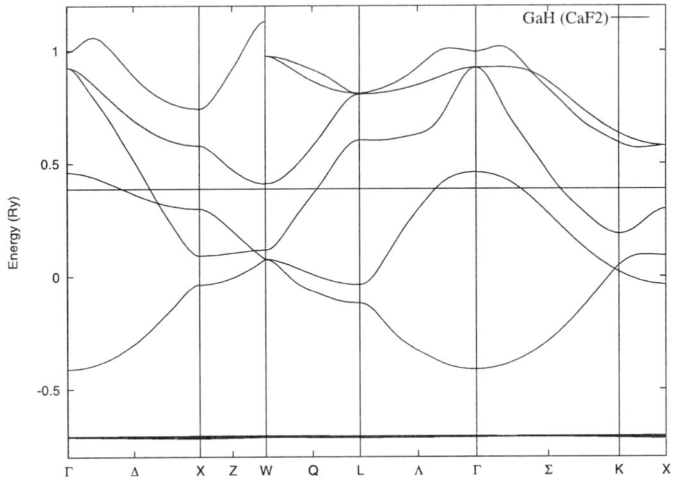

Fig. 9.16 Energy bands of GaH2 in the CaF2 structure

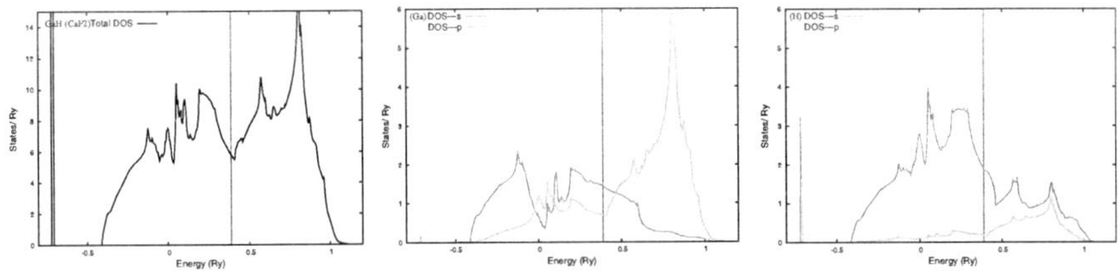

Fig. 9.17 Total, angular momentum and site decomposed densities of states of GaH2 in the CaF2 structure

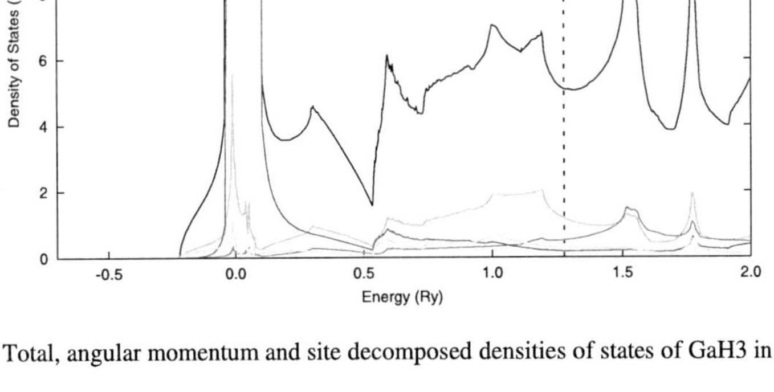

Fig. 9.18 Total, angular momentum and site decomposed densities of states of GaH3 in the Im3m structure

9.4 Indium Hydride (InH)

See Fig. 9.19.

See Tables 9.16, 9.17 and 9.18.

See Figs. 9.20 and 9.21.

See Tables 9.19 and 9.20.

See Figs. 9.22, 9.23 and 9.24.

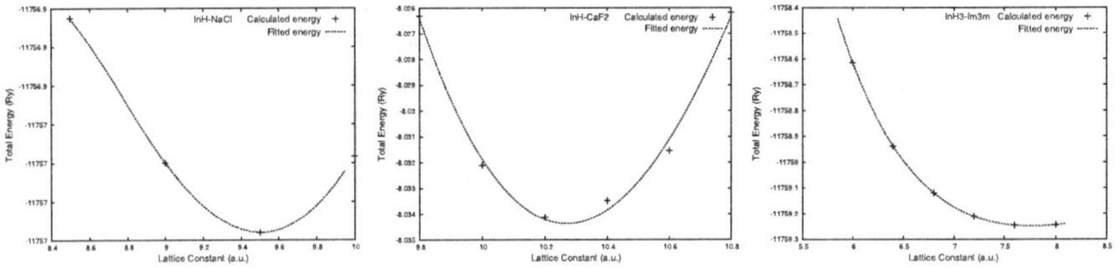

Fig. 9.19 Total energy versus lattice constant of InH in the NaCl, CaF2 and Im3m structures

Table 9.16 Lattice constant, bulk modulus, gap, total energy

Stru	a (Bohr)	B (MBar)	Gap	Total Energy (Ry)
NaCl	9.42	0.32	-	-11756.96755
CaF2	10.27	0.41	-	-11758.03476
Im3m (P=0)	7.78	0.76	-	-11759.24885
Im3m (P=2.04Mbar)	6.20	7.18	-	-11758.79874

Table 9.17 Birch fit coefficients

	A1	A2	A3	A4
NaCl	-1.175431E+04	-2.531941E+02	7.939568E+03	-8.128611E+04
CaF2	-6.948573E+00	-1.006066E+02	2.715601E+03	-1.7029579E+04
Im3m	-1.175791E+04	-1.013561E+02	1.890221E+03	1.0340139E+03

Table 9.18 DOS at Ef, Hopfield parameter, Stoner criterion

InH a=9.42 Bohr NaCl

Fermi Energy (Ry)	Total DOS (States/Ry)	Decomposed DOS inside the muffin tins s	p	eg	t2g	f
In 0.485	7.992	1.161	1.192	0.071	0.098	0.034
H 0.485	7.992	1.588	0.722	0.054	0.035	0.007

Fermi-Velocity (cm/s): 1.30150 x10E8
Plasmon Energy (eV) : 9.16356
Electron-ion interaction (Hopfield parameter) (eV/A^2) In:0.260 H:1.397

In MUFFIN-TIN RADIUS and CHARGE = 2.4728 47.0047
H MUFFIN-TIN RADIUS and CHARGE = 2.2372 1.4042
In STONER I = 0.0037
H STONER I = 0.0114
STONER PARAMETER (Ry) I = 0.0153
STONER CRITERION N*I = 0.1222

InH2 a=10.27 Bohr CaF2

Fermi Energy (Ry)	Total DOS (States/Ry)	Decomposed DOS inside the muffin tins s	p	eg	t2g	f
In 0.303	6.853	1.740	0.648	0.013	0.066	0.086
H 0.303	6.853	2.489	0.221	0.006	0.027	0.002

Fermi-Velocity (cm/s): 1.13368 x10E8
Plasmon Energy (eV) : 6.49371
Electron-ion interaction (Hopfield parameter) (eV/A^2) In:0.250 H2:2.928

In MUFFIN-TIN RADIUS and CHARGE = 2.6680 47.3320
H MUFFIN-TIN RADIUS and CHARGE = 1.7787 0.9941
In STONER I = 0.0012
H STONER I = 0.0081
STONER PARAMETER (Ry) I = 0.0174
STONER CRITERION N*I = 0.1193

InH3 a=6.2 Bohr Im3m

Fermi Energy (Ry)	Total DOS (States/Ry)	Decomposed DOS inside the muffin tins s	p	d	f
In 1.340	4.730	0.167	0.230	0.296	0.031
H 1.340	4.730	0.345	0.036	0.002	0.000

Fermi-Velocity (cm/s): 1.57210 x10E8
Plasmon Energy (eV) : 11.27696
Electron-ion interaction (Hopfield parameter) (eV/A^2) In:1.937 H3:2.690

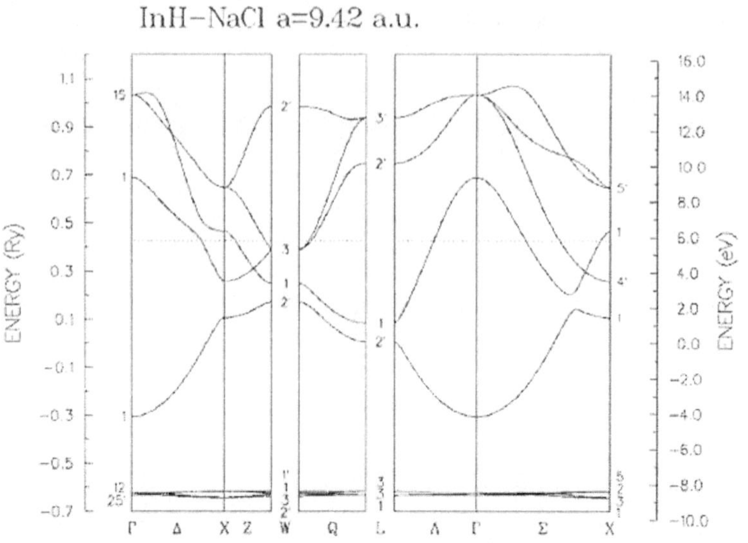

Fig. 9.20 Energy bands of InH in the NaCl structure (tight binding)

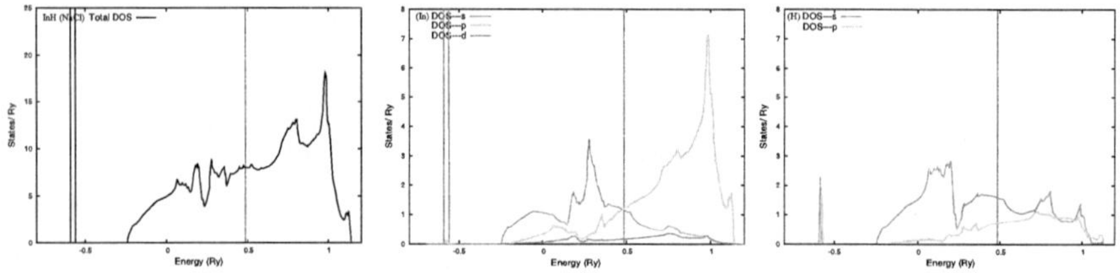

Fig. 9.21 Total, angular momentum and site decomposed densities of states of InH in the NaCl structure

Table 9.19 InH (NaCl) a = 9.42 Bohr Slater–Koster 3-center parameters

	ORTHOGONAL	NON-ORTHOGONAL	
	ENERGY INTEGRALS Ry	ENERGY INTEGRALS Ry	OVERLAP INTEGRALS
ON SITE In-In			
s,s(000)	0.19888	0.16660	
x,x(000)	0.74405	0.66271	
xy,xy(000)	-0.62483	-0.62496	
d2,d2(000)	-0.58196	-0.62589	
FIRST NEIGHBOR In-In			
s,s(110)	-0.00347	-0.01017	-0.00663
s,x(110)	-0.01415	-0.00698	0.02135
s,xy(110)	0.00833	-0.02151	0.02208
s,d2(110)	0.01183	-0.00005	0.00069
x,x(110)	0.04831	0.00926	-0.07548
x,x(011)	0.00033	-0.02270	0.01041
x,y(110)	0.02260	-0.03047	-0.05798
x,xy(110)	-0.00644	0.03117	-0.03805
x,xy(011)	0.00174	0.00522	-0.01080
z,d2(011)	0.00665	-0.00442	0.00723
z,d1(011)	-0.00263	0.00051	-0.00154
xy,xy(110)	-0.00297	0.00236	-0.00844
xy,xy(011)	0.00066	0.00323	-0.00410
xy,xz(011)	0.00059	0.00503	-0.00690
xy,d2(110)	-0.00159	-0.00323	0.00271
d2,d2(110)	0.00619	0.00067	-0.00287
d1,d1(110)	-0.02020	-0.01004	0.01858
SECOND NEIGHBOR In-In			
s,s(200)	0.01997	0.01778	-0.02304
s,x(200)	0.03612	0.02971	-0.03210
s,d2(002)	-0.02036	0.00703	-0.01277
x,x(200)	-0.07837	-0.05226	0.04249
y,y(200)	0.01383	0.01815	0.01803
x,xy(020)	0.01009	-0.01578	0.02393
z,d2(002)	0.00537	-0.00538	0.01101
xy,xy(200)	-0.00001	-0.00270	0.00438
xy,xy(002)	-0.00003	0.00129	-0.00210
d2,d2(002)	0.01416	0.00313	-0.00518
d1,d1(002)	0.00016	-0.00390	0.00622
ON SITE H-H			
s,s(000)	0.05430	0.30306	
FIRST NEIGHBOR H-H			
s,s(110)	0.00696	0.00432	-0.00530
s,x(110)	0.00000	0.00000	0.00000
x,x(110)	0.00000	0.00000	0.00000
x,x(011)	0.00000	0.00000	0.00000
x,y(110)	0.00000	0.00000	0.00000
SECOND NEIGHBOR H-H			
s,s(200)	-0.00541	-0.00111	0.00004
s,x(200)	0.00000	0.00000	0.00000
x,x(200)	0.00000	0.00000	0.00000
y,y(200)	0.00000	0.00000	0.00000
FIRST NEIGHBOR In-H			
s,s(100)	-0.07960	-0.08242	-0.03891
s,x(100)	0.00000	0.00000	0.00000
x,s(100)	0.06764	0.15414	0.15549
x,x(100)	0.00000	0.00000	0.00000
y,y(100)	0.00000	0.00000	0.00000
xy,x(010)	0.00000	0.00000	0.00000
d2,s(001)	0.09769	-0.00703	-0.00872
d2,z(001)	0.00000	0.00000	0.00000

SECOND NEIGHBOR In-H

s,s(111)	-0.00162	0.00566	0.00931
x,s(111)	-0.00153	-0.00584	-0.00782
s,x(111)	0.00000	0.00000	0.00000
x,x(111)	0.00000	0.00000	0.00000
x,y(111)	0.00000	0.00000	0.00000
xy,s(111)	0.00022	-0.00454	0.01143
xy,x(111)	0.00000	0.00000	0.00000
yz,x(111)	0.00000	0.00000	0.00000
d1,x(111)	0.00000	0.00000	0.00000

InH

BAND	ORTHOGONAL RMS ERROR mRy	MAXIMUM DEVIATION k	mRy	NON-ORTHOGONAL RMS ERROR mRy	MAXIMUM DEVIATION k	mRy
1	0.5	(008)	1.3	0.1	(044)	0.2
2	0.5	(005)	1.6	0.1	(008)	0.2
3	0.5	(066)	1.5	0.1	(044)	0.3
4	0.5	(062)	1.2	0.1	(044)	0.4
5	0.5	(042)	1.0	0.1	(044)	0.2
6	2.5	(226)	5.3	0.2	(264)	0.4
7	3.2	(007)	11.5	0.3	(226)	0.7
1-7	1.6			0.1		

ENERGY VALUES IN Ry AT SELECTED k-POINTS

	ORTHOGONAL	APW	NON-ORTHOGONAL
GAMMA 1	-0.30683	-0.30587	-0.30586
GAMMA 1	0.68926	0.68924	0.68929
GAMMA 12	-0.62301	-0.62383	-0.62370
GAMMA 15	1.03043	1.03228	1.03227
GAMMA 25'	-0.63153	-0.63157	-0.63148
X1 (008)	-0.64455	-0.64324	-0.64327
X1 (008)	0.10179	0.10599	0.10632
X1 (008)	0.46395	0.46485	0.46508
X2 (008)	-0.61653	-0.61677	-0.61679
X3 (008)	-0.64210	-0.64184	-0.64166
X5 (008)	-0.61307	-0.61378	-0.61370
X4' (008)	0.25751	0.25875	0.25843
X5' (008)	0.64134	0.64699	0.64692
L1 (444)	-0.63415	-0.63398	-0.63389
L1 (444)	0.08373	0.08347	0.08357
L3 (444)	-0.63152	-0.63171	-0.63164
L3 (444)	-0.61577	-0.61591	-0.61578
L2' (444)	0.00368	0.00757	0.00731
L2' (444)	0.74770	0.74813	0.74810
L3' (444)	0.93585	0.93741	0.93740
W1 (048)	-0.62174	-0.62153	-0.62143
W1 (048)	0.25402	0.24942	0.24949
W2' (048)	-0.63643	-0.63663	-0.63649
W2' (048)	0.16672	0.16976	0.16996
W2' (048)	0.98134	0.98469	0.98474
W1' (048)	-0.61294	-0.61375	-0.61371
W3 (048)	-0.62859	-0.62853	-0.62853
W3 (048)	0.39514	0.39046	0.39074
ODD (224)	-0.62805	-0.62808	-0.62800
ODD (224)	-0.61750	-0.61714	-0.61715
EVEN(224)	-0.63346	-0.63385	-0.63383
EVEN(224)	-0.63149	-0.63051	-0.63066
EVEN(224)	-0.62153	-0.62194	-0.62204
EVEN(224)	-0.12969	-0.13091	-0.13084
EVEN(224)	0.32010	0.31842	0.31839

Table 9.20 InH (NaCl) a = 9.42 Bohr a.u. Slater–Koster 2-center parameters

	ORTHOGONAL	NON-ORTHOGONAL	
	ENERGY INTEGRALS Ry	ENERGY INTEGRALS Ry	OVERLAP INTEGRALS
ON SITE In-In			
s	0.09604	0.08304	
p	0.68220	0.60043	
t2g	-0.62295	-0.62517	
eg	-0.62203	-0.62556	
FIRST NEIGHBOR In-In			
(sss)	-0.03615	-0.03239	0.02062
(pps)	0.09647	0.06740	-0.09216
(ppp)	-0.00176	0.03196	0.05955
(dds)	-0.00391	-0.00349	-0.00046
(ddp)	0.00038	-0.00147	0.00462
(ddd)	0.00050	0.00027	-0.00071
(sps)	0.06099	0.03780	-0.03526
(sds)	-0.01609	0.02140	-0.02351
(pds)	-0.01962	0.00857	0.00032
(pdp)	0.00209	0.01623	-0.03189
SECOND NEIGHBOR In-In			
(sss)	0.00474	0.00394	-0.00801
(pps)	-0.04847	0.02345	0.04631
(ppp)	0.01362	-0.00556	-0.02147
(dds)	0.00043	-0.00381	0.00605
(ddp)	-0.00051	-0.00120	0.00196
(ddd)	0.00003	-0.00124	0.00197
(sps)	-0.00735	0.00006	0.02198
(sds)	0.00440	-0.00122	0.00202
(pds)	-0.02821	0.00532	-0.00775
(pdp)	0.00514	0.00274	-0.00504
ON SITE H-H			
s	0.40430	0.07077	
FIRST NEIGHBOR H-H			
(sss)	0.02480	-0.00798	0.01325
(sps)	0.00000	0.00000	0.00000
(pps)	0.00000	0.00000	0.00000
(ppp)	0.00000	0.00000	0.00000
SECOND NEIGHBOR H-H			
(sss)	-0.00285	-0.00892	0.03131
(sps)	0.00000	0.00000	0.00000
(pps)	0.00000	0.00000	0.00000
(ppp)	0.00000	0.00000	0.00000
FIRST NEIGHBOR In-H			
(sss)	0.02303	0.05774	-0.14571
(sps)	0.00000	0.00000	0.00000
(pss)	0.10669	0.06333	-0.14969
(pps)	0.00000	0.00000	0.00000
(ppp)	0.00000	0.00000	0.00000
(dpp)	0.00000	0.00000	0.00000
(dss)	0.02376	0.06154	-0.12375
(dps)	0.00000	0.00000	0.00000
SECOND NEIGHBOR In-H			
(sss)	-0.01766	0.01187	-0.00231
(pss)	-0.01988	0.02232	0.01193
(sps)	0.00000	0.00000	0.00000
(pps)	0.00000	0.00000	0.00000
(ppp)	0.00000	0.00000	0.00000
(dss)	0.00594	0.00542	-0.01024
(dps)	0.00000	0.00000	0.00000
(dpp)	0.00000	0.00000	0.00000

InH

BAND	ORTHOGONAL RMS ERROR mRy	ORTHOGONAL MAXIMUM DEVIATION k	ORTHOGONAL MAXIMUM DEVIATION mRy	NON-ORTHOGONAL RMS ERROR mRy	NON-ORTHOGONAL MAXIMUM DEVIATION k	NON-ORTHOGONAL MAXIMUM DEVIATION mRy
1	1.9	(226)	6.6	0.1	(044)	0.1
2	1.7	(004)	4.3	0.1	(008)	0.2
3	0.7	(042)	1.9	0.1	(044)	0.2
4	1.2	(444)	3.0	0.1	(044)	0.4
5	1.1	(444)	3.0	0.1	(048)	0.1
6	3.2	(007)	7.4	0.1	(007)	0.3
7	8.1	(007)	27.7	0.2	(222)	0.6
1-7	3.5			0.1		

ENERGY VALUES IN Ry AT SELECTED k-POINTS

	ORTHOGONAL	APW	NON-ORTHOGONAL
GAMMA 1	-0.30928	-0.30587	-0.30595
GAMMA 1	0.68485	0.68924	0.68948
GAMMA 12	-0.62200	-0.62383	-0.62372
GAMMA 15	1.01153	1.03228	1.03221
GAMMA 25'	-0.63266	-0.63157	-0.63149
X1 (008)	-0.64183	-0.64324	-0.64323
X1 (008)	0.10813	0.10599	0.10600
X1 (008)	0.46667	0.46485	0.46495
X2 (008)	-0.61628	-0.61677	-0.61670
X3 (008)	-0.63969	-0.64184	-0.64164
X5 (008)	-0.61369	-0.61378	-0.61374
X4' (008)	0.25385	0.25875	0.25845
X5' (008)	0.64680	0.64699	0.64699
L1 (444)	-0.63378	-0.63398	-0.63400
L1 (444)	0.08088	0.08347	0.08381
L3 (444)	-0.63174	-0.63171	-0.63168
L3 (444)	-0.61288	-0.61591	-0.61584
L2' (444)	0.00426	0.00757	0.00754
L2' (444)	0.74882	0.74813	0.74813
L3' (444)	0.92113	0.93741	0.93740
W1 (048)	-0.62280	-0.62153	-0.62148
W1 (048)	0.25295	0.24942	0.24937
W2' (048)	-0.63429	-0.63663	-0.63659
W2' (048)	0.17399	0.16976	0.16986
W2' (048)	0.97811	0.98469	0.98466
W1' (048)	-0.61380	-0.61375	-0.61362
W3 (048)	-0.62750	-0.62853	-0.62856
W3 (048)	0.39868	0.39046	0.39056
ODD (224)	-0.62843	-0.62808	-0.62802
ODD (224)	-0.61756	-0.61714	-0.61720
EVEN(224)	-0.63312	-0.63385	-0.63382
EVEN(224)	-0.63102	-0.63051	-0.63056
EVEN(224)	-0.62299	-0.62194	-0.62196
EVEN(224)	-0.12903	-0.13091	-0.13108
EVEN(224)	0.32420	0.31842	0.31848

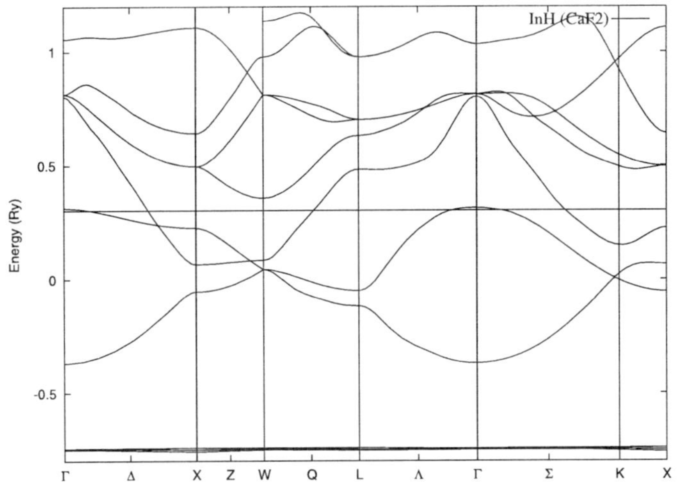

Fig. 9.22 Energy bands of InH2 in the CaF2 structure

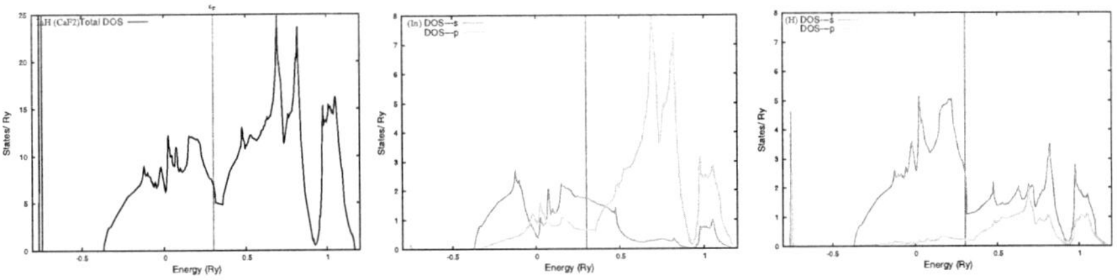

Fig. 9.23 Total, angular momentum and site decomposed densities of states of InH2 in the CaF2 structure

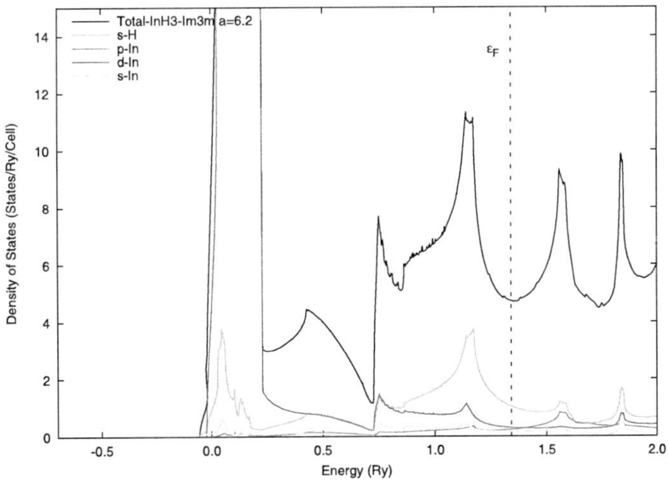

Fig. 9.24 Total, angular momentum and site decomposed densities of states of InH3 in the Im3m structure

9.5 Thallium Hydride (TlH)

See Fig. 9.25.

See Tables 9.21, 9.22 and 9.23.

See Figs. 9.26, 9.27 and 9.28.

See Tables 9.24 and 9.25.

See Fig. 9.29.

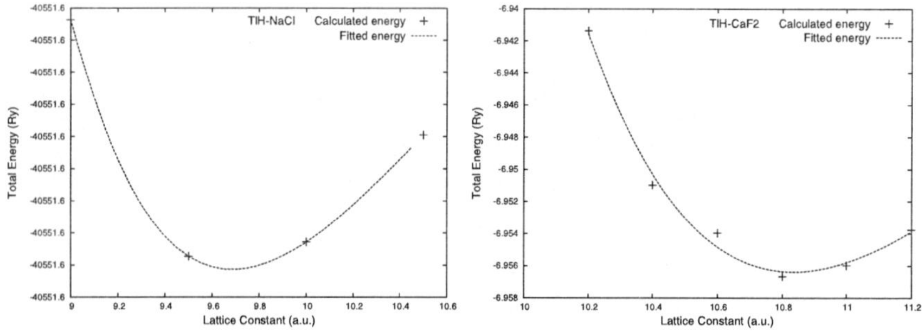

Fig. 9.25 Total energy versus lattice constant of TlH in the NaCl, and CaF2 structures

Table 9.21 Lattice constant, bulk modulus, gap, total energy

Stru	a (Bohr)	B (MBar)	Gap	Total Energy (Ry)
NaCl	9.70	0.29	-	-40555.90796
CaF2	10.83	0.29	-	-40556.95565

Table 9.22 Birch fit coefficients

	A1	A2	A3	A4
NaCl	-4.055187E+04	5.208141E+01	-2.625957E+03	4.111274E+04
CaF2	-7.790331E+00	1.493626E+02	-8.486581E+03	1.555165E+05

Table 9.23 DOS at Ef, Hopfield parameter, Stoner criterion

TlH a=9.70 Bohr NaCl

Fermi Energy (Ry)		Total DOS (States/Ry)	Decomposed DOS inside the muffin tins s	p	eg	t2g	f
Tl	0.430	6.699	0.813	1.013	0.088	0.086	0.027
H	0.430	6.699	1.829	0.541	0.035	0.020	0.007

Fermi-Velocity (cm/s): 1.32368 x10E8
Plasmon Energy (eV) : 8.16636
Electron-ion interaction (Hopfield parameter) (eV/A^2) Tl:0.468 H:1.226

Tl	MUFFIN-TIN RADIUS and CHARGE =	2.5463	79.0029
H	MUFFIN-TIN RADIUS and CHARGE =	2.3037	1.4180
Tl	STONER I =	0.0023	
H	STONER I =	0.0170	
STONER PARAMETER (Ry)	I =	0.0194	
STONER CRITERION	N*I =	0.1301	

TlH2 a=10.83 Bohr CaF2

Fermi Energy (Ry)		Total DOS (States/Ry)	Decomposed DOS inside the muffin tins s	p	eg	t2g	f
Tl	0.186	11.022	1.465	0.765	0.005	0.365	0.121
H	0.186	11.022	6.016	0.143	0.006	0.025	0.003

Fermi-Velocity (cm/s): 1.07667 x10E8
Plasmon Energy (eV) : 7.21727
Electron-ion interaction (Hopfield parameter) (eV/A^2) Tl:0.470 H2:2.333

Tl	MUFFIN-TIN RADIUS and CHARGE =	2.8148	79.4929
H	MUFFIN-TIN RADIUS and CHARGE =	1.8766	0.9685
Tl	STONER I =	0.0004	
H	STONER I =	0.0160	
STONER PARAMETER (Ry)	I =	0.0325	
STONER CRITERION	N*I =	0.3580	

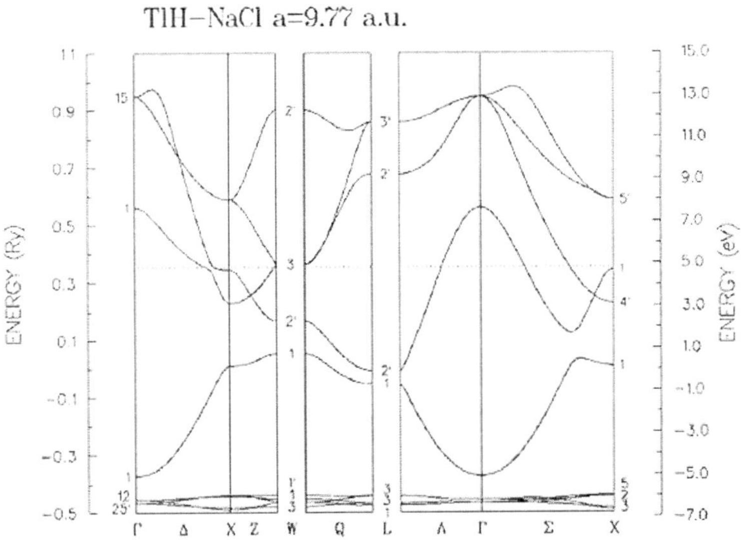

Fig. 9.26 Energy bands of TlH in the NaCl structure (tight-binding)

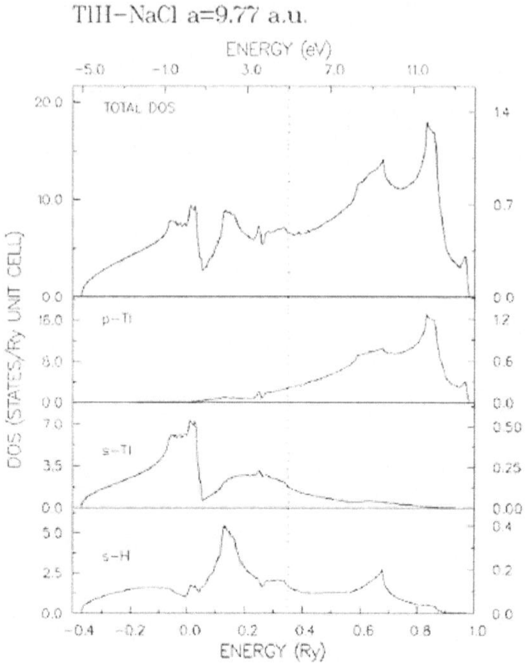

Fig. 9.27 Total, angular momentum and site decomposed densities of states of TlH in the NaCl structure (tight-binding)

Table 9.24 TlH (NaCl) a = 9.77 Bohr Slater–Koster 3-center parameters

	ORTHOGONAL	NON-ORTHOGONAL	
	ENERGY INTEGRALS Ry	ENERGY INTEGRALS Ry	OVERLAP INTEGRALS
ON SITE Tl-Tl			
s,s(000)	0.00474	-0.01527	
x,x(000)	0.62626	0.39828	
xy,xy(000)	-0.45395	-0.45604	
d2,d2(000)	-0.39572	-0.44970	
FIRST NEIGHBOR Tl-Tl			
s,s(110)	-0.00064	-0.00926	0.03412
s,x(110)	0.00193	-0.01671	0.08366
s,xy(110)	-0.01561	0.01047	-0.03411
s,d2(110)	0.00042	0.00256	0.02288
x,x(110)	0.04593	-0.03398	-0.18200
x,x(011)	-0.00063	-0.01056	0.09305
x,y(110)	0.03763	-0.02453	-0.12172
x,xy(110)	0.00628	-0.02778	0.03536
x,xy(011)	-0.01198	0.01511	-0.02757
z,d2(011)	-0.00796	0.00972	-0.02688
z,d1(011)	-0.02196	-0.01649	0.00270
xy,xy(110)	-0.00460	0.00859	-0.03022
xy,xy(011)	0.00092	0.00884	-0.01642
xy,xz(011)	-0.00003	-0.01338	0.03113
xy,d2(110)	0.00228	-0.00455	0.00448
d2,d2(110)	0.00800	0.00492	-0.01250
d1,d1(110)	-0.02843	-0.02387	0.05052
SECOND NEIGHBOR Tl-Tl			
s,s(200)	0.01777	0.01032	-0.03410
s,x(200)	0.01392	0.00312	-0.05282
s,d2(002)	0.01216	0.00307	-0.01210
x,x(200)	-0.04371	0.04873	0.11626
y,y(200)	0.01013	-0.00438	0.00606
x,xy(020)	-0.00719	0.00515	0.00361
z,d2(002)	0.00563	-0.01351	0.04696
xy,xy(200)	-0.00075	-0.01574	0.03520
xy,xy(002)	-0.00002	0.00155	-0.00397
d2,d2(002)	0.02175	0.00994	-0.01820
d1,d1(002)	-0.00143	-0.00032	0.00099
ON SITE H-H			
s,s(000)	0.10245	0.11951	
FIRST NEIGHBOR H-H			
s,s(110)	0.01075	-0.00837	0.03800
s,x(110)	0.00000	0.00000	0.00000
x,x(110)	0.00000	0.00000	0.00000
x,x(011)	0.00000	0.00000	0.00000
x,y(110)	0.00000	0.00000	0.00000
SECOND NEIGHBOR H-H			
s,s(200)	-0.02908	-0.00020	0.00866
s,x(200)	0.00000	0.00000	0.00000
x,x(200)	0.00000	0.00000	0.00000
y,y(200)	0.00000	0.00000	0.00000
FIRST NEIGHBOR Tl-H			
s,s(100)	-0.08759	-0.08941	0.04375
s,x(100)	0.00000	0.00000	0.00000
x,s(100)	0.09695	0.07558	-0.13625
x,x(100)	0.00000	0.00000	0.00000
y,y(100)	0.00000	0.00000	0.00000
xy,x(010)	0.00000	0.00000	0.00000
d2,s(001)	-0.07839	-0.02016	0.02485
d2,z(001)	0.00000	0.00000	0.00000
SECOND NEIGHBOR Tl-H			
s,s(111)	0.00793	-0.00748	0.00894
x,s(111)	-0.01389	0.00704	-0.02570
s,x(111)	0.00000	0.00000	0.00000
x,x(111)	0.00000	0.00000	0.00000
x,y(111)	0.00000	0.00000	0.00000
xy,s(111)	0.00595	0.00518	0.00541
xy,x(111)	0.00000	0.00000	0.00000
yz,x(111)	0.00000	0.00000	0.00000
d1,x(111)	0.00000	0.00000	0.00000

TlH

BAND	ORTHOGONAL RMS ERROR mRy	MAXIMUM DEVIATION k	mRy	NON-ORTHOGONAL RMS ERROR mRy	MAXIMUM DEVIATION k	mRy
1	3.8	(066)	9.7	1.2	(066)	4.1
2	5.6	(005)	21.3	0.7	(042)	2.0
3	2.0	(007)	4.3	0.6	(174)	1.9
4	2.9	(048)	7.3	0.5	(066)	1.6
5	2.5	(005)	6.5	0.4	(044)	1.2
6	9.2	(006)	20.9	0.5	(333)	1.3
7	12.0	(333)	22.9	0.7	(007)	1.9
8	16.3	(005)	51.6	1.1	(111)	3.9
9	16.5	(444)	35.9	1.0	(022)	2.9
10	23.3	(004)	46.6	0.8	(003)	1.9
1-7	11.7			0.8		

ENERGY VALUES IN Ry AT SELECTED k-POINTS

	ORTHOGONAL	APW	NON-ORTHOGONAL
GAMMA 1	-0.38236	-0.37290	-0.37246
GAMMA 1	0.54305	0.56227	0.56165
GAMMA 12	-0.45735	-0.45439	-0.45434
GAMMA 15	0.94427	0.94949	0.94905
GAMMA 25'	-0.46801	-0.46661	-0.46642
X1 (008)	-0.49474	-0.48988	-0.49075
X1 (008)	0.00790	0.01065	0.01095
X1 (008)	0.33746	0.34267	0.34386
X2 (008)	-0.43961	-0.44345	-0.44300
X3 (008)	-0.48273	-0.48316	-0.48304
X5 (008)	-0.43861	-0.43886	-0.43888
X4' (008)	0.20933	0.22919	0.22946
X5' (008)	0.58187	0.58924	0.58840
L1 (444)	-0.48146	-0.47465	-0.47291
L1 (444)	-0.07108	-0.05259	-0.05238
L3 (444)	-0.46540	-0.46704	-0.46610
L3 (444)	-0.44232	-0.44219	-0.44196
L2' (444)	-0.01468	-0.00579	-0.00674
L2' (444)	0.66375	0.67803	0.67736
L3' (444)	0.82372	0.85964	0.85891
W1 (048)	-0.45880	-0.45150	-0.45246
W1 (048)	0.04788	0.05410	0.05427
W2' (048)	-0.48261	-0.47792	-0.47882
W2' (048)	0.15502	0.16992	0.16918
W2' (048)	0.87024	0.90131	0.90040
W1' (048)	-0.43851	-0.43882	-0.43828
W3 (048)	-0.46420	-0.46285	-0.46352
W3 (048)	0.36161	0.36268	0.36261
ODD (224)	-0.45828	-0.46139	-0.46173
ODD (224)	-0.44344	-0.44410	-0.44380
ODD (224)	0.77446	0.77717	0.77878
EVEN(224)	-0.47091	-0.47207	-0.47354
EVEN(224)	-0.46310	-0.46581	-0.46715
EVEN(224)	-0.45211	-0.45295	-0.45255
EVEN(224)	-0.18923	-0.19948	-0.20022
EVEN(224)	0.22120	0.22750	0.22748
EVEN(224)	0.67629	0.67618	0.67718
EVEN(224)	0.82607	0.78119	0.78119

Table 9.25 TlH (NaCl) a = 9.77 Bohr Slater–Koster 2-center parameters

	ORTHOGONAL	NON-ORTHOGONAL	
	ENERGY INTEGRALS Ry	ENERGY INTEGRALS Ry	OVERLAP INTEGRALS
ON SITE Tl-Tl			
s	-0.08166	-0.04445	
p	0.60290	0.34234	
t2g	-0.45218	-0.45768	
eg	-0.44803	-0.45947	
FIRST NEIGHBOR Tl-Tl			
(sss)	-0.02628	-0.01662	0.11460
(pps)	0.09390	-0.03400	-0.31742
(ppp)	-0.00147	-0.00272	0.07033
(dds)	-0.00622	-0.01533	0.02030
(ddp)	-0.00040	-0.00084	0.00813
(ddd)	0.00141	0.00287	-0.00731
(sps)	0.03889	0.01787	-0.18620
(sds)	-0.01804	-0.01902	0.03396
(pds)	-0.02553	-0.03391	0.07763
(pdp)	-0.00063	0.01127	-0.04168
SECOND NEIGHBOR Tl-Tl			
(sss)	0.00313	0.00466	0.02318
(pps)	-0.03331	0.01909	-0.03397
(ppp)	0.01016	-0.00225	0.00003
(dds)	0.00144	0.00946	-0.02184
(ddp)	-0.00104	-0.00759	0.01656
(ddd)	-0.00001	0.00612	-0.01314
(sps)	0.01254	-0.00457	-0.04235
(sds)	0.01165	-0.00706	0.01373
(pds)	-0.03151	0.00076	0.01817
(pdp)	0.00769	0.00156	-0.00594
ON SITE H-H			
s	0.35822	0.34083	
FIRST NEIGHBOR H-H			
(sss)	0.01768	0.00165	-0.02966
(sps)	0.00000	0.00000	0.00000
(pps)	0.00000	0.00000	0.00000
(ppp)	0.00000	0.00000	0.00000
SECOND NEIGHBOR H-H			
(sss)	-0.00544	-0.00123	0.02531
(sps)	0.00000	0.00000	0.00000
(pps)	0.00000	0.00000	0.00000
(ppp)	0.00000	0.00000	0.00000
FIRST NEIGHBOR Tl-H			
(sss)	0.00890	0.08792	0.04221
(sps)	0.00000	0.00000	0.00000
(pss)	0.09514	0.11602	-0.06268
(pps)	0.00000	0.00000	0.00000
(ppp)	0.00000	0.00000	0.00000
(dpp)	0.00000	0.00000	0.00000
(dss)	0.03609	0.00058	0.02061
(dps)	0.00000	0.00000	0.00000
SECOND NEIGHBOR Tl-H			
(sss)	-0.01611	0.00532	0.00705
(pss)	-0.02244	0.02694	0.04984
(sps)	0.00000	0.00000	0.00000
(pps)	0.00000	0.00000	0.00000
(ppp)	0.00000	0.00000	0.00000
(dss)	0.01351	-0.01050	0.00436
(dps)	0.00000	0.00000	0.00000
(dpp)	0.00000	0.00000	0.00000

↑ TlH

BAND	ORTHOGONAL RMS ERROR mRy	ORTHOGONAL MAXIMUM DEVIATION k	ORTHOGONAL MAXIMUM DEVIATION mRy	NON-ORTHOGONAL RMS ERROR mRy	NON-ORTHOGONAL MAXIMUM DEVIATION k	NON-ORTHOGONAL MAXIMUM DEVIATION mRy
1	5.2	(066)	13.2	0.6	(044)	1.4
2	4.9	(005)	13.7	0.7	(033)	1.5
3	1.8	(048)	4.1	0.5	(444)	1.1
4	2.8	(048)	7.3	0.4	(055)	1.0
5	3.1	(033)	7.2	0.5	(048)	1.2
6	12.6	(005)	27.7	0.8	(062)	2.2
7	13.0	(008)	27.9	1.3	(062)	3.3
8	18.9	(005)	59.0	2.0	(111)	4.9
9	18.1	(444)	40.0	2.3	(064)	6.1
10	24.4	(004)	52.4	2.3	(003)	5.9
1-10	13.0			1.4		

ENERGY VALUES IN Ry AT SELECTED k-POINTS

	ORTHOGONAL	APW	NON-ORTHOGONAL
GAMMA 1	-0.38441	-0.37290	-0.37340
GAMMA 1	0.54395	0.56227	0.56405
GAMMA 12	-0.44911	-0.45439	-0.45446
GAMMA 15	0.94070	0.94949	0.94951
GAMMA 25'	-0.46955	-0.46661	-0.46570
X1 (008)	-0.49041	-0.48988	-0.48929
X1 (008)	0.00302	0.01065	0.01156
X1 (008)	0.34207	0.34267	0.33914
X2 (008)	-0.44157	-0.44345	-0.44249
X3 (008)	-0.47769	-0.48316	-0.48341
X5 (008)	-0.43910	-0.43886	-0.43868
X4' (008)	0.20132	0.22919	0.23050
X5' (008)	0.58280	0.58924	0.58751
L1 (444)	-0.47227	-0.47465	-0.47382
L1 (444)	-0.06893	-0.05259	-0.05108
L3 (444)	-0.46817	-0.46704	-0.46599
L3 (444)	-0.43576	-0.44219	-0.44194
L2' (444)	-0.01815	-0.00579	-0.00606
L2' (444)	0.65644	0.67803	0.67681
L3' (444)	0.81964	0.85964	0.85729
W1 (048)	-0.46144	-0.45150	-0.45110
W1 (048)	0.04362	0.05410	0.05365
W2' (048)	-0.47412	-0.47792	-0.47763
W2' (048)	0.16137	0.16992	0.17009
W2' (048)	0.86328	0.90131	0.89819
W1' (048)	-0.43904	-0.43882	-0.44005
W3 (048)	-0.45880	-0.46285	-0.46312
W3 (048)	0.35599	0.36268	0.36325
ODD (224)	-0.46176	-0.46139	-0.46149
ODD (224)	-0.44469	-0.44410	-0.44413
ODD (224)	0.77714	0.77717	0.78239
EVEN(224)	-0.46951	-0.47207	-0.47210
EVEN(224)	-0.46775	-0.46581	-0.46641
EVEN(224)	-0.45386	-0.45295	-0.45287
EVEN(224)	-0.19603	-0.19948	-0.20069
EVEN(224)	0.23869	0.22750	0.22750
EVEN(224)	0.67225	0.67618	0.68059
EVEN(224)	0.82645	0.78119	0.78049

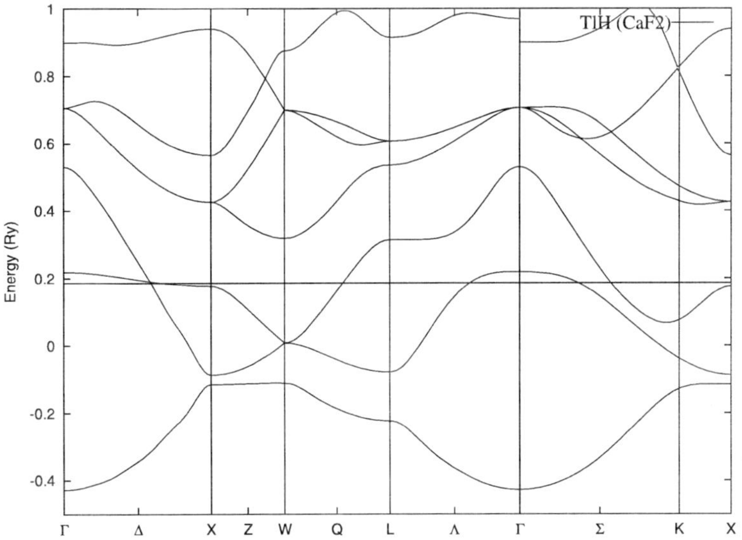

Fig. 9.28 Energy bands of TlH2 in the CaF2 structure

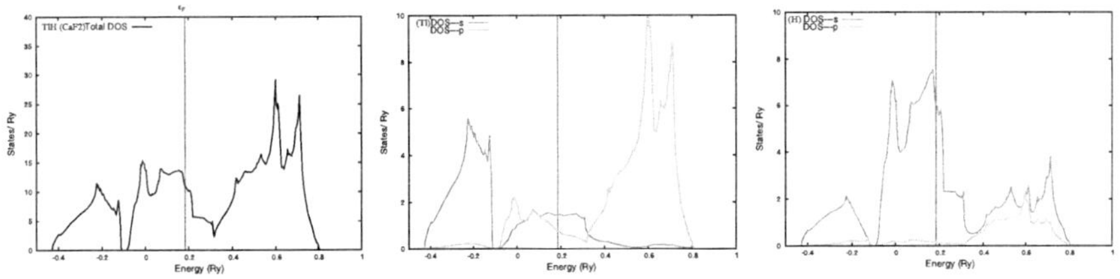

Fig. 9.29 Total, angular momentum and site decomposed densities of states of TlH2 in the CaF2 structure

Reference

1. Wikipedia

Chapter 10
Group 14 Hydrides

This chapter covers the Group 14 column of the periodic table hydrides from CH to PbH [1]. Results are presented for the crystal structures NaCl (B1), CaF2 (C1) and Im3m. These structures are not found experimentally and are presented here for the purpose of comparison through the periodic Table and to establish various trends. This group contains the classic semiconductor elements Si and Ge which with the addition of hydrogen become metallic. CH is found in a large number of polymers and clusters of the form CnH (2n+2). SiH exists as a molecule (Silane) and GeH forms as a linear chain [1]. Under consideration, at high pressures, is the Im3m structure [2]. The lattice constants, as expected, are larger than in the single elements because of the expansion of the lattice upon hydrogenation. Comparing the energy bands of the CaF2 structure to the bands of the NaCl structure we note the additional antibonding band above the Fermi level due to the second hydrogen in the CaF2 structure. Examining the densities of states figures we observe that for both crystal structures the lowest band has a mixture of s–H hydrogen and s-character of the other element. At the Fermi level the states are of s- and p-character. There are significant differences in the DOS among different crystal structures. In GeH, SnH and PbH we show d-states appearing below the GAM1 point in the energy band diagrams. Tight-binding (TB) parameters are given in the NaCl structure based on both orthogonal and non-orthogonal Hamiltonians using three- and two-center integrals. Especially, for CH the TB parameters would be very useful in building a model to describe molecules and polymers.

D. A. Papaconstantopoulos, *Band Structure of Cubic Hydrides*,
https://doi.org/10.1007/978-3-031-06878-2_10

10.1 Carbon Hydride (CH)

See Fig. 10.1 and Tables 10.1, 10.2, 10.3.

See Figs. 10.2, 10.3 and Tables 10.4, 10.5.

See Figs. 10.4, 10.5, 10.6.

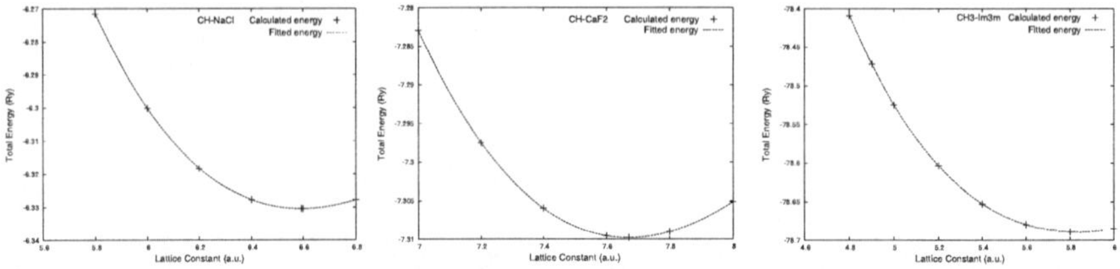

Fig. 10.1 Total energy versus lattice constant of CH in the NaCl, CaF2 and Im3m structures

Table 10.1 Lattice constant, Bulk modulus, Gap, Total energy

Stru	a (Bohr)	B (MBar)	Gap	Total Energy (Ry)
NaCl	6.59	1.34	-	-76.33041
CaF2	7.67	0.81	-	-77.30976
Im3m (P=0)	5.83	1.80	-	-78.68917
Im3m(P=1.88 MBar)	5.00	7.29	-	-78.52465

Table 10.2 Birch Fit coefficients

	A1	A2	A3	A4
NaCl	-5.441721E+00	-3.329745E+01	3.554844E+02	-7.857096E+02
CaF2	-6.454388E+00	-4.362279E+01	6.381242E+02	-2.005875E+03
Im3m	-7.705044E+01	-7.601079E+01	1.000500E+03	-2.662450E+03

Table 10.3 DOS at Ef, Hopfield parameter, Stoner criterion

```
      CH     a=6.59 Bohr          NaCl
-------------------------------------------------------------------------------
Fermi Energy    Total DOS        Decomposed DOS inside the muffin tins
   ( Ry )      (States/Ry)       s         p         eg         t2g      f
-------------------------------------------------------------------------------
C  0.742         7.424       0.437     3.920      0.025      0.033    0.007
H  0.742         7.424       0.415     0.509      0.071      0.037    0.023
-------------------------------------------------------------------------------
Fermi-Velocity (cm/s): 1.49209 x10E8
Plasmon Energy (eV)   : 17.29502
Electron-ion interaction (Hopfield parameter) (eV/A^2) C:5.582  H:1.079
-------------------------------------------------
   C     MUFFIN-TIN RADIUS and CHARGE =    1.6481    4.4553
   H     MUFFIN-TIN RADIUS and CHARGE =    1.6481    1.0645
                  C       STONER I =  0.0360
                  H       STONER I =  0.0037
         STONER PARAMETER (Ry)    I =  0.0401
         STONER CRITERION       N*I =  0.2978
-------------------------------------------------
      CH2         a=7.57 Bohr          CaF2
-------------------------------------------------------------------------------
Fermi Energy    Total DOS        Decomposed DOS inside the muffin tins
   ( Ry )      (States/Ry)       s         p         eg         t2g      f
-------------------------------------------------------------------------------
C  0.502        12.686       0.365     6.496      0.013      0.025    0.004
H  0.502        12.686       0.870     1.260      0.038      0.135    0.037
-------------------------------------------------------------------------------
Fermi-Velocity (cm/s): 1.16660 x10E8
Plasmon Energy (eV)   : 14.08463
Electron-ion interaction (Hopfield parameter) (eV/A^2) C:3.872  H2:3.995
-------------------------------------------------
   C     MUFFIN-TIN RADIUS and CHARGE =    1.6607    4.5184
   H     MUFFIN-TIN RADIUS and CHARGE =    1.6607    0.9348
                  C       STONER I =  0.0295
                  H       STONER I =  0.0017
         STONER PARAMETER (Ry)    I =  0.0332
         STONER CRITERION       N*I =  0.4208
-------------------------------------------------------------------------------
      CH3   a=5.0 Bohr               Im3m
-------------------------------------------------------------------------------
Fermi Energy    Total DOS        Decomposed DOS inside the muffin tins
   ( Ry )      (States/Ry)       s       p        d        f
C  1.288        12.045       0.206   5.499    0.044    0.012
H  1.288        12.045       0.466   0.137    0.021    0.002
-------------------------------------------------------------------------------
Fermi-Velocity (cm/s): 1.12561 x10E8
Plasmon Energy (eV)   : 17.79109
Electron-ion interaction (Hopfield parameter) (eV/A^2) C:7.786  H3:5.652
-------------------------------------------------
   C     MUFFIN-TIN RADIUS and CHARGE =    1.40    4.2706
   H     MUFFIN-TIN RADIUS and CHARGE =    1.00    0.5330
```

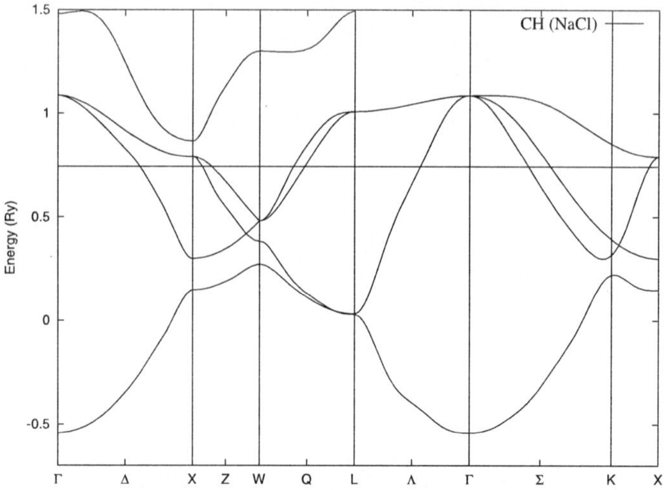

Fig. 10.2 Energy bands of CH in the NaCl structure

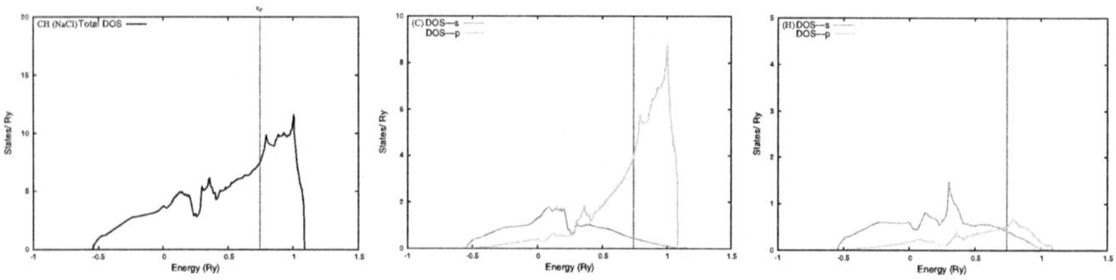

Fig. 10.3 Total, angular momentum and site decomposed densities of states of CH in the NaCl structure

Table 10.4 CH (NaCl) a = 6.59 Bohr Slater–Koster 3-center parameters

	ORTHOGONAL	NON-ORTHOGONAL	
	ENERGY INTEGRALS Ry	ENERGY INTEGRALS Ry	OVERLAP INTEGRALS
ON SITE C-C			
s,s(000)	0.24966	-0.05574	
x,x(000)	0.87514	0.56787	
FIRST NEIGHBOR C-C			
s,s(110)	0.01289	-0.06926	0.07101
s,x(110)	0.00313	0.04722	-0.07374
s,xy(110)	0.00000	0.00000	0.00000
s,d2(110)	0.00000	0.00000	0.00000
x,x(110)	0.05159	0.00671	-0.06761
x,x(011)	-0.01281	-0.01492	0.03110
x,y(110)	-0.01043	0.03357	-0.09314
x,xy(110)	0.00000	0.00000	0.00000
x,xy(011)	0.00000	0.00000	0.00000
z,d2(011)	0.00000	0.00000	0.00000
z,d1(011)	0.00000	0.00000	0.00000
xy,xy(110)	0.00000	0.00000	0.00000
xy,xy(011)	0.00000	0.00000	0.00000
xy,xz(011)	0.00000	0.00000	0.00000
xy,d2(110)	0.00000	0.00000	0.00000
d2,d2(110)	0.00000	0.00000	0.00000
d1,d1(110)	0.00000	0.00000	0.00000
SECOND NEIGHBOR C-C			
s,s(200)	0.03684	-0.01496	-0.00561
s,x(200)	-0.05257	0.02238	-0.00064
s,d2(002)	0.00000	0.00000	0.00000
x,x(200)	-0.08650	0.03033	-0.00743
y,y(200)	0.00588	-0.00589	-0.00439
x,xy(020)	0.00000	0.00000	0.00000
z,d2(002)	0.00000	0.00000	0.00000
xy,xy(200)	0.00000	0.00000	0.00000
xy,xy(002)	0.00000	0.00000	0.00000
d2,d2(002)	0.00000	0.00000	0.00000
d1,d1(002)	0.00000	0.00000	0.00000
ON SITE H-H			
s,s(000)	0.44302	0.12632	
FIRST NEIGHBOR H-H			
s,s(110)	-0.01741	-0.06550	0.03754
s,x(110)	0.00000	0.00000	0.00000
x,x(110)	0.00000	0.00000	0.00000
x,x(011)	0.00000	0.00000	0.00000
x,y(110)	0.00000	0.00000	0.00000
SECOND NEIGHBOR H-H			
s,s(200)	0.01225	0.00381	0.00243
s,x(200)	0.00000	0.00000	0.00000
x,x(200)	0.00000	0.00000	0.00000
y,y(200)	0.00000	0.00000	0.00000

FIRST NEIGHBOR C-H			
s,s(100)	-0.16826	0.11580	-0.24761
s,x(100)	0.00000	0.00000	0.00000
x,s(100)	-0.17616	0.02743	-0.33131
x,x(100)	0.00000	0.00000	0.00000
y,y(100)	0.00000	0.00000	0.00000
xy,x(010)	0.00000	0.00000	0.00000
d2,s(001)	0.00000	0.00000	0.00000
d2,z(001)	0.00000	0.00000	0.00000
SECOND NEIGHBOR C-H			
s,s(111)	0.00127	0.02492	-0.00969
x,s(111)	0.00080	0.01939	-0.01063
s,x(111)	0.00000	0.00000	0.00000
x,x(111)	0.00000	0.00000	0.00000
x,y(111)	0.00000	0.00000	0.00000
xy,s(111)	0.00000	0.00000	0.00000
xy,x(111)	0.00000	0.00000	0.00000
yz,x(111)	0.00000	0.00000	0.00000
d1,x(111)	0.00000	0.00000	0.00000

CH

	ORTHOGONAL			NON-ORTHOGONAL		
BAND	RMS ERROR	MAXIMUM DEVIATION		RMS ERROR	MAXIMUM	DEVIATION
	mRy	k	mRy	mRy	k	mRy
1	6.6	(055)	17.3	0.5	(055)	1.4
2	15.8	(008)	38.0	0.8	(264)	1.6
3	13.7	(044)	28.1	1.0	(008)	2.5
4	12.2	(444)	26.0	1.2	(062)	3.1
5	17.9	(004)	47.3	0.9	(264)	2.9
1-5	13.8			0.9		

ENERGY VALUES IN Ry AT SELECTED k-POINTS

	ORTHOGONAL	APW	NON-ORTHOGONAL
GAMMA 1	-0.54548	-0.54326	-0.54377
GAMMA 1	1.47844	1.48107	1.48116
GAMMA 15	1.08716	1.08667	1.08684
X1 (008)	0.14611	0.14765	0.14735
X1 (008)	0.85921	0.86621	0.86657
X4' (008)	0.26167	0.29971	0.30105
X5' (008)	0.77693	0.79078	0.78827
L1 (444)	0.02862	0.03248	0.03294
L2' (444)	0.02551	0.02928	0.02794
L2' (444)	1.45204	1.49364	1.49242
L3' (444)	0.98288	1.00892	1.00791
W1 (048)	0.27180	0.27145	0.27156
W2' (048)	0.36696	0.38328	0.38224
W2' (048)	1.29312	1.29941	1.30065
W3 (048)	0.49576	0.48067	0.47921
ODD (224)	0.94569	0.95970	0.96063
EVEN(224)	-0.25223	-0.25152	-0.25117
EVEN(224)	0.47015	0.46509	0.46618
EVEN(224)	0.98151	0.97043	0.97096
EVEN(224)	1.39594	1.38593	1.38750

Table 10.5 CH (NaCl) a = 6.59 Bohr Slater–Koster 2-center parameters

	ORTHOGONAL	NON-ORTHOGONAL	
	ENERGY INTEGRALS Ry	ENERGY INTEGRALS Ry	OVERLAP INTEGRALS
ON SITE C-C			
s	0.05391	-0.10283	
p	0.72716	0.57257	
FIRST NEIGHBOR C-C			
(sss)	-0.05079	-0.07739	0.10053
(pps)	0.11480	0.02263	-0.17631
(ppp)	-0.01260	-0.01560	0.03249
(dds)	0.00000	0.00000	0.00000
(ddp)	0.00000	0.00000	0.00000
(ddd)	0.00000	0.00000	0.00000
(sps)	0.07722	0.06106	-0.12688
(sds)	0.00000	0.00000	0.00000
(pds)	0.00000	0.00000	0.00000
(pdp)	0.00000	0.00000	0.00000
SECOND NEIGHBOR C-C			
(sss)	0.00534	-0.02175	-0.01152
(pps)	-0.01270	0.02466	-0.00978
(ppp)	0.00610	-0.00758	-0.00556
(dds)	0.00000	0.00000	0.00000
(ddp)	0.00000	0.00000	0.00000
(ddd)	0.00000	0.00000	0.00000
(sps)	0.00292	0.03598	0.00674
(sds)	0.00000	0.00000	0.00000
(pds)	0.00000	0.00000	0.00000
(pdp)	0.00000	0.00000	0.00000
ON SITE H-H			
s	1.08271	0.06791	
FIRST NEIGHBOR H-H			
(sss)	0.04704	-0.06825	0.04480
(sps)	0.00000	0.00000	0.00000
(pps)	0.00000	0.00000	0.00000
(ppp)	0.00000	0.00000	0.00000
SECOND NEIGHBOR H-H			
(sss)	-0.03050	0.00249	0.00081
(sps)	0.00000	0.00000	0.00000
(pps)	0.00000	0.00000	0.00000
(ppp)	0.00000	0.00000	0.00000
FIRST NEIGHBOR C-H			
(sss)	-0.00298	-0.11437	0.28420
(sps)	0.00000	0.00000	0.00000
(pss)	-0.14897	-0.02091	0.34462
(pps)	0.00000	0.00000	0.00000
(ppp)	0.00000	0.00000	0.00000
(dpp)	0.00000	0.00000	0.00000
(dss)	0.00000	0.00000	0.00000
(dps)	0.00000	0.00000	0.00000
SECOND NEIGHBOR C-H			
(sss)	0.03501	-0.03432	0.01090
(pss)	0.03068	-0.04066	0.01780
(sps)	0.00000	0.00000	0.00000
(pps)	0.00000	0.00000	0.00000
(ppp)	0.00000	0.00000	0.00000
(dss)	0.00000	0.00000	0.00000
(dps)	0.00000	0.00000	0.00000
(dpp)	0.00000	0.00000	0.00000

CH

BAND	ORTHOGONAL RMS ERROR mRy	MAXIMUM DEVIATION k	mRy	NON-ORTHOGONAL RMS ERROR mRy	MAXIMUM DEVIATION k	mRy
1	10.6	(005)	24.8	4.2	(048)	12.5
2	14.9	(008)	32.8	3.2	(055)	8.2
3	14.2	(044)	26.8	1.8	(280)	5.5
4	14.9	(064)	33.7	1.8	(174)	4.0
5	18.3	(055)	35.6	2.2	(354)	4.3
1-5	14.8			2.8		

ENERGY VALUES IN Ry AT SELECTED k-POINTS

	ORTHOGONAL	APW	NON-ORTHOGONAL
GAMMA 1	-0.55749	-0.54326	-0.54482
GAMMA 1	1.49817	1.48107	1.47947
GAMMA 15	1.08456	1.08667	1.08660
X1 (008)	0.14476	0.14765	0.14048
X1 (008)	0.85592	0.86621	0.86947
X4' (008)	0.26694	0.29971	0.30119
X5' (008)	0.77654	0.79078	0.78954
L1 (444)	0.02184	0.03248	0.02587
L2' (444)	0.00701	0.02928	0.02418
L2' (444)	1.47728	1.49364	1.49173
L3' (444)	0.98299	1.00892	1.00937
W1 (048)	0.26775	0.27145	0.28400
W2' (048)	0.39082	0.38328	0.38111
W2' (048)	1.27008	1.29941	1.29915
W3 (048)	0.49736	0.48067	0.47932
ODD (224)	0.94458	0.95970	0.96128
EVEN(224)	-0.24906	-0.25152	-0.25144
EVEN(224)	0.46798	0.46509	0.46520
EVEN(224)	0.98006	0.97043	0.97055
EVEN(224)	1.39759	1.38593	1.38782

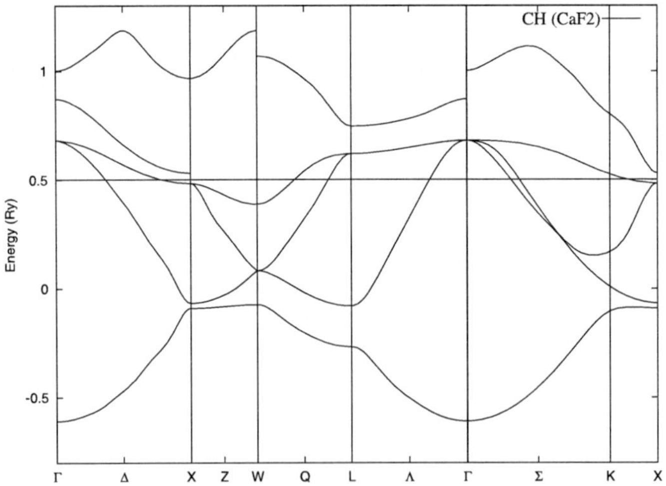

Fig. 10.4 Energy bands of CH2 in the CaF2 structure

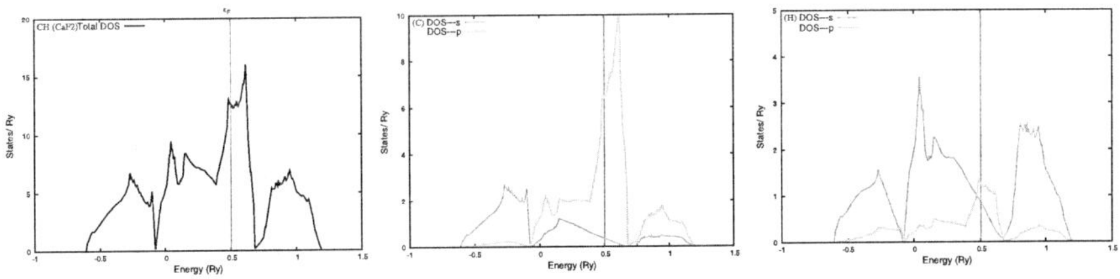

Fig. 10.5 Total, angular momentum and site decomposed densities of states of CH2 in the CaF2 structure

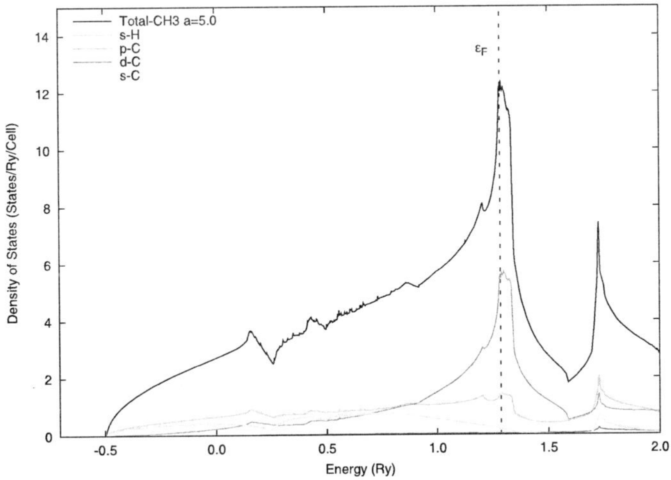

Fig. 10.6 Total, angular momentum and site decomposed densities of states of CH3 in the Im3m structure

10.2 Silicon Hydride (SiH)

See Fig. 10.7 and Tables 10.6, 10.7, 10.8.

See Figs. 10.8, 10.9 and Tables 10.9, 10.10.

See Figs. 10.10, 10.11, 10.12.

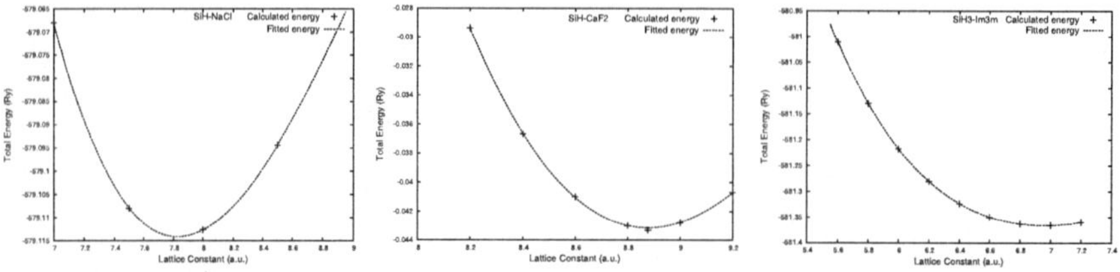

Fig. 10.7 Total energy versus lattice constant of SiH in the NaCl, CaF2 and Im3m structures

Table 10.6 Lattice constant, Bulk modulus, Gap, Total energy

Stru	a (Bohr)	B (MBar)	Gap	Total Energy (Ry)
NaCl	7.83	0.88	-	-579.07346
CaF2	8.88	0.37	-	-580.04331
Im3m (P=0)	6.96	1.04	-	-581.36579
Im3m(P=2.13 MBar)	5.60	6.90	-	-581.01080

Table 10.7 Birch Fit coefficients

	A1	A2	A3	A4
NaCl	-5.780169E+02	-6.031704E+01	9.872472E+02	-4.118370E+03
CaF2	6.587137E-01	-5.020053E+01	1.080652E+03	-6.165179E+03
Im3m	-5.797906E+02	-1.033143E+02	1.908249E+03	-6.784449E+03

Table 10.8 DOS at Ef, Hopfield parameter, Stoner criterion

```
       SiH     a=7.83 Bohr              NaCl
 ------------------------------------------------------------------------------
 Fermi Energy   Total DOS        Decomposed DOS inside the muffin tins
    ( Ry )     (States/Ry)       s          p          eg          t2g      f
 ------------------------------------------------------------------------------
 Si  0.836         7.021       0.748      1.402      0.099      0.155    0.018
 H   0.836         7.021       0.732      0.791      0.090      0.088    0.016
 ------------------------------------------------------------------------------
Fermi-Velocity (cm/s): 1.88112 x10E8
Plasmon Energy (eV)   : 16.46351
Electron-ion interaction (Hopfield parameter) (eV/A^2) Si:0.710  H:1.279
-------------------------------------------------
   Si      MUFFIN-TIN RADIUS and CHARGE =     1.9510    11.5061
    H      MUFFIN-TIN RADIUS and CHARGE =     1.9510     1.5321
                     Si         STONER I =  0.0115
                     H          STONER I =  0.0073
           STONER PARAMETER (Ry)       I =  0.0193
           STONER CRITERION          N*I =  0.1355
-------------------------------------------------
       SiH2    a=8.88 Bohr           CaF2
 ------------------------------------------------------------------------------
 Fermi Energy   Total DOS        Decomposed DOS inside the muffin tins
    ( Ry )     (States/Ry)       s          p          eg          t2g      f
 ------------------------------------------------------------------------------
 Si  0.577         8.615       0.951      1.342      0.106      0.104    0.015
 H   0.577         8.615       1.519      0.899      0.029      0.106    0.013
 ------------------------------------------------------------------------------
Fermi-Velocity (cm/s): 1.62910 x10E8
Plasmon Energy (eV)   : 13.01850
Electron-ion interaction (Hopfield parameter) (eV/A^2) Si:0.716  H2:4.412
 ------------------------------------------------------------------------------
   Si      MUFFIN-TIN RADIUS and CHARGE =     2.0180    11.5870
    H      MUFFIN-TIN RADIUS and CHARGE =     1.8258     1.2168
                     Si         STONER I =    0.0068
                     H          STONER I =    0.0035
           STONER PARAMETER (Ry)       I =    0.0140
           STONER CRITERION          N*I =    0.1206
-------------------------------------------------
        SiH3  a=5.6 Bohr           Im3m
 ------------------------------------------------------------------------------
 Fermi Energy   Total DOS        Decomposed DOS inside the muffin tins
       ( Ry )     (States/Ry)      s       p         d        f
   Si  1.363         6.457       0.390   1.407     0.297    0.036
    H  1.363         6.457       0.329   0.083     0.006    0.000
   ------------------------------------------------------------------------------
  Fermi-Velocity (cm/s): 1.74084 x10E8
  Plasmon Energy (eV)   : 16.99657
  Electron-ion interaction (Hopfield parameter) (eV/A^2) Si:1.301  H3:4.683
  -------------------------------------------------
     Si      MUFFIN-TIN RADIUS and CHARGE =     1.80    11.6999
      H      MUFFIN-TIN RADIUS and CHARGE =     1.00     0.6031
```

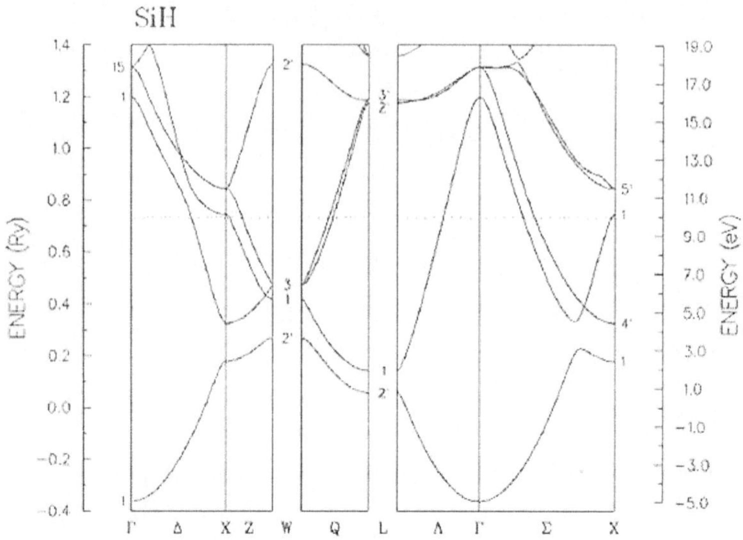

Fig. 10.8 Energy bands of SiH in the NaCl structure (tight-binding)

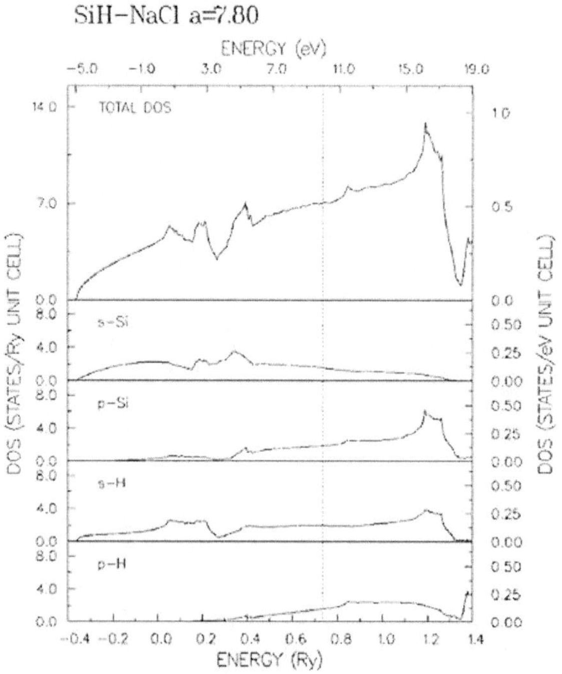

Fig. 10.9 Total, angular momentum and site decomposed densities of states of SiH in the NaCl structure (tight-binding)

Table 10.9 SiH (NaCl) a = 7.80 Bohr Slater–Koster 3-center parameters

	ORTHOGONAL	NON-ORTHOGONAL	
	ENERGY INTEGRALS Ry	ENERGY INTEGRALS Ry	OVERLAP INTEGRALS
ON SITE Si-Si			
s,s(000)	0.28748	0.02770	
x,x(000)	0.96217	0.47442	
xy,xy(000)	1.68403	10.00000	
d2,d2(000)	2.77037	10.00000	
FIRST NEIGHBOR Si-Si			
s,s(110)	-0.06361	-0.06069	0.10938
s,x(110)	0.00502	0.02622	-0.13078
s,xy(110)	0.01044	0.00000	0.00000
s,d2(110)	-0.01190	0.00000	0.00000
x,x(110)	0.06172	-0.03886	-0.13934
x,x(011)	-0.00334	-0.00696	0.07611
x,y(110)	0.02491	-0.00842	-0.17888
x,xy(110)	-0.07617	0.00000	0.00000
x,xy(011)	0.00750	0.00000	0.00000
z,d2(011)	0.03658	0.00000	0.00000
z,d1(011)	-0.03038	0.00000	0.00000
xy,xy(110)	-0.07624	0.00000	0.00000
xy,xy(011)	0.03090	0.00000	0.00000
xy,xz(011)	0.06621	0.00000	0.00000
xy,d2(110)	-0.10513	0.00000	0.00000
d2,d2(110)	0.12578	0.00000	0.00000
d1,d1(110)	0.10052	0.00000	0.00000
SECOND NEIGHBOR Si-Si			
s,s(200)	0.02385	-0.02037	-0.00889
s,x(200)	0.09529	0.02310	-0.00863
s,d2(002)	0.13088	0.00000	0.00000
x,x(200)	-0.06185	0.04505	-0.01384
y,y(200)	-0.00026	-0.01425	-0.00840
x,xy(020)	-0.00104	0.00000	0.00000
z,d2(002)	-0.03742	0.00000	0.00000
xy,xy(200)	-0.02000	0.00000	0.00000
xy,xy(002)	-0.03957	0.00000	0.00000
d2,d2(002)	0.14834	0.00000	0.00000
d1,d1(002)	-0.14739	0.00000	0.00000
ON SITE H-H			
s,s(000)	1.55959	0.08963	
FIRST NEIGHBOR H-H			
s,s(110)	-0.13182	-0.02796	0.05563
s,x(110)	0.00000	0.00000	0.00000
x,x(110)	0.00000	0.00000	0.00000
x,x(011)	0.00000	0.00000	0.00000
x,y(110)	0.00000	0.00000	0.00000
SECOND NEIGHBOR H-H			
s,s(200)	0.19979	0.00555	0.00370
s,x(200)	0.00000	0.00000	0.00000
x,x(200)	0.00000	0.00000	0.00000
y,y(200)	0.00000	0.00000	0.00000
FIRST NEIGHBOR Si-H			
s,s(100)	-0.04043	0.08219	-0.27317
s,x(100)	0.00000	0.00000	0.00000
x,s(100)	-0.14388	-0.00947	-0.40696
x,x(100)	0.00000	0.00000	0.00000
y,y(100)	0.00000	0.00000	0.00000
xy,x(010)	0.00000	0.00000	0.00000
d2,s(001)	0.64439	0.00000	0.00000
d2,z(001)	0.00000	0.00000	0.00000

SECOND NEIGHBOR Si-H

s,s(111)	0.00522	0.01992	-0.02028
x,s(111)	0.00952	0.01437	-0.02486
s,x(111)	0.00000	0.00000	0.00000
x,x(111)	0.00000	0.00000	0.00000
x,y(111)	0.00000	0.00000	0.00000
xy,s(111)	0.02795	0.00000	0.00000
xy,x(111)	0.00000	0.00000	0.00000
yz,x(111)	0.00000	0.00000	0.00000
d1,x(111)	0.00000	0.00000	0.00000

SiH

	ORTHOGONAL			NON-ORTHOGONAL		
BAND	RMS ERROR	MAXIMUM DEVIATION		RMS ERROR	MAXIMUM DEVIATION	
	mRy	k	mRy	mRy	k	mRy
1	12.1	(264)	29.9	2.8	(222)	6.9
2	15.7	(033)	40.4	6.0	(005)	14.5
3	14.2	(264)	42.0	10.0	(005)	28.4
4	8.8	(224)	26.4	8.5	(224)	20.3
5	10.6	(033)	23.7	9.3	(118)	27.4
1-5	12.5			7.8		

ENERGY VALUES IN Ry AT SELECTED k-POINTS

	ORTHOGONAL	APW	NON-ORTHOGONAL
GAMMA 1	-0.35897	-0.36360	-0.36421
GAMMA 1	1.20281	1.19844	1.20135
GAMMA 15	1.31780	1.31402	1.31394
X1 (008)	0.19321	0.17786	0.17756
X1 (008)	0.73173	0.74746	0.75432
X4' (008)	0.33034	0.32824	0.33530
X5' (008)	0.85081	0.84913	0.84405
L1 (444)	0.13991	0.14267	0.14234
L2' (444)	0.06049	0.05790	0.05289
L2' (444)	1.18793	1.17905	1.16023
L3' (444)	1.18655	1.18955	1.17895
W1 (048)	0.41072	0.41872	0.42172
W2' (048)	0.25587	0.26922	0.26469
W2' (048)	1.32011	1.32779	1.33410
W3 (048)	0.47976	0.47616	0.47068
ODD (224)	1.08250	1.05329	1.07360
EVEN(224)	-0.15323	-0.13265	-0.12922
EVEN(224)	0.50174	0.47521	0.47719
EVEN(224)	1.02912	1.03953	1.05415
EVEN(224)	1.07966	1.09208	1.09917

FERMI LEVEL QUANTITIES (Non-orthogonal fit)

ENERGY	DENSITIES OF STATES						
	Total	s-Si	p-Si	t2g-Si	eg-Si	s-H	p-H
Ry	states/Ry/cell						
0.7324	7.09	1.61	1.92	0.00	0.00	1.99	1.58
	INTEGRATED DENSITIES OF STATES						
	Total	s-Si	p-Si	t2g-Si	eg-Si	s-H	p-H
	electrons						
	5.00	2.13	0.78	0.00	0.00	1.67	0.42

VELOCITY	PLASMON ENERGY	EIGENVALUE SUM
cm/s	eV	Ry
1.78x10E8	15.64	-2.1904

Table 10.10 SiH a=7.80 Bohr Slater–Koster 2-center parameters

SiH	ORTHOGONAL	NON-ORTHOGONAL	
	ENERGY INTEGRALS Ry	ENERGY INTEGRALS Ry	OVERLAP INTEGRALS
ON SITE Si-Si			
s	0.26221	0.01226	
p	0.88011	0.48785	
t2g	2.06387	10.00000	
eg	1.97284	10.00000	
FIRST NEIGHBOR Si-Si			
(sss)	-0.03580	-0.06203	0.12173
(pps)	0.12692	-0.05947	-0.32320
(ppp)	-0.00459	-0.01064	0.06590
(dds)	-0.37662	0.00000	0.00000
(ddp)	0.10279	0.00000	0.00000
(ddd)	-0.00694	0.00000	0.00000
(sps)	0.09746	0.03325	-0.19211
(sds)	0.06858	0.00000	0.00000
(pds)	0.13494	0.00000	0.00000
(pdp)	-0.00365	0.00000	0.00000
SECOND NEIGHBOR Si-Si			
(sss)	-0.00401	-0.02206	-0.00116
(pps)	-0.03483	0.02295	-0.04250
(ppp)	0.00704	-0.01182	-0.00624
(dds)	0.05708	0.00000	0.00000
(ddp)	0.10015	0.00000	0.00000
(ddd)	-0.07004	0.00000	0.00000
(sps)	0.02065	0.01774	-0.02574
(sds)	0.06712	0.00000	0.00000
(pds)	-0.03567	0.00000	0.00000
(pdp)	-0.01483	0.00000	0.00000
ON SITE H-H			
s	0.81978	0.05991	
FIRST NEIGHBOR H-H			
(sss)	0.02184	-0.03730	0.04714
(sps)	0.00000	0.00000	0.00000
(pps)	0.00000	0.00000	0.00000
(ppp)	0.00000	0.00000	0.00000
SECOND NEIGHBOR H-H			
(sss)	-0.00811	0.00159	-0.00005
(sps)	0.00000	0.00000	0.00000
(pps)	0.00000	0.00000	0.00000
(ppp)	0.00000	0.00000	0.00000
FIRST NEIGHBOR Si-H			
(sss)	0.03120	-0.07451	0.29702
(sps)	0.00000	0.00000	0.00000
(pss)	-0.14980	0.02727	0.43034
(pps)	0.00000	0.00000	0.00000
(ppp)	0.00000	0.00000	0.00000
(dpp)	0.00000	0.00000	0.00000
(dss)	0.00000	0.00000	0.00000
(dps)	0.00000	0.00000	0.00000
SECOND NEIGHBOR Si-H			
(sss)	0.03686	-0.02136	0.01827
(pss)	0.05202	-0.02940	0.03576
(sps)	0.00000	0.00000	0.00000
(pps)	0.00000	0.00000	0.00000
(ppp)	0.00000	0.00000	0.00000
(dss)	0.00000	0.00000	0.00000
(dps)	0.00000	0.00000	0.00000
(dpp)	0.00000	0.00000	0.00000

SiH

BAND	ORTHOGONAL RMS ERROR mRy	MAXIMUM DEVIATION k	mRy	NON-ORTHOGONAL RMS ERROR mRy	MAXIMUM DEVIATION k	mRy
1	9.5	(066)	20.4	4.3	(000)	8.4
2	11.4	(048)	31.4	5.2	(005)	17.9
3	18.7	(264)	44.2	10.6	(005)	29.9
4	16.8	(008)	44.9	8.7	(264)	19.6
5	27.9	(226)	112.4	8.6	(118)	27.0
1-5	18.0			7.8		

ENERGY VALUES IN Ry AT SELECTED k-POINTS

	ORTHOGONAL	APW	NON-ORTHOGONAL
GAMMA 1	-0.35844	-0.36360	-0.37200
GAMMA 1	1.20016	1.19844	1.20243
GAMMA 15	1.30958	1.31402	1.31319
X1 (008)	0.19066	0.17786	0.17361
X1 (008)	0.77257	0.74746	0.75476
X4' (008)	0.33096	0.32824	0.33184
X5' (008)	0.85698	0.84913	0.84456
L1 (444)	0.14718	0.14267	0.14361
L2' (444)	0.06176	0.05790	0.04961
L2' (444)	1.20226	1.17905	1.15842
L3' (444)	1.18461	1.18955	1.18024
W1 (048)	0.38735	0.41872	0.42342
W2' (048)	0.26892	0.26922	0.26124
W2' (048)	1.28476	1.32779	1.33409
W3 (048)	0.47350	0.47616	0.47145
ODD (224)	1.02459	1.05329	1.07226
EVEN(224)	-0.12332	-0.13265	-0.13213
EVEN(224)	0.47834	0.47521	0.48067
EVEN(224)	1.04827	1.03953	1.04974
EVEN(224)	1.13590	1.09208	1.10317

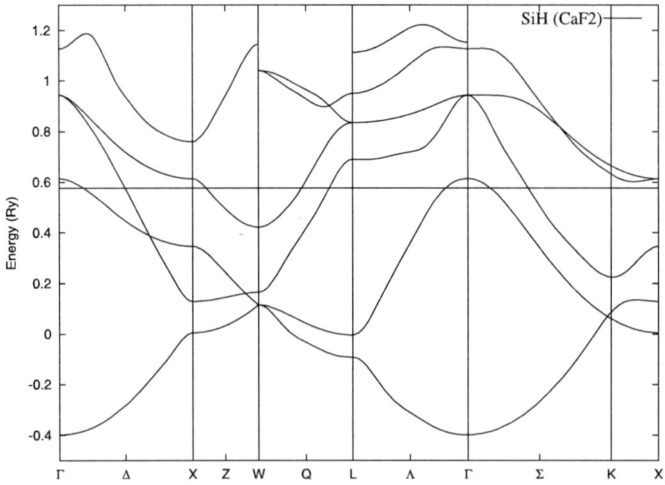

Fig. 10.10 Energy bands of SiH2 in the CaF2 structure

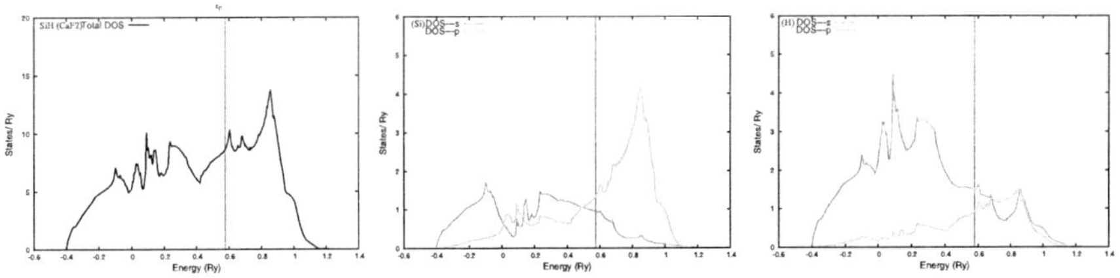

Fig. 10.11 Total, angular momentum and site decomposed densities of states of SiH2 in the CaF2 structure

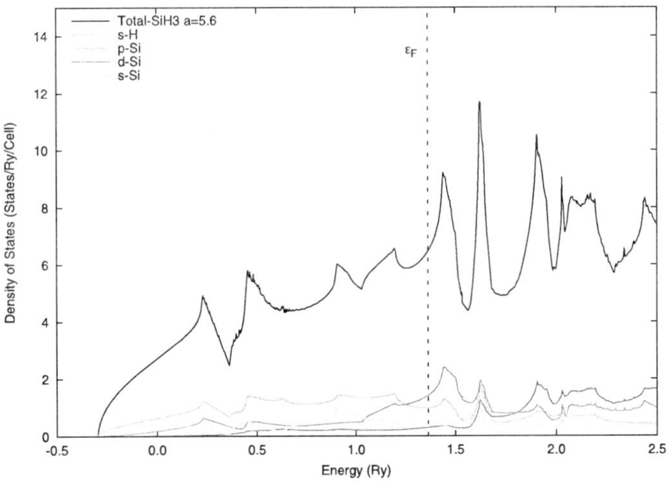

Fig. 10.12 Total, angular momentum and site decomposed densities of states of SiH3 in the Im3m structure

10.3 Germanium Hydride (GeH)

See Fig. 10.13 and Tables 10.11, 10.12, 10.13.

See Figs. 10.14, 10.15 and Tables 10.14, 10.15.

See Figs. 10.16, 10.17, 10.18.

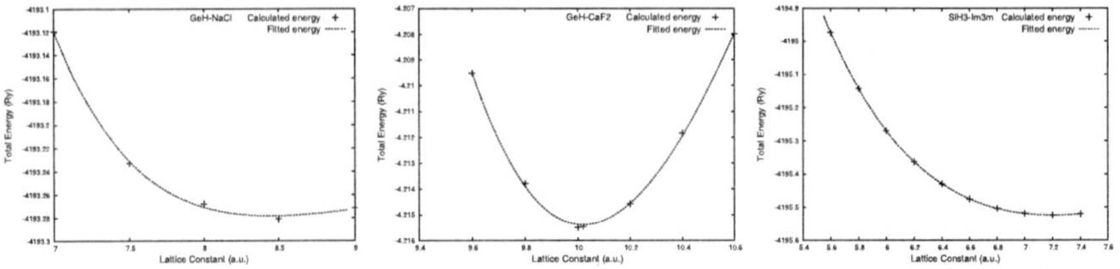

Fig. 10.13 Total energy versus lattice constant of GeH in the NaCl, CaF2 and Im3m structures

Table 10.11 Lattice constant, Bulk modulus, Gap, Total energy

Stru	a (Bohr)	B (MBar)	Gap	Total Energy (Ry)
NaCl	8.45	0.43	-	-4193.27822
CaF2	10.02	0.36	-	-4194.21546
Im3m (P=0)	7.21	0.92	-	-4195.52351
Im3m(P=2.14 MBar)	5.80	7.43	-	-4195.14266

Table 10.12 Birch fit coefficients

	A1	A2	A3	A4
NaCl	-4.193381E+03	2.262924E+01	-1.034854E+03	1.348790E+04
CaF2	-3.925230E+00	-6.774609E+00	-8.421119E+02	2.595451E+04
Im3m	-4.194144E+03	-9.248012E+01	1.618861E+03	-2.276656E+03

Table 10.13 DOS at Ef, Hopfield parameter, Stoner criterion

```
      GeH       a=8.45 Bohr            NaCl
 ---------------------------------------------------------------------------------
 Fermi Energy   Total DOS        Decomposed DOS inside the muffin tins
     ( Ry )    (States/Ry)       s          p          eg         t2g      f
 ---------------------------------------------------------------------------------
 Ge  0.632         7.049       0.460      2.098      0.070      0.112    0.022
 H   0.632         7.049       0.768      0.649      0.060      0.051    0.011
 ---------------------------------------------------------------------------------
Fermi-Velocity (cm/s): 1.70569 x10E8
Plasmon Energy (eV)   : 13.32305
Electron-ion interaction (Hopfield parameter) (eV/A^2) Ge:1.095  H:1.091
 ------------------------------------------------
   Ge      MUFFIN-TIN RADIUS and CHARGE =     2.2129   29.9377
    H      MUFFIN-TIN RADIUS and CHARGE =     2.0021    1.3448
                   Ge        STONER I =  0.0078
                   H         STONER I =  0.0063
           STONER PARAMETER (Ry)      I =  0.0144
           STONER CRITERION         N*I =  0.1013
 ------------------------------------------------
      GeH2      a=10.02 Bohr           CaF2
 ---------------------------------------------------------------------------------
 Fermi Energy   Total DOS        Decomposed DOS inside the muffin tins
     ( Ry )    (States/Ry)       s          p          eg         t2g      f
 ---------------------------------------------------------------------------------
 Ge  0.345         8.587       0.682      1.405      0.050      0.065  0.020
 H   0.345         8.587       2.612      0.720      0.009      0.050  0.016
 ---------------------------------------------------------------------------------
Fermi-Velocity (cm/s): 0.84481 x10E8
Plasmon Energy (eV)   : 5.61972
Electron-ion interaction (Hopfield parameter) (eV/A^2) Si:0.634  H2:3.656
 ---------------------------------------------------------------------------------
   Ge      MUFFIN-TIN RADIUS and CHARGE =     2.2779   29.9338
    H      MUFFIN-TIN RADIUS and CHARGE =     2.0610    1.2239
                   Ge        STONER I =  0.0033
                   H         STONER I =  0.0059
           STONER PARAMETER (Ry)      I =  0.0153
           STONER CRITERION         N*I =  0.1310
 ------------------------------------------------
       GeH3  a=5.8 Bohr                 Im3m
 ---------------------------------------------------------------------------------
 Fermi Energy   Total DOS        Decomposed DOS inside the muffin tins
      ( Ry )     (States/Ry)       s        p         d        f
   Ge  1.362       6.605        0.247    1.128     0.143    0.024
    H  1.362       6.605        0.358    0.073     0.005    0.000
   -------------------------------------------------------------------------------
   Fermi-Velocity (cm/s): 1.83843 x10E8
   Plasmon Energy (eV)   : 17.22248
   Electron-ion interaction (Hopfield parameter) (eV/A^2) Ge:4.400  H3:1.777
   ------------------------------------------------
     Ge      MUFFIN-TIN RADIUS and CHARGE =     1.700   28.9729
      H      MUFFIN-TIN RADIUS and CHARGE =     1.00     0.5755
```

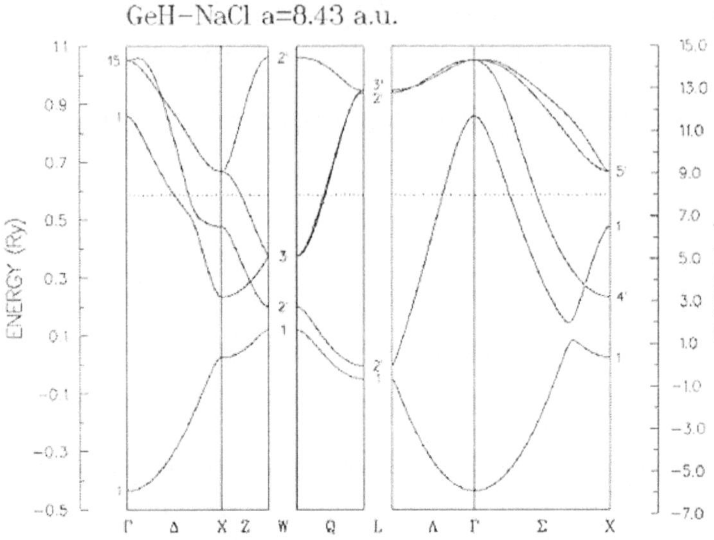

Fig. 10.14 Energy bands of GeH in the NaCl structure (tight-binding)

Fig. 10.15 Total, angular momentum and site decomposed densities of states of GeH in the NaCl structure (tight-binding)

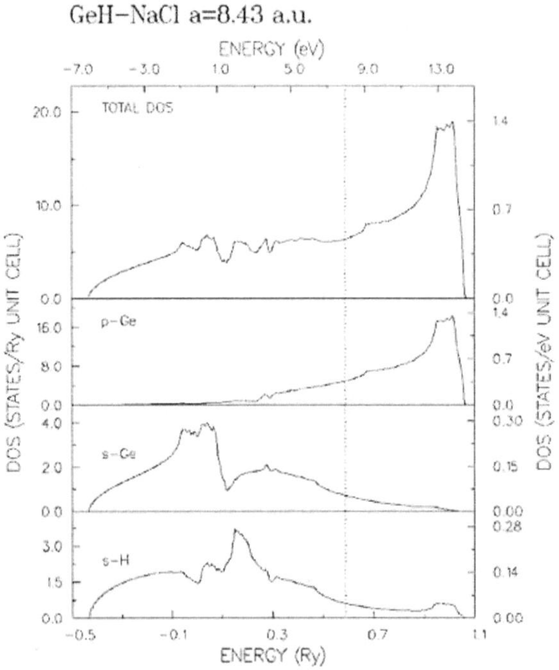

Table 10.14 GeH (NaCl) a = 8.43 Bohr Slater–Koster 3-center parameters

	ORTHOGONAL	NON-ORTHOGONAL	
	ENERGY INTEGRALS Ry	ENERGY INTEGRALS Ry	OVERLAP INTEGRALS
ON SITE Ge-Ge			
s,s(000)	0.09254	0.14257	
x,x(000)	0.77194	0.60752	
xy,xy(000)	-1.26513	-1.26647	
d2,d2(000)	-1.18581	-1.03145	
FIRST NEIGHBOR Ge-Ge			
s,s(110)	0.00341	0.02914	0.00631
s,x(110)	-0.00815	-0.00795	0.02930
s,xy(110)	-0.01450	-0.04101	0.03443
s,d2(110)	0.01752	-0.03487	-0.01602
x,x(110)	0.05115	0.01928	-0.09344
x,x(011)	-0.00114	-0.01701	0.03898
x,y(110)	0.01026	-0.03006	-0.06252
x,xy(110)	0.01309	0.05019	-0.03699
x,xy(011)	0.00052	-0.00528	0.00262
z,d2(011)	0.01889	0.00414	0.00770
z,d1(011)	0.01000	-0.00364	0.02697
xy,xy(110)	-0.00091	0.01671	-0.01380
xy,xy(011)	0.00018	0.00021	-0.00005
xy,xz(011)	-0.00039	-0.01622	0.01294
xy,d2(110)	0.00044	-0.00702	0.00519
d2,d2(110)	0.01323	0.01018	0.02272
d1,d1(110)	-0.03975	-0.05310	-0.05064
SECOND NEIGHBOR Ge-Ge			
s,s(200)	0.02457	0.04346	-0.01218
s,x(200)	0.04907	0.03550	-0.05507
s,d2(002)	-0.02571	0.03467	0.01449
x,x(200)	-0.09361	0.00640	0.10313
y,y(200)	0.01518	0.03552	0.04817
x,xy(020)	-0.01017	-0.03013	0.02394
z,d2(002)	-0.00239	-0.08257	0.03917
xy,xy(200)	-0.00027	0.00074	-0.00056
xy,xy(002)	-0.00006	0.01664	-0.01314
d2,d2(002)	0.02653	0.06770	0.00835
d1,d1(002)	0.00024	0.03664	-0.02895
ON SITE H-H			
s,s(000)	0.01009	0.27699	
FIRST NEIGHBOR H-H			
s,s(110)	0.01883	0.01231	-0.00989
s,x(110)	0.00000	0.00000	0.00000
x,x(110)	0.00000	0.00000	0.00000
x,x(011)	0.00000	0.00000	0.00000
x,y(110)	0.00000	0.00000	0.00000
SECOND NEIGHBOR H-H			
s,s(200)	-0.01622	-0.01033	0.00836
s,x(200)	0.00000	0.00000	0.00000
x,x(200)	0.00000	0.00000	0.00000
y,y(200)	0.00000	0.00000	0.00000
FIRST NEIGHBOR Ge-H			
s,s(100)	-0.10264	-0.11731	-0.09065
s,x(100)	0.00000	0.00000	0.00000
x,s(100)	0.09013	0.14104	0.01618
x,x(100)	0.00000	0.00000	0.00000
y,y(100)	0.00000	0.00000	0.00000
xy,x(010)	0.00000	0.00000	0.00000
d2,s(001)	0.17287	-0.11422	-0.18440
d2,z(001)	0.00000	0.00000	0.00000

SECOND NEIGHBOR Ge-H			
s,s(111)	-0.00340	-0.00040	-0.00351
x,s(111)	-0.00399	0.00281	-0.00778
s,x(111)	0.00000	0.00000	0.00000
x,x(111)	0.00000	0.00000	0.00000
x,y(111)	0.00000	0.00000	0.00000
xy,s(111)	0.00068	-0.00708	0.00566
xy,x(111)	0.00000	0.00000	0.00000
yz,x(111)	0.00000	0.00000	0.00000
d1,x(111)	0.00000	0.00000	0.00000

GeH

	ORTHOGONAL			NON-ORTHOGONAL		
BAND	RMS ERROR	MAXIMUM DEVIATION		RMS ERROR	MAXIMUM DEVIATION	
	mRy	k	mRy	mRy	k	mRy
1	0.7	(055)	1.5	0.1	(004)	0.2
2	1.0	(005)	3.0	0.1	(444)	0.2
3	0.6	(442)	1.5	0.1	(444)	0.2
4	0.6	(000)	1.7	0.1	(044)	0.4
5	0.7	(000)	1.7	0.1	(055)	0.2
6	3.7	(048)	10.1	0.2	(008)	0.5
7	5.5	(048)	15.2	0.3	(118)	0.9
1-7	2.6			0.2		

ENERGY VALUES IN Ry AT SELECTED k-POINTS

	ORTHOGONAL	APW	NON-ORTHOGONAL
GAMMA 1	-0.43717	-0.43555	-0.43562
GAMMA 1	0.85680	0.85775	0.85797
GAMMA 12	-1.26460	-1.26634	-1.26627
GAMMA 15	1.05006	1.05115	1.05116
GAMMA 25'	-1.26852	-1.26838	-1.26838
X1 (008)	-1.27206	-1.27025	-1.27028
X1 (008)	0.02145	0.02451	0.02504
X1 (008)	0.47382	0.47659	0.47720
X2 (008)	-1.26441	-1.26434	-1.26417
X3 (008)	-1.27138	-1.27098	-1.27098
X5 (008)	-1.26268	-1.26346	-1.26336
X4' (008)	0.23172	0.23492	0.23461
X5' (008)	0.65002	0.66832	0.66810
L1 (444)	-1.26918	-1.26831	-1.26832
L1 (444)	-0.04655	-0.04920	-0.04877
L3 (444)	-1.26799	-1.26833	-1.26813
L3 (444)	-1.26362	-1.26407	-1.26399
L2' (444)	-0.01048	-0.00373	-0.00382
L2' (444)	0.93424	0.93993	0.94055
L3' (444)	0.93946	0.94697	0.94695
W1 (048)	-1.26534	-1.26541	-1.26548
W1 (048)	0.12879	0.11871	0.11881
W2' (048)	-1.26928	-1.26866	-1.26875
W2' (048)	0.18489	0.20013	0.20040
W2' (048)	1.06373	1.06160	1.06234
W1' (048)	-1.26245	-1.26344	-1.26349
W3 (048)	-1.26750	-1.26727	-1.26727
W3 (048)	0.38166	0.37480	0.37512
ODD (224)	-1.26736	-1.26738	-1.26744
ODD (224)	-1.26553	-1.26439	-1.26430
EVEN(224)	-1.26890	-1.26863	-1.26866
EVEN(224)	-1.26736	-1.26783	-1.26781
EVEN(224)	-1.26533	-1.26544	-1.26549
EVEN(224)	-0.22932	-0.23107	-0.23105
EVEN(224)	0.30006	0.29953	0.29951

FERMI LEVEL QUANTITIES (Non-orthogonal fit)

ENERGY		DENSITIES OF STATES					
	Total	s-Ge	p-Ge	t2g-Ge	eg-Ge	s-H	p-H
Ry		states/Ry/cell					
0.5847	6.19	0.93	4.49	0.10	0.05	0.62	0.00
		INTEGRATED DENSITIES OF STATES					
	Total	s-Ge	p-Ge	t2g-Ge	eg-Ge	s-H	p-H
		electrons					
	15.00	2.01	1.39	5.79	3.88	1.92	0.00

VELOCITY	PLASMON ENERGY	EIGENVALUE SUM
cm/s	eV	Ry
1.82x10E8	13.32	-20.7372

Table 10.15 GeH (NaCl) a = 8.43 Bohr Slater–Koster 2-center parameters

	ORTHOGONAL	NON-ORTHOGONAL	
	ENERGY INTEGRALS Ry	ENERGY INTEGRALS Ry	OVERLAP INTEGRALS
ON SITE Ge-Ge			
s	-0.02653	-0.08266	
p	0.68744	0.46447	
t2g	-1.26242	-1.26650	
eg	-1.26165	-1.26660	
FIRST NEIGHBOR Ge-Ge			
(sss)	-0.03301	-0.04708	0.05236
(pps)	0.10675	0.03006	-0.20064
(ppp)	-0.00378	-0.00249	0.04854
(dds)	-0.00133	-0.00468	0.00289
(ddp)	-0.00131	-0.01096	0.00893
(ddd)	0.00110	0.01563	-0.01236
(sps)	0.06988	0.05879	-0.10760
(sds)	-0.02689	-0.01027	0.01019
(pds)	-0.03210	-0.01839	0.01714
(pdp)	0.00154	0.01369	-0.01365
SECOND NEIGHBOR Ge-Ge			
(sss)	0.00771	-0.00535	0.00459
(pps)	-0.04855	0.04065	0.01279
(ppp)	0.01515	0.00149	0.00225
(dds)	0.00073	0.01838	-0.01452
(ddp)	-0.00101	0.00590	-0.00466
(ddd)	0.00013	0.00872	-0.00688
(sps)	-0.00306	0.01202	0.00191
(sds)	0.00237	-0.00331	0.00267
(pds)	-0.03549	0.00594	-0.00435
(pdp)	0.00963	0.01708	-0.01346
ON SITE H-H			
s	0.54094	0.19801	
FIRST NEIGHBOR H-H			
(sss)	0.02891	-0.01222	-0.03838
(sps)	0.00000	0.00000	0.00000
(pps)	0.00000	0.00000	0.00000
(ppp)	0.00000	0.00000	0.00000

SECOND NEIGHBOR H-H			
(sss)	-0.01682	-0.01861	0.00711
(sps)	0.00000	0.00000	0.00000
(pps)	0.00000	0.00000	0.00000
(ppp)	0.00000	0.00000	0.00000
FIRST NEIGHBOR Ge-H			
(sss)	-0.00955	0.06975	-0.13212
(sps)	0.00000	0.00000	0.00000
(pss)	0.12661	0.08217	-0.21077
(pps)	0.00000	0.00000	0.00000
(ppp)	0.00000	0.00000	0.00000
(dpp)	0.00000	0.00000	0.00000
(dss)	0.02842	0.10780	-0.09062
(dps)	0.00000	0.00000	0.00000
SECOND NEIGHBOR Ge-H			
(sss)	-0.02911	0.00078	0.00301
(pss)	-0.02446	0.01405	0.01865
(sps)	0.00000	0.00000	0.00000
(pps)	0.00000	0.00000	0.00000
(ppp)	0.00000	0.00000	0.00000
(dss)	0.01717	-0.01203	0.01039
(dps)	0.00000	0.00000	0.00000
(dpp)	0.00000	0.00000	0.00000

GeH

	ORTHOGONAL			NON-ORTHOGONAL		
BAND	RMS ERROR	MAXIMUM DEVIATION		RMS ERROR	MAXIMUM DEVIATION	
	mRy	k	mRy	mRy	k	mRy
1	2.5	(226)	7.7	0.1	(002)	0.2
2	1.8	(264)	3.8	0.1	(444)	0.2
3	0.9	(033)	1.8	0.1	(444)	0.2
4	1.5	(444)	5.4	0.1	(044)	0.3
5	1.8	(444)	5.4	0.1	(008)	0.1
6	6.1	(004)	16.5	0.2	(264)	0.3
7	12.4	(007)	42.7	0.3	(007)	0.8
1-7	5.4			0.1		

ENERGY VALUES IN Ry AT SELECTED k-POINTS

	ORTHOGONAL	APW	NON-ORTHOGONAL
GAMMA 1	-0.44475	-0.43555	-0.43570
GAMMA 1	0.85534	0.85775	0.85788
GAMMA 12	-1.26396	-1.26634	-1.26620
GAMMA 15	1.04772	1.05115	1.05115
GAMMA 25'	-1.26990	-1.26838	-1.26842
X1 (008)	-1.26915	-1.27025	-1.27032
X1 (008)	0.02582	0.02451	0.02464
X1 (008)	0.46928	0.47659	0.47670
X2 (008)	-1.26466	-1.26434	-1.26430
X3 (008)	-1.26824	-1.27098	-1.27097
X5 (008)	-1.26330	-1.26346	-1.26332
X4' (008)	0.22391	0.23492	0.23488

X5' (008)	0.66602	0.66832	0.66833
L1 (444)	-1.27073	-1.26831	-1.26828
L1 (444)	-0.05108	-0.04920	-0.04892
L3 (444)	-1.26903	-1.26833	-1.26814
L3 (444)	-1.25869	-1.26407	-1.26409
L2' (444)	-0.01225	-0.00373	-0.00358
L2' (444)	0.93596	0.93993	0.93993
L3' (444)	0.94501	0.94697	0.94697
W1 (048)	-1.26685	-1.26541	-1.26553
W1 (048)	0.12383	0.11871	0.11879
W2' (048)	-1.26556	-1.26866	-1.26862
W2' (048)	0.20213	0.20013	0.20044
W2' (048)	1.06096	1.06160	1.06162
W1' (048)	-1.26383	-1.26344	-1.26341
W3 (048)	-1.26592	-1.26727	-1.26729
W3 (048)	0.38855	0.37480	0.37495
ODD (224)	-1.26843	-1.26738	-1.26732
ODD (224)	-1.26371	-1.26439	-1.26440
EVEN(224)	-1.27025	-1.26863	-1.26864
EVEN(224)	-1.26806	-1.26783	-1.26779
EVEN(224)	-1.26505	-1.26544	-1.26553
EVEN(224)	-0.23067	-0.23107	-0.23124
EVEN(224)	0.30826	0.29953	0.29956

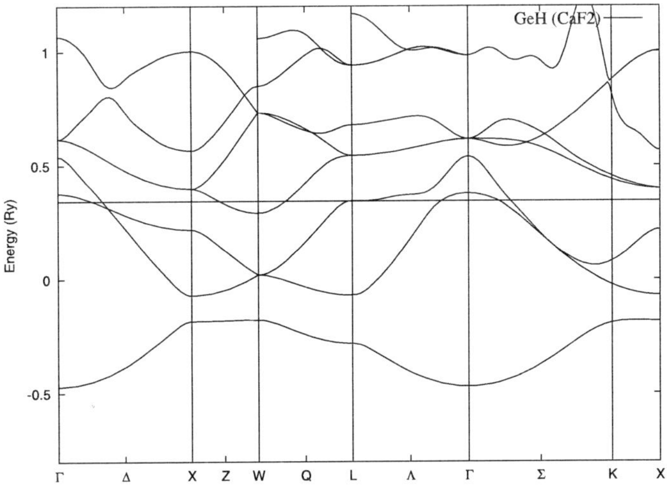

Fig. 10.16 Energy bands of GeH2 in the CaF2 structure

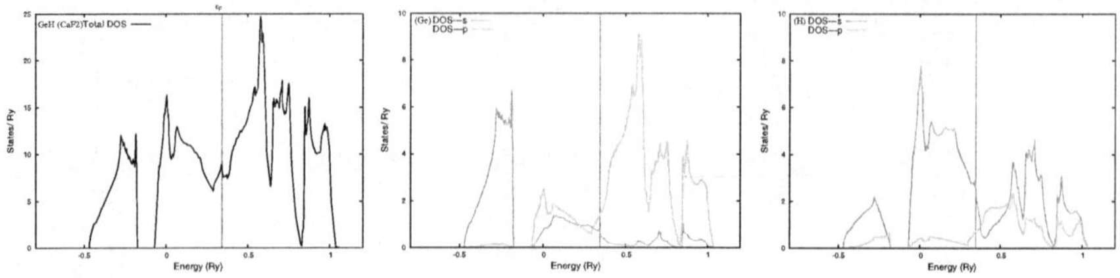

Fig. 10.17 Total, angular momentum and site decomposed densities of states of GeH2 in the CaF2 structure

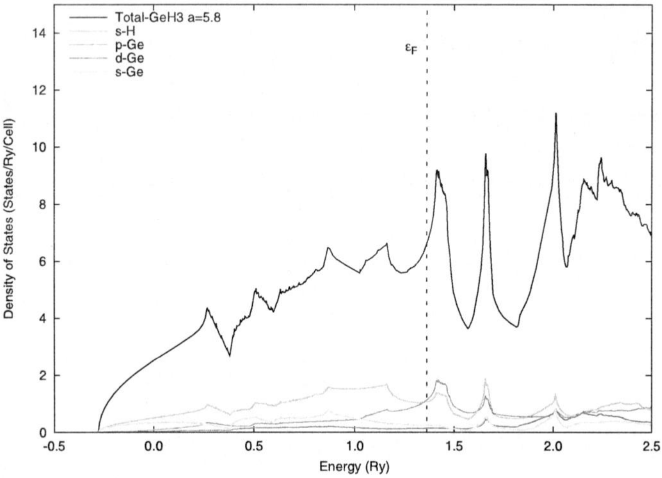

Fig. 10.18 Total, angular momentum and site decomposed densities of states of GeH3 in the Im3m structure

10.4 Tin Hydride (SnH)

See Fig. 10.19 and Tables 10.16, 10.17, 10.18.

See Figs. 10.20, 10.21 and Tables 10.19, 10.20.

See Figs. 10.22, 10.23, 10.24.

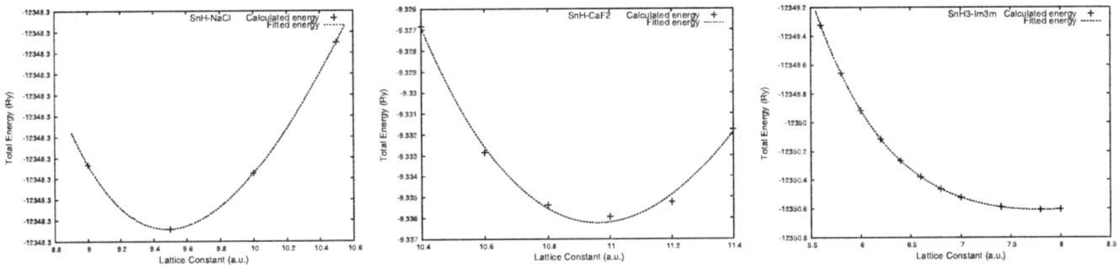

Fig. 10.19 Total energy versus lattice constant of SnH in the NaCl, CaF2 and Im3m structures

Table 10.16 Lattice constant, Bulk modulus, Gap, Total energy

Stru	a (Bohr)	B (MBar)	Gap	Total Energy (Ry)
NaCl	9.48	0.79	-	-12348.33622
CaF2	10.96	0.30	-	-12349.33641
Im3m (P=0)	7.80	0.79	-	-12350.60763
Im3m(P=2.21 MBar)	6.20	7.69		-12350.11433

Table 10.17 Birch Fit coefficients

	A1	A2	A3	A4
NaCl	-1.234693E+04	-1.016685E+02	1.986468E+03	-4.117816E+03
CaF2	-8.573971E+00	-7.274098E+01	1.738109E+03	-1.305164E+02
Im3m	-1.234918E+04	-1.089182E+02	2.063716E+03	6.629419E+02

Table 10.18 DOS at Ef, Hopfield parameter, Stoner criterion

```
       SnH       a=9.48 Bohr          NaCl
 ----------------------------------------------------------------------------------
 Fermi Energy   Total DOS       Decomposed DOS inside the muffin tins
     ( Ry )    (States/Ry)       s          p          eg          t2g      f
 ---------------------------------------------------------------------------------
 Sn  0.529        8.277       0.458      2.125      0.074      0.131    0.037
 H   0.529        8.277       1.093      0.786      0.062      0.069    0.011
 ----------------------------------------------------------------------------------
Fermi-Velocity (cm/s): 1.51302 x10E8
Plasmon Energy (eV)   : 10.73834
Electron-ion interaction (Hopfield parameter) (eV/A^2) Sn:0.626  H:0.963
------------------------------------------------
   Sn      MUFFIN-TIN RADIUS and CHARGE =     2.4885    47.7543
    H      MUFFIN-TIN RADIUS and CHARGE =     2.2515     1.4511
                     Sn        STONER I =  0.0043
                     H         STONER I =  0.0069
           STONER PARAMETER (Ry)      I =  0.0114
           STONER CRITERION         N*I =  0.0940
------------------------------------------------
       SnH2     a=10.96 Bohr              CaF2
 ----------------------------------------------------------------------------------
 Fermi Energy   Total DOS       Decomposed DOS inside the muffin tins
     ( Ry )    (States/Ry)       s          p          eg          t2g      f
 ---------------------------------------------------------------------------------
 Sn  0.289        6.676       0.531      1.393      0.067      0.021   0.012
 H   0.289        6.676       0.960      0.624      0.005      0.034   0.008
 ----------------------------------------------------------------------------------
Fermi-Velocity (cm/s): 0.84630 x10E8
Plasmon Energy (eV)   : 4.33896
Electron-ion interaction (Hopfield parameter) (eV/A^2) Sn:0.351  H2:1.315
 ----------------------------------------------------------------------------------
   Sn      MUFFIN-TIN RADIUS and CHARGE =     2.6104    47.8933
    H      MUFFIN-TIN RADIUS and CHARGE =     2.1357     1.2301
                     Sn        STONER I =  0.0020
                     H         STONER I =  0.0020
           STONER PARAMETER (Ry)      I =  0.0062
           STONER CRITERION         N*I =  0.0414
------------------------------------------------
        SnH3  a=6.2 Bohr                Im3m
 ----------------------------------------------------------------------------------
 Fermi Energy   Total DOS       Decomposed DOS inside the muffin tins
      ( Ry )     (States/Ry)       s       p         d       f
   Sn  1.442        6.853      0.188   0.624     0.197   0.040
    H  1.442        6.853      0.335   0.065     0.004   0.000
   ----------------------------------------------------------------------------------
  Fermi-Velocity (cm/s): 1.70785 x10E8
  Plasmon Energy (eV)   : 14.74607
  Electron-ion interaction (Hopfield parameter) (eV/A^2) Sn:1.571  H3:1.033
  ------------------------------------------------
     Sn      MUFFIN-TIN RADIUS and CHARGE =     1.700     45.5316
      H      MUFFIN-TIN RADIUS and CHARGE =     1.00       0.5819
```

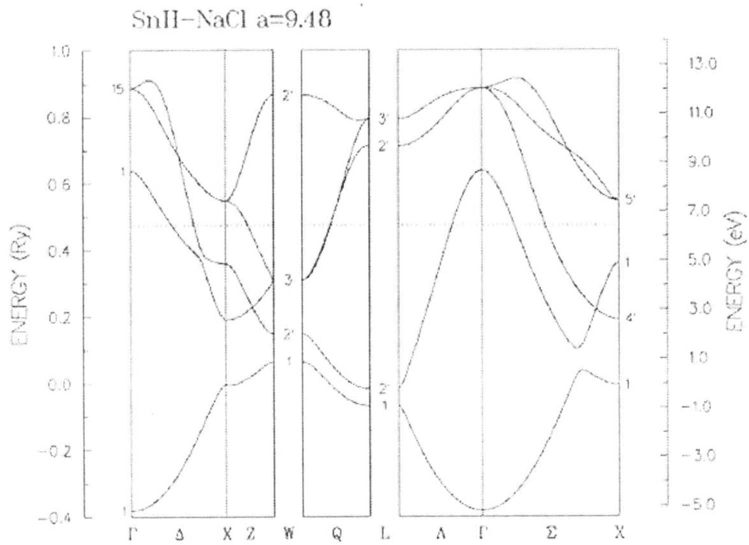

Fig. 10.20 Energy bands of SnH in the NaCl structure (tight-binding)

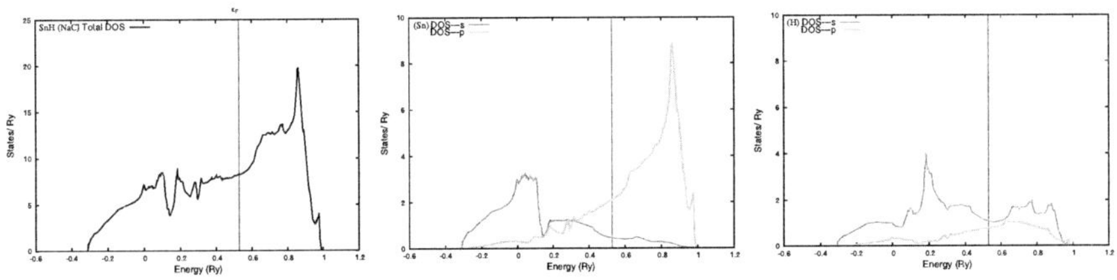

Fig. 10.21 Total, angular momentum and site decomposed densities of states of SnH in the NaCl structure

Table 10.19 SnH (NaCl) a = 9.48 Bohr Slater–Koster 3-center parameters

	ORTHOGONAL	NON-ORTHOGONAL	
	ENERGY INTEGRALS Ry	ENERGY INTEGRALS Ry	OVERLAP INTEGRALS
ON SITE Sn-Sn			
s,s(000)	0.04268	0.04361	
x,x(000)	0.63597	0.57671	
xy,xy(000)	-1.11885	-1.11890	
d2,d2(000)	-1.06172	-1.11839	
FIRST NEIGHBOR Sn-Sn			
s,s(110)	0.00283	0.00712	-0.00877
s,x(110)	-0.00675	-0.00393	0.01139
s,xy(110)	0.01015	-0.02931	0.03362
s,d2(110)	0.00912	-0.00285	-0.00061
x,x(110)	0.04362	0.01973	-0.06270
x,x(011)	-0.00028	-0.01759	0.01052
x,y(110)	0.01379	-0.02799	-0.05071
x,xy(110)	-0.00664	0.03655	-0.02464
x,xy(011)	0.00054	-0.00272	0.00198
z,d2(011)	0.01105	-0.00243	0.00496
z,d1(011)	0.00481	0.00061	0.00471
xy,xy(110)	-0.00138	0.00597	-0.00648
xy,xy(011)	0.00026	0.00549	-0.00467
xy,xz(011)	-0.00004	0.00347	-0.00300
xy,d2(110)	0.00076	-0.00628	0.00623
d2,d2(110)	0.00917	0.00164	-0.00167
d1,d1(110)	-0.02819	-0.01039	0.00913
SECOND NEIGHBOR Sn-Sn			
s,s(200)	0.01889	0.02398	-0.00210
s,x(200)	0.03660	0.03515	-0.01509
s,d2(002)	-0.01980	0.01827	-0.01491
x,x(200)	-0.07308	-0.05406	0.04154
y,y(200)	0.01195	0.02248	0.02633
x,xy(020)	0.01669	-0.02315	0.01974
z,d2(002)	0.00292	-0.02562	0.01632
xy,xy(200)	-0.00008	-0.00228	0.00191
xy,xy(002)	-0.00005	0.00519	-0.00465
d2,d2(002)	0.01906	0.00307	-0.00232
d1,d1(002)	0.00006	-0.00695	0.00624
ON SITE H-H			
s,s(000)	0.00393	0.25159	
FIRST NEIGHBOR H-H			
s,s(110)	0.01056	0.00271	-0.01647
s,x(110)	0.00000	0.00000	0.00000
x,x(110)	0.00000	0.00000	0.00000
x,x(011)	0.00000	0.00000	0.00000
x,y(110)	0.00000	0.00000	0.00000
SECOND NEIGHBOR H-H			
s,s(200)	-0.01069	-0.00102	-0.00756
s,x(200)	0.00000	0.00000	0.00000
x,x(200)	0.00000	0.00000	0.00000
y,y(200)	0.00000	0.00000	0.00000
FIRST NEIGHBOR Sn-H			
s,s(100)	-0.08415	-0.08257	-0.05837
s,x(100)	0.00000	0.00000	0.00000
x,s(100)	0.06897	0.13643	0.12867
x,x(100)	0.00000	0.00000	0.00000
y,y(100)	0.00000	0.00000	0.00000
xy,x(010)	0.00000	0.00000	0.00000
d2,s(001)	0.14093	-0.03323	0.00504
d2,z(001)	0.00000	0.00000	0.00000

SECOND NEIGHBOR Sn-H

s,s(111)	-0.00019	0.00453	0.00164
x,s(111)	-0.00392	-0.00349	-0.00406
s,x(111)	0.00000	0.00000	0.00000
x,x(111)	0.00000	0.00000	0.00000
x,y(111)	0.00000	0.00000	0.00000
xy,s(111)	0.00088	-0.00362	0.00253
xy,x(111)	0.00000	0.00000	0.00000
yz,x(111)	0.00000	0.00000	0.00000
d1,x(111)	0.00000	0.00000	0.00000

SnH

BAND	ORTHOGONAL RMS ERROR mRy	MAXIMUM DEVIATION k	mRy	NON-ORTHOGONAL RMS ERROR mRy	MAXIMUM DEVIATION k	mRy
1	0.4	(174)	1.4	0.1	(044)	0.1
2	1.0	(005)	3.4	0.1	(004)	0.2
3	0.6	(066)	1.8	0.0	(044)	0.2
4	0.7	(022)	1.5	0.1	(044)	0.3
5	0.7	(222)	1.4	0.1	(000)	0.2
6	2.4	(048)	5.9	0.1	(380)	0.3
7	3.8	(007)	14.1	0.3	(044)	0.7
1-7	1.8			0.1		

ENERGY VALUES IN Ry AT SELECTED k-POINTS

	ORTHOGONAL	APW	NON-ORTHOGONAL
GAMMA 1	-0.38199	-0.38172	-0.38171
GAMMA 1	0.63841	0.63823	0.63834
GAMMA 12	-1.11846	-1.11941	-1.11925
GAMMA 15	0.88547	0.88624	0.88625
GAMMA 25'	-1.12276	-1.12278	-1.12275
X1 (008)	-1.12726	-1.12529	-1.12533
X1 (008)	-0.00762	-0.00462	-0.00441
X1 (008)	0.36027	0.35920	0.35938
X2 (008)	-1.11577	-1.11604	-1.11599
X3 (008)	-1.12687	-1.12678	-1.12671
X5 (008)	-1.11376	-1.11465	-1.11453
X4' (008)	0.18751	0.19102	0.19069
X5' (008)	0.53877	0.54890	0.54860
L1 (444)	-1.12278	-1.12243	-1.12235
L1 (444)	-0.06596	-0.06704	-0.06686
L3 (444)	-1.12258	-1.12263	-1.12253
L3 (444)	-1.11508	-1.11559	-1.11543
L2' (444)	-0.02047	-0.01613	-0.01612
L2' (444)	0.71259	0.71301	0.71299
L3' (444)	0.78945	0.79342	0.79339
W1 (048)	-1.11881	-1.11797	-1.11793
W1 (048)	0.07069	0.06481	0.06466
W2' (048)	-1.12301	-1.12299	-1.12295
W2' (048)	0.13871	0.14778	0.14775
W2' (048)	0.86385	0.86608	0.86615
W1' (048)	-1.11354	-1.11462	-1.11453
W3 (048)	-1.12056	-1.12078	-1.12075
W3 (048)	0.31590	0.31062	0.31091
ODD (224)	-1.12183	-1.12102	-1.12103
ODD (224)	-1.11573	-1.11612	-1.11617
EVEN(224)	-1.12323	-1.12314	-1.12317
EVEN(224)	-1.12117	-1.12168	-1.12165
EVEN(224)	-1.11884	-1.11802	-1.11807
EVEN(224)	-0.21110	-0.21226	-0.21223
EVEN(224)	0.22689	0.22644	0.22626

FERMI LEVEL QUANTITIES (Non-orthogonal fit)

ENERGY		DENSITIES OF STATES					
	Total	s-Sn	p-Sn	t2g-Sn	eg-Sn	s-H	p-H
Ry		states/Ry/cell					
0.4731	7.55	0.51	6.07	0.09	0.01	0.87	0.00
		INTEGRATED DENSITIES OF STATES					
	Total	s-Sn	p-Sn	t2g-Sn	eg-Sn	s-H	p-H
		electrons					
	15.00	1.73	1.44	6.05	3.99	1.79	0.00

VELOCITY	PLASMON ENERGY	EIGENVALUE SUM
cm/s	eV	Ry
1.66x10E8	11.23	-17.7576

Table 10.20 SnH (NaCl) a = 9.48 Bohr Slater–Koster 2-center parameters

	ORTHOGONAL	NON-ORTHOGONAL	
	ENERGY INTEGRALS Ry	ENERGY INTEGRALS Ry	OVERLAP INTEGRALS
ON SITE Sn-Sn			
s	-0.05954	-0.05929	
p	0.57050	0.46992	
t2g	-1.11732	-1.11954	
eg	-1.11644	-1.11970	
FIRST NEIGHBOR Sn-Sn			
(sss)	-0.02818	-0.02753	0.01650
(pps)	0.08556	0.05817	-0.11454
(ppp)	-0.00134	0.01926	0.05636
(dds)	-0.00178	-0.01623	0.01301
(ddp)	-0.00047	-0.01886	0.01727
(ddd)	0.00056	-0.00060	0.00050
(sps)	0.05107	0.04543	-0.03215
(sds)	-0.01893	0.00315	0.00144
(pds)	-0.02108	-0.01228	0.01718
(pdp)	0.00306	0.03266	-0.03144
SECOND NEIGHBOR Sn-Sn			
(sss)	0.00359	0.00303	-0.02600
(pps)	-0.04334	0.01886	0.04068
(ppp)	0.01038	0.00223	-0.01369
(dds)	0.00046	0.00534	-0.00472
(ddp)	-0.00056	-0.00322	0.00283
(ddd)	0.00005	0.00314	-0.00279
(sps)	-0.00136	0.00746	0.02832
(sds)	0.00699	0.00185	0.00081
(pds)	-0.02921	0.01139	-0.00910
(pdp)	0.00474	0.00454	-0.00263
ON SITE H-H			
s	0.39130	0.03754	
FIRST NEIGHBOR H-H			
(sss)	0.02211	-0.00433	0.02309
(sps)	0.00000	0.00000	0.00000
(pps)	0.00000	0.00000	0.00000
(ppp)	0.00000	0.00000	0.00000

SECOND NEIGHBOR H-H			
(sss)	-0.00534	-0.01708	0.02018
(sps)	0.00000	0.00000	0.00000
(pps)	0.00000	0.00000	0.00000
(ppp)	0.00000	0.00000	0.00000
FIRST NEIGHBOR Sn-H			
(sss)	0.00711	0.06734	-0.14867
(sps)	0.00000	0.00000	0.00000
(pss)	0.10358	0.06425	-0.17989
(pps)	0.00000	0.00000	0.00000
(ppp)	0.00000	0.00000	0.00000
(dpp)	0.00000	0.00000	0.00000
(dss)	0.02265	0.10148	-0.09830
(dps)	0.00000	0.00000	0.00000
SECOND NEIGHBOR Sn-H			
(sss)	-0.01802	0.00547	-0.00114
(pss)	-0.01654	0.01312	0.00122
(sps)	0.00000	0.00000	0.00000
(pps)	0.00000	0.00000	0.00000
(ppp)	0.00000	0.00000	0.00000
(dss)	0.01163	0.00153	0.00046
(dps)	0.00000	0.00000	0.00000
(dpp)	0.00000	0.00000	0.00000

SnH

	ORTHOGONAL			NON-ORTHOGONAL		
BAND	RMS ERROR	MAXIMUM DEVIATION		RMS ERROR	MAXIMUM DEVIATION	
	mRy	k	mRy	mRy	k	mRy
1	1.5	(226)	4.3	0.1	(280)	0.3
2	1.4	(004)	3.5	0.1	(004)	0.2
3	0.6	(442)	1.6	0.1	(044)	0.3
4	1.0	(444)	3.0	0.1	(044)	0.5
5	1.0	(444)	3.0	0.1	(022)	0.2
6	3.6	(004)	9.5	0.3	(066)	0.8
7	8.4	(007)	33.1	0.4	(062)	1.0
1-7	3.6			0.2		

ENERGY VALUES IN Ry AT SELECTED k-POINTS

	ORTHOGONAL	APW	NON-ORTHOGONAL
GAMMA 1	-0.38630	-0.38172	-0.38148
GAMMA 1	0.63479	0.63823	0.63860
GAMMA 12	-1.11786	-1.11941	-1.11934
GAMMA 15	0.85690	0.88624	0.88621
GAMMA 25'	-1.12385	-1.12278	-1.12264
X1 (008)	-1.12433	-1.12529	-1.12530
X1 (008)	-0.00260	-0.00462	-0.00484
X1 (008)	0.36032	0.35920	0.35908
X2 (008)	-1.11573	-1.11604	-1.11586
X3 (008)	-1.12462	-1.12678	-1.12671
X5 (008)	-1.11470	-1.11465	-1.11453

X4' (008)	0.18310	0.19102	0.19121
X5' (008)	0.53070	0.54890	0.54863
L1 (444)	-1.12344	-1.12243	-1.12228
L1 (444)	-0.06871	-0.06704	-0.06667
L3 (444)	-1.12257	-1.12263	-1.12249
L3 (444)	-1.11262	-1.11559	-1.11556
L2' (444)	-0.02145	-0.01613	-0.01582
L2' (444)	0.71281	0.71301	0.71301
L3' (444)	0.78944	0.79342	0.79336
W1 (048)	-1.12033	-1.11797	-1.11792
W1 (048)	0.06402	0.06481	0.06449
W2' (048)	-1.12089	-1.12299	-1.12283
W2' (048)	0.15512	0.14778	0.14810
W2' (048)	0.84875	0.86608	0.86617
W1' (048)	-1.11489	-1.11462	-1.11462
W3 (048)	-1.11995	-1.12078	-1.12081
W3 (048)	0.31790	0.31062	0.31074
ODD (224)	-1.12137	-1.12102	-1.12102
ODD (224)	-1.11606	-1.11612	-1.11624
EVEN(224)	-1.12296	-1.12314	-1.12312
EVEN(224)	-1.12162	-1.12168	-1.12171
EVEN(224)	-1.11871	-1.11802	-1.11817
EVEN(224)	-0.21173	-0.21226	-0.21224
EVEN(224)	0.23131	0.22644	0.22613

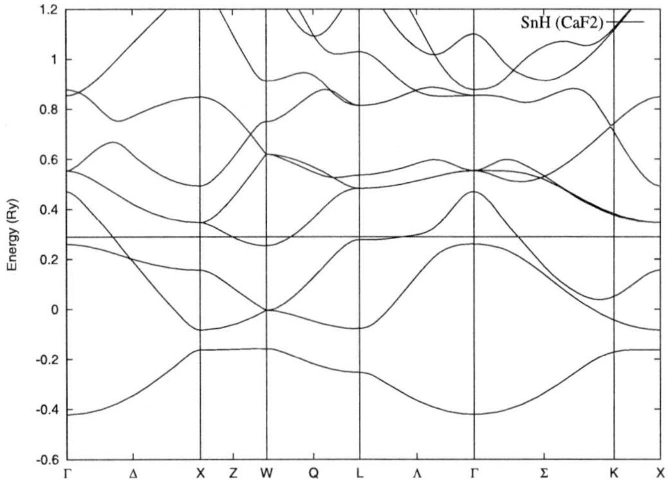

Fig. 10.22 Energy bands of SnH2 in the CaF2 structure

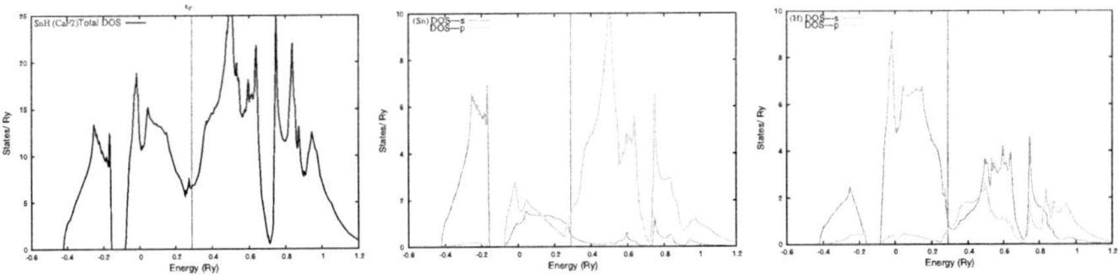

Fig. 10.23 Total, angular momentum and site decomposed densities of states of SnH2 in the CaF2 structure

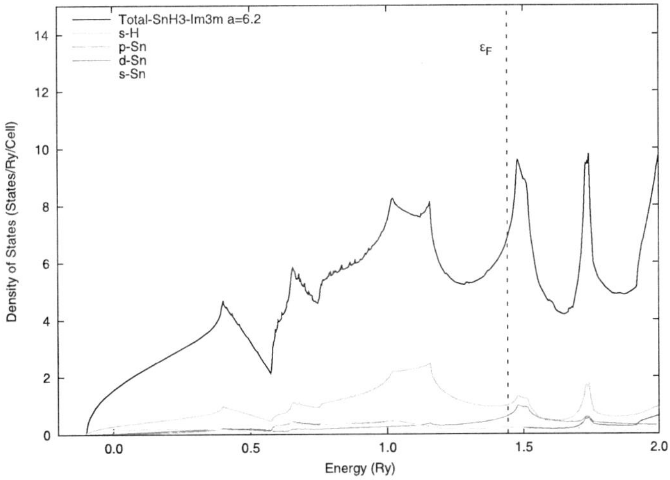

Fig. 10.24 Total, angular momentum and site decomposed densities of states of SnH3 in the Im3m structure

10.5 Lead Hydride (PbH)

See Fig. 10.25 and Tables 10.21, 10.22, 10.23.

See Figs. 10.26, 10.27 and Tables 10.24, 10.25.

See Figs. 10.28 and 10.29.

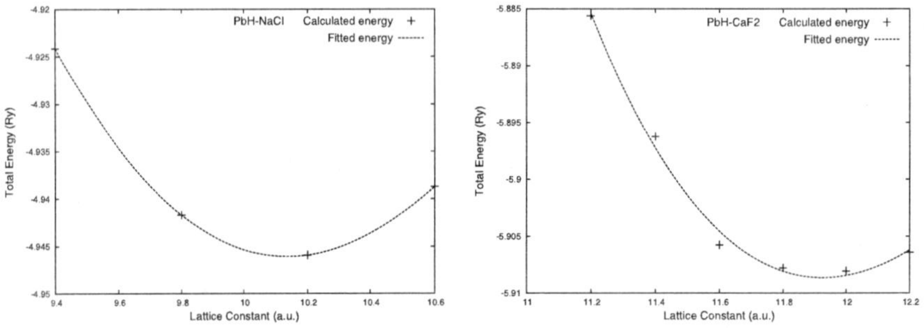

Fig. 10.25 Total energy versus lattice constant of PbH in the NaCl, and CaF2 structures

Table 10.21 Lattice constant, Bulk modulus, Gap, Total energy

Stru	a (Bohr)	B (MBar)	Gap	Total Energy (Ry)
NaCl	10.14	0.47	-	-41834.94494
CaF2 (semimetal)	11.92	0.38	-	-41835.90721

Table 10.22 Birch Fit coefficients

	A1	A2	A3	A4
NaCl	-3.276270E+00	-1.658029E+02	5.192278E+03	-4.925055E+04
CaF2	-5.152159E+00	-5.815762E+01	-6.628190E+02	8.666722E+04

Table 10.23 DOS at Ef, Hopfield parameter, Stoner criterion

```
       PbH      a=10.14 Bohr         NaCl
------------------------------------------------------------------------------
 Fermi Energy    Total DOS        Decomposed DOS inside the muffin tins
     ( Ry )     (States/Ry)       s         p         eg         t2g      f
------------------------------------------------------------------------------
 Pb  0.431         8.728       0.252     2.508     0.067      0.133    0.031
 H   0.431         8.728       1.042     0.841     0.058      0.065    0.010
------------------------------------------------------------------------------
Fermi-Velocity (cm/s): 1.22159 x10E8
Plasmon Energy (eV)   : 8.05185
Electron-ion interaction (Hopfield parameter) (eV/A^2) Pb:0.701  H:0.706
--------------------------------------------------
   Pb      MUFFIN-TIN RADIUS and CHARGE =     2.6610    79.9282
    H      MUFFIN-TIN RADIUS and CHARGE =     2.4075     1.4459
                     Pb        STONER I =  0.0032
                     H         STONER I =  0.0058
           STONER PARAMETER (Ry)      I =  0.0092
           STONER CRITERION         N*I =  0.0803
--------------------------------------------------
       PbH2     a=11.92 Bohr           CaF2
------------------------------------------------------------------------------
 Fermi Energy    Total DOS
     ( Ry )     (States/Ry)
------------------------------------------------------------------------------
 PbH  0.204         0.0     semimetal
------------------------------------------------------------------------------
   Pb      MUFFIN-TIN RADIUS and CHARGE =     2.7108   79.8689
    H      MUFFIN-TIN RADIUS and CHARGE =     2.4527    1.3511
```

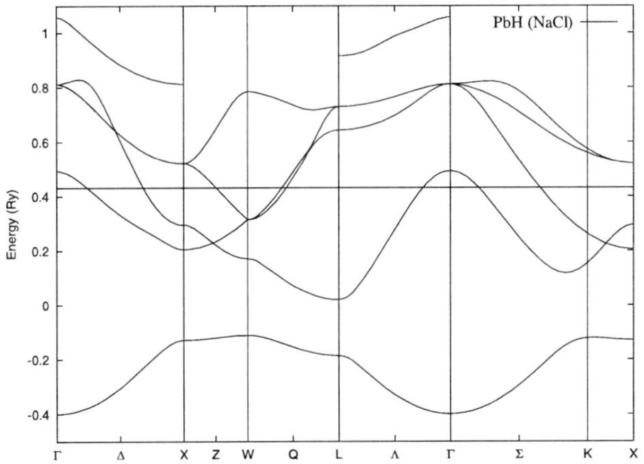

Fig. 10.26 Energy bands of PbH in the NaCl structure

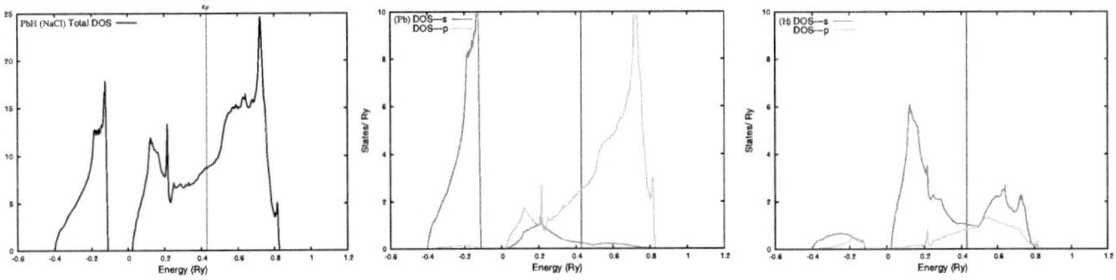

Fig. 10.27 Total, angular momentum and site decomposed densities of states of PbH in the NaCl structure

Table 10.24 PbH (NaCl) a = 10.14 Bohr Slater–Koster 3-center parameters

	ORTHOGONAL	NON-ORTHOGONAL	
	ENERGY INTEGRALS Ry	ENERGY INTEGRALS Ry	OVERLAP INTEGRALS
ON SITE Pb-Pb			
s,s(000)	0.14432	-0.20560	
x,x(000)	0.65547	0.33368	
FIRST NEIGHBOR Pb-Pb			
s,s(110)	0.01402	-0.03124	0.04983
s,x(110)	0.04026	0.03476	-0.07223
s,xy(110)	0.00000	0.00000	0.00000
s,d2(110)	0.00000	0.00000	0.00000
x,x(110)	0.00959	-0.00847	-0.09800
x,x(011)	0.01390	-0.00788	0.04441
x,y(110)	0.02571	0.00501	-0.15584
x,xy(110)	0.00000	0.00000	0.00000
x,xy(011)	0.00000	0.00000	0.00000
z,d2(011)	0.00000	0.00000	0.00000
z,d1(011)	0.00000	0.00000	0.00000
xy,xy(110)	0.00000	0.00000	0.00000
xy,xy(011)	0.00000	0.00000	0.00000
xy,xz(011)	0.00000	0.00000	0.00000
xy,d2(110)	0.00000	0.00000	0.00000
d2,d2(110)	0.00000	0.00000	0.00000
d1,d1(110)	0.00000	0.00000	0.00000
SECOND NEIGHBOR Pb-Pb			
s,s(200)	-0.01377	-0.00360	0.00065
s,x(200)	-0.00522	0.00946	-0.01062
s,d2(002)	0.00000	0.00000	0.00000
x,x(200)	-0.03172	0.02566	-0.02310
y,y(200)	0.02339	-0.00743	-0.00815
x,xy(020)	0.00000	0.00000	0.00000
z,d2(002)	0.00000	0.00000	0.00000
xy,xy(200)	0.00000	0.00000	0.00000
xy,xy(002)	0.00000	0.00000	0.00000
d2,d2(002)	0.00000	0.00000	0.00000
d1,d1(002)	0.00000	0.00000	0.00000
ON SITE H -H			
s,s(000)	0.15885	0.05740	
FIRST NEIGHBOR H -H			
s,s(110)	-0.01843	-0.02987	0.00543
s,x(110)	-0.00243	0.00000	0.00000
x,x(110)	0.03015	0.00000	0.00000
x,x(011)	0.00236	0.00000	0.00000
x,y(110)	0.03049	0.00000	0.00000
SECOND NEIGHBOR H -H			
s,s(200)	-0.01275	0.00288	0.00358
s,x(200)	-0.00877	0.00000	0.00000
x,x(200)	0.01630	0.00000	0.00000
y,y(200)	-0.00204	0.00000	0.00000
FIRST NEIGHBOR Pb-H			
s,s(100)	-0.07534	-0.07397	0.18704
s,x(100)	-0.15484	0.00000	0.00000
x,s(100)	-0.08697	-0.01358	0.35984
x,x(100)	-0.08791	0.00000	0.00000
y,y(100)	0.05671	0.00000	0.00000
xy,x(010)	0.00000	0.00000	0.00000
d2,s(001)	0.00000	0.00000	0.00000
d2,z(001)	0.00000	0.00000	0.00000

SECOND NEIGHBOR Pb-H

s,s(111)	0.00568	-0.00969	-0.00170
x,s(111)	-0.00288	-0.01240	0.00383
s,x(111)	0.00872	0.00000	0.00000
x,x(111)	-0.00055	0.00000	0.00000
x,y(111)	0.00172	0.00000	0.00000
xy,s(111)	0.00000	0.00000	0.00000
xy,x(111)	0.00000	0.00000	0.00000
yz,x(111)	0.00000	0.00000	0.00000
d1,x(111)	0.00000	0.00000	0.00000

PbH

BAND	ORTHOGONAL RMS ERROR mRy	MAXIMUM DEVIATION k	MAXIMUM DEVIATION mRy	NON-ORTHOGONAL RMS ERROR mRy	MAXIMUM DEVIATION k	MAXIMUM DEVIATION mRy
1	2.7	(006)	6.0	0.9	(006)	2.0
2	4.7	(222)	12.0	1.3	(008)	3.4
3	6.4	(111)	20.9	2.3	(174)	4.5
4	6.3	(280)	14.7	2.6	(354)	7.5
5	6.8	(033)	15.4	1.3	(066)	2.8
1-5	5.6			1.8		

ENERGY VALUES IN Ry AT SELECTED k-POINTS

	ORTHOGONAL	APW	NON-ORTHOGONAL
GAMMA 1	-0.40092	-0.39941	-0.40052
GAMMA 1	0.49205	0.49253	0.49262
GAMMA 15	0.80850	0.81218	0.81329
X1 (008)	-0.12921	-0.12902	-0.12866
X1 (008)	0.29088	0.29511	0.29476
X4' (008)	0.20052	0.20450	0.20794
X5' (008)	0.51786	0.52016	0.52020
L1 (444)	-0.19128	-0.18566	-0.18474
L1 (444)	0.91488	0.91555	10.00000
L2' (444)	0.01245	0.02080	0.01877
L2' (444)	0.64254	0.64367	0.64175
L3' (444)	0.72820	0.72982	0.72923
W1 (048)	-0.11322	-0.11078	-0.10951
W2' (048)	0.16837	0.17148	0.17051
W2' (048)	0.79560	0.78439	0.78507
W3 (048)	0.31958	0.31437	0.31123
ODD (224)	0.67044	0.66932	0.67167
EVEN(224)	-0.26086	-0.26177	-0.26160
EVEN(224)	0.18132	0.17522	0.17657
EVEN(224)	0.62079	0.62590	0.62896
EVEN(224)	0.66307	0.65016	0.65617

Table 10.25 PbH (NaCl) a = 10.14 Bohr Slater–Koster 2-center parameters

	ORTHOGONAL	NON-ORTHOGONAL	
	ENERGY INTEGRALS Ry	ENERGY INTEGRALS Ry	OVERLAP INTEGRALS
ON SITE Pb-Pb			
s	-0.14850	-0.20416	
p	0.51656	0.35488	
FIRST NEIGHBOR Pb-Pb			
(sss)	-0.00269	-0.03052	0.04917
(pps)	0.08094	-0.00845	-0.24532
(ppp)	-0.00284	-0.00877	0.04430
(dds)	0.00000	0.00000	0.00000
(ddp)	0.00000	0.00000	0.00000
(ddd)	0.00000	0.00000	0.00000
(sps)	-0.02551	0.04711	-0.09595
(sds)	0.00000	0.00000	0.00000
(pds)	0.00000	0.00000	0.00000
(pdp)	0.00000	0.00000	0.00000
SECOND NEIGHBOR Pb-Pb			
(sss)	0.00935	-0.00294	-0.00190
(pps)	-0.02421	0.02104	-0.00880
(ppp)	0.00898	-0.00737	-0.00788
(dds)	0.00000	0.00000	0.00000
(ddp)	0.00000	0.00000	0.00000
(ddd)	0.00000	0.00000	0.00000
(sps)	-0.02310	0.00865	-0.00471
(sds)	0.00000	0.00000	0.00000
(pds)	0.00000	0.00000	0.00000
(pdp)	0.00000	0.00000	0.00000
ON SITE H-H			
s	0.33331	0.03949	
FIRST NEIGHBOR H-H			
(sss)	-0.00277	-0.03122	0.01488
(sps)	0.00000	0.00000	0.00000
(pps)	0.00000	0.00000	0.00000
(ppp)	0.00000	0.00000	0.00000
SECOND NEIGHBOR H-H			
(sss)	-0.01843	0.00101	0.00026
(sps)	0.00000	0.00000	0.00000
(pps)	0.00000	0.00000	0.00000
(ppp)	0.00000	0.00000	0.00000
FIRST NEIGHBOR Pb-H			
(sss)	-0.06970	-0.07725	0.19333
(sps)	0.00000	0.00000	0.00000
(pss)	-0.07705	-0.01405	0.34160
(pps)	0.00000	0.00000	0.00000
(ppp)	0.00000	0.00000	0.00000
(dpp)	0.00000	0.00000	0.00000
(dss)	0.00000	0.00000	0.00000
(dps)	0.00000	0.00000	0.00000
SECOND NEIGHBOR Pb-H			
(sss)	0.00046	-0.01002	0.00004
(pss)	0.01983	-0.02264	0.00939
(sps)	0.00000	0.00000	0.00000
(pps)	0.00000	0.00000	0.00000
(ppp)	0.00000	0.00000	0.00000
(dss)	0.00000	0.00000	0.00000
(dps)	0.00000	0.00000	0.00000
(dpp)	0.00000	0.00000	0.00000

PbH

BAND	ORTHOGONAL RMS ERROR mRy	MAXIMUM DEVIATION k	mRy	NON-ORTHOGONAL RMS ERROR mRy	MAXIMUM DEVIATION k	mRy
1	10.1	(005)	21.0	0.7	(006)	1.5
2	13.9	(004)	32.6	1.5	(444)	3.3
3	18.1	(005)	42.1	2.6	(264)	5.3
4	16.7	(066)	35.3	2.6	(354)	7.1
5	20.5	(048)	35.4	1.4	(264)	3.4
1-5	16.3			1.9		

ENERGY VALUES IN Ry AT SELECTED k-POINTS

	ORTHOGONAL	APW	NON-ORTHOGONAL
GAMMA 1	-0.41084	-0.39941	-0.40025
GAMMA 1	0.47570	0.49253	0.49288
GAMMA 15	0.80507	0.81218	0.81329
X1 (008)	-0.13687	-0.12902	-0.12854
X1 (008)	0.28906	0.29511	0.29364
X4' (008)	0.18031	0.20450	0.20768
X5' (008)	0.51544	0.52016	0.52032
L1 (444)	-0.20462	-0.18566	-0.18444
L2' (444)	0.02420	0.02080	0.01753
L2' (444)	0.61363	0.64367	0.64111
L3' (444)	0.69662	0.72982	0.72980
W1 (048)	-0.11904	-0.11078	-0.11000
W2' (048)	0.17080	0.17148	0.17057
W2' (048)	0.74901	0.78439	0.78551
W3 (048)	0.31195	0.31437	0.31161
ODD (224)	0.66047	0.66932	0.67169
EVEN(224)	-0.25198	-0.26177	-0.26177
EVEN(224)	0.17334	0.17522	0.17666
EVEN(224)	0.62478	0.62590	0.63074
EVEN(224)	0.69812	0.65016	0.65523

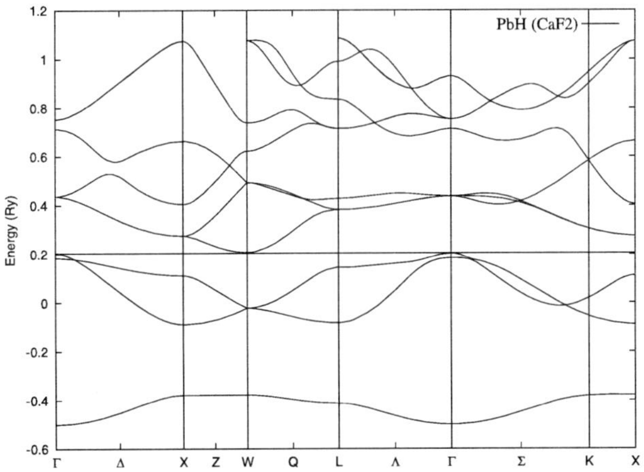

Fig. 10.28 Energy bands of PbH2 in the CaF2 structure

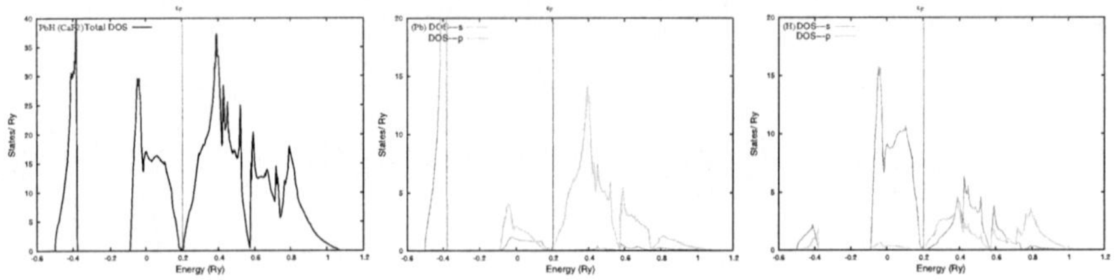

Fig. 10.29 Total, angular momentum and site decomposed densities of states of PbH2 in the CaF2 structure

References

1. Wikipedia
2. F. Fan, D.A. Papaconstantopoulos, M.J. Mehl, and B.M. Klein, High temperature superconductivity at high pressures for H3SixP(1-x), H3PxS(1-x) and H3ClxS(1-x), Journal of Physics and Chemistry of Solids, **99**, 105, (2016)

Chapter 11
Group 15 Pnictogen Hydrides

This chapter covers the Group 15 column of the periodic table hydrides from NH to BiH [1]. Results are presented for the crystal structures NaCl(B1), CsCl (B2) and Im3m. These structures are not found experimentally and are presented here for the purpose of comparison through the periodic Table and to establish various trends. The pnictogen hydrides are trihydride molecules found in a gaseous state. Following the example of SH which exists under pressure in the Im3m cubic structure we present here results for the Im3m structure exploring the possibility that some of them may form and become superconductors [2]. Examining the energy bands and densities of states figures we observe that, as in the group 14 column, the lowest band has a mixture of s–H hydrogen and s-character of the other element. At the Fermi level the states are of s- and p- character. As in the other columns of this part of the Periodic Table, the band structure is dominated by s- and p-states. The d-states are semicore states found significantly lower than the s-p states.

Gam1 point. Tight-binding(TB) parameters are given in the NaCl structure based on both orthogonal and non-orthogonal Hamiltonians using three- and two-center integrals. It could be possible to modify these TB parameters for the purpose of building a TB model to describe the trihydride molecules.

D. A. Papaconstantopoulos, *Band Structure of Cubic Hydrides*,
https://doi.org/10.1007/978-3-031-06878-2_11

11.1 Nitrogen Hydride (NH)

See Figs. 11.1 and Tables 11.1, 11.2, 11.3.

See Figs. 11.2, 11.3, Tables 11.4 and 11.5.

See Figs. 11.4, 11.5 and 11.6.

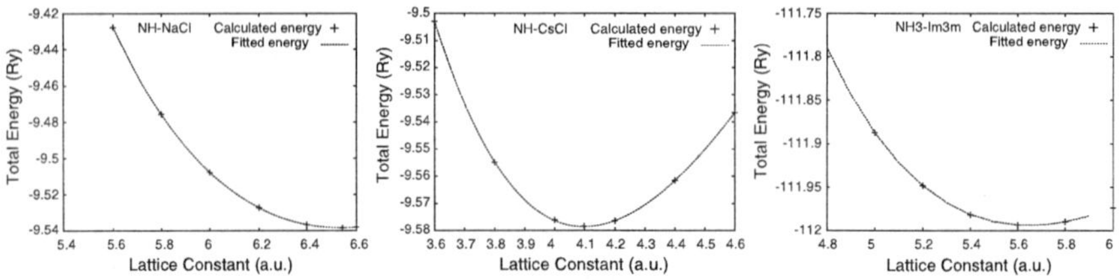

Fig. 11.1 Total energy versus lattice constant of NH in the NaCl, CsCl and Im3m structures

Table 11.1 Lattice constant, bulk modulus, gap, total energy

Stru	a (Bohr)	B (MBar)	Gap	Total Energy (Ry)
NaCl	6.54	1.62	-	-109.53836
CsCl	4.10	1.77	-	-109.57828
Im3m (P=0)	5.64	2.16	-	-111.99362
Im3m(P=1.98 MBar)	4.90	8.27	-	-111.84350

Table 11.2 Birch fit coefficients

	A1	A2	A3	A4
NaCl	-8.550441E+00	-3.557472E+01	3.531413E+02	-5.755994E+02
CsCl	-8.415721E+00	-4.289141E+01	4.569187E+02	-1.084085E+03
Im3m	-1.102955E+02	-7.248939E+01	8.582018E+02	-1.753462E+03

Table 11.3 DOS at Ef, Hopfield parameter, Stoner criterion

```
        NH          a=6.54 Bohr         NaCl
 ---------------------------------------------------------------------
---------
 Fermi Energy   Total DOS      Decomposed DOS inside the muffin   tins
     ( Ry )    (States/Ry)    s          p          eg        t2g      f
 ---------------------------------------------------------------------
--------
 N  0.604       11.761      0.150      8.501      0.019      0.024  0.006

 H  0.604       11.761      0.247      0.478      0.132      0.032  0.036
 ---------------------------------------------------------------------
---------
Fermi-Velocity (cm/s): 0.94964 x10E8
Plasmon Energy (eV)   : 14.00642
Electron-ion interaction (Hopfield parameter) (eV/A^2) N:7.033
H:0.445
------------------------------------------------
   N     MUFFIN-TIN RADIUS and CHARGE =    1.6362    5.6807
   H     MUFFIN-TIN RADIUS and CHARGE =    1.6362    0.9251
                  N       STONER I =  0.0473
                  H       STONER I =  0.0013
          STONER PARAMETER (Ry)     I =  0.0488
          STONER CRITERION        N*I =  0.5740
------------------------------------------------
        NH        a=4.10 Bohr          CsCl
 ---------------------------------------------------------------------
---------
 Fermi Energy   Total DOS      Decomposed DOS inside the muffin   tins
     ( Ry )    (States/Ry)    s          p          eg        t2g      f
 ---------------------------------------------------------------------
 ---------
 N  0.710        7.522      0.318      5.351      0.055      0.036 0.011

 H  0.710        7.522      0.588      0.217      0.018      0.101 0.039
 ---------------------------------------------------------------------
 ---------
 Fermi-Velocity (cm/s): 1.40920 x10E8
 Plasmon Energy (eV)   : 16.77679
 Electron-ion interaction (Hopfield parameter) (eV/A^2) N:9.044
 H:0.430
 ------------------------------------------------
   N     MUFFIN-TIN RADIUS and CHARGE =    1.7743    6.0682
   H     MUFFIN-TIN RADIUS and CHARGE =    1.7743    0.9598
                  N       STONER I =  0.0460
                  H       STONER I =  0.0032
          STONER PARAMETER (Ry)     I =  0.0493
          STONER CRITERION        N*I =  0.3706
 ---------------------------------------------------------------------
 ---------
        NH3  a=4.9 Bohr                 Im3m
 ---------------------------------------------------------------------
```

```
---------
Fermi Energy    Total DOS         Decomposed DOS inside the muffin tins
    ( Ry )     (States/Ry)       s        p         d        f
N  1.278          3.363        0.015    1.834     0.032    0.002
H  1.278          3.363        0.129    0.040     0.003    0.001
-----------------------------------------------------------------------
---------
Fermi-Velocity (cm/s): 1.16584 x10E8
Plasmon Energy (eV)   : 10.03672
Electron-ion interaction (Hopfield parameter) (eV/A^2) N:7.776
H3:1.683
----------------------------------------------
   N      MUFFIN-TIN RADIUS and CHARGE =     1.40     5.5240
   H      MUFFIN-TIN RADIUS and CHARGE =     1.00     0.5100
```

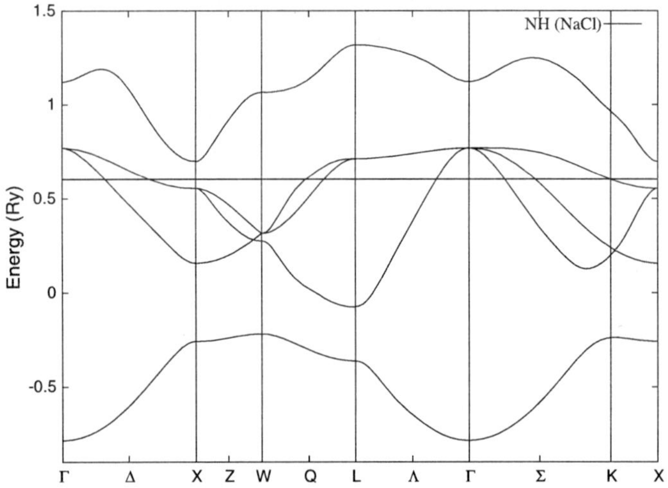

Fig. 11.2 Energy bands of NH in the NaCl structure

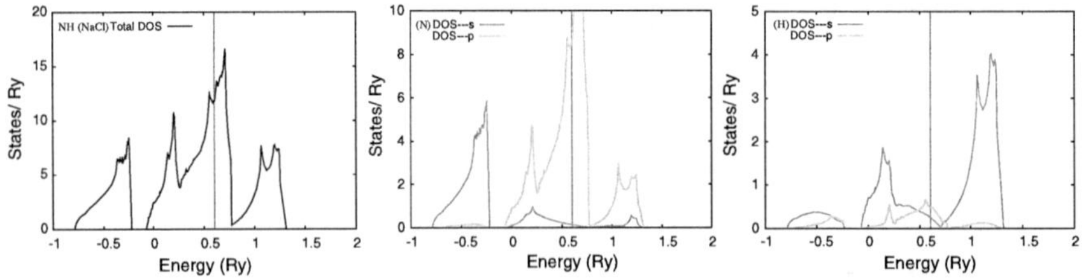

Fig. 11.3 Total, angular momentum and site decomposed densities of states of NH in the NaCl structure

Table 11.4 NH (NaCl) a = 6.54 Bohr Slatrr–Koster 3-center parameters

	ORTHOGONAL	NON-ORTHOGONAL	
	ENERGY INTEGRALS Ry	ENERGY INTEGRALS Ry	OVERLAP INTEGRALS
ON SITE N-N			
s,s(000)	-0.21901	-0.39789	
x,x(000)	0.62250	0.41116	
FIRST NEIGHBOR N-N			
s,s(110)	0.01193	-0.05591	0.03886
s,x(110)	-0.00015	0.04688	-0.04521
s,xy(110)	0.00000	0.00000	0.00000
s,d2(110)	0.00000	0.00000	0.00000
x,x(110)	0.03982	0.01854	-0.04860
x,x(011)	-0.01147	-0.01479	0.02167
x,y(110)	-0.01166	0.03972	-0.06274
x,xy(110)	0.00000	0.00000	0.00000
x,xy(011)	0.00000	0.00000	0.00000
z,d2(011)	0.00000	0.00000	0.00000
z,d1(011)	0.00000	0.00000	0.00000
xy,xy(110)	0.00000	0.00000	0.00000
xy,xy(011)	0.00000	0.00000	0.00000
xy,xz(011)	0.00000	0.00000	0.00000
xy,d2(110)	0.00000	0.00000	0.00000
d2,d2(110)	0.00000	0.00000	0.00000
d1,d1(110)	0.00000	0.00000	0.00000
SECOND NEIGHBOR N-N			
s,s(200)	0.02460	-0.00402	-0.00529
s,x(200)	-0.03142	0.01090	0.00146
s,d2(002)	0.00000	0.00000	0.00000
x,x(200)	-0.06829	0.01979	-0.00038
y,y(200)	0.00359	-0.00297	-0.00292
x,xy(020)	0.00000	0.00000	0.00000
z,d2(002)	0.00000	0.00000	0.00000
xy,xy(200)	0.00000	0.00000	0.00000
xy,xy(002)	0.00000	0.00000	0.00000
d2,d2(002)	0.00000	0.00000	0.00000
d1,d1(002)	0.00000	0.00000	0.00000
ON SITE H-H			
s,s(000)	0.42936	0.05735	
FIRST NEIGHBOR H-H			
s,s(110)	-0.01744	-0.07509	0.03941
s,x(110)	0.00000	0.00000	0.00000
x,x(110)	0.00000	0.00000	0.00000
x,x(011)	0.00000	0.00000	0.00000
x,y(110)	0.00000	0.00000	0.00000
SECOND NEIGHBOR H-H			
s,s(200)	0.00776	-0.00198	-0.00024
s,x(200)	0.00000	0.00000	0.00000
x,x(200)	0.00000	0.00000	0.00000
y,y(200)	0.00000	0.00000	0.00000
FIRST NEIGHBOR N-H			
s,s(100)	-0.16113	0.16335	-0.20940
s,x(100)	0.00000	0.00000	0.00000
x,s(100)	-0.18719	0.06759	-0.28897
x,x(100)	0.00000	0.00000	0.00000
y,y(100)	0.00000	0.00000	0.00000
xy,x(010)	0.00000	0.00000	0.00000
d2,s(001)	0.00000	0.00000	0.00000
d2,z(001)	0.00000	0.00000	0.00000
SECOND NEIGHBOR N-H			
s,s(111)	0.00234	0.01866	-0.00708
x,s(111)	-0.00092	0.01628	-0.00912
s,x(111)	0.00000	0.00000	0.00000
x,x(111)	0.00000	0.00000	0.00000
x,y(111)	0.00000	0.00000	0.00000
xy,s(111)	0.00000	0.00000	0.00000

xy,x(111)	0.00000	0.00000	0.00000
yz,x(111)	0.00000	0.00000	0.00000
d1,x(111)	0.00000	0.00000	0.00000

NH

	ORTHOGONAL			NON-ORTHOGONAL		
BAND	RMS ERROR	MAXIMUM DEVIATION		RMS ERROR	MAXIMUM DEVIATION	
	mRy	k	mRy	mRy	k	mRy
1	3.1	(055)	7.7	0.2	(055)	0.7
2	8.8	(008)	21.8	0.3	(048)	0.7
3	8.0	(226)	15.7	0.7	(008)	2.4
4	6.9	(380)	14.8	0.6	(008)	2.4
5	8.7	(055)	20.2	0.5	(055)	1.0
1-5	7.4			0.5		

ENERGY VALUES IN Ry AT SELECTED k-POINTS

	ORTHOGONAL	APW	NON-ORTHOGONAL
GAMMA 1	-0.78383	-0.78463	-0.78468
GAMMA 1	1.12223	1.12118	1.12136
GAMMA 15	0.77296	0.77048	0.77044
X1 (008)	-0.26297	-0.25942	-0.25926
X1 (008)	0.68949	0.69879	0.69893
X5 (008)	10.00000	0.00000	10.00000
X4' (008)	0.13579	0.15763	0.15814
X5' (008)	0.54617	0.55520	0.55280
L1 (444)	-0.36658	-0.36281	-0.36229
L2' (444)	-0.07683	-0.07389	-0.07420
L2' (444)	1.29762	1.31737	1.31669
L3' (444)	0.69811	0.71267	0.71223
W1 (048)	-0.21755	-0.21860	-0.21859
W2' (048)	0.26965	0.27620	0.27547
W2' (048)	1.06435	1.06513	1.06557
W3 (048)	0.32662	0.31596	0.31546
ODD (224)	0.67166	0.68116	0.68138
EVEN(224)	-0.52137	-0.51992	-0.51967
EVEN(224)	0.20943	0.20922	0.20942
EVEN(224)	0.68546	0.67754	0.67789
EVEN(224)	1.21553	1.21365	1.21334

Table 11.5 NH (NaCl) a = 6.54 Bohr Slater–Koster 2-center parameters

	ORTHOGONAL	NON-ORTHOGONAL	
	ENERGY INTEGRALS Ry	ENERGY INTEGRALS Ry	OVERLAP INTEGRALS
ON SITE N-N			
s	-0.30332	-0.40889	
p	0.49529	0.41521	
FIRST NEIGHBOR N-N			
(sss)	-0.01355	-0.06208	0.05107
(pps)	0.09170	0.05184	-0.11768
(ppp)	-0.01223	-0.01505	0.02105
(dds)	0.00000	0.00000	0.00000
(ddp)	0.00000	0.00000	0.00000
(ddd)	0.00000	0.00000	0.00000
(sps)	-0.06446	0.06632	-0.07206
(sds)	0.00000	0.00000	0.00000
(pds)	0.00000	0.00000	0.00000
(pdp)	0.00000	0.00000	0.00000
SECOND NEIGHBOR N-N			
(sss)	0.01035	-0.00439	-0.00947
(pps)	-0.00462	0.01655	-0.00316
(ppp)	0.00471	-0.00316	-0.00324
(dds)	0.00000	0.00000	0.00000
(ddp)	0.00000	0.00000	0.00000
(ddd)	0.00000	0.00000	0.00000
(sps)	-0.01666	0.01542	0.00427
(sds)	0.00000	0.00000	0.00000
(pds)	0.00000	0.00000	0.00000
(pdp)	0.00000	0.00000	0.00000
ON SITE H-H			
s	0.89533	0.08225	
FIRST NEIGHBOR H-H			
(sss)	0.00899	-0.07248	0.02994
(sps)	0.00000	0.00000	0.00000
(pps)	0.00000	0.00000	0.00000
(ppp)	0.00000	0.00000	0.00000
SECOND NEIGHBOR H-H			
(sss)	-0.04391	0.00087	0.00243
(sps)	0.00000	0.00000	0.00000
(pps)	0.00000	0.00000	0.00000
(ppp)	0.00000	0.00000	0.00000
FIRST NEIGHBOR N-H			
(sss)	-0.12524	-0.16121	0.21128
(sps)	0.00000	0.00000	0.00000
(pss)	-0.11726	-0.07705	0.27926
(pps)	0.00000	0.00000	0.00000
(ppp)	0.00000	0.00000	0.00000
(dpp)	0.00000	0.00000	0.00000
(dss)	0.00000	0.00000	0.00000
(dps)	0.00000	0.00000	0.00000
SECOND NEIGHBOR N-H			
(sss)	-0.00052	-0.02132	0.00526
(pss)	0.03122	-0.03007	0.01151
(sps)	0.00000	0.00000	0.00000
(pps)	0.00000	0.00000	0.00000
(ppp)	0.00000	0.00000	0.00000
(dss)	0.00000	0.00000	0.00000
(dps)	0.00000	0.00000	0.00000
(dpp)	0.00000	0.00000	0.00000

NH

BAND	ORTHOGONAL RMS ERROR mRy	MAXIMUM DEVIATION k	MAXIMUM DEVIATION mRy	NON-ORTHOGONAL RMS ERROR mRy	MAXIMUM DEVIATION k	MAXIMUM DEVIATION mRy
1	5.1	(062)	12.6	0.4	(055)	1.2
2	11.2	(062)	27.6	0.9	(264)	2.2
3	8.5	(444)	19.1	0.8	(008)	2.1
4	8.3	(444)	19.1	0.8	(008)	2.1
5	12.7	(008)	31.8	0.7	(055)	1.8
1-5	9.5			0.7		

ENERGY VALUES IN Ry AT SELECTED k-POINTS

	ORTHOGONAL	APW	NON-ORTHOGONAL
GAMMA 1	-0.77962	-0.78463	-0.78478
GAMMA 1	1.11551	1.12118	1.12117
GAMMA 15	0.77380	0.77048	0.77028
X1 (008)	-0.25802	-0.25942	-0.25927
X1 (008)	0.66696	0.69879	0.69903
X4' (008)	0.13809	0.15763	0.15731
X5' (008)	0.55381	0.55520	0.55314
L1 (444)	-0.36544	-0.36281	-0.36195
L2' (444)	-0.06486	-0.07389	-0.07589
L2' (444)	1.29362	1.31737	1.31565
L3' (444)	0.69356	0.71267	0.71248
W1 (048)	-0.22839	-0.21860	-0.21802
W2' (048)	0.28034	0.27620	0.27425
W2' (048)	1.06350	1.06513	1.06487
W3 (048)	0.32712	0.31596	0.31574
ODD (224)	0.66927	0.68116	0.68109
EVEN(224)	-0.51715	-0.51992	-0.51965
EVEN(224)	0.20541	0.20922	0.20972
EVEN(224)	0.68395	0.67754	0.67819
EVEN(224)	1.21925	1.21365	1.21415

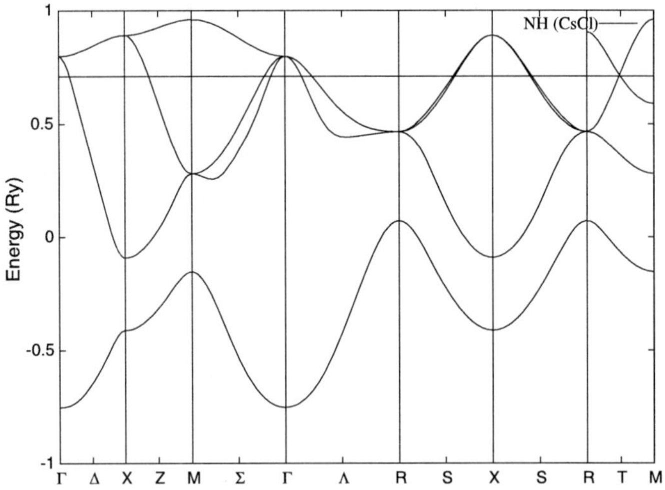

Fig. 11.4 Energy bands of NH in the CsCl structure

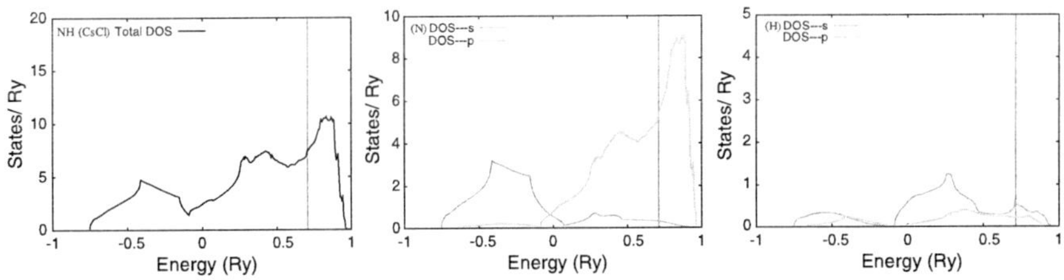

Fig. 11.5 Total, angular momentum and site decomposed densities of states of NH in the CsCl structure

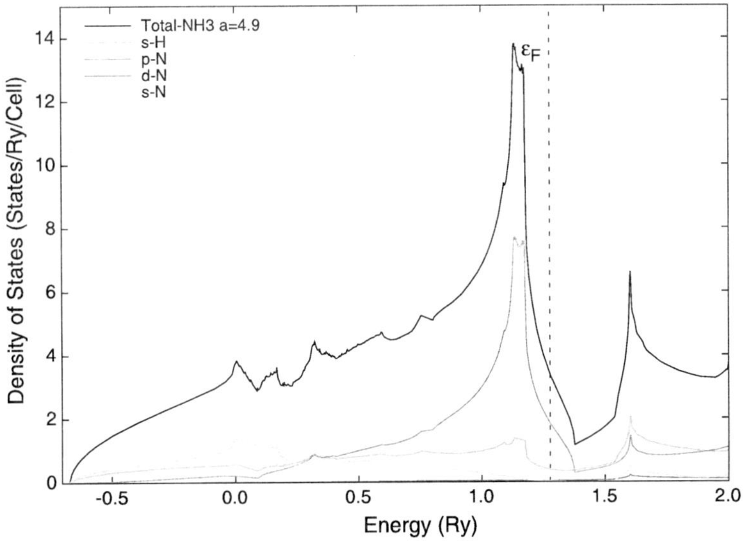

Fig. 11.6 Total, angular momentum and site decomposed densities of states of NH3 in the Im3m structure

11.2 Phosphorus Hydride (PH)

See Fig. 11.7, Tables 11.6, 11.7, and 11.8.

See Figs. 11.8, 11.9, Tables 11.9 and 11.10.

See Fig. 11.10.

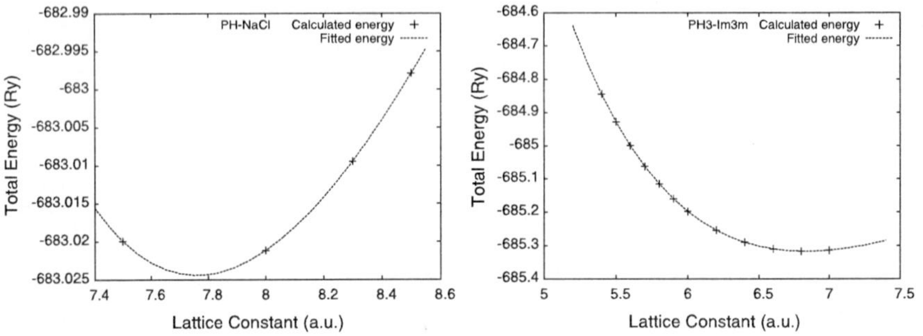

Fig. 11.7 Total energy versus lattice constant of PH in the NaCl and Im3m structures

Table 11.6 Lattice constant, bulk modulus, gap, total energy

Stru	a (Bohr)	B (MBar)	Gap	Total Energy (Ry)
NaCl	7.76	1.02	-	-683.02441
Im3m (P=0)	6.83	1.19	-	-685.31744
Im3m(P=2.09 MBar)	5.60	7.25	-	-684.99989

Table 11.7 Birch fit coefficients

	A1	A2	A3	A4
NaCl	-6.818474E+02	-6.256722E+01	9.737042E+02	-3.601103E+03
Im3m	-6.836725E+02	-1.025290E+02	1.766822E+03	-5.123713E+03

Table 11.8 DOS at Ef, Hopfield parameter, Stoner criterion

PH a=7.76 Bohr NaCl

Fermi Energy (Ry)	Total DOS (States/Ry)	Decomposed DOS inside the muffin tins s	p	eg	t2g	f
P 0.808	8.550	0.305	3.347	0.093	0.135	0.021
H 0.808	8.550	0.505	0.862	0.116	0.072	0.038

Fermi-Velocity (cm/s): 1.78326 x10E8
Plasmon Energy (eV) : 17.36874
Electron-ion interaction (Hopfield parameter) (eV/A^2) P:2.271 H:0.828

P	MUFFIN-TIN RADIUS and CHARGE =	1.9400	12.4185
H	MUFFIN-TIN RADIUS and CHARGE =	1.9400	1.4922

P STONER I = 0.0159
H STONER I = 0.0043
STONER PARAMETER (Ry) I = 0.0206
STONER CRITERION N*I = 0.1764

PH3 a=5.6 Bohr Im3m

Fermi Energy (Ry)	Total DOS (States/Ry)	Decomposed DOS inside the muffin tins s	p	d	f
P 1.305	6.817	0.135	2.384	0.282	0.030
H 1.305	6.817	0.322	0.073	0.008	0.000

Fermi-Velocity (cm/s): 1.44805 x10E8
Plasmon Energy (eV) : 14.51949
Electron-ion interaction (Hopfield parameter) (eV/A^2) P:1.898 H3:3.868

P	MUFFIN-TIN RADIUS and CHARGE =	1.80	12.5932
H	MUFFIN-TIN RADIUS and CHARGE =	1.00	0.5835

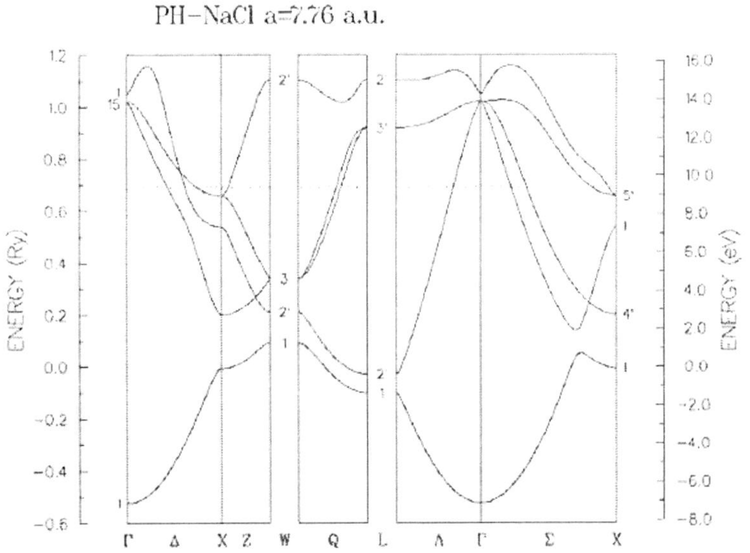

Fig. 11.8 Energy bands of PH in the NaCl structure (tight-binding)

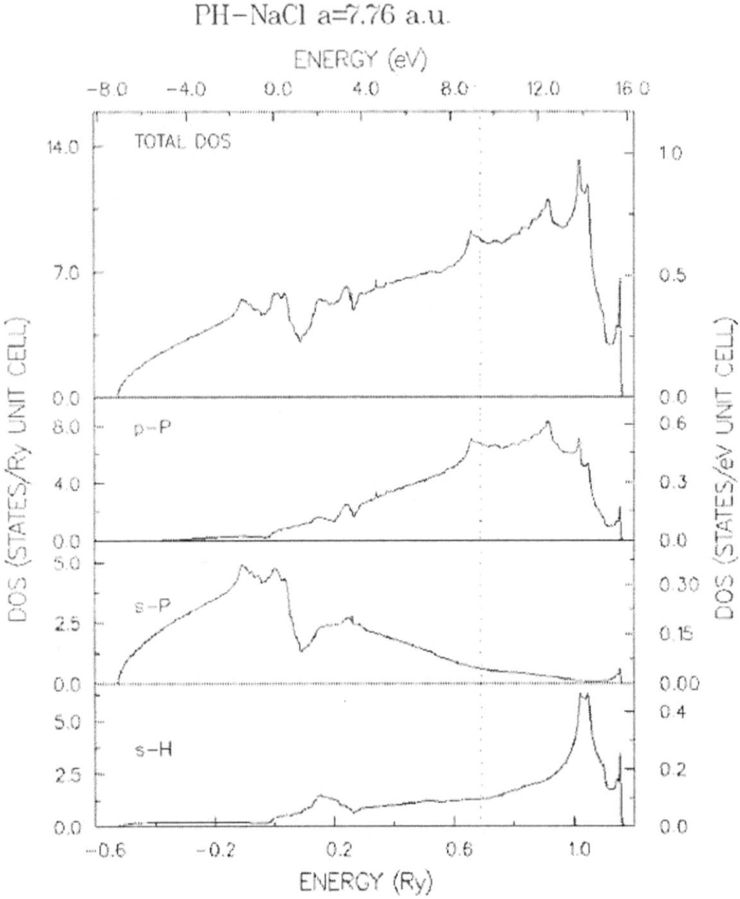

Fig. 11.9 Total, angular momentum and site decomposed densities of states of PH in the NaCl structure (tight-binding)

Table 11.9 PH (NaCl) a = 7.76 Bohr Slater–Koster 3-center parameters

	ORTHOGONAL	NON-ORTHOGONAL	
	ENERGY INTEGRALS Ry	ENERGY INTEGRALS Ry	OVERLAP INTEGRALS
ON SITE P-P			
s,s(000)	0.10083	-0.17526	
x,x(000)	1.05172	0.34125	
FIRST NEIGHBOR P-P			
s,s(110)	0.00197	-0.06478	0.08464
s,x(110)	0.00393	0.04905	-0.10647
s,xy(110)	0.00000	0.00000	0.00000
s,d2(110)	0.00000	0.00000	0.00000
x,x(110)	0.05336	-0.00943	-0.11111
x,x(011)	0.13877	-0.01509	0.05576
x,y(110)	0.00280	0.02803	-0.16134
x,xy(110)	0.00000	0.00000	0.00000
x,xy(011)	0.00000	0.00000	0.00000
z,d2(011)	0.00000	0.00000	0.00000
z,d1(011)	0.00000	0.00000	0.00000
xy,xy(110)	0.00000	0.00000	0.00000
xy,xy(011)	0.00000	0.00000	0.00000
xy,xz(011)	0.00000	0.00000	0.00000
xy,d2(110)	0.00000	0.00000	0.00000
d2,d2(110)	0.00000	0.00000	0.00000
d1,d1(110)	0.00000	0.00000	0.00000
SECOND NEIGHBOR P-P			
s,s(200)	0.02325	-0.01197	-0.00685
s,x(200)	0.00156	0.02625	-0.00157
s,d2(002)	0.00000	0.00000	0.00000
x,x(200)	-0.10065	0.04700	-0.02426
y,y(200)	0.09119	-0.00947	-0.00750
x,xy(020)	0.00000	0.00000	0.00000
z,d2(002)	0.00000	0.00000	0.00000
xy,xy(200)	0.00000	0.00000	0.00000
xy,xy(002)	0.00000	0.00000	0.00000
d2,d2(002)	0.00000	0.00000	0.00000
d1,d1(002)	0.00000	0.00000	0.00000
ON SITE H-H			
s,s(000)	0.23471	0.03090	
FIRST NEIGHBOR H-H			
s,s(110)	-0.00149	-0.04448	0.03224
s,x(110)	0.10471	0.00000	0.00000
x,x(110)	0.29555	0.00000	0.00000
x,x(011)	0.27533	0.00000	0.00000
x,y(110)	0.62181	0.00000	0.00000
SECOND NEIGHBOR H-H			
s,s(200)	0.00716	0.00583	0.00625
s,x(200)	-0.22014	0.00000	0.00000
x,x(200)	0.28304	0.00000	0.00000
y,y(200)	-0.18311	0.00000	0.00000
FIRST NEIGHBOR P-H			
s,s(100)	-0.13045	-0.11347	0.24376
s,x(100)	-0.17826	0.00000	0.00000
x,s(100)	0.12757	-0.02892	0.40548
x,x(100)	-0.03234	0.00000	0.00000
y,y(100)	-0.86146	0.00000	0.00000
xy,x(010)	0.00000	0.00000	0.00000
d2,s(001)	0.00000	0.00000	0.00000
d2,z(001)	0.00000	0.00000	0.00000
SECOND NEIGHBOR P-H			
s,s(111)	0.00017	-0.02039	0.01035
x,s(111)	0.00892	-0.01829	0.01464
s,x(111)	0.03552	0.00000	0.00000
x,x(111)	-0.03669	0.00000	0.00000

x,y(111)	-0.03613	0.00000	0.00000
xy,s(111)	0.00000	0.00000	0.00000
xy,x(111)	0.00000	0.00000	0.00000
yz,x(111)	0.00000	0.00000	0.00000
d1,x(111)	0.00000	0.00000	0.00000

PH

	ORTHOGONAL			NON-ORTHOGONAL		
BAND	RMS ERROR	MAXIMUM DEVIATION		RMS ERROR	MAXIMUM DEVIATION	
	mRy	k	mRy	mRy	k	mRy
1	6.3	(264)	16.7	0.7	(055)	1.4
2	11.2	(004)	29.9	2.3	(004)	5.9
3	14.2	(022)	35.5	2.8	(354)	6.9
4	14.6	(174)	32.6	2.8	(180)	5.5
5	15.1	(180)	30.5	3.9	(118)	10.1
1-5	12.7			2.7		

ENERGY VALUES IN Ry AT SELECTED k-POINTS

	ORTHOGONAL	APW	NON-ORTHOGONAL
GAMMA 1	-0.51950	-0.52127	-0.52149
GAMMA 1	1.04326	1.05042	1.05163
GAMMA 15	1.01879	1.02357	1.02344
X1 (008)	-0.00546	-0.00651	-0.00531
X1 (008)	0.52154	0.53911	0.54014
X4' (008)	0.20392	0.19879	0.20282
X5' (008)	0.65361	0.65747	0.65535
L1 (444)	-0.11060	-0.10011	-0.09932
L2' (444)	-0.02696	-0.02946	-0.03027
L2' (444)	1.08457	1.10647	1.10505
L3' (444)	0.89947	0.92143	0.91715
W1 (048)	0.08484	0.09262	0.09251
W2' (048)	0.21515	0.21184	0.21156
W2' (048)	1.10255	1.09885	1.10223
W3 (048)	0.33315	0.34330	0.33880
ODD (224)	0.84101	0.83882	0.84537
EVEN(224)	-0.29198	-0.29314	-0.29266
EVEN(224)	0.29628	0.29871	0.30108
EVEN(224)	0.85933	0.84047	0.84461
EVEN(224)	0.99388	0.97030	0.97534

Table 11.10 PH (NaCl) a = 7.76 Bohr Slater–Koster 2-center parameters

	ORTHOGONAL	NON-ORTHOGONAL	
	ENERGY INTEGRALS Ry	ENERGY INTEGRALS Ry	OVERLAP INTEGRALS
ON SITE P-P			
s	0.04667	-0.18244	
p	0.94650	0.38734	
FIRST NEIGHBOR P-P			
(sss)	-0.02606	-0.06671	0.08972
(pps)	0.09833	0.00372	-0.26244
(ppp)	-0.00161	-0.01808	0.04903
(dds)	0.00000	0.00000	0.00000
(ddp)	0.00000	0.00000	0.00000
(ddd)	0.00000	0.00000	0.00000
(sps)	-0.05286	0.06409	-0.14745
(sds)	0.00000	0.00000	0.00000
(pds)	0.00000	0.00000	0.00000
(pdp)	0.00000	0.00000	0.00000

SECOND NEIGHBOR P-P			
(sss)	-0.02417	-0.01314	-0.00763
(pps)	0.12035	0.02785	-0.02162
(ppp)	-0.00095	-0.00809	-0.00616
(dds)	0.00000	0.00000	0.00000
(ddp)	0.00000	0.00000	0.00000
(ddd)	0.00000	0.00000	0.00000
(sps)	-0.04288	0.02315	-0.00384
(sds)	0.00000	0.00000	0.00000
(pds)	0.00000	0.00000	0.00000
(pdp)	0.00000	0.00000	0.00000
ON SITE H-H			
s	0.69362	-0.02456	
FIRST NEIGHBOR H-H			
(sss)	0.02431	-0.05117	0.03788
(sps)	0.13913	0.00000	0.00000
(pps)	-0.24234	0.00000	0.00000
(ppp)	-0.61284	0.00000	0.00000
SECOND NEIGHBOR H-H			
(sss)	-0.00906	0.00181	0.00310
(sps)	-0.09083	0.00000	0.00000
(pps)	0.16939	0.00000	0.00000
(ppp)	0.37264	0.00000	0.00000
FIRST NEIGHBOR P-H			
(sss)	0.04024	-0.11501	0.26720
(sps)	-0.41400	0.00000	0.00000
(pss)	-0.15042	-0.02493	0.39456
(pps)	-1.22822	0.00000	0.00000
(ppp)	0.13533	0.00000	0.00000
(dpp)	0.00000	0.00000	0.00000
(dss)	0.00000	0.00000	0.00000
(dps)	0.00000	0.00000	0.00000
SECOND NEIGHBOR P-H			
(sss)	0.01952	-0.02267	0.00962
(pss)	0.02669	-0.03619	0.02279
(sps)	-0.05930	0.00000	0.00000
(pps)	-0.03031	0.00000	0.00000
(ppp)	0.06734	0.00000	0.00000
(dss)	0.00000	0.00000	0.00000
(dps)	0.00000	0.00000	0.00000
(dpp)	0.00000	0.00000	0.00000

PH

BAND	ORTHOGONAL RMS ERROR mRy	MAXIMUM DEVIATION k	mRy	NON-ORTHOGONAL RMS ERROR mRy	MAXIMUM DEVIATION k	mRy
1	7.6	(005)	20.4	1.4	(226)	3.0
2	8.2	(062)	20.4	3.5	(444)	9.6
3	11.0	(005)	34.7	3.2	(042)	6.8
4	14.5	(280)	34.7	4.1	(062)	9.4
5	16.3	(066)	36.5	5.0	(224)	16.8
1-5	12.0			3.6		

ENERGY VALUES IN Ry AT SELECTED k-POINTS

	ORTHOGONAL	APW	NON-ORTHOGONAL
GAMMA 1	-0.52006	-0.52127	-0.52300
GAMMA 1	1.03990	1.05042	1.05016
GAMMA 15	1.02108	1.02357	1.02282
X1 (008)	-0.00458	-0.00651	-0.00792
X1 (008)	0.55254	0.53911	0.53905
X4' (008)	0.20295	0.19879	0.19971
X5' (008)	0.64150	0.65747	0.65625
L1 (444)	-0.11080	-0.10011	-0.09906
L2' (444)	-0.03635	-0.02946	-0.03910

L2' (444)	1.09416	1.10647	1.09660
L3' (444)	0.90949	0.92143	0.91614
W1 (048)	0.07988	0.09262	0.09283
W2' (048)	0.21425	0.21184	0.21075
W2' (048)	1.07246	1.09885	1.09986
W3 (048)	0.35766	0.34330	0.34098
ODD (224)	0.84369	0.83882	0.84311
EVEN(224)	-0.28889	-0.29314	-0.29266
EVEN(224)	0.29746	0.29871	0.30283
EVEN(224)	0.84845	0.84047	0.84219
EVEN(224)	0.99316	0.97030	0.98706

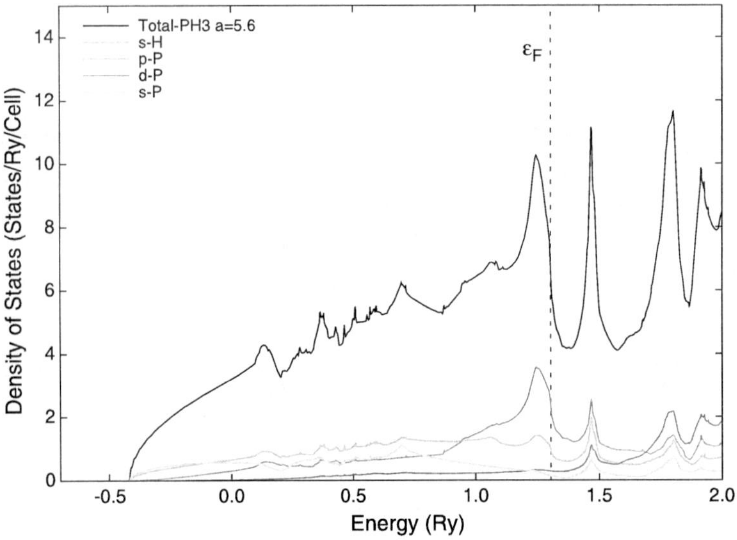

Fig. 11.10 Total, angular momentum and site decomposed densities of states of PH3 in the Im3m structure

11.3 Arsenic Hydride (AsH)

See Fig. 11.11, Tables 11.11, 11.12, and 11.13.

See Figs. 11.12, 11.13, Tables 11.14 and 11.15.

See Fig. 11.14.

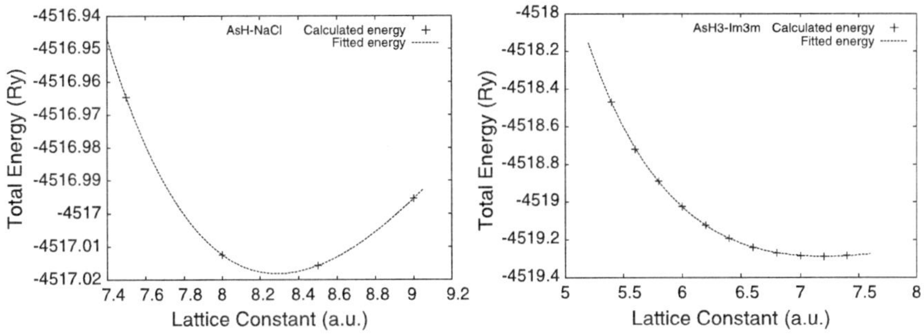

Fig. 11.11 Total energy versus lattice constant of AsH in the NaCl and Im3m structures

Table 11.11 Lattice constant, bulk modulus, gap, total energy

Stru	a (Bohr)	B (MBar)	Gap	Total Energy (Ry)
NaCl	8.29	0.94	-	-4517.01815
Im3m (P=0)	7.20	0.96	-	-4519.28865
Im3m(P=2.26 MBar)	5.80	8.03	-	-4518.89272

Table 11.12 Birch fit coefficients

	A1	A2	A3	A4
NaCl	-4.516270E+03	-3.319324E+01	1.406630E+02	5.669353E+03
Im3m	-4.517912E+03	-9.035757E+01	1.497171E+03	-4.852866E+02

Table 11.13 DOS at Ef, Hopfield parameter, Stoner criterion

```
        AsH          a=8.29 Bohr           NaCl
-----------------------------------------------------------------------------
 Fermi Energy    Total DOS         Decomposed DOS inside the muffin tins
     ( Ry )     (States/Ry)        s         p         eg         t2g      f
-----------------------------------------------------------------------------
 As  0.680        9.365          0.235     4.078     0.076     0.116    0.036
 H   0.680        9.365          0.593     0.810     0.093     0.055    0.028
-----------------------------------------------------------------------------
Fermi-Velocity (cm/s): 1.63353 x10E8
Plasmon Energy (eV)   : 15.05331
Electron-ion interaction (Hopfield parameter) (eV/A^2) As:2.262  H:0.847
-------------------------------------------------
  As      MUFFIN-TIN RADIUS and CHARGE =     2.1787    30.7060
   H      MUFFIN-TIN RADIUS and CHARGE =     1.9712     1.3564
                   As         STONER I =  0.0114
                    H         STONER I =  0.0037
           STONER PARAMETER (Ry)     I =  0.0154
           STONER CRITERION        N*I =  0.1439
-------------------------------------------------
        AsH3  a=5.8 Bohr                   Im3m
-----------------------------------------------------------------------------
 Fermi Energy    Total DOS         Decomposed DOS inside the muffin tins
     ( Ry )     (States/Ry)        s       p        d       f
 As  1.373        4.956          0.096   1.135    0.114   0.015
 H   1.373        4.956          0.239   0.053    0.004   0.000
-----------------------------------------------------------------------------
Fermi-Velocity (cm/s): 1.71974 x10E8
Plasmon Energy (eV)   : 13.95548
Electron-ion interaction (Hopfield parameter) (eV/A^2) As:1.062  H3:2.675
-------------------------------------------------
  As      MUFFIN-TIN RADIUS and CHARGE =     1.70    29.6740
   H      MUFFIN-TIN RADIUS and CHARGE =     1.00     0.5747
```

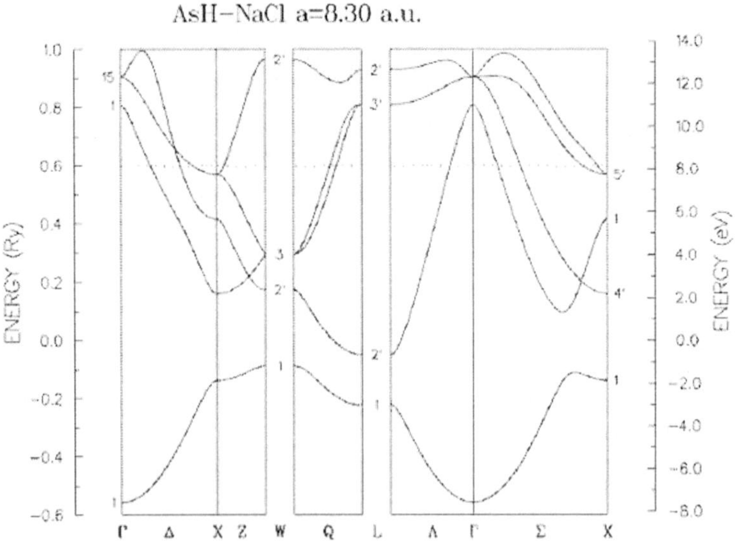

Fig. 11.12 Energy bands of AsH in the NaCl structure

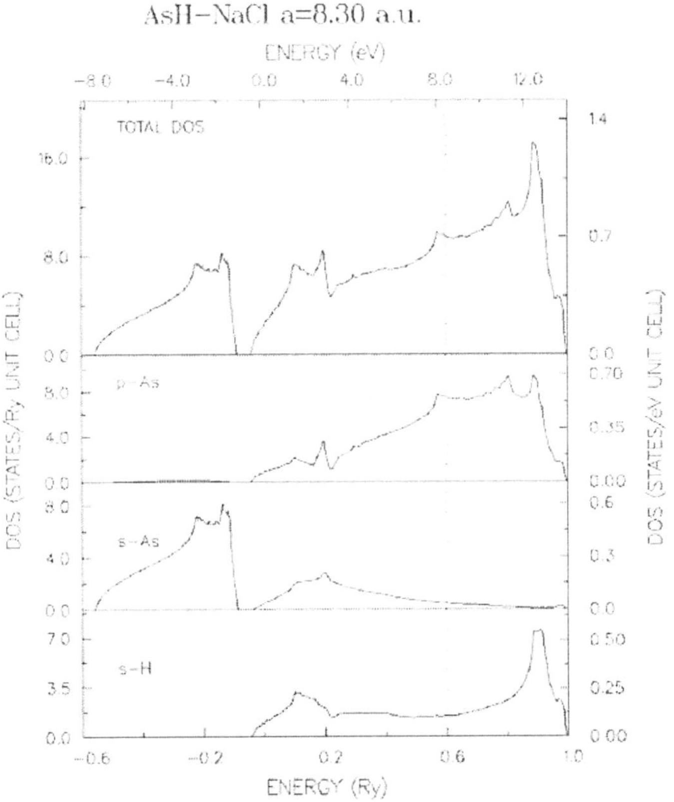

Fig. 11.13 Total, angular momentum and site decomposed densities of states of AsH in the NaCl structure

Table 11.14 AsH (NaCl) a = 8.30 Bohr Slater–Koster 3-center parameters

	ORTHOGONAL	NON-ORTHOGONAL	
	ENERGY INTEGRALS Ry	ENERGY INTEGRALS Ry	OVERLAP INTEGRALS
ON SITE As-As			
s,s(000)	0.20544	-0.27270	
x,x(000)	0.70067	0.31795	
FIRST NEIGHBOR As-As			
s,s(110)	0.02980	-0.05645	0.06970
s,x(110)	0.04387	0.04920	-0.09046
x,x(110)	0.03025	-0.00078	-0.09723
x,x(011)	0.00317	-0.01499	0.04526
x,y(110)	0.01950	0.02733	-0.15030
SECOND NEIGHBOR As-As			
s,s(200)	0.00707	-0.00794	-0.00142
s,x(200)	0.02376	0.01869	-0.01044
x,x(200)	-0.06768	0.03522	-0.02647
y,y(200)	0.02714	-0.00702	-0.00630
ON SITE H-H			
s,s(000)	0.15831	0.00198	
FIRST NEIGHBOR H-H			
s,s(110)	-0.03048	-0.04755	0.01510
s,x(110)	-0.00203	0.00000	0.00000
x,x(110)	0.05403	0.00000	0.00000
x,x(011)	0.00364	0.00000	0.00000
x,y(110)	0.04738	0.00000	0.00000
SECOND NEIGHBOR H-H			
s,s(200)	-0.02399	0.00221	0.00355
s,x(200)	-0.03898	0.00000	0.00000
x,x(200)	0.02699	0.00000	0.00000
y,y(200)	-0.00822	0.00000	0.00000
FIRST NEIGHBOR As-H			
s,s(100)	-0.09452	-0.11396	0.22900
s,x(100)	-0.15602	0.00000	0.00000
x,s(100)	-0.13628	-0.04314	0.38155
x,x(100)	-0.07195	0.00000	0.00000
y,y(100)	0.05383	0.00000	0.00000
SECOND NEIGHBOR As-H			
s,s(111)	0.01098	-0.01691	0.00227
x,s(111)	-0.00844	-0.01808	0.00744
s,x(111)	0.01307	0.00000	0.00000
x,x(111)	0.00483	0.00000	0.00000
x,y(111)	0.01011	0.00000	0.00000

	ORTHOGONAL			NON-ORTHOGONAL		
BAND	RMS ERROR	MAXIMUM DEVIATION		RMS ERROR	MAXIMUM DEVIATION	
	mRy	k	mRy	mRy	k	mRy
1	5.8	(222)	14.3	0.6	(006)	1.5
2	7.3	(033)	18.5	1.4	(007)	3.3
3	7.3	(022)	18.2	2.2	(354)	5.2
4	8.1	(280)	21.5	2.0	(062)	4.2
5	19.6	(048)	72.0	2.3	(264)	4.7
1-5	10.8			1.8		

ENERGY VALUES IN Ry AT SELECTED k-POINTS

	ORTHOGONAL	APW	NON-ORTHOGONAL
GAMMA 1	-0.55023	-0.55629	-0.55578
GAMMA 1	0.80422	0.80856	0.80810
GAMMA 15	0.90302	0.90487	0.90497
X1 (008)	-0.14443	-0.13700	-0.13700
X1 (008)	0.40938	0.41598	0.41598
X4' (008)	0.15621	0.16002	0.16300
X5' (008)	0.56596	0.56810	0.56728

L1 (444)	-0.23144	-0.22381	-0.22317
L2' (444)	-0.05403	-0.04959	-0.05108
L2' (444)	0.92774	0.93001	0.92770
L3' (444)	0.80545	0.80987	0.80702
W1 (048)	-0.09238	-0.08868	-0.08738
W2' (048)	0.17659	0.17473	0.17443
W2' (048)	0.98759	0.96128	0.96329
W3 (048)	0.30209	0.29623	0.29304

FERMI LEVEL QUANTITIES (Non-orthogonal fit)

ENERGY	DENSITIES OF STATES						
	Total	s-As	p-As	t2g-As	eg-As	s-H	p-H
Ry		states/Ry/cell					
0.5995	9.53	0.45	7.45	0.00	0.00	1.44	0.18
	INTEGRATED DENSITIES OF STATES						
	Total	s-As	p-As	t2g-As	eg-As	s-H	p-H
		electrons					
	6.00	2.74	2.13	0.00	0.00	1.09	0.04

VELOCITY	PLASMON ENERGY	EIGENVALUE SUM
cm/s	eV	Ry
1.51x10E8	14.00	-2.8307

Table 11.15 AsH (NaCl) a = 8.30 Bohr Slater–Koster 2-center parameters

	ORTHOGONAL	NON-ORTHOGONAL	
	ENERGY INTEGRALS Ry	ENERGY INTEGRALS Ry	OVERLAP INTEGRALS
ON SITE As-As			
s	-0.15451	-0.25902	
p	0.61419	0.42363	
FIRST NEIGHBOR As-As			
(sss)	-0.00782	-0.05047	0.05800
(pps)	0.09965	0.01707	-0.19489
(ppp)	-0.01274	-0.01460	0.03252
(sps)	-0.06119	0.05081	-0.04380
SECOND NEIGHBOR As-As			
(sss)	0.01174	-0.00365	-0.01508
(pps)	0.00891	-0.03445	-0.07402
(ppp)	0.00567	-0.00216	-0.00132
(sps)	-0.01356	0.00544	-0.03868
ON SITE H-H			
s	0.55387	-0.03027	
FIRST NEIGHBOR H-H			
(sss)	0.00594	-0.06719	0.02630
(sps)	0.04527	0.00000	0.00000
(pps)	-0.04720	0.00000	0.00000
(ppp)	-0.17470	0.00000	0.00000
SECOND NEIGHBOR H-H			
(sss)	-0.03315	-0.00169	0.04712
(sps)	-0.06476	0.00000	0.00000
(pps)	0.05723	0.00000	0.00000
(ppp)	0.11706	0.00000	0.00000
FIRST NEIGHBOR As-H			
(sss)	-0.09913	-0.12311	0.25647
(sps)	0.01376	0.00000	0.00000
(pss)	-0.10971	-0.05487	0.26097
(pps)	-0.24055	0.00000	0.00000
(ppp)	0.05485	0.00000	0.00000
SECOND NEIGHBOR As-H			
(sss)	-0.00105	-0.01640	0.00737
(pss)	0.03429	-0.00976	0.00371
(sps)	0.00977	0.00000	0.00000
(pps)	0.02947	0.00000	0.00000
(ppp)	0.01645	0.00000	0.00000

BAND	ORTHOGONAL RMS ERROR mRy	MAXIMUM DEVIATION k	mRy	NON-ORTHOGONAL RMS ERROR mRy	MAXIMUM DEVIATION k	mRy
1	6.7	(044)	15.1	5.6	(006)	10.4
2	9.9	(062)	25.5	12.2	(005)	33.6
3	7.4	(442)	13.2	8.4	(226)	26.2
4	12.5	(055)	25.8	12.8	(055)	34.2
5	8.4	(066)	22.4	5.9	(022)	12.9
1-5	9.2			9.5		

ENERGY VALUES IN Ry AT SELECTED k-POINTS

	ORTHOGONAL	APW	NON-ORTHOGONAL
GAMMA 1	-0.55039	-0.55629	-0.56056
GAMMA 1	0.79879	0.80856	0.80658
GAMMA 15	0.90300	0.90487	0.90942
X1 (008)	-0.13079	-0.13700	-0.12903
X1 (008)	0.40921	0.41598	0.40961
X4' (008)	0.15102	0.16002	0.17084
X5' (008)	0.55702	0.56810	0.56445
L1 (444)	-0.22683	-0.22381	-0.21745
L2' (444)	-0.05472	-0.04959	-0.07068
L2' (444)	0.93165	0.93001	0.93220
L3' (444)	0.79848	0.80987	0.80815
W1 (048)	-0.09985	-0.08868	-0.08736
W2' (048)	0.17530	0.17473	0.17367
W2' (048)	0.95854	0.96128	0.95071
W3 (048)	0.30930	0.29623	0.29726

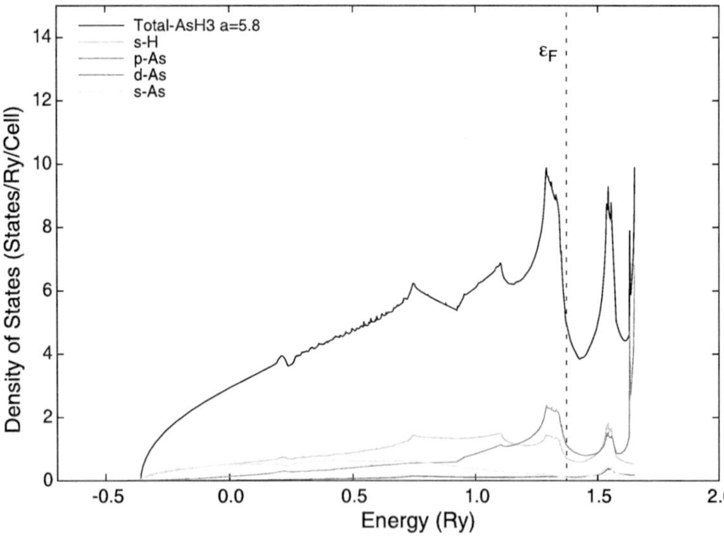

Fig. 11.14 Total, angular momentum and site decomposed densities of states of AsH3 in the Im3m structure

11.4 Antimony Hydride (SbH)

See Fig. 11.15, Tables 11.16, 11.17, and 11.18.

See Figs. 11.16, 11.17, Tables 11.19 and 11.20.

See Fig. 11.18.

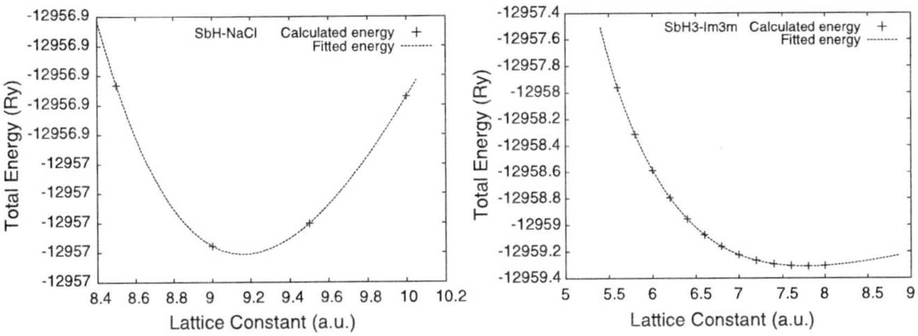

Fig. 11.15 Total energy versus lattice constant of SbH in the NaCl and Im3m structures

Table 11.16 Lattice constant, bulk modulus, gap, total energy

Stru	a (Bohr)	B (MBar)	Gap	Total Energy (Ry)
NaCl	9.16	0.73	-	-12956.96528
Im3m (P=0)	7.78	0.87	-	-12959.30920
Im3m(P=2.13 MBar)	6.25	7.55	-	-12958.83790

Table 11.17 Birch fit coefficients

	A1	A2	A3	A4
NaCl	-1.295611E+04	-5.018277E+01	4.889577E+02	7.694341E+03
Im3m	-1.295774E+04	-1.196511E+02	2.294070E+03	-3.039292E+02

Table 11.18 DOS at Ef, Hopfield parameter, Stoner criterion

SbH a=9.16 Bohr NaCl

	Fermi Energy (Ry)	Total DOS (States/Ry)	Decomposed DOS inside the muffin tins s	p	eg	t2g	f
Sb	0.633	10.383	0.269	3.773	0.078	0.130	0.071
H	0.633	10.383	0.799	1.021	0.103	0.074	0.031

Fermi-Velocity (cm/s): 1.66510 x10E8
Plasmon Energy (eV) : 13.91316
Electron-ion interaction (Hopfield parameter) (eV/A^2) Sb:1.244 H:0.819

Sb	MUFFIN-TIN RADIUS and CHARGE =	2.4071	48.3652
H	MUFFIN-TIN RADIUS and CHARGE =	2.1779	1.5078
Sb	STONER I =	0.0065	
H	STONER I =	0.0043	
STONER PARAMETER (Ry)	I =	0.0111	
STONER CRITERION	N*I =	0.1147	

SbH3 a=6.20 Bohr Im3m

	Fermi Energy (Ry)	Total DOS (States/Ry)	Decomposed DOS inside the muffin tins s	p	d	f
Sb	1.479	5.045	0.094	0.602	0.128	0.027
H	1.479	5.045	0.205	0.048	0.003	0.000

Fermi-Velocity (cm/s): 1.75650 x10E8
Plasmon Energy (eV) : 13.01175
Electron-ion interaction (Hopfield parameter) (eV/A^2) Sb:0.514 H3:1.871

Sb	MUFFIN-TIN RADIUS and CHARGE =	1.70	46.1845
H	MUFFIN-TIN RADIUS and CHARGE =	1.00	0.5889

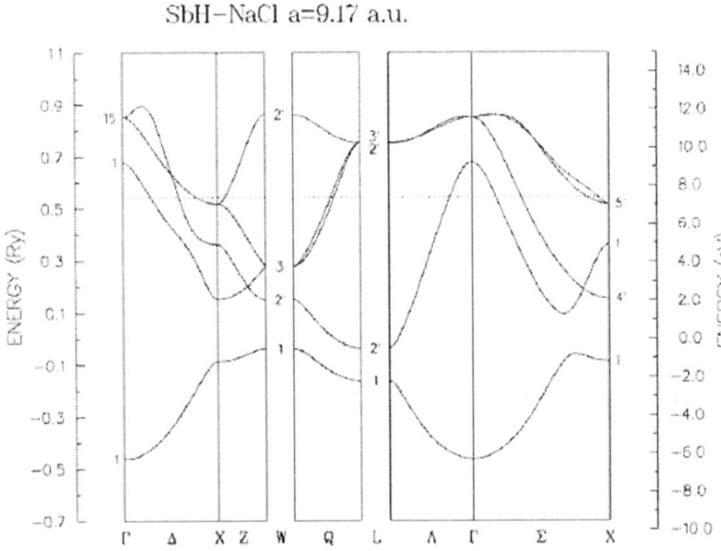

Fig. 11.16 Energy bands of SbH in the NaCl structure (tight-binding)

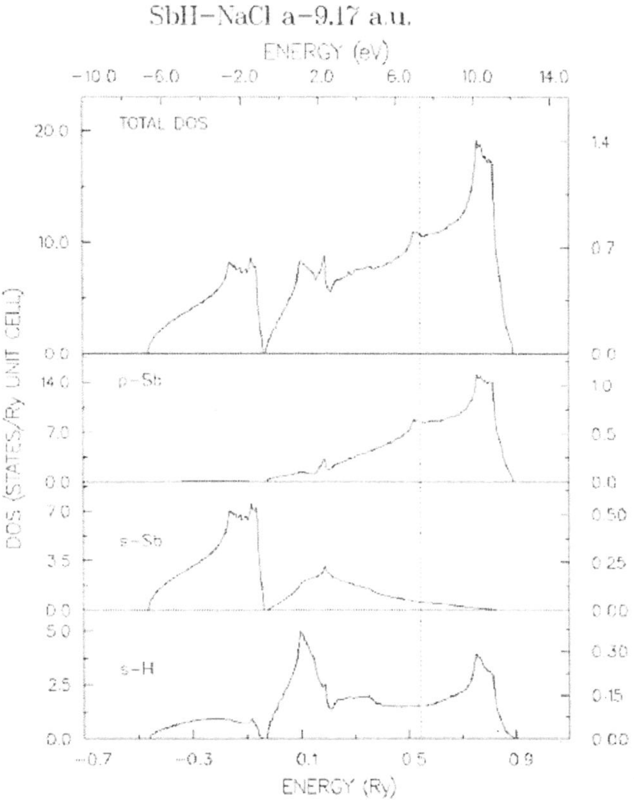

Fig. 11.17 Total, angular momentum and site decomposed densities of states of SbH in the NaCl structure (tight-binding)

Table 11.19 SbH (NaCl) a = 9.17 Bohr Slater–Koster 3-center parameters

	ORTHOGONAL	NON-ORTHOGONAL	
	ENERGY INTEGRALS Ry	ENERGY INTEGRALS Ry	OVERLAP INTEGRALS
ON SITE Sb-Sb			
s,s(000)	0.26789	-0.20282	
x,x(000)	0.66062	0.29461	
FIRST NEIGHBOR Sb-Sb			
s,s(110)	0.02147	-0.04703	0.06458
s,x(110)	0.04164	-0.04318	0.09040
s,xy(110)	0.00000	0.00000	0.00000
s,d2(110)	0.00000	0.00000	0.00000
x,x(110)	0.02239	-0.00333	-0.10262
x,x(011)	0.00656	-0.01293	0.04737
x,y(110)	0.02135	0.02182	-0.15368
x,xy(110)	0.00000	0.00000	0.00000
x,xy(011)	0.00000	0.00000	0.00000
z,d2(011)	0.00000	0.00000	0.00000
z,d1(011)	0.00000	0.00000	0.00000
xy,xy(110)	0.00000	0.00000	0.00000
xy,xy(011)	0.00000	0.00000	0.00000
xy,xz(011)	0.00000	0.00000	0.00000
xy,d2(110)	0.00000	0.00000	0.00000
d2,d2(110)	0.00000	0.00000	0.00000
d1,d1(110)	0.00000	0.00000	0.00000
SECOND NEIGHBOR Sb-Sb			
s,s(200)	-0.00550	-0.00688	0.00063
s,x(200)	0.00491	-0.01463	0.00885
s,d2(002)	0.00000	0.00000	0.00000
x,x(200)	-0.05552	0.03440	-0.01980
y,y(200)	0.02557	-0.00749	-0.00740
x,xy(020)	0.00000	0.00000	0.00000
z,d2(002)	0.00000	0.00000	0.00000
xy,xy(200)	0.00000	0.00000	0.00000
xy,xy(002)	0.00000	0.00000	0.00000
d2,d2(002)	0.00000	0.00000	0.00000
d1,d1(002)	0.00000	0.00000	0.00000
ON SITE H -H			
s,s(000)	0.11432	0.01850	
FIRST NEIGHBOR H -H			
s,s(110)	-0.02476	-0.03661	0.01302
s,x(110)	0.00322	0.00000	0.00000
x,x(110)	0.03836	0.00000	0.00000
x,x(011)	0.00448	0.00000	0.00000
x,y(110)	0.03983	0.00000	0.00000
SECOND NEIGHBOR H -H			
s,s(200)	-0.01567	0.00296	0.00362
s,x(200)	-0.02111	0.00000	0.00000
x,x(200)	0.01082	0.00000	0.00000
y,y(200)	-0.00524	0.00000	0.00000
FIRST NEIGHBOR Sb-H			
s,s(100)	-0.08054	-0.09267	0.21759
s,x(100)	-0.15658	0.00000	0.00000
x,s(100)	-0.10692	0.03614	-0.37435
x,x(100)	-0.07674	0.00000	0.00000
y,y(100)	0.04905	0.00000	0.00000
xy,x(010)	0.00000	0.00000	0.00000
d2,s(001)	0.00000	0.00000	0.00000
d2,z(001)	0.00000	0.00000	0.00000

SECOND NEIGHBOR Sb-H

s,s(111)	0.00783	-0.01295	0.00203
x,s(111)	-0.00759	0.01523	-0.00694
s,x(111)	0.01156	0.00000	0.00000
x,x(111)	0.00119	0.00000	0.00000
x,y(111)	0.00656	0.00000	0.00000
xy,s(111)	0.00000	0.00000	0.00000
xy,x(111)	0.00000	0.00000	0.00000
yz,x(111)	0.00000	0.00000	0.00000
d1,x(111)	0.00000	0.00000	0.00000

SbH

	ORTHOGONAL			NON-ORTHOGONAL		
BAND	RMS ERROR mRy	MAXIMUM DEVIATION k	mRy	RMS ERROR mRy	MAXIMUM DEVIATION k	mRy
1	3.6	(044)	7.4	0.5	(055)	1.2
2	6.3	(222)	14.0	1.4	(008)	3.3
3	6.2	(022)	21.8	2.4	(354)	6.2
4	7.8	(280)	19.5	2.1	(022)	4.1
5	11.0	(048)	40.1	1.8	(118)	3.7
1-5	7.4			1.8		

ENERGY VALUES IN Ry AT SELECTED k-POINTS

	ORTHOGONAL	APW	NON-ORTHOGONAL
GAMMA 1	-0.46211	-0.46311	-0.46327
GAMMA 1	0.67792	0.67735	0.67693
GAMMA 15	0.84664	0.85142	0.85200
X1 (008)	-0.09002	-0.08590	-0.08570
X1 (008)	0.35838	0.36033	0.36056
X4' (008)	0.15435	0.15555	0.15887
X5' (008)	0.51685	0.51969	0.51958
L1 (444)	-0.16824	-0.16306	-0.16217
L2' (444)	-0.04434	-0.03548	-0.03659
L2' (444)	0.75125	0.75433	0.75238
L3' (444)	0.75477	0.75632	0.75485
W1 (048)	-0.04478	-0.03873	-0.03831
W2' (048)	0.14848	0.15249	0.15220
W2' (048)	0.88122	0.86173	0.86314
W3 (048)	0.28498	0.27828	0.27483
ODD (224)	0.69092	0.68636	0.69009
EVEN(224)	-0.29007	-0.28790	-0.28741
EVEN(224)	0.20255	0.19387	0.19486
EVEN(224)	0.68884	0.68458	0.68910
EVEN(224)	0.70582	0.70799	0.71138

Table 11.20 SbH (NaCl) a = 9.17 Bohr Slater–Koster 2-center parameters

	ORTHOGONAL	NON-ORTHOGONAL	
	ENERGY INTEGRALS Ry	ENERGY INTEGRALS Ry	OVERLAP INTEGRALS
ON SITE Sb-Sb			
s	0.02264	-0.20663	
p	0.52925	0.32264	
FIRST NEIGHBOR Sb-Sb			
(sss)	-0.01804	-0.04847	0.07403
(pps)	0.09049	0.00841	-0.25032
(ppp)	-0.00308	-0.01441	0.04491
(dds)	0.00000	0.00000	0.00000
(ddp)	0.00000	0.00000	0.00000
(ddd)	0.00000	0.00000	0.00000
(sps)	0.02351	0.05721	-0.12718
(sds)	0.00000	0.00000	0.00000
(pds)	0.00000	0.00000	0.00000
(pdp)	0.00000	0.00000	0.00000

SECOND NEIGHBOR Sb-Sb			
(sss)	-0.01063	-0.00693	-0.00319
(pps)	-0.03764	0.02359	-0.01322
(ppp)	0.01423	-0.00728	-0.00690
(dds)	0.00000	0.00000	0.00000
(ddp)	0.00000	0.00000	0.00000
(ddd)	0.00000	0.00000	0.00000
(sps)	0.02716	0.01530	-0.00499
(sds)	0.00000	0.00000	0.00000
(pds)	0.00000	0.00000	0.00000
(pdp)	0.00000	0.00000	0.00000
ON SITE H-H			
s	0.43489	-0.01210	
FIRST NEIGHBOR H-H			
(sss)	0.01436	-0.03978	0.01852
(sps)	0.06065	0.00000	0.00000
(pps)	-0.01523	0.00000	0.00000
(ppp)	1.25288	0.00000	0.00000
SECOND NEIGHBOR H-H			
(sss)	-0.02093	0.00146	0.00128
(sps)	-0.08019	0.00000	0.00000
(pps)	-0.11601	0.00000	0.00000
(ppp)	-0.54650	0.00000	0.00000
FIRST NEIGHBOR Sb-H			
(sss)	-0.03061	0.09735	-0.23153
(sps)	0.60224	0.00000	0.00000
(pss)	0.09448	0.03628	-0.35895
(pps)	0.01181	0.00000	0.00000
(ppp)	-0.00400	0.00000	0.00000
(dpp)	0.00000	0.00000	0.00000
(dss)	0.00000	0.00000	0.00000
(dps)	0.00000	0.00000	0.00000
SECOND NEIGHBOR Sb-H			
(sss)	-0.03198	0.01535	-0.00212
(pss)	-0.02253	0.02877	-0.01179
(sps)	0.12933	0.00000	0.00000
(pps)	-0.14553	0.00000	0.00000
(ppp)	0.09167	0.00000	0.00000
(dss)	0.00000	0.00000	0.00000
(dps)	0.00000	0.00000	0.00000
(dpp)	0.00000	0.00000	0.00000

SbH

	ORTHOGONAL			NON-ORTHOGONAL		
BAND	RMS ERROR	MAXIMUM DEVIATION		RMS ERROR	MAXIMUM DEVIATION	
	mRy	k	mRy	mRy	k	mRy
1	5.9	(055)	11.0	0.6	(055)	1.4
2	13.1	(005)	27.4	1.9	(444)	4.7
3	12.4	(022)	27.6	2.6	(042)	4.8
4	16.0	(033)	39.8	2.6	(064)	5.6
5	18.4	(004)	47.0	2.1	(224)	7.3
1-5	13.8			2.1		

ENERGY VALUES IN Ry AT SELECTED k-POINTS

	ORTHOGONAL	APW	NON-ORTHOGONAL
GAMMA 1	-0.46221	-0.46311	-0.46289
GAMMA 1	0.68623	0.67735	0.67743
GAMMA 15	0.84753	0.85142	0.85204
X1 (008)	-0.08231	-0.08590	-0.08624
X1 (008)	0.36516	0.36033	0.35909
X4' (008)	0.14627	0.15555	0.15767
X5' (008)	0.52028	0.51969	0.51980
L1 (444)	-0.16014	-0.16306	-0.16195
L2' (444)	-0.01334	-0.03548	-0.04014
L2' (444)	0.74715	0.75433	0.75103
L3' (444)	0.73476	0.75632	0.75557
W1 (048)	-0.04612	-0.03873	-0.03814
W2' (048)	0.16636	0.15249	0.15173
W2' (048)	0.84297	0.86173	0.86251
W3 (048)	0.27899	0.27828	0.27580

ODD (224)	0.68177	0.68636	0.68991
EVEN(224)	-0.29709	-0.28790	-0.28769
EVEN(224)	0.20277	0.19387	0.19583
EVEN(224)	0.69776	0.68458	0.68653
EVEN(224)	0.72793	0.70799	0.71532

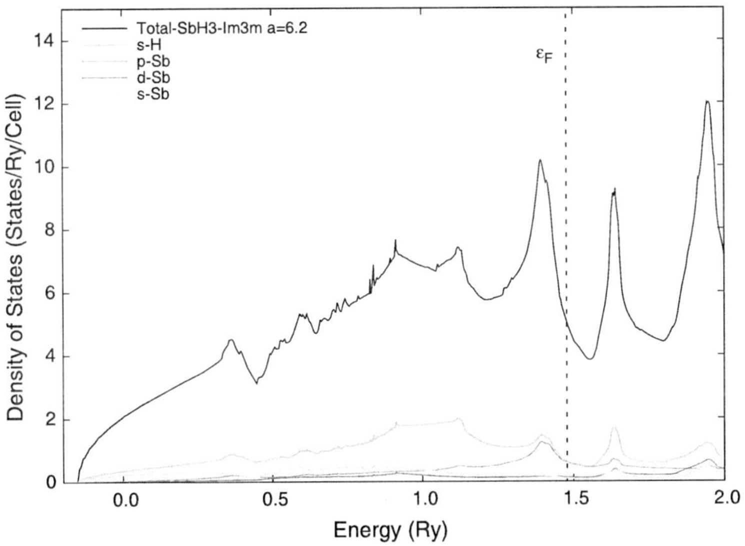

Fig. 11.18 Total, angular momentum and site decomposed densities of states of SbH3 in the Im3m structure

11.5 Bismouth Hydride (BiH)

See Fig. 11.19, Tables 11.21, 11.22, and 11.23.

See Figs. 11.20, 11.21, Tables 11.24 and 11.25.

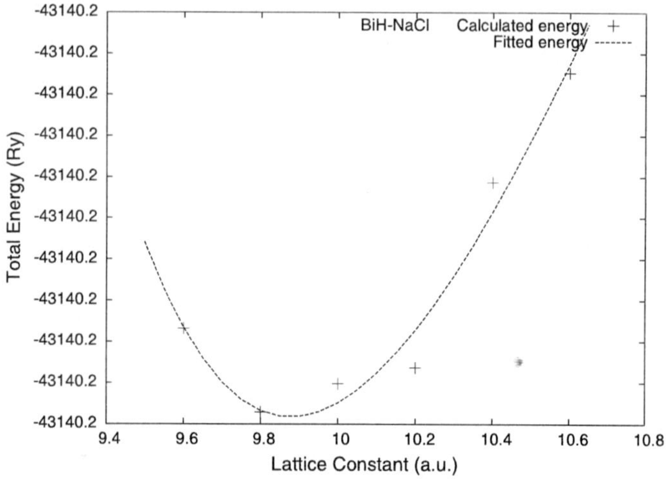

Fig. 11.19 Total energy versus lattice constant of BiH in the NaCl structure

Table 11.21 Lattice constant, bulk modulus, gap, total energy

Stru	a (Bohr)	B (MBar)	Gap	Total Energy (Ry)
NaCl	9.88	0.64	-	-43140.17772

Table 11.22 Birch fit coefficients

	A1	A2	A3	A4
NaCl	-4.314045E+04	7.678583E+01	-4.737686E+03	8.391857E+04

Table 11.23 DOS at Ef, Hopfield parameter, Stoner criterion

BiH a=9.88 Bohr NaCl

	Fermi Energy (Ry)	Total DOS (States/Ry)	Decomposed DOS inside the muffin tins s	p	eg	t2g	f
Bi	0.514	12.040	0.106	4.745	0.074	0.132	0.079
H	0.514	12.040	0.746	1.196	0.111	0.079	0.037

Fermi-Velocity (cm/s): 1.47530 x10E8
Plasmon Energy (eV) : 11.87449
Electron-ion interaction (Hopfield parameter) (eV/A^2) Bi:1.284 H:0.588

Bi	MUFFIN-TIN RADIUS and CHARGE =	2.5928	80.5326
H	MUFFIN-TIN RADIUS and CHARGE =	2.3459	1.4971
Bi	STONER I =	0.0053	
H	STONER I =	0.0035	
STONER PARAMETER (Ry)	I =	0.0090	
STONER CRITERION	N*I =	0.1078	

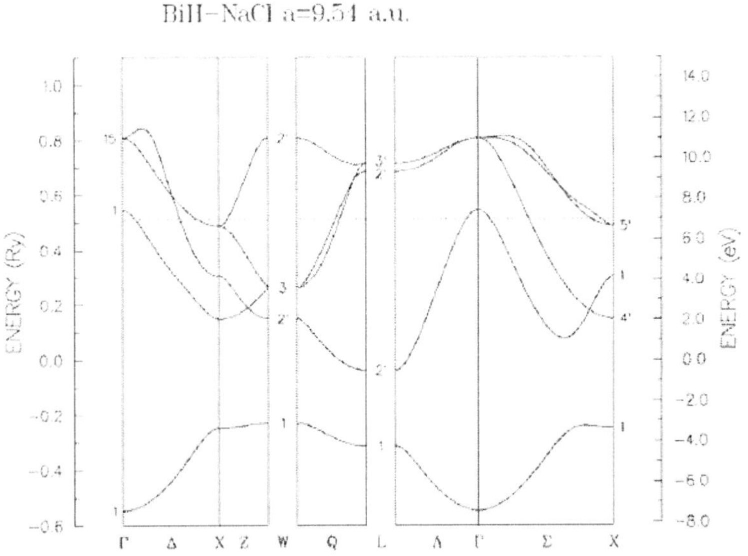

Fig. 11.20 Energy bands of BiH in the NaCl structure (tight-binding)

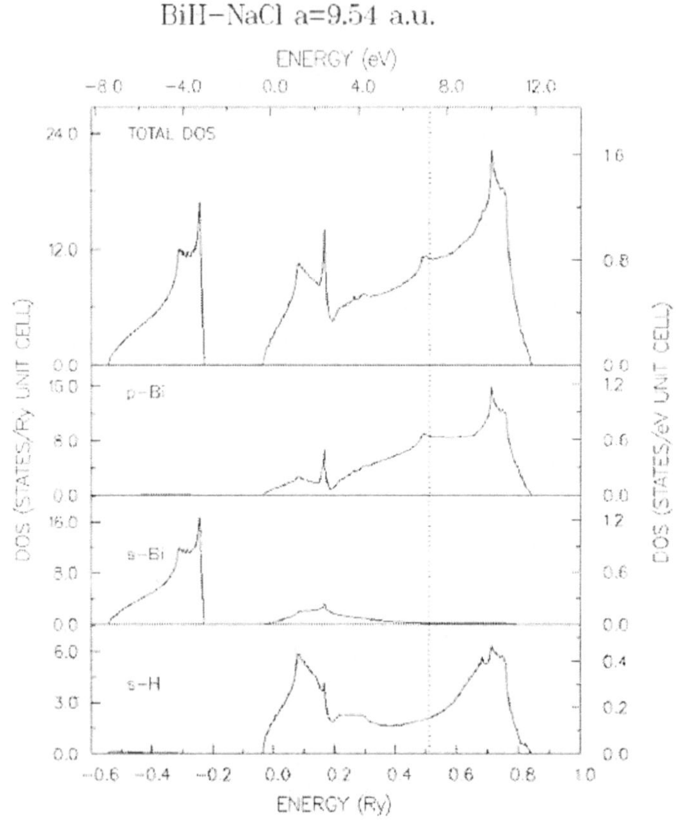

Fig. 11.21 Total, angular momentum and site decomposed densities of states of BiH in the NaCl structure (tight-binding)

Table 11.24 BiH (NaCl) a = 9.54 Bohr Slater–Koster 3-center parameters

	ORTHOGONAL	NON-ORTHOGONAL	
	ENERGY INTEGRALS Ry	ENERGY INTEGRALS Ry	OVERLAP INTEGRALS
ON SITE Bi-Bi			
s,s(000)	0.07044	-0.33530	
x,x(000)	0.62060	0.29398	
FIRST NEIGHBOR Bi-Bi			
s,s(110)	0.02014	-0.03878	0.04656
s,x(110)	0.04368	0.04570	-0.07025
x,x(110)	0.01979	0.00046	-0.09359
x,x(011)	0.00761	-0.01240	0.04057
x,y(110)	0.02408	0.02109	-0.14550
SECOND NEIGHBOR Bi-Bi			
s,s(200)	-0.01182	-0.00385	0.00095
s,x(200)	0.00827	0.01198	-0.00842
x,x(200)	-0.04531	0.03023	-0.01781
y,y(200)	0.02370	-0.00698	-0.00766
ON SITE H-H			
s,s(000)	0.12943	0.02510	
FIRST NEIGHBOR H-H			
s,s(110)	-0.02320	-0.03735	0.00105
s,x(110)	-0.00395	0.00000	0.00000
x,x(110)	0.04309	0.00000	0.00000
x,x(011)	-0.00136	0.00000	0.00000
x,y(110)	0.03579	0.00000	0.00000
SECOND NEIGHBOR H-H			
s,s(200)	-0.01494	0.00273	0.00389
s,x(200)	-0.01567	0.00000	0.00000
x,x(200)	0.02233	0.00000	0.00000
y,y(200)	-0.00195	0.00000	0.00000
FIRST NEIGHBOR Bi-H			
s,s(100)	-0.09262	-0.09756	0.19029
s,x(100)	-0.16623	0.00000	0.00000
x,s(100)	-0.10094	-0.03970	0.35232
x,x(100)	-0.07523	0.00000	0.00000
y,y(100)	0.04586	0.00000	0.00000
SECOND NEIGHBOR Bi-H			
s,s(111)	0.00837	-0.01111	-0.00469
x,s(111)	-0.00333	-0.01508	0.00155
s,x(111)	0.01197	0.00000	0.00000
x,x(111)	0.00292	0.00000	0.00000
x,y(111)	0.00354	0.00000	0.00000

BiH

	ORTHOGONAL			NON-ORTHOGONAL		
BAND	RMS ERROR mRy	MAXIMUM DEVIATION k	mRy	RMS ERROR mRy	MAXIMUM DEVIATION k	mRy
1	2.5	(044)	6.7	1.0	(006)	1.8
2	5.5	(044)	13.3	1.3	(008)	3.3
3	5.8	(022)	17.2	2.6	(354)	7.3
4	6.2	(280)	14.0	1.9	(022)	4.5
5	8.9	(048)	30.2	1.4	(004)	3.1
1-5	6.1			1.7		

ENERGY VALUES IN Ry AT SELECTED k-POINTS

	ORTHOGONAL	APW	NON-ORTHOGONAL
GAMMA 1	-0.54309	-0.54496	-0.54658
GAMMA 1	0.54576	0.54550	0.54517
GAMMA 15	0.80291	0.80677	0.80793
X1 (008)	-0.24817	-0.24857	-0.24810
X1 (008)	0.29968	0.30599	0.30571
X4' (008)	0.14061	0.14486	0.14813
X5' (008)	0.48655	0.48703	0.48755
L1 (444)	-0.31894	-0.31486	-0.31397

L2' (444)	-0.04293	-0.03689	-0.03836
L2' (444)	0.68578	0.68657	0.68549
L3' (444)	0.71274	0.71446	0.71410
W1 (048)	-0.23480	-0.23182	-0.23035
W2' (048)	0.14374	0.14998	0.14945
W2' (048)	0.82419	0.80701	0.80747
W3 (048)	0.27089	0.26631	0.26339
ODD (224)	0.65353	0.65147	0.65353
EVEN(224)	-0.39338	-0.39272	-0.39290
EVEN(224)	0.13958	0.13473	0.13619
EVEN(224)	0.64382	0.63763	0.64409
EVEN(224)	0.65490	0.65766	0.65964

FERMI LEVEL QUANTITIES (Non-orthogonal fit)

ENERGY		DENSITIES OF STATES				
	Total	s-Bi	p-Bi	t2g-Bi	eg-Bi	s-H
Ry		states/Ry/cell				
0.5109	11.06	0.25	8.73	0.00	0.00	2.09
		INTEGRATED DENSITIES OF STATES				
	Total	s-Bi	p-Bi	t2g-Bi	eg-Bi	s-H
		electrons				
	6.00	2.53	2.05	0.00	0.00	1.42

VELOCITY	PLASMON ENERGY	EIGENVALUE SUM
cm/s	eV	Ry
1.48x10E8	12.00	-2.6781

Table 11.25 BiH (NaCl) a = 9.88 Bohr Slater–Koster 2-center parameters

	ORTHOGONAL	NON-ORTHOGONAL	
	ENERGY INTEGRALS Ry	ENERGY INTEGRALS Ry	OVERLAP INTEGRALS
ON SITE Bi-Bi			
s	-0.28476	-0.33475	
p	0.63277	0.31809	
FIRST NEIGHBOR Bi-Bi			
(sss)	-0.01146	-0.03909	0.04981
(pps)	0.03171	0.01486	-0.23064
(ppp)	0.00951	-0.01356	0.03974
(dds)	0.00000	0.00000	0.00000
(ddp)	0.00000	0.00000	0.00000
(ddd)	0.00000	0.00000	0.00000
(sps)	0.00841	0.06201	-0.09376
(sds)	0.00000	0.00000	0.00000
(pds)	0.00000	0.00000	0.00000
(pdp)	0.00000	0.00000	0.00000
SECOND NEIGHBOR Bi-Bi			
(sss)	0.00580	-0.00256	-0.00310
(pps)	-0.05177	0.02344	-0.00450
(ppp)	0.01930	-0.00699	-0.00748
(dds)	0.00000	0.00000	0.00000
(ddp)	0.00000	0.00000	0.00000
(ddd)	0.00000	0.00000	0.00000
(sps)	-0.00965	0.01236	-0.00080
(sds)	0.00000	0.00000	0.00000
(pds)	0.00000	0.00000	0.00000
(pdp)	0.00000	0.00000	0.00000
ON SITE H-H			
s	0.18436	0.00228	
FIRST NEIGHBOR H-H			
(sss)	0.00829	-0.03903	0.00924
(sps)	0.04458	0.00000	0.00000
(pps)	-0.54849	0.00000	0.00000
(ppp)	0.23887	0.00000	0.00000

SECOND NEIGHBOR H-H			
(sss)	0.01746	0.00090	0.00095
(sps)	0.08777	0.00000	0.00000
(pps)	1.96622	0.00000	0.00000
(ppp)	-0.12011	0.00000	0.00000
FIRST NEIGHBOR Bi-H			
(sss)	-0.06674	-0.10276	0.19645
(sps)	0.00685	0.00000	0.00000
(pss)	-0.07451	-0.04247	0.33032
(pps)	-0.49253	0.00000	0.00000
(ppp)	0.32166	0.00000	0.00000
(dpp)	0.00000	0.00000	0.00000
(dss)	0.00000	0.00000	0.00000
(dps)	0.00000	0.00000	0.00000
SECOND NEIGHBOR Bi-H			
(sss)	0.00211	-0.01182	-0.00304
(pss)	-0.00248	-0.02768	0.00413
(sps)	0.03824	0.00000	0.00000
(pps)	-0.01940	0.00000	0.00000
(ppp)	0.01367	0.00000	0.00000
(dss)	0.00000	0.00000	0.00000
(dps)	0.00000	0.00000	0.00000
(dpp)	0.00000	0.00000	0.00000

BiH

	ORTHOGONAL			NON-ORTHOGONAL		
BAND	RMS ERROR	MAXIMUM DEVIATION		RMS ERROR	MAXIMUM DEVIATION	
	mRy	k	mRy	mRy	k	mRy
1	2.7	(055)	8.5	0.7	(044)	1.6
2	4.1	(008)	7.8	1.5	(444)	3.7
3	9.2	(022)	24.8	2.7	(354)	6.5
4	5.6	(264)	11.2	2.3	(022)	5.8
5	8.4	(003)	27.0	1.6	(224)	4.7
1-5	6.5			1.9		

ENERGY VALUES IN Ry AT SELECTED k-POINTS

	ORTHOGONAL	APW	NON-ORTHOGONAL
GAMMA 1	-0.54503	-0.54496	-0.54630
GAMMA 1	0.54614	0.54550	0.54547
GAMMA 15	0.80098	0.80677	0.80793
X1 (008)	-0.24893	-0.24857	-0.24788
X1 (008)	0.30074	0.30599	0.30442
X4' (008)	0.15268	0.14486	0.14736
X5' (008)	0.48409	0.48703	0.48777
L1 (444)	-0.31960	-0.31486	-0.31354

References

1. Wikipedia
2. P.-H. Chang, S. Silayi, D.A. Papaconstantopoulos, and M.J. Mehl, Pressure-induced high-temperature superconductivity in hypothetical H3X (X= As, Se, Br, Sb, Te and I) in the H3S structure with Im3m symmetry, J. Phys Chem Solids, **139**, 109315, (2020)

Chapter 12
Group 16 Chalcogen Hydrides

This chapter covers the Group 15 column of the periodic table hydrides from OH to PoH [1]. Results are presented for the crystal structures NaCl(B1),CaF2 (C1) and Im3m. These structures are not found experimentally and are presented here for the purpose of comparison through the periodic Table and to establish various trends. The chalcogen hydrides are usually dihydride molecules such as H2O. Of particular interest is the the Im3m cubic structure in which recently SH3 was discovered as a high temperature superconductor under 200 GPa pressure [2]. Results are presented in the Im3m structure for the other compounds of the group 16 to explore the possibility that some of them may also be superconductors. Examining the energy bands and densities of states figures we observe that, as in the group 14 and 15 columns, the lowest band has a mixture of s-H hydrogen and s-character of the other element. At the Fermi level the states are of s- and p- character. As in the other columns of this part of the Periodic Table, the band structure is dominated by s- and p-states. The d-states are semicore states found significantly lower than the s-p states.

Gam1 point. It is interesting to assess the values of the Hopfield-McMillan parameter "eta" for SH3 and OH3. In SH3 the stronger contribution comes from the H sites, while in OH3 the oxygen site contribution is dominating the electron–phonon coupling. Note that Tight-binding(TB) parameters are given in the NaCl structure based on both orthogonal and non-orthogonal Hamiltonians using three- and two-center integrals. These TB parameters could be possible to modify in order to build a TB model to describe the dihydride molecules.

D. A. Papaconstantopoulos, *Band Structure of Cubic Hydrides*,
https://doi.org/10.1007/978-3-031-06878-2_12

12.1 Oxygen Hydride (OH)

See Fig. 12.1, Tables 12.1, 12.2, and 12.3.

See Figs. 12.2, 12.3, Tables 12.4 and 12.5.

See Figs. 12.4, 12.5, and 12.6.

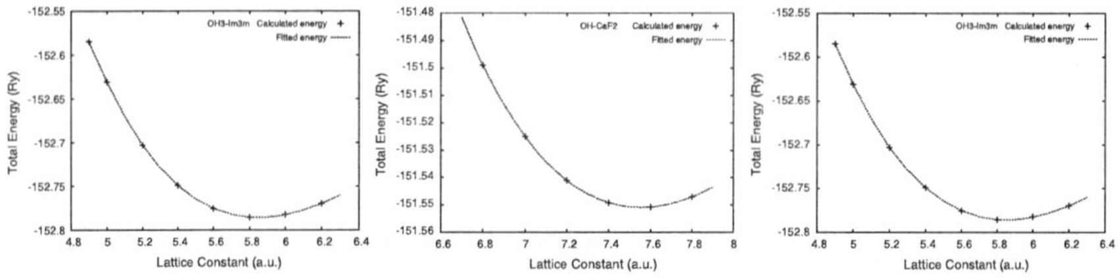

Fig. 12.1 Total energy versus lattice constant of OH in the NaCl, CaF2 and Im3m structures

Table 12.1 Lattice constant, bulk modulus, gap, total energy

Stru	a (Bohr)	B (MBar)	Gap	Total Energy (Ry)
NaCl	6.59	1.49	-	-150.44201
CaF2 (Semimetal)	7.55	1.19	-	-151.55093
Im3m (P=0)	5.85	1.69	-	-152.78580
Im3m(P=2.07 MBar)	4.90	6.84	-	-152.58480

Table 12.2 Birch fit coefficients

	A1	A2	A3	A4
NaCl	-1.496176E+02	-2.854073E+01	2.491339E+02	-3.682952E+01
CaF2	-1.504413E+02	-5.317467E+01	7.013741E+02	-1.500025E+03
Im3m	-1.510692E+02	-8.312481E+01	1.190940E+03	-4.240836E+03

Table 12.3 DOS at Ef, Hopfield parameter, Stoner criterion

OH a=6.59 NaCl

Fermi Energy (Ry)	Total DOS (States/Ry)	Decomposed DOS inside the muffin tins s	p	eg	t2g	f
O 0.409	20.761	0.032	17.561	0.009	0.012	0.004
H 0.409	20.761	0.082	0.310	0.275	0.032	0.047

Fermi-Velocity (cm/s): 0.58530 x10E8
Plasmon Energy (eV) : 11.35162
Electron-ion interaction (Hopfield parameter) (eV/A^2) O:4.247 H:0.085

```
-------------------------------------------------
  O     MUFFIN-TIN RADIUS and CHARGE =     1.6475     7.0027
  H     MUFFIN-TIN RADIUS and CHARGE =     1.6475     0.7702
                O        STONER I =  0.0557
                H        STONER I =  0.0004
        STONER PARAMETER (Ry)     I =  0.0561
        STONER CRITERION        N*I =  1.1646
-------------------------------------------------
        OH2    semimetal    a=7.55 Bohr         CaF2
-------------------------------------------------------------------------------
 Fermi Energy    Total DOS
    ( Ry )     (States/Ry)
-------------------------------------------------------------------------------
  0.302        0.0
-------------------------------------------------------------------------------
  O     MUFFIN-TIN RADIUS and CHARGE =     1.6354     7.1453
  H     MUFFIN-TIN RADIUS and CHARGE =     1.6354     0.7047
-------------------------------------------------------------------------------
        OH3  a=4.9 Bohr                Im3m
-------------------------------------------------------------------------------
 Fermi Energy    Total DOS         Decomposed DOS inside the muffin tins
    ( Ry )     (States/Ry)        s        p        d        f
O  1.483         4.770         0.349    0.920    0.145    0.005
H  1.483         4.770         0.545    0.047    0.006    0.001
-------------------------------------------------------------------------------
Fermi-Velocity (cm/s): 1.35007 x10E8
Plasmon Energy (eV)   : 13.84096
Electron-ion interaction (Hopfield parameter) (eV/A^2) O:18.838  H3:5.991
-------------------------------------------------
  O     MUFFIN-TIN RADIUS and CHARGE =     1.40     6.6740
  H     MUFFIN-TIN RADIUS and CHARGE =     1.00     0.5106
```

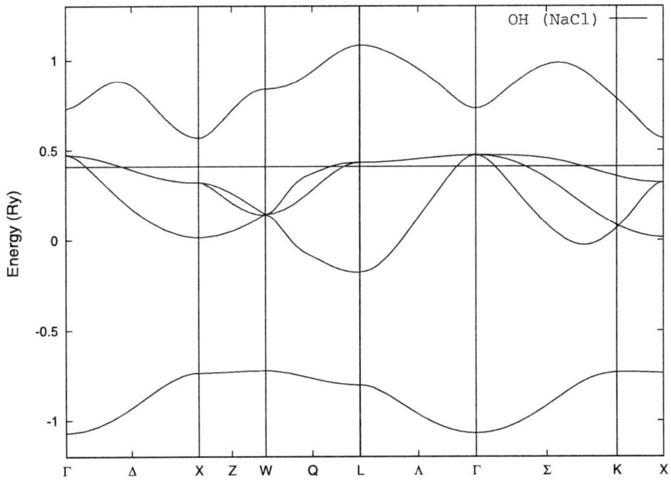

Fig. 12.2 Energy bands of OH in the NaCl structure

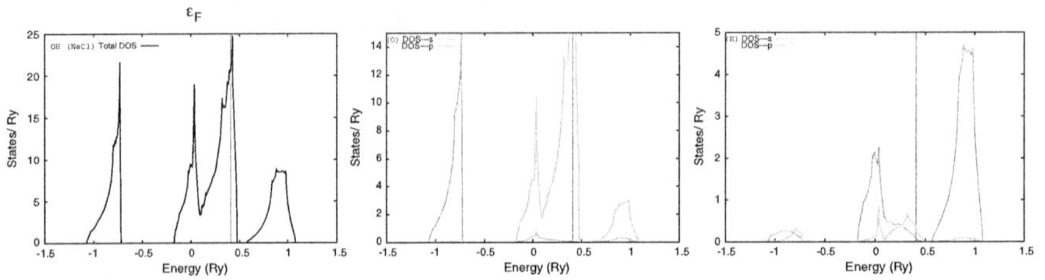

Fig. 12.3 Total, angular momentum and site decomposed densities of states of OH in the NaCl structure

Table 12.4 OH (NaCl) a = 6.59 Bohr Slater–Koster 3-center parameters

	ORTHOGONAL	NON-ORTHOGONAL	
	ENERGY INTEGRALS Ry	ENERGY INTEGRALS Ry	OVERLAP INTEGRALS
ON SITE O-O			
s,s(000)	-0.77398	-0.81303	
x,x(000)	0.36430	0.21054	
FIRST NEIGHBOR O-O			
s,s(110)	-0.01047	-0.03140	0.01155
s,x(110)	0.00737	0.03905	-0.01995
s,xy(110)	0.00000	0.00000	0.00000
s,d2(110)	0.00000	0.00000	0.00000
x,x(110)	0.02939	0.02216	-0.03287
x,x(011)	-0.00925	-0.01279	0.01426
x,y(110)	-0.00673	0.04741	-0.04098
x,xy(110)	0.00000	0.00000	0.00000
x,xy(011)	0.00000	0.00000	0.00000
z,d2(011)	0.00000	0.00000	0.00000
z,d1(011)	0.00000	0.00000	0.00000
xy,xy(110)	0.00000	0.00000	0.00000
xy,xy(011)	0.00000	0.00000	0.00000
xy,xz(011)	0.00000	0.00000	0.00000
xy,d2(110)	0.00000	0.00000	0.00000
d2,d2(110)	0.00000	0.00000	0.00000
d1,d1(110)	0.00000	0.00000	0.00000
SECOND NEIGHBOR O-O			
s,s(200)	0.00482	0.00405	-0.00717
s,x(200)	-0.00470	0.00538	0.00513
s,d2(002)	0.00000	0.00000	0.00000
x,x(200)	-0.04729	0.02172	0.00254
y,y(200)	0.00195	-0.00112	-0.00143
x,xy(020)	0.00000	0.00000	0.00000
z,d2(002)	0.00000	0.00000	0.00000
xy,xy(200)	0.00000	0.00000	0.00000
xy,xy(002)	0.00000	0.00000	0.00000
d2,d2(002)	0.00000	0.00000	0.00000
d1,d1(002)	0.00000	0.00000	0.00000
ON SITE H-H			
s,s(000)	0.45752	0.11631	
FIRST NEIGHBOR H-H			
s,s(110)	0.00051	-0.05459	0.04163
s,x(110)	0.00000	0.00000	0.00000
x,x(110)	0.00000	0.00000	0.00000
x,x(011)	0.00000	0.00000	0.00000
x,y(110)	0.00000	0.00000	0.00000

SECOND NEIGHBOR H-H			
s,s(200)	0.01170	-0.00627	-0.00465
s,x(200)	0.00000	0.00000	0.00000
x,x(200)	0.00000	0.00000	0.00000
y,y(200)	0.00000	0.00000	0.00000
FIRST NEIGHBOR O-H			
s,s(100)	-0.09656	0.17306	-0.15341
s,x(100)	0.00000	0.00000	0.00000
x,s(100)	-0.17976	0.07454	-0.26617
x,x(100)	0.00000	0.00000	0.00000
y,y(100)	0.00000	0.00000	0.00000
xy,x(010)	0.00000	0.00000	0.00000
d2,s(001)	0.00000	0.00000	0.00000
d2,z(001)	0.00000	0.00000	0.00000
SECOND NEIGHBOR O-H			
s,s(111)	0.00259	0.00791	-0.00485
x,s(111)	-0.00083	0.01042	-0.00801
s,x(111)	0.00000	0.00000	0.00000
x,x(111)	0.00000	0.00000	0.00000
x,y(111)	0.00000	0.00000	0.00000
xy,s(111)	0.00000	0.00000	0.00000
xy,x(111)	0.00000	0.00000	0.00000
yz,x(111)	0.00000	0.00000	0.00000
d1,x(111)	0.00000	0.00000	0.00000

OH

	ORTHOGONAL			NON-ORTHOGONAL		
BAND	RMS ERROR	MAXIMUM DEVIATION		RMS ERROR	MAXIMUM DEVIATION	
	mRy	k	mRy	mRy	k	mRy
1	2.1	(222)	5.0	0.2	(044)	0.4
2	4.0	(008)	10.6	0.2	(264)	0.6
3	4.8	(226)	11.6	0.5	(008)	2.4
4	4.0	(380)	7.8	0.6	(008)	2.4
5	4.3	(444)	11.6	0.4	(333)	1.5
1-5	3.9			0.4		

ENERGY VALUES IN Ry AT SELECTED k-POINTS

	ORTHOGONAL	APW	NON-ORTHOGONAL
GAMMA 1	-1.06581	-1.06937	-1.06921
GAMMA 1	0.72896	0.73190	0.73217
GAMMA 15	0.47564	0.47357	0.47343
X1 (008)	-0.73935	-0.73699	-0.73706
X1 (008)	0.56184	0.56809	0.56784
X4' (008)	0.00533	0.01591	0.01596
X5' (008)	0.31451	0.32140	0.31903
L1 (444)	-0.80289	-0.80305	-0.80278
L2' (444)	-0.17938	-0.17651	-0.17670
L2' (444)	1.07159	1.08324	1.08302
L3' (444)	0.42419	0.43190	0.43173
W1 (048)	-0.72248	-0.72336	-0.72309
W2' (048)	0.13816	0.14013	0.13993
W2' (048)	0.84441	0.84014	0.84004
W3 (048)	0.15214	0.14545	0.14546
ODD (224)	0.40574	0.41156	0.41153
EVEN(224)	-0.87914	-0.87653	-0.87635
EVEN(224)	-0.00346	-0.00425	-0.00420
EVEN(224)	0.41015	0.40517	0.40542
EVEN(224)	0.97605	0.97602	0.97534

Table 12.5 OH (NaCl) a = 6.59 Bohr Slater–Koster 2-center parameters

	ORTHOGONAL	NON-ORTHOGONAL	
	ENERGY INTEGRALS Ry	ENERGY INTEGRALS Ry	OVERLAP INTEGRALS
ON SITE O-O			
s	-0.75778	-0.81029	
p	0.27412	0.22939	
FIRST NEIGHBOR O-O			
(sss)	-0.00517	-0.03357	0.01558
(pps)	0.06767	0.05616	-0.08332
(ppp)	-0.00907	-0.01313	0.01427
(dds)	0.00000	0.00000	0.00000
(ddp)	0.00000	0.00000	0.00000
(ddd)	0.00000	0.00000	0.00000
(sps)	-0.04374	0.05519	-0.03639
(sds)	0.00000	0.00000	0.00000
(pds)	0.00000	0.00000	0.00000
(pdp)	0.00000	0.00000	0.00000
SECOND NEIGHBOR O-O			
(sss)	0.00811	0.00028	-0.00199
(pps)	-0.00179	0.01099	-0.00063
(ppp)	0.00191	-0.00153	-0.00226
(dds)	0.00000	0.00000	0.00000
(ddp)	0.00000	0.00000	0.00000
(ddd)	0.00000	0.00000	0.00000
(sps)	-0.01617	0.00296	0.00015
(sds)	0.00000	0.00000	0.00000
(pds)	0.00000	0.00000	0.00000
(pdp)	0.00000	0.00000	0.00000
ON SITE H-H			
s	0.71185	0.11047	
FIRST NEIGHBOR H-H			
(sss)	-0.00479	-0.06109	0.02915
(sps)	0.00000	0.00000	0.00000
(pps)	0.00000	0.00000	0.00000
(ppp)	0.00000	0.00000	0.00000
SECOND NEIGHBOR H-H			
(sss)	-0.03701	0.00257	0.00479
(sps)	0.00000	0.00000	0.00000
(pps)	0.00000	0.00000	0.00000
(ppp)	0.00000	0.00000	0.00000
FIRST NEIGHBOR O-H			
(sss)	-0.11573	-0.18037	0.14519
(sps)	0.00000	0.00000	0.00000
(pss)	-0.11626	-0.10487	0.23089
(pps)	0.00000	0.00000	0.00000
(ppp)	0.00000	0.00000	0.00000
(dpp)	0.00000	0.00000	0.00000
(dss)	0.00000	0.00000	0.00000
(dps)	0.00000	0.00000	0.00000
SECOND NEIGHBOR O-H			
(sss)	0.00315	-0.01052	0.00444
(pss)	0.01444	-0.02067	0.01204
(sps)	0.00000	0.00000	0.00000
(pps)	0.00000	0.00000	0.00000
(ppp)	0.00000	0.00000	0.00000
(dss)	0.00000	0.00000	0.00000
(dps)	0.00000	0.00000	0.00000
(dpp)	0.00000	0.00000	0.00000

OH

BAND	ORTHOGONAL RMS ERROR mRy	MAXIMUM DEVIATION k	mRy	NON-ORTHOGONAL RMS ERROR mRy	MAXIMUM DEVIATION k	mRy
1	1.9	(008)	4.2	0.2	(444)	0.7
2	5.3	(004)	13.3	0.5	(226)	1.3
3	5.3	(226)	11.4	0.6	(008)	2.1
4	4.4	(380)	8.5	0.6	(008)	2.1
5	5.7	(226)	13.7	0.7	(333)	2.4
1-5	4.7			0.5		

ENERGY VALUES IN Ry AT SELECTED k-POINTS

	ORTHOGONAL	APW	NON-ORTHOGONAL
GAMMA 1	-1.06933	-1.06937	-1.06918
GAMMA 1	0.73048	0.73190	0.73252
GAMMA 15	0.47630	0.47357	0.47348
X1 (008)	-0.74117	-0.73699	-0.73663
X1 (008)	0.56162	0.56809	0.56813
X4' (008)	0.00749	0.01591	0.01557
X5' (008)	0.31446	0.32140	0.31925
L1 (444)	-0.80642	-0.80305	-0.80236
L2' (444)	-0.18222	-0.17651	-0.17694
L2' (444)	1.07924	1.08324	1.08224
L3' (444)	0.42356	0.43190	0.43199
W1 (048)	-0.72090	-0.72336	-0.72341
W2' (048)	0.14630	0.14013	0.13911
W2' (048)	0.83228	0.84014	0.83999
W3 (048)	0.15335	0.14545	0.14541
ODD (224)	0.40566	0.41156	0.41149
EVEN(224)	-0.87684	-0.87653	-0.87664
EVEN(224)	-0.00811	-0.00425	-0.00401
EVEN(224)	0.41038	0.40517	0.40555
EVEN(224)	0.97827	0.97602	0.97594

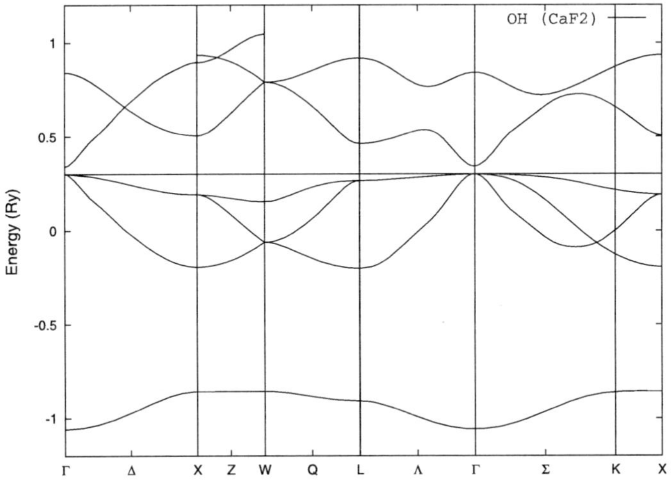

Fig. 12.4 Energy bands of OH in the CaF2 structure

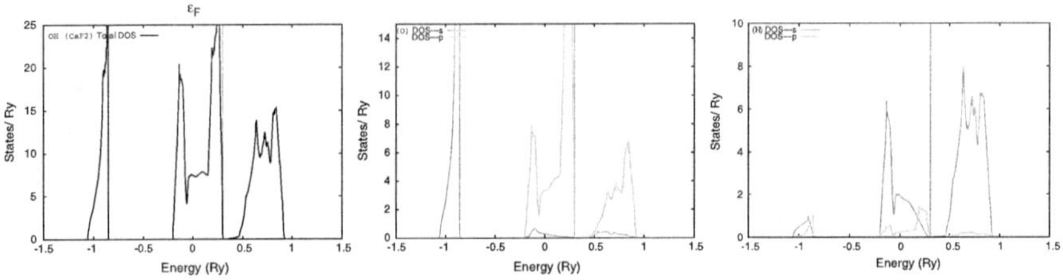

Fig. 12.5 Total, angular momentum and site decomposed densities of states of OH in the CaF2 structure

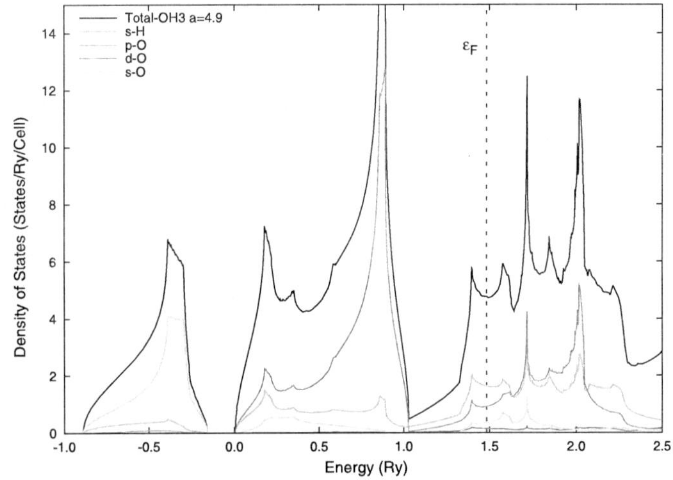

Fig. 12.6 Total, angular momentum and site decomposed densities of states of OH3 in the Im3m structure

12.2 Sulfur Hydride (SH)

See Fig. 12.7, Tables 12.6, 12.7, and 12.8.

See Figs. 12.8, 12.9, Tables 12.9, and 12.10.

See Figs. 12.10, 12.11, 12.12, 12.13, 12.14.

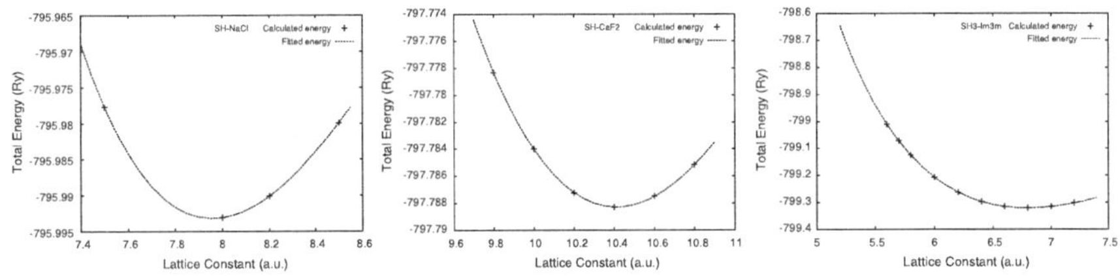

Fig. 12.7 Total energy versus lattice constant of SH in the NaCl, CaF2 and Im3m structures

Table 12.6 Lattice constant, bulk modulus, gap, total energy

Stru	a (Bohr)	B (MBar)	Gap	Total Energy (Ry)
NaCl	7.96	0.96	-	-795.99320
CaF2	10.41	0.29	-	-797.78829
Im3m (P=0)	6.79	1.24	-	-799.32074
Im3m(P=2.10 MBar)	5.60	7.45	-	-799.01059

Table 12.7 Birch fit coefficients

	A1	A2	A3	A4
NaCl	-7.955365E+02	-1.118576E+01	-3.028950E+02	7.427015E+03
CaF2	-7.970826E+02	-6.457926E+01	1.638676E+03	-7.164287E+03
Im3m	-7.976491E+02	-1.025410E+02	1.723801E+03	-4.532388E+03

Table 12.8 DOS at Ef, Hopfield parameter, Stoner criterion

```
       SH        a=7.96 Bohr           NaCl
 ----------------------------------------------------------------------------
 Fermi Energy   Total DOS         Decomposed DOS inside the muffin tins
     ( Ry )     (States/Ry)        s          p          eg          t2g     f
 ----------------------------------------------------------------------------
 S  0.512        11.790         0.123      7.041     0.090       0.077   0.012
 H  0.512        11.790         0.524      0.561     0.261       0.057   0.047
 ----------------------------------------------------------------------------
Fermi-Velocity (cm/s): 1.25468 x10E8
Plasmon Energy (eV)   : 13.81339
Electron-ion interaction (Hopfield parameter) (eV/A^2) S:3.608  H:0.407
 ---------------------------------------------
   S      MUFFIN-TIN RADIUS and CHARGE =     1.9900    13.6893
   H      MUFFIN-TIN RADIUS and CHARGE =     1.9900     1.3371
                    S        STONER I =   0.0232
                    H        STONER I =   0.0020
          STONER PARAMETER (Ry)     I =   0.0254
          STONER CRITERION        N*I =   0.2989
 ---------------------------------------------
       SH2    semimetal   a=10.41 Bohr           CaF2
 ----------------------------------------------------------------------------
 Fermi Energy   Total DOS
     ( Ry )     (States/Ry)     s          p          eg          t2g          f
 ----------------------------------------------------------------------------
 S  0.113        14.208       0.015     9.591     0.001       0.052        0.001
 H  0.113        14.208       1.131     0.406     0.026       0.418        0.057
 ----------------------------------------------------------------------------
Fermi-Velocity (cm/s): 0.35189 x10E8
Plasmon Energy (eV)   : 2.83822
Electron-ion interaction (Hopfield parameter (eV/A^2) ): S 1.080  2H 0.452
 ----------------------------------------------------------------------------
   S      MUFFIN-TIN RADIUS and CHARGE =     2.2540    14.3746
   H      MUFFIN-TIN RADIUS and CHARGE =     2.2540     1.0298
                          S        STONER I =    0.0227
                          H        STONER I =    0.0007
                  STONER PARAMETER (Ry) I =    0.0242
                  STONER CRITERION    N*I =    0.3437

 ----------------------------------------------------------------------------
       SH3   a=5.6 Bohr                   Im3m
 ----------------------------------------------------------------------------
 Fermi Energy   Total DOS         Decomposed DOS inside the muffin tins
     ( Ry )     (States/Ry)        s       p         d       f
 S  1.320        8.778          0.408    2.216     0.946   0.024
 H  1.320        8.778          0.569    0.093     0.006   0.001
 ----------------------------------------------------------------------------
Fermi-Velocity (cm/s): 1.21634 x10E8
Plasmon Energy (eV)   : 13.84612
Electron-ion interaction (Hopfield parameter) (eV/A^2) S:7.680  H3:6.855
 ---------------------------------------------
   S      MUFFIN-TIN RADIUS and CHARGE =     1.80    13.6528
   H      MUFFIN-TIN RADIUS and CHARGE =     1.00     0.5644
```

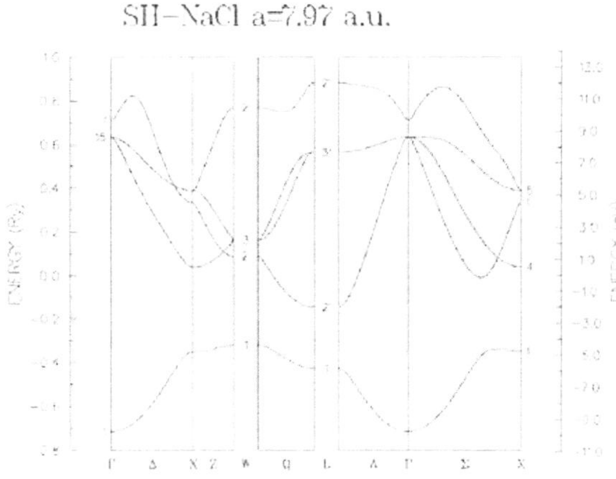

Fig. 12.8 Energy bands of SH in the NaCl structure (tight-binding)

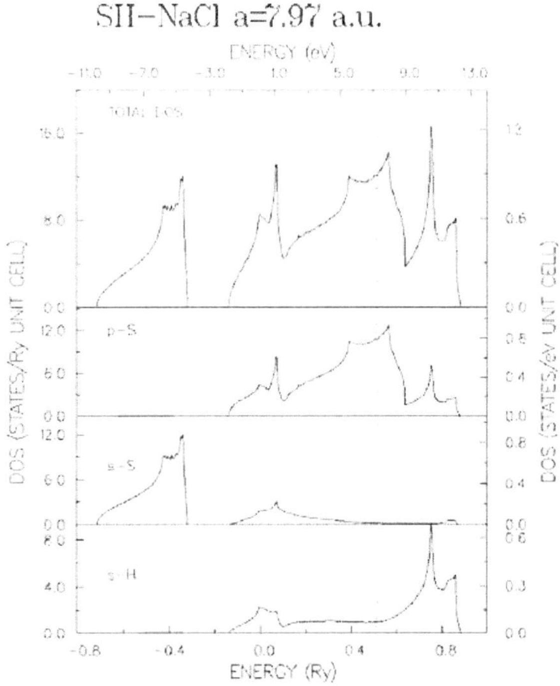

Fig. 12.9 Total, angular momentum and site decomposed densities of states of SH in the NaCl structure (tight-binding)

Table 12.9 SH (NaCl) a = 7.97 Bohr Slater–Koster 3-center parameters

	ORTHOGONAL	NON-ORTHOGONAL	
	ENERGY INTEGRALS Ry	ENERGY INTEGRALS Ry	OVERLAP INTEGRALS
ON SITE S- S			
s,s(000)	-0.29734	-0.36682	
x,x(000)	0.46628	0.21291	
FIRST NEIGHBOR S-S			
s,s(110)	0.01809	-0.00727	0.01808
s,x(110)	-0.00737	0.06137	-0.04661
x,x(110)	0.03918	0.00926	-0.08894
x,x(011)	-0.00768	-0.02818	0.01640
x,y(110)	-0.01119	0.04699	-0.09531
SECOND NEIGHBOR S-S			
s,s(200)	0.02251	0.05465	-0.10783
s,x(200)	-0.03105	0.01952	0.05806
x,x(200)	-0.06926	0.01682	-0.03089
y,y(200)	0.00550	0.00012	0.00836
ON SITE H-H			
s,s(000)	0.07780	0.22905	
FIRST NEIGHBOR H-H			
s,s(110)	-0.01708	-0.00439	-0.05529
SECOND NEIGHBOR H-H			
s,s(200)	0.01033	-0.04447	-0.02432
FIRST NEIGHBOR S-H			
s,s(100)	-0.12113	-0.05332	0.02534
x,s(100)	-0.10800	-0.09120	0.20904
SECOND NEIGHBOR S-H			
s,s(111)	0.00204	0.01104	0.00117
x,s(111)	0.00216	-0.01152	-0.03276

	ORTHOGONAL			NON-ORTHOGONAL		
BAND	RMS ERROR	MAXIMUM DEVIATION		RMS ERROR	MAXIMUM DEVIATION	
	mRy	k	mRy	mRy	k	mRy
1	5.7	(005)	16.3	7.6	(222)	17.7
2	14.2	(008)	32.9	18.8	(006)	62.1
3	13.1	(444)	27.8	13.9	(280)	48.4
4	10.9	(444)	27.8	9.6	(174)	22.6
5	20.8	(004)	58.5	12.8	(042)	30.5
1-5	13.9			13.1		

ENERGY VALUES IN Ry AT SELECTED k-POINTS

	ORTHOGONAL	APW	NON-ORTHOGONAL
GAMMA 1	-0.71816	-0.71512	-0.72024
GAMMA 1	0.70784	0.71138	0.71002
GAMMA 15	0.63248	0.63539	0.63960
X1 (008)	-0.35369	-0.34988	-0.35514
X1 (008)	0.32713	0.33463	0.35475
X4' (008)	0.00565	0.03856	0.03442
X5' (008)	0.38047	0.38956	0.39707
L1 (444)	-0.43240	-0.42777	-0.42181
L2' (444)	-0.15363	-0.14719	-0.14697
L2' (444)	0.84176	0.88518	0.87990
L3' (444)	0.53805	0.56583	0.56678
W1 (048)	-0.32469	-0.32127	-0.32081
W2' (048)	0.07357	0.09004	0.06557
W2' (048)	0.75074	0.76616	0.77782
W3 (048)	0.17105	0.16320	0.16191

FERMI LEVEL QUANTITIES (Non-orthogonal fit)

ENERGY		DENSITIES OF STATES					
	Total	s- S	p- S	t2g- S	eg- S	s-H	p-H
Ry		states/Ry/cell					
0.5122	12.10	0.21	10.71	0.00	0.00	0.97	0.21
		INTEGRATED DENSITIES OF STATES					
	Total	s- S	p- S	t2g- S	eg- S	s-H	p-H
		electrons					
	7.00	2.57	3.65	0.00	0.00	0.72	0.07

VELOCITY	PLASMON ENERGY	EIGENVALUE SUM
cm/s	eV	Ry
0.93x10E8	10.34	-3.2849

Table 12.10 SH (NaCl) a = 7.97 Bohr Slater–Koster 2-center parameters

	ORTHOGONAL	NON-ORTHOGONAL	
	ENERGY INTEGRALS Ry	ENERGY INTEGRALS Ry	OVERLAP INTEGRALS
ON SITE S- S			
s	-0.34696	-0.46113	
p	0.35259	0.22640	
FIRST NEIGHBOR S- S			
(sss)	0.00390	-0.05445	0.05339
(pps)	0.08508	0.04642	-0.19196
(ppp)	-0.00767	-0.01837	0.03244
(sps)	0.04747	0.06875	-0.09507
SECOND NEIGHBOR S- S			
(sss)	0.01552	-0.00105	-0.01105
(pps)	-0.01319	0.02097	-0.00731
(ppp)	0.00770	-0.00393	-0.00462
(sps)	0.02794	0.01377	0.00710
ON SITE H-H			
s	0.46848	-0.14050	
FIRST NEIGHBOR H-H			
(sss)	-0.00195	-0.05863	0.04246
SECOND NEIGHBOR H-H			
(sss)	-0.04315	-0.00144	0.00200
FIRST NEIGHBOR S-H			
(sss)	0.11421	-0.15425	0.23109
(pss)	-0.09532	-0.07280	0.35250
SECOND NEIGHBOR S-H			
(sss)	0.00021	-0.01715	0.00583
(pss)	0.02806	-0.03162	0.01669

	ORTHOGONAL			NON-ORTHOGONAL		
BAND	RMS ERROR	MAXIMUM DEVIATION		RMS ERROR	MAXIMUM DEVIATION	
	mRy	k	mRy	mRy	k	mRy
1	8.3	(062)	17.9	0.6	(055)	1.7
2	12.8	(004)	30.6	1.9	(444)	5.1
3	14.8	(444)	32.2	1.4	(062)	3.1
4	13.4	(444)	32.2	1.8	(064)	4.2
5	22.6	(005)	50.1	2.1	(444)	5.3
1-5	15.1			1.6		

ENERGY VALUES IN Ry AT SELECTED k-POINTS

	ORTHOGONAL	APW	NON-ORTHOGONAL
GAMMA 1	-0.72495	-0.71512	-0.71565
GAMMA 1	0.70409	0.71138	0.71074
GAMMA 15	0.63600	0.63539	0.63530
X1 (008)	-0.35872	-0.34988	-0.34913
X1 (008)	0.30663	0.33463	0.33446
X4' (008)	0.01672	0.03856	0.03857
X5' (008)	0.38769	0.38956	0.38956
L1 (444)	-0.44006	-0.42777	-0.42656
L2' (444)	-0.14744	-0.14719	-0.15233
L2' (444)	0.85201	0.88518	0.87992
L3' (444)	0.53365	0.56583	0.56438
W1 (048)	-0.33153	-0.32127	-0.32109
W2' (048)	0.09397	0.09004	0.08773
W2' (048)	0.73647	0.76616	0.76623
W3 (048)	0.17139	0.16320	0.16270

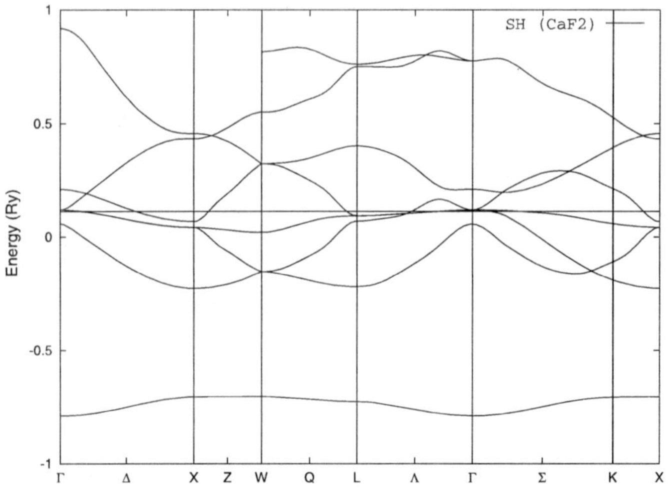

Fig. 12.10 Energy bands of SH in the CaF2 structure

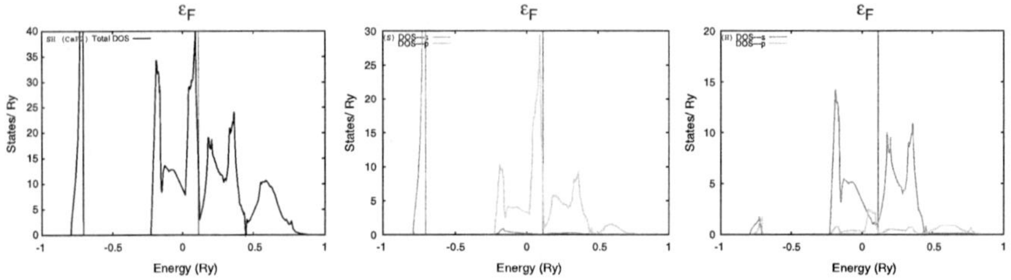

Fig. 12.11 Total, angular momentum and site decomposed densities of states of SH in the CaF2 structure

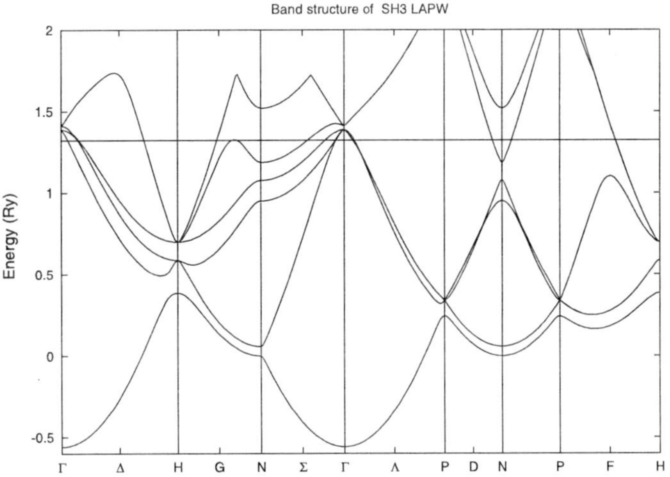

Fig. 12.12 Energy bands of SH3 in the Im3m structure

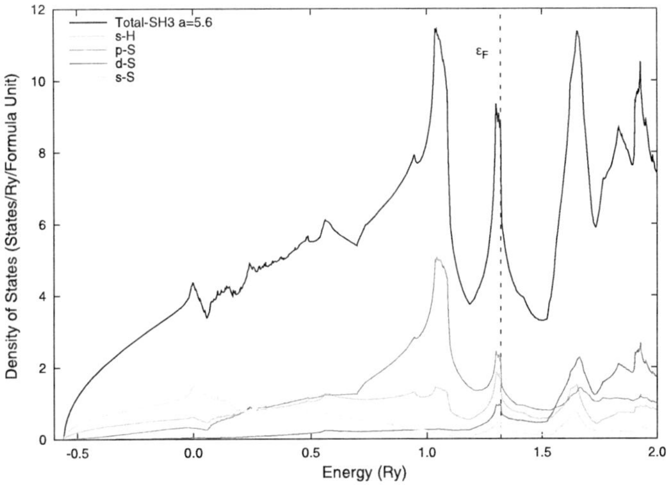

Fig. 12.13 Total, angular momentum and site decomposed densities of states of SH3 in the Im3m structure

12.3 Selenium Hydride (SeH)

See Fig. 12.14, Tables 12.11, 12.12, and 12.13.

See Figs. 12.15, 12.16, Tables 12.14, and 12.15.

See Fig. 12.17.

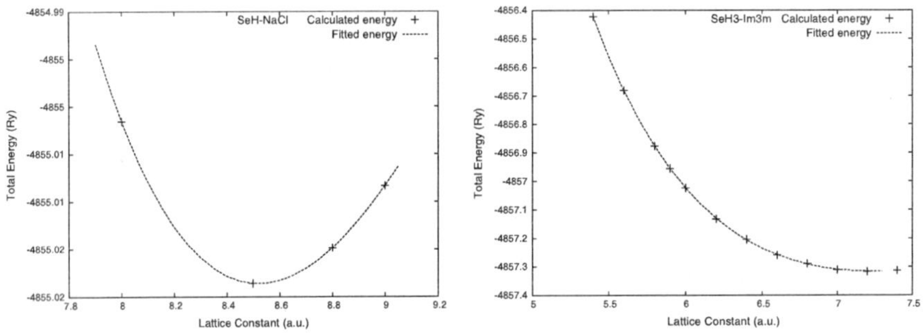

Fig. 12.14 Total energy versus lattice constant of SeH in the NaCl, and Im3m structures

Table 12.11 Lattice constant, bulk modulus, gap, total energy

Stru	a (Bohr)	B (MBar)	Gap	Total Energy (Ry)
NaCl	8.52	0.81		-4855.02363
Im3m (P=0)	7.24	0.98	-	-4857.31700
Im3m(P=2.05 MBar)	5.90	7.43	-	-4856.95527

Table 12.12 Birch fit coefficients

	A1	A2	A3	A4
NaCl	-4.854070E+03	-5.500394E+01	7.944706E+02	-4.703702E+01
Im3m	-4.855871E+03	-9.622142E+01	1.625021E+03	-7.949979E+02

Table 12.13 DOS at Ef, Hopfield parameter, Stoner criterion

SeH a=8.52 Bohr NaCl

Fermi Energy (Ry)		Total DOS (States/Ry)	Decomposed DOS inside the muffin tins s	p	eg	t2g	f
Se	0.554	12.434	0.109	7.354	0.097	0.083	0.025
H	0.554	12.434	0.724	0.566	0.209	0.050	0.034

Fermi-Velocity (cm/s): 1.29589 x10E8
Plasmon Energy (eV) : 13.23066
Electron-ion interaction (Hopfield parameter) (eV/A^2) SeH:3.302 H:0.528

Se	MUFFIN-TIN RADIUS and CHARGE =	2.2365	31.8240
H	MUFFIN-TIN RADIUS and CHARGE =	2.0235	1.2702
Se	STONER I =	0.0162	
H	STONER I =	0.0022	
STONER PARAMETER (Ry)	I =	0.0185	
STONER CRITERION	N*I =	0.2306	

SeH3 a=5.9 Bohr Im3m

Fermi Energy (Ry)		Total DOS (States/Ry)	Decomposed DOS inside the muffin tins s	p	d	f
Se	1.387	6.334	0.254	1.122	0.273	0.014
H	1.387	6.334	0.409	0.060	0.003	0.000

Fermi-Velocity (cm/s): 1.58617 x10E8
Plasmon Energy (eV) : 14.17983
Electron-ion interaction (Hopfield parameter) (eV/A^2) Se:3.876 H3:1.685

Se	MUFFIN-TIN RADIUS and CHARGE =	1.70	30.3813
H	MUFFIN-TIN RADIUS and CHARGE =	1.00	0.5579

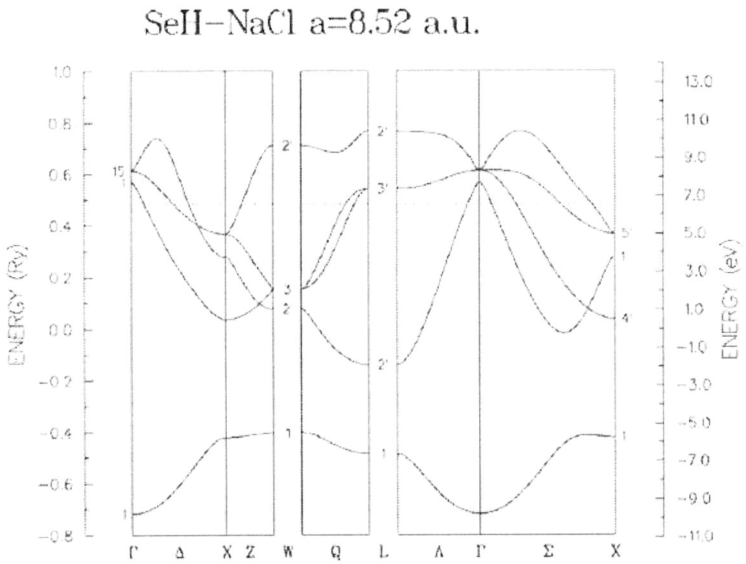

Fig. 12.15 Energy bands of SeH in the NaCl structur (tight-binding)

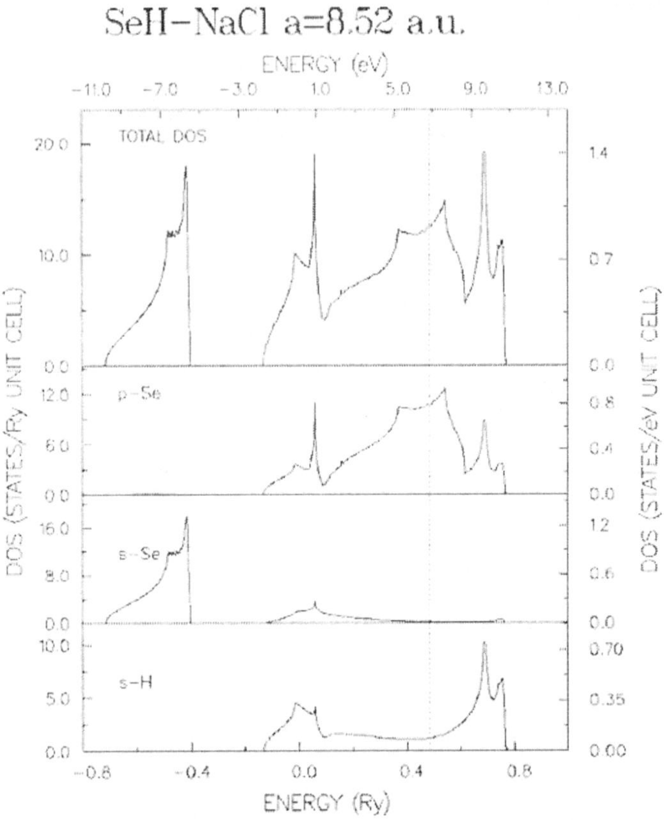

Fig. 12.16 Total, angular momentum and site decomposed densities of states of SeH in the NaCl structure (tight-binding)

Table 12.14 SeH (NaCl) a = 8.52 Bohr Slater–Koster 3-center parameters

	ORTHOGONAL	NON-ORTHOGONAL	
	ENERGY INTEGRALS Ry	ENERGY INTEGRALS Ry	OVERLAP INTEGRALS
ON SITE Se-Se			
s,s(000)	-0.15492	-0.50793	
x,x(000)	0.46028	0.19646	
FIRST NEIGHBOR Se-Se			
s,s(110)	0.03589	-0.04653	0.04613
s,x(110)	0.04616	0.04841	-0.06330
x,x(110)	0.02601	0.01300	-0.07964
x,x(011)	0.00268	-0.01579	0.03438
x,y(110)	0.00866	0.03867	-0.11816
SECOND NEIGHBOR Se-Se			
s,s(200)	0.00305	-0.00123	-0.00526
s,x(200)	0.02372	0.01162	-0.00350
x,x(200)	-0.05343	0.02639	-0.01662
y,y(200)	0.01542	-0.00385	-0.00481
ON SITE e-e			
s,s(000)	0.09792	-0.09318	
FIRST NEIGHBOR H-H			
s,s(110)	-0.03482	-0.05111	0.02188
s,x(110)	-0.01406	0.00000	0.00000
x,x(110)	0.06600	0.00000	0.00000
x,x(011)	0.00504	0.00000	0.00000
x,y(110)	0.03598	0.00000	0.00000
SECOND NEIGHBOR H-H			
s,s(200)	-0.01945	-0.00093	0.00220
s,x(200)	-0.02837	0.00000	0.00000
x,x(200)	0.04690	0.00000	0.00000
y,y(200)	0.00929	0.00000	0.00000
FIRST NEIGHBOR S-H			
s,s(100)	-0.09982	-0.14220	0.20533
s,x(100)	-0.14514	0.00000	0.00000
x,s(100)	-0.12984	-0.07242	0.35419
x,x(100)	-0.04995	0.00000	0.00000
y,y(100)	0.04650	0.00000	0.00000
SECOND NEIGHBOR S-H			
s,s(111)	0.00937	-0.01311	0.00241
x,s(111)	-0.00061	-0.01565	0.00690
s,x(111)	0.00712	0.00000	0.00000
x,x(111)	0.00426	0.00000	0.00000
x,y(111)	0.00724	0.00000	0.00000

BAND	ORTHOGONAL RMS ERROR mRy	MAXIMUM DEVIATION k	MAXIMUM DEVIATION mRy	NON-ORTHOGONAL RMS ERROR mRy	MAXIMUM DEVIATION k	MAXIMUM DEVIATION mRy
1	4.1	(222)	14.6	0.5	(006)	1.0
2	7.4	(033)	16.3	0.7	(008)	1.6
3	8.6	(007)	20.0	1.2	(354)	3.3
4	6.4	(280)	14.8	0.9	(062)	2.3
5	29.9	(444)	128.9	1.2	(264)	2.9
1-5	14.7			0.9		

ENERGY VALUES IN Ry AT SELECTED k-POINTS

	ORTHOGONAL	APW	NON-ORTHOGONAL
GAMMA 1	-0.71008	-0.71723	-0.71730
GAMMA 1	0.56752	0.56841	0.56821
GAMMA 15	0.61608	0.61299	0.61321
X1 (008)	-0.41976	-0.42108	-0.42050
X1 (008)	0.26008	0.27854	0.27837
X4' (008)	0.03063	0.03716	0.03874
X5' (008)	0.36767	0.36714	0.36665
L1 (444)	-0.48451	-0.48600	-0.48524

L2' (444)	-0.13782	-0.13839	-0.13932
L2' (444)	0.78859	0.76775	0.76658
L3' (444)	0.54012	0.54254	0.54113
W1 (048)	-0.40466	-0.40341	-0.40284
W2' (048)	0.07811	0.07713	0.07668
W2' (048)	0.73830	0.70986	0.71116
W3 (048)	0.15996	0.15512	0.15347

FERMI LEVEL QUANTITIES (Non-orthogonal fit)

ENERGY	DENSITIES OF STATES						
	Total	s-Se	p-Se	t2g-Se	eg-Se	s-H	p-H
Ry	states/Ry/cell						
0.4844	12.38	0.21	10.69	0.00	0.00	1.22	0.26
	INTEGRATED DENSITIES OF STATES						
	Total	s-Se	p-Se	t2g-Se	eg-Se	s-H	p-H
	electrons						
	7.00	2.54	3.25	0.00	0.00	1.14	0.06

VELOCITY	PLASMON ENERGY	EIGENVALUE SUM
cm/s	eV	Ry
1.01x10E8	10.25	-3.2973

Table 12.15 SeH (NaCl) a = 8.52 Bohr Slater–Koster 2-center parameters

	ORTHOGONAL	NON-ORTHOGONAL	
	ENERGY INTEGRALS Ry	ENERGY INTEGRALS Ry	OVERLAP INTEGRALS
ON SITE Se-Se			
s	-0.41546	-0.50072	
p	0.33933	0.21810	
FIRST NEIGHBOR Se-Se			
(sss)	0.00548	-0.03984	0.03632
(pps)	0.08021	0.04421	-0.19072
(ppp)	-0.00583	-0.01672	0.03225
(sps)	-0.04194	0.06396	-0.07895
SECOND NEIGHBOR Se-Se			
(sss)	0.01302	0.00023	-0.00598
(pps)	-0.01450	0.01901	-0.00643
(ppp)	0.00696	-0.00350	-0.00421
(sps)	-0.02805	0.00807	0.00342
ON SITE H-H			
s	0.38847	-0.13761	
FIRST NEIGHBOR H-H			
(sss)	-0.00542	-0.05645	0.03730
SECOND NEIGHBOR H-H			
(sss)	-0.03603	-0.00346	0.00077
FIRST NEIGHBOR Se-H			
(sss)	-0.10272	-0.14917	0.20784
(pss)	-0.09473	-0.07062	0.34364
SECOND NEIGHBOR Se-H			
(sss)	-0.00013	-0.01186	0.00384
(pss)	0.02275	-0.02904	0.01443

BAND	ORTHOGONAL RMS ERROR mRy	MAXIMUM DEVIATION k	MAXIMUM DEVIATION mRy	NON-ORTHOGONAL RMS ERROR mRy	MAXIMUM DEVIATION k	MAXIMUM DEVIATION mRy
1	6.7	(005)	13.7	0.6	(444)	1.3
2	12.8	(004)	35.6	1.5	(444)	4.2
3	14.3	(444)	30.0	1.4	(224)	2.9
4	13.0	(064)	31.1	1.5	(064)	3.6
5	19.7	(226)	42.7	1.7	(226)	4.5
1-5	14.0			1.4		

ENERGY VALUES IN Ry AT SELECTED k-POINTS

	ORTHOGONAL	APW	NON-ORTHOGONAL
GAMMA 1	-0.72795	-0.71723	-0.71706
GAMMA 1	0.56354	0.56841	0.56797
GAMMA 15	0.61239	0.61299	0.61284
X1 (008)	-0.42659	-0.42108	-0.42021
X1 (008)	0.26127	0.27854	0.27717
X4' (008)	0.01733	0.03716	0.03768
X5' (008)	0.36149	0.36714	0.36728
L1 (444)	-0.49356	-0.48600	-0.48468
L2' (444)	-0.14694	-0.13839	-0.14256
L2' (444)	0.74794	0.76775	0.76500
L3' (444)	0.51257	0.54254	0.54117
W1 (048)	-0.41133	-0.40341	-0.40451
W2' (048)	0.08146	0.07713	0.07514
W2' (048)	0.67612	0.70986	0.71011
W3 (048)	0.16156	0.15512	0.15423

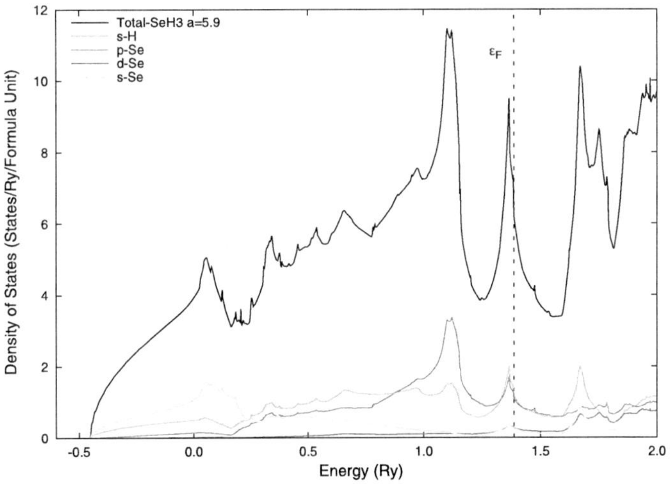

Fig. 12.17 Total, angular momentum and site decomposed densities of states of SeH3 in the Im3m structure

12.4 Tellurium Hydride (TeH)

See Fig. 12.18, Tables 12.16, 12.17, and 12.18.

See Figs. 12.19, 12.20, Tables 12.19, and 12.20.

See Fig. 12.21.

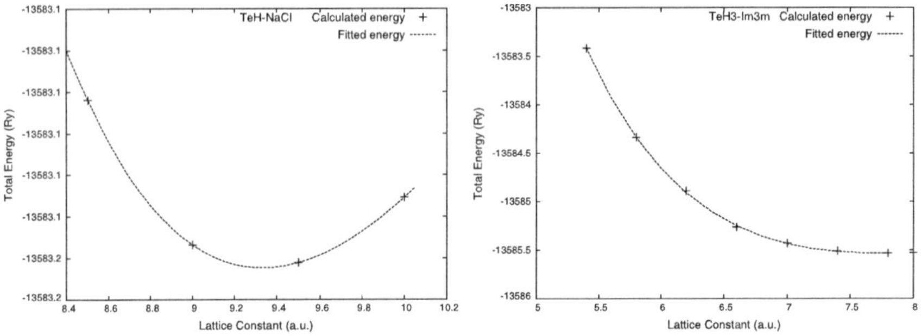

Fig. 12.18 Total energy versus lattice constant of TeH in the NaCl, and Im3m structures

Table 12.16 Lattice constant, bulk modulus, gap, total energy

Stru	a (Bohr)	B (MBar)	Gap	Total Energy (Ry)
NaCl	9.33	0.64		-13583.16234
Im3m (P=0)	7.70	1.29	-	-13585.53361
Im3m(P=2.02 MBar)	6.40	7.33	-	-13585.10315

Table 12.17 Birch fit coefficients

	A1	A2	A3	A4
NaCl	-1.358203E+04	-8.343372E+01	1.699413E+03	-5.944126E+03
Im3m	-1.358301E+04	-1.987952E+02	4.280240E+03	-1.405114E+04

Table 12.18 DOS at Ef, Hopfield parameter, Stoner criterion

TeH a=9.33 Bohr NaCl

Fermi Energy (Ry)		Total DOS (States/Ry)	Decomposed DOS inside the muffin tins s	p	eg	t2g	f
Te	0.482	12.999	0.174	6.144	0.120	0.125	0.056
H	0.482	12.999	1.154	0.774	0.219	0.078	0.037

Fermi-Velocity (cm/s): 1.45299 x10E8
Plasmon Energy (eV) : 13.25783
Electron-ion interaction (Hopfield parameter) (eV/A^2) Te:2.309 H:0.691

Te	MUFFIN-TIN RADIUS and CHARGE =	2.4465	49.3438
H	MUFFIN-TIN RADIUS and CHARGE =	2.2135	1.4722
Te	STONER I =	0.0092	
H	STONER I =	0.0035	
STONER PARAMETER (Ry)	I =	0.0128	
STONER CRITERION	N*I =	0.1664	

TeH3 a=6.4 Bohr Im3m

Fermi Energy (Ry)		Total DOS (States/Ry)	Decomposed DOS inside the muffin tins s	p	d	f
Te	1.435	6.349	0.211	0.625	0.187	0.029
H	1.435	6.349	0.332	0.051	0.002	0.000

Fermi-Velocity (cm/s): 1.67358 x10E8
Plasmon Energy (eV) : 14.19337
Electron-ion interaction (Hopfield parameter) (eV/A^2) Te:1.418 H3:0.592

Te	MUFFIN-TIN RADIUS and CHARGE =	1.70	46.7880
H	MUFFIN-TIN RADIUS and CHARGE =	1.00	0.5538

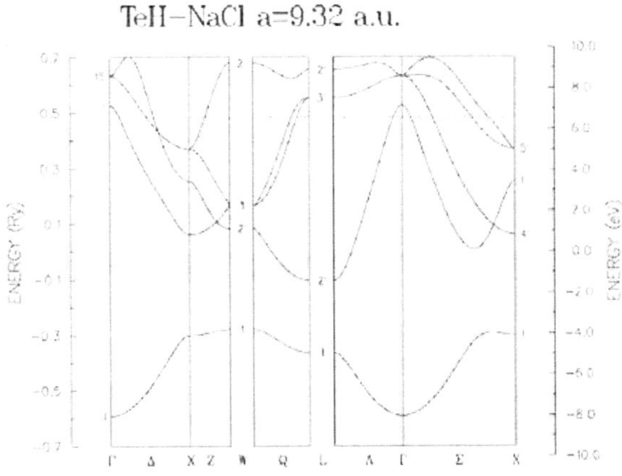

Fig. 12.19 Energy bands of TeH in the NaCl structure (tight-binding)

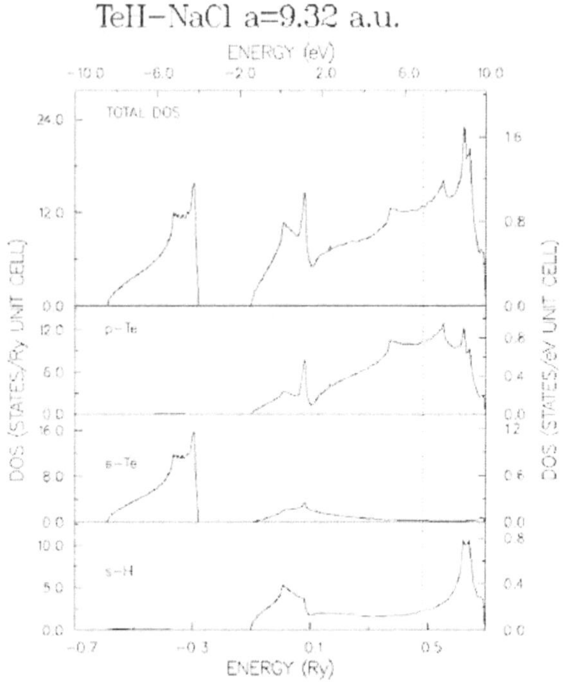

Fig. 12.20 Total, angular momentum and site decomposed densities of states of TeH in the NaCl structure (tight-binding)

Table 12.19 TeH (NaCl) a = 9.32 Bohr Slater–Koster 3-center parameters

	ORTHOGONAL	NON-ORTHOGONAL	
	ENERGY INTEGRALS Ry	ENERGY INTEGRALS Ry	OVERLAP INTEGRALS
ON SITE Te-Te			
s,s(000)	-0.04957	-0.38928	
x,x(000)	0.47014	0.19900	
FIRST NEIGHBOR Te-Te			
s,s(110)	0.02564	-0.04369	0.05112
s,x(110)	0.03739	0.04596	-0.07287
x,x(110)	0.02436	0.00861	-0.08736
x,x(011)	0.00355	-0.01417	0.03802
x,y(110)	0.01379	0.03197	-0.13376
SECOND NEIGHBOR Te-Te			
s,s(200)	0.00192	-0.00382	-0.00066
s,x(200)	0.01995	0.01227	-0.00868
x,x(200)	-0.05140	0.02627	-0.02026
y,y(200)	0.01887	-0.00452	-0.00585

ON SITE H-H			
s,s(000)	0.07134	-0.04630	
FIRST NEIGHBOR H-H			
s,s(110)	-0.02567	-0.03956	0.00838
s,x(110)	-0.00777	0.00000	0.00000
x,x(110)	0.04740	0.00000	0.00000
x,x(011)	-0.00033	0.00000	0.00000
x,y(110)	0.04273	0.00000	0.00000
SECOND NEIGHBOR H-H			
s,s(200)	-0.01562	0.00087	0.00278
s,x(200)	-0.02361	0.00000	0.00000
x,x(200)	0.02909	0.00000	0.00000
y,y(200)	-0.00421	0.00000	0.00000
FIRST NEIGHBOR Te-H			
s,s(100)	-0.08931	-0.11225	0.20467
s,x(100)	-0.12896	0.00000	0.00000
x,s(100)	-0.10510	-0.06102	0.35525
x,x(100)	-0.05188	0.00000	0.00000
y,y(100)	0.04176	0.00000	0.00000
SECOND NEIGHBOR Te-H			
s,s(111)	0.00883	-0.01118	-0.00144
x,s(111)	-0.00258	-0.01432	0.00392
s,x(111)	0.01017	0.00000	0.00000
x,x(111)	0.00414	0.00000	0.00000
x,y(111)	0.00839	0.00000	0.00000

TeH

	ORTHOGONAL			NON-ORTHOGONAL		
BAND	RMS ERROR	MAXIMUM DEVIATION		RMS ERROR	MAXIMUM DEVIATION	
	mRy	k	mRy	mRy	k	mRy
1	3.6	(222)	12.0	0.5	(006)	1.2
2	5.9	(033)	14.9	0.8	(008)	2.0
3	6.6	(118)	15.0	1.5	(354)	3.7
4	5.6	(280)	13.9	1.1	(062)	2.4
5	20.1	(444)	68.6	1.2	(264)	2.7
1-5	10.3			1.1		

ENERGY VALUES IN Ry AT SELECTED k-POINTS

	ORTHOGONAL	APW	NON-ORTHOGONAL
GAMMA 1	-0.58394	-0.58997	-0.59045
GAMMA 1	0.52321	0.52722	0.52688
GAMMA 15	0.63122	0.63058	0.63096
X1 (008)	-0.30275	-0.30046	-0.30018
X1 (008)	0.24246	0.25554	0.25521
X4' (008)	0.05228	0.05840	0.06038
X5' (008)	0.37206	0.37038	0.37010
L1 (444)	-0.36955	-0.36529	-0.36493
L2' (444)	-0.10457	-0.10384	-0.10472
L2' (444)	0.65670	0.65453	0.65373
L3' (444)	0.55262	0.55413	0.55293
W1 (048)	-0.28090	-0.28065	-0.27972
W2' (048)	0.07817	0.08169	0.08148
W2' (048)	0.69884	0.67714	0.67827
W3 (048)	0.17177	0.16785	0.16587
ODD (224)	0.50810	0.50634	0.50785
EVEN(224)	-0.44606	-0.44140	-0.44116
EVEN(224)	0.06518	0.06476	0.06557
EVEN(224)	0.49806	0.49750	0.50054
EVEN(224)	0.60173	0.60279	0.60464

FERMI LEVEL QUANTITIES (Non-orthogonal fit)

ENERGY	DENSITIES OF STATES					
	Total	s-Te	p-Te	t2g-Te	eg-Te	s-H
Ry	states/Ry/cell					
0.4816	12.80	0.28	10.29	0.00	0.00	2.23
	INTEGRATED DENSITIES OF STATES					
	Total	s-Te	p-Te	t2g-Te	eg-Te	s-H
	electrons					
	7.00	2.59	3.06	0.00	0.00	1.36

VELOCITY	PLASMON ENERGY	EIGENVALUE SUM
cm/s	eV	Ry
1.19x10E8	10.78	-2.9819

Table 12.20 TeH (NaCl) a = 9.32 Bohr Slater–Koster 2-center parameters

	ORTHOGONAL	NON-ORTHOGONAL	
	ENERGY INTEGRALS Ry	ENERGY INTEGRALS Ry	OVERLAP INTEGRALS
ON SITE Te-Te			
s	-0.30224	-0.37651	
p	0.35326	0.22535	
FIRST NEIGHBOR Te-Te			
(sss)	0.00345	-0.04054	0.05260
(pps)	0.07800	0.03102	-0.21483
(ppp)	-0.00429	-0.01356	0.03977
(sps)	-0.03589	-0.05963	0.10980
SECOND NEIGHBOR Te-Te			
(sss)	0.01261	-0.00688	0.01297
(pps)	-0.01751	0.01381	-0.01799
(ppp)	0.00785	-0.00514	-0.00754
(sps)	-0.02571	-0.00541	0.02894
ON SITE H-H			
s	0.33069	0.23564	
FIRST NEIGHBOR H-H			
(sss)	-0.00472	0.00912	-0.01762
SECOND NEIGHBOR H-H			
(sss)	-0.02898	0.00803	0.03211
FIRST NEIGHBOR Te-H			
(sss)	-0.08992	0.06576	-0.04733
(pss)	-0.08191	-0.09349	0.15196
SECOND NEIGHBOR Te-H			
(sss)	-0.00010	0.00054	0.01845
(pss)	0.02346	-0.01786	-0.03953

TeH

BAND	ORTHOGONAL RMS ERROR	MAXIMUM DEVIATION		NON-ORTHOGONAL RMS ERROR	MAXIMUM DEVIATION	
	mRy	k	mRy	mRy	k	mRy
1	8.7	(005)	17.6	0.8	(226)	2.0
2	13.0	(004)	34.7	1.4	(062)	2.9
3	14.5	(444)	32.7	2.0	(062)	4.7
4	13.9	(444)	32.7	2.0	(174)	5.0
5	19.5	(004)	42.1	3.3	(224)	7.3
1-5	14.3			2.1		

ENERGY VALUES IN Ry AT SELECTED k-POINTS

	ORTHOGONAL	APW	NON-ORTHOGONAL
GAMMA 1	-0.60129	-0.58997	-0.59013
GAMMA 1	0.51636	0.52722	0.52599
GAMMA 15	0.62735	0.63058	0.63154
X1 (008)	-0.30683	-0.30046	-0.30013
X1 (008)	0.24218	0.25554	0.25513
X4' (008)	0.03766	0.05840	0.06041
X5' (008)	0.36682	0.37038	0.36998
L1 (444)	-0.37792	-0.36529	-0.36355
L2' (444)	-0.10450	-0.10384	-0.10657
L2' (444)	0.63675	0.65453	0.65000
L3' (444)	0.52144	0.55413	0.55202
W1 (048)	-0.29082	-0.28065	-0.27967
W2' (048)	0.08253	0.08169	0.07955
W2' (048)	0.64594	0.67714	0.67510
W3 (048)	0.17083	0.16785	0.16593

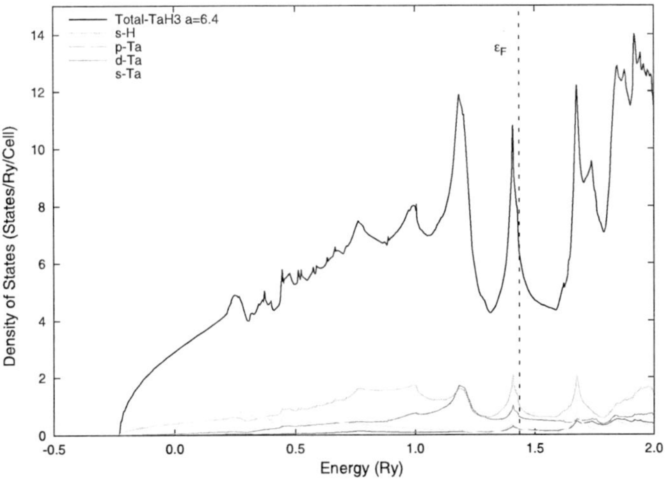

Fig. 12.21 Total, angular momentum and site decomposed densities of states of TeH3 in the Im3m structure

12.5 Polonium Hydride (PoH)

See Fig. 12.22, Tables 12.21, 12.22, and 12.23.

See Figs. 12.23, 12.24, Tables 12.24, and 12.25.

See Fig. 12.25.

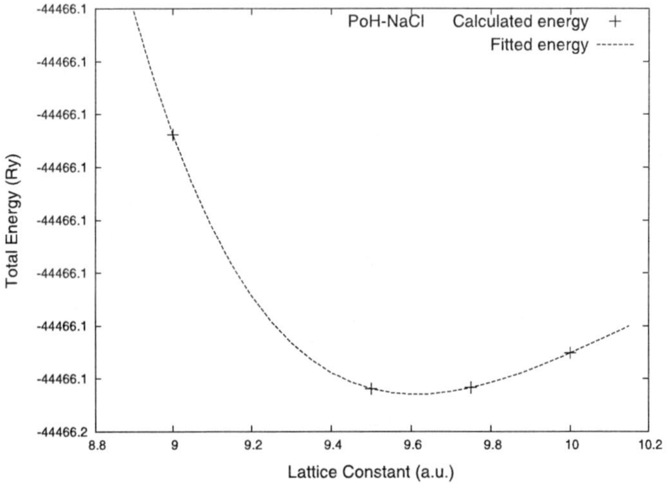

Fig. 12.22 Total energy versus lattice constant of PoH in the NaCl structure

Table 12.21 Lattice constant, bulk modulus, gap, total energy

Stru	a (Bohr)	B (MBar)	Gap	Total Energy (Ry)
NaCl	9.61	0.52		-44466.14649

Table 12.22 Birch fit coefficients

NaCl	-4.446787E+04	2.219098E+02	-9.315304E+03	1.282567E+05

Table 12.23 DOS at Ef, Hopfield parameter, Stoner criterion

PoH a=9.61 Bohr NaCl

Fermi Energy (Ry)	Total DOS (States/Ry)	Decomposed DOS inside the muffin tins s	p	eg	t2g	f
Po 0.472	13.092	0.116	5.616	0.111	0.126	0.078
H 0.472	13.092	1.311	0.882	0.209	0.089	0.038

Fermi-Velocity (cm/s): 1.52335 x10E8
Plasmon Energy (eV) : 13.34331
Electron-ion interaction (Hopfield parameter) (eV/A^2) Po:2.138 H:0.772

Po	MUFFIN-TIN RADIUS and CHARGE =	2.5200	81.2124
H	MUFFIN-TIN RADIUS and CHARGE =	2.2800	1.5344
Po	STONER I =	0.0061	
H	STONER I =	0.0041	
STONER PARAMETER (Ry)	I =	0.0104	
STONER CRITERION	N*I =	0.1355	

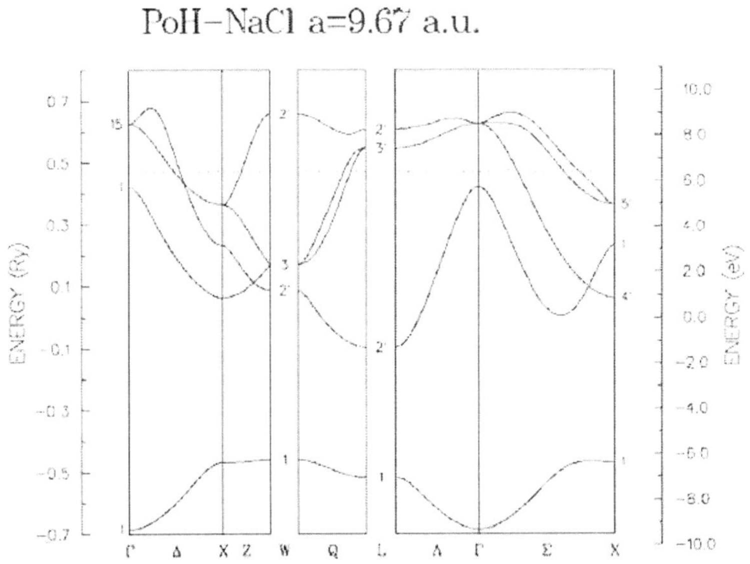

Fig. 12.23 Energy bands of TeH in the NaCl structure (tight-binding)

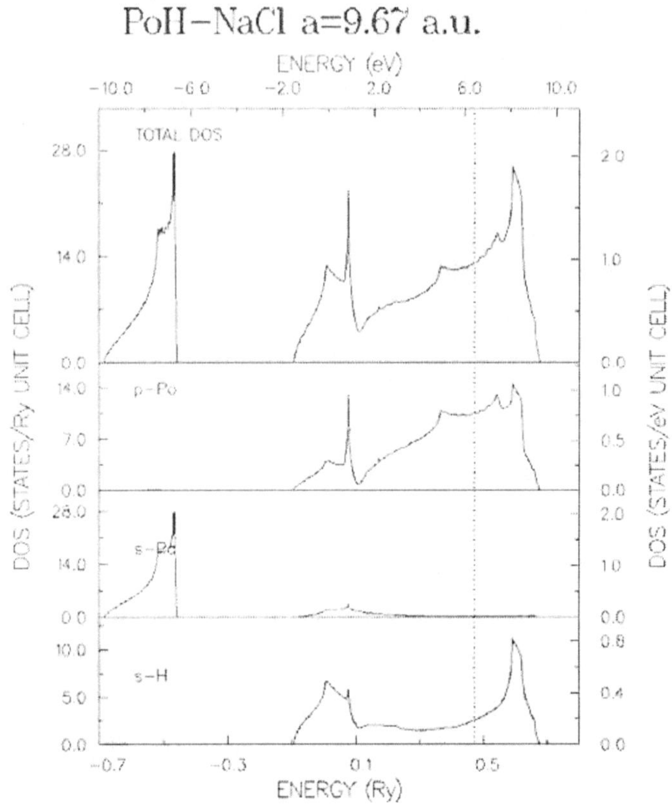

Fig. 12.24 Total, angular momentum and site decomposed densities of states of TeH in the NaCl structure (tight-binding)

Table 12.24 PoH (NaCl) a = 9.67 Bohr Slater–Koster 3-center parameters

	ORTHOGONAL	NON-ORTHOGONAL	
	ENERGY INTEGRALS Ry	ENERGY INTEGRALS Ry	OVERLAP INTEGRALS
ON SITE Po-Po			
s,s(000)	-0.23769	-0.52764	
x,x(000)	0.46549	0.20881	
FIRST NEIGHBOR Po-Po			
s,s(110)	0.02405	-0.03447	0.03597
s,x(110)	0.03862	0.04538	-0.05507
x,x(110)	0.02250	0.00896	-0.08429
x,x(011)	0.00451	-0.01323	0.03552
x,y(110)	0.01410	0.02975	-0.12809
SECOND NEIGHBOR Po-Po			
s,s(200)	-0.00453	0.00030	-0.00413
s,x(200)	0.01368	0.00965	-0.00308
x,x(200)	-0.04656	0.02454	-0.01510
y,y(200)	0.01677	-0.00448	-0.00607
ON SITE H-H			
s,s(000)	0.07535	-0.03274	
FIRST NEIGHBOR H-H			
s,s(110)	-0.02479	-0.03837	0.00515
s,x(110)	-0.00952	0.00000	0.00000
x,x(110)	0.04484	0.00000	0.00000
x,x(011)	-0.00273	0.00000	0.00000
x,y(110)	0.04287	0.00000	0.00000
SECOND NEIGHBOR H-H			
s,s(200)	-0.01027	0.00037	0.00301
s,x(200)	-0.01235	0.00000	0.00000
x,x(200)	0.03188	0.00000	0.00000
y,y(200)	-0.00051	0.00000	0.00000
FIRST NEIGHBOR Po-H			
s,s(100)	-0.09707	-0.11724	0.17483
s,x(100)	-0.13886	0.00000	0.00000
x,s(100)	-0.09581	-0.05790	0.33918
x,x(100)	-0.05646	0.00000	0.00000
y,y(100)	0.03911	0.00000	0.00000
SECOND NEIGHBOR Po-H			
s,s(111)	0.00693	-0.00896	-0.00382
x,s(111)	-0.00008	-0.01373	0.00166
s,x(111)	0.00899	0.00000	0.00000
x,x(111)	0.00331	0.00000	0.00000
x,y(111)	0.00698	0.00000	0.00000

PoH

BAND	ORTHOGONAL RMS ERROR mRy	MAXIMUM DEVIATION k	MAXIMUM DEVIATION mRy	NON-ORTHOGONAL RMS ERROR mRy	MAXIMUM DEVIATION k	MAXIMUM DEVIATION mRy
1	2.2	(222)	7.5	0.7	(062)	1.4
2	5.4	(066)	11.3	0.9	(008)	2.3
3	6.1	(007)	13.5	1.8	(354)	5.4
4	4.8	(280)	11.2	1.1	(064)	2.7
5	15.3	(444)	60.2	1.1	(022)	2.7
1-5	8.1			1.2		

ENERGY VALUES IN Ry AT SELECTED k-POINTS

	ORTHOGONAL	APW	NON-ORTHOGONAL
GAMMA 1	-0.67894	-0.68373	-0.68500
GAMMA 1	0.41886	0.42292	0.42272
GAMMA 15	0.62571	0.62532	0.62598
X1 (008)	-0.46827	-0.46786	-0.46714
X1 (008)	0.22009	0.23233	0.23204

X4' (008)	0.05733	0.06322	0.06548
X5' (008)	0.36450	0.36434	0.36453
L1 (444)	-0.51974	-0.51746	-0.51661
L2' (444)	-0.09536	-0.09374	-0.09481
L2' (444)	0.61106	0.60809	0.60805
L3' (444)	0.54791	0.54793	0.54728
W1 (048)	-0.46060	-0.46002	-0.45899
W2' (048)	0.08428	0.08819	0.08788
W2' (048)	0.67737	0.65682	0.65753
W3 (048)	0.17556	0.17190	0.16990
ODD (224)	0.50317	0.50157	0.50258
EVEN(224)	-0.56621	-0.56386	-0.56409
EVEN(224)	0.03627	0.03690	0.03801
EVEN(224)	0.48579	0.48555	0.49038
EVEN(224)	0.56950	0.57077	0.57155

FERMI LEVEL QUANTITIES (Non-orthogonal fit)

ENERGY	DENSITIES OF STATES					
	Total	s- P	p- P	t2g- P	eg- P	s-o
Ry	states/Ry/cell					
0.4716	13.16	0.16	10.48	0.00	0.00	2.51
	INTEGRATED DENSITIES OF STATES					
	Total	s- P	p- P	t2g- P	eg- P	s-o
	electrons					
	7.00	2.43	3.10	0.00	0.00	1.46

VELOCITY	PLASMON ENERGY	EIGENVALUE SUM
cm/s	eV	Ry
1.24x10E8	10.73	-3.2466

Table 12.25 PoH (NaCl) a = 9.67 Bohr Slater–Koster 2-center parameters

	ORTHOGONAL	NON-ORTHOGONAL	
	ENERGY INTEGRALS Ry	ENERGY INTEGRALS Ry	OVERLAP INTEGRALS
ON SITE Po-Po			
s	-0.45425	-0.52670	
p	0.35266	0.22791	
FIRST NEIGHBOR Po-Po			
(sss)	0.00909	-0.02836	0.02297
(pps)	0.07558	0.03205	-0.20660
(ppp)	-0.00330	-0.01351	0.03585
(sps)	-0.03545	0.06049	-0.07242
SECOND NEIGHBOR Po-Po			
(sss)	0.01149	-0.00438	0.00535
(pps)	-0.01741	0.01776	-0.00555
(ppp)	0.00685	-0.00455	-0.00632
(sps)	-0.02566	0.00473	-0.00612
ON SITE H-H			
s	0.28866	-0.07264	
FIRST NEIGHBOR H-H			
(sss)	-0.01117	-0.04338	0.02285
SECOND NEIGHBOR H-H			
(sss)	-0.02620	-0.00157	0.00187
FIRST NEIGHBOR Po-H			
(sss)	-0.08946	-0.12350	0.19351
(pss)	-0.08379	-0.05594	0.33026
SECOND NEIGHBOR Po-H			
(sss)	-0.00054	-0.00670	0.00059
(pss)	0.01875	-0.02413	0.01002

BAND	ORTHOGONAL RMS ERROR mRy	MAXIMUM DEVIATION k	MAXIMUM DEVIATION mRy	NON-ORTHOGONAL RMS ERROR mRy	MAXIMUM DEVIATION k	MAXIMUM DEVIATION mRy
1	6.5	(000)	13.7	0.4	(004)	0.9
2	15.4	(005)	41.9	1.3	(226)	3.7
3	15.0	(264)	32.1	2.0	(354)	5.2
4	13.8	(064)	32.6	1.4	(064)	4.2
5	19.1	(066)	40.8	1.6	(226)	4.4
1-5	14.6			1.4		

ENERGY VALUES IN Ry AT SELECTED k-POINTS

	ORTHOGONAL	APW	NON-ORTHOGONAL
GAMMA 1	-0.69747	-0.68373	-0.68309
GAMMA 1	0.41873	0.42292	0.42273
GAMMA 15	0.62117	0.62532	0.62589
X1 (008)	-0.46895	-0.46786	-0.46791
X1 (008)	0.22343	0.23233	0.23076
X4' (008)	0.04294	0.06322	0.06538
X5' (008)	0.35844	0.36434	0.36485
L1 (444)	-0.52320	-0.51746	-0.51703
L2' (444)	-0.10764	-0.09374	-0.09608
L2' (444)	0.59804	0.60809	0.60682
L3' (444)	0.51782	0.54793	0.54707
W1 (048)	-0.46763	-0.46002	-0.45931
W2' (048)	0.08990	0.08819	0.08674
W2' (048)	0.61909	0.65682	0.65707
W3 (048)	0.17329	0.17190	0.17013

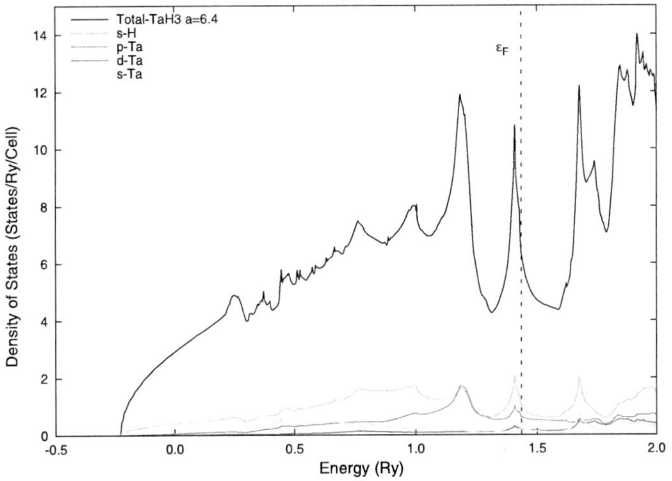

Fig. 12.25 Total, angular momentum and site decomposed densities of states of TeH3 in the Im3m structure

References

1. Wikipedia
2. D.A. Papaconstantopoulos, B.M. Klein, M.J. Mehl, W.E. Pickett, Cubic H3S around 200 Gpa: An atomic hydrogen superconductor stabilized by sulfur. Phys. Rev. B **91**,184511 (2015)
3. P.-H. Chang, S. Silayi, D.A. Papaconstantopoulos, and M.J. Mehl, Pressure-induced high-temperature superconductivity in hypothetical H3X (X= As, Se, Br, Sb, Te and I) in the H3S structure with Im3m symmetry, J. Phys Chem Solids **139**, 109315, (2020)

Chapter 13
Group 17 Hydrogen Halides

This chapter covers the Group 17 column of the periodic table hydrides from FH to AtH [1]. They are known as hydrogen halides and they are all gases at standard temperature and pressure. Results are presented for the crystal structures NaCl(B1), CsCl (B2), CaF2 (C1) and Im3m. These structures are not found experimentally and are presented here for the purpose of comparison through the periodic Table and to establish various trends. However, under high pressure at least FH3 and ClH3 have been investigated as possible high temperature superconductors [2, 3] at high pressure. The Hopfield-McMillan parameter for the F site is very large similarly to OH. Following the example of SH which exists under pressure in the Im3m cubic structure we present here results for the Im3m structure exploring the possibility that some of them may form and become superconductors. Examining the energy bands and densities of states figures we observe that, as in the group 14 column, the lowest band has a mixture of s–H hydrogen and s-character of the other element. At the Fermi level the states are of s- and p- character. As in the other columns of this part of the Periodic Table, the band structure is dominated by s- and p-states. The d-states are core states found significantly lower that the Gam1 point. Tight-binding (TB) parameters are given in the NaCl structure based on both orthogonal and non-orthogonal Hamiltonians using three- and two-center integrals. It could be possible to modify these TB parameters for the purpose of building a TB model to describe the trihydride molecules.

D. A. Papaconstantopoulos, *Band Structure of Cubic Hydrides*,
https://doi.org/10.1007/978-3-031-06878-2_13

13.1 Fluorine Hydride (FlH)

See Fig. 13.1 and Tables 13.1, 13.2 and 13.3.

See Figs. 13.2 and 13.3 and Tables 13.4 and 13.5.

See Figs. 13.4 and 13.5.

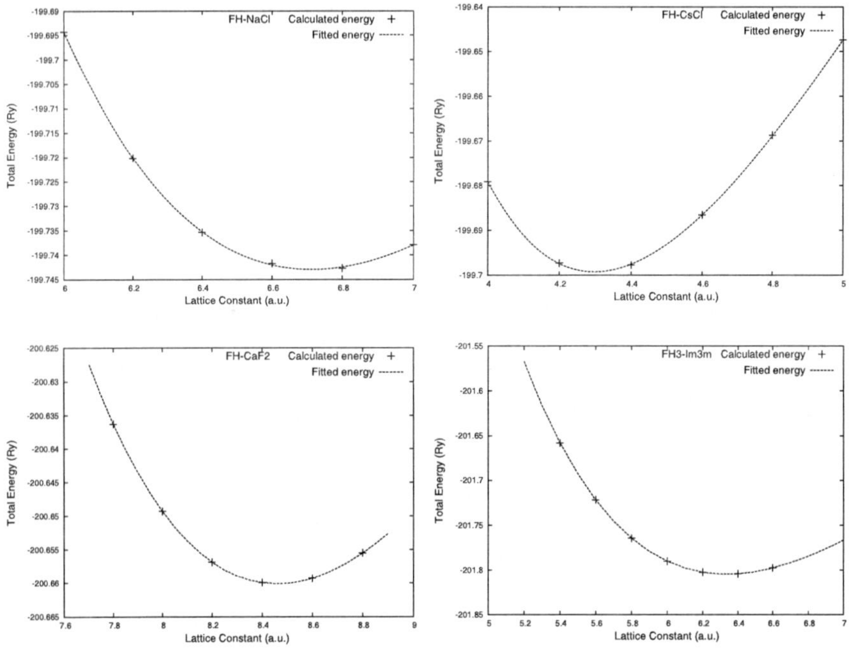

Fig. 13.1 Total energy versus lattice constant of FH in the NaCl, CsCl, CaF2 and Im3m structures

Table 13.1 Lattice constant, bulk modulus, gap, total energy

Stru	a (Bohr)	B (MBar)	Gap(Ry)	Total Energy (Ry)
NaCl	6.71	1.33	0.19	-199.74296
CsCl	4.30	1.34	-	-199.69922
CaF2	8.46	0.68	-	-200.66010
Im3m (P=0)	6.33	1.16	-	-201.80454
Im3m(P=1.29 MBar)	5.40	4.89	-	-201.65770

Table 13.2 Birch fit coefficients

	A1	A2	A3	A4
NaCl	-1.989529E+02	-2.857699E+01	2.641335E+02	-1.027122E+02
CsCl	-1.990890E+02	-1.879531E+01	6.976646E+01	1.278477E+03
CaF2	-1.997320E+02	-5.667427E+01	9.731226E+02	-3.185614E+03
Im3m	-2.004687E+02	-7.301543E+01	1.130699E+03	-3.502019E+03

Table 13.3 DOS at Ef, Hopfield parameter, Stoner criterion

```
        FH    a=6.72 Bohr    insulator    NaCl
 ---------------------------------------------------------------------------
 Fermi Energy    Total DOS
    ( Ry )     (States/Ry)
 ---------------------------------------------------------------------------
 F  0.210        0.0
 H  0.210        0.0
 ---------------------------------------------------------------------------
   F     MUFFIN-TIN RADIUS and CHARGE =    1.6782    8.3604
   H     MUFFIN-TIN RADIUS and CHARGE =    1.6782    0.6194
 ------------------------------------------------
        FH       a=4.30 Bohr             CsCl
 ---------------------------------------------------------------------------
 Fermi Energy    Total DOS
    ( Ry )     (States/Ry)     s         p         eg         t2g        f
 ---------------------------------------------------------------------------
 F  0.285         1.845      0.015     1.704     0.000      0.000      0.000
 H  0.285         1.845      0.061     0.000     0.000      0.001      0.013
 ---------------------------------------------------------------------------
Fermi-Velocity (cm/s): 0.32768 x10E8
Plasmon Energy (eV)   : 0.97436
Electron-ion interaction (Hopfield parameter (eV/A^2) ): F 0.086  H 0.000
 ---------------------------------------------------------------------------
   F     MUFFIN-TIN RADIUS and CHARGE =    1.8610   8.7123
   H     MUFFIN-TIN RADIUS and CHARGE =    1.8610   0.6289
                    F       STONER I=  0.0598
                    H       STONER I=  0.0004
             STONER PARAMETER (Ry) I = 0.0602
             STONER CRITERION    N*I = 0.1111
 ---------------------------------------------------------------------------
        FH2      a=8.46 Bohr            CaF2
 ---------------------------------------------------------------------------
 Fermi Energy    Total DOS
    ( Ry )     (States/Ry)     s         p         eg         t2g        f
 ---------------------------------------------------------------------------
 F  0.251         8.453      0.220     2.315     0.001      0.056      0.002
 H  0.251         8.453      4.487     0.135     0.004      0.010      0.016
 ---------------------------------------------------------------------------
Fermi-Velocity (cm/s): 0.59859 x10E8
Plasmon Energy (eV)   : 5.09250
Electron-ion interaction (Hopfield parameter (eV/A^2) ): F 1.469  2H 2.542
 ---------------------------------------------------------------------------
   F     MUFFIN-TIN RADIUS and CHARGE =    1.8317   8.5940
   H     MUFFIN-TIN RADIUS and CHARGE =    1.8317   0.6805
                          F       STONER I =  0.0072
                          H       STONER I =  0.0216
                   STONER PARAMETER (Ry) I =  0.0505
                   STONER CRITERION    N*I =  0.4269
 ---------------------------------------------------------------------------
        FH3  a=5.4 Bohr                Im3m
 ---------------------------------------------------------------------------
 Fermi Energy    Total DOS        Decomposed DOS inside the muffin tins
    ( Ry )     (States/Ry)       s        p        d        f
 F  1.030         6.540        0.177    2.009    0.101    0.005
 H  1.030         6.540        0.639    0.025    0.004    0.000
 ---------------------------------------------------------------------------
Fermi-Velocity (cm/s): 0.93767 x10E8
Plasmon Energy (eV)   : 9.73003
Electron-ion interaction (Hopfield parameter) (eV/A^2) F:17.486  H3:3.226
 ------------------------------------------------
   F     MUFFIN-TIN RADIUS and CHARGE =    1.50    8.0265
   H     MUFFIN-TIN RADIUS and CHARGE =    1.00    0.4163
```

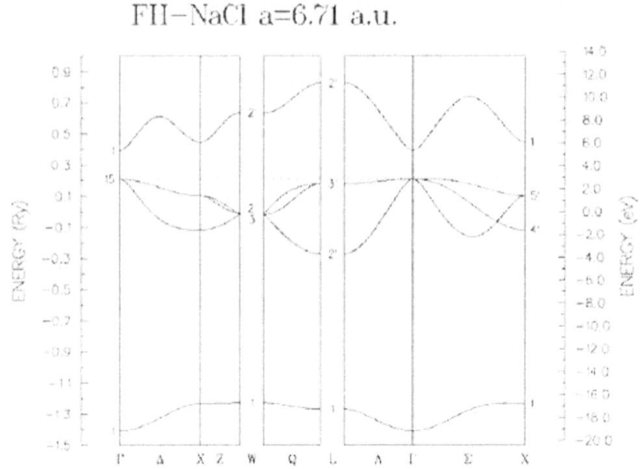

Fig. 13.2 Energy bands of FH in the NaCl structure (tight-binding)

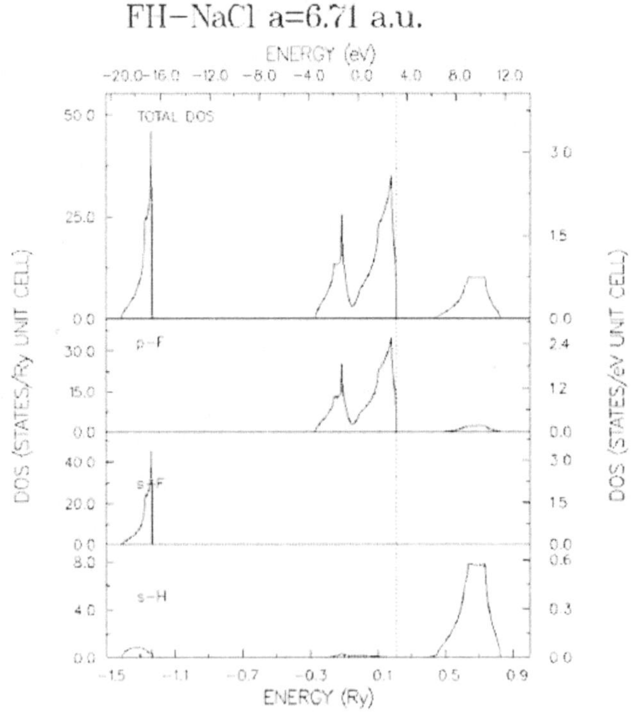

Fig. 13.3 Total, angular momentum and site decomposed densities of states of FH in the NaCl structure (tight-binding)

Table 13.4 FH (NaCl) a = 6.71 Bohr Slater-Koster 3-Center parameters

	ORTHOGONAL	NON-ORTHOGONAL	
	ENERGY INTEGRALS Ry	ENERGY INTEGRALS Ry	OVERLAP INTEGRALS
ON SITE F-F			
s,s(000)	-1.24299	-1.17593	
x,x(000)	0.11268	0.01541	
FIRST NEIGHBOR F-F			
s,s(110)	-0.00188	-0.04131	0.05024
s,x(110)	-0.00235	0.02930	-0.03738
x,x(110)	0.02074	0.01937	-0.01635
x,x(011)	-0.00702	-0.00871	0.00958
x,y(110)	0.00577	0.04541	-0.04916
SECOND NEIGHBOR F-F			
s,s(200)	0.00420	0.00117	0.01112
s,x(200)	-0.00904	0.01119	-0.02703
x,x(200)	-0.02205	0.02175	-0.02590
y,y(200)	0.00099	-0.00010	-0.00046
ON SITE H-H			
s,s(000)	0.40608	0.36864	
FIRST NEIGHBOR H-H			
s,s(110)	-0.01257	-0.01114	0.01111
SECOND NEIGHBOR H-H			
s,s(200)	-0.00621	0.00291	0.00172
FIRST NEIGHBOR F-H			
s,s(100)	0.09009	-0.02653	0.15097
x,s(100)	-0.15005	-0.05570	0.23364
SECOND NEIGHBOR F-H			
s,s(111)	-0.00242	-0.01524	0.01218
x,s(111)	-0.00059	-0.00383	0.00585

	ORTHOGONAL			NON-ORTHOGONAL		
BAND	RMS ERROR mRy	MAXIMUM DEVIATION k	MAXIMUM DEVIATION mRy	RMS ERROR mRy	MAXIMUM DEVIATION k	MAXIMUM DEVIATION mRy
1	1.5	(222)	4.1	0.3	(004)	0.6
2	1.9	(008)	4.0	0.5	(062)	1.5
3	3.0	(226)	8.7	0.5	(008)	2.4
4	2.3	(008)	4.6	0.6	(008)	2.4
5	4.4	(222)	13.0	0.9	(333)	3.2
1-5	2.8			0.6		

ENERGY VALUES IN Ry AT SELECTED k-POINTS

	ORTHOGONAL	APW	NON-ORTHOGONAL
GAMMA 1	-1.40745	-1.40936	-1.40929
GAMMA 1	0.38514	0.39126	0.39238
GAMMA 15	0.21035	0.20931	0.20919
X1 (008)	-1.23439	-1.23270	-1.23267
X1 (008)	0.44313	0.44624	0.44623
X4' (008)	-0.12148	-0.11744	-0.11771
X5' (008)	0.10060	0.10517	0.10280
L1 (444)	-1.26817	-1.26805	-1.26753
L2' (444)	-0.27135	-0.27034	-0.27097
L2' (444)	0.82139	0.83075	0.82893
L3' (444)	0.17592	0.18013	0.17963
W1 (048)	-1.22710	-1.22827	-1.22777
W2' (048)	-0.00541	-0.00688	-0.00766
W2' (048)	0.63813	0.63580	0.63520
W3 (048)	-0.01438	-0.01826	-0.01833

FERMI LEVEL QUANTITIES (Non-orthogonal fit)

ENERGY Ry	DENSITIES OF STATES states/Ry/cell Total	s-F	p-F	t2g-F	eg-F	s-H	p-H
0.2100	0.00	0.00	0.00	0.00	0.00	0.00	0.00

	INTEGRATED DENSITIES OF STATES					
Total	s-F	p-F	t2g-F	eg-F	s-H	p-H
	electrons					
8.00	1.90	5.92	0.00	0.00	0.18	0.00

VELOCITY	PLASMON ENERGY	EIGENVALUE SUM
cm/s	eV	Ry
0.00x10E8	0.00	-3.9923

Table 13.5 FH (NaCl) a = 6.71 Bohr Slater-Koster 2-Center parameters

	ORTHOGONAL	NON-ORTHOGONAL	
	ENERGY INTEGRALS Ry	ENERGY INTEGRALS Ry	OVERLAP INTEGRALS
ON SITE F-F			
s	-1.22781	-1.26271	
p	0.06870	0.05639	
FIRST NEIGHBOR F-F			
(sss)	0.00282	-0.01935	0.00930
(pps)	0.04823	0.05014	-0.03017
(ppp)	-0.00666	-0.00875	0.00615
(sps)	0.03254	-0.00731	0.01528
SECOND NEIGHBOR F-F			
(sss)	0.00711	0.01668	-0.01261
(pps)	0.00049	0.00387	0.00019
(ppp)	0.00025	0.00072	0.00417
(sps)	0.00616	0.00483	0.00046
ON SITE H-H			
s	0.52282	0.28148	
FIRST NEIGHBOR H-H			
(sss)	-0.01771	-0.07929	0.03200
SECOND NEIGHBOR H-H			
(sss)	-0.03019	-0.05575	0.01864
FIRST NEIGHBOR F-H			
(sss)	-0.11028	-0.22299	0.12891
(pss)	0.10279	-0.09429	0.01461
SECOND NEIGHBOR F-H			
(sss)	0.00333	-0.00552	0.00213
(pss)	-0.00436	0.00596	0.00430

FH

	ORTHOGONAL			NON-ORTHOGONAL		
BAND	RMS ERROR	MAXIMUM DEVIATION		RMS ERROR	MAXIMUM DEVIATION	
	mRy	k	mRy	mRy	k	mRy
1	2.3	(005)	5.0	0.5	(444)	1.1
2	4.9	(444)	11.2	0.6	(044)	1.9
3	4.1	(226)	9.2	0.7	(008)	2.4
4	3.0	(008)	7.9	0.7	(008)	2.4
5	5.1	(004)	12.3	1.0	(333)	2.9
1-5	4.0			0.7		

ENERGY VALUES IN Ry AT SELECTED k-POINTS

	ORTHOGONAL	APW	NON-ORTHOGONAL
GAMMA 1	-1.41286	-1.40936	-1.40935
GAMMA 1	0.39068	0.39126	0.39075
GAMMA 15	0.21032	0.20931	0.20888
X1 (008)	-1.23357	-1.23270	-1.23178
X1 (008)	0.44964	0.44624	0.44642
X4' (008)	-0.12227	-0.11744	-0.11738
X5' (008)	0.09732	0.10517	0.10279
L1 (444)	-1.27044	-1.26805	-1.26690
L2' (444)	-0.28152	-0.27034	-0.26970
L2' (444)	0.83262	0.83075	0.82948
L3' (444)	0.17651	0.18013	0.17969
W1 (048)	-1.22486	-1.22827	-1.22864
W2' (048)	-0.00013	-0.00688	-0.00794
W2' (048)	0.62877	0.63580	0.63548
W3 (048)	-0.01348	-0.01826	-0.01777

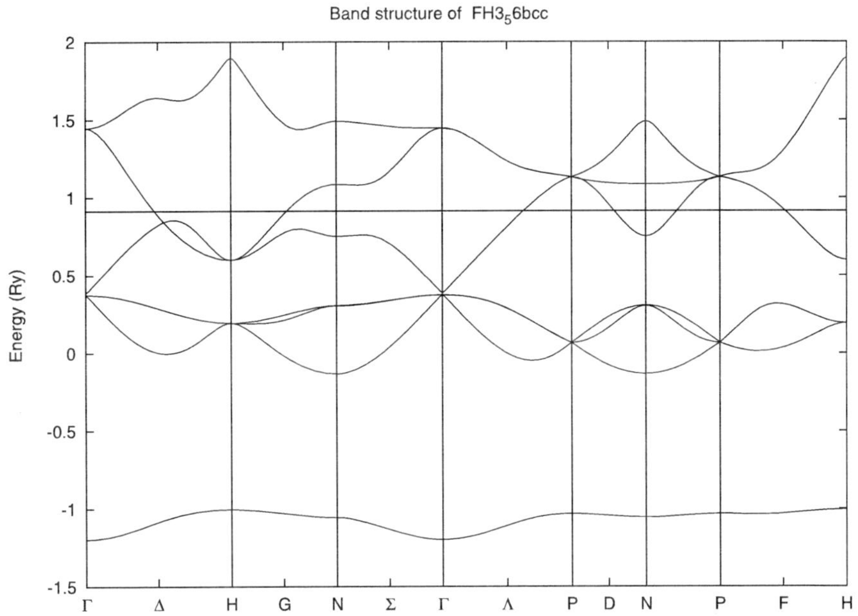

Fig. 13.4 Energy bands of FH3 in the Im3m structure

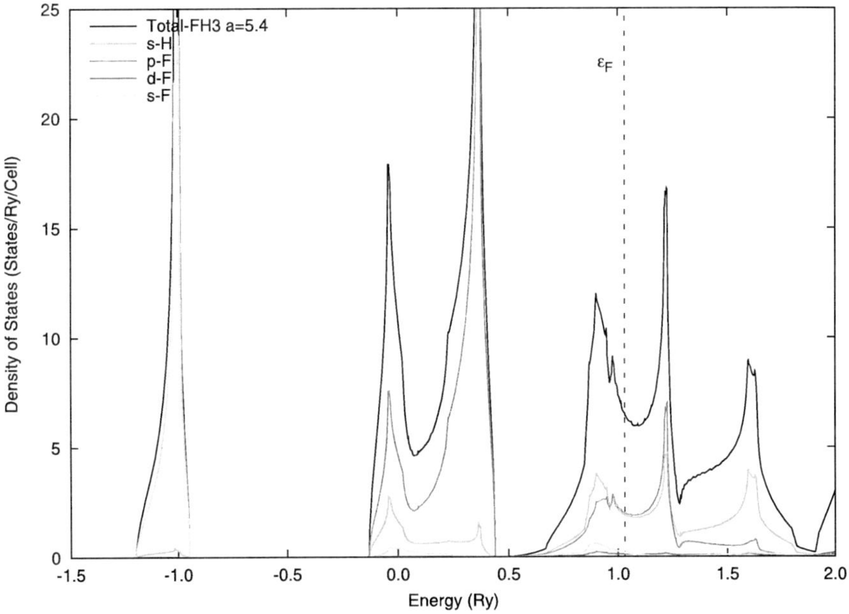

Fig. 13.5 Total, angular momentum and site decomposed densities of states of FH3 in the Im3m structure

13.2 Chlorine Hydride (ClH)

See Fig. 13.6, and Tables 13.6, 13.7 and 13.8.

See Figs. 13.7, Figs. 13.8 and Tables 13.9 and 13.10.

See Figs. 13.9 and 13.10.

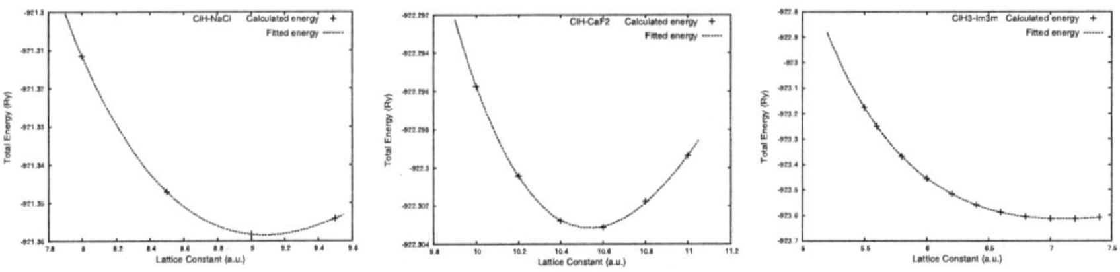

Fig. 13.6 Total energy versus lattice constant of ClH in the NaCl, CaF2 and Im3m structures

Table 13.6 Lattice constant, bulk modulus, gap, total energy

Stru	a (Bohr)	B (MBar)	Gap(Ry)	Total Energy (Ry)
NaCl	9.08	0.40	-	-921.35842
CaF2	10.54	0.26	-	-922.30317
Im3m (P=0)	7.11	0.71	-	-923.61337
Im3m(P=2.13 MBar)	5.60	7.25	-	-923.24946

Table 13.7 Birch fit coefficients

	A1	A2	A3	A4
NaCl	-9.207570E+02	-4.021882E+01	7.010542E+02	-9.474313E+02
CaF2	-9.219963E+02	-1.481238E+01	-4.815351E+02	2.372403E+04
Im3m	-9.226217E+02	-6.358253E+01	1.032147E+03	-4.160674E+02

Table 13.8 DOS at Ef, Hopfield parameter, Stoner criterion

```
        ClH      a=9.08 Bohr         NaCl
 ---------------------------------------------------------------------------
 Fermi Energy    Total DOS
     ( Ry )     (States/Ry)     s         p         eg         t2g        f
 ---------------------------------------------------------------------------
 Cl  0.141        11.500      0.012     9.260     0.028      0.002      0.002
  H  0.141        11.500      0.473     0.216     0.058      0.003      0.100
 ---------------------------------------------------------------------------
Fermi-Velocity (cm/s): 0.21425 x10E8
Plasmon Energy (eV)   : 1.90038
Electron-ion interaction (Hopfield parameter (eV/A^2) ): Cl 0.779  H 0.106
 ---------------------------------------------------------------------------

Cl      MUFFIN-TIN RADIUS and CHARGE =     2.2700   15.6747
 H      MUFFIN-TIN RADIUS and CHARGE =     2.2700    0.9675
                         Cl       STONER I =  0.0284
                          H       STONER I =  0.0008
                    STONER PARAMETER (Ry)  I =  0.0292
                    STONER CRITERION     N*I =  0.3358
 ---------------------------------------------------------------------------
       ClH2      a=10.54 Bohr          CaF2
 ---------------------------------------------------------------------------
 Fermi Energy    Total DOS
     ( Ry )     (States/Ry)     s         p         eg         t2g        f
 ---------------------------------------------------------------------------
Cl  0.096        25.221      0.299     4.753     0.004      0.685      0.038
 H  0.096        25.221     16.216     0.258     0.032      0.060      0.033
 ---------------------------------------------------------------------------
Fermi-Velocity (cm/s): 0.48909 x10E8
Plasmon Energy (eV)   : 5.17153
Electron-ion interaction (Hopfield parameter (eV/A^2) ): Cl 3.652  2H 2.829
 ---------------------------------------------------------------------------
  Cl      MUFFIN-TIN RADIUS and CHARGE =     2.3952  15.9185
   H      MUFFIN-TIN RADIUS and CHARGE =     2.1670   0.8865
                         Cl       STONER I =   0.0021
                         H        STONER I =   0.0234
                   STONER PARAMETER (Ry) I =   0.0488
                   STONER CRITERION    N*I =   1.2307
 ---------------------------------------------------------------------------
       ClH3  a=5.6 Bohr              Im3m
 ---------------------------------------------------------------------------
 Fermi Energy    Total DOS        Decomposed DOS inside the muffin tins
     ( Ry )     (States/Ry)       s       p        d       f
 Cl  1.421        3.941         0.134   0.819    0.614   0.019
  H  1.421        3.941         0.267   0.043    0.003   0.000
 ---------------------------------------------------------------------------
Fermi-Velocity (cm/s): 1.72363 x10E8
Plasmon Energy (eV)   : 13.14627
Electron-ion interaction (Hopfield parameter) (eV/A^2) Cl:5.766  H3:3.307
 -----------------------------------------------
  Cl      MUFFIN-TIN RADIUS and CHARGE =    1.80  14.7592
   H      MUFFIN-TIN RADIUS and CHARGE =    1.00   0.5594
```

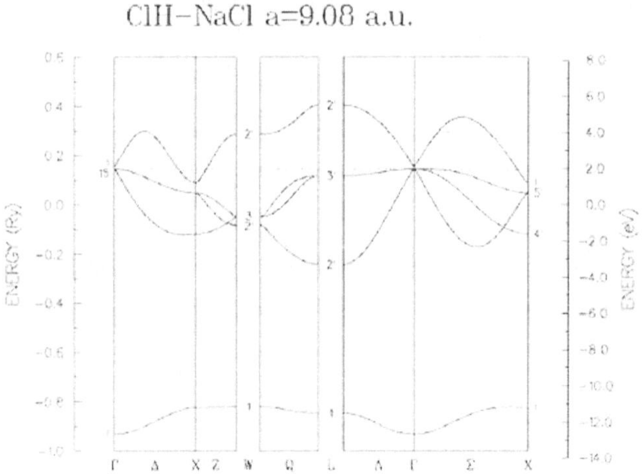

Fig. 13.7 Energy bands of ClH in the NaCl structure (tight-binding)

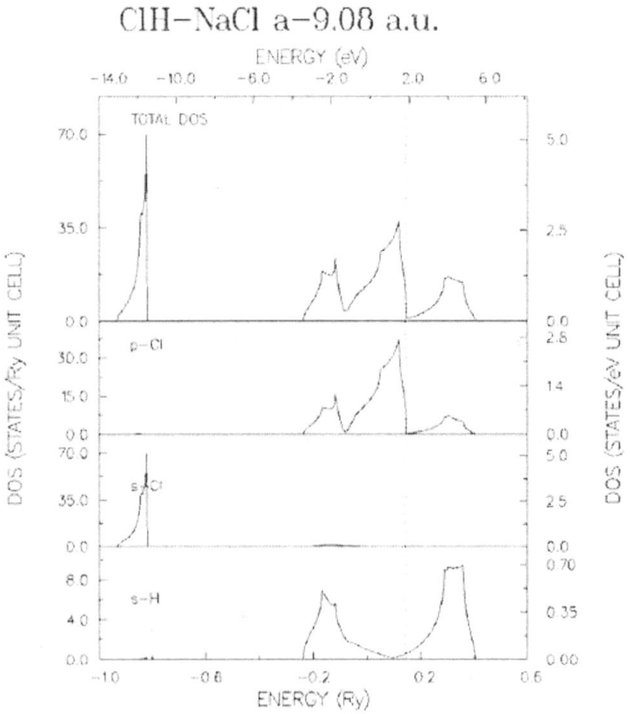

Fig. 13.8 Total, angular momentum and site decomposed densities of states of ClH in the NaCl structure (tight-binding)

Table 13.9 ClH (NaCl) a = 9.08 Bohr Slater-Koster 3-center parameters

	ORTHOGONAL	NON-ORTHOGONAL	
	ENERGY INTEGRALS Ry	ENERGY INTEGRALS Ry	OVERLAP INTEGRALS
ON SITE Cl-Cl			
s,s(000)	-0.80635	-0.83644	
x,x(000)	0.08005	-0.01829	
FIRST NEIGHBOR Cl-Cl			
s,s(110)	0.00649	0.02813	-0.03855
s,x(110)	-0.00332	0.04328	-0.02895
x,x(110)	0.01704	0.01468	-0.03407
x,x(011)	-0.00461	-0.00775	0.01479
x,y(110)	-0.00492	0.03272	-0.10134
SECOND NEIGHBOR Cl-Cl			
s,s(200)	0.00660	0.01319	-0.01384
s,x(200)	-0.00478	0.02814	-0.02664
x,x(200)	-0.02844	0.01413	-0.06547
y,y(200)	0.00134	-0.00050	-0.00162
ON SITE H-H			
s,s(000)	0.00403	-0.00884	
FIRST NEIGHBOR H-H			
s,s(110)	-0.00887	-0.01806	-0.03963
SECOND NEIGHBOR H-H			
s,s(200)	0.00262	0.00560	0.02334
FIRST NEIGHBOR Cl-H			
s,s(100)	-0.07706	-0.05699	0.05107
x,s(100)	-0.08866	-0.09995	0.21995
SECOND NEIGHBOR Cl-H			
s,s(111)	0.00129	0.00258	0.00940
x,s(111)	-0.00027	-0.00620	-0.00861

	ORTHOGONAL			NON-ORTHOGONAL		
BAND	RMS ERROR mRy	MAXIMUM DEVIATION k	mRy	RMS ERROR mRy	MAXIMUM DEVIATION k	mRy
1	0.6	(444)	1.0	0.7	(000)	1.4
2	3.1	(008)	8.3	1.3	(002)	3.1
3	3.3	(226)	9.3	0.8	(008)	2.5
4	2.6	(444)	6.1	0.9	(062)	2.8
5	3.2	(055)	7.6	1.5	(002)	3.7
1-5	2.8			1.1		

ENERGY VALUES IN Ry AT SELECTED k-POINTS

	ORTHOGONAL	APW	NON-ORTHOGONAL
GAMMA 1	-0.93093	-0.93087	-0.92949
GAMMA 1	0.15540	0.15575	0.15601
GAMMA 15	0.14645	0.14553	0.14546
X1 (008)	-0.82347	-0.82250	-0.82217
X1 (008)	0.08597	0.09096	0.08964
X4' (008)	-0.12625	-0.11796	-0.11762
X5' (008)	0.04697	0.05098	0.04848
L1 (444)	-0.84596	-0.84497	-0.84539
L2' (444)	-0.24076	-0.23866	-0.24036
L2' (444)	0.40004	0.40626	0.40619
L3' (444)	0.11187	0.11796	0.11794
W1 (048)	-0.81910	-0.81963	-0.81897
W2' (048)	-0.08181	-0.07960	-0.08117
W2' (048)	0.28727	0.28862	0.28720
W3 (048)	-0.04502	-0.04815	-0.04850

FERMI LEVEL QUANTITIES (Non-orthogonal fit)

ENERGY	DENSITIES OF STATES						
	Total	s-Cl	p-Cl	t2g-Cl	eg-Cl	s-H	p-H
Ry		states/Ry/cell					
0.1423	10.70	0.00	10.16	0.00	0.00	0.54	0.00

	INTEGRATED DENSITIES OF STATES						
	Total	s-Cl	p-Cl	t2g-Cl	eg-Cl	s-H	p-H
		electrons					
	8.00	2.20	4.98	0.00	0.00	0.82	0.00
VELOCITY	PLASMON ENERGY		EIGENVALUE SUM				
cm/s	eV		Ry				
0.24x10E8	2.06		-2.9032				

Table 13.10 ClH (NaCl) a = 9.08 Bohr Slater-Koster 2-center parameters

	ORTHOGONAL	NON-ORTHOGONAL	
	ENERGY INTEGRALS Ry	ENERGY INTEGRALS Ry	OVERLAP INTEGRALS
ON SITE Cl-Cl			
s	-0.80718	-0.84791	
p	0.02831	0.00761	
FIRST NEIGHBOR Cl-Cl			
(sss)	0.00606	-0.02756	0.02472
(pps)	0.03778	0.04114	-0.07279
(ppp)	-0.00411	-0.00780	0.01324
(sps)	0.02686	0.02101	-0.00940
SECOND NEIGHBOR Cl-Cl			
(sss)	0.00628	0.01435	-0.01785
(pps)	-0.00230	0.00315	-0.00817
(ppp)	0.00112	0.00010	0.00064
(sps)	0.01148	0.00708	0.00196
ON SITE H-H			
s	0.16006	-0.11672	
FIRST NEIGHBOR H-H			
(sss)	-0.00852	-0.05434	0.01085
SECOND NEIGHBOR H-H			
(sss)	-0.02276	-0.01924	0.02076
(sps)	0.00000	0.00000	0.00000
(pps)	0.00000	0.00000	0.00000
(ppp)	0.00000	0.00000	0.00000
FIRST NEIGHBOR Cl-H			
(sss)	0.07625	-0.15882	0.17074
(pss)	-0.07064	-0.08494	0.15752
SECOND NEIGHBOR Cl-H			
(sss)	-0.00098	-0.00785	-0.00057
(pss)	0.00595	-0.00454	0.00326

	ORTHOGONAL			NON-ORTHOGONAL		
BAND	RMS ERROR	MAXIMUM DEVIATION		RMS ERROR	MAXIMUM DEVIATION	
	mRy	k	mRy	mRy	k	mRy
1	1.1	(000)	2.5	0.4	(444)	1.3
2	5.4	(004)	13.3	0.4	(264)	0.9
3	4.3	(226)	11.0	0.6	(008)	2.2
4	3.4	(174)	7.8	0.7	(008)	2.2
5	5.5	(048)	12.1	0.5	(055)	1.6
1-5	4.3			0.5		

ENERGY VALUES IN Ry AT SELECTED k-POINTS

	ORTHOGONAL	APW	NON-ORTHOGONAL
GAMMA 1	-0.93341	-0.93087	-0.93131
GAMMA 1	0.15787	0.15575	0.15589
GAMMA 15	0.14645	0.14553	0.14535
X1 (008)	-0.82295	-0.82250	-0.82191
X1 (008)	0.08677	0.09096	0.09103
X4' (008)	-0.12293	-0.11796	-0.11764
X5' (008)	0.04465	0.05098	0.04877
L1 (444)	-0.84485	-0.84497	-0.84363
L2' (444)	-0.24891	-0.23866	-0.23879
L2' (444)	0.40638	0.40626	0.40580
L3' (444)	0.11221	0.11796	0.11739
W1 (048)	-0.81886	-0.81963	-0.81924
W2' (048)	-0.07405	-0.07960	-0.07986
W2' (048)	0.27651	0.28862	0.28831
W3 (048)	-0.04364	-0.04815	-0.04784

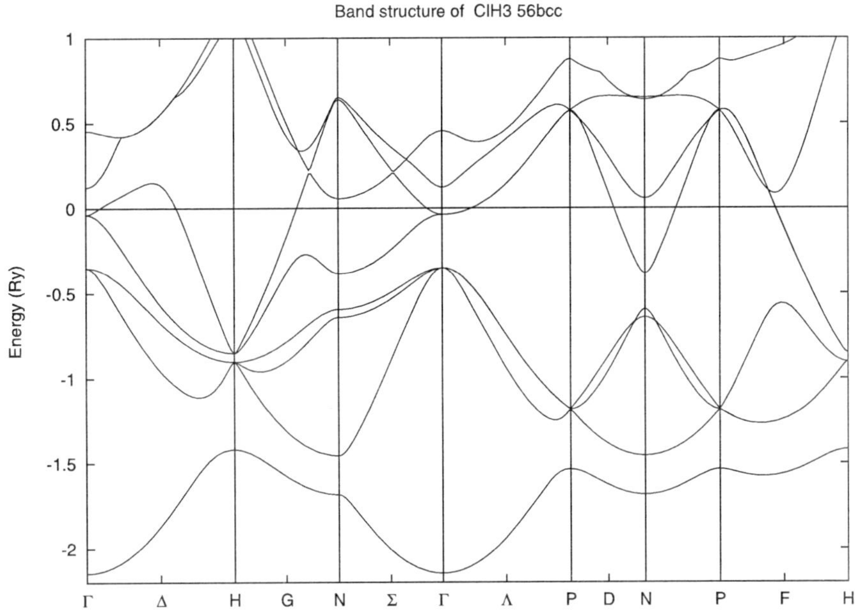

Fig. 13.9 Energy bands of ClH3 in the Im3m structure

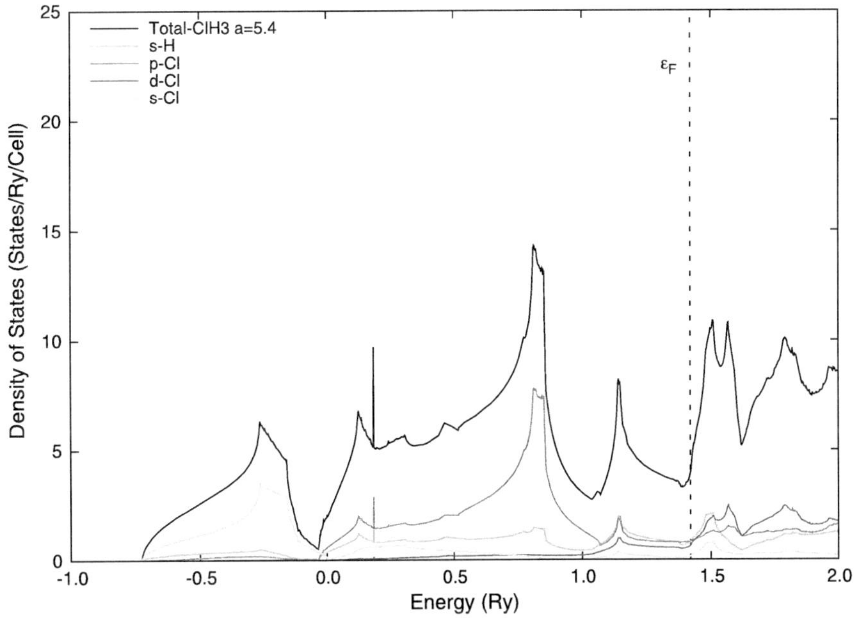

Fig. 13.10 Total, angular momentum and site decomposed densities of states of ClH3 in the Im3m structure

13.3 Bromine Hydride (BrH)

See Fig. 13.11 and Tables 13.11, 13.12, 13.13.

See Figs. 13.12, 13.13, 13.14.

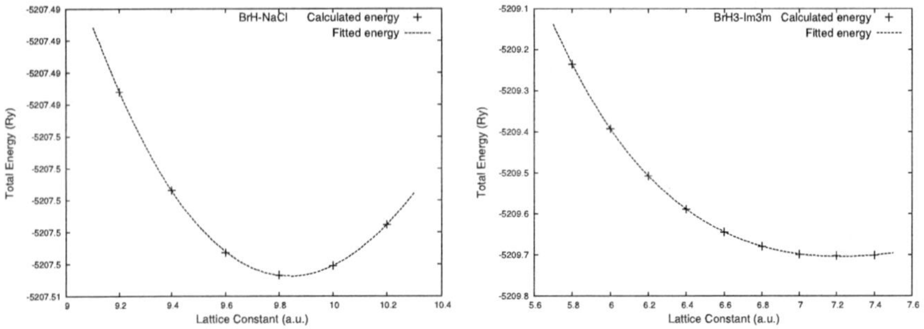

Fig. 13.11 Total energy versus lattice constant of BrH in the NaCl and Im3m structures

Table 13.11 Lattice constant, bulk modulus, gap, total energy

Stru	a (Bohr)	B (MBar)	Gap (Ry)	Total Energy (Ry)
NaCl	9.84	0.35	-	-5207.50476
Im3m (P=0)	7.21	1.10	-	-5209.70396
Im3m(P=1.83 MBar)	6.00	6.96	-	-5209.39260

Table 13.12 Birch fit coefficients

	A1	A2	A3	A4
NaCl	-5.206350E+03	-1.086817E+02	3.239656E+03	-2.950414E+04
Im3m	-5.208106E+03	-1.055146E+02	1.765857E+03	-7.831672E+02

Table 13.13 DOS at Ef, Hopfield parameter, Stoner criterion

BrH 9.84 Bohr NaCl

	Fermi Energy (Ry)	Total DOS (States/Ry)	s	p	eg	t2g	f
Br	0.140	15.513	0.015	11.915	0.038	0.005	0.003
H	0.140	15.513	0.839	0.309	0.137	0.009	0.130

Fermi-Velocity (cm/s): 0.29988 x10E8
Plasmon Energy (eV) : 2.75616
Electron-ion interaction (Hopfield parameter (eV/A^2)): Br 0.892 H 0.138

```
--------------------------------------------------------------------------
Br     MUFFIN-TIN RADIUS and CHARGE =     2.4598   33.4588
 H     MUFFIN-TIN RADIUS and CHARGE =     2.4598    1.0990
                        Br        STONER I =  0.0221
                         H        STONER I =  0.0011
                 STONER PARAMETER (Ry)  I =  0.0232
                 STONER CRITERION     N*I =  0.3602
--------------------------------------------------------------------------
      BrH3  a=6.00 Bohr                Im3m
-----------------------------------------------------------------------------
Fermi Energy   Total DOS        Decomposed DOS inside the muffin tins
   ( Ry )     (States/Ry)       s      p        d       f
Br  1.452        6.084        0.345  0.755    0.356   0.020
 H  1.452        6.084        0.427  0.054    0.003   0.000
-----------------------------------------------------------------------------
Fermi-Velocity (cm/s): 1.52303 x10E8
Plasmon Energy (eV)  : 13.01437
Electron-ion interaction (Hopfield parameter) (eV/A^2) Br:4.182  H3:3.989
-----------------------------------------------
 Br     MUFFIN-TIN RADIUS and CHARGE =    1.70  31.1579
  H     MUFFIN-TIN RADIUS and CHARGE =    1.00   0.5438
```

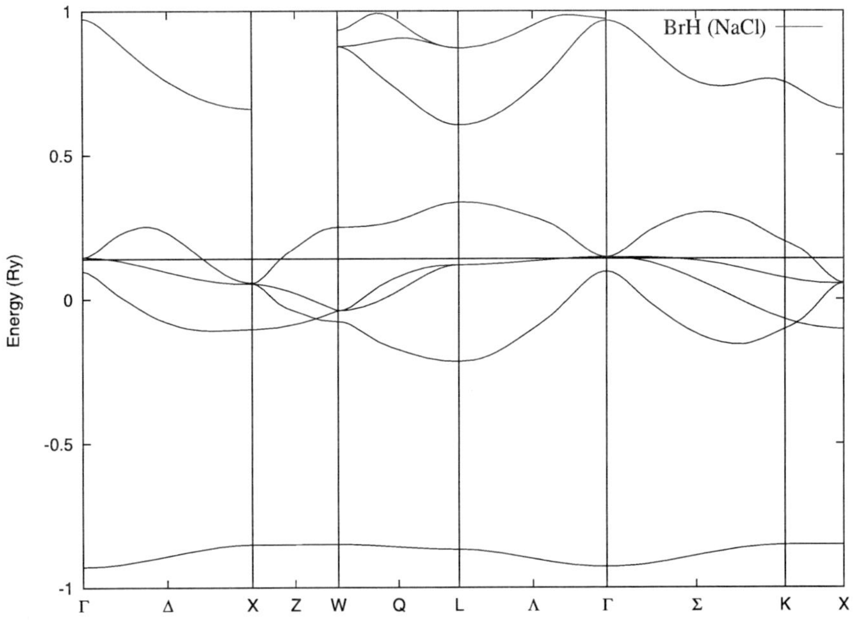

Fig. 13.12 Energy bands of BrH in the NaCl structure

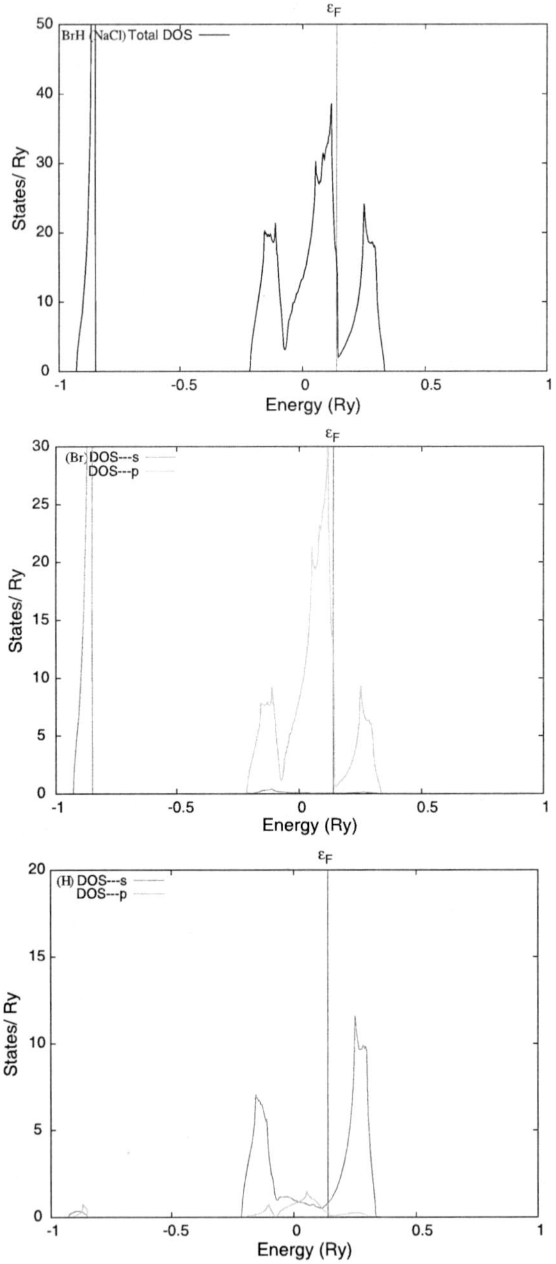

Fig. 13.13 Total, angular momentum and site decomposed densities of states of BrH in the NaCl structure

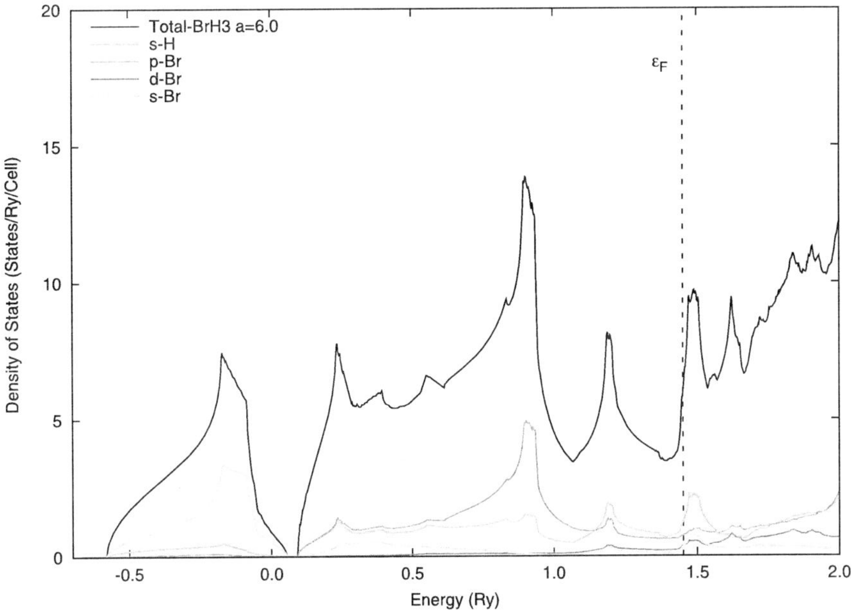

Fig. 13.14 Total, angular momentum and site decomposed densities of states of BrH3 in the Im3m structure

13.4 Iodine Hydride (IH)

See Fig. 13.15 and Tables 13.14, 13.15, 13.16.

See Figs. 13.16, 13.17, 13.18.

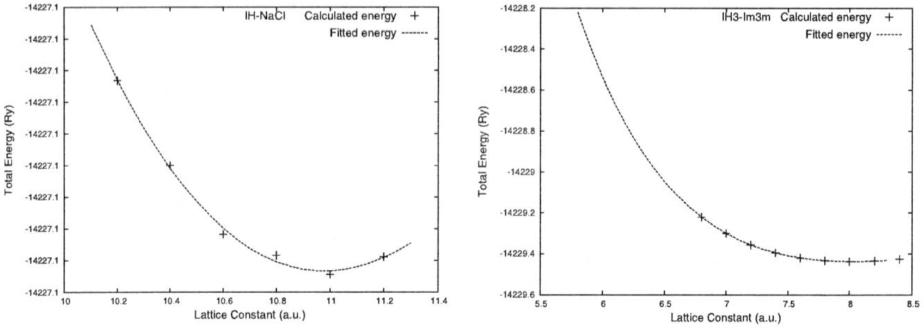

Fig. 13.15 Total energy versus lattice constant of IH in the NaCl and Im3m structures

Table 13.14 Lattice constant, bulk modulus, gap, total energy

Stru	a (Bohr)	B (MBar)	Gap(Ry)	Total Energy (Ry)
NaCl	10.98	0.21	-	-14227.11467
Im3m (P=0)	8.01	0.71	-	-14229.43745
Im3m(P=1.96 MBar)	6.40	7.04	-	-14228.97097

Table 13.15 Birch fit coefficients

	A1	A2	A3	A4
NaCl	-1.422620E+04	-1.051597E+02	3.793759E+03	-4.078147E+04
Im3m	-1.422812E+04	-1.035319E+02	1.895338E+03	5.360982E+03

Table 13.16 DOS at Ef, Hopfield parameter, Stoner criterion

```
       IH        a=10.98 Bohr        NaCl
-----------------------------------------------------------------------------
 Fermi Energy    Total DOS
    ( Ry )     (States/Ry)     s          p          eg          t2g         f
-----------------------------------------------------------------------------
  I  0.139       20.300      0.025    14.175     0.061      0.014       0.008
  H  0.139       20.300      1.632     0.453     0.269      0.025       0.155
-----------------------------------------------------------------------------
Fermi-Velocity (cm/s): 0.42948 x10E8
Plasmon Energy (eV)  : 3.83073
Electron-ion interaction (Hopfield parameter (eV/A^2) ): I 0.869  H 0.178
-----------------------------------------------------------------------------
  I     MUFFIN-TIN RADIUS and CHARGE =    2.7445   51.1230
  H     MUFFIN-TIN RADIUS and CHARGE =    2.7445    1.2782
                          I        STONER I =  0.0157
                          H        STONER I =  0.0018
                  STONER PARAMETER (Ry)  I =  0.0175
                  STONER CRITERION     N*I =  0.3559
-----------------------------------------------------------------------------
       IH3  a=6.40 Bohr                Im3m
-----------------------------------------------------------------------------
 Fermi Energy    Total DOS         Decomposed DOS inside the muffin tins
    ( Ry )     (States/Ry)       s       p        d        f
  I  1.552        5.965        0.231   0.506    0.200    0.061
  H  1.552        5.965        0.350   0.047    0.002    0.000
-----------------------------------------------------------------------------
Fermi-Velocity (cm/s): 1.61957 x10E8
Plasmon Energy (eV)  : 12.43912
Electron-ion interaction (Hopfield parameter) (eV/A^2) I:0.853  H3:2.643
-------------------------------------------------
   I     MUFFIN-TIN RADIUS and CHARGE =    1.70   47.4507
   H     MUFFIN-TIN RADIUS and CHARGE =    1.00    0.5604
```

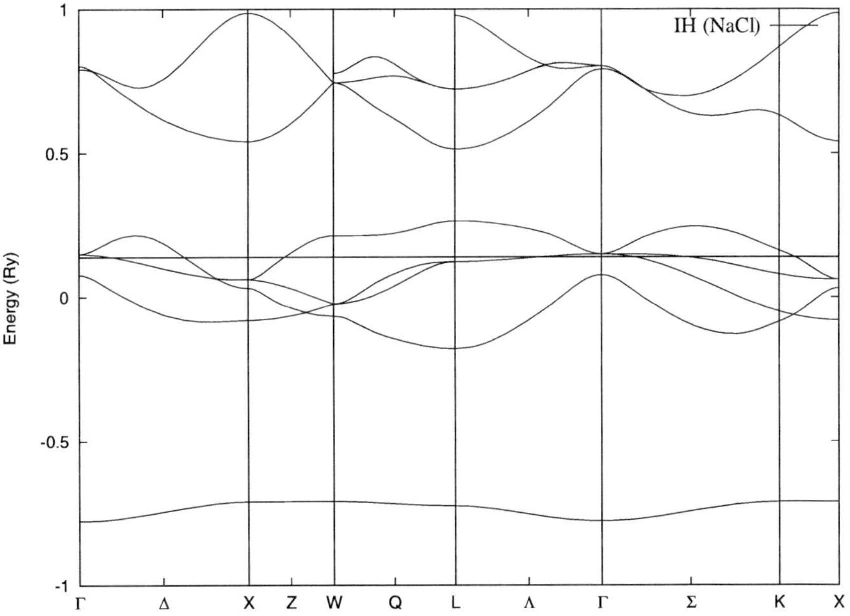

Fig. 13.16 Energy bands of IH in the NaCl structure

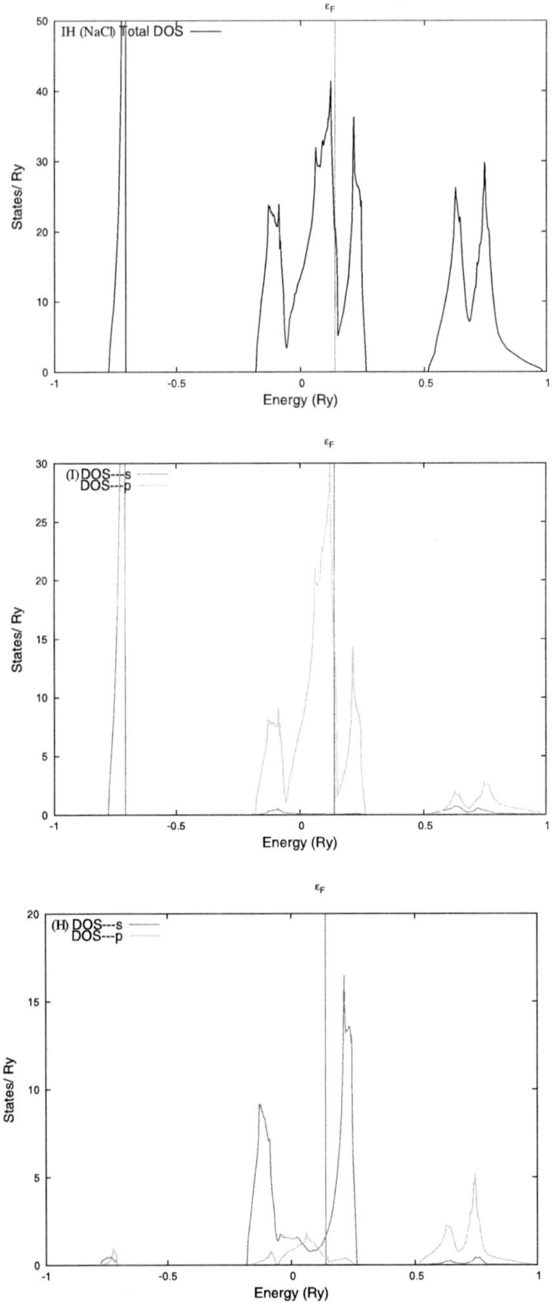

Fig. 13.17 Total, angular momentum and site decomposed densities of states of IH in the NaCl structure

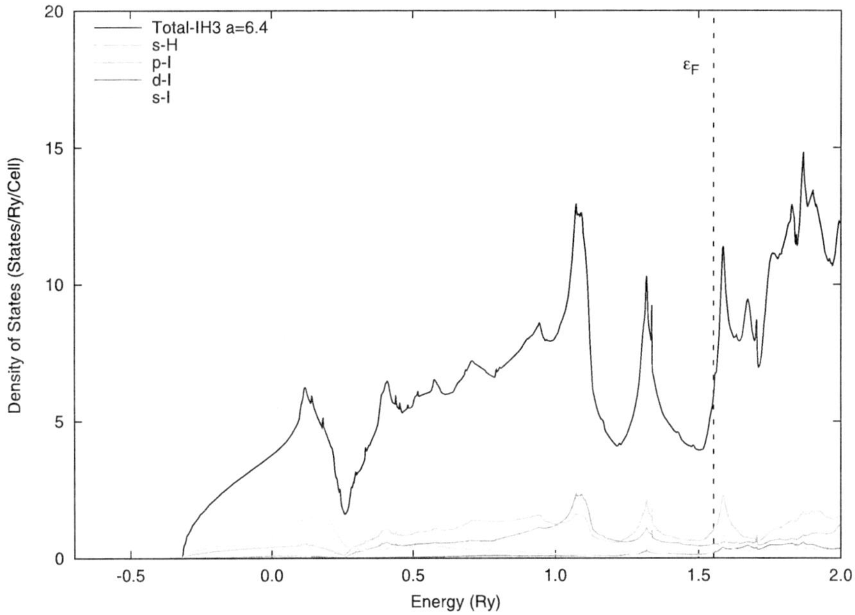

Fig. 13.18 Total, angular momentum and site decomposed densities of states of IH3 in the Im3m structure

13.5 Astatine Hydride (AtH)

See Fig. 13.19 and Tables 13.17, 13.18, 13.19.

See Figs. 13.20, 13.21.

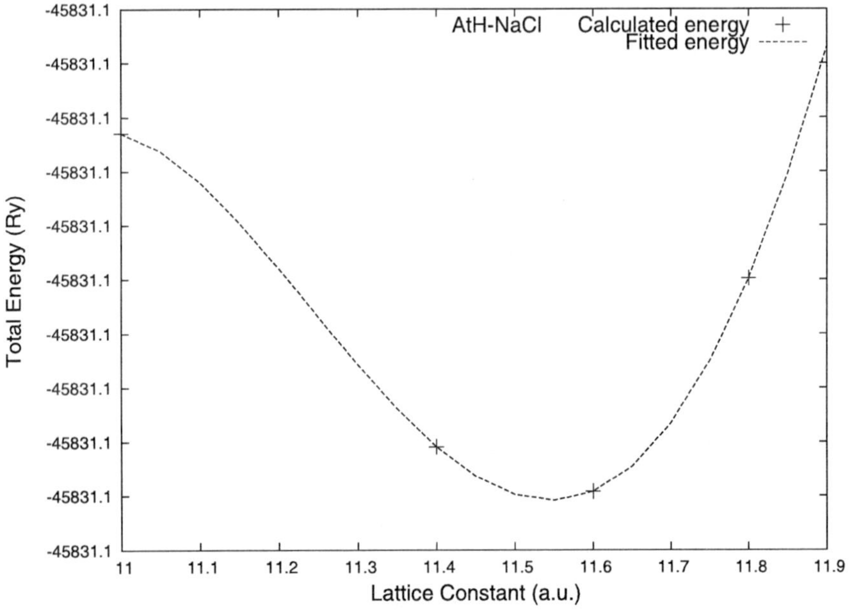

Fig. 13.19 Total energy versus lattice constant of AtH in the NaCl structure

Table 13.17 Lattice constant, bulk modulus, gap, total energy

Stru	a (Bohr)	B (MBar)	Gap(Ry)	Total Energy (Ry)
NaCl	11.55	0.30	-	-45831.07704

Table 13.18 Birch fit coefficients

	A1	A2	A3	A4
NaCl	-4.582462E+04	-9.768361E+02	4.917611E+04	-8.230783E+05

Table 13.19 DOS at Ef, Hopfield parameter, Stoner criterion

AtH a=11.55 Bohr NaCl

	Fermi Energy (Ry)	Total DOS (States/Ry)	s	p	eg	t2g	f
At	0.144	22.541	0.021	14.991	0.061	0.019	0.013
H	0.144	22.541	2.092	0.556	0.328	0.038	0.162

```
Fermi-Velocity (cm/s): 0.49221 x10E8
Plasmon Energy (eV)  : 4.28971
Electron-ion interaction (Hopfield parameter (eV/A^2) ): At 0.653  H 0.209
```

```
At      MUFFIN-TIN RADIUS and CHARGE =     2.8599    83.0296
 H      MUFFIN-TIN RADIUS and CHARGE =     2.8599     1.3416
                       At        STONER I =  0.0132
                        H        STONER I =  0.0022
                 STONER PARAMETER (Ry)  I =  0.0155
                 STONER CRITERION     N*I =  0.3488
```

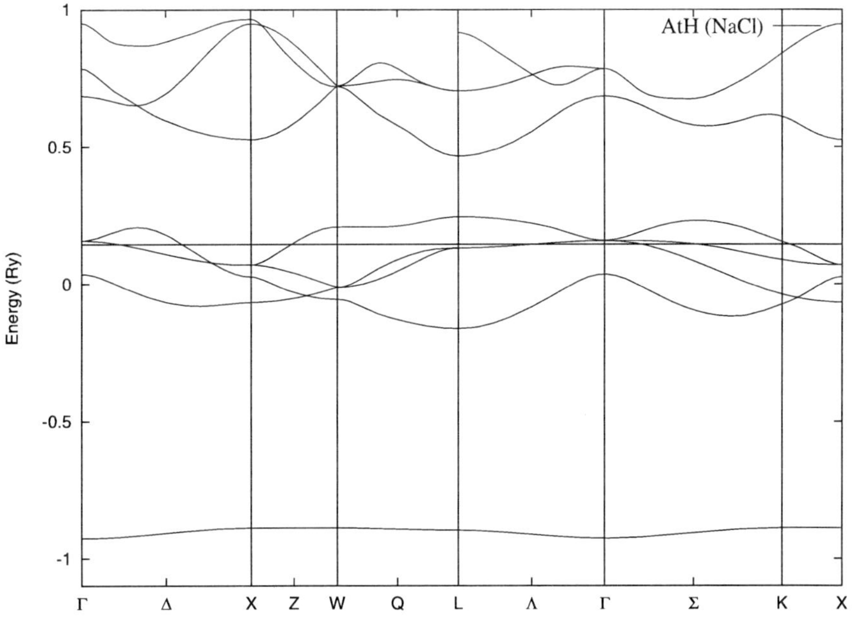

Fig. 13.20 Energy bands of AtH in the NaCl structure

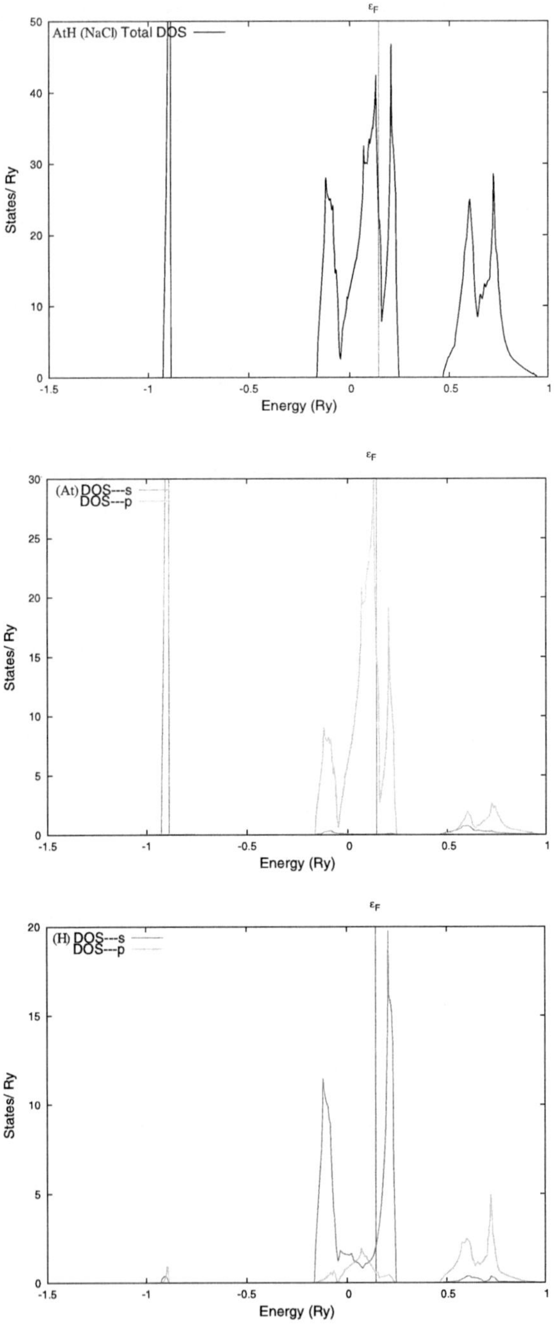

Fig. 13.21 Total, angular momentum and site decomposed densities of states of AtH in the NaCl structure

References

1. Wikipedia
2. D.A. Papaconstantopoulos, Novel Superc. Mater. **3**, 29 (2017)
3. D.A. Papaconstantopoulos, M.J. Mehl, H. Liu, Quantum Stud.: Math. Found. **5**, 23 (2018)

Chapter 14
Group 18 Hydrides

This chapter covers the Group 18 column of the periodic table hydrides from HeH to RnH [1]. These are hypothetical hydrides presented here for completeness for the NaCl structure only. It is interesting to note that all of these hypothetical hydrides satisfy the Stoner criterion showing a sharp peak in the density of states at the Fermi level with hydrogen s-like character. Examining the energy bands and densities of states figures we observe that, as in the groups (14–17 columns), the lowest band has a mixture of s–H hydrogen and s-character of the other element. At the Fermi level the states are of s- and p- character. As in the other columns of this part of the Periodic Table, the band structure is dominated by s- and p-states. The d-states are core states found significantly lower that the Gam1 point.

D. A. Papaconstantopoulos, *Band Structure of Cubic Hydrides*,
https://doi.org/10.1007/978-3-031-06878-2_14

14.1 Helium Hydride (HeH)

See Fig. 14.1 and Tables 14.1, 14.2 and 14.3.

See Fig. 14.2 and 14.3.

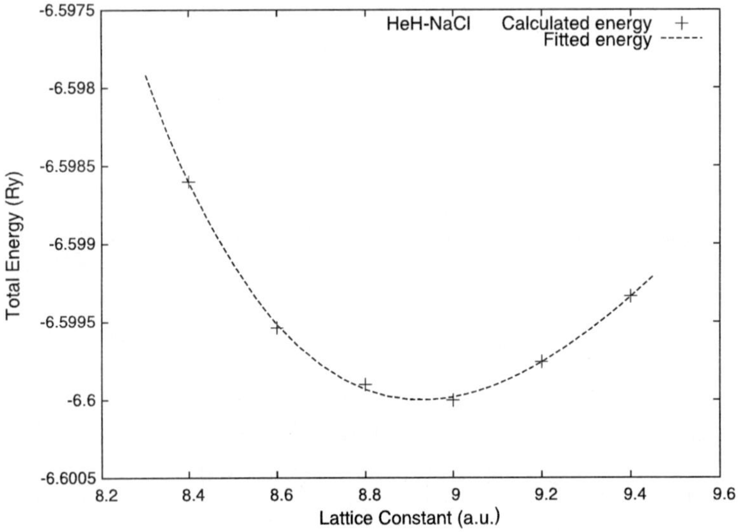

Fig. 14.1 Total energy versus lattice constant of HeH in the NaCl structure

Table 14.1 Lattice constant, bulk modulus, gap, total energy

Stru	a (Bohr)	B (MBar)	Gap (Ry)	Total Energy (Ry)
NaCl	8.93	0.06	-	-6.60003

Table 14.2 Birch fit coefficients

	A1	A2	A3	A4
NaCl	-6.583415E+00	7.999891E-01	-1.004320E+02	1.851456E+03

Table 14.3 DOS at Ef, Hopfield parameter, Stoner criterion

HeH a=8.93 Bohr NaCl

Fermi Energy (Ry)	Total DOS (States/Ry)	s	p	eg	t2g	f
He -0.089	25.623	0.509	1.007	0.131	0.002	0.028
H 0.144	25.623	19.804	0.064	0.001	0.023	0.005

Fermi-Velocity (cm/s): 0.24146 x10E8
Plasmon Energy (eV) : 3.29863

He	MUFFIN-TIN RADIUS and CHARGE =	2.2322	1.9766
H	MUFFIN-TIN RADIUS and CHARGE =	2.2322	0.7710

He STONER I = 0.0009
H STONER I = 0.1248
STONER PARAMETER (Ry) I = 0.1258
STONER CRITERION N*I = 3.2221

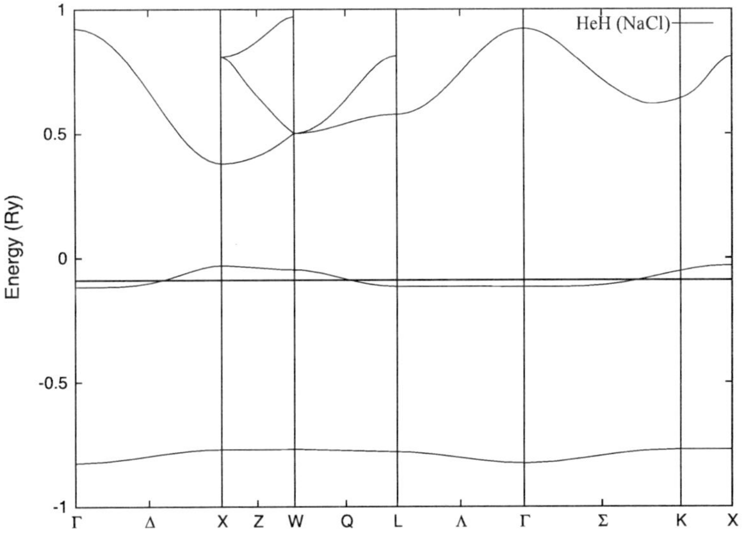

Fig. 14.2 Energy bands of HeH in the NaCl structure

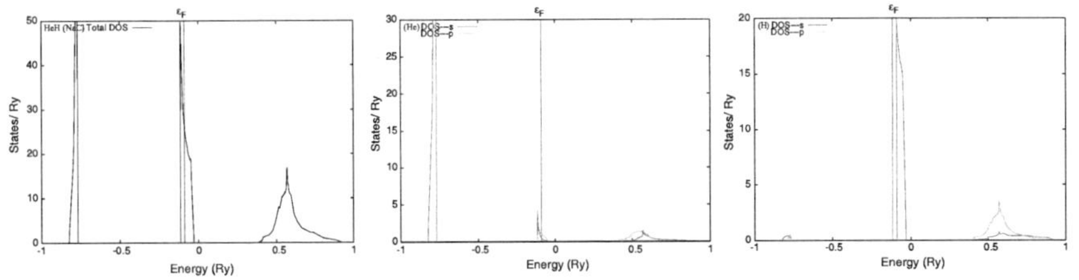

Fig. 14.3 Total, angular momentum and site decomposed densities of states of HeH in the NaCl structure

14.2 Neon Hydride (NeH)

See Fig. 14.4 and Tables 14.4, 14.5 and 14.6.

See Fig. 14.5 and 14.6

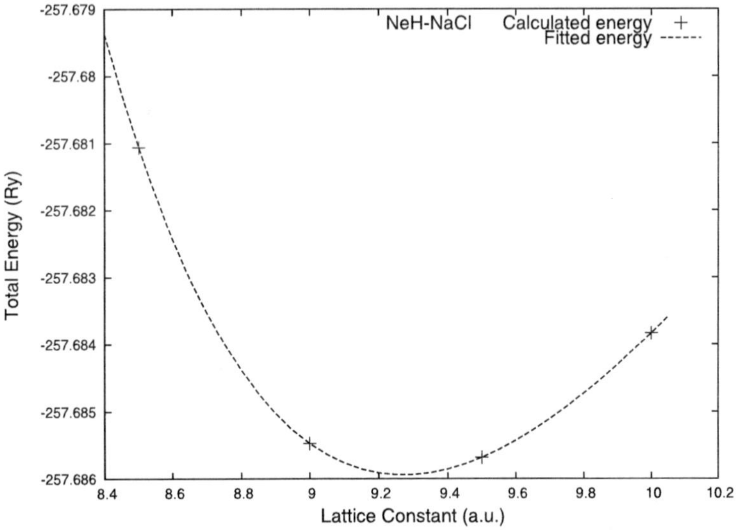

Fig. 14.4 Total energy versus lattice constant of NeH in the NaCl structure

Table 14.4 Lattice constant, bulk modulus, gap, total energy

Stru	a (Bohr)	B (MBar)	Gap(Ry)	Total Energy (Ry)
NaCl	9.27	0.08	-	-257.68594

Table 14.5 Birch fit coefficients

	A1	A2	A3	A4
NaCl	-2.576519E+02	5.791061E-01	-1.583765E+02	3.377644E+03

Table 14.6 DOS at Ef, Hopfield parameter, Stoner criterion

```
                    NeH      a=9.27 Bohr     NaCl
---------------------------------------------------------------

Fermi Energy  Total DOS    Decomposed DOS inside the muffin
tins
    ( Ry )      (States/Ry)      s         p         eg     t2g    f
---------------------------------------------------------------
    0.553       36.077          2.591    2.790   0.172  0.508 0.080
    0.553       36.077          1.934    7.652   0.038  0.143 0.040
---------------------------------------------------------------

Fermi-Velocity (cm/s): 0.31580 x10E8
Plasmon Energy (eV)   : 4.80933
Electron-ion interaction(Hopfield parameter(eV/A^2))Ne 4.12 H
7.10

-------------------------------------------------
         Ne     MUFFIN-TIN  RADIUS and CHARGE =     2.3272
9.8188
         H      MUFFIN-TIN  RADIUS and CHARGE =     2.3272
0.8219
                  Ne          STONER I = 0.0034
                  H           STONER I = 0.0144
        STONER PARAMETER  (Ry)       I = 0.0181
        STONER CRITERION           N*I = 0.6546
-------------------------------------------------
```

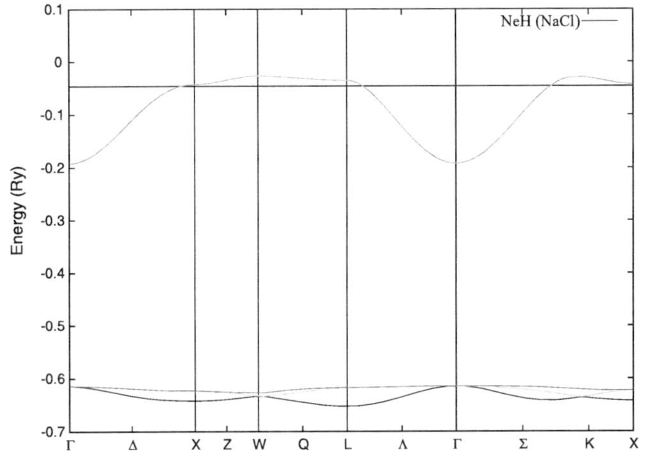

Fig. 14.5 Energy bands of NeH in the NaCl structure

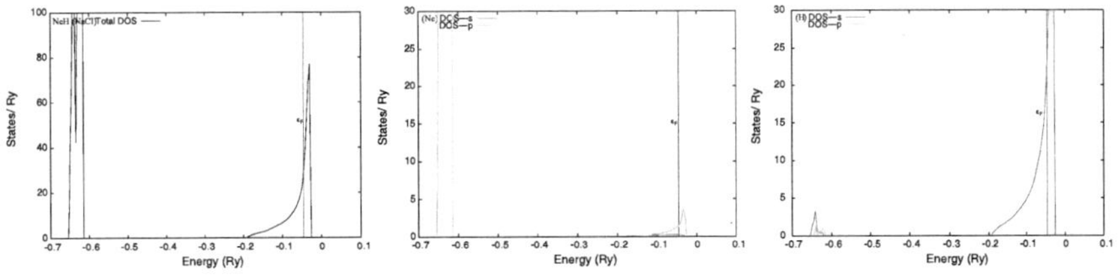

Fig. 14.6 Total, angular momentum and site decomposed densities of states of NeH in the NaCl structure

14.3 Argon Hydride (ArH)

See Fig. 14.7 and Tables 14.7, 14.8 and 14.9.

See Figs. 14.8 and 14.9.

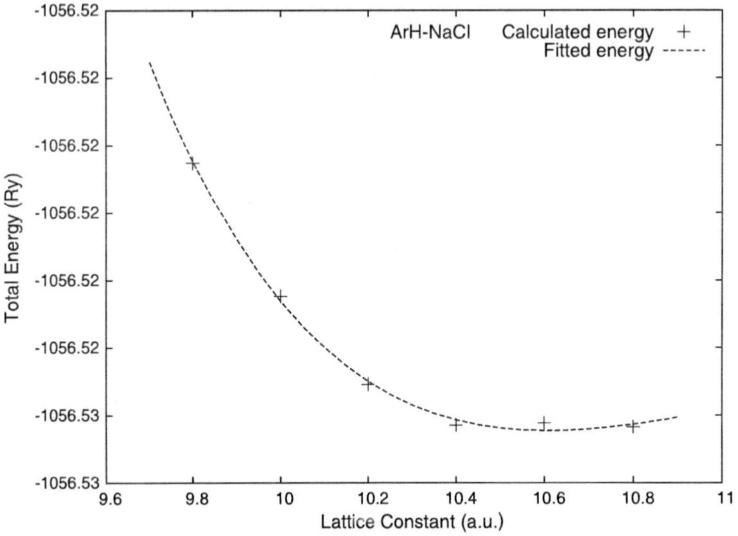

Fig. 14.7 Total energy versus lattice constant of ArH in the NaCl structure

Table 14.7 Lattice constant, bulk modulus, gap, total energy

Stru	a (Bohr)	B (MBar)	Gap(Ry)	Total Energy (Ry)
NaCl	10.61	0.04	-	-1056.52523

Table 14.8 Birch fit coefficients

	A1	A2	A3	A4
NaCl	-1.056725E+03	3.057143E+01	-1.536936E+03	2.543521E+04

Table 14.9 DOS at Ef, Hopfield parameter, Stoner criterion

ArH a=10.61 Bohr NaCl

Fermi Energy (Ry)	Total DOS (States/Ry)	Decomposed DOS inside the muffin tins s	p	eg	t2g	f
-0.021	39.313	0.303	3.829	0.783	0.004	0.078
-0.021	39.313	31.003	0.055	0.016	0.000	0.003

Fermi-Velocity (cm/s): 0.19587 x10E8
Plasmon Energy (eV) : 2.55922

Ar	MUFFIN-TIN RADIUS and CHARGE =	2.7848	17.4481
H	MUFFIN-TIN RADIUS and CHARGE =	2.5196	0.9647

Ar STONER I = 0.0005
H STONER I = 0.1095
STONER PARAMETER (Ry) I = 0.1101
STONER CRITERION N*I = 4.3271

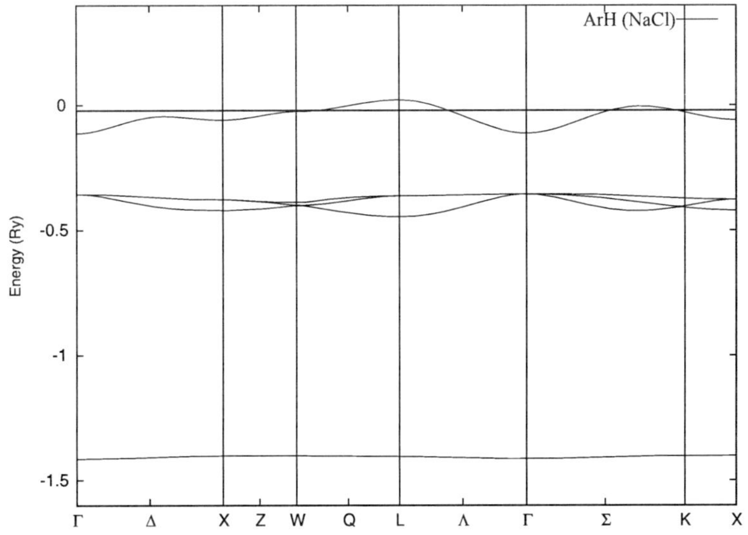

Fig. 14.8 Energy bands of ArH in the NaCl structure

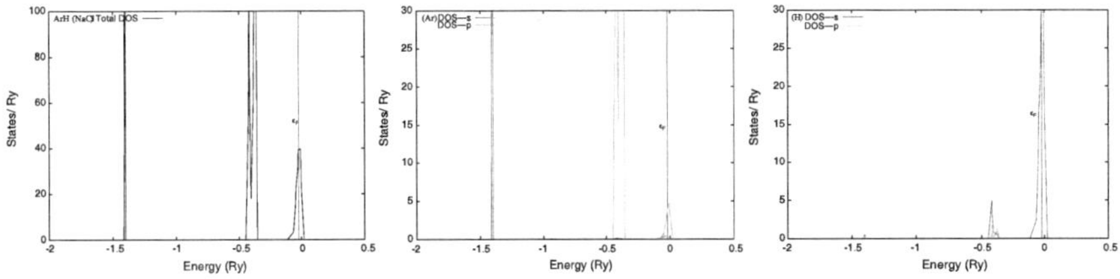

Fig. 14.9 Total, angular momentum and site decomposed densities of states of ArH in the NaCl structure

14.4 Krypton Hydride (KrH)

See Fig. 14.10 and Tables 14.10, 14.11 and 14.12.

See Figs. 14.11 and 14.12.

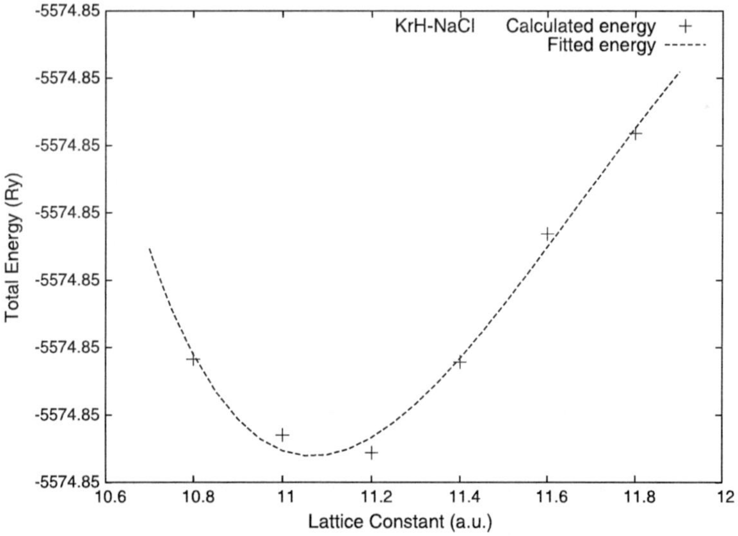

Fig. 14.10 Total energy versus lattice constant of KrH in the NaCl structure

Table 14.10 Lattice constant, bulk modulus, gap, total energy

Stru	a (Bohr)	B (MBar)	Gap(Ry)	Total Energy (Ry)
NaCl	11.99	0.08	-	-14889.16111

Table 14.11 Birch fit coefficients

	A1	A2	A3	A4
NaCl	-1.488994E+04	1.483314E+02	-9.327884E+03	1.938439E+05

Table 14.12 DOS at Ef, Hopfield parameter, Stoner criterion

XeH a=11.99 Bohr NaCl

Fermi Energy (Ry)	Total DOS (States/Ry)	Decomposed DOS inside the muffin tins s	p	eg	t2g	f
Xe 0.014	34.229	0.240	5.918	0.531	0.012	0.096
H 0.014	34.229	25.229	0.141	0.043	0.013	0.008

Fermi-Velocity (cm/s): 0.25387 x10E8
Plasmon Energy (eV) : 2.57605
Electron-ion interaction (Hopfield parameter (eV/A^2)):Xe 0.966 H 0.430

Xe	MUFFIN-TIN RADIUS and CHARGE =	3.1527	52.9078
H	MUFFIN-TIN RADIUS and CHARGE =	2.8524	1.1617
Xe	STONER I =	0.0009	
H	STONER I =	0.0850	
STONER PARAMETER (Ry)	I =	0.0860	
STONER CRITERION	N*I =	2.9429	

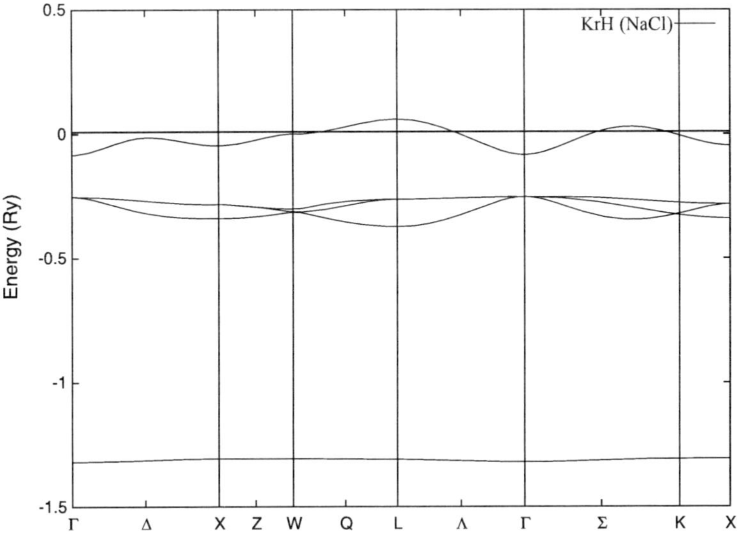

Fig. 14.11 Energy bands of KrH in the NaCl structure

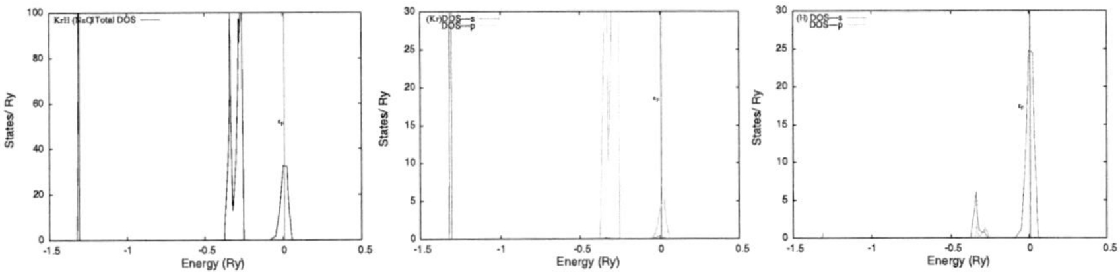

Fig. 14.12 Total, angular momentum and site decomposed densities of states of KrH in the NaCl structure

14.5 Xenon Hydride (XeH)

See Fig. 14.13 Tables 14.13, 14.14, 14.15.

See Fig. 14.14 and 14.15.

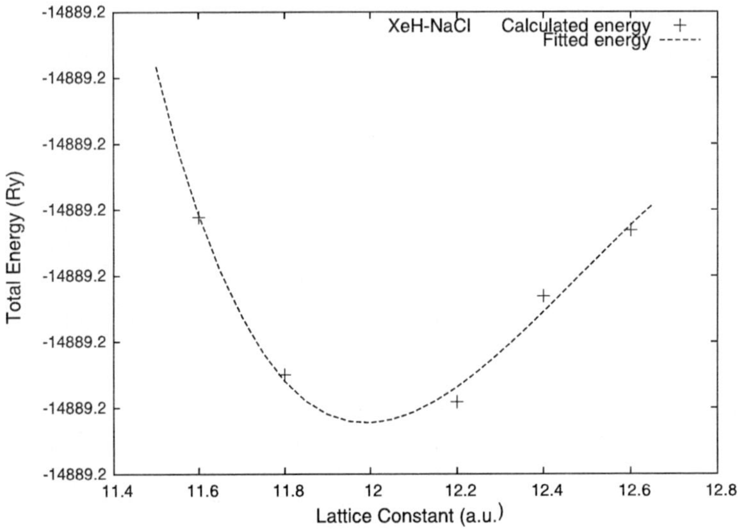

Fig. 14.13 Total energy versus lattice constant of XeH in the NaCl structure

Table 14.13 Lattice constant, bulk modulus, gap, total energy

Stru	a (Bohr)	B (MBar)	Gap(Ry)	Total Energy (Ry)
NaCl	11.99	0.08	-	-14889.16111

Table 14.14 Birch fit coefficients

	A1	A2	A3	A4
NaCl	-1.488994E+04	1.483314E+02	-9.327884E+03	1.938439E+05

Table 14.15 DOS at Ef, Hopfield parameter, Stoner criterion

```
        XeH      a=11.99 Bohr          NaCl
-----------------------------------------------------------------------------
 Fermi Energy    Total DOS          Decomposed DOS inside the muffin tins
   ( Ry )        (States/Ry)     s          p       eg        t2g      f
-----------------------------------------------------------------------------
Xe 0.014          34.229        0.240     5.918    0.531    0.012    0.096
 H 0.014          34.229       25.229     0.141    0.043    0.013    0.008
-----------------------------------------------------------------------------
Fermi-Velocity (cm/s): 0.25387 x10E8
Plasmon Energy (eV)   : 2.57605
Electron-ion interaction (Hopfield parameter (eV/A^2) ):Xe 0.966 H 0.430
-----------------------------------------------------------------------------
Xe      MUFFIN-TIN RADIUS and CHARGE =     3.1527   52.9078
 H      MUFFIN-TIN RADIUS and CHARGE =     2.8524    1.1617
                    Xe        STONER I =   0.0009
                     H        STONER I =   0.0850
          STONER PARAMETER (Ry)      I =   0.0860
          STONER CRITERION         N*I =   2.9429
```

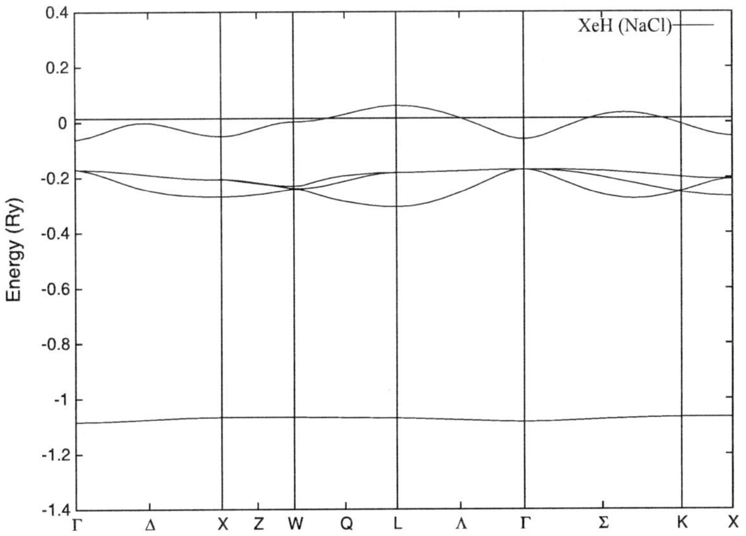

Fig. 14.14 Energy bands of XeH in the NaCl structure

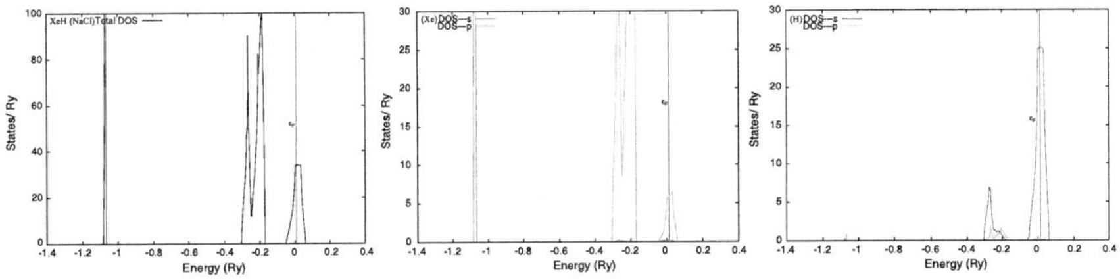

Fig. 14.15 Total, angular momentum and site decomposed densities of states of XeH in the NaCl structure

14.6 Radon Hydride (RnH)

See Fig. 14.16 and Tables 14.16, 14.17, 14.18.

See Figs. 14.17 and 14.18.

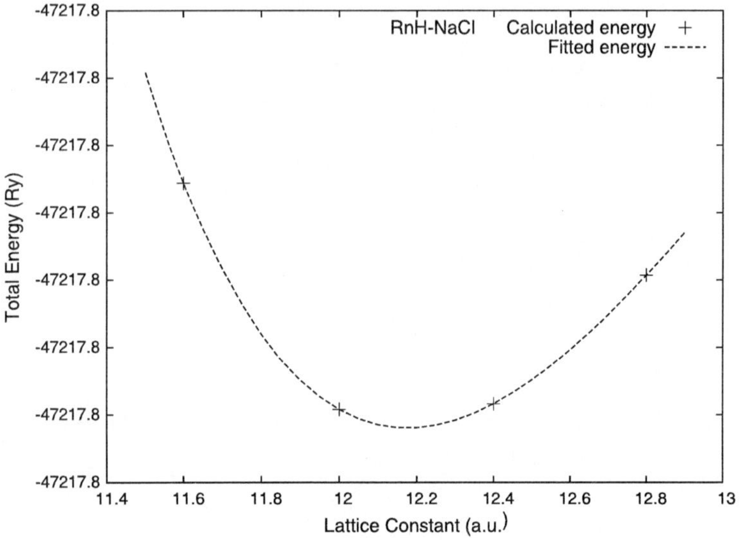

Fig. 14.16 Total energy versus lattice constant of RnH in the NaCl structure

Table 14.16 Lattice constant, bulk modulus, gap, total energy

Stru	a (Bohr)	B (MBar)	Gap(Ry)	Total Energy (Ry)
NaCl	12.18	0.09	-	-47217.76919

Table 14.17 Birch fit coefficients

	A1	A2	A3	A4
NaCl	-4.721798E+04	5.411430E+01	-4.218117E+03	1.030112E+05

Table 14.18 DOS at Ef, Hopfield parameter, Stoner criterion

```
        RnH        a=12.18 Bohr        NaCl
 ---------------------------------------------------------------------------
  Fermi Energy   Total DOS          Decomposed DOS inside the muffin tins
    ( Ry )       (States/Ry)     s        p      eg       t2g      f
 ---------------------------------------------------------------------------
 Rn 0.045         32.073        0.184    6.911   0.437    0.017    0.103
  H 0.045         32.073       22.165    0.188   0.063    0.024    0.012
 ---------------------------------------------------------------------------
Fermi-Velocity (cm/s): 0.28003 x10E8
Plasmon Energy (eV)  : 2.68759
Electron-ion interaction (Hopfield parameter (eV/A^2) ):Rn 0.995 H 0.432
 ---------------------------------------------------------------------------
 Rn      MUFFIN-TIN RADIUS and CHARGE =     3.1963  84.6926
  H      MUFFIN-TIN RADIUS and CHARGE =     2.8919   1.2174
                    Rn        STONER I =  0.0013
                     H        STONER I =  0.0746
          STONER PARAMETER (Ry)      I =  0.0758
          STONER CRITERION         N*I =  2.4317
```

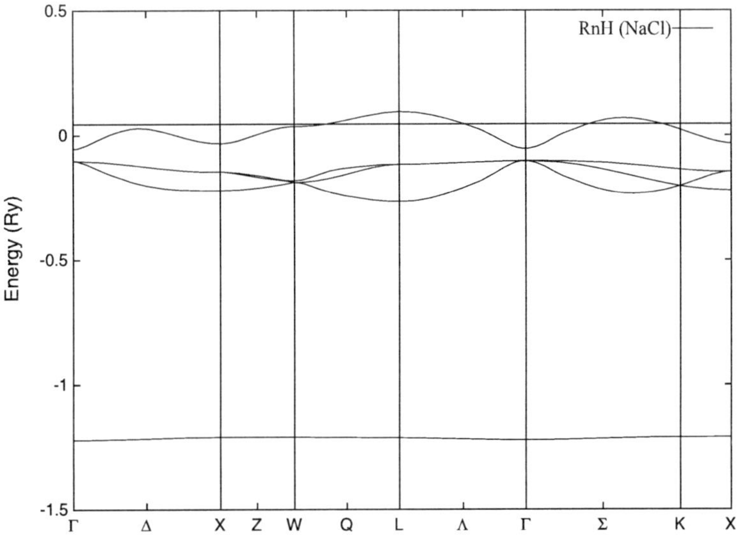

Fig. 14.17 Energy bands of RnH in the NaCl structure

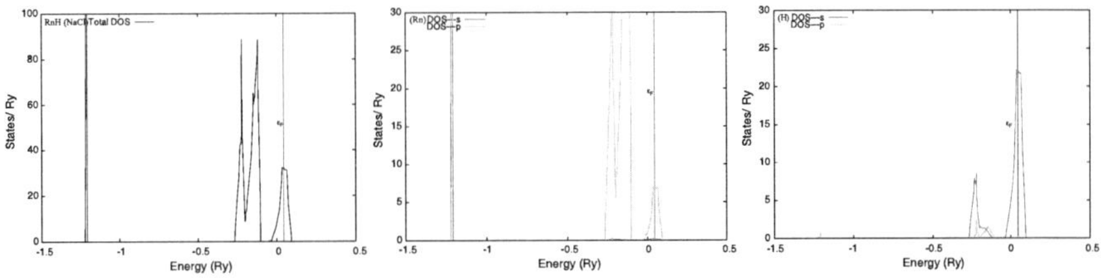

Fig. 14.18 Total, angular momentum and site decomposed densities of states of RnH in the NaCl structure

Reference

1. Wikipedia

Chapter 15
Lanthanide Hydrides

This chapter covers the lanthanide hydrides from CeH to LuH. Results are presented for the CaF2 and NaCl crystal structures [1]. Most of these systems are expected to form as antiferromagnetic dihydrides.

Examining the densities of states figures we observe that the lower occupied states have predominantly s–H character and near the Fermi level the d-2 g metal contribution is the dominant one. Comparing the energy bands of the fluorite structure to the bands of the NaCl structure we note an additional band (second band) which is due to the second hydrogen in the CaF2 structure. Characteristic of the lanthanides is that, for the paramagnetic calculations presented here, the Fermi level (Ef) falls within the f-states which results in a very large value of the DOS at Ef and also a large Stoner criterion value. Following the energy bands and DOS for both structures we find the f-states slightly above Ef for CeH and PrH. From NdH to TmH Ef falls in the middle of the f-states and for YbH and LuH the f-states are clearly below Ef.

D. A. Papaconstantopoulos, *Band Structure of Cubic Hydrides*,
https://doi.org/10.1007/978-3-031-06878-2_15

15.1 Cerium Hydride (CeH)

See Tables 15.1 and 15.2 and Figs. 15.1, 15.2, 15.3 and 15.4.

Table 15.1 Lattice constant, bulk modulus, gap, total energy

Stru	a (Bohr)	B (MBar)	Gap	Total Energy(Ry)
NaCl	9.70	0.60	-	-17718.40410
CaF2	10.55	----	-	-17719.32765
exp (CaF2)	10.55			

Table 15.2 DOS at Ef, Hopfield parameter, Stoner criterion

CeH a=9.70 Bohr NaCl

	Fermi Energy (Ry)	Total DOS (States/Ry)	Decomposed DOS inside the muffin tins s	p	eg	t2g	f
Ce	0.656	68.512	0.352	1.336	0.295	4.380	45.040
H	0.656	68.512	0.222	1.228	0.031	0.620	0.168

```
Fermi-Velocity (cm/s): 0.17096 x10E8
Plasmon Energy (eV)  : 3.37113
Electron-ion interaction (Hopfield parameter) (eV/A^2) Ce: 1.624   H: 0.031

        Ce     MUFFIN-TIN RADIUS and CHARGE =    2.5470   55.1074
        H      MUFFIN-TIN RADIUS and CHARGE =    2.3044    1.5356
                  Ce        STONER  I = 0.0114
                  H         STONER  I = 0.0002
        STONER PARAMETER (Ry)       I = 0.0116
        STONER CRITERION          N*I = 0.7978
```

CeH2 a=10.55 Bohr CaF2

	Fermi Energy (Ry)	Total DOS (States/Ry)	Decomposed DOS inside the muffin tins s	p	eg	t2g	f
Ce	0.504	47.292	0.066	0.025	0.150	0.036	41.075
H	0.504	47.292	0.079	0.873	0.054	0.071	0.055

```
Fermi-Velocity (cm/s): 0.08243 x10E8
Plasmon Energy (eV)  : 1.16552
Electron-ion interaction (Hopfield parameter) (eV/A^2) Ce: 0.009   H: 0.019

        Ce     MUFFIN-TIN RADIUS and CHARGE =    2.3983   54.6180
        H      MUFFIN-TIN RADIUS and CHARGE =    2.1699    1.4125
                  Ce        STONER I =  0.0186
                  H         STONER I =  0.0000
        STONER PARAMETER (Ry)       I = 0.0186
        STONER CRITERION          N*I = 0.8809
```

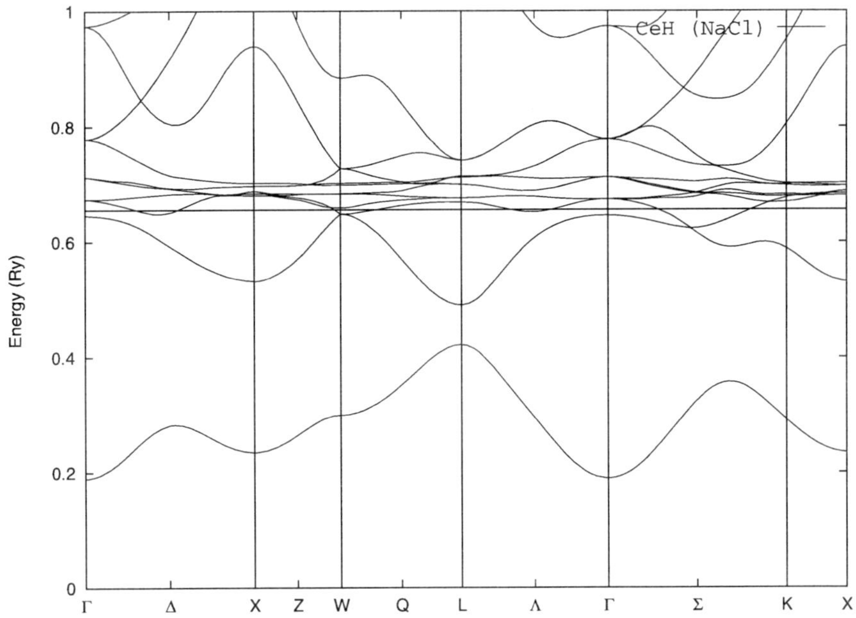

Fig. 15.1 Energy bands of CeH in the NaCl structure

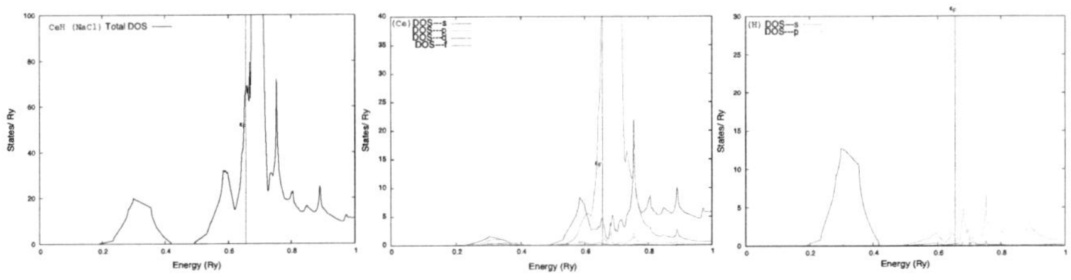

Fig. 15.2 Total, angular momentum and site decomposed densities of states of CeH in the NaCl structure

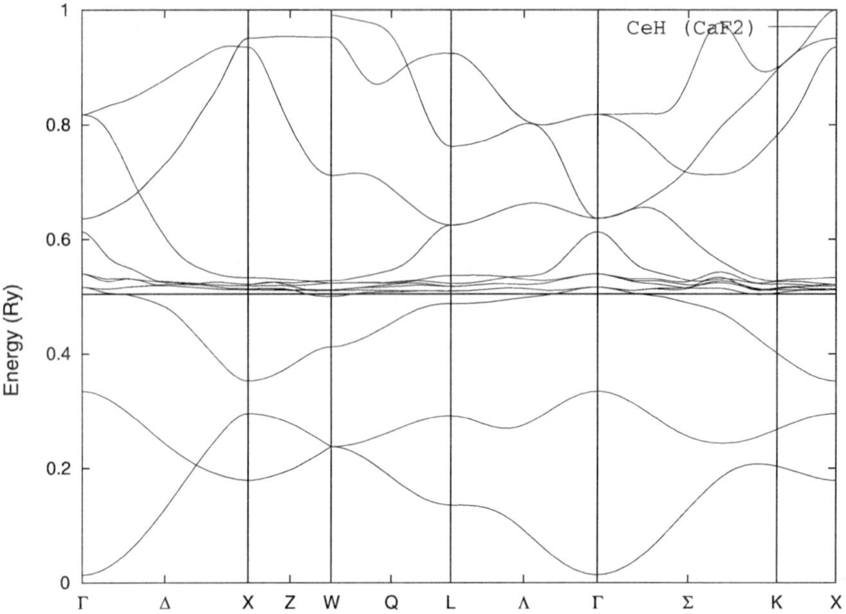

Fig. 15.3 Energy bands of CeH2 in the CaF2 structure

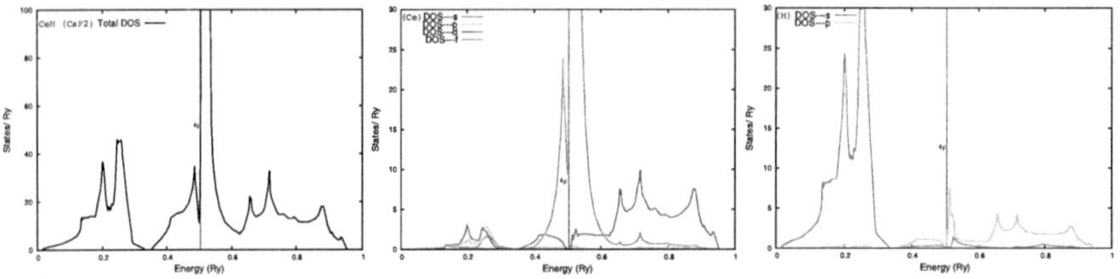

Fig. 15.4 Total, angular momentum and site decomposed densities of states of CeH2 in the CaF2 structure

15.2 Praseodymium Hydride (PrH)

See Tables 15.3, 15.4 and Figs. 15.5 and 15.6.

Table 15.3 Lattice constant, bulk modulus, gap, total energy

Stru	a (Bohr)	B (MBar)	Gap	Total Energy(Ry)
NaCl	9.58	0.58	-	-18472.85925
CaF2	10.43	----	-	-18473.75867
exp (CaF2)	10.43			

Table 15.4 DOS at Ef, Hopfield parameter, Stoner criterion

```
------------------------------------------------------------------------------
                PrH      a=9.58 Bohr                NaCl
------------------------------------------------------------------------------
   Fermi Energy   Total DOS        Decomposed DOS inside the muffin tins
     ( Ry )      (States/Ry)      s         p         eg         t2g        f
------------------------------------------------------------------------------
Pr   0.660         97.414      0.140     0.554     0.167      1.084    83.911
 H   0.660         97.414      0.095     0.626     0.022      0.398     0.128
------------------------------------------------------------------------------
Fermi-Velocity (cm/s): 0.07392 x10E8
Plasmon Energy (eV)  : 1.77104
Electron-ion interaction (Hopfield parameter) (eV/A^2) Pr: 0.152   H: 0.005
-------------------------------------------------
         Pr     MUFFIN-TIN RADIUS and CHARGE =    2.5153   56.1578
          H     MUFFIN-TIN RADIUS and CHARGE =    2.2757    1.5337
                  Pr        STONER I = 0.0180
                   H        STONER I = 0.0000
         STONER PARAMETER (Ry)      I = 0.0180
         STONER CRITERION         N*I = 1.7541
-------------------------------------------------
                PrH2     a=10.43 Bohr          CaF2
------------------------------------------------------------------------------
  Fermi Energy   Total DOS        Decomposed DOS inside the muffin tins
     (Ry)       (States/Ry)      s         p         eg         t2g        f
------------------------------------------------------------------------------
Pr  0.502       607.956       0.073     0.491     0.135      0.191    567.653
H   0.502       607.956       0.146     4.908     1.683      0.528      0.461
------------------------------------------------------------------------------
Fermi-Velocity (cm/s): 0.03564 x10E8
Plasmon Energy (eV)  : 1.84282
Electron-ion interaction (Hopfield parameter) (eV/A^2) Pr: 0.024   H: 0.035
-------------------------------------------------
         Pr     MUFFIN-TIN RADIUS and CHARGE =    2.3711   55.6865
         H      MUFFIN-TIN RADIUS and CHARGE =    2.1453    1.4041
                  Pr        STONER I =  0.0217
                   H        STONER I =  0.0000
         STONER PARAMETER (Ry)      I = 0.0218
         STONER CRITERION         N*I = 13.2204

------------------------------------------------------------------------------
```

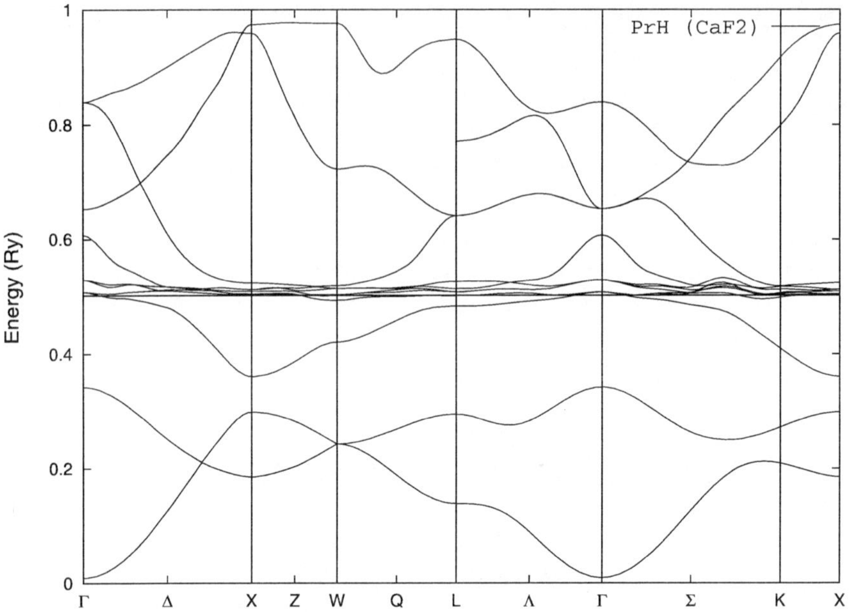

Fig. 15.5 Energy bands of PrH2 in the CaF2 structure

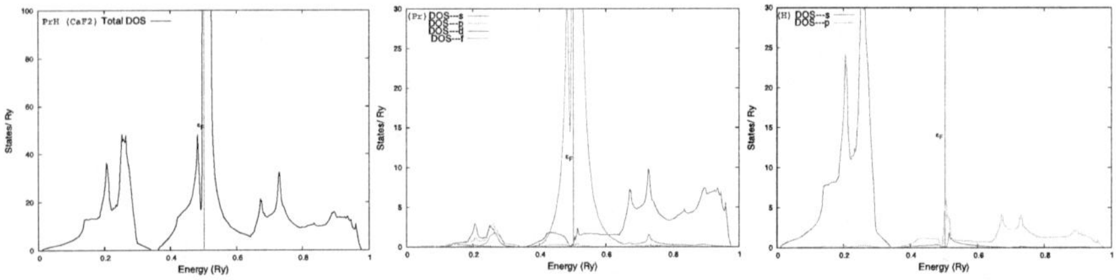

Fig. 15.6 Total, angular momentum and site decomposed densities of states of PrH2 in the CaF2 structure

15.3 Neodymium Hydride (NdH)

See Tables 15.5 and 15.6 and Figs. 15.7 and 15.8.

Table 15.5 Lattice constant, bulk modulus, gap, total energy

Stru	a (Bohr)	B (MBar)	Gap	Total Energy(Ry)
NaCl	9.54	0.58	-	-19247.00547
CaF2	10.34	----	-	-19247.90446
exp (CaF2)	10.34			

Table 15.6 DOS at Ef, Hopfield parameter, Stoner criterion

```
-----------------------------------------------------------------------------
              NdH       a=9.54 Bohr            NaCl
-----------------------------------------------------------------------------
  Fermi Energy   Total DOS        Decomposed DOS inside the muffin tins
      ( Ry )     (States/Ry)        s        p        eg         t2g        f
-----------------------------------------------------------------------------
Nd      0.651      392.949        0.045    0.856    0.433     0.866    368.613
 H      0.651      392.949        0.239    1.737    0.041     1.089      0.279
-----------------------------------------------------------------------------
Fermi-Velocity (cm/s): 0.03661 x10E8
Plasmon Energy (eV)   : 1.74521
Electron-ion interaction (Hopfield parameter (eV/A^2) ):Nd 0.055 H 0.009
------------------------------------------------
         Nd     MUFFIN-TIN RADIUS and CHARGE =    2.5032   57.2491
          H     MUFFIN-TIN RADIUS and CHARGE =    2.2648    1.5177
                   Nd        STONER I =    0.0213
                    H        STONER I =    0.0000
         STONER PARAMETER (Ry)      I =    0.0213
         STONER CRITERION         N*I =    8.3847
-----------------------------------------------------------------------------
              NdH2      a=10.34 Bohr          CaF2
-----------------------------------------------------------------------------
  Fermi Energy   Total DOS        Decomposed DOS inside the muffin tins
    (Ry)        (States/Ry)       s        p         eg        t2g         f
-----------------------------------------------------------------------------
Nd  0.493       653.887        0.048    0.449     0.253     0.337    613.163
H   0.493       653.887        0.131    4.891     1.317     0.460      0.504
-----------------------------------------------------------------------------
Fermi-Velocity (cm/s): 0.03887 x10E8
Plasmon Energy (eV)   : 2.15774
Electron-ion interaction (Hopfield parameter) (eV/A^2) Nd: 0.038   H: 0.028
------------------------------------------------
         Nd     MUFFIN-TIN RADIUS and CHARGE =    2.3506   56.7425
         H      MUFFIN-TIN RADIUS and CHARGE =    2.1267    1.4077
                  Nd        STONER I = 0.0221
                   H        STONER I = 0.0000
         STONER PARAMETER (Ry)     I = 0.0222
         STONER CRITERION        N*I = 14.4809

-----------------------------------------------------------------------------
```

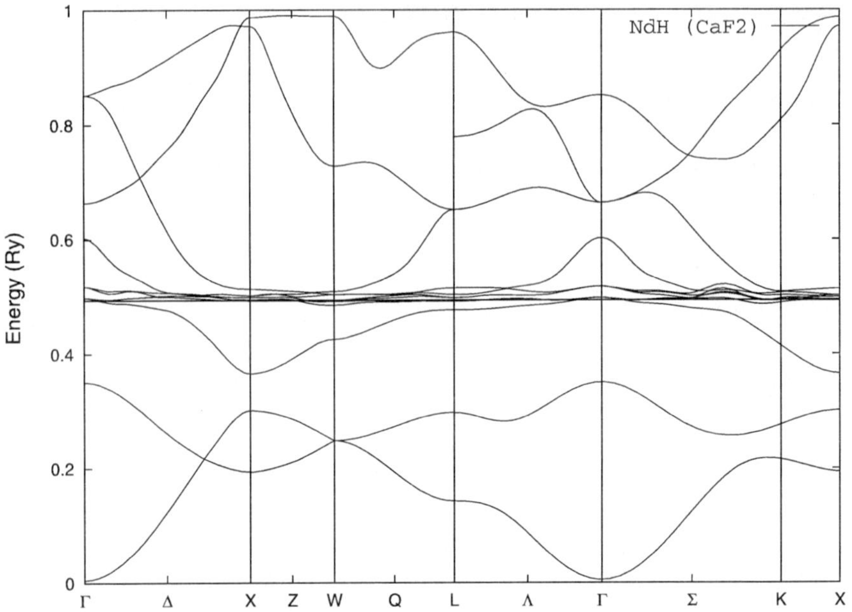

Fig. 15.7 Energy bands of NdH2 in the CaF2 structure

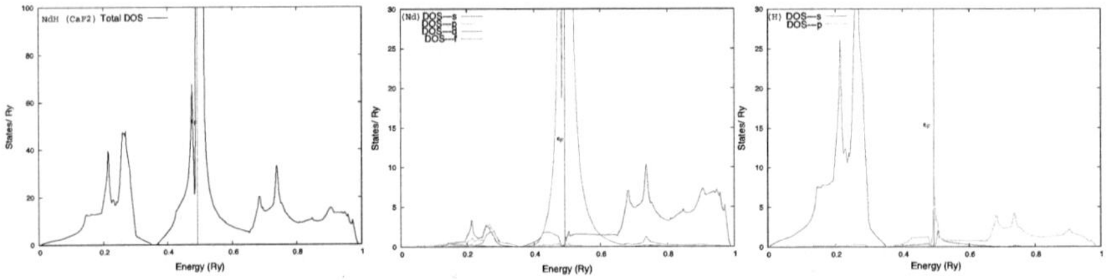

Fig. 15.8 Total, angular momentum and site decomposed densities of states of NdH2 in the CaF2 structure

15.4 Promethium Hydride (PmH)

See Tables 15.7 and 15.8 and Figs. 15.9 and 15.10.

Table 15.7 Lattice constant, bulk modulus, gap, total energy

Stru	a (Bohr)	B (MBar)	Gap	Total Energy(Ry)
NaCl	9.48	0.55	-	-20041.13129
CaF2	10.20	----	-	-20042.00809

Table 15.8 DOS at Ef, Hopfield parameter, Stoner criterion

```
-----------------------------------------------------------------------
              PmH                    NaCl
-----------------------------------------------------------------------
 Fermi Energy   Total DOS       Decomposed DOS inside the muffin tins
  ( Ry )        (States/Ry)       s        p        eg        t2g       f
-----------------------------------------------------------------------
Pm  0.643         661.846       0.023    1.567     0.927     1.160    626.671
 H  0.643         661.846       0.336    3.265     0.081     1.092      0.564
-----------------------------------------------------------------------
Fermi-Velocity (cm/s): 0.02739 x10E8
Plasmon Energy (eV)   : 1.71493
Electron-ion interaction (Hopfield parameter (eV/A^2) ): Pm 0.096  H 0.014
-----------------------------------------------------------------------
        Pm     MUFFIN-TIN RADIUS and CHARGE =    2.4898   58.3144
         H     MUFFIN-TIN RADIUS and CHARGE =    2.2527    1.5244
                 Pm       STONER I =   0.0220
                  H       STONER I =   0.0000
        STONER PARAMETER (Ry)     I =   0.0220
        STONER CRITERION        N*I =  14.5702
-----------------------------------------------------------------------
              PmH2                    CaF2
-----------------------------------------------------------------------
 Fermi Energy   Total DOS       Decomposed DOS inside the muffin tins
   (Ry)        (States/Ry)       s        p        eg        t2g       f
-----------------------------------------------------------------------
Pm  0.501      833.546       0.029    0.453     0.399     0.492    782.858
H   0.501      833.546       0.153    5.876     1.079     0.718      0.668
-----------------------------------------------------------------------
Fermi-Velocity (cm/s): 0.03846 x10E8
Plasmon Energy (eV)   : 2.45829
Electron-ion interaction (Hopfield parameter) (eV/A^2) Pm: 0.060   H: 0.031
--------------------------------------------
        Pm     MUFFIN-TIN RADIUS and CHARGE =    2.3188   57.7854
        H      MUFFIN-TIN RADIUS and CHARGE =    2.0979    1.4015
                Pm       STONER I =  0.0225
                 H       STONER I =  0.0000
        STONER PARAMETER (Ry)     I = 0.0225
        STONER CRITERION        N*I = 18.7378

-----------------------------------------------------------------------
```

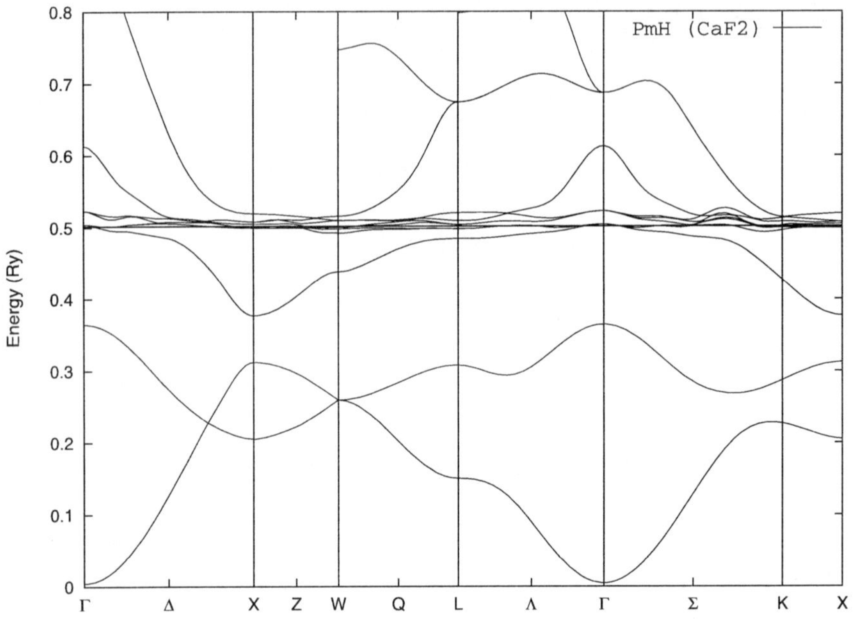

Fig. 15.9 Energy bands of PmH2 in the CaF2 structure

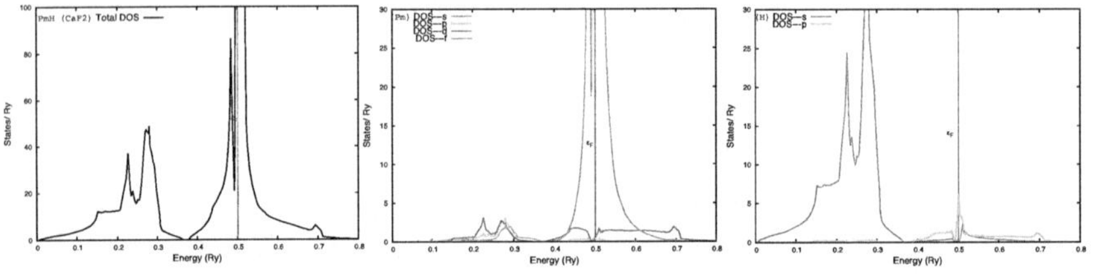

Fig. 15.10 Total, angular momentum and site decomposed densities of states of PmH2 in the CaF2 structure

15.5 Samarium Hydride (SmH)

See Tables 15.9 and 15.10 and Figs. 15.11 and 15.12.

Table 15.9 Lattice constant, bulk modulus, gap, total energy

Stru	a (Bohr)	B (MBar)	Gap	Total Energy(Ry)
NaCl	9.48	0.55	-	-20855.51324
CaF2	10.16	----	-	-20856.39524
exp (CaF2)	10.16			

Table 15.10 DOS at Ef, Hopfield parameter, Stoner criterion

```
------------------------------------------------------------------------------
                 SmH      a=9.48 Bohr          NaCl
------------------------------------------------------------------------------
Fermi Energy   Total DOS        Decomposed DOS inside the muffin tins
   ( Ry )      (States/Ry)       s         p        eg         t2g        f
------------------------------------------------------------------------------
Sm  0.631       666.294        0.009     1.490     0.911     1.389     634.292
 H  0.631       666.294        0.246     3.147     0.087     0.868       0.561
------------------------------------------------------------------------------
Fermi-Velocity (cm/s): 0.02474 x10E8
Plasmon Energy (eV)  : 1.58014
Electron-ion interaction (Hopfield parameter (eV/A^2) ): Sm 0.103 H 0.010
------------------------------------------------
         Sm     MUFFIN-TIN RADIUS and CHARGE =    2.4882   59.4038
          H     MUFFIN-TIN RADIUS and CHARGE =    2.2512    1.5049
                    Sm         STONER I =  0.0225
                     H         STONER I =  0.0000
         STONER PARAMETER (Ry)       I =  0.0225
         STONER CRITERION          N*I = 14.9875
------------------------------------------------------------------------------
                 SmH2      a=10.16 Bohr       CaF2
------------------------------------------------------------------------------
 Fermi Energy    Total DOS         Decomposed DOS inside the muffin tins
   (Ry)         (States/Ry)        s         p         eg         t2g        f
------------------------------------------------------------------------------
Sm  0.494       752.438        0.020     0.260     0.357     0.397     713.384
H   0.494       752.438        0.175     4.549     0.693     0.745       0.557
------------------------------------------------------------------------------
Fermi-Velocity (cm/s): 0.03932 x10E8
Plasmon Energy (eV)  : 2.39787
Electron-ion interaction (Hopfield parameter) (eV/A^2) Sm: 0.049   H: 0.027
-----------------------------------------------
         Sm     MUFFIN-TIN RADIUS and CHARGE =    2.3097   58.8701
         H      MUFFIN-TIN RADIUS and CHARGE =    2.0897    1.3835
                    Sm         STONER I = 0.0231
                     H         STONER I = 0.0000
         STONER PARAMETER (Ry)      I = 0.0231
         STONER CRITERION         N*I = 17.3843
------------------------------------------------------------------------------
```

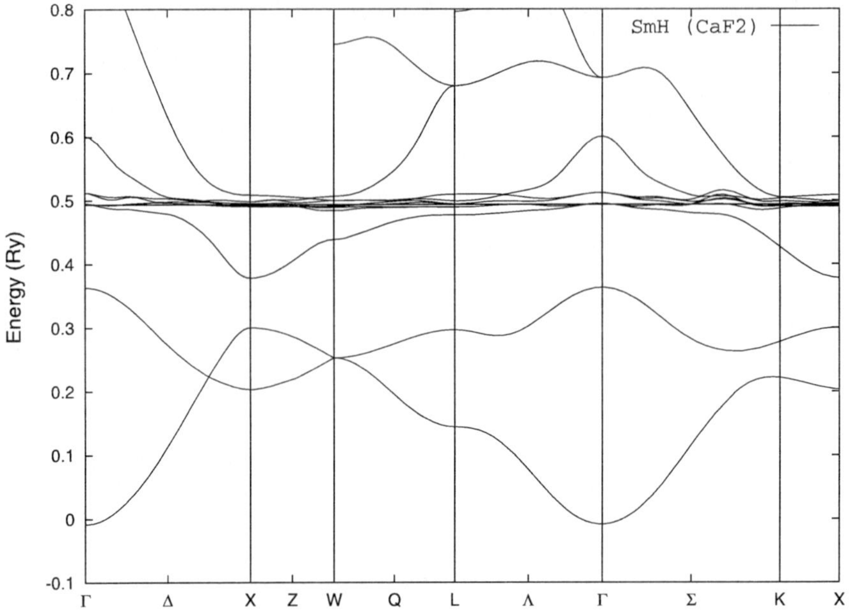

Fig. 15.11 Energy bands of SmH2 in the CaF2 structure

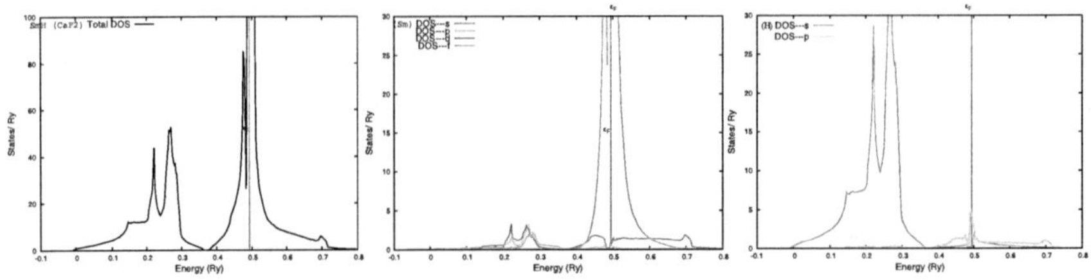

Fig. 15.12 Total, angular momentum and site decomposed densities of states of SmH2 in the CaF2 structure

15.6 Europium Hydride (EuH)

See Tables 15.11 and 15.12 and Figs. 15.13, 15.14, 15.15 and 15.16.

Table 15.11 Lattice constant, bulk modulus, gap, total energy

Stru	a (Bohr)	B (MBar)	Gap	Total Energy(Ry)
NaCl	9.46	0.55	-	-21690.43982
CaF2	10.00	----	-	-21691.30712

Table 15.12 DOS at Ef, Hopfield parameter, Stoner criterion

EuH a=9.46 Bohr NaCl

Fermi Energy (Ry)	Total DOS (States/Ry)	Decomposed DOS inside the muffin tins s	p	eg	t2g	f
Eu 0.621	611.814	0.007	1.293	1.089	0.972	585.266
H 0.621	611.814	0.153	2.903	0.107	0.423	0.556

Fermi-Velocity (cm/s): 0.02316 x10E8
Plasmon Energy (eV) : 1.31527
Electron-ion interaction (Hopfield parameter (eV/A^2)): Eu 0.092 H 0.006

Eu	MUFFIN-TIN RADIUS and CHARGE =	2.4842	60.4795
H	MUFFIN-TIN RADIUS and CHARGE =	2.2476	1.5122
Eu	STONER I =	0.0229	
H	STONER I =	0.0000	
STONER PARAMETER (Ry)	I =	0.0230	
STONER CRITERION	N*I =	14.0428	

EuH2 a=10.0 Bohr CaF2

Fermi Energy (Ry)	Total DOS (States/Ry)	Decomposed DOS inside the muffin tins s	p	eg	t2g	f
Eu 0.507	683.560	0.014	0.231	0.249	0.415	653.638
H 0.507	683.560	0.326	3.684	0.425	0.958	0.523

Fermi-Velocity (cm/s): 0.04385 x10E8
Plasmon Energy (eV) : 2.55423
Electron-ion interaction (Hopfield parameter) (eV/A^2) Eu: 0.051 H: 0.037

Eu	MUFFIN-TIN RADIUS and CHARGE =	2.2733	59.8744
H	MUFFIN-TIN RADIUS and CHARGE =	2.0568	1.3754
Eu	STONER I =	0.0238	
H	STONER I =	0.0000	
STONER PARAMETER (Ry)	I =	0.0238	
STONER CRITERION	N*I =	16.2696	

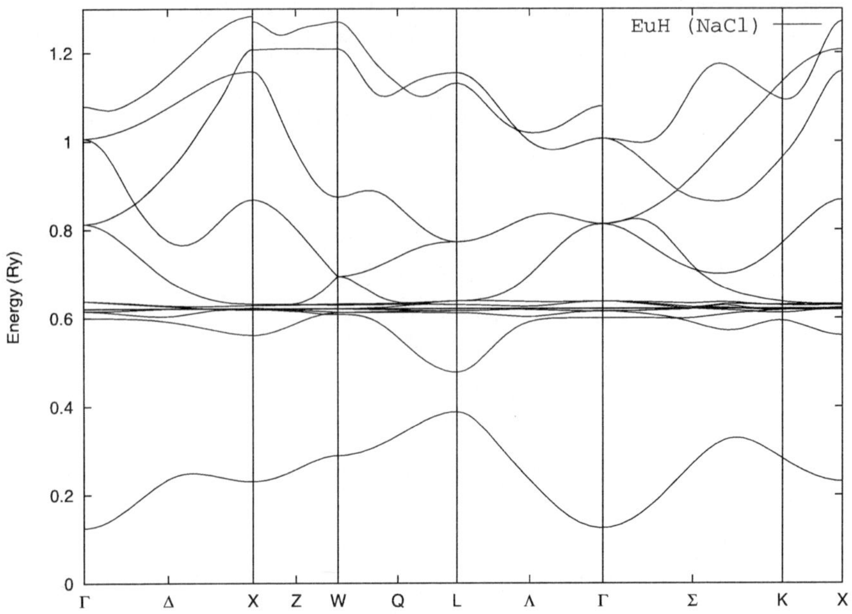

Fig. 15.13 Energy bands of EuH in the NaCl structure

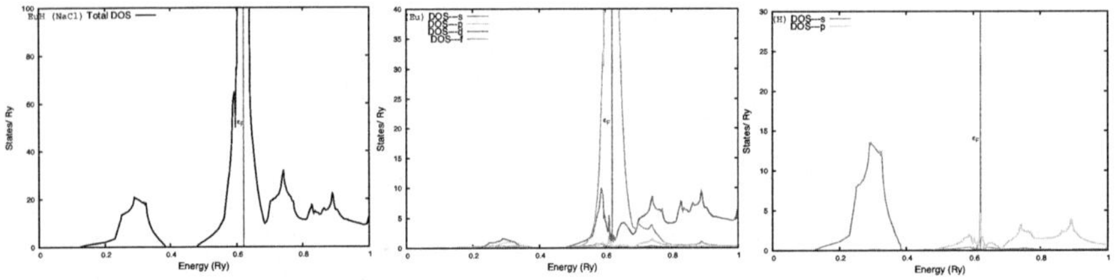

Fig. 15.14 Total, angular momentum and site decomposed densities of states of EuH in the NaCl structure

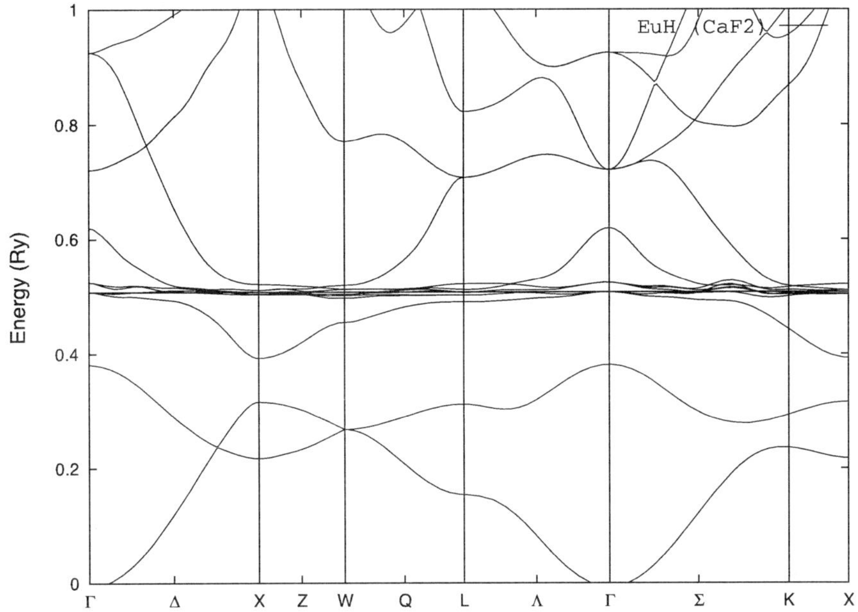

Fig. 15.15 Energy bands of EuH2 in the CaF2 structure

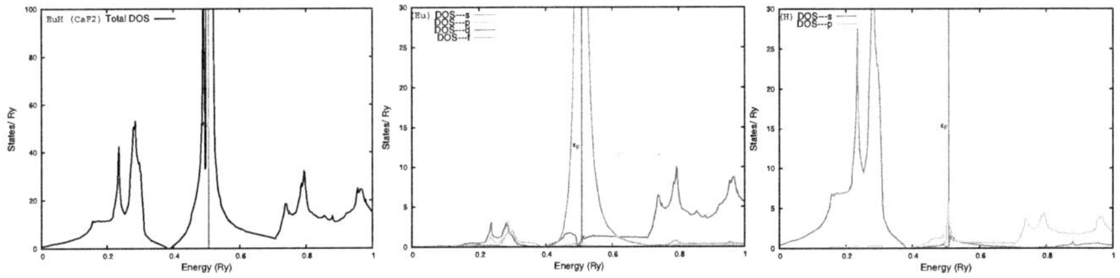

Fig. 15.16 Total, angular momentum and site decomposed densities of states of EuH2 in the CaF2 structure

15.7 Gadolinium Hydride (GdH)

See Tables 15.13 and 15.14 and Figs. 15.17 and 15.18.

Table 15.13 Lattice constant, bulk modulus, gap, total energy

Stru	a (Bohr)	B (MBar)	Gap	Total Energy(Ry)
NaCl	9.45	0.64	-	-22546.20246
CaF2	10.04	----	-	-22547.08617

Table 15.14 DOS at Ef, Hopfield parameter, Stoner criterion

GdH a=9.45 Bohr NaCl

Fermi Energy (Ry)	Total DOS (States/Ry)	Decomposed DOS inside the muffin tins s	p	eg	t2g	f
Gd 0.617	293.547	0.001	0.217	0.624	0.657	283.477
H 0.617	293.547	0.022	1.623	0.070	0.152	0.121

Fermi-Velocity (cm/s): 0.03615 x10E8
Plasmon Energy (eV) : 1.52508
Electron-ion interaction (Hopfield parameter (eV/A^2)): Gd 0.037 H 0.001

Gd	MUFFIN-TIN RADIUS and CHARGE =	2.4801	61.5505
H	MUFFIN-TIN RADIUS and CHARGE =	2.2439	1.5088
Gd	STONER I = 0.0236		
H	STONER I = 0.0000		
STONER PARAMETER (Ry)	I = 0.0237		
STONER CRITERION	N*I = 6.9446		

GdH2 a=10.04 Bohr CaF2

Fermi Energy (Ry)	Total DOS (States/Ry)	Decomposed DOS inside the muffin tins s	p	eg	t2g	f
Gd 0.492	758.875	0.009	0.244	0.147	0.582	739.323
H 0.492	758.875	0.615	2.983	0.302	1.077	0.589

Fermi-Velocity (cm/s): 0.04485 x10E8
Plasmon Energy (eV) : 2.79624
Electron-ion interaction (Hopfield parameter) (eV/A^2) Gd: 0.055 H: 0.044

Gd	MUFFIN-TIN RADIUS and CHARGE =	2.2824	60.9883
H	MUFFIN-TIN RADIUS and CHARGE =	2.0650	1.3733
Gd	STONER I = 0.0249		
H	STONER I = 0.0000		
STONER PARAMETER (Ry)	I = 0.0249		
STONER CRITERION	N*I = 18.8640		

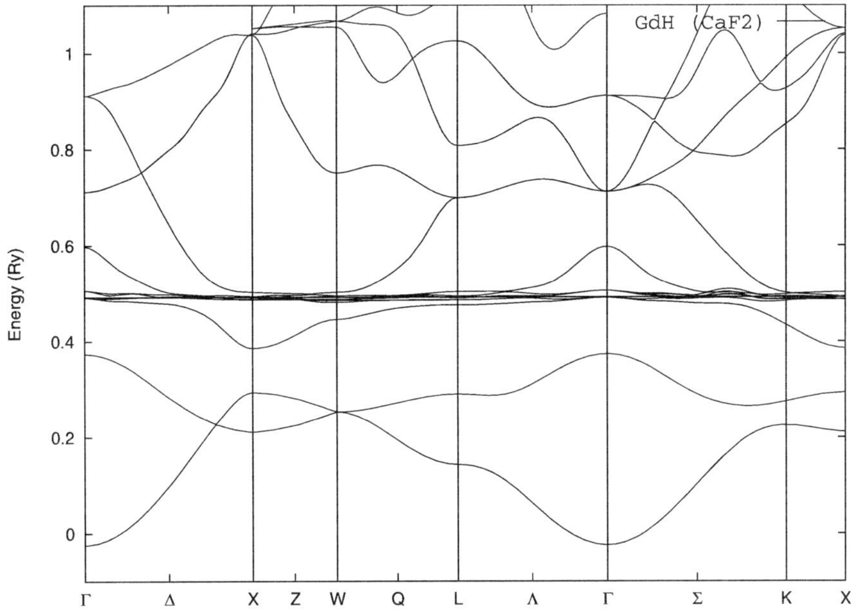

Fig. 15.17 Energy bands of GdH2 in the CaF2 structure

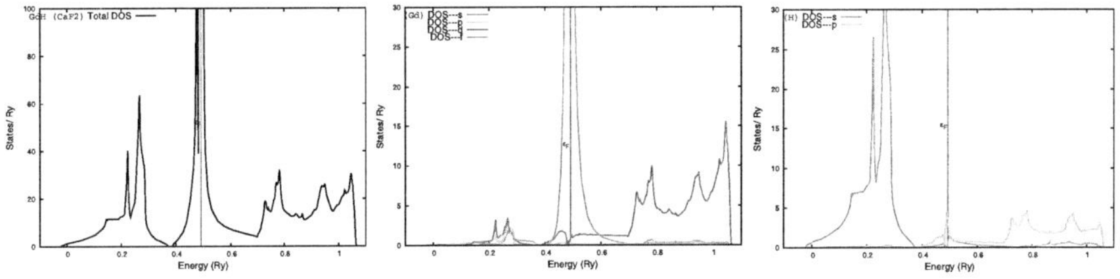

Fig. 15.18 Total, angular momentum and site decomposed densities of states of GdH2 in the CaF2 structure

15.8 Terbium Hydride (TbH)

See Tables 15.15 and 15.16 and Figs. 15.19 and 15.20.

Table 15.15 Lattice constant, bulk modulus, gap, total energy

Stru	a (Bohr)	B (MBar)	Gap	Total Energy(Ry)
NaCl	9.47	0.48	-	-23423.09441
CaF2	9.88	----	-	-23423.96421

Table 15.16 DOS at Ef, Hopfield parameter, Stoner criterion

```
-------------------------------------------------------------------------
               TbH        a=9.47 Bohr        NaCl
-------------------------------------------------------------------------
Fermi Energy   Total DOS        Decomposed DOS inside the muffin tins
   ( Ry )      (States/Ry)      s        p        eg        t2g        f
-------------------------------------------------------------------------
Tb  0.606      521.718         0.006    0.099    0.579     1.020    505.999
 H  0.606      521.718         0.039    1.472    0.216     0.318      0.204
-------------------------------------------------------------------------
Fermi-Velocity (cm/s): 0.03288 x10E8
Plasmon Energy (eV)  : 1.84364
Electron-ion interaction (Hopfield parameter (eV/A^2) ): Tb 0.042 H 0.001
---------------------------------------------
         Tb     MUFFIN-TIN RADIUS and CHARGE =    2.4854   62.6217
          H     MUFFIN-TIN RADIUS and CHARGE =    2.2487    1.4984
                   Tb         STONER I =  0.0241
                    H         STONER I =  0.0000
         STONER PARAMETER (Ry)      I =  0.0241
         STONER CRITERION         N*I = 12.5765
-------------------------------------------------------------------------
               TbH2          a=9.88 Bohr          CaF2
-------------------------------------------------------------------------
 Fermi Energy   Total DOS        Decomposed DOS inside the muffin tins
   (Ry)        (States/Ry)      s        p        eg        t2g        f
-------------------------------------------------------------------------
Tb  0.509      831.745        0.020    0.287    0.172     0.737    804.313
H   0.509      831.745        1.036    2.842    0.337     0.957      0.634
-------------------------------------------------------------------------
Fermi-Velocity (cm/s): 0.04460 x10E8
Plasmon Energy (eV)  : 2.98076
Electron-ion interaction (Hopfield parameter) (eV/A^2) Tb: 0.081   H: 0.062
---------------------------------------------
         Tb     MUFFIN-TIN RADIUS and CHARGE =    2.2460   61.9814
         H      MUFFIN-TIN RADIUS and CHARGE =    2.0321    1.3668
                  Tb         STONER I =  0.0248
                   H         STONER I =  0.0000
         STONER PARAMETER (Ry)      I =  0.0248
         STONER CRITERION         N*I = 20.6261
-------------------------------------------------------------------------
```

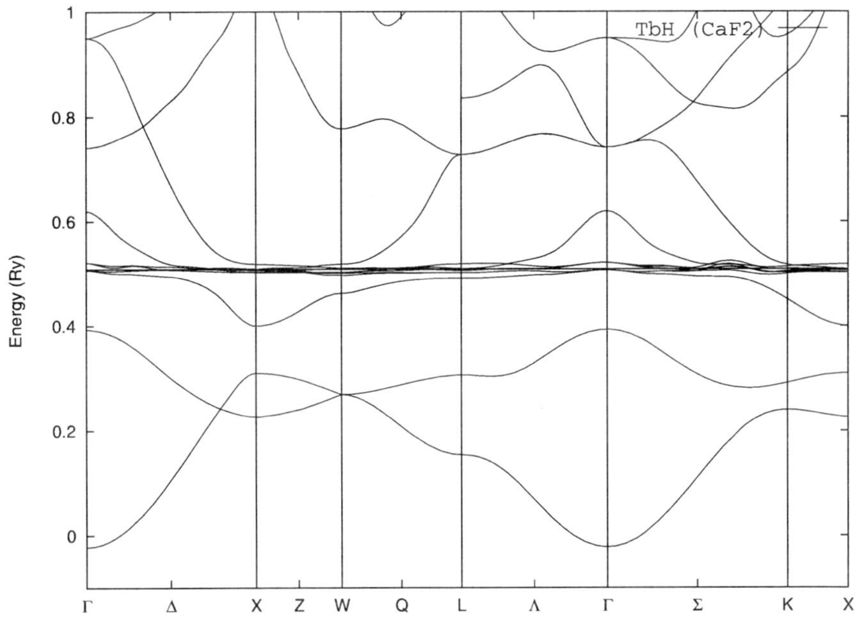

Fig. 15.19 Energy bands of TbH2 in the CaF2 structure

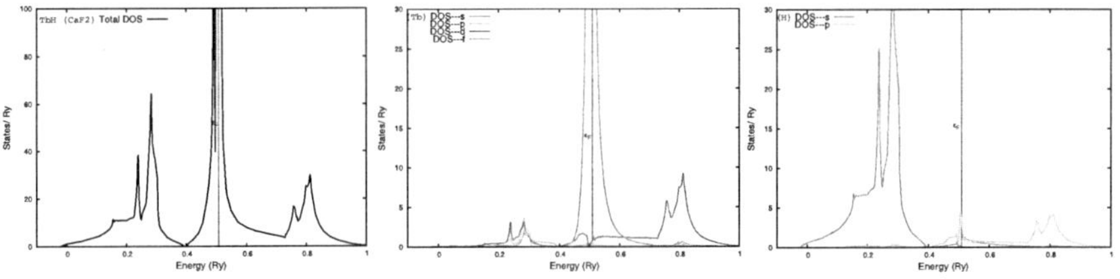

Fig. 15.20 Total, angular momentum and site decomposed densities of states of TbH2 in the CaF2 structure

15.9 Dysprosium Hydride (DyH)

See Tables 15.17 and 15.18 and Figs. 15.21 and 15.22.

Table 15.17 Lattice constant, bulk modulus, gap, total energy

Stru	a (Bohr)	B (MBar)	Gap	Total Energy(Ry)
NaCl	9.44	0.50	-	-24321.41782
CaF2	9.81	----	-	-24322.28864

Table 15.18 DOS at Ef, Hopfield parameter, Stoner criterion

```
 -----------------------------------------------------------------------------
                 DyH          a=9.44 Bohr             NaCl
 -----------------------------------------------------------------------------
 Fermi Energy    Total DOS         Decomposed DOS inside the muffin tins
    ( Ry )       (States/Ry)         s         p         eg         t2g         f
 -----------------------------------------------------------------------------
 Dy  0.605        844.423        0.019     0.084     0.414      0.828     807.723
  H  0.605        844.423        0.088     1.282     0.389      0.188       0.273
 -----------------------------------------------------------------------------
Fermi-Velocity (cm/s): 0.02426 x10E8
Plasmon Energy (eV)   : 1.69899
Electron-ion interaction (Hopfield parameter (eV/A^2) ): Dy 0.035 H 0.001
------------------------------------------------
          Dy      MUFFIN-TIN RADIUS and CHARGE =    2.4785   63.6687
           H      MUFFIN-TIN RADIUS and CHARGE =    2.2425    1.4927
                    Dy        STONER I =   0.0237
                     H        STONER I =   0.0000
          STONER PARAMETER (Ry)      I =   0.0237
          STONER CRITERION         N*I =  20.0023
 -----------------------------------------------------------------------------
                 DyH2         a=9.81 Bohr            CaF2
 -----------------------------------------------------------------------------
  Fermi Energy    Total DOS         Decomposed DOS inside the muffin tins
     (Ry)        (States/Ry)         s         p         eg         t2g         f
 -----------------------------------------------------------------------------
 Dy  0.514       869.075        0.029     0.325      0.182      0.936     837.233
 H   0.514       869.075        1.525     2.630      0.345      0.846       0.643
 -----------------------------------------------------------------------------
Fermi-Velocity (cm/s): 0.04403 x10E8
Plasmon Energy (eV)   : 3.02892
Electron-ion interaction (Hopfield parameter) (eV/A^2) Dy: 0.107   H: 0.079
------------------------------------------------
          Dy     MUFFIN-TIN RADIUS and CHARGE =    2.2301   63.0189
          H      MUFFIN-TIN RADIUS and CHARGE =    2.0177    1.3646
                   Dy       STONER I =  0.0248
                    H       STONER I =  0.0000
          STONER PARAMETER (Ry)     I =  0.0249
          STONER CRITERION        N*I = 21.5951
 -----------------------------------------------------------------------------
```

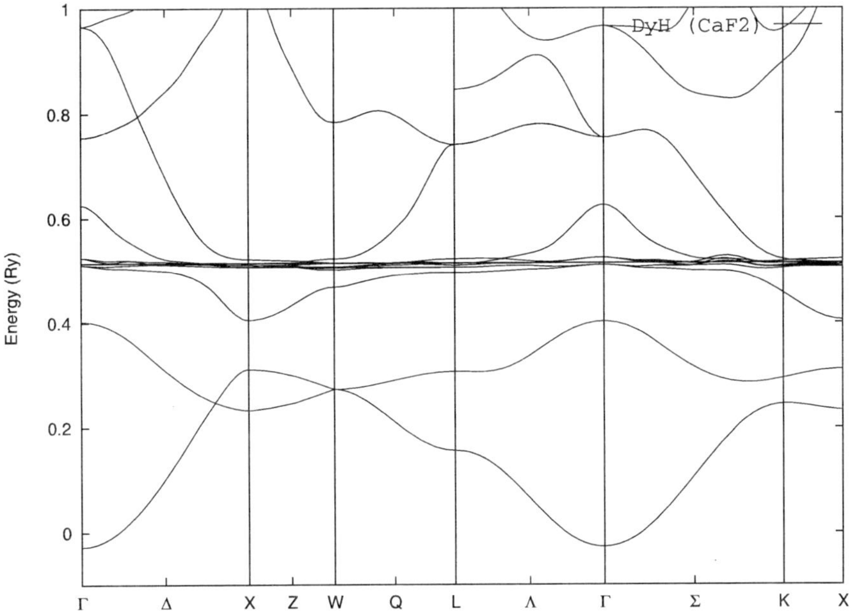

Fig. 15.21 Energy bands of DyH2 in the CaF2 structure

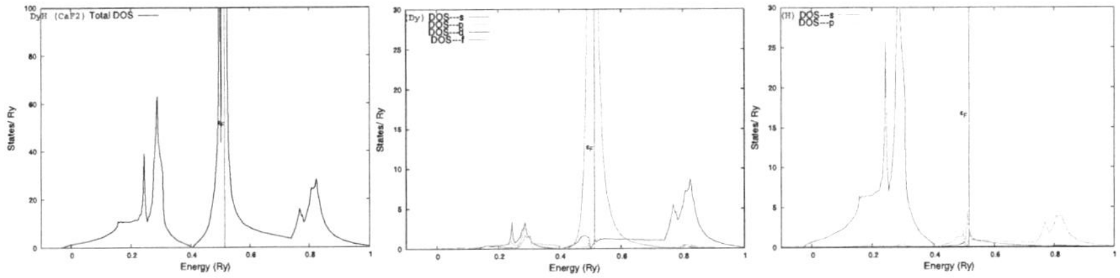

Fig. 15.22 Total, angular momentum and site decomposed densities of states of DyH2 in the CaF2 structure

15.10 Holmium Hydride (HoH)

See Tables 15.19 and 15.20 and Figs. 15.23 and 15.24.

Table 15.19 Lattice constant, bulk modulus, gap, total energy

Stru	a (Bohr)	B (MBar)	Gap	Total Energy(Ry)
NaCl	9.44	0.59	-	-25241.47982
CaF2	9.73	----	-	-25242.34828

Table 15.20 DOS at Ef, Hopfield parameter, Stoner criterion

HoH a=9.44 Bohr NaCl

Fermi Energy (Ry)		Total DOS (States/Ry)	Decomposed DOS inside the muffin tins s	p	eg	t2g	f
Ho	0.601	809.519	0.033	0.133	0.388	0.700	779.430
H	0.601	809.519	0.299	0.861	0.445	0.139	0.227

Fermi-Velocity (cm/s): 0.02089 x10E8
Plasmon Energy (eV) : 1.45633
Electron-ion interaction (Hopfield parameter (eV/A^2)): Ho 0.035 H 0.003

Ho	MUFFIN-TIN RADIUS and CHARGE =	2.4778	64.7185
H	MUFFIN-TIN RADIUS and CHARGE =	2.2418	1.4893
Ho	STONER I =	0.0242	
H	STONER I =	0.0000	
STONER PARAMETER (Ry)	I =	0.0242	
STONER CRITERION	N*I =	19.6233	

HoH2 a=9.73 Bohr CaF2

Fermi Energy (Ry)		Total DOS (States/Ry)	Decomposed DOS inside the muffin tins s	p	eg	t2g	f
Ho	0.522	756.899	0.039	0.367	0.234	1.023	722.599
H	0.522	756.899	2.007	2.267	0.294	0.646	0.562

Fermi-Velocity (cm/s): 0.04480 x10E8
Plasmon Energy (eV) : 2.92025
Electron-ion interaction (Hopfield parameter) (eV/A^2) Ho: 0.133 H: 0.103

Ho	MUFFIN-TIN RADIUS and CHARGE =	2.2119	64.0459
H	MUFFIN-TIN RADIUS and CHARGE =	2.0013	1.3614
Ho	STONER I =	0.0246	
H	STONER I =	0.0000	
STONER PARAMETER (Ry)	I =	0.0246	
STONER CRITERION	N*I =	18.6571	

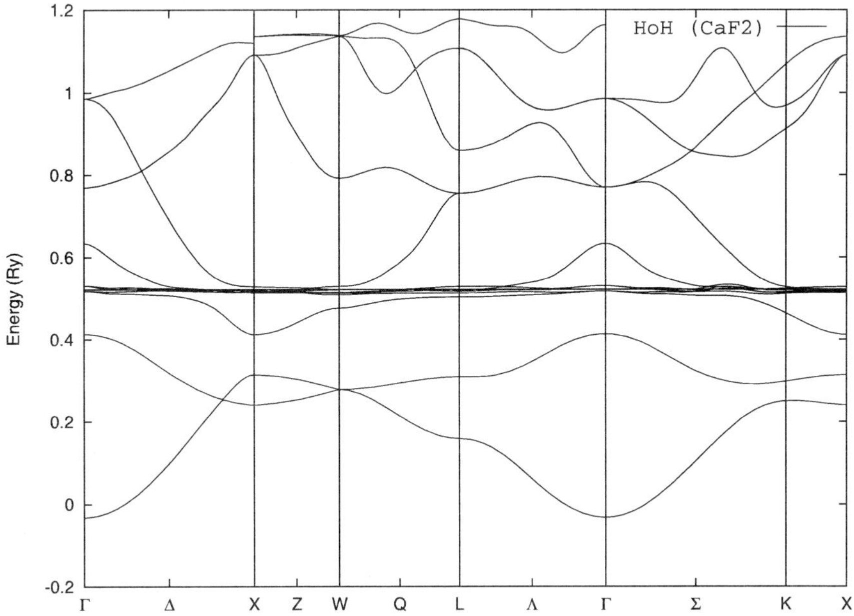

Fig. 15.23 Energy bands of HoH2 in the CaF2 structure

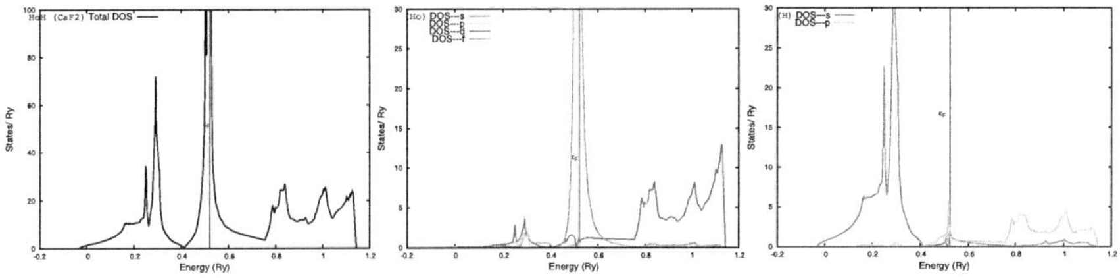

Fig. 15.24 Total, angular momentum and site decomposed densities of states of HoH2 in the CaF2 structure

15.11 Erbium Hydride (ErH)

See Tables 15.21 and 15.22 and Figs. 15.25 and 15.26.

Table 15.21 Lattice constant, bulk modulus, gap, total energy

Stru	a (Bohr)	B (MBar)	Gap	Total Energy(Ry)
NaCl	9.47	0.56	-	-26183.59339
CaF2	9.67	----	-	-26184.47024

Table 15.22 DOS at Ef, Hopfield parameter, Stoner criterion

```
--------------------------------------------------------------------------
               ErH        a=9.47 Bohr            NaCl
--------------------------------------------------------------------------
Fermi Energy    Total DOS        Decomposed DOS inside the muffin tins
   ( Ry )       (States/Ry)        s          p         eg        t2g        f
--------------------------------------------------------------------------
Er  0.591        648.674         0.070      0.122     0.314     0.860     626.912
 H  0.591        648.674         0.277      0.535     0.362     0.127       0.131
--------------------------------------------------------------------------
Fermi-Velocity (cm/s): 0.07433 x10E8
Plasmon Energy (eV)   : 4.26099
Electron-ion interaction (Hopfield parameter (eV/A^2) ): Er 0.039 H 0.002
----------------------------------------------
         Er      MUFFIN-TIN RADIUS and CHARGE =    2.4846   65.7724
          H      MUFFIN-TIN RADIUS and CHARGE =    2.2480    1.4868
                   Er        STONER I =    0.0247
                    H        STONER I =    0.0000
         STONER PARAMETER (Ry)      I =    0.0247
         STONER CRITERION         N*I =   15.9943
--------------------------------------------------------------------------
               ErH2       a=9.67 Bohr                  CaF2
--------------------------------------------------------------------------
 Fermi Energy    Total DOS        Decomposed DOS inside the muffin tins
   (Ry)          (States/Ry)       s          p         eg        t2g        f
--------------------------------------------------------------------------
Er  0.525        538.929         0.018      0.288     0.116     0.950     520.710
H   0.525        538.929         1.887      1.710     0.216     0.349       0.426
--------------------------------------------------------------------------
Fermi-Velocity (cm/s): 0.05008 x10E8
Plasmon Energy (eV)   : 2.72215
Electron-ion interaction (Hopfield parameter) (eV/A^2) Er: 0.125   H: 0.103
---------------------------------------------
         Er     MUFFIN-TIN RADIUS and CHARGE =    2.1983    65.0686
         H      MUFFIN-TIN RADIUS and CHARGE =    1.9889     1.3574
                  Er        STONER I =   0.0255
                   H        STONER I =   0.0000
         STONER PARAMETER (Ry)     I =   0.0255
         STONER CRITERION        N*I = 13.7351
--------------------------------------------------------------------------
```

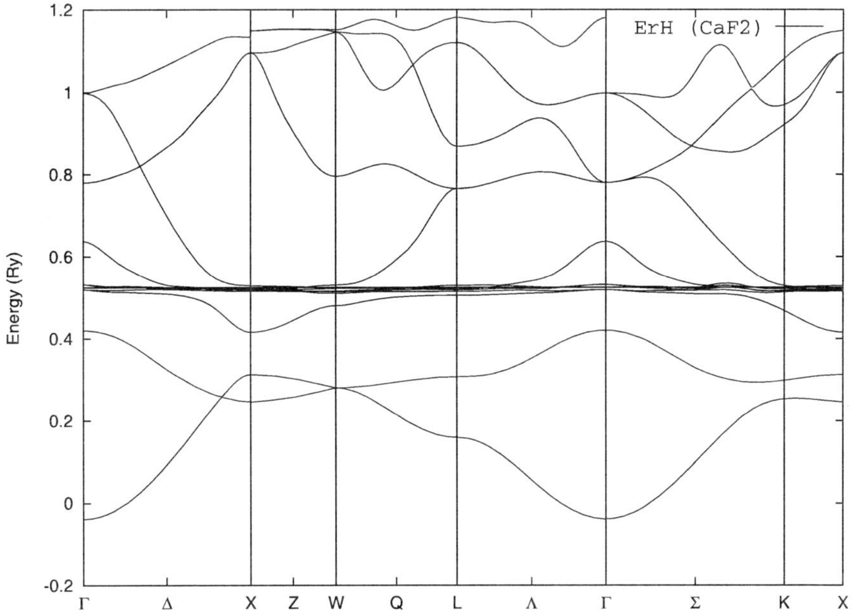

Fig. 15.25 Energy bands of ErH2 in the CaF2 structure

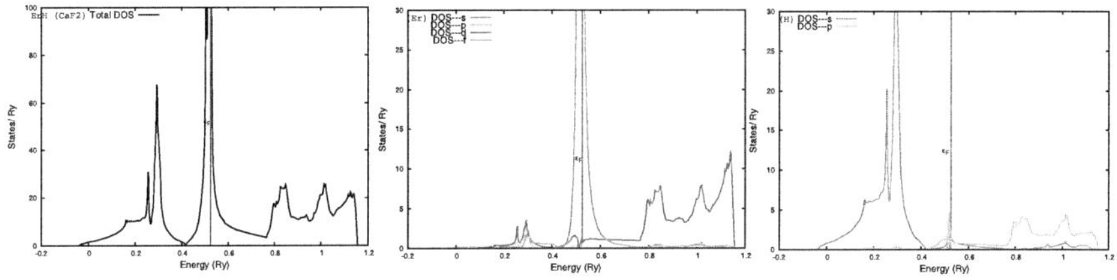

Fig. 15.26 Total, angular momentum and site decomposed densities of states of ErH2 in the CaF2 structure

15.12 Thulium Hydride (TmH)

See Tables 15.23 and 15.24 and Figs. 15.27 and 15.28.

Table 15.23 Lattice constant, bulk modulus, gap, total energy

Stru	a (Bohr)	B (MBar)	Gap	Total Energy(Ry)
NaCl	9.48	0.51	-	-27148.07583
CaF2	10.54	----	-	-27149.08738

Table 15.24 DOS at Ef, Hopfield parameter, Stoner criterion

```
---------------------------------------------------------------------------
               TmH        a=9.48 Bohr            NaCl
---------------------------------------------------------------------------
Fermi Energy   Total DOS         Decomposed DOS inside the muffin tins
   ( Ry )      (States/Ry)       s         p        eg         t2g        f
---------------------------------------------------------------------------
Tm  0.585        190.858       0.173     0.148    0.172      1.130     184.922
 H  0.585        190.858       0.322     0.427    0.040      0.127       0.028
---------------------------------------------------------------------------
Fermi-Velocity (cm/s): 0.06024 x10E8
Plasmon Energy (eV)   : 1.80434
Electron-ion interaction (Hopfield parameter (eV/A^2) ): Tm 0.066 H 0.006
-----------------------------------------------
          Tm      MUFFIN-TIN RADIUS and CHARGE =    2.4888   66.8160
           H      MUFFIN-TIN RADIUS and CHARGE =    2.2518    1.4849
                    Tm         STONER I =   0.0251
                     H         STONER I =   0.0000
          STONER PARAMETER (Ry)      I =   0.0252
          STONER CRITERION         N*I =   4.8005
---------------------------------------------------------------------------
               TmH2       a=10.54 Bohr            CaF2
---------------------------------------------------------------------------
 Fermi Energy   Total DOS         Decomposed DOS inside the muffin tins
   (Ry)         (States/Ry)       s         p         eg        t2g        f
---------------------------------------------------------------------------
Tm  0.391       589.901        0.029     0.127     0.222     0.759     577.083
H   0.391       589.901        1.495     1.256     0.144     0.192       0.368
---------------------------------------------------------------------------
Fermi-Velocity (cm/s): 0.03774 x10E8
Plasmon Energy (eV)   : 1.87551
Electron-ion interaction (Hopfield parameter) (eV/A^2) Tm: 0.048   H: 0.039
-----------------------------------------------
          Tm     MUFFIN-TIN RADIUS and CHARGE =    2.3961    66.4944
          H      MUFFIN-TIN RADIUS and CHARGE =    2.1679     1.3739
                   Tm        STONER I =  0.0257
                    H        STONER I =  0.0000
          STONER PARAMETER (Ry)     I =  0.0257
          STONER CRITERION        N*I = 15.1854
---------------------------------------------------------------------------
```

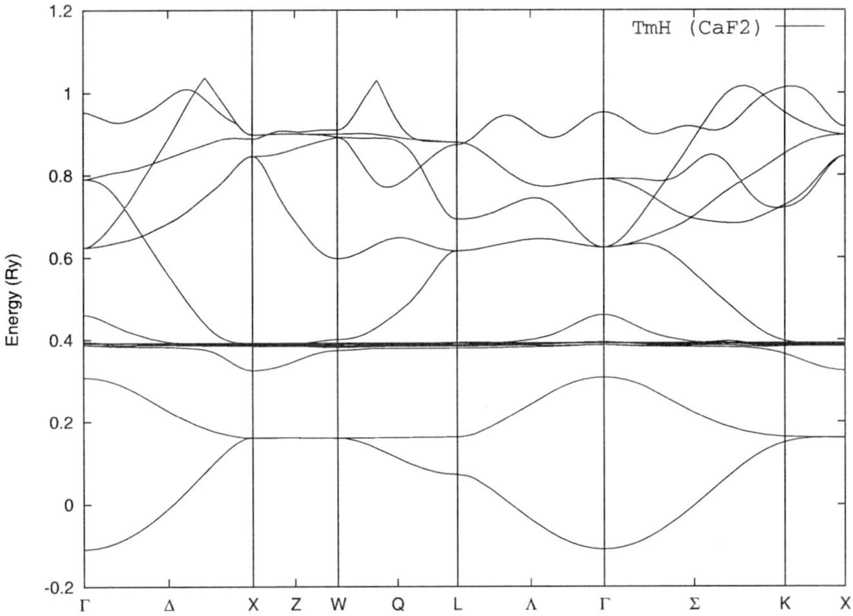

Fig. 15.27 Energy bands of TmH2 in the CaF2 structure

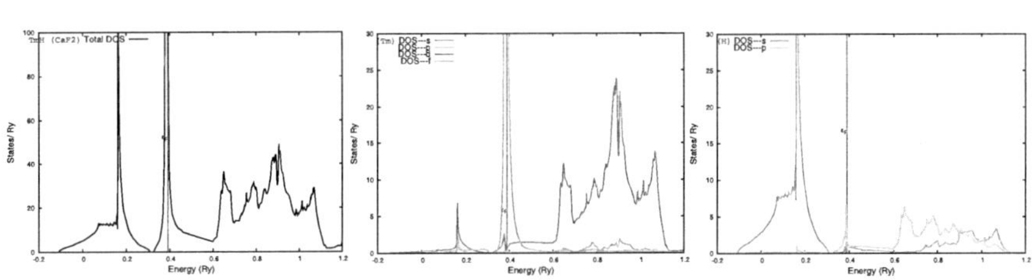

Fig. 15.28 Total, angular momentum and site decomposed densities of states of TmH2 in the CaF2 structure

15.13 Ytterbium Hydride (YbH)

See Tables 15.25 and 15.26 and Figs. 15.29, 15.30, 15.31 and 15.32.

Table 15.25 Lattice constant, bulk modulus, gap, total energy

Stru	a (Bohr)	B (MBar)	Gap	Total Energy(Ry)
NaCl	10.40	----	-	-28135.22447
CaF2	10.60	----	-	-28136.27370

Table 15.26 DOS at Ef, Hopfield parameter, Stoner criterion

```
-----------------------------------------------------------------------------
                YbH      a=10.40 Bohr          NaCl
-----------------------------------------------------------------------------
Fermi Energy   Total DOS        Decomposed DOS inside the muffin tins
   ( Ry )      (States/Ry)       s          p         eg         t2g        f
-----------------------------------------------------------------------------
Yb  0.447        20.303        0.889      0.481     0.499      2.541      6.302
 H  0.447        20.303        0.818      1.505     0.010      0.176      0.017
-----------------------------------------------------------------------------
Fermi-Velocity (cm/s): 0.41829 x10E8
Plasmon Energy (eV)   : 4.04646
Electron-ion interaction (Hopfield parameter (eV/A^2) ): Yb 0.547 H 0.393
------------------------------------------------
         Yb      MUFFIN-TIN RADIUS and CHARGE =    2.7300    68.0856
          H      MUFFIN-TIN RADIUS and CHARGE =    2.4700     1.4958
                    Yb         STONER I =    0.0046
                     H         STONER I =    0.0017
         STONER PARAMETER (Ry)       I =    0.0064
         STONER CRITERION          N*I =    0.1290
-----------------------------------------------------------------------------
                YbH2     a=10.60 Bohr          CaF2
-----------------------------------------------------------------------------
 Fermi Energy   Total DOS        Decomposed DOS inside the muffin tins
   (Ry)        (States/Ry)       s          p         eg         t2g        f
-----------------------------------------------------------------------------
Yb  0.385        24.725        0.053      0.062     0.608      0.316     18.105
H   0.385        24.725        0.948      0.732     0.016      0.055      0.014
-----------------------------------------------------------------------------
Fermi-Velocity (cm/s): 0.22356 x10E8
Plasmon Energy (eV)   : 2.31540
Electron-ion interaction (Hopfield parameter) (eV/A^2) Yb: 0.101   H: 0.328
------------------------------------------------
         Yb     MUFFIN-TIN RADIUS and CHARGE =    2.4097    67.5484
         H      MUFFIN-TIN RADIUS and CHARGE =    2.1802     1.3744
                   Yb         STONER I =   0.0151
                    H         STONER I =   0.0002
         STONER PARAMETER (Ry)       I =   0.0154
         STONER CRITERION          N*I =   0.3813
-----------------------------------------------------------------------------
```

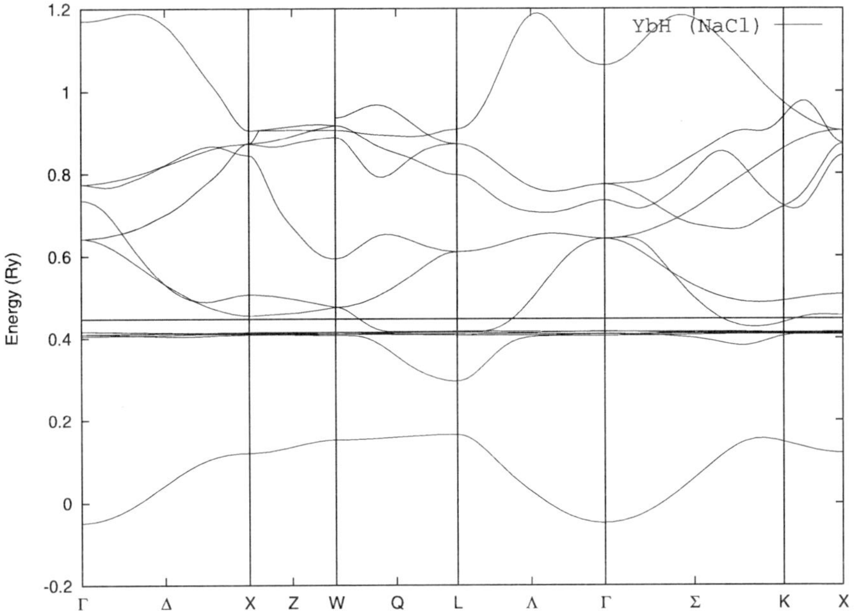

Fig. 15.29 Energy bands of YbH in the NaCl structure

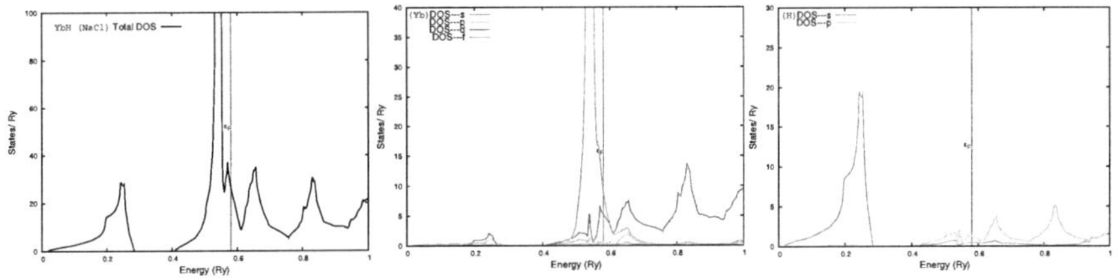

Fig. 15.30 Total, angular momentum and site decomposed densities of states of YbH in the NaCl structure

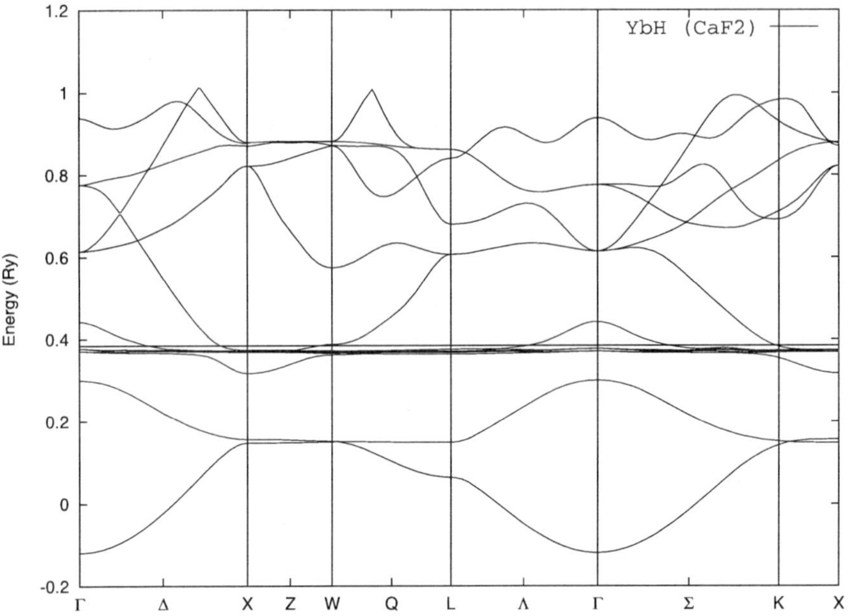

Fig. 15.31 Energy bands of YbH2 in the CaF2 structure

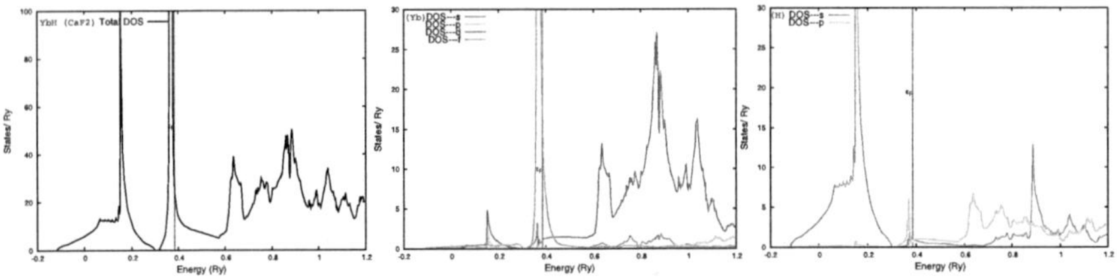

Fig. 15.32 Total, angular momentum and site decomposed densities of states of YbH2 in the CaF2 structure

15.14 Lutetium Hydride (LuH)

See Tables 15.27 and 15.28 and Figs. 15.33, 15.34, 15.35 and 15.36.

Table 15.27 Lattice constant, bulk modulus, gap, total energy

Stru	a (Bohr)	B (MBar)	Gap	Total Energy(Ry)
NaCl	9.19	0.76	-	-29145.30029
CaF2	9.43	----	-	-29146.22757

Table 15.28 DOS at Ef, Hopfield parameter, Stoner criterion

LuH a=9.19 Bohr NaCl

Fermi Energy (Ry)		Total DOS (States/Ry)	Decomposed DOS inside the muffin tins s	p	eg	t2g	f
Lu	0.662	23.156	0.306	1.112	0.714	5.744	0.222
H	0.662	23.156	0.434	1.969	0.007	0.385	0.065

Fermi-Velocity (cm/s): 0.47566 x10E8
Plasmon Energy (eV) : 5.91352
Electron-ion interaction (Hopfield parameter (eV/A^2)): Lu 4.829 H 0.368

Lu	MUFFIN-TIN RADIUS and CHARGE =	2.4130	68.5285
H	MUFFIN-TIN RADIUS and CHARGE =	2.1832	1.5077
Lu	STONER I =	0.0055	
H	STONER I =	0.0020	
STONER PARAMETER (Ry)	I =	0.0055	
STONER CRITERION	N*I =	0.1266	

LuH2 a=9.43 Bohr CaF2

Fermi Energy (Ry)		Total DOS (States/Ry)	Decomposed DOS inside the muffin tins s	p	eg	t2g	f
Lu	0.612	6.428	0.004	0.059	0.918	0.251	0.128
H	0.612	6.428	0.428	0.690	0.048	0.084	0.010

Fermi-Velocity (cm/s): 0.90166 x10E8
Plasmon Energy (eV) : 5.68434
Electron-ion interaction (Hopfield parameter) (eV/A^2) Lu: 0.715 H: 0.886

Lu	MUFFIN-TIN RADIUS and CHARGE =	2.1437	67.9012
H	MUFFIN-TIN RADIUS and CHARGE =	1.9396	1.3778
Lu	STONER I =	0.0012	
H	STONER I =	0.0014	
STONER PARAMETER (Ry)	I =	0.0044	
STONER CRITERION	N*I =	0.0283	

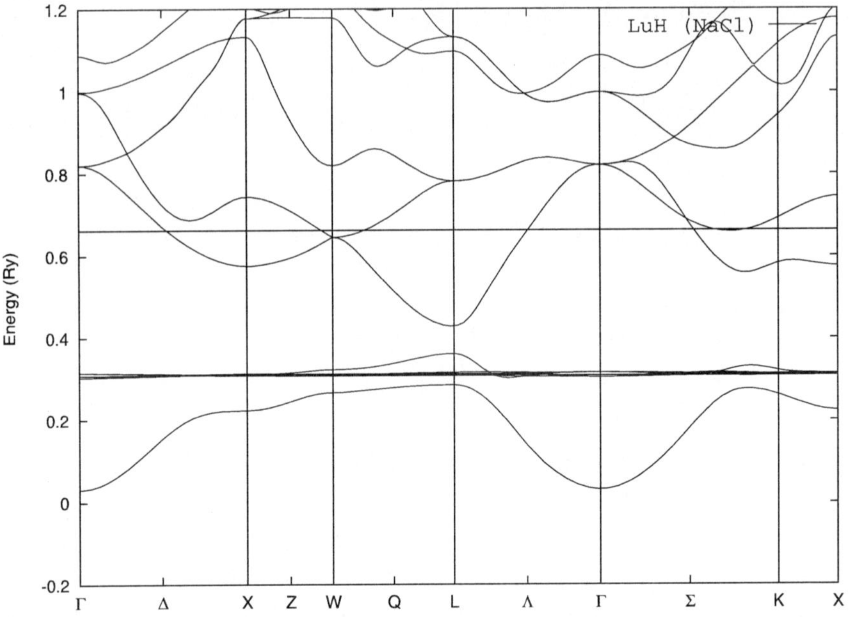

Fig. 15.33 Energy bands of LuH in the NaCl structure

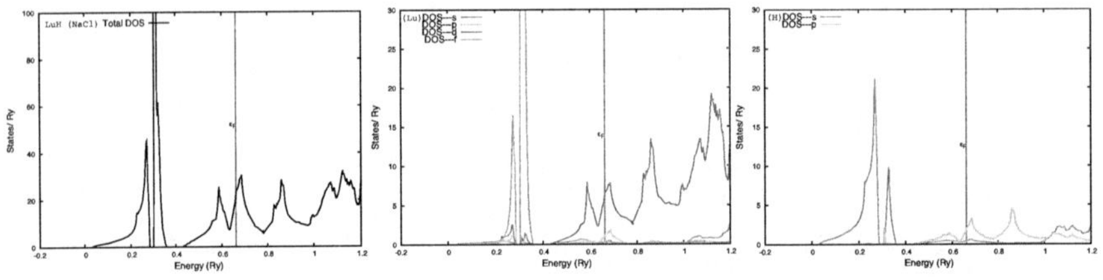

Fig. 15.34 Total, angular momentum and site decomposed densities of states of LuH in the NaCl structure

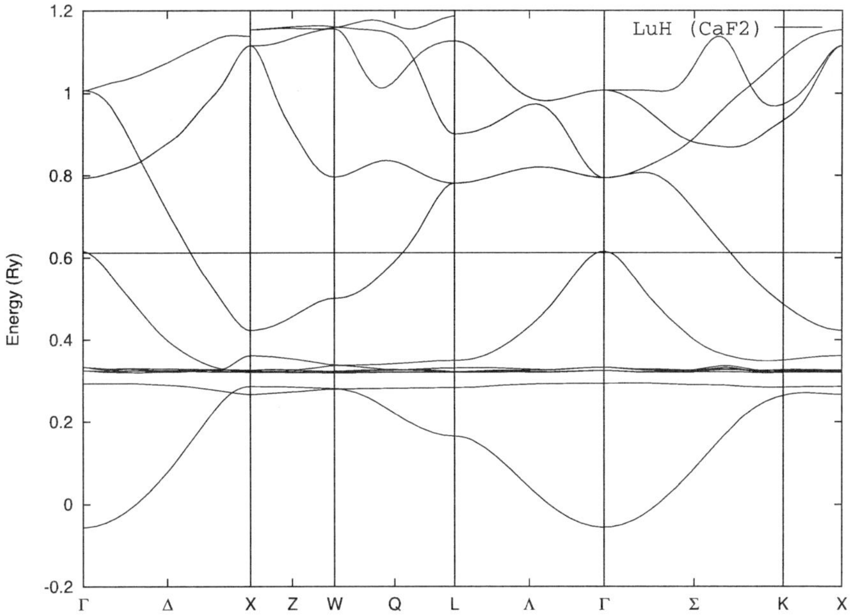

Fig. 15.35 Energy bands of LuH2 in the CaF2 structure

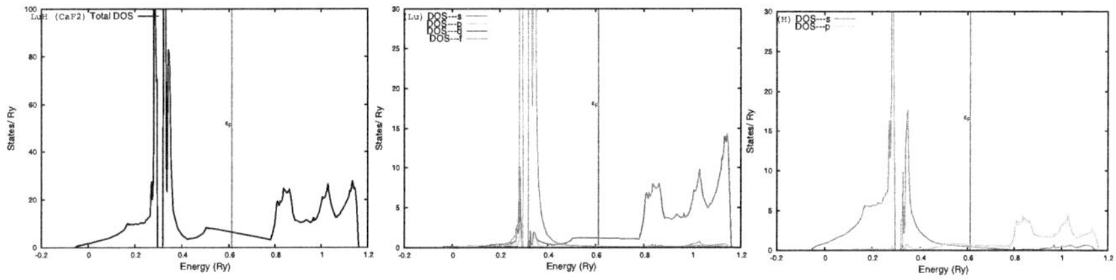

Fig. 15.36 Total, angular momentum and site decomposed densities of states of LuH2 in the CaF2 structure

Reference

1. G.G. Libowitz, The nature and properties of transition metal hydrides. J. Nucl. Mater. **2**, 1 (1960)

Chapter 16
Actinide Hydrides

This chapter covers the actinide hydrides from AcH to BkH [1].

Results are presented for the crystal structures NaCl(B1), and CaF2(C1). Most of these systems form as dihydrides in the CaF2 structure. The LDA results presented here find the equilibrium lattice constants in serious disagreement with experiment values.

Examining the densities of states figures we observe that the lower occupied states have predominantly s–H character and near the Fermi level the d-2 g metal contribution is the dominant one. Comparing the energy bands of the fluorite structure to the bands of the NaCl structure it is noted that an additional band (second band) which is due to the second hydrogen in the CaF2 structure. A similar situation as in the lanthanides.

D. A. Papaconstantopoulos, *Band Structure of Cubic Hydrides*,
https://doi.org/10.1007/978-3-031-06878-2_16

16.1 Actinium Hydride (AcH)

See Tables 16.1, 16.2, 16.3 and Figs. 16.1, 16.2.

Table 16.1 Lattice constant, Bulk modulus, Gap, Total energy

Stru	a (Bohr)	B (MBar)	Gap	Total Energy
NaCl	10.62	0.40	-	-51546.16234
CaF2	12.51	0.40	-	-51547.18580
exp (CaF2)	10.72			

Table 16.2 Birch Fit coefficients

	A1	A2	A3	A4
NaCl	-5.154424E+04	-2.172381E+02	7.909550E+03	-9.096281E+04
CaF2	-5.154803E+04	2.503469E+02	-2.136521E+04	5.629104E+05

Table 16.3 DOS at Ef, Hopfield parameter, Stoner criterion

```
                AcH     a=10.62 Bohr       NaCl
---------------------------------------------------------------------------
 Fermi Energy   Total DOS        Decomposed DOS inside the muffin tins
   (Ry)        (States/Ry)       s         p         eg         t2g        f
---------------------------------------------------------------------------
Ac 0.601        14.500         0.312     0.533     0.309      2.767      0.617
 H 0.601        14.500         0.346     1.037     0.012      0.306      0.072
---------------------------------------------------------------------------
Fermi-Velocity (cm/s): 0.54136 x10E8
Plasmon Energy (eV)   : 4.28977
Electron-ion interaction (Hopfield parameter) (eV/A^2) Ac: 1.886  H: 0.141
------------------------------------------------
  Ac      MUFFIN-TIN RADIUS and CHARGE =    2.7873    86.0403
   H      MUFFIN-TIN RADIUS and CHARGE =    2.5219     1.6102
           Ac               STONER I = 0.0039
            H               STONER I = 0.0017
          STONER PARAMETER (Ry) I =  0.0039
          STONER CRITERION     N*I =  0.0571
------------------------------------------------
                AcH2     a=12.51 Bohr       CaF2
---------------------------------------------------------------------------
 Fermi Energy   Total DOS        Decomposed DOS inside the muffin tins
   (Ry)        (States/Ry)       s         p         eg         t2g        f
---------------------------------------------------------------------------
Ac  0.335        12.538         0.019     0.147     1.474      0.251      0.209
 H  0.335        12.538         0.516     1.332     0.067      0.207      0.026
---------------------------------------------------------------------------
Fermi-Velocity (cm/s): 0.59848 x10E8
Plasmon Energy (eV)   : 3.44593
Electron-ion interaction (Hopfield parameter) (eV/A^2) Ac: 0.235   H: 0.324
------------------------------------------------
        Ac      MUFFIN-TIN RADIUS and CHARGE =    2.8451    85.9393
         H      MUFFIN-TIN RADIUS and CHARGE =    2.5741     1.5085
                  Ac          STONER I =  0.0006
                   H          STONER I =  0.0009
         STONER PARAMETER (Ry)      I = 0.0026
         STONER CRITERION         N*I = 0.0327
---------------------------------------------------------------------------
```

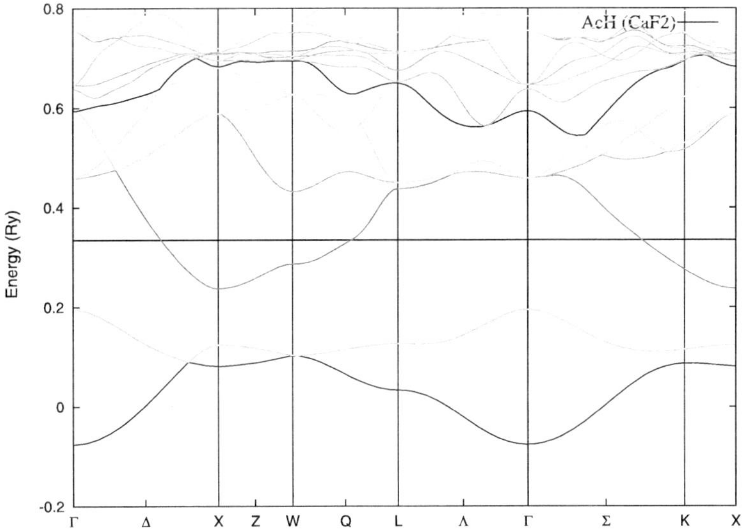

Fig. 16.1 Energy bands of AcH2 in the CaF2 structure

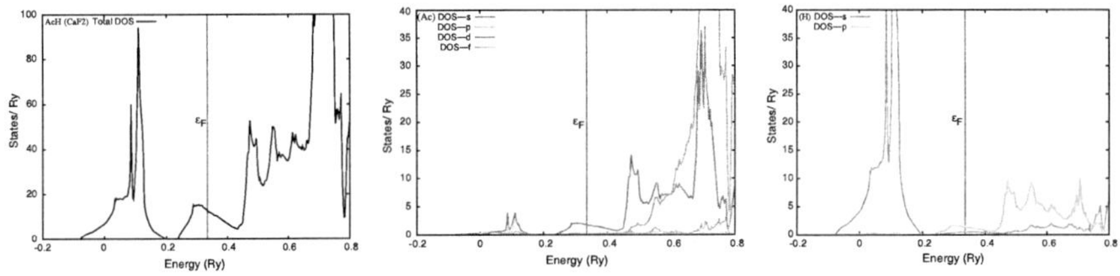

Fig. 16.2 Total, angular momentum and site decomposed densities of states of AcH2 in the CaF2 structure

16.2 Thorium Hydride (ThH)

See Tables 16.4, 16.5 and Figs. 16.3, 16.4.

Table 16.4 Lattice constant, Bulk modulus, Gap, Total energy

Stru	a (Bohr)	B (MBar)	Gap	Total Energy(Ry)
NaCl	9.96	0.76	-	-53046.62174
CaF2	11.67	0.67	-	-53047.53860
Exp (fc tetragonal)				

Table 16.5 DOS at Ef, Hopfield parameter, Stoner criterion

ThH a=9.96 Bohr NaCl

Fermi Energy (Ry)		Total DOS (States/Ry)	Decomposed DOS inside the muffin tins s	p	eg	t2g	f
Th	0.722	21.154	0.269	1.179	0.415	3.274	3.404
H	0.722	21.154	0.274	1.152	0.026	0.397	0.123

```
Fermi-Velocity (cm/s): 0.55171 x10E8
Plasmon Energy (eV)   : 5.81596
Electron-ion interaction (Hopfield parameter) (eV/A^2) Th: 4.038   H: 0.106

Th     MUFFIN-TIN RADIUS and CHARGE =    2.6136   86.4465
H      MUFFIN-TIN RADIUS and CHARGE =    2.3647    1.6450
         Th        STONER I =  0.0042
         H         STONER I =  0.0011
STONER PARAMETER (Ry)       I = 0.0042
STONER CRITERION          N*I = 0.0897
```

ThH2 a=11.67 Bohr CaF2

Fermi Energy (Ry)		Total DOS (States/Ry)	Decomposed DOS inside the muffin tins s	p	eg	t2g	f
Th	0.479	11.782	0.227	0.036	0.594	0.179	4.940
H	0.479	11.782	0.164	1.351	0.023	0.042	0.035

```
Fermi-Velocity (cm/s): 0.42749 x10E8
Plasmon Energy (eV)   : 2.64759
Electron-ion interaction (Hopfield parameter) (eV/A^2) Th: 0.041   H: 0.156

Th     MUFFIN-TIN RADIUS and CHARGE =    2.6522   86.2917
H      MUFFIN-TIN RADIUS and CHARGE =    2.3996    1.5166
         Th        STONER I =  0.0059
         H         STONER I =  0.0001
STONER PARAMETER (Ry)       I = 0.0059
STONER CRITERION          N*I = 0.0696
```

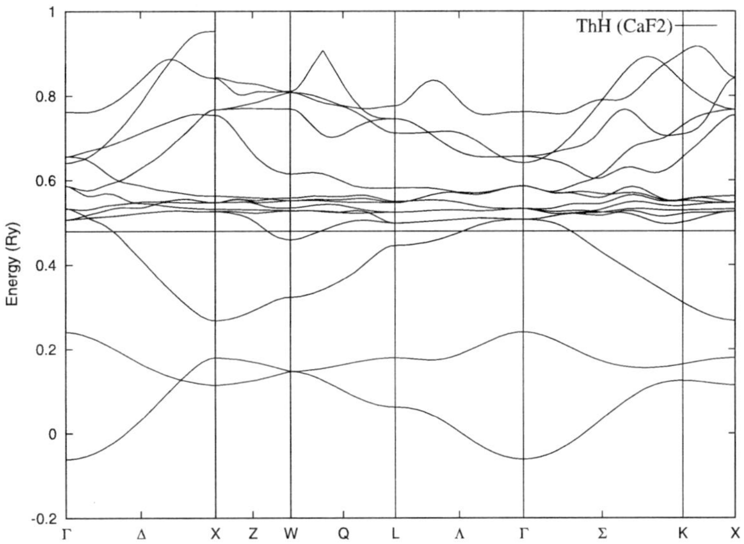

Fig. 16.3 Energy bands of ThH2 in the CaF2 structure

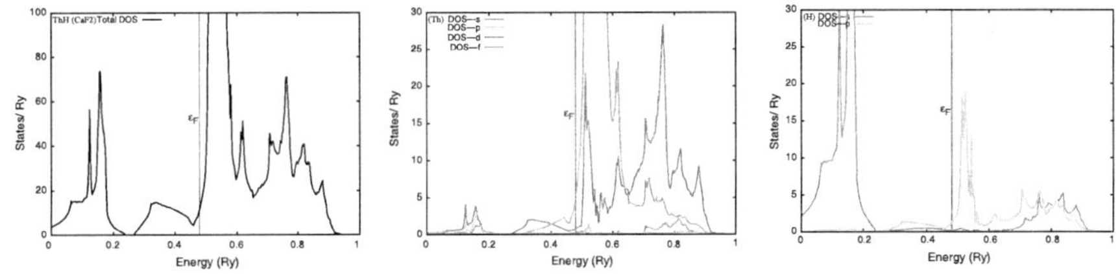

Fig. 16.4 Total, angular momentum and site decomposed densities of states of ThH2 in the CaF2 structure

16.3 Protactinium Hydride (PaH)

See Tables 16.6, 16.7 and Figs. 16.5, 16.6.

Table 16.6 Lattice constant, Bulk modulus, Gap, Total energy

Stru	a (Bohr)	B (MBar)	Gap	Total Energy(Ry)
NaCl	9.49	1.57	-	-54576.98318
CaF2	11.53	0.79	-	-54577.80490
Exp (A15)	12.57			

Table 16.7 DOS at Ef, Hopfield parameter, Stoner criterion

PaH a=9.49 Bohr NaCl

	Fermi Energy (Ry)	Total DOS (States/Ry)	s	p	eg	t2g	f
			Decomposed DOS inside the muffin tins				
Pa	0.824	19.540	0.133	0.466	0.147	0.502	12.145
H	0.824	19.540	0.110	0.560	0.016	0.210	0.064

Fermi-Velocity (cm/s): 0.32687 x10E8
Plasmon Energy (eV) : 3.55844
Electron-ion interaction (Hopfield parameter) (eV/A^2) Pa: 0.484 H: 0.027

Pa MUFFIN-TIN RADIUS and CHARGE = 2.4913 87.1726
H MUFFIN-TIN RADIUS and CHARGE = 2.2540 1.6436
Pa STONER I = 0.0089
H STONER I = 0.0003
STONER PARAMETER (Ry) I = 0.0093
STONER CRITERION N*I = 0.1819

PaH2 a=11.53 Bohr CaF2

	Fermi Energy (Ry)	Total DOS (States/Ry)	s	p	eg	t2g	f
			Decomposed DOS inside the muffin tins				
Pa	0.458	113.834	0.229	0.301	0.377	0.379	92.573
H	0.458	113.834	0.289	4.027	0.365	0.211	0.230

Fermi-Velocity (cm/s): 0.11253 x10E8
Plasmon Energy (eV) : 2.20527
Electron-ion interaction (Hopfield parameter) (eV/A^2) Pa: 0.051 H: 0.096

Pa MUFFIN-TIN RADIUS and CHARGE = 2.6210 87.3336
H MUFFIN-TIN RADIUS and CHARGE = 2.3713 1.5156
Pa STONER I = 0.0135
H STONER I = 0.0001
STONER PARAMETER (Ry) I = 0.0137
STONER CRITERION N*I = 1.5537

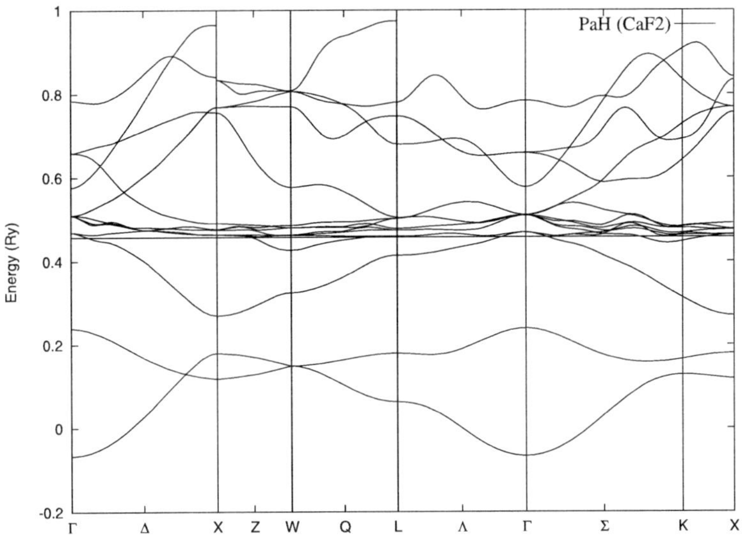

Fig. 16.5 Energy bands of PaH2 in the CaF2 structure

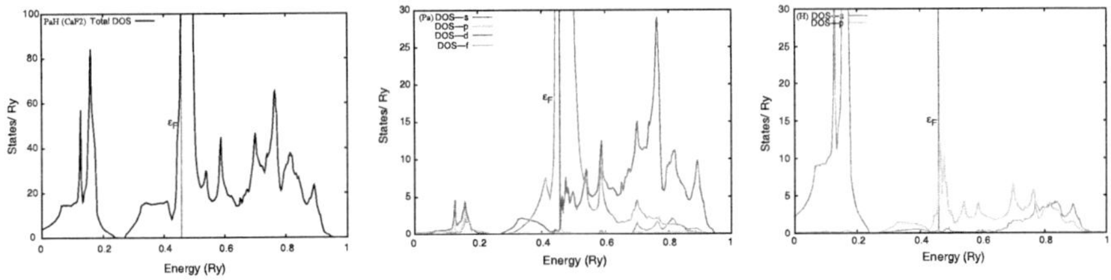

Fig. 16.6 Total, angular momentum and site decomposed densities of states of PaH2 in the CaF2 structure

16.4 Uranium Hydride (UH)

See Tables 16.8, 16.9 and Figs. 16.7, 16.8.

Table 16.8 Lattice constant, Bulk modulus, Gap, Total energy

Stru	a (Bohr)	B (MBar)	Gap	Total Energy(Ry)
NaCl	9.22	1.67	-	-56138.00550
CaF2	11.48	0.44	-	-56138.79812
Exp (A15)	12.56			

Table 16.9 DOS at Ef, Hopfield parameter, Stoner criterion

```
------------------------------------------------------------------------------
               UH         a=9.22 Bohr             NaCl
------------------------------------------------------------------------------
 Fermi Energy   Total DOS         Decomposed DOS inside the muffin tins
   (Ry)        (States/Ry)      s         p         eg         t2g        f
------------------------------------------------------------------------------
U   0.865      123.561       0.040     0.837     1.479      0.707      99.312
H   0.865      123.561       0.202     3.181     0.048      1.072       0.177
------------------------------------------------------------------------------
Fermi-Velocity (cm/s): 0.12969 x10E8
Plasmon Energy (eV)   : 3.69637
Electron-ion interaction (Hopfield parameter) (eV/A^2) U: 0.453   H: 0.049
------------------------------------------------------------------------------
          U     MUFFIN-TIN RADIUS and CHARGE =    2.4194   88.1394
          H     MUFFIN-TIN RADIUS and CHARGE =    2.1890    1.6376
                    U        STONER I = 0.0135
                    H        STONER I = 0.0002
          STONER PARAMETER (Ry)     I = 0.0137
          STONER CRITERION        N*I = 1.6970
------------------------------------------------------------------------------
               UH2             a=11.48 Bohr            CaF2
------------------------------------------------------------------------------
 Fermi Energy   Total DOS         Decomposed DOS inside the muffin tins
   (Ry)        (States/Ry)      s         p         eg         t2g        f
------------------------------------------------------------------------------
U   0.422      492.236       0.095     0.485     0.409      0.838     438.952
H   0.422      492.236       0.140     7.573     2.605      0.552       0.637
------------------------------------------------------------------------------
Fermi-Velocity (cm/s): 0.06612 x10E8
Plasmon Energy (eV)   : 2.69962
Electron-ion interaction (Hopfield parameter) (eV/A^2) U: 0.065   H: 0.052
------------------------------------------------
          U     MUFFIN-TIN RADIUS and CHARGE =    2.6090   88.4751
          H     MUFFIN-TIN RADIUS and CHARGE =    2.3605    1.4892
                    U        STONER I = 0.0159
                    H        STONER I = 0.0000
          STONER PARAMETER (Ry)     I = 0.0160
          STONER CRITERION        N*I = 7.8557

------------------------------------------------------------------------------
```

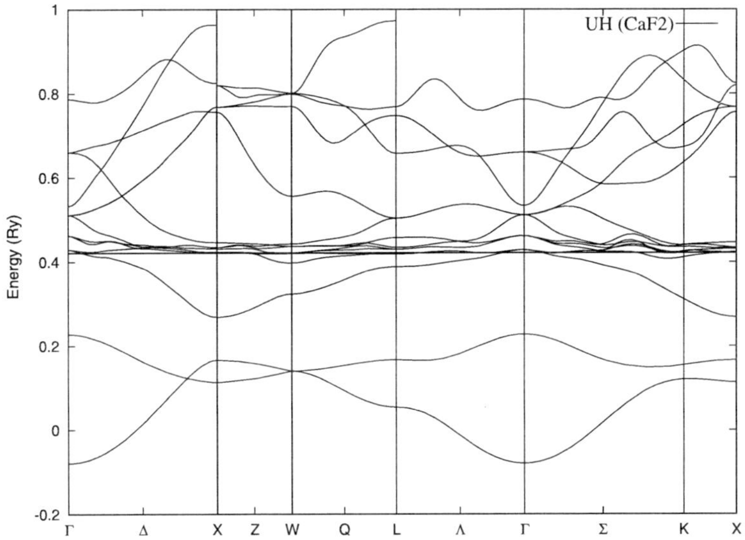

Fig. 16.7 Energy bands of UH2 in the CaF2 structure

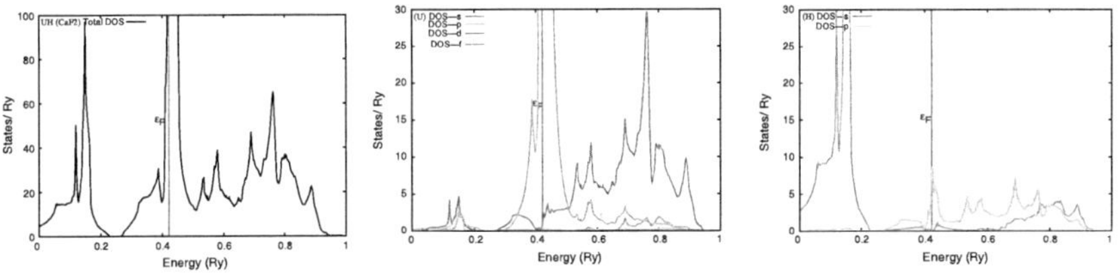

Fig. 16.8 Total, angular momentum and site decomposed densities of states of UH2 in the CaF2 structure

16.5 Neptunium Hydride (NpH)

See Tables 16.10, 16.11 and Figs. 16.9, 16.10.

Table 16.10 Lattice constant, Bulk modulus, Gap, Total energy

Stru	a (Bohr)	B (MBar)	Gap	Total Energy(Ry)
NaCl	9.05	0.77	-	-57730.36380
CaF2	11.65	0.64	-	-57731.14680

Table 16.11 DOS at Ef, Hopfield parameter, Stoner criterion

```
-------------------------------------------------------------------------
               NpH         a=9.05 Bohr           NaCl
-------------------------------------------------------------------------
 Fermi Energy   Total DOS        Decomposed DOS inside the muffin tins
   (Ry)        (States/Ry)       s        p        eg        t2g        f
-------------------------------------------------------------------------
Np   0.867      174.667       0.031    1.173     1.142     0.777    146.890
 H   0.867      174.667       0.465    2.836     0.075     1.074      0.418
-------------------------------------------------------------------------
Fermi-Velocity (cm/s): 0.09864 x10E8
Plasmon Energy (eV)  : 3.44471
Electron-ion interaction (Hopfield parameter) (eV/A^2) Np: 0.498   H: 0.074
-------------------------------------------------------------------------
         Np     MUFFIN-TIN RADIUS and CHARGE =    2.3759   89.1831
          H     MUFFIN-TIN RADIUS and CHARGE =    2.1496    1.6169
                   Np       STONER I = 0.0146
                   H        STONER I = 0.0001
         STONER PARAMETER (Ry)     I = 0.0148
         STONER CRITERION        N*I = 2.5819
-------------------------------------------------------------------------
               NpH2        a=11.65 Bohr           CaF2
-------------------------------------------------------------------------
 Fermi Energy   Total DOS        Decomposed DOS inside the muffin tins
   (Ry)        (States/Ry)       s        p        eg        t2g        f
-------------------------------------------------------------------------
Np   0.366      530.265       0.066    0.348     0.515     0.744    475.780
 H   0.366      530.265       0.110    6.939     1.312     0.653      0.693
-------------------------------------------------------------------------
Fermi-Velocity (cm/s): 0.06657 x10E8
Plasmon Energy (eV)  : 2.77572
Electron-ion interaction (Hopfield parameter) (eV/A^2) Np: 0.057   H: 0.029
---------------------------------------------
         Np     MUFFIN-TIN RADIUS and CHARGE =    2.6473   89.7300
          H     MUFFIN-TIN RADIUS and CHARGE =    2.3952    1.4715
                   Np        STONER I =  0.0159
                    H        STONER I =  0.0000
         STONER PARAMETER (Ry)      I = 0.0159
         STONER CRITERION         N*I = 8.4233

-------------------------------------------------------------------------
```

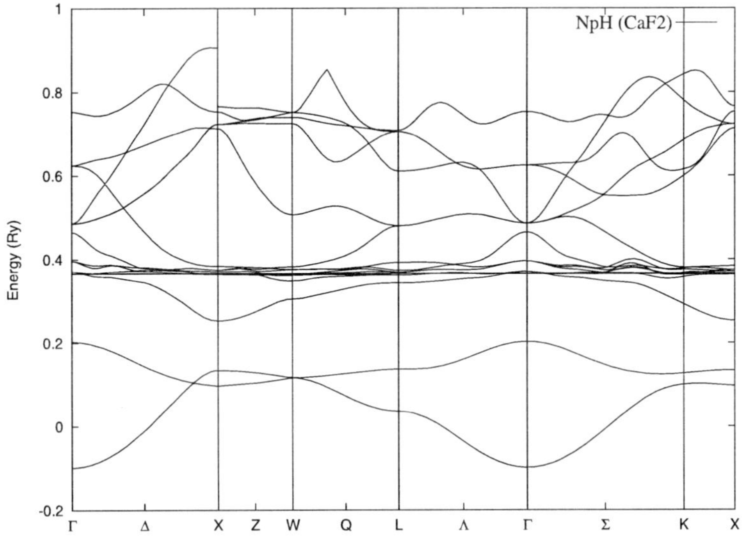

Fig. 16.9 Energy bands of NpH2 in the CaF2 structure

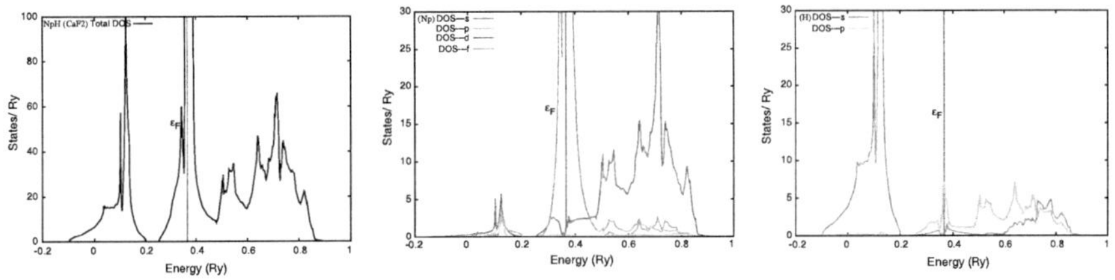

Fig. 16.10 Total, angular momentum and site decomposed densities of states of NpH2 in the CaF2 structure

16.6 Plutonium Hydride (PuH)

See Tables 16.12, 16.13 and Figs. 16.11, 16.12.

Table 16.12 Lattice constant, Bulk modulus, Gap, Total energy

Stru	a (Bohr)	B (MBar)	Gap	Total Energy(Ry)
NaCl	8.78	2.30	-	-59354.68745
CaF2	11.83	0.39	-	-59355.47874
Exp (CaF2)	10.13			

Table 16.13 DOS at Ef, Hopfield parameter, Stoner criterion

```
-----------------------------------------------------------------------------
                  PuH          a=8.78 Bohr              NaCl
-----------------------------------------------------------------------------
 Fermi Energy   Total DOS          Decomposed DOS inside the muffin tins
    (Ry)       (States/Ry)       s          p          eg         t2g        f
-----------------------------------------------------------------------------
Pu   0.917       147.549       0.035      1.838      0.266      0.846    125.265
 H   0.917       147.549       0.560      1.768      0.068      0.680      0.433
-----------------------------------------------------------------------------
Fermi-Velocity (cm/s): 0.08587 x10E8
Plasmon Energy (eV)   : 2.84802
Electron-ion interaction (Hopfield parameter) (eV/A^2) Pu: 0.657    H: 0.073
-----------------------------------------------------------------------------
          Pu      MUFFIN-TIN RADIUS and CHARGE =     2.3045    90.1373
           H      MUFFIN-TIN RADIUS and CHARGE =     2.0850     1.6014
                     Pu       STONER I = 0.0152
                     H        STONER I = 0.0001
          STONER PARAMETER (Ry)      I = 0.0153
          STONER CRITERION         N*I = 2.2582
-----------------------------------------------------------------------------
                  PuH2           a=11.83 Bohr             CaF2
-----------------------------------------------------------------------------
 Fermi Energy   Total DOS          Decomposed DOS inside the muffin tins
    (Ry)       (States/Ry)       s          p          eg         t2g        f
-----------------------------------------------------------------------------
Pu   0.323       685.940       0.060      0.243      0.521      0.673    626.668
 H   0.323       685.940       0.278      6.796      0.956      1.168      0.809
-----------------------------------------------------------------------------
Fermi-Velocity (cm/s): 0.05670 x10E8
Plasmon Energy (eV)   : 2.62400
Electron-ion interaction (Hopfield parameter) (eV/A^2) Pu: 0.046    H: 0.033
----------------------------------------------
          Pu      MUFFIN-TIN RADIUS and CHARGE =     2.6884    90.9882
           H      MUFFIN-TIN RADIUS and CHARGE =     2.4323     1.4645
                     Pu        STONER I =  0.0162
                      H        STONER I =  0.0000
          STONER PARAMETER (Ry)       I =  0.0162
          STONER CRITERION          N*I = 11.1263

-----------------------------------------------------------------------------
```

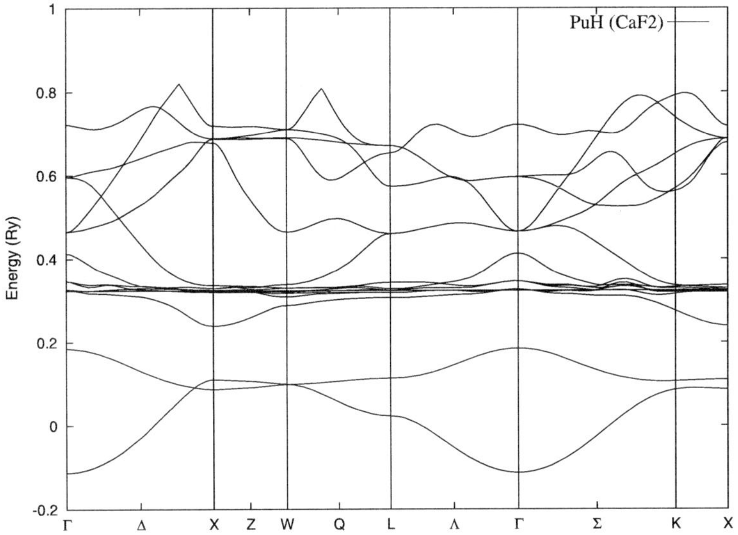

Fig. 16.11 Energy bands of PuH2 in the CaF2 structure

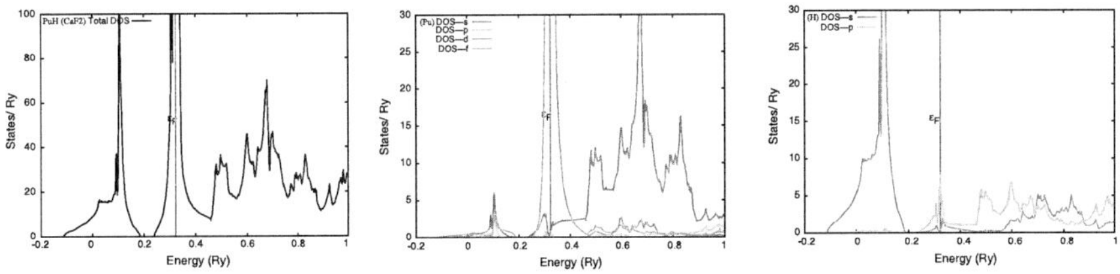

Fig. 16.12 Total, angular momentum and site decomposed densities of states of PuH2 in the CaF2 structure

16.7 Americium Hydride (AmH)

See Tables 16.14, 16.15 and Figs. 16.13, 16.14.

Table 16.14 Lattice constant, Bulk modulus, Gap, Total energy

Stru	a (Bohr)	B (MBar)	Gap	Total Energy(Ry)
NaCl	9.12	0.79	-	-61011.65492
CaF2	11.96	0.46	-	-61012.46317

Table 16.15 DOS at Ef, Hopfield parameter, Stoner criterion

```
--------------------------------------------------------------------------------
                 AmH          a=9.12 Bohr          NaCl
--------------------------------------------------------------------------------
 Fermi Energy   Total DOS          Decomposed DOS inside the muffin tins
   (Ry)        (States/Ry)       s          p          eg          t2g        f
--------------------------------------------------------------------------------
Am   0.764       302.323       0.063      3.091      0.419       1.403    266.293
 H   0.764       302.323       0.423      2.678      0.158       0.444      1.175
--------------------------------------------------------------------------------
Fermi-Velocity (cm/s): 0.03411 x10E8
Plasmon Energy (eV)   : 1.54993
Electron-ion interaction (Hopfield parameter) (eV/A^2) Am: 0.690   H: 0.035
--------------------------------------------------------------------------------
          Am     MUFFIN-TIN RADIUS and CHARGE =    2.3935   91.5834
           H     MUFFIN-TIN RADIUS and CHARGE =    2.1655    1.5755
                     Am       STONER I = 0.0159
                     H        STONER I = 0.0000
          STONER PARAMETER (Ry)     I = 0.0159
          STONER CRITERION        N*I = 4.8139
--------------------------------------------------------------------------------
                 AmH2             a=11.96 Bohr            CaF2
--------------------------------------------------------------------------------
 Fermi Energy   Total DOS          Decomposed DOS inside the muffin tins
   (Ry)        (States/Ry)       s          p          eg          t2g        f
--------------------------------------------------------------------------------
Am   0.292       662.351       0.059      0.197      0.327       0.574    625.705
 H   0.292       662.351       0.550      4.980      0.590       1.369      0.723
--------------------------------------------------------------------------------
Fermi-Velocity (cm/s): 0.05580 x10E8
Plasmon Energy (eV)   : 2.47368
Electron-ion interaction (Hopfield parameter) (eV/A^2) Am: 0.046   H: 0.033
------------------------------------------------
          Am     MUFFIN-TIN RADIUS and CHARGE =    2.7200   92.2027
           H     MUFFIN-TIN RADIUS and CHARGE =    2.4609    1.4478
                     Am        STONER I =  0.0172
                     H         STONER I =  0.0000
          STONER PARAMETER (Ry)      I =  0.0172
          STONER CRITERION         N*I = 11.3700

--------------------------------------------------------------------------------
```

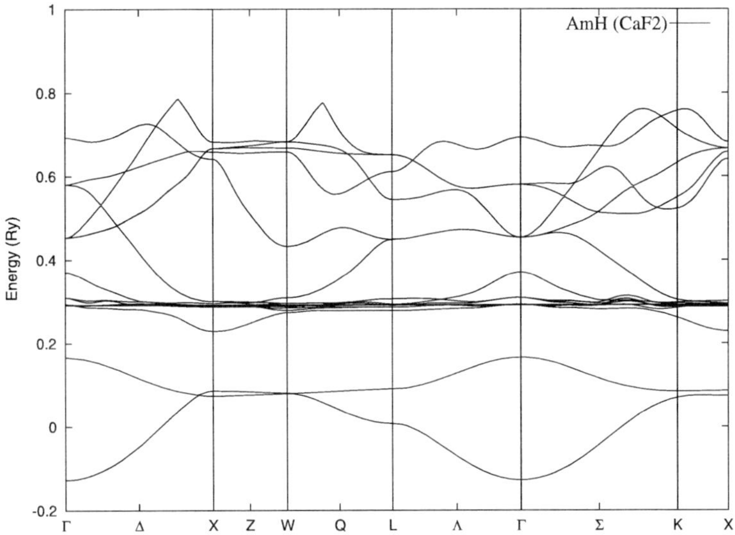

Fig. 16.13 Energy bands of AmH2 in the CaF2 structure

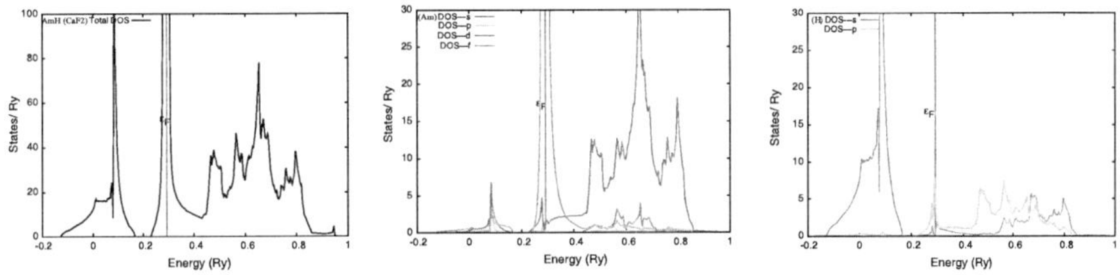

Fig. 16.14 Total, angular momentum and site decomposed densities of states of AmH2 in the CaF2 structure

16.8 Curium Hydride (CmH)

See Tables 16.16, 16.17 and Figs. 16.15, 16.16.

Table 16.16 Lattice constant, Bulk modulus, Gap, Total energy

Stru	a (Bohr)	B (MBar)	Gap	Total Energy(Ry)
NaCl	8.91	1.24	-	-62701.95489
CaF2	12.11	0.33	-	-62702.78690

Table 16.17 DOS at Ef, Hopfield parameter, Stoner criterion

```
---------------------------------------------------------------------------
                  CmH         a=8.91 Bohr              NaCl
---------------------------------------------------------------------------
 Fermi Energy   Total DOS         Decomposed DOS inside the muffin tins
   (Ry)        (States/Ry)       s         p         eg         t2g         f
---------------------------------------------------------------------------
Cm   0.797      108.581        0.017     0.555     0.546      0.625      96.859
 H   0.797      108.581        0.140     1.535     0.118      0.160       0.192
---------------------------------------------------------------------------
Fermi-Velocity (cm/s): 0.09892 x10E8
Plasmon Energy (eV)   : 2.76266
Electron-ion interaction (Hopfield parameter) (eV/A^2) Cm: 0.345   H: 0.020
---------------------------------------------------------------------------
          Cm      MUFFIN-TIN RADIUS and CHARGE =     2.3379    92.5837
           H      MUFFIN-TIN RADIUS and CHARGE =     2.1153     1.5613
                    Cm        STONER I = 0.0163
                    H         STONER I = 0.0001
          STONER PARAMETER (Ry)      I = 0.0164
          STONER CRITERION         N*I = 1.7785
---------------------------------------------------------------------------
                  CmH2             a=12.11 Bohr            CaF2
---------------------------------------------------------------------------
 Fermi Energy   Total DOS         Decomposed DOS inside the muffin tins
   (Ry)        (States/Ry)       s         p         eg         t2g         f
---------------------------------------------------------------------------
Cm   0.265      920.960        0.075     0.209     0.379      0.741     874.889
 H   0.265      920.960        0.985     4.795     0.512      1.552       0.917
---------------------------------------------------------------------------
Fermi-Velocity (cm/s): 0.04908 x10E8
Plasmon Energy (eV)   : 2.53880
Electron-ion interaction (Hopfield parameter) (eV/A^2) Cm: 0.034   H: 0.042
----------------------------------------------
          Cm      MUFFIN-TIN RADIUS and CHARGE =     2.7520    93.4072
           H      MUFFIN-TIN RADIUS and CHARGE =     2.4899     1.4381
                    Cm          STONER I =   0.0172
                    H           STONER I =   0.0000
          STONER PARAMETER (Ry)        I =   0.0172
          STONER CRITERION           N*I = 15.8706

---------------------------------------------------------------------------
```

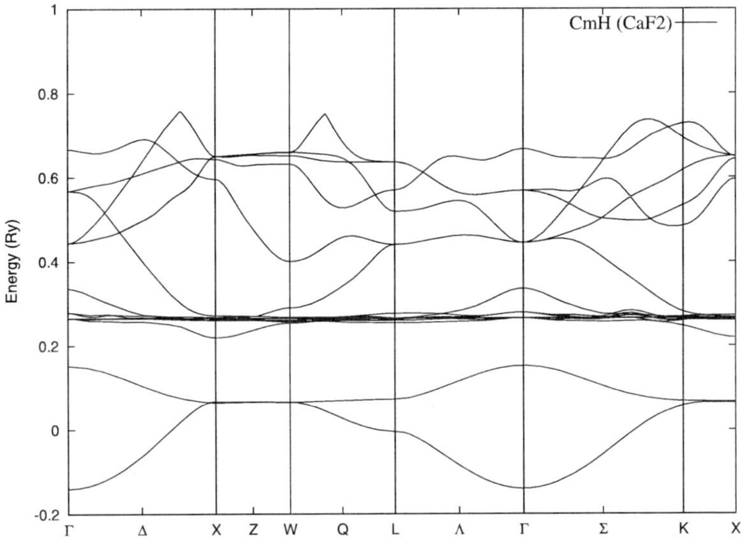

Fig. 16.15 Energy bands of CmH2 in the CaF2 structure

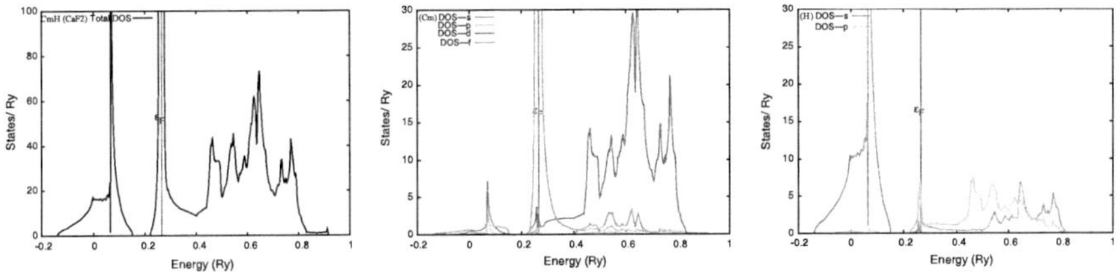

Fig. 16.16 Total, angular momentum and site decomposed densities of states of CmH2 in the CaF2 structure

16.9 Berkelium Hydride (BkH)

See Tables 16.18, 16.19 and Figs. 16.17, 16.18.

Table 16.18 Lattice constant, Bulk modulus, Gap, Total energy

Stru	a (Bohr)	B (MBar)	Gap	Total Energy(Ry)
NaCl	9.33	0.59	-	-64426.31082
CaF2	12.27	0.31	-	-64427.17260

Table 16.19 DOS at Ef, Hopfield parameter, Stoner criterion

```
--------------------------------------------------------------------------------
                 BkH          a=9.33 Bohr                  NaCl
--------------------------------------------------------------------------------
 Fermi Energy    Total DOS         Decomposed DOS inside the muffin tins
   (Ry)         (States/Ry)      s           p           eg          t2g          f
--------------------------------------------------------------------------------
Bk    0.663       209.219        0.006       0.111       0.429       0.749     197.933
 H    0.663       209.219        0.042       1.538       0.191       0.291       0.276
--------------------------------------------------------------------------------
Fermi-Velocity (cm/s): 0.08120 x10E8
Plasmon Energy (eV)   : 2.95950
Electron-ion interaction (Hopfield parameter) (eV/A^2) Bk: 0.121   H: 0.003
--------------------------------------------------------------------------------
          Bk      MUFFIN-TIN RADIUS and CHARGE =     2.4492     94.0645
           H      MUFFIN-TIN RADIUS and CHARGE =     2.2160      1.5406
                     Bk        STONER I = 0.0177
                      H        STONER I = 0.0000
          STONER PARAMETER (Ry)      I = 0.0178
          STONER CRITERION         N*I = 3.7146
--------------------------------------------------------------------------------
                 BkH2             a=12.27 Bohr                CaF2
--------------------------------------------------------------------------------
 Fermi Energy    Total DOS         Decomposed DOS inside the muffin tins
   (Ry)         (States/Ry)      s           p           eg          t2g          f
--------------------------------------------------------------------------------
Bk    0.237      1076.890        0.079       0.244       0.234       0.907    1032.170
 H    0.237      1076.890        1.816       3.683       0.494       1.467       0.985
--------------------------------------------------------------------------------
Fermi-Velocity (cm/s): 0.04556 x10E8
Plasmon Energy (eV)   : 2.50623
Electron-ion interaction (Hopfield parameter) (eV/A^2) Bk: 0.029   H: 0.044
------------------------------------------------
          Bk      MUFFIN-TIN RADIUS and CHARGE =     2.7889     94.5941
           H      MUFFIN-TIN RADIUS and CHARGE =     2.5233      1.4304
                     Bk         STONER I =   0.0174
                      H         STONER I =   0.0000
          STONER PARAMETER (Ry)       I =   0.0175
          STONER CRITERION          N*I = 18.7952

--------------------------------------------------------------------------------
```

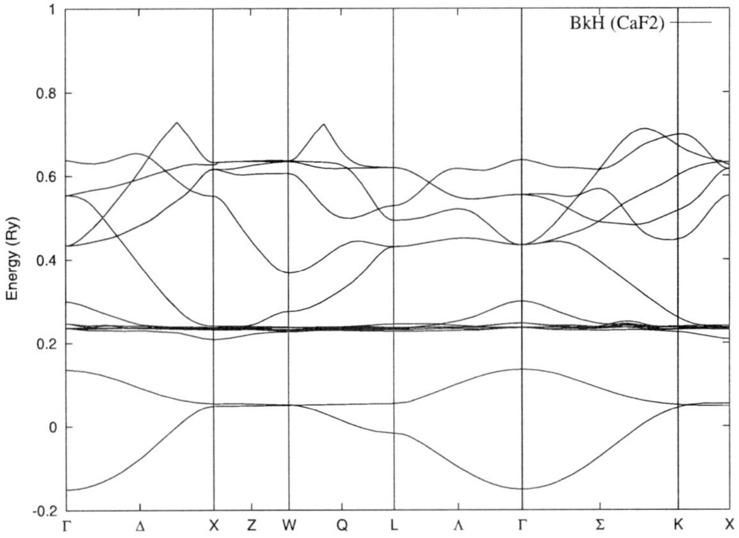

Fig. 16.17 Energy bands of BkH2 in the CaF2 structure

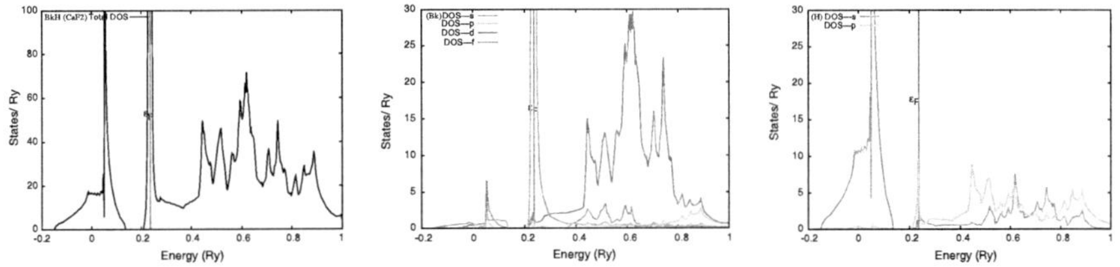

Fig. 16.18 Total, angular momentum and site decomposed densities of states of BkH2 in the CaF2 structure

16.10 Californium Hydride (CfH)

See Tables 16.20, 16.21 and Figs. 16.19, 16.20.

Table 16.20 Lattice constant, Bulk modulus, Gap, Total energy

Stru	a (Bohr)	B (MBar)	Gap	Total Energy(Ry)
NaCl	9.47	0.50	-	-66185.50032

Table 16.21 DOS at Ef, Hopfield parameter, Stoner criterion

	CfH	a=9.47 Bohr		NaCl		
Fermi Energy (Ry)	Total DOS (States/Ry)	Decomposed DOS inside the muffin tins s	p	eg	t2g	f
Cf 0.615	355.803	0.024	0.083	0.310	0.633	334.283
H 0.615	355.803	0.103	1.296	0.410	0.171	0.317

Fermi-Velocity (cm/s): 0.05659 x10E8
Plasmon Energy (eV) : 2.59640
Electron-ion interaction (Hopfield parameter) (eV/A^2) Cf: 0.081 H: 0.003

Cf	MUFFIN-TIN RADIUS and CHARGE =	2.4870	95.3005
H	MUFFIN-TIN RADIUS and CHARGE =	2.2501	1.5259
Cf	STONER I = 0.0173		
H	STONER I = 0.0000		
STONER PARAMETER (Ry)	I = 0.0173		
STONER CRITERION	N*I = 6.1686		

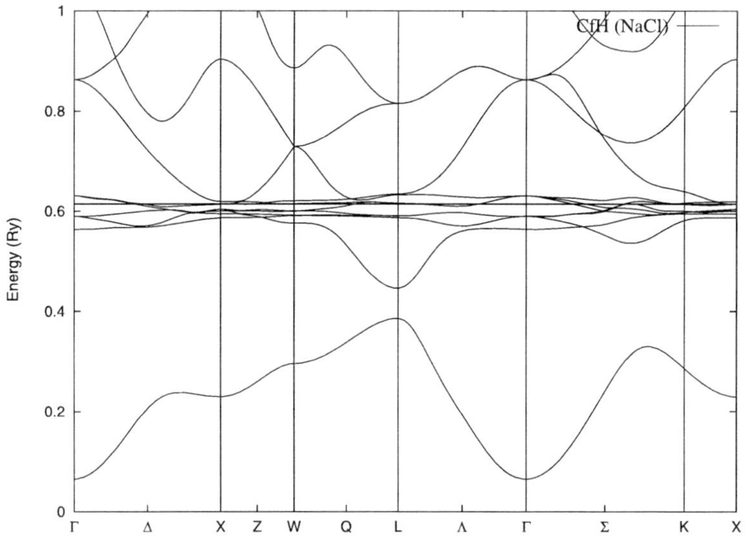

Fig. 16.19 Energy bands of CfH in the NaCl structure

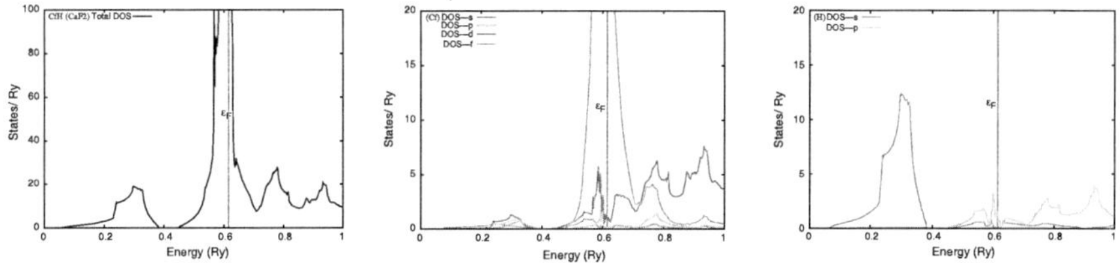

Fig. 16.20 Total, angular momentum and site decomposed densities of states of CfH in the NaCl structure

16.11 Einsteinium Hydride (EsH)

See Tables 16.22, 16.23 and Figs. 16.21, 16.22.

Table 16.22 Lattice constant, Bulk modulus, Gap, Total energy

Stru	a (Bohr)	B (MBar)	Gap	Total Energy(Ry)
NaCl	9.72	0.52	-	-67980.31317

Table 16.23 DOS at Ef, Hopfield parameter, Stoner criterion

	EsH	a=9.72 Bohr		NaCl			
	Fermi Energy (Ry)	Total DOS (States/Ry)	Decomposed DOS inside the muffin tins s	p	eg	t2g	f
Es	0.552	426.174	0.067	0.116	0.306	0.460	405.494
H	0.552	426.174	0.328	0.823	0.647	0.132	0.233

```
Fermi-Velocity (cm/s): 0.04377 x10E8
Plasmon Energy (eV)   : 2.13264
Electron-ion interaction (Hopfield parameter) (eV/A^2) Es: 0.052   H: 0.005
-----------------------------------------------------------------------------
         Es      MUFFIN-TIN RADIUS and CHARGE =     2.5521    96.5817
          H      MUFFIN-TIN RADIUS and CHARGE =     2.3091     1.5153
                   Es       STONER I = 0.0176
                   H        STONER I = 0.0000
         STONER PARAMETER (Ry)      I = 0.0176
         STONER CRITERION         N*I = 7.5046
-----------------------------------------------------------------------------
```

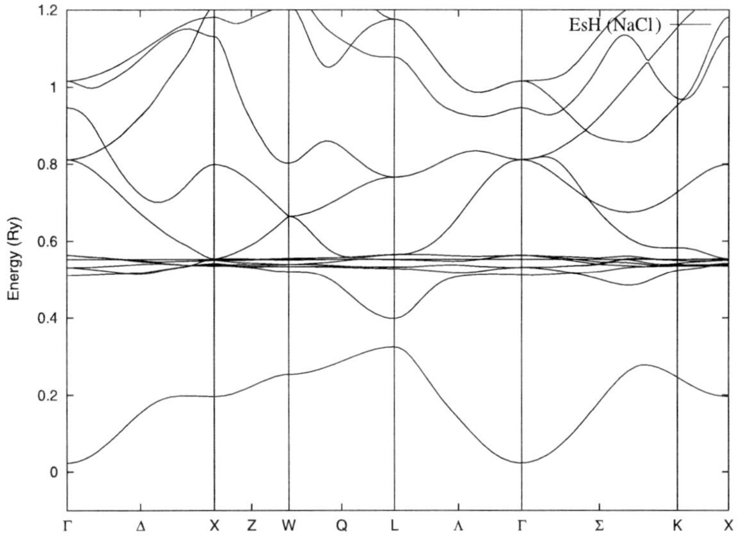

Fig. 16.21 Energy bands of EsH in the NaCl structure

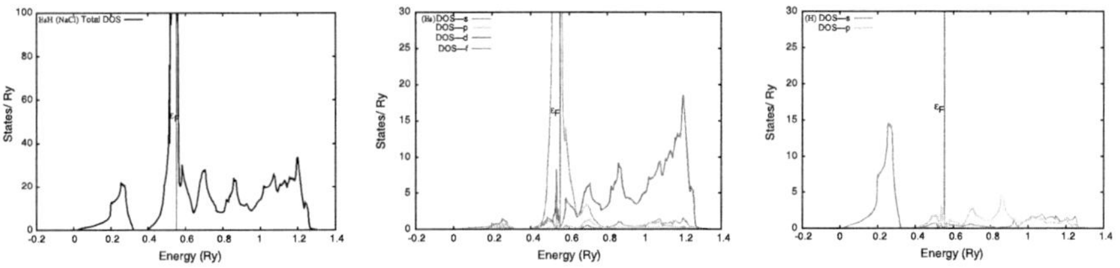

Fig. 16.22 Total, angular momentum and site decomposed densities of states of EsH in the NaCl structure

16.12 Fermium Hydride (FmH)

See Tables 16.24, 16.25 and Figs. 16.23, 16.24.

Table 16.24 Lattice constant, Bulk modulus, Gap, Total energy

Stru	a (Bohr)	B (MBar)	Gap	Total Energy(Ry)
NaCl	9.92	0.39	-	-69811.60453

Table 16.25 DOS at Ef, Hopfield parameter, Stoner criterion

	FmH	a=9.92 Bohr		NaCl		
Fermi Energy (Ry)	Total DOS (States/Ry)	Decomposed DOS inside the muffin tins				
		s	p	eg	t2g	f
Fm 0.504	301.750	0.095	0.187	0.289	0.295	290.720
H 0.504	301.750	0.657	0.355	0.346	0.097	0.121

```
Fermi-Velocity (cm/s): 0.04063 x10E8
Plasmon Energy (eV)   : 1.61477
Electron-ion interaction (Hopfield parameter) (eV/A^2) Fm: 0.037   H: 0.006

         Fm      MUFFIN-TIN RADIUS and CHARGE =    2.6045   97.7926
          H      MUFFIN-TIN RADIUS and CHARGE =    2.3565    1.4959
                    Fm       STONER I = 0.0180
                    H        STONER I = 0.0000
         STONER PARAMETER (Ry)     I = 0.0180
         STONER CRITERION        N*I = 5.4321
```

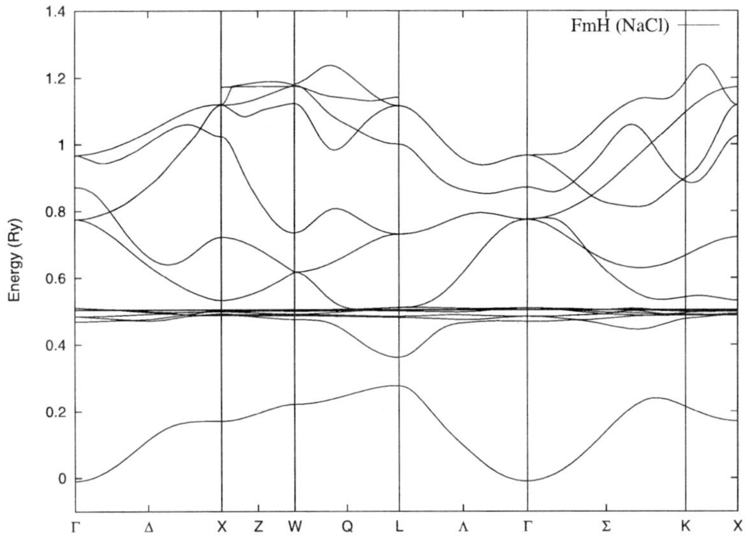

Fig. 16.23 Energy bands of FmH in the NaCl structure

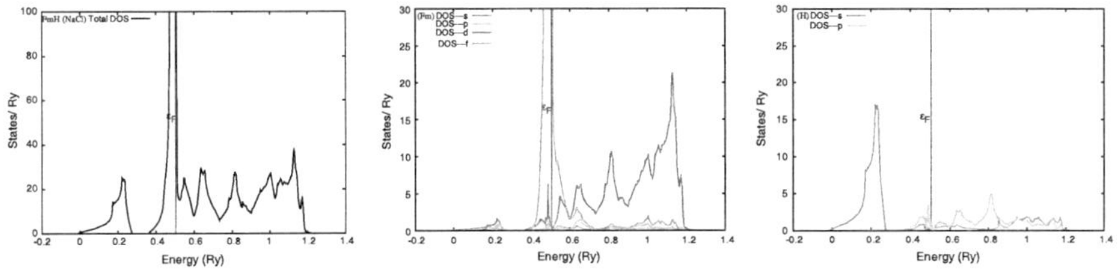

Fig. 16.24 Total, angular momentum and site decomposed densities of states of FmH in the NaCl structure

16.13 Mendelevium Hydride (MdH)

See Tables 16.26, 16.27 and Figs. 16.25, 16.26.

Table 16.26 Lattice constant, Bulk modulus, Gap, Total energy

Stru	a (Bohr)	B (MBar)	Gap	Total Energy(Ry)
NaCl	9.98	0.35	-	-71680.26139

Table 16.27 DOS at Ef, Hopfield parameter, Stoner criterion

```
--------------------------------------------------------------------------------
                MdH        a=9.98 Bohr                NaCl
--------------------------------------------------------------------------------
 Fermi Energy   Total DOS         Decomposed DOS inside the muffin tins
    (Ry)        (States/Ry)       s          p          eg         t2g         f
--------------------------------------------------------------------------------
Md   0.483        86.583        0.152      0.124      0.056      0.248     83.180
 H   0.483        86.583        0.416      0.225      0.011      0.034      0.018
--------------------------------------------------------------------------------
Fermi-Velocity (cm/s): 0.08347 x10E8
Plasmon Energy (eV)   : 1.60488
Electron-ion interaction (Hopfield parameter) (eV/A^2) Md: 0.026   H: 0.008
--------------------------------------------------------------------------------
         Md      MUFFIN-TIN RADIUS and CHARGE =    2.6201    98.9178
          H      MUFFIN-TIN RADIUS and CHARGE =    2.3706     1.4794
                    Md       STONER I = 0.0180
                    H        STONER I = 0.0000
         STONER PARAMETER (Ry)      I = 0.0181
         STONER CRITERION         N*I = 1.5631
--------------------------------------------------------------------------------
```

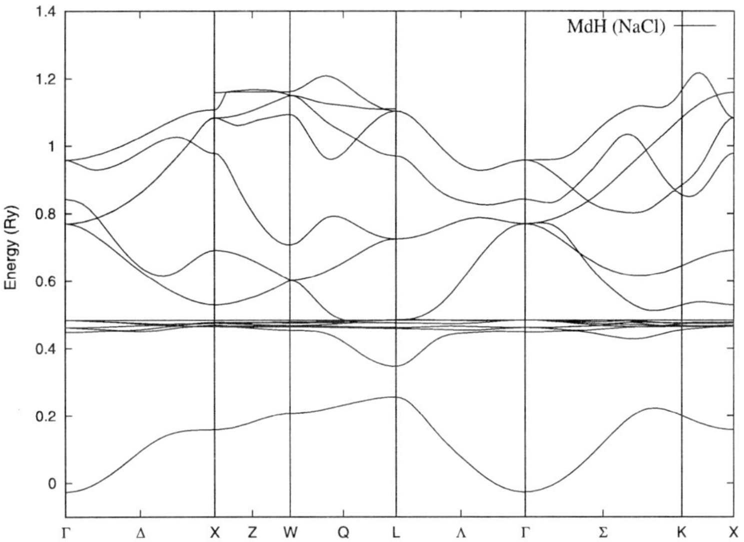

Fig. 16.25 Energy bands of MdH in the NaCl structure

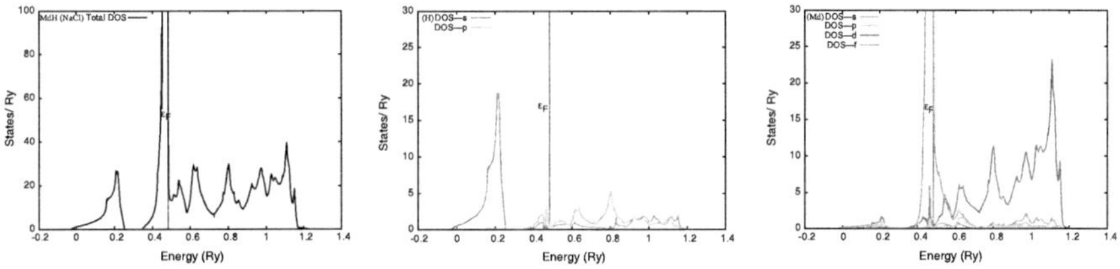

Fig. 16.26 Total, angular momentum and site decomposed densities of states of MdH in the NaCl structure

16.14 Nobelium Hydride (NoH)

See Tables 16.28, 16.29 and Figs. 16.27, 16.28.

Table 16.28 Lattice constant, Bulk modulus, Gap, Total energy

Stru	a (Bohr)	B (MBar)	Gap	Total Energy(Ry)
NaCl	10.18	0.36	-	-73587.18082

Table 16.29 DOS at Ef, Hopfield parameter, Stoner criterion

```
-----------------------------------------------------------------------------
                NoH        a=10.18 Bohr              NaCl
-----------------------------------------------------------------------------
 Fermi Energy   Total DOS        Decomposed DOS inside the muffin tins
   (Ry)        (States/Ry)        s         p         eg        t2g        f
-----------------------------------------------------------------------------
No    0.484       10.867       0.929     0.270     0.310      1.375      0.626
 H    0.484       10.867       0.894     1.165     0.021      0.094      0.013
-----------------------------------------------------------------------------
Fermi-Velocity (cm/s): 0.71592 x10E8
Plasmon Energy (eV)   : 5.23263
Electron-ion interaction (Hopfield parameter) (eV/A^2) No: 0.604   H: 0.661
-----------------------------------------------------------------------------
         No     MUFFIN-TIN RADIUS and CHARGE =    2.6720    99.9438
          H     MUFFIN-TIN RADIUS and CHARGE =    2.4175     1.4988
                  No      STONER I = 0.0019
                  H       STONER I = 0.0043
         STONER PARAMETER (Ry)    I = 0.0065
         STONER CRITERION       N*I = 0.0700
-----------------------------------------------------------------------------
```

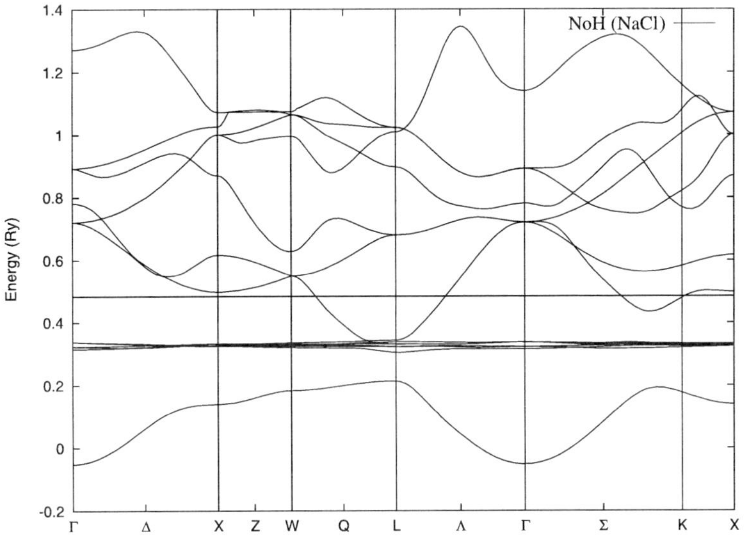

Fig. 16.27 Energy bands of NoH in the NaCl structure

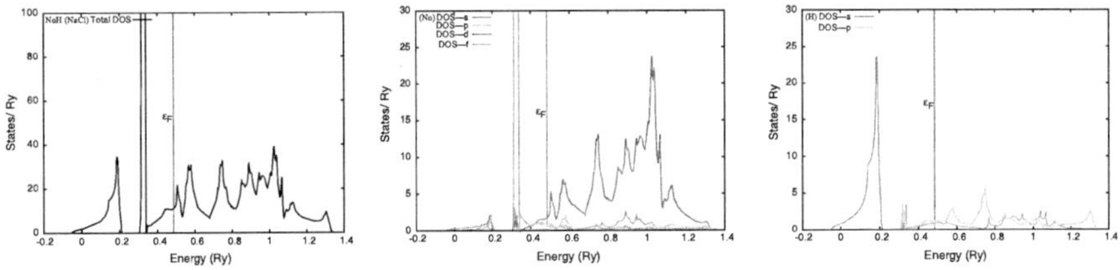

Fig. 16.28 Total, angular momentum and site decomposed densities of states of NoH in the NaCl structure

16.15 Lawrencium Hydride (LrH)

See Tables 16.30, 16.31 and Figs. 16.29, 16.30.

Table 16.30 Lattice constant, Bulk modulus, Gap, Total energy

Stru	a (Bohr)	B (MBar)	Gap	Total Energy(Ry)
NaCl	9.64	0.70	-	-75533.11642

Table 16.31 DOS at Ef, Hopfield parameter, Stoner criterion

```
--------------------------------------------------------------------------------
               LrH        a=9.64 Bohr              NaCl
--------------------------------------------------------------------------------
 Fermi Energy   Total DOS        Decomposed DOS inside the muffin tins
   (Ry)        (States/Ry)       s         p         eg        t2g        f
--------------------------------------------------------------------------------
Lr   0.610        21.872       0.618     0.866     0.851     4.686     0.206
 H   0.610        21.872       0.775     1.884     0.017     0.397     0.053
--------------------------------------------------------------------------------
Fermi-Velocity (cm/s): 0.54176 x10E8
Plasmon Energy (eV)   : 6.09243
Electron-ion interaction (Hopfield parameter) (eV/A^2) Lr: 3.464   H: 0.540
--------------------------------------------------------------------------------
         Lr     MUFFIN-TIN RADIUS and CHARGE =    2.5312   100.4194
          H     MUFFIN-TIN RADIUS and CHARGE =    2.2901     1.5494
                   Lr       STONER I = 0.0025
                   H        STONER I = 0.0023
         STONER PARAMETER (Ry)      I = 0.0052
         STONER CRITERION         N*I = 0.1126
--------------------------------------------------------------------------------
```

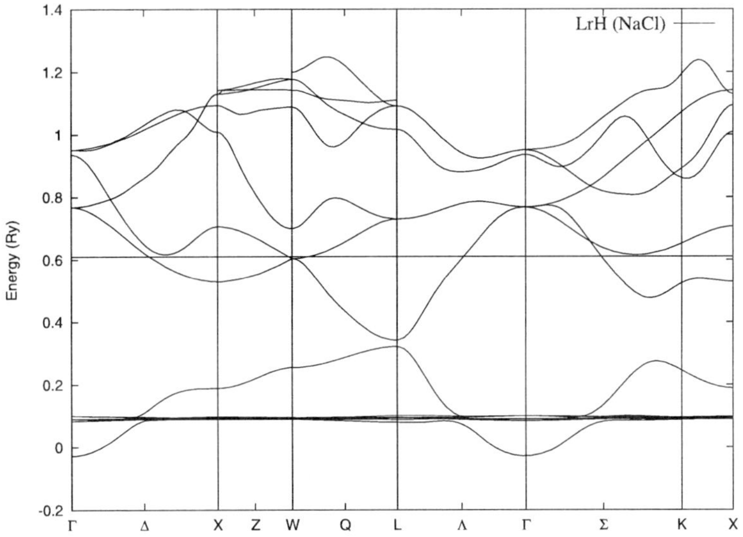

Fig. 16.29 Energy bands of LrH in the NaCl structure

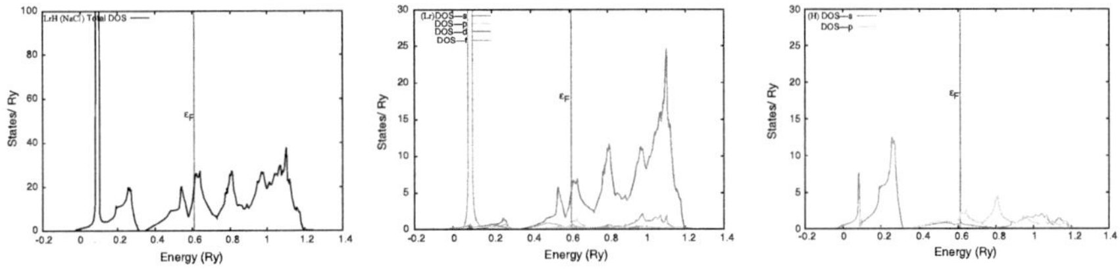

Fig. 16.30 Total, angular momentum and site decomposed densities of states of LrH in the NaCl structure

Reference

1. G.G. Libowitz, The nature and properties of transition metal hydrides. J. Nucl. Mater. **2**, 1 (1960)

Appendix A
Computer Program to Generate Tight-Binding Eigenvalues for the NaCl Structure

D. A. Papaconstantopoulos, *Band Structure of Cubic Hydrides*,
https://doi.org/10.1007/978-3-031-06878-2

```
      PROGRAM NACLSK
C    AUTHOR: D.A.PAPACONSTANTOPOULOS
C    NaCl 13x13 matrix--REAL, BUT EQUIVALENT TO COMPLEX--OVERLAP VERSION
C    CREATES INPUT TO TETRAHEDRON
C      CAUTION: THE DIMENSIONING OF Q10,Q11,Q12,Q13, SHOULD BE
C      CONSISTENT WITH THE NUMER OF K-POINTS CREATED BY THE
C      ASSIGNED VALUE TO MESH
      IMPLICIT REAL*8 (A-H,O-Z)
      CHARACTER*10 INA,INB
      CHARACTER*80 TITLE
      DIMENSION UU(13,13),T(13,13),H(13,13),U(13,13),S(13,13)
      DIMENSION TITLE(20),PARAM(114)
      DIMENSION A(3),HNEW(13,13)
      DIMENSION Q10(13,505),Q11(13,505),Q12(13,505),Q13(13,505)
     &,Q20(13,505),Q21(13,505)
      INA = 'par.in'
      OPEN(10,FILE=INA,BLANK='ZERO')
c     OPEN(11,FILE='dosdat',BLANK='ZERO')
      INB = 'nacl.out'
      OPEN(15,FILE=INB,BLANK='ZERO')
      READ (10,35) TITLE
   35 FORMAT(20A4)
c     WRITE (15,335) TITLE
  335 FORMAT (1H1,20A4)
C     ICENTER=3 THREE-CENTER APPROXIMATION
C     ICENTER=2 TWO-CENTER APPROXIMATION
C     MESH  K-POINT MESH GENERATOR
      READ (10,9999) MESH,ICENTER
c     WRITE(15,9998) MESH,ICENTER
 9999 FORMAT (5I5)
 9998 FORMAT (5X,'MESH=',I5,10X,'ICENTER=',I5)
      IF (ICENTER.EQ.3) NPAR=60
      IF (ICENTER.EQ.2) NPAR=50
      NS=13
      SR2=SQRT(2.)
      SR3=SQRT(3.)
c     IF (ICENTER.EQ.3)
c    1READ(10,55)A1,A6,A14,A19,A2,A3,A4,A5,A7,A8,A9,A10,A11,A12,A13,
c    C A15,A16,A17,A18,A20,A21,B1,B2,B3,B4,B5,B6,B7,B8,B9,B10,B11
c    C                            ,S2,S3,S4,S5,S7,S8,S9,S10,S11,S12,S13,
c    C S15,S16,S17,S18,S20,S21,T1,T2,T3,T4,T5,T6,T7,T8,T9,T10,T11
C     THE ORDER OF THE SK-PARAMETERS HERE IS THE SAME AS IN THE TABLES
c     IF (ICENTER.EQ.3)
c    1WRITE(15,999)A1,A6,A14,A19,A2,A3,A4,A5,A7,A8,A9,A10,A11,A12,A13,
c    C A15,A16,A17,A18,A20,A21,B1,B2,B3,B4,B5,B6,B7,B8,B9,B10,B11
c    C                            ,S2,S3,S4,S5,S7,S8,S9,S10,S11,S12,S13,
c    C S15,S16,S17,S18,S20,S21,T1,T2,T3,T4,T5,T6,T7,T8,T9,T10,T11
c     IF (ICENTER.EQ.2) READ(10,55)   (PARAM(I),I=1,50)
c     IF (ICENTER.EQ.2) WRITE(15,999)   (PARAM(I),I=1,50)
c     IF (ICENTER.EQ.3) GO TO 500
C      THIS SECTION IS FOR THE TWO-CENTER CALCULATIONS
  500 CONTINUE
```

```
  999 FORMAT(//10X,'VALUES OF THE SK PARAMETERS',//60(25X,F15.8/))
           DO 1010 I=1,NPAR
c      INDEX=I
c      READ(10,55) PARAM(I),IFREEZE(I),INDEX,(XLABE(I,M),M=1,5)
       READ(10,55) PARAM(I)
   55  FORMAT(F13.10)
c 55   FORMAT(F13.10,3X,I1,I3,4X,5A4,36X)
c      WRITE(15,999)PARAM(I),IFREEZE(I),INDEX,(XLABE(I,M),M=1,5),TITLE
c      IF(IFREEZE(I).EQ.1) GOTO 1010
c          J=J+1
c          PAR(J)=PARAM(I)
 1010 CONTINUE
           DO 1011 I=1,NPAR
c      WRITE(15,55) PARAM(I)
 1011 CONTINUE
       N=13
       IF (ICENTER.EQ.3) GO TO 666
       A1=PARAM(1)
       A6=PARAM(2)
       A14=PARAM(3)
       A19=PARAM(4)
       P1=PARAM(5)
       P2=PARAM(6)
       P3=PARAM(7)
       P4=PARAM(8)
       P5=PARAM(9)
       P6=PARAM(10)
       P12=PARAM(11)
       P14=PARAM(12)
       P24=PARAM(13)
       P35=PARAM(14)
       B1=PARAM(15)
       B4=PARAM(16)
       B5=PARAM(17)
       B10=PARAM(18)
       B8=PARAM(19)
       B9=PARAM(20)
       B2=PARAM(21)
       B3=PARAM(22)
       B7=PARAM(23)
       B6=PARAM(24)
       C1=PARAM(25)
       C6=PARAM(26)
       C2=PARAM(27)
       CC3=PARAM(28)
       CC7=PARAM(29)
       CC8=PARAM(30)
       EE=PARAM(31)
       EE2=PARAM(32)
       E4=PARAM(33)
       E5=PARAM(34)
       D1=PARAM(35)
```

```
      D2=PARAM(36)
      D3=PARAM(37)
      D4=PARAM(38)
      D5=PARAM(39)
      D6=PARAM(40)
      D7=PARAM(41)
      D8=PARAM(42)
C   PARAMETERS OF 2ND NN A-B INTERACIONS S-P ORBITALS H-MATRIX
      F1=PARAM(43)
      PSS=PARAM(44)
      SPS=PARAM(45)
      PPS=PARAM(46)
      PPP=PARAM(47)
C   PARAMETERS OF 2ND NN A-B INTERACIONS SP-D ORBITALS H-MATRIX
      SDS=PARAM(48)
      PDS=PARAM(49)
      PDP=PARAM(50)
      B11=B9
      A2=P1
      A3=P12/SR2
      A4=SR3*P14/2.
      A5=-P14/2.
      A7=(P2+P3)/2.
      A8=P3
      A9=(P2-P3)/2.
      A10=SR3*SR2*P24/4.
      A11=P35/SR2
      A12=(SR2*SR3*P35+P24/SR2)/4.
      A13=(-SR3*P24+2.*P35)/(4.*SR2)
      A15=(3.*P4+P6)/4.
      A16=(P5+P6)/2.
      A17=(P5-P6)/2.
      A18=SR3*(P6-P4)/4.
      A20=(P4+3.*P6)/4.
      A21=P5
      C3=CC3/SR2
      C7=(CC7+CC8)/2.
      C8=CC8
      C9=(CC7-CC8)/2.
C    THREE TO TWO CENTER RELATIONSHIPS FOR 2ND NN A-B INTERACTIONS
C    OF S AND P ORBITALS.
      F2=PSS/SR3
      F3=SPS/SR3
      F4=(PPS+2.*PPP)/3.
      F5=(PPS-PPP)/3.
C   SIMILARLY INCLUDING D ORBITALS OF ATOM A
      G1=SDS/SR3
      G2=(PDS+PDP/SR3)/3.
      G3=(PDS-2.*PDP/SR3)/3.
C    Changed June 1991
      G4=PDP/SR3
 666  CONTINUE
```

```
IF (ICENTER.EQ.2) GO TO 667
A1=PARAM(1)
A6=PARAM(2)
A14=PARAM(3)
A19=PARAM(4)
A2=PARAM(5)
A3=PARAM(6)
A4=PARAM(7)
A5=PARAM(8)
A7=PARAM(9)
A8=PARAM(10)
A9=PARAM(11)
A10=PARAM(12)
A11=PARAM(13)
A12=PARAM(14)
A13=PARAM(15)
A15=PARAM(16)
A16=PARAM(17)
A17=PARAM(18)
A18=PARAM(19)
A20=PARAM(20)
A21=PARAM(21)
B1=PARAM(22)
B2=PARAM(23)
B3=PARAM(24)
B4=PARAM(25)
B5=PARAM(26)
B6=PARAM(27)
B7=PARAM(28)
B8=PARAM(29)
B9=PARAM(30)
B10=PARAM(31)
B11=PARAM(32)
C1=PARAM(33)
C2=PARAM(34)
C3=PARAM(35)
C6=PARAM(36)
C7=PARAM(37)
C8=PARAM(38)
C9=PARAM(39)
EE=PARAM(40)
EE2=PARAM(41)
E4=PARAM(42)
E5=PARAM(43)
D1=PARAM(44)
D2=PARAM(45)
D3=PARAM(46)
D4=PARAM(47)
D5=PARAM(48)
D6=PARAM(49)
D7=PARAM(50)
D8=PARAM(51)
```

```
      F1 =PARAM(52)
      F2 =PARAM(53)
      F3 =PARAM(54)
      F4=PARAM(55)
      F5=PARAM(56)
      G1 =PARAM(57)
      G2 =PARAM(58)
      G3=PARAM(59)
      G4=PARAM(60)                                                  106,7
  667 CONTINUE
      JK=0
      IMESH=2*MESH+1
C     THE K-VECTOR UNITS ARE CHOSEN SO THAT THE POINT X HAS
C     CORDINATES (8,0,0)
      SCALE= 8./FLOAT(2*MESH)
      DO 451  IYBK=1,IMESH
      DO 451  IXBK=1,IYBK
      DO 451  IZBK=1,IXBK
      IT= IXBK+IYBK+IZBK-3
      IF(MESH*3.LT.IT) GO TO 451
      JK=JK+1
      A(1)= FLOAT(IXBK-1)
      A(2)= FLOAT(IYBK-1)
      A(3)= FLOAT(IZBK-1)
      XX =A(1)*SCALE
      YY =A(2)*SCALE
      ZZ =A(3)*SCALE
      PI = 3.14159265
      X1 = PI*XX/8.
      Y1 = PI*YY/8.
      Z1 = PI*ZZ/8.
       COSX=COS(X1)
       COSY=COS(Y1)
       COSZ=COS(Z1)
       SINX=SIN(X1)
       SINY=SIN(Y1)
       SINZ=SIN(Z1)
      COS2X=COS(2.*X1)
      COS2Y=COS(2.*Y1)
      COS2Z=COS(2.*Z1)
      SIN2X=SIN(2.*X1)
      SIN2Y=SIN(2.*Y1)
      SIN2Z=SIN(2.*Z1)
      H(1,1)=A1+4.*A2*(COSX*COSY+COSY*COSZ+COSZ*COSX)+2.*B1*(COS2X+COS2Y
     1+COS2Z)
      H(1,2)=4.*A3*SINX*(COSY+COSZ)+2.*B2*SIN2X
      H(1,3)=4.*A3*SINY*(COSZ+COSX)+2.*B2*SIN2Y
      H(1,4)=4.*A3*SINZ*(COSX+COSY)+2.*B2*SIN2Z
C
C+++++H(1,2),H(1,3),H(1,4) ARE REVERSED IN SIGN SO THAT THE REAL MATRIX IS
C     EQUIVALENT TO THE COMPLEX MATRIX
C
```

```
 H(1,2)=-H(1,2)
 H(1,3)=-H(1,3)
 H(1,4)=-H(1,4)
 H(1,5)=-4.*A4*SINX*SINY
 H(1,6)=-4.*A4*SINY*SINZ
 H(1,7)=-4.*A4*SINZ*SINX
 H(1,8)=2.*SR3*A5*COSZ*(COSY-COSX)+SR3*B3*(COS2X-COS2Y)
 H(1,9)=-2.*A5*(-2.*COSX*COSY+COSY*COSZ+COSZ*COSX)+B3*(-COS2X-COS2Y
1+2.*COS2Z)
 H(2,2)=A6+4.*A7*COSX*(COSY+COSZ)+4.*A8*COSY*COSZ+2.*B4*COS2X+2.*B5
1*(COS2Y+COS2Z)
 H(2,3)=-4.*A9*SINX*SINY
 H(2,4)=-4.*A9*SINX*SINZ
 H(2,5)=4.*SINY*(A10*COSX+A11*COSZ)+2.*B6*SIN2Y
 H(2,6)=0.
 H(2,7)=4.*SINZ*(A10*COSX+A11*COSY)+2.*B6*SIN2Z
 H(2,8)=2.*SR3*A12*SINX*(COSY+COSZ)+2.*A13*SINX*(COSY-COSZ)+SR3*B7*
1SIN2X
 H(2,9)=-2.*A12*SINX*(COSY+COSZ)+2.*SR3*A13*SINX*(COSY-COSZ)-B7*
1SIN2X
 H(3,3)=A6+4.*A7*COSY*(COSZ+COSX)+4.*A8*COSZ*COSX+2.*B4*COS2Y+
12.*B5*(COS2Z+COS2X)
 H(3,4)=-4.*A9*SINY*SINZ
 H(3,5)=4.*SINX*(A10*COSY+A11*COSZ)+2.*B6*SIN2X
 H(3,6)=4.*SINZ*(A10*COSY+A11*COSX)+2.*B6*SIN2Z
 H(3,7)=0.
 H(3,8)=-2.*SR3*A12*SINY*(COSZ+COSX)-2.*A13*SINY*(COSX-COSZ)-SR3*B7
1*SIN2Y
 H(3,9)=-2.*A12*SINY*(COSZ+COSX)-2.*SR3*A13*SINY*(COSZ-COSX)-B7*
1SIN2Y
 H(4,4)=A6+4.*A7*COSZ*(COSX+COSY)+4.*A8*COSX*COSY+2.*B4*COS2Z+2.*B5
1*(COS2X+COS2Y)
 H(4,5)=0.
 H(4,6)=4.*SINY*(A10*COSZ+A11*COSX)+2.*B6*SIN2Y
 H(4,7)=4.*SINX*(A10*COSZ+A11*COSY)+2.*B6*SIN2X
 H(4,8)=4.*A13*SINZ*(COSY-COSX)
 H(4,9)=4.*A12*SINZ*(COSX+COSY)+2.*B7*SIN2Z
 H(5,5)=A14+4.*A15*COSX*COSY+4.*A16*COSZ*(COSX+COSY)+2.*B8*(COS2X+
1COS2Y)+2.*B9*COS2Z
 H(5,6)=-4.*A17*SINX*SINZ
 H(5,7)=-4.*A17*SINY*SINZ
 H(5,8)=0.
 H(5,9)=-4.*A18*SINX*SINY
 H(6,6)=A14+4.*A15*COSY*COSZ+4.*A16*COSX*(COSY+COSZ)+2.*B8*(COS2Z
1+COS2Y)+2.*B9*COS2X
 H(6,7)=-4.*A17*SINX*SINY
 H(6,8)=-2.*SR3*A18*SINY*SINZ
 H(6,9)=2.*A18*SINY*SINZ
 H(7,7)=A14+4.*A15*COSX*COSZ+4.*A16*COSY*(COSX+COSZ)+2.*B8*(COS2X+
1COS2Z)+2.*B9*COS2Y
 H(7,8)=2.*SR3*A18*SINX*SINZ
 H(7,9)=2.*A18*SINX*SINZ
```

```
      H(8,8)=A19+3.*A20*COSZ*(COSX+COSY)+4.*A21*(COSX*COSY+COSX*COSZ/4.+
     1COSY*COSZ/4.)+(3./2.)*B10*(COS2X+COS2Y)+2.*B11*(COS2X/4.+COS2Y/4.+
     2COS2Z)
      H(8,9)=SR3*COSZ*(COSX-COSY)*(A20-A21)+(SR3/2.)*(COS2Y-COS2X)*(B10-
     1B11)
      H(9,9)=A19+4.*A20*(COSX*COSY+COSX*COSZ/4.+COSY*COSZ/4.)+3.*A21*
     1COSZ*(COSX+COSY)+2.*B10*(COS2X/4.+COS2Y/4.+COS2Z)+(3./2.)*B11*
     2(COS2X+COS2Y)
C
C     CHANGES TO INCORPORATE 2ND NN OF THE A-B INTERACTIONS
      H(1,10)=2.*D1*(COSX+COSY+COSZ)+8.*F1*COSX*COSY*COSZ
      H(2,10)=2.*D3*SINX+8.*F2*SINX*COSY*COSZ
      H(3,10)=2.*D3*SINY+8.*F2*SINY*COSZ*COSX
      H(4,10)=2.*D3*SINZ+8.*F2*SINZ*COSX*COSY
      H(5,10)=-8.*G1*SINX*SINY*COSZ
      H(6,10)=-8.*G1*COSX*SINY*SINZ
      H(7,10)=-8.*G1*SINX*COSY*SINZ
      H(8,10)=SR3*D7*(COSX-COSY)
      H(9,10)=D7*(-COSX-COSY+2.*COSZ)
      H(10,10) =
     1 C1+4.*C2*(COSX*COSY+COSY*COSZ+COSZ*COSX)+2.*EE*(COS2X+COS2Y
     1+COS2Z)
      H(1,11)=2*D2*SINX+8.*F3*SINX*COSY*COSZ
      H(1,12)=2*D2*SINY+8.*F3*SINY*COSZ*COSX
      H(1,13)=2*D2*SINZ+8.*F3*SINZ*COSX*COSY
      H(2,11)=2*D4*COSX+2*D5*(COSY+COSZ)+8.*F4*COSX*COSY*COSZ
      H(2,12)=-8.*F5*SINX*SINY*COSZ
      H(2,13)=-8.*F5*SINX*SINZ*COSY
      H(3,11)=H(2,12)
      H(3,12)=2*D4*COSY+2*D5*(COSZ+COSX)+8.*F4*COSX*COSY*COSZ
      H(3,13)=-8.*F5*SINY*SINZ*COSX
      H(4,11)=H(2,13)
      H(4,12)=H(3,13)
      H(4,13)=2*D4*COSZ+2*D5*(COSY+COSX)+8.*F4*COSX*COSY*COSZ
c    END OF 2ND NN A-B INTERACTIONS FOR S AND P ORBITALS
      H(5,11)=2.*D6*SINY+8.*G2*COSX*SINY*COSZ
      H(5,12)=2.*D6*SINX+8.*G2*SINX*COSY*COSZ
      H(5,13)=-8.*G3*SINX*SINY*SINZ
      H(6,11)=-8.*G3*SINX*SINY*SINZ
      H(6,12)=2.*D6*SINZ+8.*G2*COSX*COSY*SINZ
      H(6,13)=2.*D6*SINY+8.*G2*COSX*SINY*COSZ
      H(7,11)=2.*D6*SINZ+8.*G2*COSX*COSY*SINZ
      H(7,12)=-8.*G3*SINX*SINY*SINZ
      H(7,13)=2.*D6*SINX+8.*G2*SINX*COSY*COSZ
      H(8,11)=SR3*D8 *SINX+8.*G4*SINX*COSY*COSZ
      H(8,12)=-SR3*D8 *SINY-8.*G4*COSX*SINY*COSZ
      H(8,13)=0.0
      H(9,11)=-D8 *SINX-(8./SR3)*G4*SINX*COSY*COSZ
      H(9,12)=-D8 *SINY-(8./SR3)*G4*COSX*SINY*COSZ
      H(9,13)=2*D8 *SINZ+(16./SR3)*G4*COSX*COSY*SINZ
      H(10,11)=-4.*C3*SINX*(COSY+COSZ)-2.*EE2*SIN2X
      H(10,12)=
```

```
     1       -4.*C3*SINY*(COSZ+COSX)-2.*EE2*SIN2Y
      H(10,13)=
     1       -4.*C3*SINZ*(COSX+COSY)-2.*EE2*SIN2Z
      H(11,11)=C6+
     1           4.*C7*COSX*(COSY+COSZ)+4.*C8*COSY*COSZ+2.*E4*COS2X+
     22.*E5*(COS2Y+COS2Z)
      H(11,12)= -C9*SINX*SINY*4.
      H(11,13)= -C9*SINX*SINZ*4.
      H(12,12)=C6+
     1           4.*C7*COSY*(COSZ+COSX)+4.*C8*COSZ*COSX+2.*E4*COS2Y+
     22.*E5*(COS2Z+COS2X)
      H(12,13)= -C9*SINY*SINZ*4.
      H(13,13)=C6+
     1           4.*C7*COSZ*(COSX+COSY)+4.*C8*COSX*COSY+2.*E4*COS2Z+
     22.*E5*(COS2X+COS2Y)
C  Overlap matrix
C   Zero the overlap matrix
      S(1,1)=1.
      S(1,2)=0.0
      S(1,3)=0.0
      S(1,4)=0.0
      S(1,5)=0.0
      S(1,6)=0.0
      S(1,7)=0.0
      S(1,8)=0.0
      S(1,9)=0.0
      S(1,10)=0.0
      S(1,11)=0.0
      S(1,12)=0.0
      S(1,13)=0.0
      S(2,2)=1.
      S(2,3)=0.0
      S(2,4)=0.0
      S(2,5)=0.0
      S(2,6)=0.
      S(2,7)=0.0
      S(2,8)=0.0
      S(2,9)=0.0
      S(2,10)=0.0
      S(2,11)=0.0
      S(2,12)=0.0
      S(2,13)=0.0
      S(3,3)=1.
      S(3,4)=0.0
      S(3,5)=0.0
      S(3,6)=0.0
      S(3,7)=0.
      S(3,8)=0.0
      S(3,9)=0.0
      S(3,10)=0.0
      S(3,11)=0.0
      S(3,12)=0.0
```

```
S(3,13)=0.0
S(4,4)=1.
S(4,5)=0.
S(4,6)=0.0
S(4,7)=0.0
S(4,8)=0.0
S(4,9)=0.0
S(4,10)=0.0
S(4,11)=0.0
S(4,12)=0.0
S(4,13)=0.0
S(5,5)=1.
S(5,6)=0.0
S(5,7)=0.0
S(5,8)=0.
S(5,9)=0.0
S(5,10)=0.0
S(5,11)=0.0
S(5,12)=0.0
S(5,13)=0.0
S(6,6)=1.
S(6,7)=0.0
S(6,8)=0.0
S(6,9)=0.0
S(6,10)=0.0
S(6,11)=0.0
S(6,12)=0.0
S(6,13)=0.0
S(7,7)=1.
S(7,8)=0.0
S(7,9)=0.0
S(7,10)=0.0
S(7,11)=0.0
S(7,12)=0.0
S(7,13)=0.0
S(8,8)=1.
S(8,9)=0.0
S(8,10)=0.0
S(8,11)=0.0
S(8,12)=0.0
S(8,13)=0.0
S(9,9)=1.
S(9,10)=0.0
S(9,11)=0.0
S(9,12)=0.0
S(9,13)=0.0
S(10,10)=1.
S(10,11)=0.
S(10,12)=0.
S(10,13)=0.
S(11,11)=1.
S(11,12)=0.
```

```
      S(11,13)=0.
      S(12,12)=1.
      S(12,13)=0.
      S(13,13)=1.
      DO 310 L=1,N
      DO 310 K=1,N
      H(K,L)=H(L,K)
      S(K,L)=S(L,K)
  310 CONTINUE
c     WRITE(15,325) XX,YY,ZZ
  325 FORMAT(3(F6.2,2X))
      NR=0
      IEGEN=0
C     DIAGONALIZES THE OVERLAP MATRIX S
      CALL HDIAG(S,NS,IEGEN, U,NR,NS)
      DO 8111 M=1,NS
      DO 8111 J=1,M
      HNEW(M,J)=0.0
      DO 8200 K=1,NS
      DO 8200 L=1,NS
      HNEW(M,J)=HNEW(M,J)+U(K,M)*H(K,L)*U(L,J)
 8200 CONTINUE
c     S(M,M)=ABS(S(M,M))
c     S(J,J)=ABS(S(J,J))
      HNEW(M,J)=HNEW(M,J)/(SQRT(S(M,M))*SQRT(S(J,J)))
       HNEW(J,M)=HNEW(M,J)
 8111 CONTINUE
C     DIAGONALIZES THE EFFECTIVE HAMILTONIAN HNEW
      CALL HDIAG(HNEW,NS,IEGEN,UU,NR,NS)
      DO 850 I1=1,NS
      DO 850 I2=1,I1
      T(I1,I2)=0.0
      DO 860 J1=1,NS
c     S(J1,J1)=ABS(S(J1,J1))
      T(I1,I2)=T(I1,I2)+U(I1,J1)*UU(J1,I2)/SQRT(S(J1,J1))
  860 CONTINUE
  850 CONTINUE
      DO 851 I1=1,NS
      DO 851 I2=I1,NS
      T(I1,I2)=0.0
      DO 861 J1=1,NS
c     S(J1,J1)=ABS(S(J1,J1))
      T(I1,I2)=T(I1,I2)+U(I1,J1)*UU(J1,I2)/SQRT(S(J1,J1))
  861 CONTINUE
  851 CONTINUE
      DO 390 J=1,NS
      ANORM=0.0
      DO 865 LNORM=1,NS
      ANORM=ANORM+(UU(LNORM,J)**2)/S(LNORM,LNORM)
  865 CONTINUE
      ANORM=SQRT(ANORM)
      DO 866 LNORM=1,NS
```

```
      T(LNORM, J)=T(LNORM, J)/ANORM
      U(LNORM, J)=T(LNORM, J)
  866 CONTINUE
  390 CONTINUE
      CALL SORT(HNEW, U, NS)
C     THE Q10, Q11, Q12, Q13, Q20, Q21 ARE THE SUMS OF THE SQUARES OF THE
C     EIGENVECTORS OVER 13 BANDS. THESE ARE USED TO FIND THE
C     ANGULAR MOMENTUM CHARACTER OF EACH STATE AND THUS
C     DECOMPOSE THE DOS
      DO 729   J=1, NS
      Q10(J, JK )=U(1, J)**2
      Q11(J, JK )= U(2, J)**2 + U(3, J)**2 + U(4, J)**2
      Q12(J, JK )= U(5, J)**2 + U(6, J)**2 + U(7, J)**2
      Q13(J, JK )= U(8, J)**2 + U(9, J)**2
      Q20(J, JK )=U(10, J)**2
      Q21(J, JK )= U(11, J)**2 + U(12, J)**2 + U(13, J)**2
  729 CONTINUE
  332 FORMAT(1X, 5F10.5)
      DO 5554 J=1, NS
      WRITE(15, 330)HNEW(J, J), Q10(J, JK), Q11(J, JK), Q12(J, JK), Q13(J, JK)
     &, Q20(J, JK), Q21(J, JK)
 5554 CONTINUE
  330 FORMAT(F9.5, 3X, 6F8.5)
  451 CONTINUE
      WRITE(15, 452) JK
  452 FORMAT(1X, 'NUMBER OF K-POINTS', 2X, I5)
      STOP
      END
      SUBROUTINE SORT(H, U, N)
      implicit real*8 (A-H, O-Z)
C     THIS ROUTINE ORDERS EIGENVALUES AND EIGENVECTORS
      DIMENSION H(13, 13), U(13, 13), T(13)
      NN=N-1
      DO 50 L=1, NN
      LL=L+1
      DO 60 M=LL, N
      IF (H(L, L)-H(M, M)) 60, 60, 70
   70 TEMP=H(L, L)
      H(L, L)=H(M, M)
      H(M, M)=TEMP
      DO 100 MM=1, N
      T(MM)=U(MM, L)
      U(MM, L)=U(MM, M)
  100 U(MM, M)=T(MM)
   60 CONTINUE
   50 CONTINUE
      RETURN
      END
      SUBROUTINE HDIAG(H, N, IEGEN, U, NR, NN)
      implicit real*8 (A-H, O-Z)
C    THIS ROUTINE COMES FROM THE MIT COMPUTATIONS CENTER
C        THE JACOBI METHOD.
```

```
C     CALLING SEQUENCE FOR DIAGONALIZATION
C     CALL HDIAG(H,N,IEGEN,U,NR,NN)
C            WHERE H IS THE ARRAY TO BE DIAGONALIZED.
C     N IS THE ORDER OF THE MATRIX, H.
C
C     IEGEN MUST BE SET UNEQUAL TO ZERO IF ONLY EIGENVALUES ARE
C            TO BE COMPUTED.
C
C     A DIMENSION STATEMENT MUST BE INSERTED IN THE SUBROUTINE.
C     DIMENSION H(NN,NN),U(NN,NN),X(NN),IQ(NN)
      DIMENSION H(13,13),U(13,13)
      DIMENSION X(100), IQ(100)
      IF (IEGEN.NE.0) GO TO 15
 10   DO 14 I=1,N
      DO 14 J=1,N
      IF (I. NE.J) GO TO 12
 11   U(I,J)=1.0
      GO TO 14
 12   U(I,J)=0.
 14   CONTINUE
C
 15   NR = 0
      IF (N.LE.1) RETURN
C
C     SCAN FOR LARGEST OFF DIAGONAL ELEMENT IN EACH ROW
C     X(I) CONTAINS LARGEST ELEMENT IN ITH ROW
C     IQ(I) HOLDS SECOND SUBSCRIPT DEFINING POSITION OF ELEMENT
C
 17   NMI1=N-1
      DO 30 I=1,NMI1
      X(I) = 0.
      IPL1=I+1
      DO 30 J=IPL1,N
      IF (X(I).GT. ABS(H(I,J))) GO TO 30
 20   X(I)=ABS (H(I,J))
      IQ(I)=J
 30   CONTINUE
C
C     SET INDICATOR FOR SHUT-OFF.RAP=2**-27,NR=NO. OF ROTATIONS
      RAP=7.450580596E-6
      HDTEST=1.0E35
C
C     FIND MAXIMUM OF X(I) S FOR PIVOT ELEMENT AND
C     TEST FOR END OF PROBLEM
C
 40   DO  70  I=1,NMI1
      IF (I.GT.1.AND.XMAX.GE.X(I)) GOT 070
 60   XMAX=X(I)
      IPIV=I
      JPIV=IQ(I)
 70   CONTINUE
C
```

```
C      IS MAX. X(I) EQUAL TO ZERO, IF LESS THAN HDTEST, REVISE HDTEST
       IF (XMAX.LE.0.) RETURN
80     IF (HDTEST.GT.0..AND.XMAX.GT.HDTEST) GO TO148
 90    HDIMIN = ABS ( H(1,1) )
       DO 110  I= 2,N
       IF (HDIMIN.GT.ABS(H(I,I))) HDIMIN=ABS(H(I,I))
 110   CONTINUE
C
       HDTEST=HDIMIN*RAP
C
C      RETURN IF MAX.H(I,J)LESS THAN(2**-27)ABSF(H(K,K)-MIN)
       IF (HDTEST.GE.XMAX) RETURN
 148   NR = NR+1
C
C      COMPUTE TANGENT, SINE AND COSINE,H(I,I),H(J,J)
 150   TANG=SIGN(2.D0,(H(IPIV,IPIV)-H(JPIV,JPIV)))*H(IPIV,JPIV)/(ABS (H(I
      1PIV,IPIV)-H(JPIV,JPIV))+SQRT ((H(IPIV,IPIV)-H(JPIV,JPIV))**2+4.0*H
      2(IPIV,JPIV)**2))
       COSINE=1.0/SQRT (1.0+TANG**2)
       SINE=TANG*COSINE
       HII=H(IPIV,IPIV)
       H(IPIV,IPIV)=COSINE**2*(HII+TANG*(2.*H(IPIV,JPIV)+TANG*H(JPIV,JPIV
      1)))
       H(JPIV,JPIV)=COSINE**2*(H(JPIV,JPIV)-TANG*(2.*H(IPIV,JPIV)-TANG*H
      1II))
       H(IPIV,JPIV)=0.
C
C       PSEUDO RANK THE EIGENVALUES
C       ADJUST SINE AND COS FOR COMPUTATION OF H(IK) AND U(IK)
       IF (H(IPIV,IPIV).GE.H(JPIV,JPIV)) GO TO 153
 152   HTEMP = H(IPIV,IPIV)
       H(IPIV,IPIV) = H(JPIV,JPIV)
       H(JPIV,JPIV) = HTEMP
C        RECOMPUTE SINE AND COS
       HTEMP = SIGN  (1.0D0, -SINE) * COSINE
       COSINE = ABS  (SINE)
       SINE = HTEMP
 153   CONTINUE
C
C      INSPECT THE IQS BETWEEN I+1 AND N-1 TO DETERMINE
C      WHETHER A NEW MAXIMUM VALUE SHOULD BE COMPUTED SINCE
C      THE PRESENT MAXIMUM IS IN THE I OR J ROW.
C
       DO 350 I=1,NMI1
       IF (I.EQ.IPIV.OR.I.EQ.JPIV) GO TO 350
210    IF (IQ(I).NE.IPIV.AND.IQ(I).NE.JPIV)      GO TO 350
 240   K=IQ(I)
 250   HTEMP=H(I,K)
       H(I,K)=0.
       IPL1=I+1
       X(I) =0.
C
```

```
C     SEARCH IN DEPLETED ROW FOR NEW MAXIMUM
C
      DO 320 J=IPL1,N
      IF (X(I)  .GT.ABS(H(I,J))) GO TO 320
  300 X(I) = ABS (H(I,J))
      IQ(I)=J
  320 CONTINUE
      H(I,K)=HTEMP
  350 CONTINUE
C
      X(IPIV) =0.
      X(JPIV) =0.
C
C     CHANGE THE OTHER ELEMENTS OF H
C
      DO 530 I=1,N
C
      IF(I-IPIV)370,530,420
  370 HTEMP = H(I,IPIV)
      H(I,IPIV) = COSINE*HTEMP + SINE*H(I,JPIV)
      IF (X(I).GE.ABS(H(I,IPIV))) GO TO 390
  380 X(I) = ABS (H(I,IPIV))
      IQ(I) = IPIV
  390 H(I,JPIV) = -SINE*HTEMP + COSINE*H(I,JPIV)
      IF (X(I).GE.ABS(H(I,JPIV))) GO TO 530
      X(I)=ABS(H(I,JPIV))
      IQ(I) = JPIV
      GO TO 530
C
  420 IF(I-JPIV)430,530,480
  430 HTEMP = H(IPIV,I)
      H(IPIV,I) = COSINE*HTEMP + SINE*H(I,JPIV)
      IF (X(IPIV).GE.ABS(H(IPIV,I))) GO TO 450
  440 X(IPIV) = ABS (H(IPIV,I))
      IQ(IPIV) = I
  450 H(I,JPIV) = -SINE*HTEMP + COSINE*H(I,JPIV)
      IF (X(I).GE.ABS(H(I,JPIV))) GO TO 530
      X(I)=ABS (H(I,JPIV))
      IQ(I)=JPIV
      GO TO 530
C
  480 HTEMP = H(IPIV,I)
      H(IPIV,I) = COSINE*HTEMP + SINE*H(JPIV,I)
      IF (X(IPIV).GE.ABS(H(IPIV,I))) GO TO 500
  490 X(IPIV) = ABS (H(IPIV,I))
      IQ(IPIV) = I
  500 H(JPIV,I) = -SINE*HTEMP + COSINE*H(JPIV,I)
      IF (X(JPIV).GE.ABS(H(JPIV,I))) GOTO 530
  510 X(JPIV) = ABS (H(JPIV,I))
      IQ(JPIV) = I
  530 CONTINUE
C
```

```
C     TEST FOR COMPUTATION OF EIGENVECTORS
C
      IF (IEGEN.NE.0) GO TO 40
  540 DO 550 I=1,N
      HTEMP=U(I,IPIV)
      U(I,IPIV)=COSINE*HTEMP+SINE*U(I,JPIV)
  550 U(I,JPIV)=-SINE*HTEMP+COSINE*U(I,JPIV)
      GO TO 40
      END
```

The parameters given below are for CuH (2-center orthogonal). For another material the user has to type the TB parameters given in the Tables of the Handbook, in exactly the same order.

Caution: The hydrogen onsite "p" parameter is arbitrarily given below as 10.0 (It doesn't appear in the Tables) to keep the p-levels out of the way. It would have an exact value for a carbide calculation.

The "8" generates a k-point mesh for 505 k points in the irreducible BZ. NOT to be used for plotting energy bands.

The "2" means 2-center TB parameters. It would be "3" for 3-center params.

Remove the above to run the program. The parameters given below are for CuH.

For another material the user has to type the TB parameters given in the Tables of the Handbook, in exactly the same order.

```
CuH 2-center orthog
   8   2
 0.7400259972   0  1   s on-site              CuH
 1.5060045719   0  2   p on-site              CuH
 0.4652805924   0  3   t2g on-site            CuH
 0.4626203775   0  4   eg on-site             CuH
-0.0862571150   0  5      sss 2nd nn A-A      CuH
 0.1887781769   0  6      pps 2nd nn A-A      CuH
-0.0102373008   0  7      ppp 2nd nn A-A      CuH
-0.0228107497   0  8      dds 2nd nn A-A      CuH
 0.0099785505   0  9      ddp 2nd nn A-A      CuH
-0.0001492867   0 10      ddd 2nd nn A-A      CuH
 0.1405877769   0 11      sps 2nd nn A-A      CuH
-0.0398081429   0 12      sds 2nd nn A-A      CuH
-0.0523708202   0 13      pds 2nd nn A-A      CuH
 0.0119420076   0 14      pdp 2nd nn A-A      CuH
 0.0125410324   0 15      sss 4th nn A-A      CuH
-0.0992608070   0 16      pps 4th nn A-A      CuH
 0.0279004369   0 17      ppp 4th nn A-A      CuH
 0.0000347503   0 18      dds 4th nn A-A      CuH
-0.0002100729   0 19      ddp 4th nn A-A      CuH
-0.0001720828   0 20      ddd 4th nn A-A      CuH
-0.0242424160   0 21      sps 4th nn A-A      CuH
-0.0033862733   0 22      sds 4th nn A-A      CuH
-0.0047177281   0 23      pds 4th nn A-A      CuH
-0.0021732016   0 24      pdp 4th nn A-A      CuH
 0.9718690515   0 25   s on-site B-B          CuH
10.0000000000   1 26   p on-site B-B          CuH
```

REMOVE the above to run the program.

```
0.0496456362  0 27   sss 2nd nn B-B    CuH
0.0000000000  1 28   sps 2nd nn B-B    CuH
0.0000000000  1 29   pps 2nd nn B-B    CuH
0.0000000000  1 30   ppp 2nd nn B-B    CuH
0.0045700432  0 31   sss 4th nn B-B    CuH
0.0000000000  1 32   sps 4th nn B-B    CuH
0.0000000000  1 33   pps 4th nn B-B    CuH
0.0000000000  1 34   ppp 4th nn B-B    CuH
0.0587895960  0 35   sss 1st nn A-B    CuH
0.0000000000  1 36   sps 1st nn A-B    CuH
0.1869927794  0 37   pss 1st nn A-B    CuH
0.0000000000  1 38   pps 1st nn A-B    CuH
0.0000000000  1 39   ppp 1st nn A-B    CuH
0.0000000000  1 40   pdp 1st nn A-B    CuH
0.0713763312  0 41   sds 1st nn A-B    CuH
0.0000000000  1 42   pds 1st nn A-B    CuH
-0.0359935872 0 43   sss 3rd nn A-B    CuH
-0.0457092784 0 44   pss 3rd nn A-B    CuH
0.0000000000  1 45   sps 3rd nn A-B    CuH
0.0000000000  1 46   pps 3rd nn A-B    CuH
0.0000000000  1 47   ppp 3rd nn A-B    CuH
-0.0140490225 0 48   sds 3rd nn A-B    CuH
0.0000000000  1 49   pds 3rd nn A-B    CuH
0.0000000000  1 50   pdp 3rd nn A-B    CuH
```

The parameters given below are for CuH (3-center orthogonal). For another material the user has to type the TB parameters given in the Tables of the Handbook, in exactly the same order.

Caution: The hydrogen onsite "p" parameter is arbitrarily given below as 10.0 (It doesn't appear in the Tables) to keep the p-levels out of the way. It would have an exact value for a carbide calculation.

The "8" generates a k-point mesh for 505 k points in the irreducible BZ. NOT to be used for plotting energy bands.

The "3" means 3-center TB parameters. It would be "2" for 2-center params.

Remove the above to run the program.

```
CuH 3-center orthog
    8     3
 1.0133436918   0  1                    CuH
 1.6299730539   0  2                    CuH
 0.4669095576   0  3                    CuH
 0.5523251891   0  4                    CuH
 0.0070993169   0  5                    CuH
-0.0390431844   0  6                    CuH
-0.0347827710   0  7                    CuH
-0.0250461921   0  8                    CuH
 0.0905985832   0  9                    CuH
-0.0000801649   0 10                    CuH
 0.0442320406   0 11                    CuH
 0.0352347530   0 12                    CuH
-0.0050138300   0 13                    CuH
 0.0063122837   0 14                    CuH
 0.0193088688   0 15                    CuH
-0.0174550749   0 16                    CuH
 0.0054789842   0 17                    CuH
 0.0045968308   0 18                    CuH
 0.0090510566   0 19                    CuH
 0.0090507874   0 20                    CuH
-0.0348815545   0 21                    CuH
 0.0593079962   0 22                    CuH
 0.0829818100   0 23                    CuH
 0.0467784964   0 24                    CuH
-0.1623541564   0 25                    CuH
 0.0210846700   0 26                    CuH
 0.0027799814   0 27                    CuH
-0.0472122058   0 28                    CuH
-0.0007821462   0 29                    CuH
 0.0000663938   0 30                    CuH
 0.0302998964   0 31                    CuH
-0.0003523610   0 32                    CuH
 0.1711303145   0 33                    CuH
-0.0177616198   0 34                    CuH
 0.0000000000   1 35                    CuH
10.0000000000   1 36                    CuH
 0.0000000000   1 37                    CuH
 0.0000000000   1 38                    CuH
 0.0000000000   1 39                    CuH
-0.0050270767   0 40                    CuH
 0.0000000000   1 41                    CuH
 0.0000000000   1 42                    CuH
 0.0000000000   1 43                    CuH
-0.1016161218   0 44                    CuH
 0.0000000000   1 45                    CuH
 0.0295211356   0 46                    CuH
 0.0000000000   1 47                    CuH
 0.0000000000   1 48                    CuH
 0.0000000000   1 49                    CuH
-0.0313003585   0 50                    CuH
 0.0000000000   1 51                    CuH
 0.0141851548   0 52                    CuH
-0.0314414278   0 53                    CuH
 0.0000000000   1 54                    CuH
 0.0000000000   1 55                    CuH
 0.0000000000   1 56                    CuH
-0.0030053612   0 57                    CuH
 0.0000000000   1 58                    CuH
 0.0000000000   1 59                    CuH
 0.0000000000   1 60                    CuH
```

This file is the output of the program naclorthog.f and becomes the input to the program doste.f which calculates the DOS. The data below correspond to the 3-center TB parameters. Next line explains the different columns.

```
  Ener (Ry)   Q(s)    Q(p)    Q(t2g)  Q(eg)   Q(s-H)  last column only for
 carbides; not needed for hydrides.
Remove the above instructions to run the DOS with : doste.f
This file normally has 13 bands x 505 k-points =6565 lines
 Of course for hydrides only 10 bands are needed.
 To save space we show here the first k-point (Gamma) and the last W
 -0.21929     0.08080 0.00000 0.00000 0.00000 0.91920 0.00000
  0.43793     0.00000 0.00000 1.00000 0.00000 0.00000 0.00000
  0.43793     0.00000 0.00000 1.00000 0.00000 0.00000 0.00000
  0.43793     0.00000 0.00000 1.00000 0.00000 0.00000 0.00000
  0.48718     0.00000 0.00000 0.00000 1.00000 0.00000 0.00000
  0.48718     0.00000 0.00000 0.00000 1.00000 0.00000 0.00000
  1.60150     0.91920 0.00000 0.00000 0.00000 0.08080 0.00000
  2.11407     0.00000 1.00000 0.00000 0.00000 0.00000 0.00000
  2.11407     0.00000 1.00000 0.00000 0.00000 0.00000 0.00000
  2.11407     0.00000 1.00000 0.00000 0.00000 0.00000 0.00000
 10.00000     0.00000 0.00000 0.00000 0.00000 0.00000 1.00000
 10.00000     0.00000 0.00000 0.00000 0.00000 0.00000 1.00000
 10.00000     0.00000 0.00000 0.00000 0.00000 0.00000 1.00000
-------------------------------------------------------------

  0.16915     0.00000 0.02295 0.00000 0.01818 0.95888 0.00000
  0.39759     0.00000 0.08018 0.91982 0.00000 0.00000 0.00000
  0.39759     0.00000 0.08018 0.91982 0.00000 0.00000 0.00000
  0.47279     0.01864 0.00000 0.00000 0.98136 0.00000 0.00000
  0.53347     0.00000 0.00000 1.00000 0.00000 0.00000 0.00000
  0.77235     0.00000 0.02172 0.00000 0.96562 0.01266 0.00000
  0.99040     0.00000 0.91982 0.08018 0.00000 0.00000 0.00000
  0.99040     0.00000 0.91982 0.08018 0.00000 0.00000 0.00000
  1.11554     0.98136 0.00000 0.00000 0.01864 0.00000 0.00000
  2.11307     0.00000 0.95533 0.00000 0.01620 0.02847 0.00000
 10.00000     0.00000 0.00000 0.00000 0.00000 0.00000 1.00000
 10.00000     0.00000 0.00000 0.00000 0.00000 0.00000 1.00000
 10.00000     0.00000 0.00000 0.00000 0.00000 0.00000 1.00000
 NUMBER OF K-POINTS    505  Only 2 k-points given above.
```

Appendix B
Computer Program to Calculate Densities of States by the Tetrahedron Method

D. A. Papaconstantopoulos, *Band Structure of Cubic Hydrides*,
https://doi.org/10.1007/978-3-031-06878-2

```
C  TETRAHEDRON---FOR calculating DOS
c  Writen by L.L.Boyer and D.A.Papaconstantopoulos
c  Modified by Papa to do the interpolation at Ef  (April 6, 2019)
C  When converting from monatomic to diatomic (like TiFe), the following
C  changes should be made:
C   1) Dimensions on E in the common blocks should go from E(5,969,9) to
C      E(9,969,18).
C   2) NP,NB, and NAT should be changed as noted below in the program.
C   3) CHANGE 630 FORMAT AS IS EXPLAINED THERE.
C   4) NSAV controls the number of deviations printed at each band for
C      each mesh.
C  To convert back to diatomic, all these changes must be made in
reverse.
      IMPLICIT REAL*8(A-H,O-Z)
C     REAL*4 E
C     REAL*4 EIN(NAT,NP)
      CHARACTER*10 INA,INB,INC,IND
      COMMON/COMM/   FERV(9000),SFER(9000),OMEGA(9000),
     1  EM(9000),E(7,969,13),DOS(7,9000),ELEC(9000)
      CHARACTER*40 TITLE
C     WRITE(*,9551)
 9551 FORMAT(1X,'ENTER DATA FILE')
 9552 FORMAT(A10)
 9553 FORMAT(A10)
C INPUTS
C  dosdat.in  is an input. See attached file.
C  nacl.out  is the output of the TB program
      OPEN (10,FILE='dosdat.in',BLANK = 'ZERO')
      OPEN (11,FILE='nacl.out',BLANK = 'ZERO')
c     OPEN (9,FILE='nacl.out',BLANK = 'ZERO')
c OUTPUTS
      OPEN(19,FILE='dosdat.bin.plot',BLANK = 'ZERO')
      OPEN(15,FILE='fort.8',BLANK = 'ZERO')
      OPEN(20,FILE='dosdat.out',BLANK = 'ZERO')
      OPEN(13,FILE='dosapw.itp',BLANK = 'ZERO')
      NH=9000
C   FOR BCC AND FCC RUNS:
      NP=969
CC    NP=969  CORRESPONDS TO CUBIC MESH EQUIVALENT TO 505 FCC
      NB=13
      NAT=7
C   FOR CsCl RUNS:
C     NP=165
C     NB=18
C     NAT=9
C
      CALL TETDS(NH,NP,NB,NAT)
      STOP
      END
      SUBROUTINE CALDV(INC,NBD,MESH,NAT,NP,NB)
      IMPLICIT REAL*8(A-H,O-Z)
      DIMENSION INC(6,3,3),KT(4,3),KTS(4,3,18  ),DEV(18  )
      DIMENSION ETS(4,18  ),ET(4)
      COMMON/COMM/  FERV(9000),SFER(9000),OMEGA(9000),
```

```
     1  EM(9000),E(7,969,13),DOS(7,9000),ELEC(9000)
C      NSAV=18
       NSAV=8
       MESH2=MESH*2
       M2P1=MESH2+1
       DO 200 IB=1,NBD
       WRITE(20,1) IB
     1 FORMAT('-LARGEST DEVIATIONS FOR BAND NUMBER',I3)
       JNC=MESH
   150 IF(JNC.EQ.0) GO TO 80
       DO 160 I=1,NSAV
   160 DEV(I)=0.0
       KKK=0
       DO 100 II=1,M2P1,JNC
       KT(1,1)=II-1
       DO 100 JJ=1,II,JNC
       KT(1,2)=JJ-1
       DO 100 KK=1,JJ,JNC
       KKK=KKK+1
       KT(1,3)=KK-1
       DO 70 NN=1,6
       DO 40 L=1,3
       DO 40 M=1,3
    40 KT(L+1,M)=KT(1,M)+INC(NN,L,M)*JNC
       DO 60 L=2,4
       IF(KT(L,1).GT.MESH2) GO TO 70
       IF(KT(L,3).GT.KT(L,2)) GO TO 70
       IF(KT(L,2).GT.KT(L,1)) GO TO 70
    60 CONTINUE
       DO 65 L=1,4
       K1=KT(L,1)
       K2=KT(L,2)
       K3=KT(L,3)
       IN=K1*(K1+1)*(K1+2)/6 + K2*(K2+1)/2 + K3 + 1
    65 ET(L)=E(1,IN,IB)
       EAV=(ET(1)+ET(2)+ET(3)+ET(4))*0.25
       ADEV=ABS(ET(1)-EAV)+ABS(ET(2)-EAV)+ABS(ET(3)-EAV)+ABS(ET(4)-EAV)
       ADEV=ADEV*0.25
       IF(ADEV.LE.DEV(NSAV)) GO TO 76
       DO 75 I=2,NSAV
       J=NSAV-I+1
       IF(ADEV.LT.DEV(J)) GO TO 77
    75 CONTINUE
       J=J-1
    77 CONTINUE
       J=J+2
       IF(J.GT.NSAV) GO TO 73
       DO 78 I=J,NSAV
       N=NSAV-I+J
       DEV(N)=DEV(N-1)
       DO 79 L=1,4
       ETS(L,N)=ETS(L,N-1)
       DO 79 M=1,3
    79 KTS(L,M,N)=KTS(L,M,N-1)
```

```
   78 CONTINUE
   73 CONTINUE
      DEV(J-1)=ADEV
      DO 74 L=1,4
      ETS(L,J-1)=ET(L)
      DO 74 M=1,3
   74 KTS(L,M,J-1)=KT(L,M)
   76 CONTINUE
   70 CONTINUE
  100 CONTINUE
      WRITE(20,2) KKK
    2 FORMAT(1X,I5,' POINT MESH')
      WRITE(20,3)
    3 FORMAT(4X,'DEV',12X,'E1',6X,'K1',14X,'E2',6X,'K2',14X,'E3',6X,'K3'
     1,14X,'E4',6X,'K4')
      DO 85 I=1,NSAV
      WRITE(20,4) DEV(I),(ETS(L,I),(KTS(L,M,I),M=1,3),L=1,4)
    4 FORMAT(1X,F9.5,4(5X,F10.5,3I3))
   85 CONTINUE
      JNC=JNC/2
      GO TO 150
   80 CONTINUE
  200 CONTINUE
      RETURN
      END
      SUBROUTINE TETDS(NH,NP,NB,NAT)
      IMPLICIT REAL*8(A-H,O-Z)
c     PARAMETER NNH=2000
c     REAL*4 E
c     REAL*4 EIN(NAT,NP)
c     REAL*4 TITLE(20)
         REAL*8 INTERP,NUMELEC
      COMMON/COMM/  FERV(9000),SFER(9000),OMEGA(9000),
     1  EM(9000),E(7,969,13),DOS(7,9000),ELEC(9000)
      DIMENSION DENSFL(15),DOS1(9000),DOS2(9000),DOS3(9000),DOS4(9000),
     1  DOS5(9000),DOS6(9000),DOS7(9000),DOS8(9000),DOS9(9000),
     2  DOS10(9000),DOS11(9000),dos12(9000),dos13(9000),dos14(9000),
     3  DOS15(9000),DOS16(9000)
      CHARACTER*40 TITLE
      DIMENSION IJX(60),IJN(60)
      DIMENSION KT(4,3)
      DIMENSION INC(6,3,3),IND(4),EO(16,4),S0(16),S1(16),S2(16),S3(16)
      DIMENSION ST(35),DL(35)
      DIMENSION ELC(9000)
      DATA INC/1,1,1,1,1,1,1,1,0,0,1,1,1,0,0,0,1,0,1,1,1,1,1,1,1,1,1,1,0
     A,0,0,1,1,0,0,0,1,1,1,1,1,1,0,0,1,1,1,1,0,0,0,1,0,1/
      NNH=9000
      IF(NH.LE.NNH) GO TO 256
      WRITE(20,257)
  257 FORMAT(' NH TOO BIG -- STOPPING')
      STOP
  256 CONTINUE
      E01=0.0
      DO 4 I=1,NH
```

```
      ELC(I)=0.0
      FERV(I)=0.0
      SFER(I)=0.0
    4 OMEGA(I)=0.0
      DO 5 I=1,NAT
      DO 5 J=1,NP
      DO 5 K=1,NB
    5 E(I,J,K)=0.0
      DO 6 I=1,NAT
      DO 6 J=1,NH
    6 DOS(I,J)=0.0
      PI=3.1415926
      READ(10,500)TITLE,IATOM
      READ(10,501) NUMELEC
      WRITE(*,551) NUMELEC
  500 FORMAT(A40,I5)
  501 FORMAT(F6.2)
  551 FORMAT(1X,'NUMEL',F6.2)
      WRITE(*,600)TITLE
      WRITE(20,600)TITLE
  600 FORMAT(///,20X,A40,///)
      WRITE(20,601)
  601 FORMAT(20X,'****************************************',///)
C     ISTR =1    BCC     ISTR =2   FCC
C     ISTR =3   CUBIC (INCLUDING CsCl)
      READ(10,141)ISTR,ISTRU
 141  FORMAT (2I4)
      IF((ISTR.GT.3).OR.(ISTR.LT.1)) THEN
         WRITE(20,1611)
 1611    FORMAT(' ERROR IN ISTR------STOPPING')
         STOP 12
      ENDIF
      IF (ISTR .EQ.1) WRITE(20,161)
 161  FORMAT (3X,'STRUCTURE BCC')
      IF (ISTR .EQ.2) WRITE(20,162)
 162  FORMAT (3X,'STRUCTURE FCC')
      IF (ISTR .EQ.3) WRITE(20,163)
 163  FORMAT (3X,'STRUCTURE CUBIC')
      READ(10,142)PIA
 142  FORMAT (F10.5)
      ALAT=PI/PIA
      WRITE(20,143)ALAT
 143  FORMAT (3X,'ALAT=',F10.5)
      READ(10,1) MESH,NBD,IAT,NRG,ISD
      WRITE(20,1)MESH,NBD,IAT,NRG,ISD
    1 FORMAT(5I5)
      MESH2=MESH*2
      M2P1=MESH2+1
      NRP1=NRG+1
      READ(10,2) (ST(I),DL(I),I=1,NRP1)
      WRITE(20,2)(ST(I),DL(I),I=1,NRP1)
    2 FORMAT(8F10.5)
      NPTS=1
      EM(1)=ST(1)
```

```
      DO 15 I=1,NRG
   16 NPTS=NPTS+1
      EM(NPTS)=EM(NPTS-1)+DL(I)
      IF(NPTS.GE.NH) GO TO 17
      IF(EM(NPTS).LE.ST(I+1)) GO TO 16
      NPTS=NPTS-1
      EM(NPTS)=ST(I+1)
   15 CONTINUE
   17 CONTINUE
      IF (ISTR.EQ.1) INBZ=MESH2+2
      IF (ISTR.EQ.2) INBZ=MESH*3+3
      IF (ISTR.EQ.3)  GO TO 1000
      DO 210 I=1,M2P1
      DO 210 J=1,I
      DO 210 K=1,J
      IF (ISTR.EQ.2) GO TO 213
      IF (I+J   .GT.INBZ) GO TO 210
      IF (ISTR.EQ.1) GO TO 214
 213  IF (I+J+K.GT.INBZ) GO TO 210
 214  NID=I*(I-1)*(I+1)/6+J*(J-1)/2+K
c     WRITE (51,*) NBD,IAT
      DO 215 IB=1,NBD
c     WRITE (51,*) NID,IB
      READ (11,211) (E(L,NID,IB),L=1,IAT)
C*************CAUTION THIS IS THE EMPIRICAL SHIFT FOR
FE*****************
C     E(1,NID,IB)=E(1,NID,IB)+0.01611
 211  FORMAT (F9.5,3X,6F8.5)
C    FREE ELECTRON TEST
      XESH2=MESH2
C     E(1,NID,IB)=((I-1)/XESH2)**2+((J-1)/XESH2)**2+((K-1)/XESH2)**2
C
c     IF (IAT.LE.6) GO TO 215
c     READ (11,212)              (E(L,NID,IB),L=6,IAT)
 212  FORMAT (12X,5F8.5)
 215  CONTINUE
      IF (ISTR.EQ.1) GO TO 220
      II=IABS(I-1-MESH2)+1
      JJ=IABS(J-1-MESH2)+1
      KK=IABS(K-1-MESH2)+1
      IDN=KK*(KK-1)*(KK+1)/6+JJ*(JJ-1)/2+II
C     WRITE(20,217) I,J,K,NID,IDN
 217  FORMAT (1X,5I5)
      DO 216 IB=1,NBD
      DO 216  L=1,IAT
      E(L,IDN,IB)=E(L,NID,IB)
 216  CONTINUE
      IF (ISTR.EQ.2) GO TO 210
 220  DO 218 IJK=1,3
      II=IABS(I-1-MESH2)+1
      JJ=IABS(J-1-MESH2)+1
      KK=IABS(K-1-MESH2)+1
      IF (IJK.EQ.1) IDN=JJ*(JJ-1)*(JJ+1)/6+II*(II-1)/2+K
      IF (IJK.EQ.2) IDN=KK*(KK-1)*(KK+1)/6+JJ*(JJ-1)/2+I
```

```
      IF (IJK.EQ.3) IDN=KK*(KK-1)*(KK+1)/6+II*(II-1)/2+J
C     WRITE(20,217) I,J,K,NID,IDN
      DO 219 IB=1,NBD
C     PRINT 712, E(1,NID,IB)
      DO 219  L=1,IAT
 712  FORMAT (1X,F10.5)
      E(L,IDN,IB)=E(L,NID,IB)
 219  CONTINUE
 218  CONTINUE
 210  CONTINUE
CC    WRITE(20,711) (E(1,J,1),J=1,NP)
CC    WRITE(20,711) (E(1,J,2),J=1,NP)
C     WRITE (7,711) (E(1,J,5),J=1,NP)
C     WRITE (7,711) (E(1,J,6),J=1,NP)
 711  FORMAT (10F8.6)
 1000 I=0
      IF (ISTR.NE.3) GO TO 21
C    FREE ELECTRON TEST
C     DO 310 I=1,M2P1
C     DO 310 J=1,I
C     DO 310 K=1,J
C     NID=I*(I-1)*(I+1)/6+J*(J-1)/2+K
C     DO 315 IB=1,NBD
C     XESH2=MESH2
C     E(1,NID,IB)=((I-1)/XESH2)**2+((J-1)/XESH2)**2+((K-1)/XESH2)**2
      WRITE(20,217) I,J,K,NID
C     PRINT 712, E(1,NID,IB)
C315  CONTINUE
C310  CONTINUE
C
      DO 222 NID=1,NP
      DO 20 J=1,NBD
      READ(11,211) (E(L,NID,J),L=1,5)
      READ(11,211) DUM,(E(L,NID,J),L=6,9)
  20  CONTINUE
 222  CONTINUE
  21  CONTINUE
      MAX=NP
      IJN(1)=1
      DO 140 J=1,NBD
      EJX=E(1,1,J)
      EJN=E(1,1,J)
      DO 160 I=2,MAX
      IF(E(1,I,J).GT.EJX) EJX=E(1,I,J)
      IF(E(1,I,J).LT.EJN) EJN=E(1,I,J)
  160 CONTINUE
      IMIN=IJN(J)
      DO 170 I=IMIN,NPTS
      II=I
      IF(EM(I).LT.EJN) IJN(J)=I
      IF(EM(I).GT.EJX) GO TO 171
  170 CONTINUE
  171 IJX(J)=II
      IJN(J+1)=IJN(J)
```

```
 140 CONTINUE
     WRITE(20,901) NPTS
 901 FORMAT(1X,10I5)
     WRITE(20,901) (IJN(I),I=1,NBD)
     WRITE(20,901) (IJX(I),I=1,NBD)
 902 FORMAT(1X,10F10.5)
     FILL=MESH2**3
     DO 230 JB=1,NBD
     IMAX=IJX(JB)+1
     IF(IMAX.GT.NPTS) GO TO 230
     DO 240 ID=IMAX,NPTS
 240 ELC(ID)=ELC(ID)+FILL
 230 CONTINUE
     IF(ISD.LE.0) CALL CALDV(INC,NBD,MESH,NAT,NP,NB)
     IF(ISD.LT.0) STOP
     DO 100 II=1,M2P1
     KT(1,1)=II-1
     DO 100 JJ=1,II
     KT(1,2)=JJ-1
     DO 100 KK=1,JJ
     KT(1,3)=KK-1
     DO 70 N=1,6
     DO 40 L=1,3
     DO 40 M=1,3
  40 KT(L+1,M)=KT(1,M)+INC(N,L,M)
     DO 60 L=2,4
     IF(KT(L,1).GT.MESH2) GO TO 70
     IF(KT(L,3).GT.KT(L,2)) GO TO 70
     IF(KT(L,2).GT.KT(L,1)) GO TO 70
  60 CONTINUE
     DO 65 L=1,4
     K1=KT(L,1)
     K2=KT(L,2)
     K3=KT(L,3)
     IND(L)=(K1*(K1+1)*(K1+2))/6 + K2*(K2+1)/2 + K3 + 1
  65 CONTINUE
     DO 80 JB=1,NBD
     DO 85 L=1,4
     ID=IND(L)
     DO 85 M=1,IAT
     EO(M,L)=E(M,ID,JB)
  85 CONTINUE
     ED21=EO(1,2)-EO(1,1)
     ED31=EO(1,3)-EO(1,1)
     ED41=EO(1,4)-EO(1,1)
     PLAS=ED21**2+2.0*ED31**2+2.0*ED41**2-2.0*ED41*ED31-2.0*ED21*ED31
     IF (ISTR.EQ.1)PLAS=PLAS*MESH **2/PIA**2
     IF (ISTR.EQ.2)PLAS=PLAS*MESH **2/PIA**2
     IF (ISTR.EQ.3)PLAS=PLAS*MESH2**2/PIA**2
     SRPLAS=SQRT(PLAS)
C    PRINT 1001, ((KT(L,M),M=1,3),L=1,4),(EO(1,L),L=1,4),PLAS
1001 FORMAT (1X,4(3I3,2X),4F10.5,E16.7)
  52 LAST=0
     DO 50 L=2,4
```

```
      LM1=L-1
      IF(EO(1,L).GE.EO(1,LM1)) GO TO 50
      DO 51 M=1,IAT
      HELP=EO(M,L)
      EO(M,L)=EO(M,LM1)
      EO(M,LM1)=HELP
   51 CONTINUE
      LAST=1
   50 CONTINUE
      IF(LAST.EQ.1) GO TO 52
      I1EI2=0
      IF(EO(1,1).EQ.EO(1,2)) GO TO 71
      E01=1.0/(EO(1,2)-EO(1,1))
      E02=1.0/(EO(1,3)-EO(1,1))
      E03=1.0/(EO(1,4)-EO(1,1))
      C0=E01*E02*E03
      DO 74 L=2,IAT
      S0(L)=(E01*(EO(L,2)-EO(L,1))+E02*(EO(L,3)-EO(L,1))+E03*(EO(L,4)
     A-EO(L,1)))/3.0
   74 CONTINUE
      GO TO 77
   71 I1EI2=1
   77 CONTINUE
      IF(EO(1,3).EQ.EO(1,2)) GO TO 72
      E02=1.0/(EO(1,3)-EO(1,1))
      E03=1.0/(EO(1,4)-EO(1,1))
      E12=1.0/(EO(1,3)-EO(1,2))
      E13=1.0/(EO(1,4)-EO(1,2))
      C1=E02*E03*E12
      C2=E03*E12*E13
      DO 75 L=2,IAT
      H1=(EO(L,3)-EO(L,1))*E02+(EO(L,4)-EO(L,1))*E03
      S1(L)=E12*(EO(L,2)-EO(L,1)+(EO(1,2)-EO(1,1))*H1)/3.0
      H2=3.0*(EO(L,4)-EO(L,1))+EO(L,3)-EO(L,4)
      H1=(EO(L,2)-EO(L,4))*E13+(EO(L,1)-EO(L,4))*EO3
      H1=H1*(EO(1,4)-EO(1,3))
      S2(L)=E12*(H1+H2)/3.0
C     S1(L)=E03*(EO(L,4)-EO(L,1))+E02*(EO(L,3)-EO(L,1))*0.5
C     S2(L)=E13*(EO(L,4)-EO(L,2))+E12*(EO(L,3)-EO(L,2))*0.5
   75 CONTINUE
      V0=C0
      V1=-E01*E12*E13
      V2=0.0
      IF(I1EI2.EQ.0) GO TO 78
      V0=-E02*E02*E03
      V1=-E02*E03*E03
      V2=3.0*E02*E03
   78 CONTINUE
   72 CONTINUE
      IF(EO(1,4).EQ.EO(1,3)) GO TO 73
      E03=1.0/(EO(1,4)-EO(1,1))
      E13=1.0/(EO(1,4)-EO(1,2))
      E23=1.0/(EO(1,4)-EO(1,3))
      C3=E03*E13*E23
```

```
      DO 76 L=2,IAT
      S3(L)=(E03*(EO(L,4)-EO(L,1))+E13*(EO(L,4)-EO(L,2))+E23*(EO(L,4)
     A-EO(L,3)))/3.0
C       WRITE(15,1517) S3(L),L
 1517 FORMAT(1X,'S3(L),L',E12.6,5X,I5)
   76 CONTINUE
   73 CONTINUE
      IMIN=IJN(JB)
      IMAX=IJX(JB)
      DO 55 ID=IMIN,IMAX
      EX=EM(ID)
      IF(EX.GE.EO(1,4)) GO TO 56
      IF(EX.LE.EO(1,1)) GO TO 55
      IF(EX.LE.EO(1,2)) GO TO 57
      IF(EX.LT.EO(1,3)) GO TO 58
      FAC=C3*(EX-EO(1,4))**2
      VAC=FAC*(EO(1,4)-EX)
      ELC(ID)=ELC(ID)+1.0-VAC
      FERV(ID)=FERV(ID) +FAC*PLAS
      SFER(ID)=SFER(ID)+FAC*SRPLAS
      DOS(1,ID)=DOS(1,ID)+FAC
      DO 93 L=2,IAT
      UPD=EO(L,4)+(EX-EO(1,4))*S3(L)
      IF(UPD.GE.0.0) GO TO 593
      WRITE(6,693) II,JJ,KK,N,L,UPD
      WRITE(6,793) (EO(1,LL),LL=1,4),EX
      WRITE(6,793) (EO(L,LL),LL=1,4),S3(L)
  693 FORMAT(' E3<E<E4',5I3,F10.5)
  793 FORMAT(1X,6F9.5)
  593 CONTINUE
      DOS(L,ID)=DOS(L,ID)+FAC*UPD
   93 CONTINUE
      GO TO 55
   58 CONTINUE
      VAC2=(EX-EO(1,1))**2
      VAC0=VAC2*(EX-EO(1,1))
      VAC1=(EX-EO(1,2))**3
      VAC=V0*VAC0+V1*VAC1+V2*VAC2
      ELC(ID)=ELC(ID)+VAC
      FAC=(EO(1,3)-EX)*(EX-EO(1,1))*C1 + (EO(1,4)-EX)*(EX-EO(1,2))*C2
      FERV(ID)=FERV(ID) +FAC*PLAS
      SFER(ID)=SFER(ID)+FAC*SRPLAS
      DOS(1,ID)=DOS(1,ID)+FAC
      DO 92 L=2,IAT
      ASBAR=(EO(1,3)-EX)*S1(L)+(EX-EO(1,2))*S2(L)
      UPD=EO(L,1)+ASBAR
      IF(UPD.GE.0.0) GO TO 592
      WRITE(6,692) II,JJ,KK,N,L,UPD,ASBAR
      WRITE(6,793) (EO(1,LL),LL=1,4),EX
      WRITE(6,793) (EO(L,LL),LL=1,4),S1(L),S2(L)
  692 FORMAT(' E2<E<E3',5I3,2F10.5)
  592 CONTINUE
      DOS(L,ID)=DOS(L,ID)+FAC*UPD
   92 CONTINUE
```

```
      GO TO 55
   57 CONTINUE
      FAC=C0*(EX-EO(1,1))**2
      VAC=FAC*(EX-EO(1,1))
      ELC(ID)=ELC(ID)+VAC
      FERV(ID)=FERV(ID)  +FAC*PLAS
      SFER(ID)=SFER(ID)+FAC*SRPLAS
      DOS(1,ID)=DOS(1,ID)+FAC
      DO 91 L=2,IAT
      UPD=EO(L,1)+(EX-EO(1,1))*S0(L)
      IF(UPD.GE.0.0) GO TO 591
      WRITE(6,691) II,JJ,KK,N,L,UPD
      WRITE(6,793)  (EO(1,LL),LL=1,4),EX
      WRITE(6,793)  (EO(L,LL),LL=1,4),S0(L)
  691 FORMAT(' E1<E<E2',5I3,F10.5)
      STOP
  591 CONTINUE
      DOS(L,ID)=DOS(L,ID)+FAC*UPD
   91 CONTINUE
      GO TO 55
   56 CONTINUE
      ELC(ID)=ELC(ID)+1.0
   55 CONTINUE
   80 CONTINUE
   70 CONTINUE
  100 CONTINUE
      IF (ISTR .EQ.1) V=ALAT**3/2.0
      IF (ISTR .EQ.2) V=ALAT**3/4.0
      IF (ISTR .EQ.3) V=ALAT**3
      DO 120 I=1,NPTS
C     CONVERTS  TO CM/SEC
      IF (DOS(1,I).GT.0.0D+0) FERV(I)=FERV(I)*1.1964/DOS(1,I)
      FERV(I)=SQRT(FERV(I))
      IF (DOS(1,I).GT.0.0D+0) SFER(I)=SFER(I)*1.0938/DOS(1,I)
 120  CONTINUE
      FAC=6.0/(MESH2**3)
      DO 110 J=1,IAT
      DO 110 I=1,NPTS
      DOS(J,I)=FAC*DOS(J,I)
      OMEGA(I)=50.916714*FERV(I)*(DOS(1,I)/(2.0*V))**0.5
  110 CONTINUE
      FAC= 2.0/(MESH2**3)
      DO 111 J=1,NPTS
  111 ELC(J)=FAC*ELC(J)
      ELEC(1)=0.0
      DO 115 I=2,NPTS
      ELEC(I)=ELEC(I-1)+0.5*(EM(I)-EM(I-1))*(DOS(1,I-1)+DOS(1,I))
  115 CONTINUE
      WRITE(15,623) TITLE
      NG= NPTS/55 + 1
      DO 200  I=1,NG
      WRITE(20,623)TITLE
  623 FORMAT(A40)
      WRITE(20,620)
```

```
  620 FORMAT(        1X,'CELL',5X,'ENERGY',5X,'S1',9X,'P1',9X,'D1',
     $ 9X,'F1',9X,'S2',9X,'P2',9X,'D2',9X,'F2',6X,'DENSITY',3X,
     $ 'TOTAL NUMBER',/)
      NTOP= I*55
      NBOT= NTOP - 54
      DO 200  J=NBOT,NTOP
      IF(J.GT.NPTS) GO TO 200
      WRITE(20,630)J,EM(J),(DOS(L,J),L=2,NAT),DOS(1,J),ELEC(J),ELC(J)

C     FOR DIATOMIC MATERIALS CHANGE 4E11.4 TO 8E11.4
  630 FORMAT(1X,I4,F8.4,6E11.4,E13.6,2F7.3)
      WRITE(15,640) J,EM(J), DOS(1,J),ELC(J),FERV(J),SFER(J),OMEGA(J)
  640 FORMAT(I5,6F10.5)
      WRITE(15,645) (DOS(L,J),L=2,NAT)
  645 FORMAT(6F10.5)
      WRITE(19,650) J,EM(J),ELC(J),DOS(1,J),(DOS(L,J),L=2,IAT)
  650 FORMAT(I5,20F10.4)
  200 CONTINUE
      DO 201  I=1,NG
      WRITE(20,621)
  621 FORMAT(        1X,'CELL',5X,'ENERGY',5X,'FERV'9X,'SFER',9X,'OMEGA'
     1,6X,'DENSITY',3X,'TOTAL NUMBER',/)
      NTOP= I*55
      NBOT= NTOP - 54
      DO 201  J=NBOT,NTOP
      IF(J.GT.NPTS) GO TO 201
      WRITE(20,631) J,EM(J),FERV(J),SFER(J),OMEGA(J),DOS(1,J),ELC(J)
  631 FORMAT(1X,I5,F11.4,5E13.6)
  201 CONTINUE
C INTERPOLATE INPUT DATA TO FIND FERMI ENERGY,TOTAL DOS AT EF, AND
C   COMPONENT ENERGIES AT EF
 1989 DO 4090 L=1,J
        DOS1(L)=DOS(1,L)
        DOS2(L)=DOS(2,L)
        DOS3(L)=DOS(3,L)
        DOS4(L)=DOS(4,L)
        DOS5(L)=DOS(5,L)
        DOS6(L)=DOS(6,L)
        DOS7(L)=DOS(7,L)
c       DOS8(L)=DOS(8,L)
c       DOS9(L)=DOS(9,L)
c       DOS10(L)=DOS(10,L)
c       DOS11(L)=DOS(11,L)
c       DOS12(L)=DOS(12,L)
c       DOS13(L)=DOS(13,L)
c       DOS14(L)=DOS(14,L)
c       DOS15(L)=DOS(15,L)
c       DOS16(L)=DOS(16,L)
 4090 CONTINUE
      MCOUNT=0
      DO 4080 L=1,(J-1)
        IF (ELC(L).GE.NUMELEC) GOTO 178
        GOTO 4080
 178    IF (MCOUNT.NE.0) GOTO 4080
```

```
        MCOUNT=L
 4080 CONTINUE
C     LINEAR INTERPOLATION OF THE FERMI LEVEL
      S=NUMELEC
      MCM = MCOUNT - 1
C     WRITE(*,*)'S=',S,' MCM=',MCM,' ELC=',ELC,' EM=',EM
      EFL = INTERP(S,ELC,EM,MCM,2)
C     WRITE(*,*)'S=',S,' MCM=',MCM,' ELC=',ELC,' EM=',EM
      DNORFL=INTERP(S,ELC,DOS1,MCM,2)
c  Fermi Velocity
      DFERV=INTERP(S,ELC,FERV,MCM,2)
c     DFERV=INTERP(S,ELC,SFER,MCM,2)
c  Plasmon Energy
      DOMEG=INTERP(S,ELC,OMEGA,MCM,2)
      DENSFL(1)=INTERP(S,ELC,DOS2,MCM,2)
      DENSFL(2)=INTERP(S,ELC,DOS3,MCM,2)
      DENSFL(3)=INTERP(S,ELC,DOS4,MCM,2)
      DENSFL(4)=INTERP(S,ELC,DOS5,MCM,2)
      DENSFL(5)=INTERP(S,ELC,DOS6,MCM,2)
      DENSFL(6)=INTERP(S,ELC,DOS7,MCM,2)
      DENSFL(7)=INTERP(S,ELC,DOS8,MCM,2)
      DENSFL(8)=INTERP(S,ELC,DOS9,MCM,2)
      DENSFL(9)=INTERP(S,ELC,DOS10,MCM,2)
      DENSFL(10)=INTERP(S,ELC,DOS11,MCM,2)
      DENSFL(11)=INTERP(S,ELC,DOS12,MCM,2)
      DENSFL(12)=INTERP(S,ELC,DOS13,MCM,2)
      DENSFL(13)=INTERP(S,ELC,DOS14,MCM,2)
      DENSFL(14)=INTERP(S,ELC,DOS15,MCM,2)
      DENSFL(15)=INTERP(S,ELC,DOS16,MCM,2)
      WRITE(13,500) TITLE,IATOM
c  Writes Fermi velocity and Plasmon energy   9 Aug 2011
      WRITE(13,971) EFL,DNORFL,(DENSFL(I),I=1,5)
      WRITE(14,972) EFL,DNORFL,(DENSFL(I),I=1,5),DFERV,DOMEG
      IF (IAT1.GT.10)
     1WRITE(13,971) EFL,DNORFL,(DENSFL(I),I=6,10)
      IF (IAT1.GT.10)
     1WRITE(14,972) EFL,DNORFL,(DENSFL(I),I=6,10),DFERV,DOMEG
      IF (IAT1.EQ.16)
     1WRITE(13,971) EFL,DNORFL,(DENSFL(I),I=11,15)
      IF (IAT1.EQ.16)
     1WRITE(14,972) EFL,DNORFL,(DENSFL(I),I=11,15),DFERV,DOMEG
  971 FORMAT(7F10.5)
  972 FORMAT(9F10.5)
      RETURN
      END
      REAL*8 FUNCTION INTERP(XX,X,F,J1,N)
      IMPLICIT REAL*8(A-H,O-Z)
      dimension X(9000),F(9000)
c     REAL X(700),F(700)
c     DOUBLE PRECISION FX,P
      FX=0.
      ISTART=J1-N+N/2+1
      J2=ISTART+N-1
      DO 10 J=ISTART,J2
```

```
          P=F(J)
          DO 5 I=ISTART,J2
            IF(I.EQ.J) GOTO 5
            P=P*(XX-X(I))/(X(J)-X(I))
5         CONTINUE
10      FX=FX+P
        INTERP = FX
        RETURN
        END
```

```
CuH 3-Center                                        2
12.0                          !Valence electrons
   2   2                     ! (1=BCC,2=FCC,3=CUBIC[INCLUDING CsCl])
0.43737                       ! PI/lat.con.
    8    13     7     1     0 !(k-MESH;NO. OF BANDS;IAT=5 FOR 1 ATOM,7 FOR
2;No of energy ranges
  -0.30000   0.00100   1.50000   0.00000   0.00000   0.00000   0.00000
0.00000
```

This is the output file with the DOS. The different columns are labeled as shown below. The p(H) is zero all the way through. To save space we have cut off the file and show values only near EF.

Index	Energy	No of elec	Tot-DOS	s (M)	p (M)	t2g (M)	eg (M)	s (H)	p(H)
1075	0.7740	11.9994	0.2816	0.0056	0.0697	0.0119	0.1929	0.0015	0.0000
1076	0.7750	11.9996	0.2159	0.0044	0.0533	0.0089	0.1481	0.0011	0.0000
1077	0.7760	11.9998	0.1488	0.0032	0.0340	0.0056	0.1053	0.0008	0.0000
1078	0.7770	11.9999	0.0839	0.0021	0.0156	0.0026	0.0632	0.0005	0.0000
1079	0.7780	12.0000	0.0397	0.0012	0.0036	0.0005	0.0342	0.0002	0.0000
1080	0.7790	12.0000	0.0169	0.0005	0.0007	0.0001	0.0154	0.0001	0.0000
1081	0.7800	12.0000	0.0043	0.0001	0.0001	0.0000	0.0040	0.0000	0.0000
1082	0.7810	12.0000	0.0000	0.0000	0.0000	0.0000	0.0000	0.0000	0.0000
1083	0.7820	12.0000	0.0000	0.0000	0.0000	0.0000	0.0000	0.0000	0.0000
1084	0.7830	12.0000	0.0041	0.0002	0.0003	0.0000	0.0036	0.0000	0.0000
1085	0.7840	12.0000	0.0306	0.0015	0.0032	0.0005	0.0254	0.0001	0.0000
1086	0.7850	12.0001	0.0833	0.0040	0.0116	0.0019	0.0656	0.0002	0.0000
1087	0.7860	12.0002	0.1530	0.0076	0.0212	0.0033	0.1205	0.0004	0.0000
1088	0.7870	12.0004	0.2270	0.0114	0.0320	0.0046	0.1783	0.0006	0.0000
1089	0.7880	12.0006	0.3029	0.0157	0.0434	0.0060	0.2373	0.0007	0.0000
1090	0.7890	12.0010	0.3803	0.0203	0.0537	0.0072	0.2983	0.0009	0.0000
1091	0.7900	12.0014	0.4620	0.0259	0.0610	0.0083	0.3658	0.0010	0.0000
1092	0.7910	12.0019	0.5433	0.0308	0.0769	0.0099	0.4246	0.0011	0.0000
1093	0.7920	12.0025	0.6201	0.0357	0.0937	0.0116	0.4780	0.0012	0.0000
1094	0.7930	12.0031	0.7006	0.0412	0.1068	0.0131	0.5383	0.0013	0.0000
1095	0.7940	12.0039	0.7865	0.0488	0.1154	0.0145	0.6064	0.0013	0.0000
1096	0.7950	12.0047	0.8656	0.0559	0.1218	0.0152	0.6712	0.0014	0.0000
1097	0.7960	12.0056	0.9332	0.0635	0.1234	0.0160	0.7289	0.0014	0.0000
1098	0.7970	12.0066	0.9400	0.0622	0.1333	0.0169	0.7262	0.0015	0.0000
1099	0.7980	12.0075	0.9469	0.0609	0.1436	0.0178	0.7232	0.0015	0.0000
1100	0.7990	12.0085	0.9539	0.0595	0.1542	0.0188	0.7199	0.0015	0.0000
1101	0.8000	12.0094	0.9609	0.0582	0.1651	0.0199	0.7162	0.0015	0.0000

```
CuH 3-Center                                        2
    0.78200   0.00000   0.00000   0.00000   0.00000   0.00000   0.00000

  Ef=0.782  These are the DOS at Ef. They are zero because CuH is a
semimetal.
```

Appendix C
Computer Program Applying the Coherent Potential Approximation

1st line Label

2nd line NPARM = 60 number of 3-Center TB parameters.
NDIM = 9 size of the metal unperturbed Hamiltonian.
NKS = 8 sets a k mesh for 505 points.
NUELEC=11.95 number of electrons with 5% H vacancies.

Lines 3–62 TB parameters for and orthogonal 3-center calculation.

Line 63 CB = 0.05 amount of H vacancies.

EMIN and EMAX lower and higher energies of the DOS window.

DELE energy step. EPIV pivot energy chosen somewhere in the middle.

SAGR and SAGI initial values for the real and imaginary parts of the self-energy.

CONV convergence criterion.

Line 64 EPS small imaginary part added to the energy; controls the disorder broadening of the DOS.

D. A. Papaconstantopoulos, *Band Structure of Cubic Hydrides*,
https://doi.org/10.1007/978-3-031-06878-2

```
      PROGRAM CPA
      IMPLICIT REAL*8 (A-H,O-Z)
      COMMON /A/ RR(50,9,9),RI(50,9,9),XR(50,9,9),XI(50,9,9)
      COMMON/B/QQ,WEIGHT,TOTVOL,NS
      DIMENSION R(9,9),WK(110)
C     DIMENSION RR(50,9,9),RI(50,9,9),XR(50,9,9),XI(50,9,9)
      DIMENSION H(9,9),UPARM(60),X(9,9),V(9)
      DIMENSION A(20)
      DIMENSION AR( 5300),AI( 5300),HMZ(9,9, 5300),VST(9, 5300)
      DIMENSION HHYD( 5300)
      DIMENSION WEIGHT( 5300),QQ (5300,3)
      DIMENSION DUM(4000),EDUM(4000),ANUMEL(4000)
      DIMENSION RES(8,4000)
      DIMENSION          DUMM(4000), DUMP2(4000),DUMS2(4000)
      DIMENSION DUMS1(4000),DUMP1(4000),DUMD1(4000),DUMD2(4000)
      DIMENSION DENSFL(6),PAR(60)
      REAL*8 INTERP
      REAL*8 NUELEC
      OPEN(5,FILE='cpa.in',BLANK='ZERO')
      OPEN(6,FILE='cpa.out',BLANK='ZERO')
      OPEN(20,FILE='cpa.dos',BLANK='ZERO')
      PI=3.141593
      NEN=4000
 1000 FORMAT (3I5,F10.5)
 1001 FORMAT(8F10.4)
 1002 FORMAT(I5,3F10.5)
 1003 FORMAT(6D15.5)
 2002 FORMAT(20A4)
 9000 FORMAT(1H1)
      READ(5,2002) (A(I),I=1,20)
      READ(5,1000) NPARM,NDIM,NKS,NUELEC
      WRITE(6,1000) NPARM,NDIM,NKS,NUELEC
      DO 8600 J=1,NPARM
      READ(5,1040) UPARM(J),NNP
 1040 FORMAT(F13.10,4X,I3)
      WRITE(6,1006) UPARM(J),NNP
 1006 FORMAT(10X,F15.8,I10)
      IF(NNP.EQ.J) GO TO 8600
      WRITE(6,1045)
 1045 FORMAT(//10X,37H PARAMETERS OUT OF ORDER --- STOPPING)
      STOP
 8600 CONTINUE
      DO 555 I=1,NEN
  555 ANUMEL(I)=0.0
c   This allows for reading a second set of TB parameters for another hydride
c   and then average them to treat a ternary in the virtual crystal approximation.
c     DO 8601 J=1,NPARM
c     READ(5,1040) PAR(J),NNP
c     WRITE(6,1006) PAR(J),NNP
c     IF(NNP.EQ.J) GO TO 8601
c     WRITE(6,1045)
c     STOP
```

```
 8601 CONTINUE
      CALL BZONE(NKS)
 9623 CONTINUE
      READ(5,1001) CB,EMIN,EMAX,DELE,EPIV,SAGR,SAGI,CONV
c     DO 8602 J=1,NPARM
c8602 UPARM(J)=(1.0-CB)*UPARM(J)+CB*PAR(J)
      WRITE(6,9000)
      WRITE(6,2002) (A(I),I=1,20)
      WRITE(6,1005) (UPARM(I),I=1,NPARM)
 1005 FORMAT(10X,F15.8)
      WRITE(6,1001) CB,EMIN,EMAX,DELE,EPIV,SAGR,SAGI
      READ(5,500) EPS,SKIP
  500 FORMAT(2F15.8)
      WRITE(6,600) EPS
  600 FORMAT (///50X,' EPS= ',F15.8)
      EH=UPARM(33)
      E=EMIN
      DO 19  I=1,NS
      XK1=QQ(I,1)
      XK2=QQ(I,2)
      XK3=QQ(I,3)
      CALL MATS(XK1,XK2,XK3,H,UPARM,V,NDIM,HYD)
      DO 19  K=1,NDIM
      VST(K,I)= V(K)
      DO 19  L=1,NDIM
      HMZ (K,L,I)= H(K,L)
      HHYD(I)= HYD
   19 CONTINUE
      N50= NS/50 + 1
      M50= 50*(N50 - 1)
      DO9999IXYZ=1,NEN
C     ITIME=ICLOCK(DUM)
      DO 810  I1=1,N50
      J50= (I1-1)*50
      I50= 50
      IF((I1*50).GT.NS) I50= NS - M50
      IF(I50.LE.0) GO TO 810
      DO 820  I2=1,I50
      I= J50 + I2
      DO7K=1,NDIM
      DO7L=1,NDIM
      H(K,L)=  HMZ(K,L,I)
      IF(K.EQ.L)GOTO79
      RR(I2,K,L)= H(K,L)
      RI(I2,K,L)= 0.0
      XR(I2,K,L)= 0.0
      XI(I2,K,L)= 0.0
      GOTO7
   79  RR(I2,K,K)= H(K,K) - E
      RI(I2,K,K)= - EPS
      XR(I2,K,K)= 1.0
      XI(I2,K,K)= 0.0
```

```
 7     CONTINUE
  820 CONTINUE
        CALL CLSS(9,9,9,9,I50)
C      CALL CLSS(9,9,RR,RI,9,XR,XI,9,I50)
       DO 9  I2=1,I50
       I= J50 + I2
       QR= 0.0
       QI= 0.0
       DO77K=1,NDIM
       DO77L=1,NDIM
       QR= QR + VST(K,I)*XR(I2,K,L)*VST(L,I)
       QI= QI + VST(K,I)*XI(I2,K,L)*VST(L,I)
   77 CONTINUE
       C= QR
       D= QI
       AR(I)= E + C - HHYD(I)
       AI(I)=EPS+D
 9     CONTINUE
  810 CONTINUE
       SIGR= SAGR
       SIGI= SAGI
       CHKR= SIGR
       CHKI= SIGI
       DO999N=1,500
       GR= 0.0
       GI= 0.0
       MCHK=0
       DO99I=1,NS
       W=AR(I)-SIGR
        Y=AI(I)-SIGI
       DEN=W*W+Y*Y
       GR = GR + W*WEIGHT(I)/DEN
       GI= GI - Y*WEIGHT(I)/DEN
 99    CONTINUE
       GR= GR/TOTVOL
       GI= GI/TOTVOL
       GDEN= GR*GR + GI*GI
       SIGR= EH - GR*CB/GDEN
       SIGI= CB*GI/GDEN
       IF(( ABS(SIGR-CHKR).LT.CONV   ).AND.( ABS(SIGI-CHKI).LT.CONV   ))
      1MCHK=1
       CHKR= SIGR
       CHKI= SIGI
       IF(MCHK.EQ.1)GOTO9991
 999   CONTINUE
       PRINT1004,E
 1004 FORMAT('    SIGMA DIDN T CONVERGE FOR E=',F10.4)
       GOTO9999
 9991 CONTINUE
       PRINT1002,N,SIGR,SIGI
C      EL=DCMPLX(E,EPS)-SIG
       ELR= E - SIGR
```

```
      ELI= EPS - SIGI
      ELMAG= ELR**2 + ELI**2
      ELRMAG= ELR/ELMAG
      ELIMAG= ELI/ELMAG
      GGI= 0.0
      PDS= 0.0
      PDP= 0.0
      PDD1=0.0
      PDD2=0.0
      DO 860  I1=1,N50
      J50= (I1-1)*50
      I50= 50
      IF((I1*50).GT.NS) I50= NS - M50
      IF(I50.LE.0) GO TO 860
      DO 870  I2=1,I50
      I= J50 + I2
      DO87K=1,NDIM
      DO87L=1,NDIM
      HMZKL= HMZ(K,L,I)
      RR(I2,K,L)=HMZKL+VST(K,I)*VST(L,I)*ELRMAG
      RI(I2,K,L)=     - VST(K,I)*VST(L,I)*ELIMAG
      XR(I2,K,L)= 0.0
      XI(I2,K,L)=0.0
      IF(K.NE.L) GO TO 87
      XR(I2,K,K)= 1.0
      RR(I2,K,K)= RR(I2,K,K) - E
      RI(I2,K,K)= RI(I2,K,K) - EPS
87    CONTINUE
 870  CONTINUE
       CALL CLSS(9,9,9,9,I50)
      DO 880  I2=1,I50
      I= J50 + I2
      X1= XI(I2,1,1)
      X2= XI(I2,2,2)+XI(I2,3,3)+XI(I2,4,4)
      X3=XI(I2,5,5)+XI(I2,6,6)+XI(I2,7,7)
      X4=                                     XI(I2,8,8)+XI(I2,9,9)
      PDS= PDS +      (X1)*WEIGHT(I)
      PDP= PDP +      (X2)*WEIGHT(I)
      PDD1=PDD1+      (X3)*WEIGHT(I)
      PDD2=PDD2+      (X4)*WEIGHT(I)
 880  CONTINUE
 860  CONTINUE
      ROH= -GI/PI
      PDS= PDS/(PI*TOTVOL)
      PDP= PDP/(PI*TOTVOL)
      PDD1=PDD1/(PI*TOTVOL)
      PDD2=PDD2/(PI*TOTVOL)
      ROPD= PDS + PDP + PDD1+PDD2
      RO=ROH+ROPD
      ANUMEL(IXYZ)=ANUMEL(IXYZ)+RO*2.0*DELE
      PRINT1002,IXYZ,CB
      RES(1,IXYZ)= E
```

```
      RES(2,IXYZ)= ROH
      RES(3,IXYZ)= PDS
      RES(4,IXYZ)= PDP
      RES(5,IXYZ)= PDD1
      RES(6,IXYZ)= PDD2
      RES(7,IXYZ)= ROPD
      RES(8,IXYZ)= RO
      PRINT 610,E,ROH,PDS,PDP,PDD1,PDD2,ROPD,RO
  610 FORMAT(/ 5X,8F12.6)
      SAGR=SIGR
      SAGI=SIGI
      E=E+DELE
      IF (E.GE.EMAX) GO TO 1234
 9999 CONTINUE
 1234 CONTINUE
      DO 3422  L=1,IXYZ
      EDUM(L)= RES(1,L)
      DUM(L)= RES(8,L)*2.0
      DUMS1(L )=RES(3,L)*2.0
      DUMP1(L )=RES(4,L)*2.0
      DUMD1(L )=RES(5,L)*2.0
      DUMD2(L )=RES(6,L)*2.0
      DUMS2(L )=RES(2,L) *2.0
      DUMP2(L )=0.0
 3422 CONTINUE
      JNUM= IXYZ - 1
      CALL SIMP(EDUM,DUM,ANUMEL,JNUM)
      ANUMEL(1)= 0.0
      ANUMEL(IXYZ)=ANUMEL(IXYZ-1) + 2.0*DELE*RES(8,IXYZ)
      PRINT 6423
 6423 FORMAT(1H1)
      WRITE(6,194)
      WRITE(20,194)
  194 FORMAT(2X,"ENERGY    Hydrogen-s  METAL-s  METAL-p METAL-t2g METAL
     &-eg MET-TOT-d TOTAL-DOS ELECTRONS"/)
      DO 3425  J=1,IXYZ
      WRITE (20,612) (RES(I,J),I=1,8),ANUMEL(J)
 612  FORMAT(9F10.5)
      PRINT 611, (RES(I,J),I=1,8),ANUMEL(J)
  611 FORMAT (5X,F10.4,10X,9F12.5)
 3425 CONTINUE
      MCOUNT=0
      DO 4080 L=1,IXYZ
      IF(ANUMEL(L).GT.NUELEC) GO TO 178
      GO TO 4080
  178 IF(MCOUNT .NE. 0) GO TO 4080
      MCOUNT=L
 4080 CONTINUE
C     LAGRANGIAN INTERPOLATION OF THE FERMI LEVEL
      MC=MCOUNT
      MCM=MC-1
      S=NUELEC
```

```
      EFL=INTERP(S,ANUMEL    ,EDUM(1) ,MCM,2)
      DNORFL=INTERP(S,ANUMEL    ,DUM (1)    ,MCM,2)/2.0
      WRITE(6,841)
  841 FORMAT(//65X,11H (PER SPIN)  )
      DENSFL(1)=INTERP(S,ANUMEL    ,DUMS1(1),MCM,2)/2.0
      DENSFL(2)=INTERP(S,ANUMEL    ,DUMP1(1),MCM,2)/2.0
      DENSFL(3)=INTERP(S,ANUMEL    ,DUMD1(1),MCM,2)/2.0
      DENSFL(4)=INTERP(S,  ANUMEL  ,DUMD2(1),MCM,2)/2.0
      DENSFL(5)=INTERP(S,ANUMEL    ,DUMS2(1)  ,MCM,2)/2.0
      DENSFL(6)=INTERP(S,ANUMEL    ,DUMP2(1)  ,MCM,2)/2.0
      WRITE(6,193)
      WRITE(20,193)
  193 FORMAT(1X,12HFERMI ENERGY,3X,9HELECTRONS,3X,9HTOTAL DOS,7X,1HS,7X,
     &1HP,7X,4HDT2G,7X,3HDEG,7X,1HS,6X,1HP //)
      WRITE(6,839) EFL,NUELEC,DNORFL,(DENSFL(I),I=1,6)
      WRITE(20,839) EFL,NUELEC,DNORFL,(DENSFL(I),I=1,6)
  839  FORMAT(2F10.5    ,3X,7F10.5//)
      STOP
      END
      SUBROUTINE BZONE(KK)
C  DIVIDES THE IRREDUCIBLE 1/48TH OF THE BRILLIOUIN
C  ZONE OF A CUBIC CRYSTAL INTO A MESH OF BOXES.
C      KK=DETERMINES MESH SIZE
C      NSFB=1,2,3 FOR SIMPLE CUBIC,FCC,AND BCC, RESPECTIVELY.
C      IMAX=MAXIMUM VALUE OF RUNNING INDEX
C      IMIN=MINIMUM VALUE OF RUNNING INDEX
C
      IMPLICIT REAL*8 (A-H,O-Z)
      DIMENSION Q(3)
      DIMENSION WEIGHT( 5300),QQ (5300,3)
      COMMON/B/QQ,WEIGHT,TOTVOL,NS
  111 FORMAT(///'   (NSFB,KK,IMIN,IMAX)=',10I4)
  101 FORMAT(//)
  701 FORMAT('   NS=',I5)
      IMAX=-1
      NSFB=2
      IMIN=1
C***********************************************************************
      NS = 0
C  NS IS TO BE THE MESH POINT SEQUENCE NUMBER
C      BREAK UP IRREDUCIBLE SECTOR OF BZ.
      GO TO (230,235,240),NSFB
  230 JIMAX = KK
      PP = 0.5
      GO TO 250
  235 JIMAX = 4*KK
      PP = 1.0
      GO TO 250
  240 JIMAX = 2*KK
      PP = 1.0
  250 CUBE = PP/JIMAX
      IF(IMAX)260,260,270
```

```
  260 IMAX = JIMAX
  270 IF(IMIN)280,280,290
  280 IMIN = 3
  290 PRINT 111, NSFB,KK,IMIN,IMAX
C  BRILLOUIN ZONE NOW BROKEN UP INTO THREE PARTS (KM = 1,3)
      TOTVOL=0.0D0
      DO 800 KM = 1,3
      AHALF = CUBE/2.0
      VOLUME = 8.0*AHALF**3
      ASQ = AHALF**2
C**  AHALF IS HALF THE LENGTH OF A LITTLE CUBE EDGE  .
      DO 700 I = IMIN,IMAX
      GO TO (410,415,420),NSFB
  410 JMAX = I
      GO TO 430
  415 JMAX = 3*KK+0.5001-ABS (3*KK+0.5-I)
      KFCC = 6*KK + 1
      GO TO 430
  420 JMAX = KK+0.5001-ABS (KK+0.5-I)
  430 DO 700 J=1,JMAX
      DO 700 K=1,J
      GO TO (450,440,450),NSFB
  440 IF(I+J+K-KFCC) 450,450,700
  450 Q(1) = (I-0.5)*CUBE
      Q(2) = (J-0.5)*CUBE
      Q(3) = (K-0.5)*CUBE
C  THE *LOGICAL IF* FORM OF THE STATEMENT(S) WHICH FOLLOW(S), IS
C*    IF(JMAX.NE.J.AND.J.NE.K) 490,460
      IF((JMAX-J)*(J-K)) 490,460,490
  460 IF(K-JMAX) 480,470,470
  470 U=1.0
      GO TO 500
  480 U = 3.0
      GO TO 500
  490 U = 6.0
  500 NS = NS + 1
C*******************************************************************
      WEIGHT(NS)=U*VOLUME
      TOTVOL=TOTVOL+WEIGHT(NS)
      QQ(NS,1)=Q(1)
      QQ(NS,2)=Q(2)
      QQ(NS,3)=Q(3)
C**********************************************************************
  700 CONTINUE
      PRINT 701, NS
      PRINT 101
      KK = 3*KK
      CUBE = CUBE/3.0
      IF(IMIN-1) 900,900,720
  720 CONTINUE
      IMAX = (IMIN-1)*3
      GO TO (760,740,800), KM
```

```
  740 IMIN = 1
      GO TO 800
  760 IMIN = 4
  800 CONTINUE
  900 CONTINUE
      RETURN
      END
      SUBROUTINE CLSS(N,M,IA,IB,NMAT)
      IMPLICIT REAL*8 (A-H,O-Z)
      COMMON /A/ AR(50,9,9),AI(50,9,9),BR(50,9,9),BI(50,9,9)
C     DIMENSION AR(50,9,9),AI(50,9,9),BR(50,9,9),BI(50,9,9),X(50),T(50)
      DIMENSION X(50),T(50)
      DIMENSION L(50),SR1(50),SI1(50),SR2(50),SI2(50),VR(50),VI(50)
      DIMENSION HR(50),HI(50)
      ID=1
      IF(N.GT.IA.OR.N.GT.IB.OR.N.LE.0.OR.M.LE.0) GO TO 17
      DO 11 I=1,N
      IM1=I-1
      DO 110 II=1,NMAT
      TT=AR(II,I,I)**2+AI(II,I,I)**2
      T(II)=SQRT(TT)
  110 CONTINUE
      IF(I.EQ.N) GO TO 130
      DO 109 II=1,NMAT
  109 L(II)=I
      IP1=I+1
      DO 1 J=IP1,N
      DO 111 II=1,NMAT
      XX=AR(II,J,I)**2+AI(II,J,I)**2
      X(II)=SQRT(XX)
      IF(X(II).LE.T(II)) GO TO 111
      T(II)=X(II)
      L(II)=J
  111 CONTINUE
    1 CONTINUE
      DO 112 II=1,NMAT
      IF(T(II).EQ.0.0) GO TO 15
      IF(T(II).LT.1.0E-15) PRINT 80
  112 CONTINUE
   80 FORMAT('  SYSTEM NEAR SINGULAR --- CALCULATION CONTINUING')
      DO 113 II=1,NMAT
      LL=L(II)
      IF(LL.EQ.I) GO TO 4
      DO 2 J=1,N
      TEMPR=AR(II,I,J)
      TEMPI=AI(II,I,J)
      AR(II,I,J)=AR(II,LL,J)
      AI(II,I,J)=AI(II,LL,J)
      AR(II,LL,J)=TEMPR
      AI(II,LL,J)=TEMPI
    2 CONTINUE
      DO 3 J=1,M
```

```
      TEMPR=BR(II,I,J)
      TEMPI=BI(II,I,J)
      BR(II,I,J)=BR(II,LL,J)
      BI(II,I,J)=BI(II,LL,J)
      BR(II,LL,J)=TEMPR
      BI(II,LL,J)=TEMPI
    3 CONTINUE
    4 CONTINUE
  113 CONTINUE
  130 CONTINUE
      DO 114 II=1,NMAT
      SR1(II)=AR(II,I,I)**2+AI(II,I,I)**2
  114 CONTINUE
      DO 115 II=1,NMAT
      VR(II)=AR(II,I,I)/SR1(II)
      VI(II)=AI(II,I,I)/SR1(II)
  115 CONTINUE
      IF(I.EQ.N) GO TO 9
      DO 8 J=IP1,N
      DO 116 II=1,NMAT
      SR1(II)=0.0
      SI1(II)=0.0
      SR2(II)=0.0
      SI2(II)=0.0
  116 CONTINUE
      IF(IM1.EQ.0) GO TO 6
      DO 5 K=1,IM1
      DO 5 II=1,NMAT
      SR1(II)=SR1(II)+AR(II,I,K)*AR(II,K,J)-AI(II,I,K)*AI(II,K,J)
      SI1(II)=SI1(II)+AR(II,I,K)*AI(II,K,J)+AR(II,K,J)*AI(II,I,K)
    5 CONTINUE
    6 CONTINUE
      DO 117 II=1,NMAT
      HR(II)=AR(II,I,J)-SR1(II)
      HI(II)=AI(II,I,J)-SI1(II)
  117 CONTINUE
      DO 118 II=1,NMAT
      AR(II,I,J)=HR(II)*VR(II)+HI(II)*VI(II)
      AI(II,I,J)=HI(II)*VR(II)-HR(II)*VI(II)
  118 CONTINUE
      DO 7 K=1,I
      DO 7 II=1,NMAT
      SR2(II)=SR2(II)+AR(II,J,K)*AR(II,K,IP1)-AI(II,J,K)*AI(II,K,IP1)
      SI2(II)=SI2(II)+AR(II,J,K)*AI(II,K,IP1)+AI(II,J,K)*AR(II,K,IP1)
    7 CONTINUE
      DO 119 II=1,NMAT
      AR(II,J,IP1) = AR(II,J,IP1)-SR2(II)
      AI(II,J,IP1) = AI(II,J,IP1)-SI2(II)
  119 CONTINUE
    8 CONTINUE
    9 CONTINUE
      DO 120 II=1,NMAT
```

```
      IF(T(II).EQ.0.0) GO TO 15
  120 CONTINUE
      DO 11 J=1,M
      DO 121 II=1,NMAT
      SR1(II)=0.0
  121 SI1(II)=0.0
      IF(IM1.EQ.0) GO TO 21
      DO 10 K=1,IM1
      DO 10 II=1,NMAT
      SR1(II)=SR1(II)+AR(II,I,K)*BR(II,K,J)-AI(II,I,K)*BI(II,K,J)
      SI1(II)=SI1(II)+AI(II,I,K)*BR(II,K,J)+AR(II,I,K)*BI(II,K,J)
   10 CONTINUE
   21 CONTINUE
      DO 122 II=1,NMAT
      HR(II)=BR(II,I,J)-SR1(II)
      HI(II)=BI(II,I,J)-SI1(II)
  122 CONTINUE
      DO 123 II=1,NMAT
      BR(II,I,J)=HR(II)*VR(II)+HI(II)*VI(II)
      BI(II,I,J)=HI(II)*VR(II)-HR(II)*VI(II)
  123 CONTINUE
   11 CONTINUE
      IF(N.EQ.1) RETURN
      NM1=N-1
      DO 13 I=1,M
      DO 13 J=1,NM1
      NMJ=N-J
      DO 124 II=1,NMAT
      SR1(II)=0.0
  124 SI1(II)=0.0
      NMJP1=NMJ+1
      DO 12 K=NMJP1,N
      DO 12 II=1,NMAT
      SR1(II)=SR1(II)+AR(II,NMJ,K)*BR(II,K,I)-AI(II,NMJ,K)*BI(II,K,I)
      SI1(II)=SI1(II)+AR(II,NMJ,K)*BI(II,K,I)+AI(II,NMJ,K)*BR(II,K,I)
   12 CONTINUE
      DO 125 II=1,NMAT
      BR(II,NMJ,I)=BR(II,NMJ,I)-SR1(II)
      BI(II,NMJ,I)=BI(II,NMJ,I)-SI1(II)
  125 CONTINUE
   13 CONTINUE
      RETURN
   15 ID=2
   17 PRINT 85
   85 FORMAT(' DIMENSIONING PROBLEM IN CLSS')
      RETURN
      END
      SUBROUTINE SIMP(XX,FX,AX,NX)
      IMPLICIT REAL*8 (A-H,O-Z)
      DIMENSION XX(4000),FX(4000),AX(4000)
      DELX= XX(2) - XX(1)
      AX(1)= 0.0
```

```
      DO 10  IX=2,NX,2
      IF(NX-IX)2,3,2
    2 AX(IX)= DELX*(5.0*FX(IX-1)+8.0*FX(IX)-FX(IX+1))/12.0+AX(IX-1)
      AX(IX+1)=DELX*(FX(IX-1)+4.0*FX(IX)+FX(IX+1))/3.0+AX(IX-1)
   10 CONTINUE
      GO TO 4
    3 AX(NX)=DELX*(-FX(NX-2)+8.0*FX(NX-1)+5.0*FX(NX))/12.0+AX(NX-1)
    4 RETURN
      END
      REAL*8 FUNCTION INTERP   (XX,X,F,J1,N)
      IMPLICIT REAL*8 (A-H,O-Z)
      DIMENSION   X(1),F(1)
      FX=0.
      ISTART=J1-N+N/2+1
      J2=ISTART+N-1
      DO 10 J=ISTART,J2
      P=F(J)
      DO 5 I=ISTART,J2
      IF (I.EQ.J) GO TO 5
      P=P*(XX-X(I))/(X(J)-X(I))
    5 CONTINUE
   10 FX=FX+P
      INTERP=FX
      RETURN
      END
      SUBROUTINE MATS(X1,X2,X3,WW,PAR,V,MAXM1,HYD)
      IMPLICIT REAL*8 (A-H,O-Z)
      DIMENSION H(10,10),PAR(60)
      DIMENSION WW(9,9),V(9)
      YY=3.1415926536
      MAX=MAXM1+1
      A1 =PAR(1)
      A6 =PAR(2)
      A14 =PAR(3)
      A19 =PAR(4)
      A2 =PAR(5)
      A3 =PAR(6)
      A4 =PAR(7)
      A5=PAR(8)
      A7=PAR(9)
      A8=PAR(10)
      A9=PAR(11)
      A10=PAR(12)
      A11=PAR(13)
      A12=PAR(14)
      A13=PAR(15)
      A15=PAR(16)
      A16=PAR(17)
      A17=PAR(18)
      A18=PAR(19)
      A20=PAR(20)
      A21=PAR(21)
```

```
      B1=PAR(22)
      B2=PAR(23)
      B3=PAR(24)
      B4=PAR(25)
      B5=PAR(26)
      B6=PAR(27)
      B7=PAR(28)
      B8=PAR(29)
      B9=PAR(30)
      B10=PAR(31)
      B11=PAR(32)
      C1=PAR(33)
      C2=PAR(34)
      C3=PAR(35)
      C6=PAR(36)
      C7=PAR(37)
      C8=PAR(38)
        C9=PAR(39)
        EE=PAR(40)
        EE2=PAR(41)
        E4=PAR(42)
        E5=PAR(43)
        D1=PAR(44)
        D2=PAR(45)
        D3=PAR(46)
        D4=PAR(47)
        D5=PAR(48)
        D6=PAR(49)
        D7=PAR(50)
        D8=PAR(51)
      F1=PAR(52)
      F2=PAR(53)
      F3=PAR(54)
      F4=PAR(55)
      F5=PAR(56)
      G1=PAR(57)
      G2=PAR(58)
      G3=PAR(59)
      G4=PAR(60)
      SR2=SQRT(2.)
      SR3=SQRT(3.)
      DO9I=1,MAX
      DO9J=1,MAX
9         H(I,J)=0.0
      X1=YY*X1
      Y1=YY*X2
      Z1=YY*X3
      COSX= COS(X1)
      COSY= COS(Y1)
      COSZ= COS(Z1)
      SINX= SIN(X1)
      SINY= SIN(Y1)
```

```
      SINZ= SIN(Z1)
      COS2X=COS(2.*X1)
      COS2Y=COS(2.*Y1)
      COS2Z=COS(2.*Z1)
      SIN2X=SIN(2.*X1)
      SIN2Y=SIN(2.*Y1)
      SIN2Z=SIN(2.*Z1)
      H(1,1)=A1+4.*A2*(COSX*COSY+COSY*COSZ+COSZ*COSX)+2.*B1*(COS2X+
     1COS2Z)
C
C+++++H(1,2),H(1,3),H(1,4) ARE REVERSED IN SIGN SO THAT THE REAL MATRIX
C     EQUIVALENT TO COMPLEX MATRIX
C
      H(1,2)=4.*A3*SINX*(COSY+COSZ)+2.*B2*SIN2X
      H(1,3)=4.*A3*SINY*(COSZ+COSX)+2.*B2*SIN2Y
      H(1,4)=4.*A3*SINZ*(COSX+COSY)+2.*B2*SIN2Z
      H(1,2)=-H(1,2)
      H(1,3)=-H(1,3)
      H(1,4)=-H(1,4)
      H(1,5)=-4.*A4*SINX*SINY
      H(1,6)=-4.*A4*SINY*SINZ
      H(1,7)=-4.*A4*SINZ*SINX
      H(1,8)=2.*SR3*A5*COSZ*(COSY-COSX)+SR3*B3*(COS2X-COS2Y)
      H(1,9)=-2.*A5*(-2.*COSX*COSY+COSY*COSZ+COSZ*COSX)+B3*(-COS2X-COS2Y
     1+2.*COS2Z)
      H(2,2)=A6+4.*A7*COSX*(COSY+COSZ)+4.*A8*COSY*COSZ+2.*B4*COS2X+2.*B5
     1*(COS2Y+COS2Z)
      H(2,3)=-4.*A9*SINX*SINY
      H(2,4)=-4.*A9*SINX*SINZ
      H(2,5)=4.*SINY*(A10*COSX+A11*COSZ)+2.*B6*SIN2Y
      H(2,6)=0.
      H(2,7)=4.*SINZ*(A10*COSX+A11*COSY)+2.*B6*SIN2Z
      H(2,8)=2.*SR3*A12*SINX*(COSY+COSZ)+2.*A13*SINX*(COSY-COSZ)+SR3*B7*
     1SIN2X
      H(2,9)=-2.*A12*SINX*(COSY+COSZ)+2.*SR3*A13*SINX*(COSY-COSZ)-B7*
     1SIN2X
      H(3,3)=A6+4.*A7*COSY*(COSZ+COSX)+4.*A8*COSZ*COSX+2.*B4*COS2Y+
     12.*B5*(COS2Z+COS2X)
      H(3,4)=-4.*A9*SINY*SINZ
      H(3,5)=4.*SINX*(A10*COSY+A11*COSZ)+2.*B6*SIN2X
      H(3,6)=4.*SINZ*(A10*COSY+A11*COSX)+2.*B6*SIN2Z
      H(3,7)=0.
      H(3,8)=-2.*SR3*A12*SINY*(COSZ+COSX)-2.*A13*SINY*(COSX-COSZ)-SR3*B7
     1*SIN2Y
      H(3,9)=-2.*A12*SINY*(COSZ+COSX)-2.*SR3*A13*SINY*(COSZ-COSX)-B7*
     1SIN2Y
      H(4,4)=A6+4.*A7*COSZ*(COSX+COSY)+4.*A8*COSX*COSY+2.*B4*COS2Z+2.*B5
     1*(COS2X+COS2Y)
      H(4,5)=0.
      H(4,6)=4.*SINY*(A10*COSZ+A11*COSX)+2.*B6*SIN2Y
      H(4,7)=4.*SINX*(A10*COSZ+A11*COSY)+2.*B6*SIN2X
      H(4,8)=4.*A13*SINZ*(COSY-COSX)
```

```
      H(4,9)=4.*A12*SINZ*(COSX+COSY)+2.*B7*SIN2Z
      H(5,5)=A14+4.*A15*COSX*COSY+4.*A16*COSZ*(COSX+COSY)+2.*B8*(COS2X+
     1COS2Y)+2.*B9*COS2Z
      H(5,6)=-4.*A17*SINX*SINZ
      H(5,7)=-4.*A17*SINY*SINZ
      H(5,8)=0.
      H(5,9)=-4.*A18*SINX*SINY
      H(6,6)=A14+4.*A15*COSY*COSZ+4.*A16*COSX*(COSY+COSZ)+2.*B8*(COS2Z
     1+COS2Y)+2.*B9*COS2X
      H(6,7)=-4.*A17*SINX*SINY
      H(6,8)=-2.*SR3*A18*SINY*SINZ
      H(6,9)=2.*A18*SINY*SINZ
      H(7,7)=A14+4.*A15*COSX*COSZ+4.*A16*COSY*(COSX+COSZ)+2.*B8*(COS2X+
     1COS2Z)+2.*B9*COS2Y
      H(7,8)=2.*SR3*A18*SINX*SINZ
      H(7,9)=2.*A18*SINX*SINZ
      H(8,8)=A19+3.*A20*COSZ*(COSX+COSY)+4.*A21*(COSX*COSY+COSX*COSZ/4.+
     1COSY*COSZ/4.)+(3./2.)*B10*(COS2X+COS2Y)+2.*B11*(COS2X/4.+COS2Y/4.+
     2COS2Z)
      H(8,9)=SR3*COSZ*(COSX-COSY)*(A20-A21)+(SR3/2.)*(COS2Y-COS2X)*(B10-
     1B11)
      H(9,9)=A19+4.*A20*(COSX*COSY+COSX*COSZ/4.+COSY*COSZ/4.)+3.*A21*
     1COSZ*(COSX+COSY)+2.*B10*(COS2X/4.+COS2Y/4.+COS2Z)+(3./2.)*B11*
     2(COS2X+COS2Y)
C     CHANGES TO INCORPORATE 2ND NN OF THE A-B INTERACTIONS
      H(1,10)=2.*D1*(COSX+COSY+COSZ)+8.*F1*COSX*COSY*COSZ
      H(2,10)=2.*D3*SINX+8.*F2*SINX*COSY*COSZ
      H(3,10)=2.*D3*SINY+8.*F2*SINY*COSZ*COSX
      H(4,10)=2.*D3*SINZ+8.*F2*SINZ*COSX*COSY
      H(5,10)=-8.*G1*SINX*SINY*COSZ
      H(6,10)=-8.*G1*COSX*SINY*SINZ
      H(7,10)=-8.*G1*SINX*COSY*SINZ
      H(8,10)=SR3*D7*(COSX-COSY)
      H(9,10)=D7*(-COSX-COSY+2.*COSZ)
      H(10,10) =
     1 C1+4.*C2*(COSX*COSY+COSY*COSZ+COSZ*COSX)+2.*EE*(COS2X+COS2Y
     1+COS2Z)
C    END OF 2ND NN A-B INTERACTIONS FOR S AND P ORBITALS
      DO99J=1,MAX
      DO99I=1,J
 99   H(J,I)=H(I,J)
      DO8I=1,MAXM1
      V(I)=H(I,MAX)
      DO8J=1,MAXM1
 8    WW(I,J)=H(I,J)
      HYD= H(10,10) - C1
      RETURN
      END
```

```
 CPA FOR Cu-H X=0.95
  60     9     8  11.95
 1.0133436918   0  1          CuH
 1.6299730539   0  2          CuH
 0.4669095576   0  3          CuH
 0.5523251891   0  4          CuH
 0.0070993169   0  5          CuH
-0.0390431844   0  6          CuH
-0.0347827710   0  7          CuH
-0.0250461921   0  8          CuH
 0.0905985832   0  9          CuH
-0.0000801649   0 10          CuH
 0.0442320406   0 11          CuH
 0.0352347530   0 12          CuH
-0.0050138300   0 13          CuH
 0.0063122837   0 14          CuH
 0.0193088688   0 15          CuH
-0.0174550749   0 16          CuH
 0.0054789842   0 17          CuH
 0.0045968308   0 18          CuH
 0.0090510566   0 19          CuH
 0.0090507874   0 20          CuH
-0.0348815545   0 21          CuH
 0.0593079962   0 22          CuH
 0.0829818100   0 23          CuH
 0.0467784964   0 24          CuH
-0.1623541564   0 25          CuH
 0.0210846700   0 26          CuH
 0.0027799814   0 27          CuH
-0.0472122058   0 28          CuH
-0.0007821462   0 29          CuH
 0.0000663938   0 30          CuH
 0.0302998964   0 31          CuH
-0.0003523610   0 32          CuH
 0.1711303145   0 33          CuH
-0.0177616198   0 34          CuH
 0.0000000000   1 35          CuH
10.0000000000   1 36          CuH
 0.0000000000   1 37          CuH
 0.0000000000   1 38          CuH
 0.0000000000   1 39          CuH
-0.0050270767   0 40          CuH
 0.0000000000   1 41          CuH
 0.0000000000   1 42          CuH
 0.0000000000   1 43          CuH
-0.1016161218   0 44          CuH
 0.0000000000   1 45          CuH
 0.0295211356   0 46          CuH
 0.0000000000   1 47          CuH
 0.0000000000   1 48          CuH
 0.0000000000   1 49          CuH
-0.0313003585   0 50          CuH
```



```
0.0000000000    1 51                        CuH
0.0141851548    0 52                        CuH
-0.0314414278   0 53                        CuH
0.0000000000    1 54                        CuH
0.0000000000    1 55                        CuH
0.0000000000    1 56                        CuH
-0.0030053612   0 57                        CuH
0.0000000000    1 58                        CuH
0.0000000000    1 59                        CuH
0.0000000000    1 60                        CuH
0.05      -1.50      1.00      0.002     0.15      0.4       -0.01     0.001
0.002
```

ENERGY	Hydrogen-s	METAL-s	METAL-p	METAL-t2g	METAL -eg	MET-TOT-d	TOTAL-DOS	ELECTRONS
cut off to save space								
-1.50000	0.00023	0.00011	0.00021	0.00050	0.00031	0.00113	0.00136	0.00000
-1.49800	0.00023	0.00011	0.00021	0.00050	0.00031	0.00113	0.00136	0.00001
-1.49600	0.00023	0.00011	0.00021	0.00050	0.00031	0.00113	0.00136	0.00001
-1.49400	0.00023	0.00011	0.00021	0.00050	0.00031	0.00113	0.00136	0.00002
-1.49200	0.00023	0.00011	0.00021	0.00050	0.00031	0.00113	0.00137	0.00002
-1.49000	0.00023	0.00011	0.00021	0.00050	0.00031	0.00114	0.00137	0.00003
-1.48800	0.00023	0.00011	0.00021	0.00050	0.00031	0.00114	0.00137	0.00003
-1.48600	0.00023	0.00011	0.00021	0.00051	0.00031	0.00114	0.00138	0.00004
-1.48400	0.00024	0.00011	0.00021	0.00051	0.00031	0.00114	0.00138	0.00004
-1.48200	0.00024	0.00011	0.00021	0.00051	0.00031	0.00114	0.00138	0.00005
-1.48000	0.00024	0.00011	0.00021	0.00051	0.00031	0.00115	0.00138	0.00005
-1.47800	0.00024	0.00011	0.00021	0.00051	0.00031	0.00115	0.00139	0.00006
-1.47600	0.00024	0.00011	0.00021	0.00051	0.00032	0.00115	0.00139	0.00007
-1.47400	0.00024	0.00011	0.00021	0.00051	0.00032	0.00115	0.00139	0.00007
-1.47200	0.00024	0.00011	0.00021	0.00051	0.00032	0.00116	0.00139	0.00008
-1.47000	0.00024	0.00011	0.00021	0.00051	0.00032	0.00116	0.00140	0.00008
-1.46800	0.00024	0.00011	0.00021	0.00052	0.00032	0.00116	0.00140	0.00009
-1.46600	0.00024	0.00011	0.00022	0.00052	0.00032	0.00116	0.00140	0.00009
-1.46400	0.00024	0.00011	0.00022	0.00052	0.00032	0.00116	0.00141	0.00010
-1.46200	0.00024	0.00011	0.00022	0.00052	0.00032	0.00117	0.00141	0.00011
-1.46000	0.00024	0.00011	0.00022	0.00052	0.00032	0.00117	0.00141	0.00011
-1.45800	0.00024	0.00011	0.00022	0.00052	0.00032	0.00117	0.00141	0.00012
-1.45600	0.00024	0.00011	0.00022	0.00052	0.00032	0.00117	0.00142	0.00012
-1.45400	0.00024	0.00011	0.00022	0.00052	0.00032	0.00118	0.00142	0.00013
-1.45200	0.00024	0.00011	0.00022	0.00052	0.00032	0.00118	0.00142	0.00013
-1.45000	0.00025	0.00011	0.00022	0.00053	0.00032	0.00118	0.00143	0.00014
-1.44800	0.00025	0.00011	0.00022	0.00053	0.00032	0.00118	0.00143	0.00014
-1.44600	0.00025	0.00011	0.00022	0.00053	0.00033	0.00119	0.00143	0.00015
-1.44400	0.00025	0.00011	0.00022	0.00053	0.00033	0.00119	0.00144	0.00016
-1.44200	0.00025	0.00011	0.00022	0.00053	0.00033	0.00119	0.00144	0.00016
-1.44000	0.00025	0.00011	0.00022	0.00053	0.00033	0.00119	0.00144	0.00017
-1.43800	0.00025	0.00011	0.00022	0.00053	0.00033	0.00119	0.00144	0.00017
-1.43600	0.00025	0.00012	0.00022	0.00053	0.00033	0.00120	0.00145	0.00018
-1.43400	0.00025	0.00012	0.00022	0.00053	0.00033	0.00120	0.00145	0.00019
-1.43200	0.00025	0.00012	0.00022	0.00054	0.00033	0.00120	0.00145	0.00019
-1.43000	0.00025	0.00012	0.00022	0.00054	0.00033	0.00120	0.00146	0.00020
-1.42800	0.00025	0.00012	0.00022	0.00054	0.00033	0.00121	0.00146	0.00020
-1.42600	0.00025	0.00012	0.00022	0.00054	0.00033	0.00121	0.00146	0.00021
-1.42400	0.00025	0.00012	0.00022	0.00054	0.00033	0.00121	0.00146	0.00021
-1.42200	0.00025	0.00012	0.00022	0.00054	0.00033	0.00121	0.00147	0.00022
-1.42000	0.00026	0.00012	0.00022	0.00054	0.00033	0.00122	0.00147	0.00023
-1.41800	0.00026	0.00012	0.00022	0.00054	0.00033	0.00122	0.00147	0.00023
-1.41600	0.00026	0.00012	0.00022	0.00054	0.00034	0.00122	0.00148	0.00024
-1.41400	0.00026	0.00012	0.00022	0.00055	0.00034	0.00122	0.00148	0.00024
-1.41200	0.00026	0.00012	0.00022	0.00055	0.00034	0.00123	0.00148	0.00025
-1.41000	0.00026	0.00012	0.00022	0.00055	0.00034	0.00123	0.00149	0.00026
-1.40800	0.00026	0.00012	0.00022	0.00055	0.00034	0.00123	0.00149	0.00026
-1.40600	0.00026	0.00012	0.00022	0.00055	0.00034	0.00123	0.00149	0.00027
-1.40400	0.00026	0.00012	0.00022	0.00055	0.00034	0.00123	0.00150	0.00027

```
 -1.40200   0.00026   0.00012   0.00023   0.00055   0.00034   0.00124   0.00150   0.00028
 -1.40000   0.00026   0.00012   0.00023   0.00055   0.00034   0.00124   0.00150   0.00029
 -1.39800   0.00026   0.00012   0.00023   0.00056   0.00034   0.00124   0.00151   0.00029
 -1.39600   0.00026   0.00012   0.00023   0.00056   0.00034   0.00124   0.00151   0.00030
 -1.39400   0.00026   0.00012   0.00023   0.00056   0.00034   0.00125   0.00151   0.00030
 -1.39200   0.00027   0.00012   0.00023   0.00056   0.00034   0.00125   0.00151   0.00031
 -1.39000   0.00027   0.00012   0.00023   0.00056   0.00034   0.00125   0.00152   0.00032
 -1.38800   0.00027   0.00012   0.00023   0.00056   0.00035   0.00125   0.00152   0.00032
 -1.38600   0.00027   0.00012   0.00023   0.00056   0.00035   0.00126   0.00152   0.00033
 -1.38400   0.00027   0.00012   0.00023   0.00056   0.00035   0.00126   0.00153   0.00033
 -1.38200   0.00027   0.00012   0.00023   0.00057   0.00035   0.00126   0.00153   0.00034
 -1.38000   0.00027   0.00012   0.00023   0.00057   0.00035   0.00126   0.00153   0.00035
 -1.37800   0.00027   0.00012   0.00023   0.00057   0.00035   0.00127   0.00154   0.00035
 -1.37600   0.00027   0.00012   0.00023   0.00057   0.00035   0.00127   0.00154   0.00036
 -1.37400   0.00027   0.00012   0.00023   0.00057   0.00035   0.00127   0.00154   0.00036
 -1.37200   0.00027   0.00012   0.00023   0.00057   0.00035   0.00127   0.00155   0.00037
 -1.37000   0.00027   0.00012   0.00023   0.00057   0.00035   0.00128   0.00155   0.00038
 -1.36800   0.00027   0.00012   0.00023   0.00057   0.00035   0.00128   0.00155   0.00038
 -1.36600   0.00027   0.00012   0.00023   0.00058   0.00035   0.00128   0.00156   0.00039
 -1.36400   0.00028   0.00012   0.00023   0.00058   0.00035   0.00129   0.00156   0.00040
FERMI ENERGY   ELECTRONS   TOTAL DOS      S        P        DT2G        DEG        S      P

   0.84271   11.95000      0.54488   0.04313   0.31152   0.06293   0.12454   0.00276   0.00000
```

Index

D. A. Papaconstantopoulos, *Band Structure of Cubic Hydrides*,
https://doi.org/10.1007/978-3-031-06878-2